T0211183

GRAPH SPECTRA FOR COMPLEX NETWORKS

This concise and self-contained introduction builds up the spectral theory of graphs from scratch, with linear algebra and the theory of polynomials developed in the later parts. The book focuses on properties and bounds for the eigenvalues of the adjacency, Laplacian and effective resistance matrices of a graph. The goal of the book is to collect spectral properties that may help to understand the behavior or main characteristics of real-world networks. The chapter on spectra of complex networks illustrates how the theory may be applied to deduce insights into real-world networks.

The second edition contains new chapters on topics in linear algebra and on the effective resistance matrix, and treats the pseudoinverse of the Laplacian. The latter two matrices and the Laplacian describe linear processes, such as the flow of current, on a graph. The concepts of spectral sparsification and graph neural networks are included.

PIET VAN MIEGHEM is Professor at the Delft University of Technology. His research interests lie in network science: the modeling and analysis of complex networks such as infrastructural networks (for example telecommunication, power grids and transportation) as well as biological, brain, social and economic networks.

GRAPH SPECTRA FOR COMPLEX NETWORKS

NETWORKS

Second Edition

PIET VAN MIEGHEM

Delft University of Technology

CAMBRIDGE
UNIVERSITY PRESS

Shaftesbury Road, Cambridge CB2 8EA, United Kingdom

One Liberty Plaza, 20th Floor, New York, NY 10006, USA

477 Williamstown Road, Port Melbourne, VIC 3207, Australia

314–321, 3rd Floor, Plot 3, Splendor Forum, Jasola District Centre,
New Delhi – 110025, India

103 Penang Road, #05–06/07, Visioncrest Commercial, Singapore 238467

Cambridge University Press is part of Cambridge University Press & Assessment,
a department of the University of Cambridge.

We share the University's mission to contribute to society through the pursuit of
education, learning and research at the highest international levels of excellence.

www.cambridge.org
Information on this title: www.cambridge.org/9781009366809

DOI: 10.1017/9781009366793

First edition © Cambridge University Press 2011
Second edition © Piet Van Mieghem 2023

This publication is in copyright. Subject to statutory exception and to the provisions
of relevant collective licensing agreements, no reproduction of any part may take
place without the written permission of Cambridge University Press & Assessment.

First published 2011
Second edition 2023

A catalogue record for this publication is available from the British Library

A Cataloging-in-Publication data record for this book is available from the Library of Congress

ISBN 978-1-009-36680-9 Paperback

Cambridge University Press & Assessment has no responsibility for the persistence
or accuracy of URLs for external or third-party internet websites referred to in this
publication and does not guarantee that any content on such websites is, or will
remain, accurate or appropriate.

in memory of Saskia and Nathan

to Huijuan

Contents

Preface to the second edition

There is no place for ugly mathematics (G. H. Hardy)

After more than a decade, a new edition was felt needed. The interest in and the role of networks is still increasing, although the landscape of graph spectra is not dramatically changed, but is slowly evolving. New theory or theory that I have missed in the first edition is added. For example, I include the matrix theory of linear processes on a graph, whose dynamics is proportional to the underlying topology, such as fluids flowing in a network of pipes or electrical current in a resistor network. The vector of the injected current at nodes is connected to the vector of potentials at those nodes by a weighted Laplacian as explained in **art**. 14, from which the pseudoinverse of the Laplacian naturally arises. The physics and meaning of the diagonal elements of the pseudoinverse as well as the effective resistance matrix of a graph are treated in Chapter 5.

The computation of graph spectra, eigenvalues and eigenvectors requires the theory of linear algebra and polynomials. In the first edition, the book was divided into two parts, where the second part originated from exploded appendices. This second edition consists of three parts. The core of the book is Part I on Spectra of Graphs, which consumes more than half of the pages and seven chapters. The main theory on the eigenvalue equation (1.3), that comprises matrix and determinant operations in linear algebra, is summarized in the Eigensystem in Part II. The theory of polynomials, which also belongs to function theory, is contained in Part III. Those two last parts contain the general theory, which is applied to graphs in Part I. The reason for the separation is the inclusion of many nice results that make those two last parts self-contained. Parts II and III can be read independently of Part I.

Apart from the correction of errors and the deletion of a few articles (**art**.) in the first edition, many additions have been included in this second edition. Some additions are new and not published before. The list of new material in this second edition is:

in Chapter 2: **art**. 12, 13, 14 to 16, 28, 29, 33, 35, 38 to 40, 17, 19, 21 to 24;

in Chapter 3: **art**. 43, 44, 52 to 58, 61, 64, 70, 71, 81, 87 to 91, 93 98;

in Chapter 4: **art**. 118, 120, 128 to 132, 139, 160, 161;

Chapter 5 on the effective resistance matrix;

in Chapter 6: Sections 6.4.3, 6.11 and 6.12;

in Chapter 7: **art**. 172, 182 and Section 7.5.3;

in Chapter 8: Sections 8.8 to 8.11;

Chapter 9 contains matrix transformations and properties of the determinant;

in Chapter 10: **art.** 240, 241, 248, 249, 253, 256, 257, 258, 259, 271, 281 and Sections 10.2, 10.7 and 10.10;

in Chapter 11: **art.** 293, 304, 305, 308, 312, 334, 336 and Section 11.6;

and in Chapter 12: Section 12.7.

Just as in the first edition, the main focus is on undirected graphs, whose graph-related matrices as the adjacency matrix and Laplacian are symmetric. For symmetric matrices, the eigenvalue decomposition is effective, simple and beautiful. Asymmetic matrices such as the non-backtracking matrix and the Markovian transition probability matrix specifying the directed Markov graph are not treated. Another omission concerns eigenvectors of graph-related matrices. Apart from their computation, relatively little is understood about eigenvectors, although we expect that progress will occur in the near future. A reason for this belief is the discovery of the geometric simplex representation of an undirected graph, which is a third equivalent representation besides the topology and the spectral domain, explained in the Preface to the first edition below. Any undirected graph, possibly weighted, on N nodes is a simplex – a generalization of a triangle in higher dimensions than two – in the $N-1$ dimensional Euclidean space, as first deduced by Fiedler (2009) and rediscovered by us (Devriendt and Van Mieghem, 2019a) while studying electrical resistor networks. That simplex is intimately related to eigenvectors of the Laplacian matrix, but we omit the simplex geometry of a graph and simplicial complexes. A last omission is specific topics in the relatively new field of graph signal processing, for which we refer to the recent book by Ortega (2022) and Section 8.11 for the concepts of graph neural networks. Graph signal processing analyzes data generated by processes on graphs and its aim is similar to that of Network Science; roughly the same topics are treated, only the approach and nomenclature differs somewhat. Here, we follow the network science terminology. While this book contains inequalities for eigenvalues of graph-related matrices, Stanić (2015) devotes an entire book on eigenvalue inequalities, which complements ours.

Finally, I hope that this new edition is easier to read: cross-referencing between articles **art.** is greatly improved and I have tried to fabricate many **art.**'s as more independent blocks that can stand on their own. To increase the readability, the equation labels in Part II and III contain as first indicator A and B, respectively, instead of the chapter number that is maintained in the core Part I.

July 2023   Piet Van Mieghem

Preface to the first edition

During the first years of the third millennium, considerable interest arose in complex networks such as the Internet, the world-wide web, biological networks, utility infrastructures (for transport of energy, waste, water, trains, cars and aircrafts), social networks, human brain networks, and so on. It was realized that complex networks are omnipresent and of crucial importance to humanity, whose still augmenting living standards increasingly depend on complex networks. Around the beginning of the new era, general laws such as "preferential attachment" and the "power law of the degree" were observed in many, totally different complex networks. This fascinating coincidence gave birth to an area of new research that is still continuing today. But, as is often the case in science, deeper investigations lead to more questions and to the conclusion that so little is understood of (large) networks. For example, the rather simple but highly relevant question "What is a robust network?" seems beyond the realm of present understanding. The most natural way to embark on solving the question consists of proposing a set of metrics that tend to specify and quantify "robustness". Soon one discovers that there is no universal set of metrics, and that the metrics of any set are dependent on each other and on the structure of the network.

Any complex network can be represented by a graph. Any graph can be represented by an adjacency matrix, from which other matrices such as the Laplacian are derived. These graph related matrices are defined in Chapter 2. One of the most beautiful aspects of linear algebra is the notion that, to each matrix, a set of eigenvalues with corresponding eigenvectors can be associated. The physical meaning of an "eigen" system is best understood by regarding the matrix as a geometric transformation of "points" in a space. Those "points" define a vector: a line segment from an origin that ends in the particular point and that is directed from origin to end. The transformation (rotation, translation, scaling) of the vector is again a vector in the same space, but generally different from the original vector. The vector that after the transformation turns out to be proportional with itself is called an eigenvector and the proportionality strength or the scaling factor is the eigenvalue. The Dutch and German adjective "eigen" means something that is inherent to itself, a characteristic or fundamental property. Thus, knowing that each graph is represented by a matrix, it is natural to investigate the "eigensystem", the set of all eigenvalues with corresponding eigenvectors because the "eigensystem" characterizes the graph. Stronger even, since both the adjacency and Laplacian matrix are symmetric, there is a one-to-one correspondence between the matrix and the "eigensystem", established in **art.** 247.

In a broader context, transformations have proved very fruitful in science. The

xiii

most prominent is undoubtedly the Fourier (or Laplace) transform. Many branches of science ranging from mathematics, physics and engineering abound with examples that show the power and beauty of the Fourier transform. The general principle of such transforms is that one may study the problem in either of two domains: in the original one and in the domain after transformation, and that there exists a one-to-one correspondence between both domains. For example, a signal is a continuous function of time that may represent a message or some information produced over time. Some properties of the signal are more appropriately studied in the time-domain, while others are in the transformed domain, the frequency domain. This analogy motivates us to investigate some properties of a graph in the topology domain, represented by a graph consisting of a set of nodes connected by a set of links, while other properties may be more conveniently dealt with in the spectral domain, specified by the set of eigenvalues and eigenvectors.

The duality between topology and spectral domain is, of course, not new and has been studied in the field of mathematics called *algebraic graph theory*. Several books on the topic, for example by Cvetković *et al.* (1995); Biggs (1996); Godsil and Royle (2001) and recently by Cvetković *et al.* (2009), have already appeared. Notwithstanding these books, the present one is different in a few aspects. First, I have tried to build-up the theory as a connected set of basic building blocks, called articles, which are abbreviated by **art.** The presentation in article-style was inspired by great scholars in past, such as Gauss (1801) in his great treatise *Disquisitiones Arithmeticae*, Titchmarsh (1964) in his *Theory of Functions*, and Hardy and Wright (2008) in their splendid *Introduction to the Theory of Numbers*, and many others that cannot be mentioned all. To some extent, it is a turning back to the past, where books were written for peers, and without exercise sections, which currently seem standard in almost all books. Thus, this book does not contain exercises. Second, the book focuses on general theory that applies to all graphs, and much less to particular graphs with special properties, of which the Petersen graph, shown in Fig. 2.3, is perhaps the champion among all. In that aspect, the book does not deal with a zoo of special graphs and their properties, but confines itself to a few classes of graphs that depend at least on a single parameter, such as the number of nodes, that can be varied. Complex networks all differ and vary in at least some parameters. Less justifiable is the omission of multigraphs, directed graphs and weighted graphs. Third, I have attempted to make the book as self-contained as possible and, as a peculiar consequence, the original appendices consumed about half of the book! Thus, I decided to create two parts, the main Part I on the spectra, while Part II overviews interesting results in linear algebra and the theory of polynomials that are used in Part I. Since each chapter in Part II discusses a wide area in mathematics, in fact, separate books on each topic are required. Hence, only the basic theory is discussed, while advanced topics are not covered, because the goal to include Part II was to support Part I. Beside being supportive, Part II contains interesting theory that opens possibilities to advance spectral results. For example, Laguerre's beautiful Theorem 91 may once be applied to the characteristic polynomials of a class of graphs with the same number of negative, positive and zero eigenvalues of the adjacency matrix.

A drawback is that the book does not contain a detailed list of references pointing to the original, first published papers: it was not my intention to survey the literature on the spectra of graphs, but rather to write a cohesive manuscript on results and on methodology. Sometimes, different methods or new proofs of a same result are presented. The monograph by Cvetković *et al.* (1995), complemented by Cvetković *et al.* (2009), still remains the invaluable source for references and tables of graph spectra.

I would like to thank Huijuan Wang, for her general interest, input and help in pointing me to interesting articles. Further, I am most grateful to Fernando Kuipers for proofreading the first version of the manuscript, to Roeloef Koekoek for reviewing Chapter 12 on orthogonal polynomials, and to Jasmina Omic for the numerical evaluation of bounds on the largest eigenvalue of the adjacency matrix. Javier Martin Hernandez, Dajie Liu and Xin Ge have provided me with many nice pictures of graphs and plots of spectra. Stojan Trajanovski has helped me with the m-dimensional lattice and **art. 153**. Wynand Winterbach showed that the assortativity of regular graphs is not necessarily equal to one, by pointing to the example of the complete graph minus one link (Section 8.5.1.1). Rob Kooij has constructed Fig. 4.1 as a counter example for the common belief that Fiedler's algebraic connectivity is always an adequate metric for network robustness with respect to graph disconnectivity. As in my previous book (Van Mieghem, 2006), David Hemsley has suggested a number of valuable textual improvements.

The book is a temporal reflection of the current state of the art: during the process of writing, progress is being made. In particular, the many bounds that typify the field are continuously improved. The obvious expectation is that future progress will increasingly shape and fine-tune the field into – hopefully – maturity. Hence, the book will surely need to be updated and all input is most welcome. Finally, I hope that the book may be of use to others and that it may stimulate and excite people to dive into the fascinating world of complex networks with the rigorous devices of algebraic graph theory offered here.

Ars mathematicae

October 2010   PIET VAN MIEGHEM

Acknowledgements

I would like to thank Massimo Achterberg and Xiangrong Wang for numerous comments and corrections. Hale Cetinay has contributed to the pseudoinverse of the Laplacian (Section 4.2). Karel Devriendt has informed me about several results in the literature that were unknown to me and he has derived new results in Section 5.2 on the effective resistance. Moreover, Karel also continued to show me his own new inventions, about eigenvectors of the Laplacian and about effective resistances, after he left our group NAS at Delft University of Technology. Further, Albert Cenen Cerda, Bastian Prasse, Qiang Liu and Jaron Sanders have read chapters together during several weeks around the end of 2018 and beginning of 2019 and provided me with a list of errors and suggested improvements. Ivan Jokić has derived the Hadamard product identity in **art.** 96.

I am very pleased that Massimo Achterberg, Scott Dalhgren, Clare Dennison, Karel Devriendt, Johan Dubbeldam, Ivan Jokić, Yingyue Ke, Rob Kooij, Geert Leus, Rogier Noldus, Bastian Prasse and Dragan Stevanović have read several chapters, pointed me to errors and provided input on the final version, just before publication.

Symbols

Only when explicitly mentioned, will we deviate from the standard notation and symbols outlined here.

Random variables and matrices are written with capital letters, while complex, real, integer, etc., variables are in lower case. For example, X refers to a random variable, A to a matrix, whereas x is a real number and z is a complex number. Also the element a_{ij} of a matrix A is written with a small letter. Usually, i, j, k, l, m, n are integers. Operations on random variables are denoted by $[.]$, whereas $(.)$ is used for real or complex variables. A set of elements is embraced by $\{.\}$. The largest integer smaller than or equal to x is denoted by $\lfloor x \rfloor$, whereas $\lceil x \rceil$ equals the smallest integer larger than or equal to x.

Linear Algebra

A $n \times m$ matrix $\begin{bmatrix} a_{11} & \cdots & a_{1m} \\ \vdots & & \\ a_{n1} & \cdots & a_{nm} \end{bmatrix}$

$\det A \quad = \begin{vmatrix} a_{11} & \cdots & a_{1n} \\ \vdots & & \\ a_{n1} & \cdots & a_{nn} \end{vmatrix}$: determinant of a square matrix A

$\mathrm{trace}(A) \quad = \sum_{j=1}^{n} a_{jj}$: sum of diagonal elements of A

$\mathrm{diag}(a) \quad = \mathrm{diag}(a_1, a_2, \ldots, a_n)$: diagonal matrix with diagonal elements equal to the components of the vector $a = (a_1, a_2, \ldots, a_n)$ while all off-diagonal elements are zero

$A^T \quad$ transpose of a matrix, the rows of A are the columns of A^T

$A^* \quad$ matrix in which each element is the complex conjugate of the corresponding element in A

$A^H \quad = (A^*)^T$: Hermitian of matrix A

$c_A(x) \quad = \det(A - xI)$: characteristic polynomial of A

$\mathrm{adj} A \quad = A^{-1} \det A$: adjugate of A

AB	matrix product of $n \times m$ matrix A and $m \times l$ matrix B with element $(AB)_{ij} = \sum_{k=1}^{m} a_{ik}b_{kj}$
$A \circ B$	Hadamard product of $n \times n$ matrix A and $n \times n$ matrix B with element $(A \circ B)_{ij} = a_{ij}b_{ij}$
J	all-one matrix
u	all-one vector
I	$\mathrm{diag}(u)$, identity matrix
$Q(\lambda)$	$= \frac{c_A(\lambda)}{\lambda I - A}$: adjoint of A
e_j	basic vector: all components are zero, except component j is 1
δ_{kj}	Kronecker delta, $\delta_{kj} = 1$ if $k = j$, else $\delta_{kj} = 0$

Probability theory

$\Pr[X]$	probability of the event X
$E[X]$	$= \mu$: expectation of the random variable X
$\mathrm{Var}[X]$	$= \sigma_X^2$: variance of the random variable X
$f_X(x)$	$= \frac{dF_X(x)}{dx}$: probability density function of X
$F_X(x)$	probability distribution function of X
$\varphi_X(z)$	probability generating function of X
	$\varphi_X(z) = E[z^X]$ when X is a discrete r.v.
	$\varphi_X(z) = E[e^{-zX}]$ when X is a continuous r.v.
$\{X_k\}_{1 \le k \le m}$	$= \{X_1, X_2, \ldots, X_m\}$
$X_{(k)}$	k-th order statistics, k-th smallest value in the set $\{X_k\}_{1 \le k \le m}$
P	transition probability matrix (Markov process)
$1_{\{x\}}$	indicator function: $1_{\{x\}} = 1$ if the event or condition $\{x\}$ is true, else $1_{\{x\}} = 0$. For example, $\delta_{kj} = 1_{\{k=j\}}$

Graph theory

$\mathcal{L}$	set of links in graph G
$\mathcal{N}$	set of nodes in graph G
L	$= |\mathcal{L}|$: number of links in graph G
N	$= |\mathcal{N}|$: number of nodes in graph G
A	adjacency matrix of graph G
B	incidence matrix of graph G
Q	$= BB^T$ Laplacian matrix of graph G
$Q^\dagger$	pseudoinverse of the Laplacian matrix of graph G
Ω	effective resistance matrix of graph G
H	hopcount in a graph (random variable) or hopcount matrix
$l(G)$	line graph of graph G
Δ	$= \mathrm{diag}(d)$: diagonal matrix of the nodal degrees
d	degree vector of a graph G
d_j	degree of node j
$d_{(j)}$	the j-th largest degree of node in graph G

$d_{\max}$	maximum degree in graph G
$d_{\min}$	minimum degree in graph G
D	degree (random variable) in graph G
$\kappa_{\mathcal{N}}(G)$	vertex (node) connectivity of graph G
$\kappa_{\mathcal{L}}(G)$	edge (link) connectivity of graph G
R_G	effective graph resistance
$\blacktriangle_G$	the number of triangles in graph G
$\{\lambda_k\}_{1 \leq k \leq N}$	set of eigenvalues of A ordered as $\lambda_1 \geq \lambda_2 \geq \cdots \geq \lambda_N$
$\{\mu_k\}_{1 \leq k \leq N}$	set of eigenvalues of Q ordered as $\mu_1 \geq \mu_2 \geq \cdots \geq \mu_N$
N_k	total number of walks with length k
W_k	number of closed walks with length k
ρ	diameter of graph G
ρ_D	degree assortativity of graph G
ω	clique number of graph G
K_N	the complete graph with N nodes
$K_{n,m}$	the complete bi-partite graph with $N = n + m$
P_N	path on N nodes

1

Introduction

Despite the fact that complex networks are the driving force behind the investigation of the spectra of graphs, it is not the purpose of this book to dwell on complex networks. A generally accepted, all-encompassing definition of a complex network does not seem to be available. Instead, complex networks are understood by instantiation: the Internet, transportation (car, train, airplane) and infrastructural (electricity, gas, water, sewer) networks, biological molecules, the human brain network, social networks, software dependency networks, are examples of complex networks. There is such a large literature about complex networks, predominantly in the physics community, that providing a detailed survey is a daunting task. We content ourselves here with referring to some review articles by Strogatz (2001); Newman *et al.* (2001); Albert and Barabási (2002); Newman (2003b), and to books in the field by Watts (1999); Barabási (2002); Dorogovtsev and Mendes (2003); Barrat *et al.* (2008); Dehmer and Emmert-Streib (2009); Newman (2010), and to references in these works. Application of spectral graph theory to chemistry and physics are found in Cvetković *et al.* (1995, Chapter 8).

A few years ago, the study of complex networks has been called *Network Science* Barabási (2016); Newman (2018). Networks consists of two main ingredients: (a) a dynamic process, such as transport of items from node a to node b and (b) an underlying topology or graph, over which the process evolves over time. In general, the graph of the network is not fixed, but can change over time steered by some second process. In time-varying networks, there are thus at least two processes, which may be either independent or coupled by a third interaction process. The best example, as experienced during the Covid pandemic, is epidemic spread on a human contact graph: (a) the epidemic is governed by a viral infection process and (b) the human mobility process creates the contact graph. Both processes may be coupled by a third process, when viral awareness information is distributed and humans can change contacts depending on whether people in their surrounding are infected or not. Usually, the process on a graph specifies the directions of links, while the graph itself reflects only link existence and is undirected.

In summary, most networks contain dynamic processes beside the graph. Network science studies the duality between process and graph and thus encompasses graph theory.

1.1 Graph of a network

The graph of a network, denoted by G, consists of a set $\mathcal{N}$ of N nodes connected by a set $\mathcal{L}$ of L links. Sometimes, nodes and links are called vertices and edges, respectively, and are correspondingly denoted by the set V and E. Here and in my book on *Performance Analysis* (Van Mieghem, 2014), a graph is denoted by $G(\mathcal{N}, \mathcal{L})$ or $G(N, L)$ to avoid conflicts with the expectation operator E in probability theory. There is no universal notation of a graph, although in graph theory $G = (V, E)$ often occurs, while in network theory and other applied fields, nodes and links are used and the notation $G(\mathcal{N}, \mathcal{L})$ or $G(N, L)$ appears. None of these notations is ideal nor optimized, but fortunately in most cases, the notation G for a graph seems sufficient. As explained in Devriendt and Van Mieghem (2019a) and mentioned in the preface, any undirected, possibly weighted graph on N nodes can be represented in the $N - 1$-dimensional Euclidean space by a simplex, whose vertices represent the nodes of the graph G, but the edges of the simplex differ from the links! Therefore, we adhere to nodes and link in the *topology domain* and we talk about vertices, edges, angles and faces in the *geometric domain*. Besides the graph and geometric domain, the third domain is the *spectral domain*, which is the main focus of this book. Between these three different representations of a graph G, there is a one-to-one correspondence for undirected graphs, implying that all information about the graph in one domain is preserved in another domain.

Graphs, in turn, can be represented by a matrix (**art. 1**). The simplest among these graph-related matrices is the adjacency matrix A, whose entries or elements are

$$a_{ij} = 1_{\{\text{node } i \text{ is connected to node } j\}} \tag{1.1}$$

where 1_x is the indicator function and equal to one if the event or condition x is true, else it is zero. All elements a_{ij} of the adjacency matrix are thus either 1 or 0 and A is symmetric for undirected graphs. Unless mentioned otherwise, we assume in this book that the graph is undirected and that A and other graph-related matrices are symmetric.

1.2 Eigenvalues and eigenvectors of a graph

If the graph consists of N nodes and L links, then **art.** 247 demonstrates that the $N \times N$ *symmetric* adjacency matrix can be written as

$$A = X\Lambda X^T \tag{1.2}$$

where the $N \times N$ orthogonal matrix X contains as columns the normalized eigenvectors $x_1, x_2, ..., x_N$ of A belonging to the real eigenvalues $\lambda_1 \geq \lambda_2 \geq ... \geq \lambda_N$, represented by the eigenvalue vector $\lambda = (\lambda_1, \lambda_2, \ldots, \lambda_N)$, and where the matrix $\Lambda = \text{diag}(\lambda)$. The basic relation (1.2) is an instance of the general eigenvalue problem (**art.** 235) for an arbitrary square matrix A with eigenvalue ξ, where A is not necessarily an adjacency matrix,

$$Ax = \xi x \qquad (1.3)$$

Assuming that the matrix A has N linearly independent eigenvectors, which implies that the matrix A is not defective nor has a Jordan form (Meyer, 2000), then the eigenvalue equation (1.3) can be written for each solution $Ax_k = \lambda_k x_k$ in terms of the orthogonal matrix $X = \begin{bmatrix} x_1 & x_2 & \cdots & x_N \end{bmatrix}$ as

$$AX = X\Lambda$$

The assumption of N linearly independent eigenvectors also means that $\text{rank}(X) = N$ and that the inverse matrix X^{-1} exists. Right-multiplying both sides by X^{-1} yields

$$A = X\Lambda X^{-1}$$

Art. 247 shows that symmetric matrices possess orthogonal eigenvectors, implying that $X^{-1} = X^T$, which brings us to (1.2). The eigenvalue equation (1.3) and its specific form for symmetric matrices (1.2) form the cornerstone of this book. Usually, although other definitions occur, the spectrum of a graph refers to the set of eigenvalues $\{\lambda_j\}_{1 \le j \le N}$ of a graph-related matrix and an eigenmode of an operator or matrix is the eigenvector belonging to an eigenvalue.

This basic relation (1.2) equates the *topology domain*, represented by the adjacency matrix, to the *spectral domain* of the graph, represented by the eigensystem in terms of the orthogonal matrix X of eigenvectors and the diagonal matrix Λ with corresponding eigenvalues. The major difficulty lies in the map from topology to spectral domain, $A \to X\Lambda X^T$, because the inverse map from spectral to topology domain, $X\Lambda X^T \to A$, consists of straightforward matrix multiplications. Thus, most of the efforts in this book lie in computing or deducing properties of X and Λ, given A. Even more confining, most endeavors are devoted to the diagonal matrix Λ of eigenvalues and the distribution and properties of the eigenvalues $\{\lambda_j\}_{1 \le j \le N}$ of A and of other graph-related matrices. It is fair to say that not too much is known about the eigenvectors and the distribution and properties of eigenvector components. A state of the current art is presented by Cvetković *et al.* (1997).

1.3 Interpretation and contemplation

One of the most studied eigenvalue problems is the stationary Schrödinger equation in quantum mechanics (see, e.g., Cohen-Tannoudji *et al.* (1977)),

$$H\varphi(r) = E\varphi(r)$$

where $\varphi(r)$ is the wave function, E is the energy eigenvalue of the Hamiltonian (linear) differential operator

$$H = -\frac{\hbar^2}{2m}\Delta + V(r)$$

in which the Laplacian operator is $\Delta = \frac{\partial^2}{\partial x^2} + \frac{\partial^2}{\partial y^2} + \frac{\partial^2}{\partial z^2}$, $\hbar = \frac{h}{2\pi}$ and $h \simeq 6.62 \times 10^{-34}$Js is Planck's constant, m is the mass of an object subject to a potential field $V(r)$ and r is a three-dimensional location vector. The wave function $\varphi(r)$ is generally complex, but $|\varphi(r)|^2$ represents the density function of the probability that the object is found at position r. The mathematical theory of second-order linear differential operators is treated, for instance, by Titchmarsh (1962, 1958).

While the interpretation of the eigenfunction $\varphi(r)$ of the Hamiltonian H, the continuous counterpart of an eigenvector with discrete components, and its corresponding energy eigenvalue E is well understood, the meaning of an eigenvector of a graph is rather vague and not satisfactory. An attempt is as follows. The basic equation (1.3) of the eigenvalue problem, combined with the zero-one nature of the adjacency matrix A, states that the j-th component of the eigenvector x_k belonging to eigenvalue λ_k can be written as

$$\lambda_k(x_k)_j = (Ax_k)_j = \sum_{l=1}^{N} a_{jl}(x_k)_l = \sum_{l \in \text{ neighbors}(j)} (x_k)_l \qquad (1.4)$$

where neighbors$(j) = \{l \in \mathcal{N} : a_{jl} = 1\}$ denotes the set of all direct neighbors of node j. In a simple graph (**art. 1**), there are no self-loops, i.e. $a_{jj} = 0$, and the eigenvector component $(x_k)_j$ multiplied by the eigenvalue λ_k equals the sum of the *other* eigenvector components $(x_k)_l$ over all direct neighbors l of node j. Since all eigenvectors of the adjacency matrix A are orthogonal[1] (**art. 247**), each eigenvector can be interpreted as describing a different inherent property of the graph. The precise meaning of that property depends upon the graph-related matrix viewed as an operator that acts upon a vector or points in the N-dimensional space. The eigenvalue basic equation (1.2) says that there are only N such inherent properties and the orthogonality of X or of the eigenvectors tells us that these inherent properties are independent. The above component equation (1.4) then expresses that the value $(x_k)_j$ of the inherent property k, belonging to the eigenvalue λ_k and specified by the eigenvector x_k, at each node j equals a weighted sum of those values $(x_k)_l$ over all its direct neighbors l and each such sum has a same weight λ_k^{-1} (provided $\lambda_k \neq 0$, else the average over all direct neighbors of those values $(x_k)_l$ is zero). Since both sides of the basic equation (1.3) can be multiplied by some non-zero number or quantity, we may interpret that the value of property k is expressed in own "physical" units. Perhaps, depending on the nature of the complex network, some of these units can be determined or discovered, but the pure mathematical

---

[1] Mathematically, the eigenvectors form an orthogonal basis that spans the entire N-dimensional space. Each eigenvector "adds" or specifies one dimension or one axis (orthogonal to all others) in that N-dimensional coordinate frame.

description (1.3) of the eigenvalue problem does not contain this information. Although the focus here is on eigenvectors, equation (1.4) also provides interesting information about the eigenvalues, for which we refer to **art. 273**.

Equation (1.4) reflects a *local* property with value $(x_k)_j$ that only depends on the corresponding values $(x_k)_l$ of direct neighbors. But this local property for node j holds *globally* for any node j, with a same strength or factor λ_k. This local and global aspect of the eigenstructure is another fascinating observation, that is conserved after "self-replication". Indeed, using (1.4) with index $j = l$ into (1.4) yields

$$\lambda_k^2 (x_k)_j = \sum_{l_1=1}^{N} a_{jl_1} \sum_{l_2=1}^{N} a_{l_1 l_2} (x_k)_{l_2} = \sum_{l_2=1}^{N} (A^2)_{jl_2} (x_k)_{l_2}$$

$$= d_j (x_k)_j + \sum_{l_2 \text{ is a second hop neighbor of } j} (x_k)_{l_2}$$

because **art. 19** shows that $(A^2)_{jj} = d_j$, where d_j is the degree, i.e. the number of neighbors, of node j. The idea can be continued and a subsequent substitution of (1.4) leads to an expression that involves a sum over all three hops nodes away from node j. Subsequent iterations relate the expansion of the graph around node j in the number m of hops, further elaborated in **art. 6** and **art. 65**, to the eigenvalue structure as

$$\left\{ \lambda_k^m - (A^m)_{jj} \right\} (x_k)_j = \sum_{l_m \text{ is an } m\text{-th hop neighbor of } j} (x_k)_{l_m} \tag{1.5}$$

The larger m, the more globally the environment around node j is extended.

The alternative representation (A.138) of $A = X \Lambda X^T$,

$$A = \sum_{k=1}^{N} \lambda_k x_k x_k^T$$

shows that there is a hierarchy in importance of the properties, specified by the absolute value of the eigenvalues, because all eigenvectors are scaled to have equal unit norm. In particular, possible zero eigenvalues contain properties that the graph does not possess, because the corresponding eigenvectors do not contribute to the structure – the adjacency matrix A – of the graph. In contrast, the properties belonging to the largest (in absolute value) eigenvalues have a definite and strong influence on the graph structure.

Another observation[2] is that the definition of the adjacency matrix A is somewhat arbitrary. Indeed, we may agree to assign the value α to the existence of a link and β otherwise, where α and $\beta \neq \alpha$ can be any complex number. Clearly, the graph is then equally well described by a new adjacency matrix $A(\alpha, \beta) = (\alpha - \beta) A + \beta J$, where J is the all-one matrix. Unless $\alpha = 1$ and $\beta = 0$, the eigenvalues and eigenvectors of $A(\alpha, \beta)$ are different from those of A. This implies that an entirely

---

[2] Communicated to me by Dajie Liu.

different, but still consistent theory of the spectra of graphs can be built. We have
not pursued this track here, although we believe that for certain problems a more
appropriate choice of α and β than $\alpha = 1$ and $\beta = 0$ may simplify the solution.

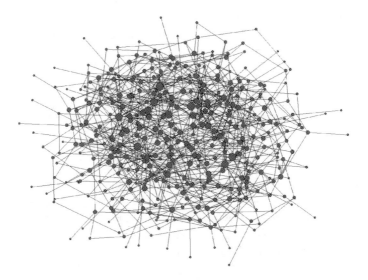

Fig. 1.1. A realization of an Erdős-Rényi random graph $G_p(N)$ with $N = 400$ nodes,
$L = 793$ links and average degree $\frac{2L}{N}$ of about 4. The link density $p \simeq 10^{-2}$ equals the
probability to have a link between two arbitrary chosen nodes in $G_p(N)$. The size of a
node is drawn proportional to its degree.

When encountering the subject for the first time, one may be wondering where all
the energy is spent, because the problem of finding the eigenvalues of A, reviewed in
Chapter 10, basically boils down to solving the zeros of the associated characteristic
polynomial (**art. 235**). In addition, we know (**art. 1**), due to symmetry of A, that
all zeros are real (**art. 247**), a fact that considerably simplifies matters as shown
in Chapter 11. For, nearly all of the polynomials with real coefficients possess
complex zeros and only a very small subset has zeros that are all real. This suggests
that there must be something special about these eigenvalues and characteristic
polynomials of A. Orthogonal polynomials form a fascinating class of polynomials
with real coefficients whose zeros are all real, which are studied in Chapter 12 and
which are related to orthogonal eigenvectors.

Much of the research in the spectral analysis of graphs is devoted to understand
properties of the graph by inspecting the spectra of mainly two matrices, the adjacency matrix A and the Laplacian Q, defined in **art. 4**. For example, how does
the spectrum, the set of all eigenvalues, show that a graph is connected? What
is the physical meaning of the largest and smallest eigenvalue, how large or small
can they be? How are eigenvalues changing when nodes and/or links are added

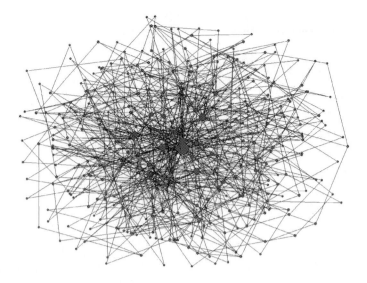

Fig. 1.2. An instance of a Barabási-Albert graph with $N = 400$ nodes and $L = 780$ links, which is about the same as in Fig. 1.1. The size of a node is drawn proportional to its degree.

to the graph? Deeper questions are, "Is Λ alone, without X in (1.2), sufficient to characterize a graph?", "How are the spacings, the differences between consecutive eigenvalues, distributed and what do spacings physically mean?", or, extremal problems as "What is the class of graphs on N nodes and L links that achieves the largest second smallest eigenvalue of the Laplacian?", and so on.

1.4 Outline of the book

Chapter 2 introduces some definitions and concepts of algebraic graph theory, which are needed in Part I. We embark on the spectrum in Chapter 3, that focuses on the eigenvalues of the adjacency matrix A. In Chapter 4, we continue with the investigation of the spectrum of the Laplacian Q. As argued by Mohar, the theory of the Laplacian spectrum is richer and contains more beautiful achievements than that of the adjacency matrix. Mohar's view is supported by the effective resistance matrix Ω in Chapter 5, that is closely related to the Laplacian matrix Q. In Chapter 6, we compute the entire adjacency spectrum and sometimes also the Laplacian spectrum of special types of classes containing at least one variable parameter such as the number of nodes N or/and the number of links L. Chapter 6 thus illustrates the theory of Chapter 3 and Chapter 4 by useful examples. In fact, the book originated from Chapter 6 and it was a goal to collect all spectra of graphs (with at least one parameter) that can be computed analytically. The underlying thought was to explain the spectrum of a complex network by features

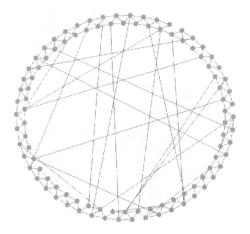

Fig. 1.3. The Watts-Strogatz small-world graph on $N = 100$ nodes and with nodal degree $D = 4$ (or $k = 2$ as explained in Section 6.2) and rewiring probability $p_r = \frac{1}{100}$.

appearing in "known spectra". Chapter 7 complements Chapter 6 asymptotically when graphs grow large, $N \to \infty$. For large graphs, the density or distribution of the eigenvalues (as nearly continuous variables) is more appealing and informative than the long list of eigenvalues. Apart from the three marvelous scaling laws by Wigner, Marčenko-Pastur and McKay, we did not find many explicit results on densities of eigenvalues of graphs. Finally, Chapter 8, the last chapter of Part I, applies the spectral knowledge of the previous chapters to gain physical insight into the nature of complex networks.

As mentioned in the Preface (first edition), the results derived in Part I have been built on the general theory of linear algebra and of polynomials with real coefficients, summarized in Part II and Part III, respectively.

1.5 Classes of graphs

The main classes of graphs in the study of complex networks are: the class of Erdős-Rényi random graphs (Fig. 1.1), whose fascinating properties are derived in Bollobás (2001); the class of Watts-Strogatz small-world graphs (Fig. 1.3) first explored in Watts (1999); the class of Barabási-Albert power law graphs (Fig. 1.2 and Fig. 1.4) introduced by Barabási and Albert (1999); and the regular hyper-lattices in several dimensions.

The Erdős-Rényi random graph is the simplest random model for a network. Its analytic tractability in a wide range of graph problems has resulted in the richest and most beautiful theory among classes of graphs. In many cases, the Erdős-Rényi random graph serves as a basic model that provides a fast benchmark for first order estimates and behaviors in real networks. Usually, if a graph problem cannot be

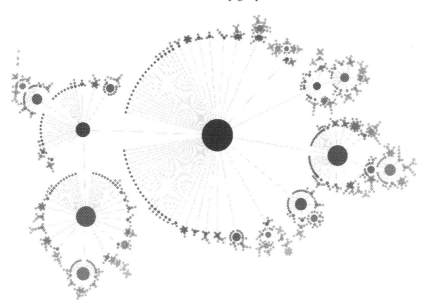

Fig. 1.4. A Barabási "fractal-like" tree with $N = 1000$ nodes, grown by adding at each step one new node to nodes already in the tree and proportional to their degree.

solved analytically for the Erdős-Rényi random graph or for hyper-lattices, little hope exists that other classes of (random) graphs may have a solution. However, in particular the degree distribution of complex networks does not match well with the binomial degree distribution of Erdős-Rényi random graphs (drawn in Fig. 1.5) and this observation has spurred the search for "more realistic models".

After random rewiring of links, the Watts-Strogatz small-world graphs in Section 6.2 possess a relatively high clustering and short hopcount. The probability p_r that a link is rewired is a powerful tool in Watts-Strogatz small-world graphs to balance between "long hopcounts" (p_r is small) and "small-worlds" ($p_r \to 1$).

The most distinguishing property of large Barabási-Albert power law graphs is the power law degree distribution, $\Pr[D = k] \approx ck^{-\tau}$ where[3] $c = \left(\sum_{k=1}^{N-1} k^{-\tau}\right)^{-1} \approx \frac{1}{\zeta(\tau)}$ for large N, which is observed as a major characteristic in many real-world complex networks. Fig. 1.5 compares the degree distribution of the Erdős-Rényi random graph shown in Fig. 1.1 and of the Barabási-Albert power law graph in Fig. 1.2, both with the same number of nodes ($N = 400$) and almost the same average degree ($E[D] = 4$). The insert illustrates the characteristic power law of the Barabási-Albert graph, recognized by a straight line in a log-log plot. Most nodes in the Barabási-Albert power law graph have small degree, while a few nodes have degree larger than 10 (which is the maximum degree in the realization here of

---

[3] The Dirichlet series $\zeta(s) = \sum_{k=1}^{\infty} \frac{1}{k^s}$ defines the Riemann-Zeta function (Titchmarsh and Heath-Brown, 1986) for complex numbers s with $\text{Re}(s) > 1$.

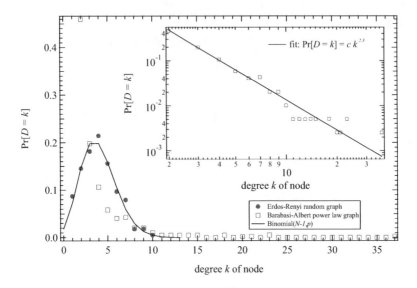

Fig. 1.5. The probability density function (pdf) of the nodal degree in the Erdős-Rényi random graph shown in Fig. 1.1 and in the Barabási-Albert power law graph in Fig. 1.2.

the Erdős-Rényi random graph with the same number of nodes and links), and even one node has 36 neighbors. A power law graph is often called a "scale-free graph", meaning that there is no typical scale for the degree. Thus, the standard deviation $\sigma_D = \sqrt{\mathrm{Var}\,[D]}$ is usually larger than the average $E\,[D]$, such that the latter is not a good estimate for the random variable D of the degree, in contrast to Gaussian or binomial distributions, where the bell-shape is centered around the mean with, usually, small variance. Physically, power law behavior can be explained by the notion of long-range dependence, heavy correlations over large spacial or temporal intervals and of self-similarity. A property is self-similar if on various scales in time or space or aggregation levels (e.g., hierarchical structuring of nodes in a network) about the same behavior is observed. The result is that a local property is magnified or scaled-up towards a global extent. Mathematically, if $\Pr\,[D = \alpha k] = c\alpha^{-\tau}k^{-\tau}$, then $\Pr\,[\alpha^{-1}D = k] = \alpha^{-\tau}\Pr\,[D = k]$: scaling a property – here, the degree D – by a factor α^{-1} leads to precisely the same distribution, apart from a proportionality constant $\alpha^{-\tau}$. Thus, on different scales, the behavior "looks" similar.

There is also a large number of more dedicated classes, such as Ramanujan graphs and the Kautz graphs, shown in Fig. 1.6, that possess interesting extremal properties. We will not further elaborate on the different properties of these classes; we have merely included some of them here to illustrate that complex networks are studied by comparing observed characteristics to those of "classes of graphs with known properties".

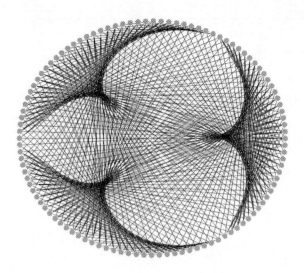

Fig. 1.6. The Kautz graph of degree $d = 3$ and of dimension $n = 3$ has $(d + 1) \, d^n$ nodes and $(d + 1) \, d^{n+1}$ links. The Kautz graph has the smallest diameter of any, possibly directed, graph with N nodes and degree d.

1.6 Outlook

I believe that we still do not understand "networks" sufficiently well. For example, if the adjacency matrix of a large graph is given, it seems quite complex to tell without visualization of the graph by computing graph metrics only, what the properties of the network are. A large number of topological metrics may be listed such as hopcount, eccentricity, diameter, girth, expansion, betweenness, distortion, degree, assortativity, coreness, clique number, clustering coefficient, vertex and edge connectivity and others. We humans see a pile of numbers, but often miss the overall picture and understanding.

The spectrum, that is for a sufficiently large graph a unique fingerprint as conjectured in van Dam and Haemers (2003), may reveal much more. First, graph or topology metrics are generally correlated and dependent. In contrast, eigenvalues weigh the importance of eigenvectors, that are all orthogonal, which makes the spectrum a more desirable device. Second, earlier research on photoluminescence spectra (Borghs *et al.*, 1989) provided useful and precise information about the structural properties of doped GaAs substrates. By inspecting carefully the differences in peaks and valleys, in gaps and in the broadness of the distribution of eigenvalues, that physically represented energy levels in the solid described by Schrödinger's equation in Section 1.3, insight gradually arose. A similar track may be followed to understand real-world networks. We hope that the mathematical properties of spectra, presented here, may help in achieving this goal.

Part I

Spectra of graphs

2

Algebraic graph theory

The elementary basics of the matrix theory for graphs $G(N, L)$ is outlined. The books by Cvetković *et al.* (1995) and Biggs (1996) are standard works on algebraic graph theory.

2.1 Graph related matrices

1. *Adjacency matrix A.* The adjacency matrix A of a graph G with N nodes is an $N \times N$ matrix with elements $a_{ij} = 1$ only if the pair of nodes (i, j) is connected by a link l of G, otherwise $a_{ij} = 0$. If the graph is undirected, the existence of the link l implies that $a_{ij} = a_{ji}$ and the adjacency matrix $A = A^T$ is a symmetric, zero-one matrix. It is assumed further in this book that the graph G does not contain self-loops ($a_{ii} = 0$) nor multiple links between two nodes. Graphs without self-loops and without multiple links between two nodes are called *simple*.

The complement G^c of the graph G consists of the same set of nodes but with a link l between (i, j) if there is *no* link $l = (i, j)$ in G and vice versa. Thus, $(G^c)^c = G$ and the adjacency matrix A^c of the complement G^c is $A^c = J - I - A$, where J is the all-one matrix ($(J)_{ij} = 1$) and I is the identity matrix. The links in

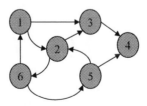

Fig. 2.1. A directed graph with $N = 6$ and $L = 9$. The links are lexicographically ordered, $l_1 = 1 \rightarrow 2, l_2 = 1 \rightarrow 3, l_3 = 1 \longleftarrow 6, l_4 = 2 \rightarrow 3$, etc.

a graph can be numbered in some way, for example, lexicographically as illustrated in Fig. 2.1. Due to different node labeling, the same graph structure can possess many different adjacency matrices (see Section 2.5 below).

15

2. *Incidence matrix B.* Information about the direction of the links is specified by the *incidence matrix B*, an $N \times L$ matrix with elements

$$b_{il} = \begin{cases} 1 & \text{if link } l = i \longrightarrow j \\ -1 & \text{if link } l = i \longleftarrow j \\ 0 & \text{otherwise} \end{cases}$$

If e_k is the k-th $N \times 1$ basic vector of the N-dimensional space with $(e_k)_j = 1$ if $k = j$ and otherwise $(e_k)_j = 0$, then the l-th column vector of B, associated to link $l = i \longrightarrow j$, equals $e_i - e_j$. Each column in B has only two non-zero elements. The adjacency matrix and incidence matrix of the graph in Fig. 2.1 are

$$A = \begin{bmatrix} 0 & 1 & 1 & 0 & 0 & 1 \\ 1 & 0 & 1 & 0 & 1 & 1 \\ 1 & 1 & 0 & 1 & 0 & 0 \\ 0 & 0 & 1 & 0 & 1 & 0 \\ 0 & 1 & 0 & 1 & 0 & 1 \\ 1 & 1 & 0 & 0 & 1 & 0 \end{bmatrix}, \quad B = \begin{bmatrix} 1 & 1 & -1 & 0 & 0 & 0 & 0 & 0 & 0 \\ -1 & 0 & 0 & 1 & -1 & 1 & 0 & 0 & 0 \\ 0 & -1 & 0 & -1 & 0 & 0 & 1 & 0 & 0 \\ 0 & 0 & 0 & 0 & 0 & 0 & -1 & -1 & 0 \\ 0 & 0 & 0 & 0 & 1 & 0 & 0 & 1 & -1 \\ 0 & 0 & 1 & 0 & 0 & -1 & 0 & 0 & 1 \end{bmatrix}$$

An important property of the incidence matrix B is that the sum of each column equals zero,

$$u^T B = 0 \tag{2.1}$$

where $u = (1, 1, \ldots, 1)$ is the all-one vector, also written as an $N \times 1$ matrix $u = \begin{bmatrix} 1 & 1 & \cdots & 1 \end{bmatrix}^T$.

An undirected graph can be represented by an $N \times (2L)$ incidence matrix B, where each link (i, j) is counted twice, once for the direction $i \to j$ and once for the direction $j \to i$. In that case, the degree of each node is just doubled. A link $l = (i, j)$ between node i and j in an undirected graph is also denoted as $l = i \sim j$ or $l = i \leftrightarrows j$. Instead of using the incidence matrix, the unsigned incidence matrix R, defined in art. 25, is more appropriate for an undirected graph.

3. *Degree of a node.* By the definition of the adjacency matrix A, the row sum i of A equals the degree d_i of node i,

$$d_i = \sum_{k=1}^{N} a_{ik} \tag{2.2}$$

A neighbor j of a node i is a node in the graph G connected by a link to node i, thus obeying $a_{ij} = 1$. The degree d_i is the number of neighbors of node i and $0 \leq d_i \leq N - 1$. However, only $N - 1$ degree values are possible in a simple graph, because the existence of $d_k = 0$ for some node k excludes the existence of a degree equal to $N - 1$ and vice versa. Consequently, in any graph G with N nodes, there are at least two nodes with the same degree.

Since $\sum_{i=1}^{N} \sum_{k=1}^{N} a_{ik} = 2L$, where L is the number of links in the graph G, the basic law for the degree follows as

$$\sum_{i=1}^{N} d_i = 2L \tag{2.3}$$

Probabilistically, when considering an arbitrary nodal degree[1] D, the basic law for the degree becomes

$$E[D] = \frac{2L}{N}$$

meaning that the average degree or expectation of D in a graph G is twice the ratio of the number L of links over the number N of nodes. Especially in large real-world networks, a probabilistic approach is adequate as illustrated in Chapter 8.

The basic law of the degree (2.3) implies that any graph G possesses an *even* (possibly zero) number of nodes with *odd* degree. Indeed, the sum in (2.3) can be split over nodes with even and odd degree so that

$$\sum_{i=1}^{N} d_i^{(o)} = 2L - \sum_{i=1}^{N} d_i^{(e)}$$

where $d_i^{(o)}$ is an odd integer if the degree of node i is odd, otherwise $d_i^{(o)} = 0$ (and similarly for the even degree $d_i^{(e)}$). The right-hand side is always even, which implies that each simple graph must contain an even number of odd degree nodes.

Let us define the degree vector $d = \begin{bmatrix} d_1 & d_2 & \cdots & d_N \end{bmatrix}^T$, then both (2.2) and (2.3) have a compact vector presentation as

$$Au = d \tag{2.4}$$

and

$$u^T A u = u^T d = d^T u = 2L \tag{2.5}$$

For a directed graph, the in-degree d_i^{in} and out-degree d_i^{out} of node i are defined as the number of links entering and leaving, respectively, node i. From the incidence matrix B, the number of "1" elements in row i equals d_i^{out}, while the number of "-1" elements in row i equals d_i^{in}. From an asymmetric adjacency matrix A (where $a_{ij} = 1$ only if there is link from node $i \longrightarrow j$, otherwise $a_{ij} = 0$), we find that

$$Au = d^{out} \text{ and } u^T A = \left(d^{in}\right)^T$$

If A is symmetric, then $u^T A = u^T A^T = (Au)^T$ and $d^{out} = d^{in} = d$.

4. *Laplacian matrix* Q. The relation between adjacency and incidence matrix is given by the *admittance* matrix or *Laplacian* Q,

$$Q = BB^T = \Delta - A \tag{2.6}$$

where $\Delta = \mathrm{diag}(d_1, d_2, \ldots, d_N)$ is the degree matrix. Indeed, if $i \neq j$ and recalling that each column in the incidence matrix B has precisely two non-zero elements,

$$q_{ij} = \left(BB^T\right)_{ij} = \sum_{k=1}^{L} b_{ik} b_{jk} = \begin{cases} -1 & \text{if } (i, j) \text{ are linked} \\ 0 & \text{if } (i, j) \text{ are not linked} \end{cases}$$

---

[1] The random variable D of the degree in a graph G is equal to one of the possible realizations or outcomes $d_1, d_2, \ldots, d_N$ of the degrees in G.

from which the "link decomposition" of the Laplacian, derived in (4.5), follows as

$$Q = \sum_{(i,j)\in\mathcal{L}} (e_i - e_j)(e_i - e_j)^T$$

If $i = j$, then $\sum_{k=1}^{L} b_{ik}^2 = d_i$ in (2.6) is the number of links that have node i in common. If self-loops are allowed in a graph, then the right-hand side of definition (2.6) shows that self-loops do not influence the Laplacian Q.

The basic property $u^T B = 0$ in (2.1) of the incidence matrix B leads in (2.6) to

$$Qu = 0$$

Consequently, each row sum $\sum_{j=1}^{N} q_{ij} = 0$, which shows that Q is singular, implying that $\det Q = 0$.

Since BB^T is symmetric, so is Q and A. Hence, although the incidence matrix B specifies the direction of links in the graph, (2.6) loses information about directions and A in (2.6) only reflects the existence of links between a pair of nodes, corresponding to an undirected graph. Consequently, if A is asymmetric and specifies, just like B, the direction of links in a directed graph, then (2.6) does not hold. Moreover, the asymmetric matrix $\Delta - A$ does not define an asymmetric Laplacian, because the row sum of $\Delta - A$ is not everywhere zero. By replacing the degree in Δ by the in-degree or out-degree, either the column sum or the row sum of $\Delta - A$ is zero, so that we may define two different asymmetric "Laplacian" matrices. The arguments illustrate that, generally, directed graphs possess less elegant properties[2] than undirected graphs and give rise to a more complicated analysis.

The Laplace matrix Q can be viewed as a discrete operator acting on a vector. The relation with its continuous counterpart, the Laplacian differential operator, is explained by Merris (1994) for a lattice graph.

5. *Matrices of weighted graphs.* Weighted graphs often appear in practice, where a link between node i and node j in the graph G is specified by one or more real numbers that reflect e.g. a delay, a monetary cost when using the link, the energy needed when traveling over that link, a performance loss, a geographic distance, a quality of service metric in telecommunication networks, like packet loss, jitter, etc.. We call any such real number, that specifies a link characteristic, a weight w_{ij} of the link between node i and j and the $N \times N$ weighted matrix W represents the weights between all pairs (i, j) of adjacent nodes. In most cases, analyses are limited

---

[2] Perhaps the major disadvantage of directed graphs is that the eigenvalues are not necessarily real (since **art. 247** does not apply). Even worse, the asymmetric adjacency matrix A may not be diagonalizable and may possess a Jordan canonical form (**art. 239**).

From a physical point of view, flows in networks (**art. 14**) can propagate in either direction, depending on the driving force or potential difference; the incidence matrix B specifies the direction of the flow in the link, while the adjacency matrix $A = A^T$ determines the existence of a link. If the adjaceny matrix is asymmetric, then some links only allow propagation of flows in one direction and forbid the flow in the other direction. Physically, such an asymmetric situation requires non-linear elements (such as diodes in an electrical network or water tubes with directional shutters), which seriously complicate "linear" theory. Nevertheless, asymmetry naturally occurs in www-links, social relations and the Markov graph of a Markov process.

to one link weight, but multiple-parameter routing explained in Van Mieghem and Kuipers (2004) is an example where each entry in the matrix W is a vector, rather than a single real number. The *link weight structure*, the set of all link weights of graph G, is usually specified by a process or a function on the network, so that link weight w_{ij} may depend upon link weight w_{kl}. Since a process on a graph typically introduces directions, W is generally not a symmetric matrix.

We will denote graph matrices of a weighted graph by a tilde to distinguish them from graph matrices of the unweighted graph. For example, the element $\tilde{a}_{ij}$ of the weighted adjacency matrix $\tilde{A}$ represents the weight w_{ij} of a link between node i and j and $\tilde{a}_{jj} = 0$ for all $1 \le j \le N$. Using the Hadamard[3] product $\circ$, the weighted adjacency matrix $\tilde{A}$ equals $\tilde{A} = W \circ A$, where $\tilde{a}_{ij} = w_{ij} a_{ij}$ and a_{ij} is an element of the adjacency matrix A. Hence, the unweighted case can be regarded as a special case where the weighted matrix $W = J$ is the all-one matrix.

A particular class of weighted graphs are undirected weighted graphs, where the corresponding weighted adjacency matrix is symmetric, $\tilde{A} = \tilde{A}^T$. The weighted degree of node i is $\tilde{d}_i = \sum_{j=1}^N \tilde{a}_{ij}$, while the degree vector is $\tilde{d} = \tilde{A}u$. Similarly, the corresponding weighted Laplacian can be defined as $\tilde{Q} = \operatorname{diag}\left(\tilde{d}\right) - \tilde{A} = \tilde{\Delta} - \tilde{A}$, thus $\tilde{q}_{ij} = -\tilde{a}_{ij}$ if $i \ne j$, else, $\tilde{q}_{jj} = -\sum_{i=1; i \ne j}^N \tilde{q}_{ji}$ and $\tilde{Q} = \tilde{Q}^T$.

6. *Walk, path and cycle.* A *walk* of length k from node i to node j is a succession of k links (arcs) or k hops of the form $(r_0 \to r_1)(r_1 \to r_2) \cdots (r_{k-1} \to r_k)$, where node label $r_0 = i$ and $r_k = j$. A *closed walk* of length k is a walk that starts in node $r_0 = i$ and returns, after k hops, to that same node $r_k = i$. A *path* is a walk in which all nodes are different, i.e. $r_l \ne r_m$ for all $0 \le l \ne m \le k$. A *cycle* of length k is a closed walk with different intermediate nodes, i.e. $r_l \ne r_m$ for all $0 \le l \ne m < k$. For an undirected walk, path or cycle, we replace the directed link $r_i \to r_j$ by the undirected link $r_i \sim r_j$. An Eulerian walk (circuit) is a closed walk containing each link of the graph G once, while a Hamiltonian cycle contains each node of G exactly once.

7. *A shortest path.* We consider only additive link weights such that the weight of a path $\mathcal{P}$ is $w(\mathcal{P}) = \sum_{l \in \mathcal{P}} w_l$, i.e., $w(P)$ equals the sum of the weights of the constituent links of the path $\mathcal{P}$. The shortest path $\mathcal{P}^*_{a \to b}$ from node a to node b is the path with minimal weight, thus, $w\left(\mathcal{P}^*_{a \to b}\right) \le w\left(\mathcal{P}_{a \to b}\right)$ for all paths $\mathcal{P}_{a \to b}$. The shortest path weight matrix S has elements $s_{ij} = w\left(\mathcal{P}^*_{i \to j}\right)$. If all link weights are equal to $w_{ij} = 1$ as in an unweighted graph, shortest paths are shortest hop paths and $w\left(P^*_{i \to j}\right) = h_{ij}$ is the hopcount, i.e. the length in hops or links of the shortest path between node i and node j, also called the *distance* between nodes i and j, or sometimes, the length of $\mathcal{P}^*_{i \to j}$. In weighted graphs, the hopcount h_{ij} is generally different from the weight $s_{ij} = w\left(\mathcal{P}^*_{ij}\right)$ of a shortest path.

In man-made infrastructures, two major types of transport exist: either a packet

---

[3] The Hadamard product (Horn and Johnson, 1991) is the entrywise product of two matrices: $(A \circ B)_{ij} = A_{ij} B_{ij}$. If A and B are both diagonal matrices, then $A.B = A \circ B$.

(e.g. car, parcel, IP-packet, container) or a flow (e.g. electric current, water, gas). Transport is either flow-based or path-based. Packets follow a single path from source to destination, whereas a flow spreads over all possible paths. Generally, packets in a weighted network follow shortest paths. The flow analogon of the shortest path weight matrix S is the effective resistance matrix Ω in Chapter 5.

There exist many routing algorithms to compute shortest paths in networks. The most important of these routing algorithms are explained, for example, in Van Mieghem (2010) and Cormen *et al.* (1991).

8. *Graph matrices and distance matrices.* Many other graph-related matrices, in short graph matrices, can be defined and we mention only a few. The effective resistance matrix Ω is studied in Chapter 5. The modularity matrix M is defined and discussed in **art.** 151. The probability transfer matrix $P = \Delta^{-1}A$ of a random walk on a graph is a stochastic matrix, because all elements of P lie in the interval $[0, 1]$ and each row sum is 1. Graph matrices can be scaled or normalized, e.g., normalized Laplacians are $\Delta^{-1}Q$ or $\Delta^{-\frac{1}{2}}Q\Delta^{-\frac{1}{2}}$.

A *distance matrix* D is a non-negative matrix, where element d_{ij} specifies a distance measure between node i and j in a graph. For example, if the distance measure is equal to the hopcount h_{ij}, then $h_{ii} = 0$. Thus, distance matrices possess a zero diagonal and contain the distances between each pair (i, j) of nodes in a graph. Any element of a distance matrix obeys the triangle inequality (**art.** 201): $0 \leq d_{ij} \leq d_{il} + d_{lj}$. The spectrum of distance matrices is reviewed by Aouchiche and Hansen (2014). Both H, S and Ω are distance matrices.

The hopcount matrix H of the directed graph in Fig. 2.1,

$$H = \begin{bmatrix} 0 & 1 & 1 & 2 & 3 & 2 \\ 2 & 0 & 1 & 2 & 2 & 1 \\ \times & \times & 0 & 1 & \times & \times \\ \times & \times & \times & 0 & \times & \times \\ 3 & 1 & 2 & 1 & 0 & 2 \\ 1 & 2 & 2 & 2 & 1 & 0 \end{bmatrix}$$

illustrates asymmetry in directed graphs as well as the possibility of the non-existence, marked by $\times$ in the above matrix, of a path between two nodes, although the graph is connected. For these reasons, we usually confine to undirected, connected graphs. Since $h_{ij} = h_{ji}$ in an undirected, connected graph, the corresponding distance matrix H is symmetric, with positive integer off-diagonal elements and with zero elements on the diagonal.

2.2 The incidence matrix B

The $N \times L$ incidence matrix B in **art.** 2 transforms an $L \times 1$ vector y of the "link"-space to an $N \times 1$ vector x of the "nodal" space by $x = By$. Physically, this transformation is best understood when y is a flow or current vector through links in a network, while x is the externally injected current in nodes of the graph G as discussed in **art**. 14 below. We first concentrate on mathematical properties of the incidence matrix B.

9. *Rank of the incidence matrix* B.

Theorem 1 *If the graph G is connected, then rank$(B) = N - 1$.*

Proof: The basic property $u^T B = 0$ in (2.1) implies that rank$(B) \leq N - 1$. Suppose that there exists a non-zero vector $x \neq \alpha u$ for any real number α such that $x^T B = 0$. Under that assumption, the vector u and x are independent and the kernel (or zero space of B) consisting of all vectors v such that $v^T B = 0$ has at least rank 2, and consequently rank$(B) \leq N - 2$. We will show that x is not independent, but proportional to u. Consider row j in B corresponding to the non-zero component x_j. All non-zero elements in the row vector $(B)_j$ are links incident to node j. Since each column of B only consists of two elements (with opposite signs), for each link l incident to node j, there is precisely one other row k in B with a non-zero element in column l. In order for the linear relation $x^T B = 0$ to hold, we thus conclude that $x_j = x_k$, and this observation holds for all nodal indices j and k because G is connected. This implies that $x^T B = \alpha u^T B$, which shows that the rank of the incidence matrix cannot be lower than $N - 1$. $\qquad\square$

An immediate consequence is that rank$(B) = N - k$ if the graph has k disjoint but connected components, because then (see also **art**. 116) there exists a relabeling of the nodes such that B can be partitioned as

$$
B = \begin{bmatrix} B_1 & O & \cdots & O \\ O & B_2 & & \vdots \\ \vdots & & \ddots & \\ O & & \cdots & B_k \end{bmatrix}
$$

10. *The cycle-space and cut-space of a graph G.* The cycle-space of a graph G consists of all possible cycles in that graph. A cycle (**art**. 6) can have two cycle orientations. This means that the orientation of links in a cycle either coincides with the cycle orientation or that it is the reverse of the cycle orientation. For example, the cycle $(1 - 2)(2 - 6)(6 - 1)$ in Fig. 2.1 corresponds to the links (columns in B) $1, 6$ and 3 and all links are oriented in the same direction along the cycle. When adding columns $1, 3$ and 6, the sum is zero, which is equivalent to $By = 0$ with $y = (1, 0, 1, 0, 0, 1, 0, 0, 0)$. On the other hand, the triplet $(1 - 2)(2 - 3)(3 - 1)$, corresponding to the links $1, 4$ and 2, is not a cycle, because not all links are oriented in the same direction such that $y = (1, -1, 0, 1, 0, 0, 0, 0, 0)$ has now negative sign components.

In general, if $By = 0$, then the non-zero components of the vector y are links of a cycle. Indeed, consider the j-th row $(By)_j = x_j$. If node j is not incident with links of the cycle, then $x_j = 0$. If node j is incident with some links of the cycle, then it is incident with precisely two links, with opposite sign such that x_j is again zero.

Since the rank of B is $N - k$, where k is the number of connected components, the rank of the kernel (or null space) of B is $L - N + k$. Hence, the dimension of the cycle-space of a graph equals the rank of the kernel of B, which is $L - N + k$. The orthogonal complement of the cycle-space is called the cut-space, with dimension $N - k$. Thus, the cut-space is the space consisting of all vectors y for which $By = x \neq 0$. Since $u^T x = 0$ by (2.1), the non-negative components of x are the nodes belonging to one partition and the negative components define the other partition. These two disjoint sets of nodes thus define a cut in the graph, a set of links whose removal separates the graph G in two disjoint subgraphs. For example in Fig. 2.1, $By = \begin{bmatrix} 1 & 0 & -1 & -2 & 1 & 1 \end{bmatrix}$ defines a cut that separates nodes 3 and 4 from the rest. Section 4.4 further investigates the partitioning of a graph.

11. *Cycles and cuts in a connected graph G.* A spanning tree T in the graph G is a connected subgraph of G that contains all N nodes of G. Any tree on N nodes has $N - 1$ links, whose set is denoted by $\mathcal{T} \subset \mathcal{L}$, and a tree does not contain a cycle.

The definition of a spanning tree $\mathcal{T}$ of the graph G leads to an interesting property: If a link $l \in \mathcal{L}$, but $l \notin \mathcal{T}$, is added to the spanning tree T, then there is a unique cycle in the graph $\mathcal{T} \cup \{l\}$. Indeed, let l be a link between node i and j. Since l does not belong to the spanning tree T, the nodes i and j are not directly connected, but there is a path from node i to node j in spanning tree T, because T is connected. The addition of l to T results in two different paths from node i to node j. By the definition of a cycle, the graph $\mathcal{T} \cup \{l\}$ contains one cycle $cyc\,(T, l)$, which is unique by construction and to which we can associate a vector y_l obeying $By_l = 0$ by **art. 10**. The length of that cycle contains at most N links, because the longest shortest path in the spanning tree has at most $N - 1$ links.

The companion property is: if a link $h \in \mathcal{T}$ (clearly, $h \in \mathcal{L}$) is removed, then there is a unique cut $cut\,(T, h)$, that contains link h and links $e \in \mathcal{L}$, but $e \notin \mathcal{T}$. Similarly, we can associate a vector y_h to the cut $cut\,(T, h)$ that obeys $By_h \neq 0$.

Since there are $L - N + 1$ links of G that do not belong to the spanning tree T, we can construct $L - N + 1$ cycles and the set of cycles $\{cyc\,(T, l)\}_{l \in \mathcal{L} \backslash \mathcal{T}}$ forms an independent set, because a link l belongs to a cycle $cyc\,(T, l)$, but not to another cycle $cyc\,(T, g)$ for $g \neq l$. Moreover, $L - N + 1$ is the dimension of the cycle-space of G (**art. 10**) and the set of vectors y_l, obeying $By_l = 0$, for $l \in \mathcal{L} \backslash \mathcal{T}$ represents a basis for the cycle-subspace of G. Analogously, the set of cuts $\{cut\,(T, h)\}_{h \in \mathcal{T}}$ with associated set of vectors y_h, obeying $By_h \neq 0$, for $h \in \mathcal{T}$ represents a basis for the cut-subspace of G.

12. *Spanning trees and the incidence matrix B.* Consider the incidence matrix B of a graph G and remove an arbitrary row in B, corresponding to a node n. Let M_n be one of the $\binom{L}{N-1}$ square $(N - 1) \times (N - 1)$ submatrices of B without row n and let G_n denote the subgraph of G on $N - 1$ nodes formed by the links in the columns of M_n. Since there are $N - 1$ columns in M_n, the subgraph G_n has precisely $N - 1$ links, where some links may start or end at node n, outside the node set of G_n. We will now investigate det M_n.

(a) Suppose first that there is no node with degree 1 in G, except possibly for n, in which case G_n is not a tree spanning $N - 1$ nodes. Since the number of links is $L(G_n) = N - 1$, the basic law of the degree (2.3) shows that there must be a zero degree node in G_n. If the zero degree node is not n, then G_n has a zero row and det $M_n = 0$. If n is the zero degree node, then each column of M_n contains a 1 and -1. Thus, each row sum of M_n is zero and det $M_n = 0$.

(b) In the other case, G_n has a node i with degree 1. Then, the i-th row in G_n only has one non-zero element, either 1 or -1. After expanding det M_n by this i-th row, we obtain a new $(N - 2) \times (N - 2)$ determinant $M_{n;i}$ corresponding to the graph $G_{n;i}$, formed by the links in the columns of $M_{n;i}$. For det $M_{n;i}$, we can repeat the analysis: either $G_{n;i}$ is not a tree spanning the $N - 2$ nodes of G except for nodes n and i, in which case det $M_{n;i} = 0$ or det $M_{n;i} = \pm$ det $M_{n;i;k}$.

Iterating this process shows that the determinant of any square submatrix M of B is either 0, when the corresponding graph formed by the links, corresponding to the columns in M is not a spanning tree, or ± 1, when that corresponding graph is a spanning tree. Thus, we have shown:

Theorem 2 (Poincaré) *The determinant of any square submatrix of the incidence matrix B is either 0, 1, or -1.*

If the determinant of any square submatrix of a matrix is 0, 1, or -1, then that matrix is said to be *totally unimodular*. Hence, the incidence matrix B is totally unimodular.

13. *The matrix C representing cycles in G.* **Art.** 11 suggests to write the incidence matrix B of the graph G as

$$B = \begin{bmatrix} B_T & B_{G \setminus T} \\ & b_N \end{bmatrix} \tag{2.7}$$

where the $(N - 1) \times (N - 1)$ square matrix B_T has as columns the (partial[4]) links of the spanning tree T of G, the $(N - 1) \times (L - N + 1)$ matrix $B_{G \setminus T}$ contains the remaining links of G not belonging to T and the $1 \times L$ vector b_N is linearly dependent on the $N - 1$ first rows of B, because rank$(B) = N - 1$ by Theorem 1. The $L \times (L - N + 1)$ cycle matrix C, in which a column represents a cycle of G, is defined by

$$C = \begin{bmatrix} C_T \\ I_{L-N+1} \end{bmatrix}$$

where the $(N - 1) \times (L - N + 1)$ matrix C_T contains elements of the vectors y_l, obeying $By_l = 0$, for $l \in \mathcal{L} \setminus T$. The basic property $By_l = 0$ of a cycle y_l translates to

---

[4] The row N, corresponding to node N, is not included in B_T and links to or from node N in the columns of B_T only contain a 1 or -1.

the matrix equation $BC = 0$, from which $B_T C_T + B_{G \setminus T} = 0$. **Art.** 12 demonstrates that $\det B_T = \pm 1$, implying that the inverse of B_T exists, thus

$$C_T = -B_T^{-1} B_{G \setminus T} \tag{2.8}$$

Analogously for the cut-subspace of G, the $L \times (N - 1)$ matrix F whose columns contain the $N - 1$ vectors y_h, obeying $B y_h \neq 0$, for $h \in T$,

$$F = \begin{bmatrix} I_{N-1} \\ F_T \end{bmatrix}$$

Since each column of F belongs to the orthogonal complement of the cycle-subspace of G, it holds that $C^T F = 0$, from which $C_T^T + F_T = 0$ and, with (2.8),

$$F_T = -C_T^T = \left(B_T^{-1} B_{G \setminus T} \right)^T \tag{2.9}$$

In summary, the basic cycle matrix C_T in (2.8) and the basic cut matrix F_T in (2.9) can be expressed in terms of the incidence matrix B for each spanning tree T in G. The idea to concentrate on a spanning tree T of G originates from Kirchhoff (1847), who found the solution of the current-voltage relations in a resistor network in terms of T.

14. *Electrical resistor network.* The importance of the incidence matrix B and the Laplacian matrix Q of a graph G is nicely illustrated by the current-voltage relations in a resistor network. The flows of currents in a network, steered by forces created by potential differences between nodes, is an example of a linear process, where the dynamic process is proportional to the network's graph. Other examples of processes, that are "linear" in the graph, are water (or fluid or gas) networks, where water flows through pipes and the potential of a node corresponds with its height, heat diffusion in a network, where the nodal potential is its temperature, and mechanical networks where springs connect nodes and nodal displacements are related to potentials.

The $L \times 1$ flow vector y possesses a component $y_l = y_{ij} = -y_{ji}$, which denotes the electrical current flowing through the link $l = i \sim j$ from node i to node j. Kirchhoff's current law

$$x = By \tag{2.10}$$

is a conservation law. The j-th row in (2.10), $x_j = (By)_j = \sum_{k=1}^{L} B_{jk} y_k$, states that, at each node j in the network G, the current x_j leaving ($x_j \leq 0$) or entering ($x_j \geq 0$) must equal the sum of currents over links incident to j. If current $x_j \geq 0$ is injected at node j, the flow conservation at node j is also written as

$$x_j = \sum_{i \in \text{ neighbors}(j)} y_{ji} = \sum_{i=1}^{N} a_{ij} y_{ji} \tag{2.11}$$

Thus, if no current ($x_j = 0$) is injected nor leaving the node j, then the net current flow, the sum of the flows over links incident at node j, is zero. If $By = 0$, then

art. 10 shows that the non-zero components of y form a cycle. Left-multiplying both sides of $x = By$ in Kirchhoff's current law (2.10) by u^T and using (2.1) yields $u^T x = 0$, which means that the net flow, influx plus outflow, in the network is zero. Thus, $By = x$ reflects a *conservation law*: the demand x_j offered at node j in the network is balanced by the sum of currents or flows at node j and the net demand of influx and outflow to the network is zero.

Each link $l = i \sim j$ between node i and node j contains a resistor with resistance $r_l = r_{ij}$. A flow y_{ij} is said to be physical if there is an associated potential function v on the nodes of the network such that

$$v_i - v_j = r_{ij} y_{ij} \tag{2.12}$$

In electrical networks, the potential function is called the "voltage", whereas in hydraulic networks, it is called the "pressure". The relation (2.12), known as the law of Ohm, reflects that the potential difference $v_i - v_j$ generates a force that drives the current y_{ij} from node i to node j (if $v_i - v_j > 0$, else in the opposite direction) and that the potential difference is proportional to the current y_{ij}. The proportionality constant equals the resistance[5] $r_{ij} > 0$ between node i and j. For other electrical network elements such as capacitors and inductances, the relations between potential and current are more complicated than Ohm's law (2.12) and can be derived from the laws of Maxwell (see e.g. Feynman *et al.* (1963)). We rewrite Ohm's law (2.12) in terms of the current $y_{ij} = \frac{1}{r_{ij}}(v_i - v_j)$ flowing through the link $l = (i, j)$, which becomes in matrix form

$$y_{L \times 1} = \text{diag}\left(\frac{1}{r_{ij}}\right)_{L \times L} \left(B^T\right)_{L \times N} v_{N \times 1} \tag{2.13}$$

where the $N \times 1$ vector v contains as elements the voltage v_j at each node j in G and $\text{diag}\left(\frac{1}{r_{ij}}\right)$ has diagonal elements $\left(\frac{1}{r_1}, \ldots, \frac{1}{r_l}, \ldots, \frac{1}{r_L}\right)$ where $r_l = r_{ij}$ is the resistance of link $l = (i, j)$. Substituting Ohm's law (2.13) into Kirchhoff's conservation law (2.10) yields

$$x = B\text{diag}\left(\frac{1}{r_{ij}}\right) B^T v$$

Similar to the unweighted Laplacian decomposition $Q = BB^T$ in (2.6), we define the $N \times N$ weighted, symmetric Laplacian matrix[6]

$$\widetilde{Q} = B\text{diag}\left(\frac{1}{r_{ij}}\right) B^T \tag{2.14}$$

---

[5] If $r_{ij} = 0$, then the potential v_i of node i and v_j of node j are the same by Ohm's law (2.12). From an electrical point of view, both nodes cannot be differentiated and we can merge node i and j in the graph to one node. Therefore, we further assume that $r_{ij} > 0$ in the graph G.

[6] Since $r_{ij} > 0$, we can write $\widetilde{Q} = B\text{diag}\left(\frac{1}{r_{ij}}\right) B^T = B\text{diag}\left(\frac{1}{\sqrt{r_{ij}}}\right)\left(B\text{diag}\left(\frac{1}{\sqrt{r_{ij}}}\right)\right)^T$ and may consider the $N \times L$ matrix $\widetilde{B} = B\text{diag}\left(\frac{1}{\sqrt{r_{ij}}}\right)$ as a "weighted incidence" matrix and the unit of the element $\widetilde{B}_{ij}$ is $\frac{1}{\sqrt{\text{Ohm}}}$. The law of Ohm in (2.13) transforms to $y = \text{diag}\left(\frac{1}{\sqrt{r_{ij}}}\right)\widetilde{B}^T v$, so that $\widetilde{B}$ apparently lacks a physical interpretation.

The weighted Laplacian $\widetilde{Q}$ also generalizes the definition (2.6) of the Laplacian $Q = \Delta - A$ to $\widetilde{Q} = \widetilde{\Delta} - \widetilde{A}$, where the $N \times N$ weighted, symmetric adjacency matrix $\widetilde{A}$ with elements $\widetilde{a}_{ij} = \frac{a_{ij}}{r_{ij}}$ possesses a corresponding weighted degree diagonal matrix $\widetilde{\Delta} = \text{diag}\left(\widetilde{d}_1, \widetilde{d}_2, \ldots, \widetilde{d}_N\right)$ with $\widetilde{d}_j = \left(\widetilde{A}u\right)_j$ introduced in **art.** 8. Alternatively, substitution of Ohm's law $y_{ji} = \frac{1}{r_{ij}}(v_j - v_i)$ into the nodal conservation law (2.11) for node j yields

$$x_j = \sum_{i=1}^{N} \frac{a_{ij}}{r_{ij}}(v_j - v_i) = v_j \sum_{i=1}^{N} \frac{a_{ij}}{r_{ij}} - \sum_{i=1}^{N} \frac{a_{ij}}{r_{ij}}v_i = v_j \sum_{i=1}^{N} \widetilde{a}_{ij} - \sum_{i=1}^{N} \widetilde{a}_{ij}v_i$$

which is, in matrix form, $x = \left(\widetilde{\Delta} - \widetilde{A}\right)v = \widetilde{Q}v$, where the weighted degree is $\widetilde{d}_j = \sum_{j=1}^{N} \widetilde{a}_{ij}$. While link $l = i \sim j$ contains a resistor with resistance $r_l = r_{ij}$, the link weight is $w_l = w_{ij} = \frac{1}{r_l}$.

In summary, we arrive at the fundamental relation between the $N \times 1$ injected current flow vector x into nodes of the network and the $N \times 1$ voltage vector v at the nodes

$$x = \widetilde{Q}v \tag{2.15}$$

Clearly, if all resistances equal $r_{ij} = 1$ Ohm, then the unweighted case with the standard matrices A, B and Q is retrieved. Most properties transfer to the weighted graph related matrices: the weighted Laplacian $\widetilde{Q} = B\text{diag}\left(\frac{1}{r_{ij}}\right)B^T = \widetilde{B}\widetilde{B}^T$ is positive semidefinite (as follows from **art.** 101) and the conservation of total injected flows $u^T x = u^T \widetilde{Q} v = 0$, due to the basic property (2.1) of the incidence matrix B. The power, the energy per unit time (in watts), dissipated in a resistor network is the sum of power dissipated in each resistor, which equals $\mathcal{P} = v^T x$. The fundamental relation (2.15) leads to the quadratic form $\mathcal{P} = v^T \widetilde{Q}v = \sum_{l \in \mathcal{L}} \left(\frac{v_{l+} - v_{l-}}{\sqrt{r_l}}\right)^2$, which will allow us in **art.** 103 to relate the power $\mathcal{P}$ to eigenvalues of the weighted Laplacian $\widetilde{Q}$.

15. *Harmonic functions.* The continuous description of $x = \widetilde{Q}v$ in (2.15) is the Poisson equation $\nabla^2 \phi(r) = -\frac{\rho(r)}{\epsilon_0}$, where the potential $\phi(r)$ is a continuous function of the position $r = (r_1, r_2, \ldots, r_m)$ of a point in an m-dimensional space, the Laplace operator is $\nabla^2 = \frac{\partial^2}{\partial r_1^2} + \frac{\partial^2}{\partial r_2^2} + \cdots + \frac{\partial^2}{\partial r_m^2}$, the charge density $\rho(r)$ specifies the location of electrical charges and the permittivity constant ϵ_0 balances the physical units at the left- and right-hand side. The Poisson equation is related to Gauss's divergence law of the electrical field, that appears as the first Maxwell equation (see, e.g., Feynman *et al.* (1963), Morse and Feshbach (1978)). If the potential $\phi(r)$ is defined at some boundary or surface S that encloses a volume without charges inside, then $\nabla^2 \phi(r) = 0$ for $r \notin S$ and the solution $\phi(r)$ of the Laplace differential equation is called a harmonic function. Harmonic functions possess many nice properties and are the fundamental corner stone, via the Riemann-Cauchy equations, of analytic functions in the complex plane (Titchmarsh, 1964). In the discrete setting, the

Laplace operator ∇^2 in a continuous space is replaced by a Laplacian matrix $\widetilde{Q}$ on a graph and this powerful association results in more properties of and deeper insight in the Laplacian than the adjacency matrix.

If the current x is injected in some nodes $\mathcal{S} \subset \mathcal{N}$, equivalent with the boundary S, while $x_j = 0$ if $j \notin \mathcal{S}$, then $\left(\widetilde{Q}v\right)_j = 0$ and $v_j = \frac{1}{d_j}\sum_{i=1}^N \widetilde{a}_{ij}v_i$ is a weighted average of the potential of its direct neighbors. The voltage vector v in $x = \widetilde{Q}v$ is called a harmonic at node j if $\left(\widetilde{Q}v\right)_j = 0$. Similar to the continuous setting, known as Dirichlet's boundary problem, Doyle and Snell (1984) prove that a harmonic function $v(j)$, defined on the nodes $j \in \mathcal{N}$ of the graph, achieves its maximum and minimum value at the boundary S. This important property of harmonic functions follows physically from the voltages as potentials in electrical networks (see also Section 5.3.2).

If $x = 0$, then (2.15) indicates that $\widetilde{Q}v = 0$, which is an eigenvalue equation. If the graph G is connected (see **art.** 116), the (weighted) Laplacian has one zero eigenvalue belonging to eigenvector proportional to the all-one vector u, so that the potential or voltage vector $v = \alpha u$, for a non-zero real α. The law of Ohm (2.13) and the basic property (2.1) of the incidence matrix B then show that $y = 0$, thus all currents are zero. Another consequence of the basic property (2.1) of the incidence matrix B is that $\det \widetilde{Q} = 0$ and that the general relation (2.15) *cannot* be directly inverted as $v = \widetilde{Q}^{-1}x$. In Section 4.2, the inversion problem is analyzed and a general method based on the pseudoinverse $\widetilde{Q}^\dagger$ of the Laplacian matrix $\widetilde{Q}$ is presented.

16. *Electrical resistor network revisited.* Kirchhoff (1847) considered a variant of the setting in **art.** 14, where the external current vector x is replaced by an external voltage difference vector δv_{ext} over links of G. The law of Ohm in (2.13) becomes $\delta v = \text{diag}(r_{ij})\, y + \delta v_{\text{ext}}$, where the link potential difference vector is $\delta v = B^T v$. If $By = 0$, then **art.** 10 shows that the non-zero components of y form a cycle. Kirchhoff (1847) demonstrated[7] that $C^T \delta v = 0$: the sum of the voltage differences over a cycle is zero, which is Kirchhoff's voltage law.

Considering a spanning tree T as explained in **art.** 10 and 13, we write the link current vector y and potential difference vector δv as

$$y = \begin{bmatrix} y_T \\ y_{G\backslash T} \end{bmatrix} \quad \text{and} \quad \delta v = \begin{bmatrix} \delta v_T \\ \delta v_{G\backslash T} \end{bmatrix}$$

Since there are no external currents, i.e. $x = 0$ and $By = 0$, the link current vector y with (2.7) obeys

$$\begin{bmatrix} B_T & B_{G\backslash T} \\ & b_N \end{bmatrix}\begin{bmatrix} y_T \\ y_{G\backslash T} \end{bmatrix} = B_T y_T + B_{G\backslash T}\, y_{G\backslash T} = 0$$

---

[7] More generally, if the magnetic field is time-invariant (see, e.g., Feynman *et al.* (1963)), the Maxwell equation $\nabla \times \overrightarrow{E} = 0$, where $\overrightarrow{E}$ is the electric field vector, and Stokes' theorem then state that $\oint \overrightarrow{E}\, d\overrightarrow{s} = 0$, implying that any closed contour over the electric field is zero.

and, invoking (2.8),

$$y_T = -B_T^{-1} B_{G\backslash T} y_{G\backslash T} = C_T y_{G\backslash T}$$

Thus,

$$y = \begin{bmatrix} y_T \\ y_{G\backslash T} \end{bmatrix} = \begin{bmatrix} C_T \\ I_{L-N+1} \end{bmatrix} y_{G\backslash T} = C y_{G\backslash T}$$

illustrating that the whole current vector only depends on those current vector components, associated with links that are not in the spanning tree T. Similar, $C^T \delta v = 0$ leads to $C^T \mathrm{diag}(r_{ij}) y = -C^T \delta v_{\mathrm{ext}}$. Substituting $y = C y_{G\backslash T}$ then yields

$$\left(C^T \mathrm{diag}\left(r_{ij} \right) C \right) y_{G\backslash T} = -C^T \left(\delta v \right)_{\mathrm{ext}}$$

Finally, the $(L - N + 1) \times (L - N + 1)$ matrix $C^T \mathrm{diag}(r_{ij}) C$ has rank $L - N + 1$ and is invertible,

$$y_{G\backslash T} = -\left(C^T \mathrm{diag}\left(r_{ij} \right) C \right)^{-1} C^T \delta v_{\mathrm{ext}}$$

which is Kirchhoff's solution. In fact, Kirchhoff (1847) evaluates the solution further in terms of all spanning trees, reviewed without proof by Schnakenberg (1976). Section 5.6 expresses the effective resistance in terms of spanning trees.

2.3 Connectivity, walks and paths

17. *Connectivity of a graph.* A graph G is connected if there exists a walk (**art. 6**) between each pair of nodes in G.

Theorem 3 *If a graph G is disconnected, then its complement G^c is connected.*

Proof: Since a graph G is disconnected, G possesses at least two connected components G_1 and G_2. There are two situations: (a) If node $i \in G_1$ and node $j \in G_2$, then no link in G connects them. By the definition of the complement of a graph (**art. 1**), there will be a link $i \sim j$ in G^c. (b) If node i and j are in the same connected component in G, then consider any node m in a different connected component. The argument in situation (a) shows that the link $i \sim m$ and the link $j \sim m$ exist in G^c. Consequently, i and j are connected by the path $P = i \sim m \sim j$. Combining the two possible situations demonstrates that any two nodes are reachable in G^c, implying that the graph G^c is connected. $\quad\square$

The converse of Theorem 3, "If G is connected, then its complement G^c is disconnected" is not always true. For example, if G is a tree (except for the star $K_{1,N-1}$), then G^c is connected. Section 4.1.1 gives additional properties of a graph's connectivity.

18. *The number of k-hops walks.* **Art.** 6 has defined a walk. Due to its importance, Lemma 1 is proved in two ways.

Lemma 1 *The number of walks of length k from node i to node j is equal to the element $\left(A^k\right)_{ij}$.*

Proof by induction: For $k = 1$, the number of walks of length 1 between node i and node j equals the number of direct links between i and j, which is by definition the element a_{ij} in the adjacency matrix A. Suppose the lemma holds for $k - 1$. A walk of length k consists of a walk of length $k - 1$ from i to some node r which is adjacent to j. By the induction hypothesis, the number of walks of length $k - 1$ from i to r is $\left(A^{k-1}\right)_{ir}$ and the number of walks with length 1 from r to j equals a_{rj}. The total number of walks from i to j with length k then equals $\sum_{r=1}^{N} \left(A^{k-1}\right)_{ir} a_{rj} = \left(A^k\right)_{ij}$ (by the rules of matrix multiplication). $\qquad\square$

Proof by direct computation: After q iterations in k of the matrix multiplication rule $\left(M^k\right)_{ij} = \sum_{r_{k-1}=1}^{N} \left(M^{k-1}\right)_{ir_{k-1}} m_{r_{k-1}j}$ for any matrix M, we obtain

$$\left(M^k\right)_{ij} = \sum_{r_{k-1}=1}^{N} \sum_{r_{k-2}=1}^{N} \cdots \sum_{r_{k-q}=1}^{N} \left(M^{k-q}\right)_{ir_{k-q}} m_{r_{k-q}r_{k-(q-1)}} \cdots m_{r_{k-2}r_{k-1}} m_{r_{k-1}j}$$

When $q = k - 1$, then $\left(M^{k-q}\right)_{ir_{k-q}} = m_{ir_1}$ and it holds for any matrix M that

$$\left(M^k\right)_{ij} = \sum_{r_1=1}^{N} \sum_{r_2=1}^{N} \cdots \sum_{r_{k-1}=1}^{N} m_{ir_1} m_{r_1r_2} \cdots m_{r_{k-2}r_{k-1}} m_{r_{k-1}j}$$

and applied to the adjacency matrix A,

$$\left(A^k\right)_{ij} = \sum_{r_1=1}^{N} \sum_{r_2=1}^{N} \cdots \sum_{r_{k-1}=1}^{N} a_{ir_1} a_{r_1r_2} \cdots a_{r_{k-2}r_{k-1}} a_{r_{k-1}j} \qquad (2.16)$$

With the convention $r_0 = i$ and $r_k = j$, (2.16) can be written as

$$\left(A^k\right)_{ij} = \sum_{r_1=1}^{N} \sum_{r_2=1}^{N} \cdots \sum_{r_{k-1}=1}^{N} \prod_{l=0}^{k-1} a_{r_l r_{l+1}} \qquad (2.17)$$

where the indicator function $\prod_{l=0}^{k-1} a_{r_l r_{l+1}} = a_{ir_1} a_{r_1r_2} \cdots a_{r_{k-2}r_{k-1}} a_{r_{k-1}j}$ is one if and only if all links in the walk $(i = r_0 \to r_1)(r_2 \to r_3) \cdots (r_{k-1} \to r_k = j)$ exist (i.e. $a_{r_l r_{l+1}} = 1$ for all values of l in $[0, k - 1]$), otherwise it is zero. The $(k - 1)$-fold multiple summation in the explicit expressions (2.16) and (2.17) ranges over all possible, directed walks $(i = r_0 \to r_1)(r_2 \to r_3) \cdots (r_{k-1} \to r_k = j)$ with k hops (**art.** 6) between node i and j and enumerates, out of all possible walks, the existing walks in the graph, reflected by $\prod_{l=0}^{k-1} a_{r_l r_{l+1}} = 1$. $\qquad\square$

The maximum possible number of walks with k hops between two nodes in a graph with N nodes is attained in the complete graph K_N, whose adjacency matrix is $A_{K_N} = J - I$, and equals $(J - I)_{ij}^{k}$. Invoking Newton's binomium, which is allowed because J and I commute, we have

$$(J - I)^k = \sum_{m=0}^{k} \binom{k}{m} J^m (-I)^{k-m}$$

Since $J^m = N^{m-1}J$ for $m > 0$, then $(J - I)^k = (-1)^k I + \sum_{m=1}^{k} \binom{k}{m} N^{m-1}(-1)^{k-m} J$. The binomium gives $(J - I)^k = (-1)^k I + \frac{1}{N}\left((N-1)^k - (-1)^k\right) J$, from which the maximum possible number of walks with k hops between node i and node j in any graph follows as

$$(J - I)_{ij}^k = \begin{cases} \frac{1}{N}\left((N-1)^k - (-1)^k\right) & \text{for } i \neq j \\ \frac{1}{N}\left((N-1)^k - (-1)^k\right) + (-1)^k & \text{for } i = j \end{cases} \tag{2.18}$$

19. *Lower bounds for* $\left(A^k\right)_{ij}$. For any integer $0 \leq n \leq k$, the matrix multiplication form

$$\left(A^k\right)_{ij} = \sum_{q=1}^{N} \left(A^{k-n}\right)_{iq} \left(A^n\right)_{qj} \tag{2.19}$$

reduces, for $n = 1$ and taking into account the absence of self-loops, i.e. $a_{jj} = 0$, to

$$\left(A^k\right)_{jj} = \sum_{q=1;q\neq j}^{N} \left(A^{k-1}\right)_{jq} a_{qj}$$

illustrating for each node j that $\left(A^k\right)_{jj}$ does not depend on $\left(A^{k-1}\right)_{jj}$. For $n = 2$, symmetry in the adjacency matrix, $A = A^T$, yields

$$\left(A^2\right)_{jj} = \sum_{k=1}^{N} a_{jk}a_{kj} = \sum_{k=1}^{N} a_{jk}^2 = \sum_{k=1}^{N} a_{jk} = d_j \tag{2.20}$$

The off-diagonal element $\left(A^2\right)_{ij} = \sum_{k=1}^{N} a_{ik}a_{jk}$ counts the number of nodes k that have a link to both node i and node j; i.e. the number of joint neighbors of node i and node j, so that $0 \leq \left(A^2\right)_{ij} \leq \min(d_i, d_j)$. Hence, $\left(A^2\right)_{ij}$ obeys both (A.185) and (A.186) in **art. 279**, because of the basic inequality between the arithmetic and geometric mean of two non-negative real numbers x and y: $\min(x, y) \leq \sqrt{xy} \leq \frac{x+y}{2}$. For $n = 2$ and $k > 2$ in (2.19), we find

$$\left(A^k\right)_{ij} = \left(A^{k-2}\right)_{ii}\left(A^2\right)_{ij} + \left(A^{k-2}\right)_{ij} d_j + \sum_{q=1;q\neq\{i,j\}}^{N} \left(A^{k-2}\right)_{iq}\left(A^2\right)_{qj} \quad \text{for } i \neq j$$

$$\left(A^k\right)_{jj} = \left(A^{k-2}\right)_{jj} d_j + \sum_{q=1;q\neq j}^{N} \left(A^{k-2}\right)_{jq}\left(A^2\right)_{qj} \quad \text{for } i = j$$

The last equation leads to the recursion inequality $\left(A^k\right)_{jj} \geq \left(A^{k-2}\right)_{jj} d_j$ for $k \geq 2$, that, after iteration, results for even $k = 2m$ into

$$\left(A^{2m}\right)_{jj} \geq d_j^m \tag{2.21}$$

but, for odd $k = 2m + 1$, we can only deduce $\left(A^{2m+1}\right)_{jj} \geq 0$ and equality can occur, e.g. in the path graph, studied in Section 6.4. Similarly, the first equation for $i \neq j$ when $n = 2$ leads, for $k \geq 2$, to the recursion inequality

$$\left(A^k\right)_{ij} \geq \left(A^{k-2}\right)_{ii} \left(A^2\right)_{ij} + d_j \left(A^{k-2}\right)_{ij}$$

After p iterations, we have

$$\left(A^k\right)_{ij} \geq \left(A^2\right)_{ij} \left\{\sum_{q=0}^{p} \left(A^{k-2(q+1)}\right)_{ii} d_j^q\right\} + d_j^{p+1} \left(A^{k-2(p+1)}\right)_{ij}$$

For odd $k = 2m + 1$ and $p = m - 1$, we can conclude from the lower bound

$$\left(A^{2m+1}\right)_{ij} \geq \left(A^2\right)_{ij} \left\{\sum_{q=0}^{m-1} \left(A^{2(m-(q+1)+1)}\right)_{ii} d_j^q\right\} + d_j^m a_{ij}$$

that

$$\left(A^{2m+1}\right)_{ij} \geq d_j^m a_{ij}$$

Even $k = 2m$ and $p = m - 1$ give us $\left(A^{2m}\right)_{ij} \geq \left(A^2\right)_{ij} \sum_{q=0}^{m-1} d_j^q \left(A^{2(m-1-q)}\right)_{ii}$. Invoking the lower bound (2.21) yields

$$\left(A^{2m}\right)_{ij} \geq \left(A^2\right)_{ij} \sum_{q=0}^{m-1} d_j^q d_i^{m-1-q} = \left(A^2\right)_{ij} \frac{d_j^m - d_i^m}{d_j - d_i}$$

In conclusion, the properties in the number $\left(A^k\right)_{ij}$ of walks from node i to node j with odd and even length k differ quite significantly, as will be supported by the spectral investigations in **art. 58**. The reason is that A^{2m} is a positive semidefinite matrix (**art. 278**), while A^{2m+1} is not.

20. *The number of k-hops paths.* The number of paths with k hops between node i and node j follows from (2.16) by excluding possible same nodes in the walk,

$$X_k(i, j; N) = \sum_{r_1 \neq \{i,j\}} \sum_{r_2 \neq \{i,r_1,j\}} \cdots \sum_{r_{k-1} \neq \{i,r_1,\ldots,r_{k-2},j\}} a_{ir_1} a_{r_1 r_2} \cdots a_{r_{k-1}j}$$

valid for $k > 1$ and $N > 2$, while the number of paths with $k = 1$ hop between the node pair (i, j) is $X_1(i, j; N) = a_{ij}$. Symmetry of the adjacency matrix A implies that $X_k(i, j; N) = X_k(j, i; N)$. The definition of a path restricts the first index r_1 to $N - 2$ possible values, the second r_2 to $N - 3$, etc., such that the maximum number of k-hop paths, which is attained in the complete graph K_N, where $a_{ij} = 1$ for each link (i, j), equals

$$\prod_{l=1}^{k-1} (N - 1 - l) = \frac{(N-2)!}{(N-k-1)!}$$

whereas the total possible number of walks with k hops is given in (2.18). If we allow self-loops ($a_{jj} \neq 0$), then (2.16) with $\prod_{l=0}^{k-1} a_{r_l r_{l+1}} = 1$ leads to the total possible number of walks with k hops equal to N^{k-1}.

The total number M_N of paths between two nodes in the complete graph is

$$M_N = \sum_{j=1}^{N-1} \frac{(N-2)!}{(N-j-1)!} = (N-2)! \sum_{k=0}^{N-2} \frac{1}{k!} = (N-2)!e - R$$

where the remainder

$$R = (N-2)! \sum_{j=N-1}^{\infty} \frac{1}{j!} = \sum_{j=0}^{\infty} \frac{(N-2)!}{(N-1+j)!}$$

$$= \frac{1}{N-1} + \frac{1}{(N-1)N} + \frac{1}{(N-1)N(N+1)} + \cdots$$

$$< \sum_{j=1}^{\infty} \left(\frac{1}{N-1}\right)^j = \frac{1}{N-2}$$

implying that for $N \geq 3$, the remainder $R < 1$. But M_N is an integer. Hence, the total number of paths in K_N is exactly equal to

$$M_N = [e(N-2)!] \tag{2.22}$$

where $e = 2.718\,281...$ and $[x]$ denotes the largest integer smaller than or equal to x. Since any graph is a subgraph of the complete graph, the maximum total number of paths between two nodes in any graph is upper bounded by $[e(N-2)!]$.

21. *Hopcount h_{ij} in a connected graph.* A graph G is connected if there exists a walk between each pair of nodes in G. Lemma 1 shows that connectivity is equivalent to the existence of some integer $k > 0$ for which $\left(A^k\right)_{ij} \neq 0$ for each nodal pair (i,j). The lowest integer $k = h_{ij}$, where $i \neq j$, for which $\left(A^k\right)_{ij} \neq 0$, but $\left(A^m\right)_{ij} = 0$, for all $0 \leq m < k$, equals the number of hops in the shortest walk – which is then a path – from node i to node j. Thus, for $i \neq j$, the vector $\left(A_{ij}, \left(A^2\right)_{ij}, \ldots, \left(A^{k-1}\right)_{ij}, \left(A^k\right)_{ij}\right)$ with $k = h_{ij}$ components equals $\left(A^k\right)_{ij} e_k$, where e_k is the k-th basic vector of the k-th dimensional space. If $i = j$, then we define the hopcount of the shortest path to be $h_{ii} = 0$. Hence, the element h_{ij} in the distance matrix H, defined in **art.** 8, equals $h_{ij} = k1_{\left\{\min_k : (A^k)_{ij} \neq 0\right\}}$ for $i \neq j$ and $h_{ii} = 0$. The hopcount h_{ij} of the shortest path P_{ij}^* between node i and node j is a unique integer, although there can be multiple shortest paths between node i and node j, so that $\left(A^{h_{ij}}\right)_{ij} \geq 1$.

Each off-diagonal $(i \neq j)$ element in the hopcount matrix H obeys

$$h_{ij} = \min_{1 \leq r \leq N} \left(\frac{1}{a_{ir}} + h_{rj}\right) \tag{2.23}$$

Indeed, if node r is a direct neighbor of node i, then $a_{ir} = 1$ and the hopcount of the remaining path from node r to node j equals h_{rj}. The minimum-hop (or shortest) path travels over that neighbor r of node i with the minimum remaining hops to the destination node j. If $r = j$ and $a_{ij} = 1$, then we find, with $h_{jj} = 0$, hopcount 1 for the direct neighbor path. If r is not a neighbor of i, then $\frac{1}{a_{ir}} = \infty$, which removes the index $q = r$ entry $\frac{1}{a_{ir}} + h_{rj}$ from the minimal set $\left\{\frac{1}{a_{iq}} + h_{qj}\right\}_{1\leq q\leq N}$ in (2.23). Since G is connected, thus excluding isolated nodes, there is at least one element $a_{ir} = 1$ in that minimal set. The non-linear recursion (2.23) can also be written as

$$h_{ij} = 1 + \min_{r\in \text{ neighbors}(i)} h_{rj}$$

22. *Diameter of a graph.* The *diameter* of the graph G, denoted by ρ and sometimes by ρ_G or $\rho(G)$, is the number of hops in the longest shortest path in G and equals $\rho = \max_{1\leq i\leq N;1\leq j\leq N} h_{ij}$. In a connected graph, the diameter is upper bounded by $\rho \leq N - 1$, the hopcount $N - 1$ of the longest possible shortest path in any connected graph on N nodes. The maximal diameter $\rho = N - 1$ occurs in a path on N nodes. The diameter of a connected graph G is lower bounded by $\rho \geq 1$ and the minimal diameter $\rho = 1$ only occurs in the complete graph K_N. If G is disconnected, the diameter is not defined, but sometimes put as $\rho > N$ or $\rho \to \infty$ or, even $\rho = 0$; in principle, any integer outside the interval $[1, N - 1]$ can serve as an indication of the non-existence of the diameter. We remark that $A^\rho - J$ is not necessarily a non-negative matrix, because $\left(A^{h_{ij}+1}\right)_{ij}$ can be zero[8], even though $\left(A^{h_{ij}}\right)_{ij} \geq 1$.

Lemma 2 *Let $f_k > 0$ for any $k \geq 0$ and A be the adjacency matrix of a connected graph G, then all elements of the matrix $\sum_{k=0}^{m} f_k A^k$ are positive for $m \geq \rho$. If $m < \rho$, the non-negative matrix $\sum_{k=0}^{m} f_k A^k$ contains at least one zero element.*

Proof: The definition of the diameter implies that, for each node pair (i, j) in a connected graph G, there exists a path with hopcount at most equal to ρ. This means that $\left(A^k\right)_{ij}$ is non-zero for at least one integer $k \in [0, \rho]$. In addition, there exists a pair (r, q), separated by the longest shortest path in G, for which $\left(A^k\right)_{rq} = 0$ for all $k < \rho$. Since each coefficient $f_k > 0$, it follows that $\sum_{k=0}^{\rho} f_k \left(A^k\right)_{ij} > 0$ for each node pair (i, j), but $\sum_{k=0}^{m} f_k A^k$ with $m < \rho$ contains at least one zero element, namely $\sum_{k=0}^{m} f_k \left(A^k\right)_{rq} = 0$. $\qquad\square$

When $f_k = \alpha^{\rho-k}\binom{\rho}{k}$ with $\alpha > 0$, then $\sum_{k=0}^{\rho} f_k A^k = \sum_{k=0}^{\rho} \binom{\rho}{k}\alpha^{\rho-k} A^k = (\alpha I + A)^\rho$, which leads to the known result that the diameter ρ is the smallest integer for which the matrix $(I + A)^\rho$ has positive elements. Since $\left(A^k\right)_{ij}$ are integers, it also follows that $(I + A)^\rho - J$ is a non-negative matrix (see Section 10.6). We infer from Lemma 2 that, for each node pair (i, j), at least one of the matrices in the sequence $\{A^m\}_{0\leq m\leq \rho} = \{I, A, A^2, \ldots, A^\rho\}$ contains a non-zero (i, j) element,

---

[8] For example, in a path graph, studied in Section 6.4, with $N = 3$ and adjacency matrix $A = \begin{bmatrix} 0 & 1 & 0 \\ 1 & 0 & 1 \\ 0 & 1 & 0 \end{bmatrix}$, there is not a walk with length 2 (nor any even number) between node 1 and node 2 (i.e. $\left(A^{2k}\right)_{12} = 0$ for $k > 1$), while there is a walk of odd length, thus $\left(A^{2k+1}\right)_{12} > 0$ for $k > 0$. The diameter $\rho = 2$, but $A^2 - J$ contains negative elements.

while there is at least one node pair (r, q), corresponding to the longest shortest path in G with ρ hops, whose entries in the sequence $\{A^m\}_{0 \leq m < \rho}$ are zero. The next Lemma generalizes this observation.

Lemma 3 *For any diagonal matrix M and for each node pair (i, j), at least one of the matrices in the sequence $\{(A + M)^m\}_{0 \leq m \leq \rho}$ contains a non-zero (i, j) element.*

Proof: Let G and G' denote the graph represented by the adjacency matrix A without self-loops ($a_{jj} = 0$ for any node j) and the same graph with weighted self-loops (equal to m_{jj} for node j), respectively. As explained in **art. 21**, the smallest integer $k = h_{ij}$, where $i \neq j$, for which $\left(A^k\right)_{ij} \neq 0$, but $\left(A^m\right)_{ij} = 0$, for all $0 \leq m < k$, is the hopcount of the shortest *path* in G from node i to node j. The expression (2.16) indicates that $\left(A^{h_{ij}}\right)_{ij} \neq 0$ does not depend on any diagonal element of A, because a path is a walk with all nodes different. This means that $\left((A + M)^{h_{ij}}\right)_{ij} = \left(A^{h_{ij}}\right)_{ij} \neq 0$. In addition, for $m < h_{ij}$, there is no path in G from i to j with m hops. Since a diagonal element, associated to a self-loop in G', cannot help to reach node j from i if there is no path from i to j in the graph G and thus also not in G', there also holds that $((A + M)^m)_{ij} = (A^m)_{ij} = 0$ for $m < h_{ij}$. These facts demonstrate Lemma 3. Only when $m > h_{ij}$, then $(A + M)^m_{ij}$ can differ from $(A^m)_{ij}$.   □

An interesting consequence of Lemma 3 is that, also for the Laplacian $Q = \Delta - A$, one of the matrices in the sequence $\{Q^m\}_{0 \leq m \leq \rho}$ contains a non-zero (i, j) element. Finally, combining Lemma 2 and 3 leads to the statement that there exists a matrix polynomial $p_m (A + M)$ of degree $m \in [0, \rho]$, whose (i, j)-th element is non-zero.

23. *h-hops adjacency matrix.* Analogous to Estrada (2012), who defines a path-Laplacian, we define the h-hops graph $_hG$ on N nodes as the graph that contains a link between i and j if their distance in an original graph G is h hops. The corresponding h-hops adjacency matrix $_hA$ has elements

$$(_hA)_{ij} = 1_{\{h_{ij}=h\}} \tag{2.24}$$

We define $_0A = I$ and, clearly, $_1A = A$. **Art.** 21 shows that a walk with $h = \min_k \left\{ (A^k)_{ij} \neq 0 \right\}$ is also the shortest path between i and j and that, for $i \neq j$,

$$(_hA)_{ij} = 1_{\left\{ \{\forall m \in [1,h):(A^m)_{ij}=0\} \cap \{(A^h)_{ij}>0\} \right\}} \tag{2.25}$$

while the diagonal elements $(_hA)_{jj}$ in (2.24) are zero for $h > 0$. **Art.** 22 illustrates that the composed event $\left\{ \forall m \in [1, h) : (A^m)_{ij} = 0 \right\}$ is also equal to the event $\left\{ \sum_{m=1}^{h-1} (A^m)_{ij} = 0 \right\}$, because all elements in A^k are non-negative. For the same reason, the last event is also equal to the event $\left\{ \sum_{m=1}^{h-1} c_m (A^m)_{ij} = 0 \right\}$, where $c_m > 0$ for each index m. Hence, the number of conditions to be checked in (2.25) is reduced to two in

$$(_hA)_{ij} = 1_{\{\{\sum_{m=1}^{h-1} c_m(A^m)_{ij}=0\}\cap\{(A^h)_{ij}>0\}\}} = 1_{\{\sum_{m=1}^{h-1} c_m(A^m)_{ij}=0\}} 1_{\{(A^h)_{ij}>0\}} \tag{2.26}$$

Finally, we can choose $c_m = \binom{h}{m}$ so that $\sum_{m=1}^{h} c_m(A^m)_{ij} = \left((A+I)^h\right)_{ij}$ for $i \neq j$ and (2.26) simplifies to

$$(_hA)_{ij} = 1_{\{((A+I)^{h-1})_{ij}=0\}} 1_{\{(A^h)_{ij}>0\}}$$

Lemma 2 states that $\sum_{m=0}^{h} c_m(A^m)_{ij} > 0$ for all $h \geq \rho$ and, consequently, (2.26) implies that $_hA = O$ for all $h > \rho$ as well as

$$\sum_{h=0}^{N-1} {}_hA = \sum_{h=0}^{\rho} {}_hA = J$$

The relation with the distance matrix H in **art.** 8 and **art.** 21 is

$$H = \sum_{h=1}^{\rho} h\, {}_hA$$

The number of links $\frac{1}{2}u^T\, {}_hAu$ in the graph $_hG$ equals the number of node pairs connected by an h-hop shortest path.

The sequence of h-hops adjacency matrices $\{_hA\}_{1\leq h\leq\rho} = \{_1A, _2A, \ldots, _\rho A\}$ defines a multi-layer network where, in each xy-plane, the graph $_hG$ is depicted and along the z-axis, the number h of hops is varied. Such multi-layer network may visualize how the links $(_hA)_{ij}$ around node i to any other node j in G vary with hop h and it allows to construct the levelset (Van Mieghem, 2014, Sec. 16.2.2), the set containing the number of nodes $(_hAu)_i$ at each level h in a shortest path tree rooted at node i of G and depicted in Fig. 6.4.

24. *Effects of link removals on the diameter.* Schoone et al. (1987) have derived bounds for the maximum diameter of a still connected graph G_k, obtained from an original graph G with diameter ρ after the removal of k links. For undirected graphs G, Schoone et al. (1987) prove an upper bound for the diameter in G_k of $(k+1)\rho$ and a lower bound of $(k+1)\rho - k$, for even ρ, and of $(k+1)\rho - 2k + 2$, for odd $\rho \geq 3$. For the special cases of $k = 2$ and $k = 3$, the exact bounds are $\rho(G_2) \leq 3\rho - 1$ and $\rho(G_3) \leq 4\rho - 2$, respectively. In addition, Schoone et al. (1987) prove that the problem of finding G_k by removing k links in G so that $\rho(G_k)$ is at least m as well as the related problem of finding the graph G_l by adding l links to G so that $\rho(G_l) \leq m$ is NP-complete.

2.4 The line graph

25. The line graph $l(G)$ of the graph $G(N, L)$ has as set of nodes the links of G and two nodes in the line graph $l(G)$ are adjacent if and only if they have, as links in G, one node of G in common. Given the graph G, the definition thus specifies the line graph operator $l\,(.)$. The line graph $l\,(G)$ of G is sometimes called the "dual" or "interchanged" or "derived" graph of G. For example, the line graph of the star $K_{1,n}$ is the complete graph K_n and the line graph of the example graph in Fig. 2.1 is drawn in Fig. 2.2. When G is connected, then also $l\,(G)$ is connected as follows from the definition[9] of the line graph $l\,(G)$.

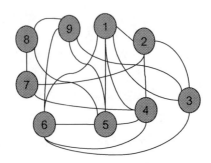

Fig. 2.2. The line graph of the undirected variant of the graph drawn in Fig. 2.1.

We denote by R the absolute value of the incidence matrix B, i.e., $R_{ij} = |b_{ij}|$. In other words, $R_{ij} = 1$ if node i and link j are incident, otherwise $R_{ij} = 0$. Hence, the unsigned incidence matrix R ignores the direction of links in the graph, in contrast to the incidence matrix B. Analogously to the definition of the Laplacian in **art.** 4, we may verify that the $N \times N$ adjacency matrix A of the graph G is written in terms of the unsigned $N \times L$ node-link incidence matrix R as

$$A = RR^T - \Delta \qquad (2.27)$$

The $L \times L$ adjacency matrix of the line graph $l\,(G)$ is similarly written in terms of R as

$$A_{l(G)} = R^T R - 2I \qquad (2.28)$$

The matrix $B^T B$ is generally a $(-1, 0, 1)$-matrix. Taking the absolute value of its entries equals $R^T R$, whereas the Laplacian matrix $Q = 2\Delta - RR^T = BB^T$.

In a graph G, where multiple links with the same direction between two nodes are excluded, we consider

$$\left(B^T B\right)_{ij} = \sum_{n=1}^{N} b_{ni} b_{nj} = \begin{cases} 1 & \text{if both link } i \text{ and } j \text{ either start or end in node } n \\ -1 & \text{if either link } i \text{ or } j \text{ starts or ends in node } n \\ -2 & \text{if link } i \text{ and } j \text{ have two nodes in common} \end{cases}$$

The latter case, where $\left(B^T B\right)_{ij} = -2$, occurs for a bidirectional link between two nodes. If the links at each node of the graph G either all start or all end, then

---

[9] In a connected graph G, each node is reachable from any other node via a path (a sequence of adjacent links, **art.** 6). Similarly, in the dual setting corresponding to the line graph, each link in G is reachable from any other link via a path (a sequence of adjacent nodes or neighbors).

we observe that $\left(B^T B\right)_{ij} = 1$ for all links i and j and, in that case, it holds that $B^T B = R^T R$. An interesting example of such a graph is the general bipartite graph, studied in Section 6.8, where the direction of the links is the same for each node in the set $\mathcal{M}$ to each node in the other set $\mathcal{N} \backslash \mathcal{M}$.

26. *Basic properties of the line graph.* The number of nodes in the line graph $l\left(G\right)$ equals the number L of links in G. The number of links in the line graph $l\left(G\right)$ is computed from the basic law of the degree (2.5) and (2.28) with the $L \times 1$ all-one vector u as

$$L_{l(G)} = \frac{1}{2} u^T A_{l(G)} u = \frac{1}{2} u^T R^T R u - u^T u$$
$$= \frac{1}{2} \|Ru\|_2^2 - L$$

It follows from the definition of the unsigned incidence matrix R that $u_{1 \times N}^T R = 2u_{L \times 1}^T$ or

$$R^T u = 2u \tag{2.29}$$

which is the companion of (2.1), and that

$$Ru = d \tag{2.30}$$

because the row sum of $\sum_{l=1}^{L} R_{il} = d_i$, the number of links in G incident to node i. Hence, we find that the number of links in the line graph $l\left(G\right)$ equals

$$L_{l(G)} = \frac{1}{2} d^T d - L = \frac{1}{2} \sum_{i=1}^{N} d_i^2 - L \tag{2.31}$$

Alternatively, each node i in G with degree d_i generates in the line graph $l\left(G\right)$ precisely d_i nodes that are all connected to each other as a clique, corresponding to $\binom{d_i}{2}$ links. The number of links in $l\left(G\right)$ is thus also

$$L_{l(G)} = \sum_{i=1}^{N} \binom{d_i}{2}$$

Art. 4 indicates that the average degree of a node in the line graph $l\left(G\right)$ is

$$E\left[D_{l(G)}\right] = \frac{2L_{l(G)}}{N_{l(G)}} = \frac{1}{L} \sum_{i=1}^{N} d_i^2 - 2$$

The degree vector of the line graph $l\left(G\right)$ follows from (2.4) as

$$d_{l(G)} = A_{l(G)} u_{L \times 1} = R^T R u - 2u$$
$$= R^T d - 2u$$

Each column of R (as in the incidence matrix B) contains only two non-zero elements and the vector component $\left(R^T d\right)_l = d_{l+} + d_{l-}$, where l_+ denotes the node at the start and l_- the node at the end of the link l. Hence, the maximum (and similarly minimum) degree of the line graph $l\left(G\right)$ equals

$$\max d_{l(G)} = \max_{1 \leq l \leq L} \left(d_{l+} + d_{l-} - 2\right) \leq d_{(1)} + d_{(2)} - 2$$

where $d_{(k)}$ denotes the k-th largest degree in G and $d_{(k-1)} \geq d_{(k)}$ for $2 \leq k \leq N$.

Example The degree vector of a regular graph with degree r is $d = r u_{N \times 1}$. The degree vector of the corresponding line graph is $d_{l(G)} = R^T d - 2u = r R^T u - 2u$ and with (2.29), we find $d_{l(G)} = 2\left(r - 1\right) u_{L \times 1}$. The line graph of a regular graph with degree r is also a regular graph with degree $2\left(r - 1\right)$. The total number of links follows from $L_{l(G)} = \sum_{i=1}^{N} \binom{d_i}{2} = N \frac{r(r-1)}{2}$ or from the basic law of the degree (2.5), $L_{l(G)} = \frac{1}{2} d_{l(G)}^T u = \left(r - 1\right) L = \left(r - 1\right) \frac{r}{2} N$.

The sum of all off-diagonal elements in A^2 equals

$$\sum_{i=1}^{N} \sum_{j=1;j\neq i}^{N} \left(A^2\right)_{ij} = \sum_{i=1}^{N} \sum_{j=1;j\neq i}^{N} \sum_{k=1}^{N} a_{ik} a_{kj} = \sum_{k=1}^{N} \sum_{i=1}^{N} a_{ki} \sum_{j=1;j\neq i}^{N} a_{kj}$$

$$= \sum_{k=1}^{N} \sum_{i=1}^{N} a_{ki} \left(d_k - a_{ki}\right) = \sum_{k=1}^{N} \left(d_k \sum_{i=1}^{N} a_{ki} - \sum_{i=1}^{N} a_{ki}\right)$$

and, thus

$$\sum_{i=1}^{N} \sum_{j=1;j\neq i}^{N} \left(A^2\right)_{ij} = \sum_{k=1}^{N} d_k \left(d_k - 1\right) = 2L_{l(G)} \tag{2.32}$$

where the last equality follows from (2.31) and $\sum_{i=1}^{N} \sum_{j=1;j\neq i}^{N} \left(A^2\right)_{ij}$ equals twice the total number of two-hop walks with different source and destination nodes. In other words, the total number of connected triplets of nodes in G, which is half of (2.32), equals the number of links in the line graph $l\left(G\right)$.

The $L \times L$ Laplacian matrix $Q_{l(G)}$ of the line graph $l\left(G\right)$ is, by definition (2.6),

$$Q_{l(G)} = \mathrm{diag}\left(d_{l(G)}\right) - A_{l(G)}$$
$$= \mathrm{diag}\left(R^T d\right) - R^T R$$

which illustrates that the relation between the Laplacian Q of the graph G and the Laplacian $Q_{l(G)}$ of its line graph $l\left(G\right)$ is less obvious.

27. Since $R^T R$ is a Gram matrix (**art. 280**), all eigenvalues of $R^T R$ are non-negative. Hence, it follows from (2.28) that the eigenvalues of the adjacency matrix of the line graph $l\left(G\right)$ are not smaller than -2.

The adjacency spectra of the line graph $l\left(G\right)$ and of G are related by Lemma 11 in **art. 284** since

$$\det\left(\left(R^T R\right)_{L \times L} - \lambda I\right) = \lambda^{L-N} \det\left(\left(R R^T\right)_{N \times N} - \lambda I\right)$$

Using the definitions (2.28) and (2.27) in **art.** 25 yields

$$\det\left(A_{l(G)} - (\lambda - 2)I\right) = \lambda^{L-N}\det\left(\Delta + A - \lambda I\right)$$

or

$$\det\left(A_{l(G)} - \lambda I\right) = (\lambda + 2)^{L-N}\det\left(\Delta + A - (\lambda + 2)I\right) \tag{2.33}$$

The eigenvalues of the adjacency matrix of the line graph $l(G)$ are those of the unsigned Laplacian $\Delta + A$ in **art.** 30 shifted over -2 and an eigenvalue at -2 with multiplicity $L - N$.

If $B^T B = R^T R$, then Lemma 11 indicates that

$$\det\left((B^T B)_{L\times L} - \lambda I\right) = \lambda^{L-N}\det\left((BB^T)_{N\times N} - \lambda I\right)$$

from which

$$\det(Q - \lambda I) = \lambda^{N-L}\det\left(A_{l(G)} - (\lambda - 2)I\right)$$

or

$$\det\left(A_{l(G)} - \lambda I\right) = (\lambda + 2)^{L-N}\det\left(Q - (\lambda + 2)I\right) \tag{2.34}$$

In graphs G, where $B^T B = R^T R$, the eigenvalues of the adjacency matrix of the line graph $l(G)$ are those of the Laplacian $Q = \Delta - A$ shifted over -2 and an eigenvalue at -2 with multiplicity $L - N$.

The restriction, that all eigenvalues of an adjacency matrix are not less than -2, is not sufficient to characterize line graphs (Biggs, 1996, p. 18). The state-of-the-art knowledge about line graphs is reviewed by Cvetković *et al.* (2004), who treat the characterization of line graphs in detail. Referring for proofs to Cvetković *et al.* (1995, 2004), we mention here only:

Theorem 4 (Krausz) *A graph is a line graph if and only if its set of links can be partitioned into "non-trivial" cliques, namely (i) two cliques have at most one node in common and (ii) each node belongs to at most two cliques.*

Theorem 5 (Van Rooij and Wilf) *A graph is a line graph if and only if (i) it does not contain the star $K_{1,3}$ as an induced subgraph and (ii) the remaining (or opposite) nodes in any two triangles with a common link must be adjacent and each of such triangles must be connected to at least one other node in the graph by an odd number of links.*

28. *Inverse line graph.* Given a line graph $l(G)$, it is possible to reconstruct the original graph G by the inverse line graph operation $l^{-1}(.)$, so that $l^{-1}(l(G)) = G$ returns the original graph G.

Each link l in G connects two nodes i and j and is transformed in the line graph $l(G)$ to a node l that belongs to two cliques $\widehat{K}_{d_i}$ and $\widehat{K}_{d_j}$, where a clique, denoted by $\widehat{K}_n$, contains the complete graph K_n and additional links to other nodes outside the complete graph K_n. If a line graph $l(G)$ can be partitioned into cliques (Krausz' Theorem 4), then the number of those cliques equals the number N of nodes in G and each node l in $l(G)$, belonging to two cliques i and j, corresponds to a link l in G between two nodes i and j. Apart from the line graph $l(G) = K_3$, that has two original graphs, the triangle K_3 and the star $K_{1,3}$ on four nodes, the reconstruction or inverse line graph $l^{-1}(G)$ is unique by a theorem of Whitney (1932).

Algorithms to compute the original graph G from the line graph $l(G)$ are presented by Lehot (1974) and Roussopoulos (1973). Our inverse line graph algorithm ILIGRA complements and has advantages over Lehot's and Roussopoulos' algorithm, as explained in Liu *et al.* (2015).

29. *Repeated line graph transformations.* The Cauchy-Schwarz inequality (A.72), $\left(\sum_{i=1}^{N} d_i\right)^2 \leq N \sum_{i=1}^{N} d_i^2$ with equality only for regular graphs where $d_j = r$ for each node j, the basic law of the degree (2.3) and (2.31) indicate that

$$L_{l(G)} \geq L\left(\frac{2L}{N} - 1\right) \qquad (2.35)$$

The number $L_{l(G)}$ of links in the line graph can only be equal to the number L of links in the original graph if the average degree $\frac{2L}{N} = 2$ and the graph is regular. Hence, the line graph of a cycle C_N on N nodes is again the cycle C_N, i.e. $l(C_N) = C_N$.

For $k \geq 1$, van Rooij and Wilf (1965) have constructed the sequence $G_0, G_1, \ldots, G_k$ of graphs, where the graph $G_k = l(G_{k-1})$ has N_k nodes and L_k links and where the original graph G_0 is possibly the only non-line graph. The k-th line graph iterate $\underbrace{l.l \ldots l}_{k \text{ times}}(G_0)$ is denoted by $G_k = l^k(G_0)$. The line graph of the path P_N on N nodes is $l(P_N) = P_{N-1}$. Hence, the k-th iterate $l^k(P_N) = P_{N-k}$ becomes the empty graph for $k = N - 1$, while the cycle, obeying $l^k(C_N) = C_N$, is invariant under a line graph transformation.

The basic property (**art. 26**) of the line graph shows that $N_k = L_{k-1}$ and (2.35) becomes

$$\frac{N_{k+1}}{N_k} \geq 2\frac{N_k}{N_{k-1}} - 1$$

Let $\nu_k = \frac{N_{k+1}}{N_k}$, then $\nu_{k+1} \geq 2\nu_k - 1$, equivalent to $\nu_{k+1} - 1 \geq 2(\nu_k - 1)$ and after p iterations, we obtain

$$\nu_k - 1 \geq 2(\nu_{k-1} - 1) \geq 2^2(\nu_{k-2} - 1) \geq \ldots \geq 2^p(\nu_{k-p} - 1)$$

If $k - p = 0$, then, with $\nu_0 = \frac{L}{N}$, we find that $\nu_k = \frac{N_{k+1}}{N_k} \geq 2^k\left(\frac{L}{N} - 1\right) + 1$. With $\xi = \frac{L}{N} - 1$, iterating $N_{k+1} = \nu_k N_k$ downwards yields

$$N_k \geq N \prod_{j=0}^{k-1}\left(1 + 2^j\xi\right) = N\xi^k 2^{\frac{k(k-1)}{2}} \prod_{j=0}^{k-1}\left(1 + \frac{1}{2^j\xi}\right) \qquad (2.36)$$

If G_j is regular, then all G_k with $k > j$ are also regular graphs (**art.** 26), in which case the equality sign in (2.36) holds. Hence, if G_0 is a regular graph with degree r, then $\xi = \frac{r}{2} - 1$ and equality holds in (2.36) so that $N_k = N \prod_{j=0}^{k-1} \left(1 + 2^j \left(\frac{r}{2} - 1\right)\right)$. Since the degree of a node in any graph with $N > 3$ is smaller than or equal to $r = N - 1$ in the complete graph, we find an upper bound

$$N_k \leq N \prod_{j=0}^{k-1} \left(1 + 2^j \left(\frac{N-1}{2} - 1\right)\right)$$

In summary, for any graph with $\xi = \frac{L}{N} - 1 > 0$, $N > 3$ (but excluding the star $K_{1,3}$, because $l(K_{1,3}) = K_3$) and at least one nodal degree $d_i \geq 3$, the number N_k of nodes in G_k is increasing in k rapidly[10] as $O\left(N\xi^k 2^{\frac{k(k-1)}{2}}\right)$.

Xiong (2001) has shown for a connected graph G_0 different from a path that $l^n(G_0)$ is Hamiltonian if $n \leq \rho - 1$, where ρ is the diameter (**art.** 22) of G_0, while Harary and Nash-Williams (1965) prove that, if G_0 is Eulerian (**art.** 6), then $l^3(G_0)$ is Hamiltonian and conversely.

30. *Unsigned Laplacian.* The unsigned or signless Laplacian $\overline{Q} = \Delta + A$, studied by Cvetković *et al.* (2007), possesses a number of interesting properties. The definition (2.27) shows that $\overline{Q} = RR^T$ is a positive semidefinite matrix and all its eigenvalues are non-negative (**art.** 27). The smallest eigenvalue of $\overline{Q}$ of a connected graph is only equal to zero if the graph is bipartite. Indeed, $\overline{Q}x = 0$ implies that $Rx = 0$, which is only possible if $x_i = -x_j$ for every link $l = i \sim j$ in the graph, i.e. only if G is bipartite (**art.** 25). Cvetković *et al.* (2007) show that this zero eigenvalue is simple in a connected graph and that the multiplicity of the zero eigenvalue of $\overline{Q}$ in any graph equals the number of bipartite components. The smallest eigenvalue of the signless Laplacian can be regarded as a measure of the non-bipartiteness of a graph. Stanić (2015) devotes a chapter on inequalities of the signless Laplacian.

---

[10] The fundamental cornerstone in the theory of Gaussian polynomials, defined as

$$\begin{bmatrix} k \\ l \end{bmatrix}(q) = \frac{\prod_{j=1}^{k}(1-q^j)}{\prod_{j=1}^{l}(1-q^j)\,\prod_{j=1}^{k-l}(1-q^j)} = \prod_{j=1}^{l}\frac{(1-q^{k-j+1})}{(1-q^j)} \qquad (2.37)$$

is

$$Q_k(z,x) = \prod_{m=0}^{k-1}(x+q^m z) = \sum_{m=0}^{k} \begin{bmatrix} k \\ m \end{bmatrix}(q)\, q^{m(m-1)/2}\, z^m\, x^{k-m} \qquad (2.38)$$

which bears a striking resemblance to Newton's binomium (Rademacher, 1973; Goulden and Jackson, 1983) for $q = 1$ so that $\begin{bmatrix} k \\ l \end{bmatrix}(1) = \binom{k}{l}$. We define $Q_k(-x,x) = \delta_{0k}$ in correspondence to the first factor for $m = 0$ in the product. The so-called q-analog (2.38) of Newton's binomium is derived via induction from the recursion $Q_k(z,x) = (x + q^{k-1} z)\, Q_{k-1}(z,x)$ for $k > 0$ and $Q_0(x,z) = 1$. When k tends to infinity, (2.38) leads for $|q| < 1$ to

$$\prod_{m=0}^{\infty}(1+q^m z) = \sum_{m=0}^{\infty} \frac{q^{m(m-1)/2}}{\prod_{j=1}^{m}(1-q^j)}\, z^m \qquad (2.39)$$

2.5 Permutations, partitions and the quotient graph

31. *Permutation matrix* P. Consider the set $\mathcal{N} = \{n_1, n_2, \ldots, n_N\}$ of nodes of G, where n_j is the label of node j. The most straightforward way is the labeling $n_j = j$. Suppose that the nodes in G are relabeled. This means that there is a permutation, often denoted by π, that rearranges the node identifiers n_j as $n_i = \pi(n_j)$. The corresponding permutation matrix P has, on row i, element $p_{ij} = 1$ if $n_i = \pi(n_j)$, and $p_{ij} = 0$ otherwise. Thus, in each row there is precisely one non-zero element equal to 1 and, consequently, it holds that

$$Pu = u$$

For example, the set of nodes $\{1, 2, 3, 4\}$ is permuted to the set $\{2, 4, 1, 3\}$ by the permutation matrix

$$P = \begin{bmatrix} 0 & 1 & 0 & 0 \\ 0 & 0 & 0 & 1 \\ 1 & 0 & 0 & 0 \\ 0 & 0 & 1 & 0 \end{bmatrix}$$

If the vector $v = (1, 2, 3, 4)$, then the permuted vector $w = Pv = (2, 4, 1, 3)$. Next, $z = Pw = P^2 v = (4, 3, 2, 1)$, then $y = Pz = P^3 v = (3, 1, 4, 2)$, and, finally, $Py = P^4 v = v$. Thus, $P^4 = I$. The observation $P^N = I$ holds in general for each $N \times N$ permutation matrix P: each node can be relabeled to one of the $\{n_1, n_2, \ldots, n_N\}$ possible labels and the permutation matrix maps each time a label $n_j \to \pi(n_j) = n_i$, where, generally, $n_i \neq n_j$, else certain elements are not permuted[11]. After N relabelings, we arrive again at the initial labeling and $P^N = I$. The definition (A.27) of the determinant shows that $\det P = \pm 1$, because in each row there is precisely one non-zero element equal to 1.

Another example of a permutation matrix is the unit-shift relabeling transformation in Section 6.2.1.

32. *A permutation matrix P is an orthogonal matrix.* Since a permutation matrix P relabels a vector v to a vector $w = Pv$, both vectors v and w contain the same components, but in a different order (provided $P \neq I$), such that their norms (**art.** 201) are equal, $\|v\| = \|w\|$. Using the Euclidean norm $\|x\|_2^2 = x^T x$, the equality $v^T v = w^T w$ implies that $P^T P = I$, such that P is an orthogonal matrix (**art.** 247).

If G_1 and G_2 are two directed graphs on the same set of nodes, then they are called *isomorphic*[12] if and only if there is a permutation matrix P such that $P^T A_{G_1} P = A_{G_2}$. Since permutation matrices are orthogonal, $P^{-1} = P^T$, the spectra of G_1 and G_2 are identical (**art.** 247) : the spectrum (set of eigenvalues) is an invariant of the isomorphism class of a graph. However, the converse "if the spectrum (set of eigenvalues) is the same, then the graph is isomorphic" is not true

---

[11] The special permutation $P = I$ does not, in fact, relabel nodes.

[12] The word "isomorphism" stems from $\iota\sigma o\varsigma$ (isos: same) and $\mu o\rho\varphi\eta$ (morphei: form).

in general. There exist nonisomorphic graphs that have precisely the same set of eigenvalues and such graphs are called *cospectral graphs*.

33. *A permutation matrix P is a doubly-stochastic matrix.* Left-multiplying both sides of $Pu = u$ with P^T and using $P^T P = I$ in **art.** 32 leads to $P^T u = u$. Since each element $p_{ij} \in [0, 1]$ and the row sum of P equals 1, i.e. $Pu = u$, we conclude that P is a stochastic matrix and property $P^T u = u$ makes P a doubly-stochastic matrix.

34. *Automorphism.* We investigate the effect of a permutation π of the nodal set $\mathcal{N}$ of a graph on the structure of the adjacency matrix A. Suppose that $n_i = \pi(n_j)$ and $n_k = \pi(n_l)$, then we have with the definition of P in **art.** 31,

$$(PA)_{il} = \sum_{m=1}^{N} p_{im} a_{ml} = a_{jl}$$

$$(AP)_{il} = \sum_{m=1}^{N} a_{im} p_{ml} = a_{ik}$$

In order for A and P to commute, i.e. $PA = AP$, we observe that, between each node pair (n_j, n_l) and its permutation $(\pi(n_j), \pi(n_l))$ there must be a link such that $a_{jl} = 1 = a_{ik}$. An *automorphism* of a graph is a permutation π of the nodal set $\mathcal{N}$ such that (n_i, n_j) is a link of G if and only if $(\pi(n_i), \pi(n_j))$ is a link of G. Hence, if the permutation π is an automorphism, then A and P commute. In fact, an automorphism is an isomorphism of the graph G to itself and represents a form of symmetry that maps the graph onto itself. A classical example is the Peterson graph in Fig. 2.3: by rotating the five nodes (both inner as outer ring) over 72 degrees, we obtain again a Peterson graph. All possible such permutations, that preserve all details of its structure, constitute the automorphism group of a graph G, denoted by $\text{Aut}(G)$. A graph is called *symmetric* if there are non-trivial, i.e. excluding $P = I$, automorphisms ($|\text{Aut}(G)| > 1$), and *asymmetric* if the trivial permutation $P = I$ is the only automorphism ($|\text{Aut}(G)| = 1$). Determining $\text{Aut}(G)$ or testing whether a graph has a non-trivial automorphism is a "hard" problem, likely NP-complete, but its hardness class is still unknown, just as the graph isomorphism problem (**art.** 38).

The consequences of the commutation $PA = AP$ for the spectrum of the adjacency matrix A are interesting. Suppose that x is an eigenvector of A belonging to the eigenvalue λ, then

$$APx = PAx = P\lambda x = \lambda Px$$

which implies that Px is also an eigenvector of A belonging to eigenvalue λ. If x and Px are linearly independent, then λ cannot be a simple eigenvalue. Thus, an automorphism produces multiple eigenvectors belonging to a same eigenvalue.

35. *Enumeration of graphs.* The total number of undirected graphs $G(N, L)$ with

N nodes and L links equals

$$n_{G(N,L)} = \binom{\binom{N}{2}}{L} \qquad (2.40)$$

which is the number of ways that we can distribute the L ones, corresponding to the L links, in the upper (or lower) triangular part of an $N \times N$ symmetric adjacency matrix, containing $\binom{N}{2}$ possible positions. The total number of undirected graphs with N nodes then follows by summing (2.40) over all possible number of links, $0 \le L \le \binom{N}{2}$, as

$$n_{G(N)} = 2^{\binom{N}{2}} \qquad (2.41)$$

The enumeration has implicitly assumed that all nodes are distinguishable. For example, each node has a certain characteristic property (i.e. a label, a color, a size, etc.). In many cases, the nodes of a graph are all of the same type and indistinguishable, which means that, if we relabel two nodes, the resulting graph is still the same or isomorphic to the former. The number of ways in which we can relabel the N nodes is $N!$. However, the number of graphs isomorphic to a given graph G is $N!/|Aut(G)|$. Therefore, for any class C of graphs closed under isomorphism (e.g. all graphs, or all regular graphs), the number of isomorphism classes is $\overline{|Aut(C)|}/N!$, where $\overline{|Aut(G)|}$ is the average size of the automorphism group of a graph in G. Hence, the total number of undirected, nonisomorphic graphs is

$$n_{\text{nonisomorphic } G(N)} = \frac{2^{\binom{N}{2}}}{N!} \overline{|Aut(G(N))|} \qquad (2.42)$$

where $\overline{|Aut(G(N))|}$ is the average number of automorphisms among all graphs on N nodes and the complicating factor in (2.42).

In some cases, the enumeration of graph properties (such as the number of walks (**art.** 59), the number triangles in (3.8) and spanning trees (**art.** 117)) can be efficiently computed from the spectrum of the graph, while in other cases, enumeration leads to a challenging combinatorial problem (such as the number of regular or cospectral graphs (**art.** 40)). Techniques for enumeration of graph properties, including a proof of (2.42), are discussed in depth in the book by Harary and Palmer (1973).

36. *Partitions.* A generalization of a permutation is a partition that separates the nodal set $\mathcal{N}$ of a graph in disjoint, non-empty subsets of $\mathcal{N}$, whose union is $\mathcal{N}$. The $k \in \{1, 2, \ldots, N\}$ disjoint, non-empty subsets generated by a partition are sometimes called cells, and denoted by $\{C_1, C_2, \ldots, C_k\}$. If $k = N$, the partition reduces to a permutation. We also denote a partition by π.

Let $\{C_1, C_2, \ldots, C_k\}$ be a partition of the set $\mathcal{N} = \{1, 2, \ldots, N\}$ of nodes and let

A be a symmetric matrix, that is partitioned as

$$A = \begin{bmatrix} A_{1,1} & \cdots & A_{1,k} \\ \vdots & & \vdots \\ A_{k,1} & \cdots & A_{k,k} \end{bmatrix}$$

where the block matrix $A_{i,j}$ is the submatrix of A formed by the rows in C_i and the columns in C_j. For example, the partition $C_1 = \{1, 3\}$, $C_2 = \{2, 4, 6\}$ and $C_3 = \{5\}$ of the nodes in Fig. 2.1 leads to the partitioned adjacency matrix

$$\pi A = \begin{bmatrix} \begin{bmatrix} 0 & 1 \\ 1 & 0 \end{bmatrix} & \begin{bmatrix} 1 & 0 & 1 \\ 1 & 1 & 0 \end{bmatrix} & \begin{bmatrix} 0 \\ 0 \end{bmatrix} \\ \begin{bmatrix} 1 & 1 \\ 0 & 1 \\ 1 & 0 \end{bmatrix} & \begin{bmatrix} 0 & 0 & 1 \\ 0 & 0 & 0 \\ 1 & 0 & 0 \end{bmatrix} & \begin{bmatrix} 1 \\ 1 \\ 1 \end{bmatrix} \\ \begin{bmatrix} 0 & 0 \end{bmatrix} & \begin{bmatrix} 1 & 1 & 1 \end{bmatrix} & [0] \end{bmatrix}$$

which is obtained from the matrix A on p. 16 by relabeling nodes according to $1 = \pi(1), 2 = \pi(3), 3 = \pi(2), 4 = \pi(4), 5 = \pi(6), 6 = \pi(5)$. The characteristic matrix S of the partition, also called the community matrix S, is the $N \times k$ matrix whose columns are the vectors C_k labeled in accordance with πA. Thus, in the example, the partition $C_1 = \{1, 3\}$, $C_2 = \{2, 4, 6\}$ and $C_3 = \{5\}$, translates after relabeling into $\pi C_1 = \{\pi(1) = 1, \pi(3) = 2\}$, $\pi C_2 = \{\pi(2) = 3, \pi(4) = 4, \pi(6) = 5\}$ and $\pi C_3 = \{\pi(5) = 6\}$, respectively, with corresponding matrix S

$$S = \begin{bmatrix} 1 & 0 & 0 \\ 1 & 0 & 0 \\ 0 & 1 & 0 \\ 0 & 1 & 0 \\ 0 & 1 & 0 \\ 0 & 0 & 1 \end{bmatrix} = \begin{bmatrix} u_2 & 0 & 0 \\ 0 & u_3 & 0 \\ 0 & 0 & u_1 \end{bmatrix}$$

where u_j is the all one vector of dimension j. Clearly, $S^T S = \text{diag}(2, 3, 1)$.

In general, $S^T S = \text{diag}(|C_1|, |C_2|, \ldots, |C_k|)$, where $|C_k|$ equals the number of elements in the set C_k. Each row of S only contains one non-zero element, which follows from the definition of a partition: a node can only belong to one cell or community of the partition and the union of all cells is again the complete set $\mathcal{N}$ of nodes. Thus, the elements of the $N \times k$ community matrix S after relabeling are

$$S_{ij} = \begin{cases} 1 & \text{if node } i \text{ belongs to the community } \pi C_j \\ 0 & \text{otherwise} \end{cases}$$

or compactly, $S_{ij} = 1_{\{\pi^{-1}(i) \in C_j\}}$. The columns of S are orthogonal and $\text{trace}(S^T S) = N$.

37. *Quotient matrix.* The quotient matrix corresponding to the partition specified by $\{C_1, C_2, \ldots, C_k\}$ is defined as the $k \times k$ matrix

$$A^\pi = \left(S^T S\right)^{-1} S^T \left(_\pi A\right) S \tag{2.43}$$

where $\left(S^T S\right)^{-1} = \mathrm{diag}\left(\frac{1}{|C_1|}, \frac{1}{|C_2|}, \ldots, \frac{1}{|C_k|}\right)$. The quotient matrix of the matrix A of the example in **art.** 36 is

$$A^\pi = \begin{bmatrix} 1 & 2 & 0 \\ \frac{4}{3} & \frac{2}{3} & 1 \\ 0 & 3 & 0 \end{bmatrix}$$

We can verify that $(A^\pi)_{ij}$ denotes the average row sum of the block matrix $(_\pi A)_{i,j}$. An example of the quotient matrix Q^π of a Laplacian Q is given in Section 6.13.

If the row sum of each block matrix $A_{i,j}$ is equal to the same constant, then the partition π is called *equitable* or *regular*. In that case, $A_{i,j} u = (_\pi A)_{i,j} u$ or $_\pi A S = S A^\pi$. Also, a partition π is equitable if, for any i and j, the number of neighbors that a node in C_i has in the cell C_j does not depend on the choice of a node in C_i.

For example, consider a node v in the Petersen graph shown in Fig. 2.3 and construct the three cell partitions as $C_1 = \{v\}$, C_2 is the set of the neighbors of v and C_3 is the set of nodes two hops away from v. The number of neighbors of v in

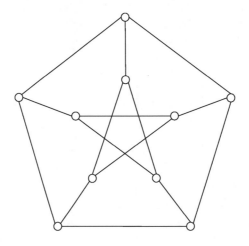

Fig. 2.3. The Petersen graph.

C_2 is three and zero in C_3, while the number of neighbors of a node in C_2 with C_3 is two such that

$$A^\pi = \begin{bmatrix} 0 & 3 & 0 \\ 1 & 0 & 2 \\ 0 & 1 & 2 \end{bmatrix}$$

A distance partition with respect to node v is the partition of $\mathcal{N}$ into the sets of nodes in G at distance r from a node v. A distance partition is, in general, not equitable.

If v is an eigenvector of A^{π} belonging to the eigenvalue λ, then Sv is an eigenvector of ${}_{\pi}A$ belonging to the same eigenvalue λ. Indeed, left-multiplication of the eigenvalue equation $A^{\pi} v = \lambda v$ by S yields

$$\lambda S v = S A^{\pi} v = ({}_{\pi}A) S v$$

This property makes equitable partitions powerful.

For example, the adjacency matrix of the complete bipartite graph $K_{m,n}$ (see Section 6.7) has an equitable partition with $k = 2$. The corresponding quotient matrix is $A^{\pi} = \begin{bmatrix} 0 & m \\ n & 0 \end{bmatrix}$ whose eigenvalues are $\pm\sqrt{mn}$, which are the non-zero eigenvalues of $K_{m,n}$. The quotient matrix of the complete multipartite graph is derived in Section 6.9. Exact solutions of the epidemic mean-field equations in Prasse *et al.* (2021) rely on equitable partitions.

The quotient graph of an equitable partition, denoted by G^{π}, is the directed graph with the cells of the partition π as its nodes and with $(A^{\pi})_{ij}$ links going from cell/node C_i to node C_j. Thus, $(A^{\pi})_{ij}$ equals the number of links that join a node in the cell C_i to the nodes in cell C_j. In general, the quotient graph contains multiple links and self-loops. The subgraph induced by each cell in an equitable partition is necessarily a regular graph because each node in cell C_i has the same number of neighbors in cell C_j.

2.6 Cospectral graphs

Cospectral graphs are nonisomorphic graphs that possess the same set of eigenvalues, as earlier defined in **art. 32**. Since the spectrum of graphs is the main theme in this book, we cannot avoid devoting some attention to cospectral graphs.

38. Checking whether two graphs have the same adjacency eigenvalues is a polynomial, thus "easy" problem. However, determining whether two cospectral graphs are isomorphic can be non-polynomial, thus "hard", but it is currently unknown (McKay and Piperno, 2014) whether the graph isomorphism problem is NP-hard.

Almost all non-star-like trees are *not* determined by the spectrum of the adjacency matrix (van Dam and Haemers, 2003). Godsil and Royle (2001) start by the remark that the spectrum of a graph does not determine the degrees, nor whether the graph is planar and that there are many graphs that are cospectral, i.e., although graphs are different (nonisomorphic), their spectrum is the same. Cvetković *et al.* (2009) devote a whole chapter on the characterization of graphs by their spectrum. They list theorems on graphs that are determined by their spectrum such as regular graphs with degree $r = 2$ and complete bipartite graphs, but they also present counter examples. Finally, van Dam and Haemers (2003) conjecture that sufficiently large graphs are determined by their spectrum, roughly speaking because the probability of having cospectral graphs becomes vanishingly small when the number of nodes N increases. A major tool to construct cospectral graphs is Godsil-McKay switching.

39. *Godsil-McKay switching for cospectral graph construction.* Godsil and McKay (1982) have invented an ingenious way to construct cospectral graphs by using a certain partitioning π of a graph and by rewiring a specific set of links, which is called "switching". They start by proposing the partition $\pi = \{C_1, C_2, \ldots, C_k, F\}$, where (a) any two nodes in C_i have the same number of neighors in C_j, for $1 \leq i, j \leq k$ and i can be the same as j; (b) a node $v \in F$ has either zero, $n_i/2$ or n_i neighbors in C_i, where the number of nodes in C_i is $n_i = |C_i|$. Any graph G with N nodes can be partitioned in this way, in particular, if $C_j = \{j\}$ and $F = \{N\}$. Of course, the interest lies in finding non-trivial partitions where $n_i > 1$, for at least some i. The adjacency matrix corresponding to this partition π is denoted as a block matrix

$$\pi A = \begin{bmatrix} A_{1,1} & A_{1,2} & \cdots & A_{1,k} & A_{F_1} \\ A_{1,2}^T & A_{2,2} & \cdots & A_{2,k} & A_{F_2} \\ \vdots & \vdots & \ddots & \vdots & \vdots \\ A_{1,k}^T & A_{2,k}^T & \cdots & A_{k,k} & A_{F_k} \\ A_{F_1}^T & A_{F_2}^T & \cdots & A_{F_k}^T & A_F \end{bmatrix}$$

where $A_{j,j}$ is the $n_j \times n_j$ adjacency matrix of the set of nodes belonging to C_j and the adjacency matrix $A_{i,j}$ and A_{F_j} describe the interlinking between the sets C_i and C_j and between the sets C_j and F, respectively. By construction, the row sum of each block matrix $A_{i,j}$ is constant, thus $A_{i,j}u = d_{ij}u$, where d_{ij} denotes the number of neighbors in C_j that each node in C_i has. The row sum of A_{F_j}, i.e. $A_{F_j}u = f_j u$, where f_j is either 0, $n_j/2$ or n_j. Since all block matrices of πA are adjacency matrices and symmetric, the column sums are constant as well. Next, Godsil and McKay (1982) introduce the $m \times m$ matrix $V_m = \frac{2}{m}J_{m \times m} - I_m$, where the all-one $r \times m$ matrix $J_{r \times m} = u_r.u_m^T$. The matrix V_m features interesting properties, because V_m is a Householder reflection (see art. 197). First, using $J_{l \times q}J_{q \times k} = u_l u_q^T u_q u_k^T$ and $u_q^T u_q = q$ so that $J_{l \times q}J_{q \times k} = qJ_{l \times k}$, we find that

$$V_m^2 = I_m \tag{2.44}$$

Next, for an $m \times n$ matrix X with constant row sum r and column sum c, it holds that

$$V_m X V_n = X \tag{2.45}$$

Indeed, $V_m X = \frac{2}{m}u_m u_m^T X - X = \frac{2c}{m}u_m u_n^T - X$ from which

$$V_m X V_n = \left(\frac{2c}{m}u_m u_n^T - X\right)\left(\frac{2}{n}u_n u_n^T - I_n\right)$$

$$= \left(\frac{2c}{m} - \frac{2r}{n}\right)u_m u_n^T + X$$

The sum of all elements in X equals $rm = cn$, from which $\frac{c}{m} = \frac{r}{n}$, demonstrating (2.45). Finally, if the $2m \times 1$ vector x contains m zero elements and m one elements, then the definition $V_m = \frac{2}{m} u_m u_m^T - I_m$ directly shows that

$$V_{2m} x = u_{2m} - x \tag{2.46}$$

This last property (2.46) motivates the Godsil-McKay construction of the graph G^* obtained from G with adjacency matrix $_\pi A$ as follows. For all those sets C_i, where each node $v \in F$ is connected to $n_i/2$ nodes in C_i, these $n_i/2$ links are deleted and each node $v \in F$ is reconnected to the other $n_i/2$ nodes in the set C_i. The fascinating relation between G and G^* is that G^* and G, as well as their complements G^{*c} and G^c, have the same adjacency eigenvalues. Hence, G and G^* are cospectral with cospectral complement. The proof is surprisingly easy, the adjacency matrix of G^* satisfies

$$_\pi A^* = V \left(_\pi A \right) V \tag{2.47}$$

where the block-diagonal matrix $V = \mathrm{diag}\left(V_{n_1}, V_{n_2}, \ldots, I_{|F|} \right)$. Property (2.45) illustrates that $_\pi A^*$ is the same as $_\pi A$, except for the last block row and block column. Property (2.46) switches in A_{F_j} all zero entries into one and vice versa. Finally, left-multiplying both sides of (2.47) by V and invoking property (2.44) shows that the eigenvalue equation $_\pi A^* y = \lambda y$ is equivalent to $_\pi A (Vy) = \lambda (Vy)$. Hence, $_\pi A^*$ and $_\pi A$ possess the same eigenvalues with the corresponding eigenvectors y and Vy. Since the adjacency matrix of the complement G^c is also a block matrix with constant row and column sums and of similar block structure as $_\pi A$, the same arguments also demonstrate that G^{*c} and G^c are cospectral.

The Godsil-McKay construction of the cospectral graph G^* illustrates that the main difficulty lies in finding a non-trivial Godsil-MacKay partition π with corresponding adjacency matrix $_\pi A$. The useful properties of V_m for cospectral graph constructions result from the fact that the labeling of nodes in any cell C_i and F does not influence a sum as $u_m^T z_m = \sum_{k=1}^m z_k$, so that only constant row (and column) sums are required in the Godsil-McKay construction.

40. Although cospectral graphs are not easy to construct, they should not be ignored. The following theorem, due to Brendan McKay, implies that the probability to draw a regular graph (**art. 55**) out of the set of all nonisomorphic graphs with N nodes is substantially lower than randomly choosing a cospectral graph.

Theorem 6 (McKay) *For sufficiently large N, the number of cospectral graphs exceeds the number of regular graphs.*

Proof[13]**:** The number of pairs of cospectral graphs, conjectured by Godsil and McKay (1982) and proved by Haemers and Spence (2004, Theorem 3), is at least $\left(\frac{1}{24} - o(1) \right) N^3 g_{N-1}$, where $g_N = n_{\mathrm{nonisomorphic}\ G(N)}$ in (2.42) is the number of

---
[13] Private communication with Brendan McKay.

nonisomorphic graphs with N nodes (**art.** 35). Since pairs of cospectral graphs are a subset of all cospectral graphs and since most graphs are asymmetric (**art.** 34), we find with (2.42), for large N, that the number of cospectral graphs is lower bounded by $n_{\text{cospectral graph}} \geq cN^4 \frac{2^{\binom{N-1}{2}}}{N!}$, where c is a constant. The total number of regular graphs was determined, for large N, by McKay and Wormald (1990, Corollary 1),

$$n_{\text{regular graphs}} \sim \sqrt{2e}\beta_N \frac{2^{\frac{N^2}{2}}}{(\pi N)^{\frac{N}{2}}}$$

where β_N, specified in McKay and Wormald (1990, Corollary 1), has a different value depending on whether N is even, $1 \bmod 4$ or $3 \bmod 4$. Let the constant $b = \max_N \sqrt{2e}\beta_N$ so that $b \approx 4.2$. Most regular graphs are shown in Krivelevich *et al.* (2001) to be asymmetric. The total number of nonisomorphic regular graphs is, for large N, at most

$$n_{\text{nonisomorphic regular graphs}} \leq b' \frac{2^{\frac{N^2}{2}}}{(\pi N)^{\frac{N}{2}} N!}$$

where the constant b' is slightly larger than b. The ratio

$$\frac{n_{\text{nonisomorphic regular graphs}}}{n_{\text{cospectral graph}}} \leq \frac{b'}{c} \frac{1}{(\pi N)^{\frac{N}{2}} N^4 2^{-\frac{3}{2}N+1}} = O\left(\frac{e^{-\frac{N}{2}(\ln N + \ln \pi - 3\ln 2)}}{N^4}\right)$$

rapidly tends to zero with N. $\qquad\square$

3

Eigenvalues of the adjacency matrix

Only general results of the adjacency eigenvalue spectrum of an undirected graph G are treated. The spectrum of special types of graphs is computed in Chapter 6.

3.1 General properties

41. For an $N \times N$ symmetric, possibly weighted, adjacency matrix A, **art.** 247 shows that A has N real eigenvalues, which we order as $\lambda_N \leq \lambda_{N-1} \leq \cdots \leq \lambda_1$. Apart from a similarity transform (**art.** 239), the set of eigenvalues $\{\lambda_1, \lambda_2, \ldots, \lambda_N\}$ with corresponding set of eigenvectors $\{x_1, x_2, \ldots, x_N\}$ is unique. A relabeling of the nodes in the graph, which is a permutation discussed in Section 2.5 and a special type of similarity transform, obviously does not alter the structure of the graph, but merely expresses the eigenvectors in a different base.

The classical Perron-Frobenius Theorem 75 in **art.** 269 for non-negative, irreducible matrices states that the largest eigenvalue λ_1 is a simple and non-negative root of the characteristic polynomial in (A.95) possessing the only eigenvector of A with non-negative components. The largest eigenvalue λ_1 is also called the *spectral radius* of the graph.

42. *Range of eigenvalues of the adjacency matrix.* Gerschgorin's Theorem 65 applied to the adjacency matrix states that any eigenvalue of A lies in the interval $[-d_{\max}, d_{\max}]$, where $d_{\max}$ is the maximum degree in the graph G. Hence, $\lambda_1 \leq N - 1$ and this maximum is attained in the complete graph (see Section 6.1).

Theorem 109, with $m = 1$ and using (3.7) below, indicates that all the eigenvalues of A are contained in the interval $\left[-\sqrt{\frac{2L}{N}(N-1)}, \sqrt{\frac{2L}{N}(N-1)}\right]$. In terms of the average degree $d_{av} = \frac{2L}{N}$ and combining both Theorems 65 and 109, any adjacency eigenvalue of an undirected graph obeys

$$\lambda_k \in \left[-\min\left(d_{\max}, \sqrt{d_{av}(N-1)}\right), \min\left(d_{\max}, \sqrt{d_{av}(N-1)}\right)\right]$$

43. *Fundamental weights.* Left-multiplying the eigenvalue equation $Ax_k = \lambda_k x_k$ in

51

(1.3) by the all-one vector u and invoking $u^T A = d^T$ in (2.4) yields

$$\lambda_k = \frac{d^T x_k}{u^T x_k} = \sum_{j=1}^{N} d_j \frac{(x_k)_j}{\sum_{m=1}^{N} (x_k)_m} \tag{3.1}$$

which expresses the k-th eigenvalue as a weighted sum of the nodal degrees. Due to its appearance in many spectral relations,

$$w_k = u^T x_k = \sum_{j=1}^{N} (x_k)_j \tag{3.2}$$

is called the k-th fundamental weight of the graph G. Fundamental weights are related to graph angles (3.31) and can be regarded as graph metrics.

44. Let x_k denote the eigenvector of A belonging to the eigenvalue λ_k that satisfies the normalization $x_k^T x_k = 1$. **Art.** 251 shows that $\lambda_k = x_k^T A x_k$. Writing out the quadratic form yields

$$\lambda_k = x_k^T A x_k = \sum_{i=1}^{N} \sum_{j=1}^{N} a_{ij} (x_k)_i (x_k)_j = 2 \sum_{i=1}^{N-1} \sum_{j=i+1}^{N} a_{ij} (x_k)_i (x_k)_j$$

which can be written as

$$\lambda_k = 2 \sum_{l=1}^{L} (x_k)_{l+} (x_k)_{l-} \tag{3.3}$$

where a link $l \in \mathcal{L}$ joins the nodes l^+ and l^-. The expression (3.3) shows that any eigenvalue λ_k of the adjacency matrix A can be written as a sum of products of eigenvector components over all links in the graph G. An analogous representation for the Laplacian is given in **art.** 103. In particular, for the largest eigenvalue λ_1, all terms in (3.3) are non-negative (**art.** 41).

By invoking $0 \le (z_i - z_j)^2$, we observe that $2 z_i z_j \le z_i^2 + z_j^2 \le \sum_{i=1}^{N} z_i^2 = z^T z$. Hence, when considering normalized vectors such that $x_k^T x_k = \|x_k\|_2^2 = 1$, any term in (3.3) is bounded by one, $2 (x_k)_{l+} (x_k)_{l-} \le 1$. Moreover, the equality $2 (x_k)_{l+} (x_k)_{l-} = 1$ is only possible if and only if $(x_k)_{l+}^2 + (x_k)_{l+}^2 = 1$, in which case $(x_k)_{l+} = (x_k)_{l-} = \frac{1}{\sqrt{2}}$ and all the other eigenvector components are zero. For the largest eigenvector x_1, this situation can only occur for a graph G consisting of K_2 and $N - 2$ disjoint nodes and (3.3) indicates that $\lambda_1 = 1$. In summary, for any connected graph G, the Perron-Frobenius Theorem 75 in **art.** 269 shows, for $N > 2$, that $0 < 2 (x_1)_{l+} (x_1)_{l-} < 1$.

3.2 Characteristic polynomial $c_A(\lambda)$ of the adjacency matrix A

This section applies the general theory in Chapter 10 to the adjacency matrix A and mainly investigates the coefficients $c_k (A)$ of the characteristic polynomial $c_A(\lambda)$.

45. *Eigenvalues are either integer or irrational.* The characteristic polynomial

$$c_A(\lambda) = \det(A - \lambda I) = \sum_{k=0}^{N} c_k(A)\,\lambda^k \tag{3.4}$$

defined in **art. 235**, has integer coefficients $c_k(A)$ and $c_N(A) = (-1)^N$. **Art. 292** demonstrates that the only rational zeros of $c_A(\lambda)$, i.e., zeros belonging to $\mathbb{Q}$, are integers. This property also holds for the Laplacian matrix Q. For example, $\frac{3}{4}$ is never an eigenvalue of A nor Q.

Art. 293 gives additional methods to check from the integer coefficients $c_k(A)$ in (3.4) whether $c_A(\lambda)$ can be factored into two lower degree polynomials with integer coefficients.

46. Since $a_{ii} = 0$, we have that $\operatorname{trace}(A) = 0$. From (A.99), the coefficient $c_{N-1}(A)$ of the characteristic polynomial $c_A(\lambda)$ is

$$c_{N-1}(A) = \sum_{k=1}^{N} \lambda_k = 0 \tag{3.5}$$

47. *Newton identities.* Applying the Newton identities (B.4) or (B.8) in **art. 294** to the characteristic polynomial (A.95) and (A.97) of the adjacency matrix with $z_k = \lambda_k$, $a_k = c_k(A)$ and using $c_{N-1}(A) = 0$ from (3.5) yields for the first few values,

$$(-1)^N c_{N-2}(A) = -\frac{1}{2} \sum_{k=1}^{N} \lambda_k^2$$

$$(-1)^N c_{N-3}(A) = -\frac{1}{3} \sum_{k=1}^{N} \lambda_k^3$$

$$(-1)^N c_{N-4}(A) = \frac{1}{8} \left(\left(\sum_{k=1}^{N} \lambda_k^2\right)^2 - 2 \sum_{k=1}^{N} \lambda_k^4 \right)$$

48. The coefficient $c_0(A)$ follows from (A.98) as $c_0(A) = \det A = \prod_{k=1}^{N} \lambda_k$. Applying the Hadamard inequality (A.78) for the determinant of a matrix yields, with (2.2),

$$|\det A| \leq \prod_{j=1}^{N} \left(\sum_{i=1}^{N} a_{ji}^2 \right)^{\frac{1}{2}} = \prod_{j=1}^{N} \left(\sum_{i=1}^{N} a_{ji} \right)^{\frac{1}{2}} = \prod_{j=1}^{N} \sqrt{d_j}$$

Hence, with $\det A = \prod_{k=1}^{N} \lambda_k$ in (A.98), we find

$$(\det A)^2 = \prod_{k=1}^{N} \lambda_k^2 \leq \prod_{j=1}^{N} d_j \tag{3.6}$$

49. The coefficient $c_{N-2}(A)$ follows from (A.96) as $c_{N-2}(A) = (-1)^N \sum_{\text{all}} M_2$ and is explicitly given as $c_{N-2}(A) = (-1)^N \sum_{i=1}^{N} \sum_{j=i+1}^{N} (a_{ii}a_{jj} - a_{ij}a_{ji})$ in **art.** 210. Since $a_{ii} = 0$, the double sum is $-\frac{1}{2}u^T A u$ and (2.5) leads to

$$(-1)^N c_{N-2}(A) = -L$$

The Newton identities in **art.** 47 show that the number of links L equals

$$L = \frac{1}{2} \sum_{k=1}^{N} \lambda_k^2 \tag{3.7}$$

Since $E[\lambda] = \frac{1}{N} \sum_{k=1}^{N} \lambda_k = 0$ in **art.** 46, the variance $\text{Var}[\lambda] = \frac{1}{N} \sum_{k=1}^{N} (\lambda_k - E[\lambda])^2$ of the adjacency eigenvalues equals, invoking the basic law of the degree (2.3),

$$\text{Var}[\lambda] = \frac{1}{N} \sum_{k=1}^{N} \lambda_k^2 = \frac{2L}{N} = E[D]$$

This stochastic interpretation is helpful to understand the density function of the adjacency eigenvalues in Section 8.

50. Each principal submatrix $M_{3\times3}$ of the adjacency matrix A is of the form

$$M_{3\times3} = \begin{bmatrix} 0 & x & z \\ x & 0 & y \\ z & y & 0 \end{bmatrix}$$

and the corresponding minor $M_3 = \det M_{3\times3} = 2xyz$ is only non-zero for $x = y = z = 1$. That form of $M_{3\times3}$ corresponds with a subgraph of three different nodes that are fully connected in a triangle. Since (A.96) reduces to $(-1)^N c_{N-3}(A) = -\sum_{\text{all}} M_3$ and $c_{N-3}(A) = -2\times$ the number $\blacktriangle_G$ of triangles in G. From **art.** 47, it follows that the number of triangles in G is

$$\blacktriangle_G = \frac{1}{6} \sum_{k=1}^{N} \lambda_k^3 \tag{3.8}$$

51. *Coefficient c_k of the characteristic polynomial* $c_A(\lambda) = \sum_{k=0}^{n} c_k \lambda^k$. From (A.96) and by identifying the structure of a minor M_k of the adjacency matrix A of an *undirected* graph, any coefficient $c_{N-k}(A)$ can be expressed in terms of graph characteristics,

$$(-1)^N c_{N-k}(A) = \sum_{\mathcal{G} \in G_k} (-1)^{cycles(\mathcal{G})} \tag{3.9}$$

where G_k is the set of all subgraphs of G with exactly k nodes and $cycles(\mathcal{G})$ is the number of cycles in a subgraph $\mathcal{G} \in G_k$. The minor M_k is a determinant of the $M_{k\times k}$ submatrix of A and defined (see **art.** 208) as

$$M_k = \sum_p (-1)^{\sigma(p)} a_{1p_1} a_{2p_2} \cdots a_{kp_k}$$

where the sum is over all $k!$ permutations $p = (p_1, p_2, \ldots, p_k)$ of $(1, 2, \ldots, k)$ and $\sigma(p)$ is the number of interchanges of $(1, 2, \ldots, k)$ to obtain $(p_1, p_2, \ldots, p_k)$. Only if all the links $(1, p_1), (2, p_2), \ldots, (k, p_k)$ are contained in G, then $a_{1p_1} a_{2p_2} \ldots a_{kp_k}$ is non-zero. Since $a_{jj} = 0$, the sequence of contributing links $(1, p_1), (2, p_2), \ldots, (k, p_k)$ is a set of disjoint cycles such that each node in G_k belongs to exactly one of these cycles and $\sigma(p)$ depends on the number of those disjoint cycles. The minor M_k is constructed from a specific set $\mathcal{G} \in G_k$ of k out of N nodes and in total there are $\binom{N}{k}$ such sets in G_k, which is rewritten as (3.9).

Directed graphs. Harary (1962) discusses the determinant $\det A$ of a *directed* graph, from which another expression than (3.9) for the coefficients $c_k(A)$ of the characteristic polynomial $c_A(\lambda)$ can be derived. An elementary subgraph $\mathcal{H}$ of G on k nodes is a graph in which each component is either a link between two distinct nodes or a cycle. Here, a cycle is thus of at least length 3, possessing at least three nodes or links. Harary observes that, in the determinant of the adjacency matrix A (or in each of its minors) of a *directed* graph, each directed cycle of even (odd) length contributes negatively (positively) to $\det A$. Let e_c denote the number of even components in an elementary subgraph, i.e. containing an even number of nodes. Each cycle in an undirected graph corresponds to the two directions in its directed companion. Harary (1962) shows that the coefficient of the characteristic polynomial $c_A(\lambda)$ of the adjacency matrix of a *directed* graph can be written as

$$(-1)^N c_{N-k}(A) = \sum_{\mathcal{H} \in H_k} (-1)^{e_c(\mathcal{H})} 2^{|c(\mathcal{H})|}$$

where $c(\mathcal{H})$ is the set of components that are cycles in $\mathcal{H}$ and H_k denotes the set of all elementary subgraphs $\mathcal{H}$ of G with k nodes.

Undirected trees. Confining to trees, where the only cycles are directed cycles of length 2 corresponding to the links of the tree, Mowshowitz (1972) further explores (3.9) and shows for the characteristic polynomial $c_{A_T}(\lambda) = \det(A_T - \lambda I) = \sum_{k=0}^{N} c_k(A_T) \lambda^k$ of a tree T that the coefficient $a_k = (-1)^k c_{N-k}(A_T) = (-1)^r h_r(T)$ if $k = 2r$ is even, $a_k = 0$ if k is odd and $a_0 = 1$, where $h_r(T)$ obeys the recursion, as a special case of (3.106) below,

$$h_r(T + l_{uv}) = h_{r-1}(T) + h_{r-1}(T_{\backslash \{u\}})$$

where $T + l_{uv}$ is the tree T to which a link l_{uv} with end point v is added at node u and $T_{\backslash \{u\}}$ is the tree from which a node u is removed. Mowshowitz (1972) mentions that $|a_k|$ is the number of sets consisting of k pairwise non-incident links of T (i.e. links that do not share common nodes), which equals the number of independent sets of links of size k in T, also called the number of matchings of size k in tree T.

52. Characteristic polynomials of graphs with one node removed. If $G_{\backslash \{j\}}$ is the graph obtained from the graph G after the removal of node j, then its adjacency characteristic polynomial (**art. 45**) is

$$c_{A\setminus\{j\}}(\lambda) = \det\left(A_{\setminus\{j\}} - \lambda I\right) = \sum_{k=0}^{N} c_k\left(A_{\setminus\{j\}}\right)\lambda^k$$

After equating corresponding powers in λ in formula $\frac{dc_A(\lambda)}{d\lambda} = -\sum_{j=1}^{N} c_{A\setminus\{j\}}(\lambda)$ in (A.46) in **art. 213**, we obtain, for $0 \le k \le N-1$,

$$c_{k+1}(A) = -\frac{1}{k+1}\sum_{j=1}^{N} c_k\left(A_{\setminus\{j\}}\right) \tag{3.10}$$

The case $k = N-1$ and $N-2$ are identities. For $k = N-3$, we obtain with $c_{N-2}(A) = -L$ in **art. 49** the relation $L_G = \frac{1}{N-2}\sum_{j=1}^{N} L_{G\setminus\{j\}}$ between the number of links in G and $G_{\setminus\{j\}}$. Similarly, for $k = N-4$ and invoking **art. 50**, we find that the corresponding relation $\blacktriangle_G = \frac{1}{N-3}\sum_{j=1}^{N}\blacktriangle_{G\setminus\{j\}}$ for the number of triangles. Hence, the average number of links $\frac{1}{N}\sum_{j=1}^{N} L_{G\setminus\{j\}}$ and of triangles $\frac{1}{N}\sum_{j=1}^{N}\blacktriangle_{G\setminus\{j\}}$ is always smaller than the number of links L_G and triangles $\blacktriangle_G$ in G, respectively.

The relation (3.10) can be understood by considering two ensembles of graphs. The first $S_1 = \{G, G, \ldots, G\}$ contains N times the original graph G, while the second ensemble is $S_2 = \{G_{\setminus\{1\}}, G_{\setminus\{2\}}, \ldots, G_{\setminus\{N\}}\}$. An enumeration, such as the number of links or triangles, in S_1 simply equals N times that enumeration, while in S_2 each link is affected twice, a triangle three times, etc. by removing in total N (different) nodes. Hence, we find that $NL_G - \sum_{j=1}^{N} L_{G\setminus\{j\}} = 2L_G$ and $N\blacktriangle_G - \sum_{j=1}^{N}\blacktriangle_{G\setminus\{j\}} = 3\blacktriangle_G$.

53. *Newton's inequalities.* Since the characteristic polynomial $c_A(x)$ has real coefficients and real zeros, Newton's Theorem 97 in **art. 327** provides the inequality, for $1 \le k \le N-1$,

$$c_k^2(A) \ge c_{k+1}(A)\, c_{k-1}(A)\,\frac{k+1}{k}\frac{N-k+1}{N-k} \tag{3.11}$$

Because $c_{N-1}(A) = 0$ in **art. 46**, the Newton inequality (3.11) for $k = N-1$ and $N-2$ does not yield a useful bound. For $k = N-3$, on the other hand, we have, using $c_{N-2}(A) = -L$ in **art. 49** and $c_{N-3}(A) = -2\blacktriangle_G$ in **art. 50**, that

$$\frac{3(N-3)}{(N-2)L}(\blacktriangle_G)^2 \ge -c_{N-4}(A)$$

Finally, with $c_{N-4}(A) = \frac{1}{8}\left(4L^2 - 2\sum_{k=1}^{N}\lambda_k^4\right)$ from **art. 47** and **art. 49**, we find the inequality

$$\sum_{k=1}^{N}\lambda_k^4 \le \frac{12(N-3)}{(N-2)L}(\blacktriangle_G)^2 + 2L^2 \tag{3.12}$$

54. *Eigenvalue bounds for triangles.* From the inequality (A.14) for Hölder q-norms, we find for $p > q > 0$ that, if $\sum_{k=1}^{N}|\lambda_k|^q < \Lambda^q$, then $\sum_{k=1}^{N}|\lambda_k|^p < \Lambda^p$.

Since $\sum_{k=1}^{N} \lambda_k = 0$, not all λ_k can be positive and combined with $\left| \sum_{k=1}^{N} \lambda_k^p \right| \leq \sum_{k=1}^{N} |\lambda_k|^p$, we also have that $\left| \sum_{k=1}^{N} \lambda_k^p \right| < \Lambda^p$. Applied to the case where $q = 2$ and $p = 3$ gives the following implication: if $\sum_{k=2}^{N} \lambda_k^2 < \lambda_1^2$ then $\left| \sum_{k=2}^{N} \lambda_k^3 \right| < \lambda_1^3$. In that case, the number of triangles given in (3.8) is

$$\blacktriangle_G = \frac{1}{6} \lambda_1^3 + \frac{1}{6} \sum_{k=2}^{N} \lambda_k^3 \geq \frac{1}{6} \lambda_1^3 - \frac{1}{6} \left| \sum_{k=2}^{N} \lambda_k^3 \right| > 0$$

Invoking (3.7) to the implication: if $2L = \sum_{k=2}^{N} \lambda_k^2 + \lambda_1^2 < 2\lambda_1^2$, then the number of triangles $\blacktriangle_G$ in G is at least one. In summary[1], if $\lambda_1 > \sqrt{L}$, then the graph G contains at least one triangle.

Theorem 7 (Mantel) *A graph G with N nodes and more than $\left[\frac{N^2}{4} \right]$ links contains at least one triangle.*

Proof: If $L > \frac{N^2}{4} \geq \left[\frac{N^2}{4} \right]$, which is equivalent to $N < 2\sqrt{L}$, then the classical lower bound on the largest eigenvalue (3.63) in **art. 72** is $\lambda_1 \geq \frac{2L}{N} > \frac{2L}{2\sqrt{L}} = \sqrt{L}$ and $\lambda_1 > \sqrt{L}$ is precisely the condition above to have at least one triangle. $\square$

Mantel's Theorem 7 is best possible, because the complete bipartite graph $K_{\frac{N}{2}, \frac{N}{2}}$, with N even, contains $L = \frac{N^2}{4}$ links, but no triangle. Its generalization due to Turán for any clique size is stated in Section 6.9, where the Turán graph is studied. Nikiforov (2021) proves a related, but more complicated result: if $\lambda_1 \geq \sqrt{L}$, then the maximum number of triangles with a common edge in the graph G, called the booksize $bk(G)$ of G, is $bk(G) > \frac{1}{12} \sqrt[4]{L}$, unless G is a complete bipartite graph with possibly some isolated nodes. Nikiforov (2021) also proves the instance $r = 2$ of his conjecture with Bollobás: If a graph G with L links and $N \geq r + 1$ nodes does not contain a clique K_{r+1}, then $\lambda_1^2 + \lambda_2^2 \leq 2 \left(1 - \frac{1}{r} \right) L$.

3.3 Regular graphs

The class of regular graphs possesses a lot of specific and remarkable properties that justify the discussion of some spectrum related properties here.

55. *Regular graphs.* Every node j in a regular graph has the same degree $d_j = r$ and relation (2.2) indicates that each row sum of A equals r. The basic law of the degree (2.3) reduces for regular graphs to $2L = Nr$, implying that, if the degree r is odd, then the number of nodes N must be even.

---

[1] Nikiforov (2021) remarks that Eva Nosal in her master thesis at the University of Calgary in 1970 has proved this elegant result. Mantel's Theorem 7 of 1907 has been greatly extended by Turán in 1941 as mentioned in Bollobás (1998, p. 6), who gives a non-spectral proof.

Theorem 8 *The maximum degree $d_{\max} = \max_{1 \leq j \leq N} d_j$ is the largest eigenvalue of the adjacency matrix A of a connected graph G if and only if the corresponding graph is regular, i.e. $d_j = d_{\max} = r$ for all j.*

Proof: If x is an eigenvector of A belonging to eigenvalue $\lambda = d_{\max}$ so is each vector kx for each complex k (**art. 235**). Thus, we can scale the eigenvector x such that the maximum component, say $x_m = 1$, and $x_k \leq 1$ for all k. The eigenvalue equation $Ax = d_{\max}x$ for that maximum component x_m is

$$d_{\max}x_m = d_{\max} = \sum_{j=1}^{N} a_{mj}x_j = \sum_{j \in \text{ neighbor}(m)} x_j$$

which implies that all $x_j = 1$ whenever $a_{mj} = 1$, i.e., when the node j is adjacent to node m. Hence, the degree of node m is $d_m = d_{\max}$. For any node j adjacent to m for which the component $x_j = 1$, a same eigenvalue relation holds and thus $d_j = d_{\max}$. Proceeding with this process shows that every node $k \in G$ has same degree $d_k = d_{\max}$ because G is connected. Hence, $x = u$ where $u^T = [1 \ 1 \ \cdots \ 1]$ and the Perron-Frobenius Theorem 75 shows that u is the eigenvector belonging to the largest eigenvalue of A. Conversely, if G is connected and regular, then $\sum_{j=1}^{N} a_{mj} = d_{\max} = r$ for each m such that u is the eigenvector belonging to eigenvalue $\lambda = d_{\max}$, and the only possible eigenvector (**art. 41**). Hence, there is only one eigenvalue $d_{\max} = r$. □

Theorem 8 shows that, for a regular graph, $Au = ru$, and, thus, $AJ = rJ$. After taking the transpose, $(AJ)^T = JA = rJ$, we see that $AJ = JA$. Thus, A and J commute if G is regular.

Theorem 9 (Hoffman) *A graph G is regular and connected if and only if there exists a polynomial p such that $J = p(A)$.*

Proof: (a) If $J = p(A)$, then J and A commute and, hence, G is regular. (b) Since the largest eigenvalue r is simple (**art. 41**), the Laplacian $Q = rI - A$ has a zero eigenvalue with multiplicity 1. Theorem 21 then states that a regular graph G is connected. Conversely, let G be connected and regular. We can diagonalize the adjacency matrix A of G by using an orthogonal matrix formed by its eigenvectors (**art. 247**). This basis of eigenvectors of A also diagonalizes J as $\text{diag}(N, 0, \ldots, 0)$, because J and A commute (**art. 284**). Consider the polynomial

$$q(x) = \frac{c_A(x)}{x - r} = \prod_{j=2}^{N} (\lambda_k(A) - x)$$

where $c_A(x)$ is the characteristic polynomial of A, then $J = N\frac{q(A)}{q(r)}$, because the projections on the basis vectors are $q(A)x_j = 0$ if $x_j \neq u$ and $q(A)u = q(r)u$, while $Ju = Nu$. Thus, the polynomial $p(x) = N\frac{q(x)}{q(r)}$ satisfies the requirement. □

The proof shows that, if $m_{c_A}(x)$ is the minimal polynomial (**art.** 228) associated to the characteristic polynomial $c_A(x)$ and $q_{m_c}(x) = \frac{m_{c_A}(x)}{x-r}$, the polynomial $p_{m_c}(x) = N \frac{q_{m_c}(x)}{q_{m_c}(r)}$ of possibly lower degree can be found (see **art.** 229).

56. *Strongly regular graphs.* Following Cvetković *et al.* (1995), we first define $\varpi(v,w)$ as the number of nodes adjacent to *both* node v and node $w \neq v$. In other words, $\varpi(v,w)$ is the number of common neighbors of both v and w. A regular graph G of degree $r > 0$, different from the complete graph K_N, is called strongly regular if $\varpi(v,w) = n_1$ for each pair (v,w) of adjacent nodes and $\varpi(v,w) = n_2$ for each pair (v,w) of non-adjacent nodes. A strongly regular graph is completely defined by the parameters (N, r, n_1, n_2).

Examples The Petersen graph in Fig. 2.3 is a strongly regular graph with parameters $(10, 3, 0, 1)$. Cvetković *et al.* (2009) show how many strongly regular graphs can be constructed from line graphs. The line graph $l(K_N)$ of the complete graph is strongly regular with parameters $\left(\frac{N(N-1)}{2}, 2N-4, N-2, 4\right)$ for $N > 3$. The corresponding eigenvalues of the $\binom{N}{2} \times \binom{N}{2}$ adjacency matrix of $l(K_N)$ are $r = 2N-4$, $[(-2)]^{\binom{N-1}{2}-1}$ and $[(N-4)]^{N-1}$, for $N > 3$. Another example is the class of Paley graphs P_q, whose nodes belong to the finite field $\mathbb{F}_q$ of order q, where q is a prime power congruent to 1 modulo 4, and whose links (i, j) are present if and only if $i - j$ is a quadratic residue (see Hardy and Wright (2008)). The Paley graph P_q is strongly regular with parameters $\left(|\mathbb{F}_q|, \frac{q-1}{2}, \frac{q-5}{4}, \frac{q-1}{4}\right)$. Bollobás (2001, Chapter 13) discusses properties of the Paley graph and its generalizations, the Cayley graphs and conference graphs.

The number of common neighbors of two different nodes i and j is equal to the number of 2-hop walks between i and j. Thus, Lemma 1 states that $\varpi(i, j) = \left(A^2\right)_{ij}$ if $i \neq j$. **Art.** 19 shows that $\left(A^2\right)_{ii} = d_i = r$. The condition for strong regularity states that, for different nodes i and j, $\left(A^2\right)_{ij} = n_1 a_{ij} + n_2(1 - a_{ij})$, because $\varpi(i, j) = n_1$ if node i and j are neighbors, hence, $a_{ij} = 1$ and $\varpi(i, j) = n_2$, if they are not, i.e. $a_{ij} = 0$. Adding the two mutual exclusive conditions together with $\varpi(i, j) = \left(A^2\right)_{ij}$ demonstrates the relation. Combining all entries into a matrix form yields

$$A^2 = n_1 A + n_2 A^c + rI$$

Finally, using $A^c = J - I - A$ in **art.** 1, we obtain the matrix relation that characterizes strong regularity,

$$A^2 = (n_1 - n_2) A + n_2 J + (r - n_2) I$$

from which $J = \frac{1}{n_2}\left(A^2 + (n_2 - n_1) A + (n_2 - r) I\right)$. Hence, the polynomial $J = p(A)$ in Hoffman's Theorem 9 is the quadratic polynomial

$$p_2(z) = \frac{1}{n_2}\left(z^2 + (n_2 - n_1) z + (n_2 - r)\right)$$

from which we deduce that the minimal polynomial $m_{c_A}(x) = \frac{q_{m_c}(r)}{N}(x-r)p_2(x)$ is of degree 3. The definition of a minimal polynomial in **art.** 310 implies that the adjacency matrix A of G possesses precisely three distinct eigenvalues $\lambda_1 = r$, λ_2 and λ_3, where λ_2 and λ_3 are zeros of $p_2(x)$, related by $n_1 - n_2 = \lambda_2 + \lambda_3$ and $n_2 - r = \lambda_2\lambda_3$. The property that strongly regular graphs have three different eigenvalues explains why the complete graph K_N must be excluded in the definition above. In summary, we have proved:

Theorem 10 *A connected graph G is strongly regular with degree $r > 0$ if and only if its adjacency matrix A has three distinct eigenvalues $\lambda_1 = r$, λ_2 and λ_3, which satisfy*

$$n_1 = r + \lambda_2 + \lambda_3 + \lambda_2\lambda_3$$
$$n_2 = r + \lambda_2\lambda_3$$

where n_1 and n_2 are the number of common neighbors of adjacent and non-adjacent nodes, respectively.

3.4 Powers of the adjacency matrix

Before concentrating on the total number of walks in a graph in Section 3.5, we review the eigenvalue equation $A^k x = \lambda^k x$ in Section 1.1, which reads in matrix form, $A^k X = X\Lambda^k$, where the orthogonal matrix X, satisfying $X^T X = XX^T = I$, contains the eigenvectors in its columns.

57. *Eigenvalue relation $\Lambda^k = X^T A^k X$.* Formula $\Lambda^k = X^T A^k X$ allows us to express the eigenvalue λ_m^k in terms of the elements of the matrix A^k as

$$\lambda_m^k \delta_{ml} = \sum_{i=1}^{N}\sum_{j=1}^{N} \left(A^k\right)_{ij} (x_m)_i (x_l)_j \tag{3.13}$$

which is equivalent to

$$\begin{cases} \lambda_m^k = \sum_{i=1}^{N}\sum_{j=1}^{N} \left(A^k\right)_{ij}(x_m)_i(x_m)_j \\ 0 = \sum_{i=1}^{N}\sum_{j=1}^{N} \left(A^k\right)_{ij}(x_m)_i(x_l)_j & \text{if } m \neq l \end{cases}$$

Only if node i and j are connected by a k-hop walk (**art.** 18) and $\left(A^k\right)_{ij} > 0$, then (3.13) shows that their corresponding components of the m-th eigenvector ($m = l$) contribute in (3.13) to λ_m^k. Thus, (3.13) is another representation of (1.5). For $m = 1$, we directly find (3.3) again in **art.** 44. We rewrite (3.13) as $\lambda_m^k \delta_{ml} = \sum_{i=1}^{N}\sum_{j=1}^{i} \left(A^k\right)_{ij}(x_m)_i(x_l)_j + \sum_{i=1}^{N}\sum_{j=i+1}^{N} \left(A^k\right)_{ij}(x_m)_i(x_l)_j$, reverse the summations in the first term and split off the $i = j$ terms for a symmetric matrix $A = A^T$ to obtain

$$\sum_{i=1}^{N} \left(A^k\right)_{ii}(x_m)_i^2 = \lambda_m^k - 2\sum_{i=1}^{N}\sum_{j=i+1}^{N} \left(A^k\right)_{ij}(x_m)_j(x_m)_i \tag{3.14}$$

and for $m \neq l$

$$\sum_{i=1}^{N} \left(A^k\right)_{ii} (x_m)_i (x_l)_i = - \sum_{i=1}^{N} \sum_{j=i+1}^{N} \left(A^k\right)_{ij} \left\{ (x_m)_j (x_l)_i + (x_m)_i (x_l)_j \right\}$$

These relations reduce for $k = 0$ with $A^0 = I$ to the orthogonality conditions of eigenvectors in (A.124). For $m = 1$, the Perron-Frobenius Theorem 75 tells us that $(x_1)_j \geq 0$ and (3.14) leads, for all integer $k \geq 0$, to the bound

$$\sum_{i=1}^{N} \left(A^k\right)_{ii} (x_1)_i^2 \leq \lambda_1^k$$

(a) Substituting $ab = \frac{(a^2+b^2)-(a-b)^2}{2}$ in (3.13) and, denoting $\sum_{i=1}^{N} \left(A^n\right)_{ij} = \left(A^n u\right)_j$, gives us

$$\lambda_m^k = \sum_{j=1}^{N} \left(A^k u\right)_j (x_m)_j^2 - \frac{1}{2} \sum_{i=1}^{N} \sum_{j=1}^{N} \left(A^k\right)_{ij} \left((x_m)_i - (x_m)_j\right)^2 \qquad (3.15)$$

where each term in a sum is non-negative, while (b) $ab = \frac{(a+b)^2-(a^2+b^2)}{2}$ results in

$$\lambda_m^k = \frac{1}{2} \sum_{i=1}^{N} \sum_{j=1}^{N} \left(A^k\right)_{ij} \left((x_m)_i + (x_m)_j\right)^2 - \sum_{j=1}^{N} \left(A^k u\right)_j (x_m)_j^2$$

which reduces for $k = 1$ to

$$\lambda_m = \sum_{l \in \mathcal{L}} \left((x_m)_{l+} + (x_m)_{l-}\right)^2 - \sum_{j=1}^{N} d_j (x_m)_j^2 \qquad (3.16)$$

Relation (3.16) is also obtained from the unsigned incidence matrix R in **art. 25**, that obeys $A = RR^T - \Delta$,

$$\lambda_m = x_m^T A x_m = x_m^T R R^T x_m - x_m^T \Delta x_m = \left\| R^T x_m \right\|_2^2 - x_m^T \Delta x_m$$

(c) Invoking $ab = \frac{(a+b)^2-(a-b)^2}{4}$ in (3.13) yields

$$4\lambda_m^k = \sum_{i=1}^{N} \sum_{j=1}^{N} \left(A^k\right)_{ij} \left((x_m)_i + (x_m)_j\right)^2 - \sum_{i=1}^{N} \sum_{j=1}^{N} \left(A^k\right)_{ij} \left((x_m)_i - (x_m)_j\right)^2$$

which reduces for $k = 1$ to

$$\lambda_m = \frac{1}{2} \sum_{l \in \mathcal{L}} \left((x_m)_{l+} + (x_m)_{l-}\right)^2 - \frac{1}{2} \sum_{l \in \mathcal{L}} \left((x_m)_{l+} - (x_m)_{l-}\right)^2 \qquad (3.17)$$

and expresses an eigenvalue of the adjacency matrix as a difference of two sums of squares.

58. *Eigenvalue relation $A^k = X\Lambda^k X^T$.* The reverse $A^k = X\Lambda^k X^T$ of $\Lambda^k =$

$X^T A^k X$ in **art.** 57 expresses the number of walks $(A^k)_{ij}$ with k hops (**art.** 18) in terms of the eigenvalues $\lambda_1^k, \lambda_2^k, \ldots, \lambda_N^k$ of A^k as

$$(A^k)_{ij} = \sum_{m=1}^{N} \lambda_m^k (x_m)_i (x_m)_j \tag{3.18}$$

whose matrix form is

$$A^k = \sum_{m=1}^{N} \lambda_m^k x_m x_m^T \tag{3.19}$$

The expression (3.18) can be generalized to a function f of the matrix A, via Taylor series in **art.** 231, resulting in $(f(A))_{ij} = \sum_{k=1}^{N} f(\lambda_k)(x_k)_i (x_k)_j$. When applying **art.** 257 to $f(x) = x^k$, then (A.142), (A.143) and (A.144) translate to, respectively,

$$(A^k)_{ij} = \frac{1}{2} \sum_{m=1}^{N} \lambda_m^k \left((x_m)_i + (x_m)_j\right)^2 - \frac{(A^k)_{ii} + (A^k)_{jj}}{2}$$

$$(A^k)_{ij} = \frac{(A^k)_{ii} + (A^k)_{jj}}{2} - \frac{1}{2} \sum_{m=1}^{N} \lambda_m^k \left((x_m)_i - (x_m)_j\right)^2 \tag{3.20}$$

$$(A^k)_{ij} = \frac{1}{4} \sum_{m=1}^{N} \lambda_m^k \left((x_m)_i + (x_m)_j\right)^2 - \frac{1}{4} \sum_{m=1}^{N} \lambda_m^k \left((x_m)_i - (x_m)_j\right)^2$$

The corresponding bounds (A.145) and (A.146) are

$$(1 + \delta_{ij}) \min_{1 \leq m \leq N} \lambda_m^k \leq \frac{(A^k)_{ii} + (A^k)_{jj}}{2} + (A^k)_{ij} \leq (1 + \delta_{ij}) \max_{1 \leq m \leq N} \lambda_m^k \tag{3.21}$$

$$(1 - \delta_{ij}) \min_{1 \leq m \leq N} \lambda_m^k \leq \frac{(A^k)_{ii} + (A^k)_{jj}}{2} - (A^k)_{ij} \leq (1 - \delta_{ij}) \max_{1 \leq m \leq N} \lambda_m^k \tag{3.22}$$

while (A.147) leads to the bound

$$\left|(A^k)_{ij} - \frac{\delta_{ij}}{2} \left(\min_{1 \leq m \leq N} \lambda_m^k + \max_{1 \leq m \leq N} \lambda_m^k\right)\right| \leq \frac{1}{2} \left(\max_{1 \leq m \leq N} \lambda_m^k - \min_{1 \leq m \leq N} \lambda_m^k\right) \tag{3.23}$$

Since the row vectors (A.125) in X are orthogonal (A.126) and excluding the empty graph, (3.20) for $k = 2n$ and $i \neq j$ leads to the strict inequality

$$(A^{2n})_{ij} < \frac{(A^{2n})_{ii} + (A^{2n})_{jj}}{2}$$

which is a general property of positive semidefinite and symmetric matrix (**art.** 279). The bound (3.23) indicates that each non-diagonal element is bounded by

$$0 \leq (A^k)_{ij} \leq \frac{1}{2} \left(\lambda_1^k - \min_{1 \leq m \leq N} \lambda_m^k\right)$$

while each diagonal element is bounded by

$$\max\left(0, \min_{1\leq m\leq N} \lambda_m^k\right) \leq \left(A^k\right)_{jj} \leq \lambda_1^k$$

Alternatively, the Perron-Frobenius theorem, as explained in **art.** 41, states that all eigenvector components are non-negative, which leads in (1.5) to the same bound

$$\left(A^k\right)_{jj} \leq \lambda_1^k \tag{3.24}$$

Writing (3.18) as $\left(A^k\right)_{ij} = \lambda_1^k \left(x_1\right)_i \left(x_1\right)_j \left(1 + \sum_{m=2}^{N} \left(\frac{\lambda_m}{\lambda_1}\right)^k \frac{(x_m)_i (x_m)_j}{(x_1)_i (x_1)_j}\right)$ leads, for large k and provided $\lambda_1 > |\lambda_N|$, thus excluding bipartite graphs in Section 6.7, to the asymptotic form

$$\left(A^k\right)_{ij} \sim \lambda_1^k \left(x_1\right)_i \left(x_1\right)_j \text{ for } k \to \infty \tag{3.25}$$

which also means $\lim_{k\to\infty} \frac{(A^k)_{ij}}{\lambda_1^k} = (x_1)_i (x_1)_j$. The diagonal elements of A^k for even $k = 2n$ follow from (3.18) as

$$\left(A^{2n}\right)_{jj} = \sum_{m=1}^{N} (x_m)_j^2 \lambda_m^{2n}$$

consists of all non-negative terms, illustrating that $\left(A^{2n}\right)_{jj}$ is always positive in a connected graph because $(x_1)_j > 0$, in agreement with **art.** 19. **Art.** 257 indicates that $\left(A^k\right)_{ij} \leq \frac{1}{2}\left(\lambda_1^k - \min_{1\leq m\leq N}\lambda_m^k\right)$ for $i \neq j$, illustrating sharper upper bounds for even $k = 2n$, $\left(A^{2n}\right)_{ij} \leq \frac{1}{2}\lambda_1^{2n}$, than for odd $k = 2n + 1$, $\left(A^{2n+1}\right)_{ij} \leq \lambda_1^{2n+1}$, but diagonal elements are bounded by $\left(A^k\right)_{jj} \leq \lambda_1^k$. Thus, apart from $\left(A^{2n}\right)_{jj} \sim \lambda_1^{2n} (x_1)_j^2$ in (3.25) for large n, it follows for any finite n that

$$(x_1)_j^2 \lambda_1^{2n} < \left(A^{2n}\right)_{jj} < \lambda_1^{2n}$$

For $k = 2n + 1$, (3.18) is

$$\left(A^{2n+1}\right)_{jj} = \sum_{m=1;\lambda_m>0}^{N} (x_m)_j^2 \lambda_m^{2n+1} - \sum_{l=1;\lambda_l<0}^{N} (x_l)_j^2 \left|\lambda_l^{2n+1}\right|$$

which can be zero as demonstrated in **art.** 19.

3.5 The number of walks

59. The total number N_k of walks of length k in a graph follows from Lemma 1 as

$$N_k = \sum_{i=1}^{N} \sum_{j=1}^{N} (A^k)_{ij} = u^T A^k u \tag{3.26}$$

where u is the all-one vector. For example, $N_0 = N$ and $N_1 = 2L$. If the graph is undirected, i.e. $A = A^T$, each walk $(A^k)_{ij}$ with $i \neq j$ is counted twice in (3.26). Invoking (2.4) and $A^T = A$, we can write

$$N_k = u^T A^T A^{k-2} A u = d^T A^{k-2} d$$

If $k = 2$, we obtain

$$N_2 = d^T d = \sum_{k=1}^{N} d_k^2 = N \left((E[D])^2 + \text{Var}[D] \right) \tag{3.27}$$

The number of walks $\{N_0, N_1, \ldots, N_{10}\}$ in the graph of Fig. 2.1 ignoring directions and up to $k = 10$ is $\{6, 18, 56, 174, 542, 1688, 5258, 16378, 51016, 158910, 494990\}$.

Substituting $A^k = \sum_{n=1}^{N} \lambda_n^k x_n x_n^T$ in (3.19) into (3.26) expresses the total number $N_k = u^T A^k u$ of walks of length k in terms of the eigenvalues of A as

$$N_k = \sum_{n=1}^{N} \left(x_n^T u \right)^2 \lambda_n^k \tag{3.28}$$

where $x_n^T u = \sum_{j=1}^{N} (x_n)_j$ is the n-th fundamental weight in **art.** 43. When the normalized eigenvector $x_1 = \frac{u}{\sqrt{N}}$ as in regular graphs (**art.** 55) of degree r, where $\lambda_1 = r$, the number of all walks with k hops in (3.28) simplifies to

$$N_{k;\text{regular graph}} = N r^k \tag{3.29}$$

due to orthogonality (A.124) of eigenvectors.

Since $N_k = u^T A^m A^{k-m} u = (A^m u)^T (A^{k-m} u)$, the Cauchy-Schwarz inequality (A.12) shows that

$$\left| (A^m u)^T (A^{k-m} u) \right|^2 \leq \left(u^T A^{2m} u \right) \left(u^T A^{2(k-m)} u \right)$$

from which we obtain, for integers $0 \leq m \leq k$, the inequality

$$N_k^2 \leq N_{2m} N_{2k-2m} \tag{3.30}$$

Equality only holds for regular graphs. In particular for $m = 0$, it holds that $N_k^2 \leq N N_{2k}$.

60. *Graph angles.* Geometrically, the scalar product $x_n^T u = \sum_{j=1}^{N} (x_n)_j$ is the projection of the eigenvector x_n onto the vector u,

$$x_n^T u = \|x_n\|_2 \|u\|_2 \cos \gamma_n = \sqrt{N} \cos \gamma_n \tag{3.31}$$

where γ_n is the angle between the eigenvector x_n and the all-one vector u. The total number N_k of walks of length k, written in terms of the "graph angles" as coined by Cvetković *et al.* (1997), is

$$N_k = N \sum_{n=1}^{N} \lambda_n^k \cos^2 \gamma_n \tag{3.32}$$

Since $N_0 = N$, (3.28) becomes in terms of fundamental weights $w_k = u^T x_k$ in (3.2)

$$N = \sum_{n=1}^{N} \left(x_n^T u\right)^2 = \sum_{n=1}^{N} w_n^2 \tag{3.33}$$

which is equivalent with (3.32) for graph angles to

$$\sum_{n=1}^{N} \cos^2 \gamma_n = 1$$

61. *Probabilistic interpretation.* Besides graph angles in **art. 60**, we add a probabilistic approach, based on the general property (see e.g. Van Mieghem (2014, p. 13)) of the expectation operator $E[.]$ on a function g of a discrete random variable Λ, that can have N possible outcomes $\lambda_1, \lambda_2, \ldots, \lambda_N$,

$$E\left[g\left(\Lambda\right)\right] = \sum_{j=1}^{N} g\left(\lambda_j\right) \Pr\left[\Lambda = \lambda_j\right]$$

Comparison of $N_k = \sum_{n=1}^{N} \left(x_n^T u\right)^2 \lambda_n^k$ in (3.28) and (3.33) then suggest us to define $\Pr\left[\Lambda = \lambda_j\right] = \frac{\left(x_n^T u\right)^2}{N}$ so that $E\left[\Lambda^k\right] = \frac{N_k}{N}$. The corresponding probability generating function is

$$\varphi_\Lambda\left(z\right) = E\left[z^\Lambda\right] = \sum_{j=1}^{N} \Pr\left[\Lambda = \lambda_j\right] z^{\lambda_j}$$

which holds for any symmetric matrix M, where Λ reflects an arbitrary eigenvalue of M.

If the function g is convex, then Jensen's inequality (see e.g. Van Mieghem (2014, Section 5.2)) tells us that

$$g\left(E\left[\Lambda\right]\right) \le E\left[g\left(\Lambda\right)\right] \tag{3.34}$$

Since $g\left(x\right) = x^{2p}$ is convex for any real x and real p, whereas $g\left(x\right) = x^p$ is convex for non-negative real x, Jensen's inequality (3.34) translates to

$$\left(E\left[\Lambda\right]\right)^{2k} = \left(\frac{N_1}{N}\right)^{2k} = d_{av}^{2k} \le E\left[\Lambda^{2k}\right] = \frac{N_{2k}}{N}$$

Only for a *positive semidefinite* matrix M (see Section 10.8), it holds for any integer k that

$$N_k \ge \frac{N_1^k}{N^{k-1}} = N d_{av}^k$$

For even $k = 2m$ number of hops, equality in $d_{av}^{2m} \le \frac{N_{2m}}{N}$ is reached for regular graphs with average degree $d_{av} = r$ as shown in (3.29). Thus, the total number N_k of walks with even length $k = 2m$ in any graph is at least as large as in a regular graph with the same (integer) average degree.

62. *The generating function $N_G(z)$.* The generating function of the total number of walks in a graph G is defined as

$$N_G(z) = \sum_{k=0}^{\infty} N_k z^k \tag{3.35}$$

The two different expressions in **art. 59** result in two different expressions for $N_G(z)$. First, substituting the definition (3.26) into (3.35) yields

$$N_G(z) = u^T \left(\sum_{k=0}^{\infty} A^k z^k \right) u = u^T (I - zA)^{-1} u \tag{3.36}$$

where $|z| < \frac{1}{\lambda_1}$ in order for the infinite series to converge (**art. 231**). Since A is symmetric, there holds for any analytic function $f(z)$, possessing a power series expansion around some point, that $f(A) = (f(A))^T$. Thus, we have that

$$u^T (I - zA)^{-1} u = u^T \left((I - zA)^{-\frac{1}{2}} \right)^T (I - zA)^{-\frac{1}{2}} u = \left\| (I - zA)^{-\frac{1}{2}} u \right\|_2^2$$

which shows that

$$N_G(z) = \left\| (I - zA)^{-\frac{1}{2}} u \right\|_2^2 \geq 0$$

for all real z obeying $|z| < \frac{1}{\lambda_1}$. The zeros of $N_G(z)$ are simple and lie in between two consecutive eigenvalues of A as follows from interlacing in **art. 263**.

Second, invoking (3.28) gives, for $|z| < \frac{1}{\lambda_1}$,

$$N_G(z) = \sum_{n=1}^{N} \left(x_n^T u \right)^2 \sum_{k=0}^{\infty} \lambda_n^k z^k = \sum_{n=1}^{N} \frac{\left(x_n^T u \right)^2}{1 - \lambda_n z} \tag{3.37}$$

For regular graphs (**art. 55**), where $x_1 = \frac{u}{\sqrt{N}}$ is the eigenvector belonging to $\lambda_1 = r$, the generating function (3.37) of the total number of walks simplifies to

$$N_{\text{regular graph}}(z) = \frac{N}{1 - rz} \tag{3.38}$$

Cvetković *et al.* (1995, p. 45) have found an elegant formula[2] for $N_G(z)$ by rewriting $u^T (I - zA)^{-1} u$ using (A.65). Indeed, for $k = 1$ in (A.65) and $C_{N \times 1} = xu$ and

---

[2] The characteristic polynomial of the complement G^c is

$$\det (A^c - \lambda I) = \det (J - A - (\lambda + 1) I)$$
$$= (-1)^N \det \left((A + (\lambda + 1) I) \left(I - (A + (\lambda + 1) I)^{-1} J \right) \right)$$
$$= (-1)^N \det ((A + (\lambda + 1) I)) \det \left(I - (A + (\lambda + 1) I)^{-1} u.u^T \right)$$

where we have used that $J = u.u^T$. Using the "rank 1 update" formula (A.66), we find

$$\det (A^c - \lambda I) = (-1)^N \left(1 - u^T (A + (\lambda + 1) I)^{-1} u \right) \det (A + (\lambda + 1) I) \tag{3.39}$$

With the definition $N_G(z) = u^T (I - zA)^{-1} u$ in (3.36) of the generating function $N_G(z)$, we arrive again at Cvetkovic's formula (3.40).

$D_{N \times 1} = u$, we obtain with $J = u.u^T$,

$$\det(A + zJ) = \det A \det(1 + zu^T A^{-1} u) = (1 + zu^T A^{-1} u) \det A$$

Replacing $A \to I - zA$ results in

$$u^T (I - zA)^{-1} u = \frac{1}{z} \left(\frac{\det(I + z(J - A))}{\det(I - zA)} - 1 \right)$$

The right-hand side can be written in terms of the complement $A^c = J - I - A$ as

$$u^T (I - zA)^{-1} u = \frac{1}{z} \left((-1)^N \frac{\det\left(A^c + \frac{z+1}{z}I\right)}{\det\left(A - \frac{1}{z}I\right)} - 1 \right)$$

Finally, using the characteristic polynomial $c_A(z) = \det(A - zI)$ of a matrix A, we arrive at Cvetkovic's formula, for $|z| < \frac{1}{\lambda_1}$,

$$N_G(z) = \frac{1}{z} \left((-1)^N \frac{c_{A^c}\left(-\frac{1}{z} - 1\right)}{c_A\left(\frac{1}{z}\right)} - 1 \right) \tag{3.40}$$

which shows that $zN_G(z) + 1$ is a ratio of two real polynomials, both with real zeros and of degree at most N.

Combining (3.37) and (3.40) yields, with $\lambda = \frac{1}{z}$,

$$(-1)^N \frac{c_{A^c}(-\lambda - 1)}{c_A(\lambda)} = 1 + \sum_{n=1}^{N} \frac{\left(x_n^T u\right)^2}{\lambda - \lambda_n}$$

The right-hand side can be written as a fraction of two polynomials, in which the denominator polynomial has only simple zeros. From this observation, Cvetković *et al.* (1995) deduced that, if $c_A(\lambda)$ has an eigenvalue λ with multiplicity $p > 1$, then the characteristic polynomial of the complement $c_{A^c}(\lambda)$ contains an eigenvalue $-\lambda - 1$ with multiplicity $p - 1 \le q \le p + 1$.

63. *The total number of walks N_k and the sum of degree powers.* Fiol and Garriga (2009) have proven the inequality

$$N_k \le \sum_{j=1}^{N} d_j^k \tag{3.41}$$

Equality in (3.41) for all $k \ge 0$ is only achieved for regular graphs, because $N_{k;\text{regular graph}} = Nr^k$ in (3.29). For $k \le 2$, equality in (3.41) holds in general, because $N_0 = N = \sum_{j=1}^{N} d_j^0$, $N_1 = 2L = \sum_{j=1}^{N} d_j$ and $N_2 = d^T d = \sum_{j=1}^{N} d_j^2$.
Proof of (3.41): For $k > 2$, the total number N_k of walks of length k is

$$N_k = u^T A^T A^{k-2} Au = d^T A^{k-2} d = \sum_{i=1}^{N} \sum_{j=1}^{N} d_i \left(A^{k-2}\right)_{ij} d_j$$

$$= \sum_{i=1}^{N} \left(A^{k-2}\right)_{ii} d_i^2 + 2 \sum_{i=1}^{N} \sum_{j=i+1}^{N} \left(A^{k-2}\right)_{ij} d_i d_j$$

where the last sum holds by symmetry of $A = A^T$. From $0 \leq (a-b)^2 = a^2 + b^2 - 2ab$, we bound as $2 \sum_{i=1}^{N} \sum_{j=i+1}^{N} (A^{k-2})_{ij} d_i d_j \leq \sum_{i=1}^{N} \sum_{j=i+1}^{N} (A^{k-2})_{ij} \{d_i^2 + d_j^2\}$ and

$$N_k \leq \sum_{i=1}^{N} (A^{k-2})_{ii} d_i^2 + \sum_{i=1}^{N} \sum_{j=i+1}^{N} (A^{k-2})_{ij} \{d_i^2 + d_j^2\} = \sum_{i=1}^{N} \sum_{j=1}^{N} (A^{k-2})_{ij} d_j^2$$

$$= u^T A^{k-2} d^2$$

where the vector $d^j = \left(d_1^j, d_2^j, \ldots, d_N^j\right)$. This derivation suggests the induction argument

$$N_k \leq u^T A^{k-m} d^m \leq u^T A^{k-m-1} d^{m+1} \tag{3.42}$$

which has been demonstrated already for $m = 0, 1$ and 2. Assume now that it holds for $m = \nu \geq 0$, then the induction inequality (3.42) is proved when we can show that it also holds for $m = \nu + 1$. Using $Au = d$ in (2.4) and $A = A^T$,

$$u^T A^{k-\nu} d^\nu = u^T A^T A^{k-\nu-1} d^\nu = d^T A^{k-\nu-1} d^\nu = \sum_{i=1}^{N} \sum_{j=1}^{N} d_i \left(A^{k-\nu-1}\right)_{ij} d_j^\nu$$

$$= \sum_{i=1}^{N} \left(A^{k-\nu-1}\right)_{ii} d_i^{\nu+1} + \sum_{i=1}^{N} \sum_{j=i+1}^{N} \left(A^{k-\nu-1}\right)_{ij} \left(d_i d_j^\nu + d_j d_i^\nu\right)$$

Fiol and Garriga (2009) cleverly use the inequality for positive numbers a and b,

$$a^k b + a b^k = a^{k+1} + b^{k+1} - \left(a^k - b^k\right)(a - b) \leq a^{k+1} + b^{k+1}$$

with equality if and only if $a = b$, and obtain

$$u^T A^{k-\nu} d^\nu \leq \sum_{i=1}^{N} \left(A^{k-\nu-1}\right)_{ii} d_i^{\nu+1} + \sum_{i=1}^{N} \sum_{j=i+1}^{N} \left(A^{k-\nu-1}\right)_{ij} \left(d_j^{\nu+1} + d_i^{\nu+1}\right)$$

$$= \sum_{i=1}^{N} \sum_{j=1}^{N} \left(A^{k-\nu-1}\right)_{ij} d_j^{\nu+1} = u^T A^{k-(\nu+1)} d^{\nu+1}$$

which establishes the induction inequality (3.42) and proves (3.41). $\qquad \square$

The fundamental form of the Laplacian (4.3) in **art. 101**, applied to $x = d$,

$$d^T Q d = \sum_{l \in \mathcal{L}} (d_{l+} - d_{l-})^2$$

and $d^T Q d = d^T (\Delta - A) d = \sum_{j=1}^{N} d_j^3 - d^T A d$ lead to

$$\sum_{j=1}^{N} d_j^3 - N_3 = \sum_{l \in \mathcal{L}} (d_{l+} - d_{l-})^2 = \frac{1}{2} \sum_{i=1}^{N} \sum_{j=1}^{N} (d_i - d_j)^2 a_{ij} \tag{3.43}$$

where the right-hand side sums, over all links l in the graph, the square of the difference between the degrees at both sides of the link l. Section 8.5 relates this expression to the linear correlation coefficient of the degrees in a graph and to the (dis)assortative property of a graph.

64. *Inequalities for the sum of degree powers.* Cioabă (2006) writes the sum of degree powers in terms of a node m of the graph G as

$$\sum_{j=1}^{N} d_j^x = d_m^x + \sum_{j=1}^{N} a_{mj} d_j^x + \sum_{j=1}^{N} (A^c)_{mj} d_j^x$$

where x is a real number and where $(A^c)_{mj}$ is the element (m,j) of the complement $A^c = J - I - A$ of the adjacency matrix A. The first sum contains the x-th powers of the degrees of direct neighbors of node m, while the second sum contains the x-th powers of the degrees of nodes that are not adjacent to m. With the definitions of the averages

$$M_x(m) = \frac{1}{d_m} \sum_{j=1}^{N} a_{mj} d_j^x \leq (d_{\max})^x \qquad (3.44)$$

and

$$R_x(m) = \frac{1}{N - 1 - d_m} \sum_{j=1}^{N} (A^c)_{mj} d_j^x \geq (d_{\min})^x \qquad (3.45)$$

where the inequalities hold for $x > 0$ and imply the bound

$$M_x(m) - R_x(m) \leq (d_{\max})^x - (d_{\min})^x$$

and $E[D^x] = \frac{1}{N} \sum_{j=1}^{N} d_j^x$, the above relation can be recast as

$$d_m^x + (N-1) M_x(m) = NE[D^x] + (N - 1 - d_m)(M_x(m) - R_x(m))$$
$$\leq NE[D^x] + (N - 1 - d_m)((d_{\max})^x - (d_{\min})^x)$$

and produces for $x > 0$ the bound of Cioabă (2006)

$$\frac{d_m^x}{N-1} + M_x(m) \leq \frac{N}{N-1} E[D^x] + \left(1 - \frac{d_m}{N-1}\right)((d_{\max})^x - (d_{\min})^x)$$

Furthermore, summing $d_m M_x(m)$ in (3.44) over all m yields

$$\sum_{m=1}^{N} d_m M_x(m) = \sum_{m=1}^{N} \sum_{j=1}^{N} a_{mj} d_j^x = \sum_{j=1}^{N} \left(\sum_{m=1}^{N} a_{mj}\right) d_j^x = \sum_{j=1}^{N} d_j^{x+1} \qquad (3.46)$$

Multiplying both sides in Cioabă's bound by d_m and denoting $y = ((d_{\max})^x - (d_{\min})^x)$,

$$\frac{d_m^{x+1}}{N-1} + d_m M_x(m) \leq d_m \left\{\frac{1}{N-1} \sum_{j=1}^{N} d_j^x + y\right\} - \frac{y}{N-1} d_m^2$$

and summing over all nodes m yields for $x > 0$, after invoking (3.46) and (2.3) and after multiplying both sides by $\frac{N-1}{N}$, Cioabă's recursive inequality in $\sum_{j=1}^{N} d_j^x$ is

$$\sum_{j=1}^{N} d_j^{x+1} \leq \frac{2L}{N} \sum_{j=1}^{N} d_j^x + \frac{\left((d_{\max})^x - (d_{\min})^x\right)}{N} \left\{ 2L(N-1) - \sum_{j=1}^{N} d_j^2 \right\} \qquad (3.47)$$

If $x < 0$ and the graph is connected (i.e. $d_{\min} \geq 1$), then $M_x(m) - R_x(m) \geq (d_{\max})^x - (d_{\min})^x$ and

$$\sum_{j=1}^{N} d_j^{x+1} \geq \frac{2L}{N} \sum_{j=1}^{N} d_j^x - \frac{\left((d_{\min})^x - (d_{\max})^x\right)}{N} \left\{ 2L(N-1) - \sum_{j=1}^{N} d_j^2 \right\} \qquad (3.48)$$

Cioabă (2006) shows that equality in (3.47) is obtained for regular graphs and for connected graphs with exactly t nodes of degree $N-1$ and the remaining $N-t$ nodes form an independent set[3] for $1 \leq t \leq N$.

If two sequences are non-increasing, $x_1 \geq x_2 \geq \cdots \geq x_n$ and $y_1 \geq y_2 \geq \cdots \geq y_n$, then the Chebyshev's sum inequality, proved in Van Mieghem (2014), is

$$\frac{1}{n} \sum_{j=1}^{n} x_j y_j \geq \left(\frac{1}{n} \sum_{j=1}^{n} x_j \right) \left(\frac{1}{n} \sum_{j=1}^{n} y_j \right) \qquad (3.49)$$

If one sequence is non-increasing and the other is non-decreasing, then the opposite inequality sign holds. Application of Chebyshev's sum inequality (3.49) results in the lower bound

$$\sum_{j=1}^{N} d_j^{x+1} \geq \frac{2L}{N} \sum_{j=1}^{N} d_j^x \qquad (3.50)$$

Comparing the Chebyshev lower bound (3.50) with Cioabă's upper bound (3.47), indeed, illustrates that equality holds for regular graphs.

When $x = 1$ in (3.47) and in (3.50), we find the bounds

$$\frac{(2L)^2}{N} \leq \sum_{j=1}^{N} d_j^2 \leq 2L \frac{2L + (d_{\max} - d_{\min})(N-1)}{N + d_{\max} - d_{\min}}$$

For $x = -1$ in (3.48), we obtain

$$\sum_{j=1}^{N} \frac{1}{d_j} \leq \frac{N^2}{2L} + \left(\frac{1}{d_{\min}} - \frac{1}{d_{\max}} \right) \left\{ (N-1) - \frac{\sum_{j=1}^{N} d_j^2}{2L} \right\}$$

$$\leq \frac{N^2}{2L} + \left(\frac{1}{d_{\min}} - \frac{1}{d_{\max}} \right) \left\{ (N-1) - \frac{2L}{N} \right\}$$

---

[3] There are no links between the nodes of an independent set.

while the Chebyshev lower bound[4] (3.50) yields

$$\frac{N^2}{2L} \le \sum_{j=1}^{N} \frac{1}{d_j} \qquad (3.51)$$

Hence, the harmonic mean $E\left[\frac{1}{D}\right] = \frac{1}{N}\sum_{j=1}^{N} \frac{1}{d_j}$ is bounded by

$$\frac{1}{E[D]} \le E\left[\frac{1}{D}\right] \le \frac{1}{E[D]} + \frac{1}{N}\left(\frac{1}{d_{\min}} - \frac{1}{d_{\max}}\right) E[D^c]$$

where $E[D^c] = N - 1 - E[D]$ is the average degree in the complementary graph. The bound (3.51) is generalized, for graphs with degree $d_j \ge 1$, by the Hölder inequality (A.10), for $x_j = d_j^r$, $y_j = \frac{1}{d_j^r}$, $p > 1$ and $r > 0$, to the lower bound

$$\frac{N^p}{\left(\sum_{j=1}^{N} d_j^{\frac{rp}{p-1}}\right)^{p-1}} \le \sum_{j=1}^{N} \frac{1}{d_j^{rp}}$$

65. *Number of closed walks W_k.* The number of closed walks W_k of length k in graph G is defined in **art. 6**; Lemma 1 and **art. 243** show that

$$W_k = \sum_{j=1}^{N}(A^k)_{jj} = \operatorname{trace}\left(A^k\right) = \sum_{j=1}^{N} \lambda_j^k \qquad (3.52)$$

Art. 46 shows that the mean $E[\lambda] = \frac{1}{N}\sum_{j=1}^{N} \lambda_j = 0$. The definition (3.52) demonstrates that all centered moments of the adjacency eigenvalues are non-negative and equal to

$$E\left[(\lambda - E[\lambda])^k\right] = \frac{W_k}{N}$$

Hence, the centered k-th moment is equal to the number of closed walks of length k per node. The special case for $k = 2$ is $\operatorname{Var}[\lambda] = \frac{2L}{N}$, which is deduced in **art. 49**. If $k = 3$, then (3.8) indicates that

$$E\left[(\lambda - E[\lambda])^3\right] = \frac{6\blacktriangle_G}{N}$$

The skewness s_λ, that measures the lack of symmetry of the distribution around the mean, is defined as the normalized third moment,

$$s_\lambda = \frac{E\left[(\lambda - E[\lambda])^3\right]}{\left(E\left[(\lambda - E[\lambda])^2\right]\right)^{3/2}} = \frac{3\blacktriangle_G}{L\sqrt{E[D]}}$$

Since a tree does not have triangles, $\blacktriangle_G = 0$, the minimum possible skewness, $s_\lambda = 0$, in the distribution of adjacency eigenvalues is achieved for a tree. In

---

[4] Similarly, (3.51) follows from the Cauchy-Schwarz inequality (A.72) with $x_j = \sqrt{d_j}$ and $y_j = \frac{1}{\sqrt{d_j}}$.

Section 6.8, we will indeed show that only the adjacency spectrum of a tree is symmetric around the mean (or origin $\lambda = 0$). For $k = 4$, the inequality (3.12) bounds the number W_4 of closed walks of length 4.

66. *Generating function of the number of closed walks W_k.* The number of closed walks W_k of length k in graph G has a nice generating function, which is derived from Jacobi's general identity (**art. 215**). Using the Taylor series (**art. 231**) of $(I - zA)^{-1}$, convergent for $|z| < \frac{1}{\lambda_1}$, into Jacobi's trace formula (A.53) yields

$$\frac{1}{z} \sum_{k=1}^{\infty} \text{trace}\left(A^k\right) z^k = \frac{d}{dz} \log \det \left(I - zA\right)$$

With $W_k = \text{trace}(A^k)$ and $W_0 = N$, the generating function of the number of closed walks W_k in G and convergent for $|z| < \frac{1}{\lambda_1}$ is

$$W_G(z) = \sum_{k=0}^{\infty} W_k z^k = N + z \frac{d}{dz} \log \det \left(I - zA\right) \tag{3.53}$$

Substitution of the last equality in (3.52) into the generating function (3.53) yields, for $|z| < \frac{1}{\lambda_1}$,

$$W_G(z) = \sum_{j=1}^{N} \sum_{k=0}^{\infty} \lambda_j^k z^k = \sum_{j=1}^{N} \frac{1}{1 - \lambda_j z} \tag{3.54}$$

In terms of the characteristic polynomial $c_A(\lambda) = \sum_{k=0}^{N} c_k \lambda^k$ of A, which is $c_A(\lambda) = \det(A - \lambda I) = (-\lambda)^N \det\left(I - \frac{1}{\lambda}A\right)$, we have

$$\det(I - zA) = (-z)^N c_A\left(z^{-1}\right) = \sum_{k=0}^{N} (-1)^N c_{N-k} z^k$$

with $(-1)^N c_N = 1$. Then, we deduce from (3.53) that

$$\sum_{k=0}^{N} (-1)^N c_{N-k} z^k = \exp\left(\sum_{k=1}^{\infty} W_k \frac{z^k}{k}\right)$$

from which, by Taylor's theorem,

$$(-1)^N c_{N-k} = \frac{1}{k!} \frac{d^k}{dz^k} \exp\left(\sum_{k=1}^{\infty} \frac{W_k}{k} z^k\right)\Bigg|_{z=0} \tag{3.55}$$

Relation (3.55) is equivalent to the Newton identities (**art. 47**). By applying our characteristic coefficients defined in (B.10) in **art. 47** or in Van Mieghem (2007), the above derivatives can be explicitly computed for any finite k.

67. The generating function of the number of closed walks of length k that start and terminate at node j (**art. 6**), is defined as

$$W_G(z; j) = \sum_{k=0}^{\infty} \left(A^k\right)_{jj} z^k \tag{3.56}$$

Substituting the j-th diagonal element of (3.19) into (3.56) yields, for $|z| < \frac{1}{\lambda_1}$,

$$W_G(z; j) = \sum_{n=1}^{N} \left(x_n x_n^T\right)_{jj} \sum_{k=0}^{\infty} \lambda_n^k z^k = \sum_{n=1}^{N} \frac{\left(x_n x_n^T\right)_{jj}}{1 - \lambda_n z}$$

Art. 255 indicates that $\left(x_k x_k^T\right)_{jj} = \left((x_k)_j\right)^2$, such that

$$W_G(z; j) = \sum_{n=1}^{N} \frac{\left((x_n)_j\right)^2}{1 - \lambda_n z} \tag{3.57}$$

By definition, we have that $W_G(z) = \sum_{j=1}^{N} W_G(z; j)$.

Combining (A.52) and (A.162) in **art. 262** yields

$$\frac{\det\left(zI - A_{\backslash\{j\}}\right)}{\det(zI - A)} = \sum_{n=1}^{N} \frac{\left(x_n x_n^T\right)_{jj}}{z - \lambda_n}$$

where $A_{\backslash\{j\}}$ is the $(N-1) \times (N-1)$ adjacency matrix obtained from A by deleting the j-th row and column. Thus, $A_{\backslash\{j\}}$ is the adjacency matrix of the subgraph $G \backslash \{j\}$ of G obtained from the graph G by deleting node j and all its incident links. Hence,

$$\frac{\det\left(zI - A_{\backslash\{j\}}\right)}{\det(zI - A)} = \frac{1}{z} W_G\left(\frac{1}{z}; j\right)$$

and written in terms of the characteristic polynomial of a matrix A, $c_A(z) = \det(A - zI)$, we obtain

$$W_G(z; j) = -\frac{c_{A_{\backslash\{j\}}}\left(\frac{1}{z}\right)}{z c_A\left(\frac{1}{z}\right)}$$

The relation between the characteristic polynomials $c_{A_{\backslash\{j\}}}(z)$ and $c_A(z)$ is further studied in **art. 85**.

68. *Relations between N_k and W_k.* Let m be the maximizer of the fundamental weights $w_k = x_k^T u$ in **art. 43** over all $1 \le k \le N$ eigenvectors such that $x_m^T u \ge x_k^T u$ for any $1 \le k \ne m \le N$. Geometrically, the "graph angle" representation in (3.31), $x_k^T \frac{u}{\sqrt{N}} = \cos(\gamma_k)$, reflects that all orthogonal eigenvectors $x_1, x_2, \ldots, x_N$ start at the origin and end on an N-dimensional unit sphere centered at the origin. The graph angle between x_k and $\frac{u}{\sqrt{N}}$ is largest for x_1, by the Perron-Frobenius Theorem 75 in **art. 269**, because x_1 and $\frac{u}{\sqrt{N}}$ lie in the same N-dimensional "quadrant" as both their components are non-negative. Any other vector x_k must be orthogonal to x_1, implying that x_k cannot lie in the "opposite" N-dimensional "quadrant", where all components or coordinates are negative, and in which the resulting $\cos(\alpha_k)$ also can be large. Another, though less transparent, argument follows from the Cauchy identity (A.71),

$$w_k^2 = \left(u^T x_k\right)^2 = N - \frac{1}{2}\sum_{j=1}^{N}\sum_{l=1}^{N}\left((x_k)_j - (x_k)_l\right)^2 = N - \sum_{j=1}^{N}\sum_{l=1}^{j-1}\left((x_k)_j - (x_k)_l\right)^2$$

which illustrates that the maximizer over all $w_k^2 = \left(u^T x_k\right)^2$ has minimum difference between its components. Thus, it is the eigenvector x_m that is as close as possible to the vector $\frac{1}{\sqrt{N}}u$ with all components exactly the same. In conclusion, $m = 1$ and $w_1 > w_k$ for all $1 < k \leq N$. **Art.** 60 demonstrates that $w_1 \geq 1$. This result also follows from **art.** 203, because $w_1 = x_1^T u = \sum_{j=1}^{N}(x_1)_j = \sum_{j=1}^{N}\left|(x_1)_j\right| = \|x_1\|_1$ and $\|x_1\|_1 \geq \|x_1\|_2 = 1$. A much sharper lower bound (3.114) for w_1 is derived in **art.** 93 as a consequence of the Motzkin-Straus Theorem 17.

Likewise, let q be the index that minimizes $\left(x_k^T u\right)^2 \geq \left(x_q^T u\right)^2$ for any $1 \leq k \neq q \leq N$. Recall that $x_q^T u = 0$ for a regular graph. Then, the total number of walks N_k in (3.28) is lower and upper bounded for even k as

$$\left(x_q^T u\right)^2 \sum_{n=1}^{N}\lambda_n^k \leq \sum_{n=1}^{N}\left(x_n^T u\right)^2 \lambda_n^k \leq \left(x_1^T u\right)^2 \sum_{n=1}^{N}\lambda_n^k$$

Invoking the number of closed walks W_k of length k in graph G (**art.** 65), $W_k = \sum_{n=1}^{N}\lambda_n^k$, and the total number N_k of walks (3.28), leads to the inequality (only for even k)

$$\left(x_q^T u\right)^2 W_k \leq N_k \leq \left(x_1^T u\right)^2 W_k \leq NW_k$$

where the last inequality follows from (3.33), with equality for regular graphs.

The $N \times 1$ total walk vector $\mathbf{N} = (N, N_1, N_2, \ldots, N_{N-1})$ can be written with (3.28) as

$$
\begin{bmatrix} N \\ N_1 \\ \vdots \\ N_{N-2} \\ N_{N-1} \end{bmatrix} =
\begin{bmatrix}
1 & 1 & \cdots & 1 & 1 \\
\lambda_1 & \lambda_2 & \vdots & \lambda_{N-1} & \lambda_N \\
\vdots & \vdots & \cdots & \vdots & \vdots \\
\lambda_1^{N-2} & \lambda_2^{N-2} & \cdots & \lambda_{N-1}^{N-2} & \lambda_N^{N-2} \\
\lambda_1^{N-1} & \lambda_2^{N-1} & \cdots & \lambda_{N-1}^{N-1} & \lambda_N^{N-1}
\end{bmatrix}
\cdot
\begin{bmatrix}
\left(u^T x_1\right)^2 \\
\left(u^T x_2\right)^2 \\
\vdots \\
\left(u^T x_{N-1}\right)^2 \\
\left(u^T x_N\right)^2
\end{bmatrix}
$$

and, in matrix notation,

$$\mathbf{N} = V_N(\lambda)\, t_x$$

where $V_N(\lambda)$ is the Vandermonde matrix (A.75) in **art.** 224 and where the $N \times 1$ vector t_x has $w_k^2 = \left(u^T x_k\right)^2$ as its k-th component. Similarly, the closed walk vector $\mathbf{W} = (W_0, W_1, W_2, \ldots, W_{N-1})$ is written as

$$\mathbf{W} = V_N(\lambda)\, u$$

3.6 Diameter of a graph

Lemma 1 implies that $\left(A^k\right)_{ij}$ is non-zero if and only if node i and j can be joined in the graph by a walk of length k. Thus, if the shortest path from node i to j consists of h hops or links, then $\left(A^h\right)_{ij} \neq 0$, while $\left(A^k\right)_{ij} = 0$ for $1 \leq k < h$. **Art.** 22 defines the *diameter* ρ of the graph G as the number of hops $h = \rho$ in the longest shortest path in G and equals $\rho = \max_{1 \leq i \leq N; 1 \leq j \leq N} h_{ij}$, where h_{ij} is the hopcount between node i and j in a connected graph (**art. 21**).

69. *Diameter of a graph.*

Theorem 11 *The number of distinct eigenvalues of the adjacency matrix A is at least equal to $\rho + 1$, where ρ is the diameter of the graph.*

First proof: Art. 21 and 22 indicate that the matrix A^h cannot be written as a linear combination of $I, A, A^2, \ldots, A^{h-1}$. By definition of the diameter ρ as the longest shortest path, we thus conclude that the matrices $I, A, A^2, \ldots, A^\rho$ are linearly independent. **Art.** 254 shows that the matrix E_k, that represents the orthogonal projection onto the eigenspace of λ_k, is a polynomial in A. Thus, the vector space spanned by $I, A, A^2, \ldots, A^\rho$ is also spanned by a corresponding set of matrices E_k, which obey $E_k E_m = 1_{\{k=m\}}$. Let $Y = \sum_{k=1}^{\rho+1} c_k E_k$, then $c_j = E_j Y$ is only zero if all E_k are linearly independent. The matrices E_k and E_m are only linearly independent if they belong to a distinct eigenvalue of A. The linear independence of the set $I, A, A^2, \ldots, A^\rho$ thus implies that at least $\rho + 1$ eigenvalues of A must be distinct. $\qquad \square$

We may rephrase Theorem 11 as: "The diameter ρ of a graph G obeys $\rho \leq l - 1$, where l is the number of different eigenvalues of A". The second proof may be found easier and more elegant.

Second proof: Suppose that the adjacency matrix A has precisely l distinct eigenvalues. **Art.** 228 shows that A obeys $m_{c_A}(A) = O$, where the minimal polynomial $m_{c_A}(z) = \sum_{k=0}^{l} b_k z^k$ has degree l. Hence, we may write

$$\left(A^l\right)_{ij} = -\frac{1}{b_l} \sum_{k=0}^{l-1} b_k \left(A^k\right)_{ij} \tag{3.58}$$

which shows that the diameter $\rho \leq l - 1$. For, assume that $\rho > l - 1$, then there is at least one pair (i, j) for which $\left(A^k\right)_{ij} = 0$ for $0 \leq k \leq l - 1$. But, the minimal polynomial in (3.58) then shows that also $\left(A^l\right)_{ij} = 0$ and, further any $\left(A^{l+q}\right)_{ij} = 0$ for any integer $q \geq 0$, because $m_{c_A}(A) = O$ implies that any power r of A higher than $r \geq l$ can be written as a linear combination of powers A^k with k not exceeding l. The definition of the diameter, equivalent to $(A^\rho)_{ij} \neq 0$, while $\left(A^k\right)_{ij} = 0$ for $1 \leq k < \rho$, contradicts that $\rho > l - 1$. $\qquad \square$

Third proof (limited to a regular graph): Let us denote the l distinct eigenvalues of the adjacency matrix A of a regular graph G with degree r by $r =$

$\alpha_1 > \alpha_2 > \cdots > \alpha_l$. The Lagrange polynomial $p_{l-1}(x) = \prod_{k=2}^{l} \frac{x-\alpha_k}{r-\alpha_k}$ of degree $l-1$ in **art.** 303 passes through the points $(\alpha_k, 0)$ for $2 \le k \le l$ and $(r, 1)$. If the $v = |\mathcal{V}|$ nodes in the set $\mathcal{V}$ and the $w = |\mathcal{W}|$ nodes in the set $\mathcal{W}$ in a regular graph G with degree r are separated by at least l hops, the van Dam-Haemers inequality (4.103) in **art.** 161 shows that

$$\frac{vw}{(N-v)(N-w)} \le 0$$

implying that there are no nodes in G at a distance of l hops from each other, which is equivalent to a diameter $\rho < l$. $\qquad\qquad\qquad\qquad\qquad\qquad\qquad \Box$

As an example, consider the complete graph K_N whose adjacency matrix has precisely $l = 2$ distinct eigenvalues, $\lambda_1 = N - 1$ and $\lambda_2 = -1$, as computed in Section 6.1. Theorem 11 states that the diameter is at most $\rho = l - 1 = 1$. Since the diameter is at least equal to $\rho = 1$, we conclude from Theorem 11 that the diameter in the complete graph equals $\rho = 1$, as anticipated. If there is only $l = 1$ eigenvalue, then the diameter is at most equal to $\rho = 0$ and this situation (e.g. from (3.5)) corresponds to the empty graph only, where each eigenvalue $\lambda_k = 0$.

It follows from Lemma 3 that Theorem 11 also holds for the Laplacian matrix Q: If a connected graph G has l distinct adjacency or Laplacian eigenvalues, then its diameter ρ is at most $l - 1$, i.e. $\rho \le l - 1$.

70. *Spectral upper bound for the diameter of a graph.* Chung (1989) has proven:

Theorem 12 (Chung) *Let* $y = \min_{1 \le k \le N}(x_1)_k$, *then the diameter* ρ *of a graph, in which* $|\lambda_2| \ge |\lambda_N|$, *is upper bounded by*

$$\rho \le \left\lceil \frac{\log\left(\frac{1}{y^2} - 1\right)}{\log \lambda_1 - \log|\lambda_2|} \right\rceil \qquad\qquad (3.59)$$

Proof: We bound $\left(A^k\right)_{ij} = \sum_{n=1}^{N} \lambda_n^k (x_n)_i (x_n)_j$ in (3.18) as

$$\left(A^k\right)_{ij} \ge \lambda_1^k (x_1)_i (x_1)_j - \left|\sum_{n=2}^{N} \lambda_n^k (x_n)_i (x_n)_j\right|$$

because the largest eigenvalue is always positive and the Perron-Frobenius Theorem 75 in **art.** 269 states that the eigenvector components of x_1 are non-negative. Further, we have that

$$\left|\sum_{n=2}^{N} \lambda_n^k (x_n)_i (x_n)_j\right| \le \sum_{n=2}^{N} |\lambda_n^k| \left|(x_n)_i (x_n)_j\right| \le \max_{2 \le n \le N} |\lambda_n^k| \sum_{n=2}^{N} \left|(x_n)_i (x_n)_j\right|$$

where $\max_{2 \leq n \leq N} |\lambda_n^k| = \max \left(|\lambda_2^k|, |\lambda_N^k| \right) = |\lambda_2|^k$, because the graph features $|\lambda_2| \geq |\lambda_N|$. Invoking the Cauchy-Schwarz inequality (**art. 222**),

$$\sum_{n=2}^{N} \left| (x_n)_i \, (x_n)_j \right| \leq \sqrt{\sum_{n=2}^{N} \left| (x_n)_i \right|^2 \sum_{n=2}^{N} \left| (x_n)_j \right|^2}$$

and the eigenvector normalization $x_k^T x_k = 1$ in (A.124), such that $\sum_{n=2}^{N} |(x_n)_i|^2 = 1 - (x_1)_i^2$, leads to $\left| \sum_{n=2}^{N} \lambda_n^k (x_n)_i (x_n)_j \right| \leq |\lambda_2|^k \sqrt{\left(1 - (x_1)_i^2 \right) \left(1 - (x_1)_j^2 \right)}$ so that $\left(A^k \right)_{ij} \geq \lambda_1^k (x_1)_i (x_1)_j - |\lambda_2|^k \sqrt{\left(1 - (x_1)_i^2 \right) \left(1 - (x_1)_j^2 \right)}$. The diameter ρ is the smallest value of k for which $\left(A^k \right)_{ij} > 0$ for each element in A^k. Requiring that the above inequality is strictly larger than zero amounts to

$$\left(\frac{\lambda_1}{|\lambda_2|} \right)^k > \frac{\sqrt{\left(1 - (x_1)_i^2 \right) \left(1 - (x_1)_j^2 \right)}}{(x_1)_i (x_1)_j} \geq \frac{1 - y^2}{y^2}$$

which proves the theorem. $\qquad\qquad\qquad\qquad\qquad\qquad\qquad\qquad\qquad\qquad$ □

Theorem 12 shows that, when $\frac{\lambda_1}{|\lambda_2|}$ is large implying that the spectral gap $\lambda_1 - \lambda_2$ of the graph is large, the diameter ρ is small. Equality in (3.59) is reached for the complete graph, where $y = \frac{1}{\sqrt{N}}$, $\lambda_1 = N - 1$ and $|\lambda_2| = |\lambda_N| = 1$. Theorem 12 does not apply to the complete bipartite graph $K_{m,n}$ (Section 6.7), where $\lambda_N = -\lambda_1 = \sqrt{mn}$ and all other eigenvalues are zero.

71. *The spectral radius λ_1 and the diameter ρ.* Communications networks are designed to possess a small diameter, that results in efficient transport of packets with low end-to-end delay and packet loss. In order to be less vulnerable to epidemic malware, the spectral radius of the graph should be minimal, which corresponds to a high epidemic threshold, as demonstrated in Van Mieghem *et al.* (2009). Inspired by those requirements, van Dam and Kooij (2007) proposed to find those graphs with minimum spectral radius λ_1 given the diameter ρ of the graph. A few years later, Cioabă *et al.* (2010) proved

Theorem 13 *If ρ is the diameter of a graph G with N nodes, then the spectral radius is lower bounded by*

$$\lambda_1 \geq (N - 1)^{\frac{1}{\rho}} \qquad\qquad\qquad\qquad\qquad (3.60)$$

Moreover, Cioabă *et al.* (2010) showed that equality in (3.60) holds for $\rho = 1$ if and only if G is the complete graph K_N and for $\rho = 2$ if and only if G is the star $K_{1,N-1}$, the pentagon, the Petersen graph (Fig. 2.3), the Hoffman–Singleton graph, or a putative 57-regular graph on $3250 = 57^2 - 1$ nodes. These cases for $\rho = 1$ and $\rho = 2$ were earlier found in van Dam and Kooij (2007).

Proof of (3.60): The diameter equals the number of hops in the longest shortest path in the graph G and a finite diameter implies that G is connected (**art. 22**). This means that each node j is reachable from another node i in G by a shortest path possessing k hops, with $k \leq \rho$, of the form[5] $\mathcal{P}^*_{i \sim j} = n_0 \sim n_1 \sim n_2 \sim \cdots \sim n_{k-1} \sim n_k$, where $n_0 = i$ and $n_k = j$ (**art. 6**) and where each node n_m in the path is different. If $k = \rho$, then the shortest path $\mathcal{P}^*_{i \sim j}$ is also a walk with ρ hops. If $k < \rho$, then there exists a walk $\mathcal{W}_{i \sim j; \rho} = n_0 \sim n_1 \sim n_2 \sim \cdots \sim n_{k-1} \sim n_k = j \sim n_{k-1} \sim n_{k-2} \sim \cdots$ with ρ hops, in which the walk segment from node i up to node j is unique, because it is the shortest path between i and j. Hence, for each node i, we can reach each other node j by a walk $\mathcal{W}_{i \sim j; \rho}$ with ρ hops. Thus, there are at least $N - 1$ different walks with ρ hops from i to another node j in G. This holds for each source node i, so that the total number of walks with ρ hops obeys $N_\rho \geq N(N - 1)$. The lower bound (3.60) then follows from (3.65). □

Although Theorem 13 indicates that $\frac{\lambda_1}{(N-1)^{\frac{1}{\rho}}} \geq 1$ for any graph, Cioabă *et al.* (2010) made the interesting claim that, for any graph class with $\rho > 1$ in which the number N of nodes can grow unboundedly, there holds that

$$\lim_{N \to \infty} \frac{\lambda_1}{(N - 1)^{\frac{1}{\rho}}} = 1 \qquad (3.61)$$

They showed that their claim (3.61) is related to the degree-diameter problem that asks for the graph with a maximum number $N_{(d_{\max}, \rho)}$ of nodes, given the maximum degree $d_{\max}$ and the diameter ρ. Bollobás (2004) has conjectured, for $\rho > 3$, that

$$\lim_{N \to \infty} \inf \frac{N_{(d_{\max}, \rho)}}{d_{\max}^\rho} = 1 \qquad (3.62)$$

Cioabă *et al.* (2010) demonstrated that (3.61) is true if the conjecture (3.62) is true.

3.7 The spectral radius λ_1

The largest eigenvalue λ_1 of the adjacency matrix A, also called the spectral radius of a graph G, appears in many applications. In dynamic processes on graphs, the inverse of the largest eigenvalue λ_1 characterizes the threshold of the phase transition of both virus spread (Van Mieghem *et al.*, 2009) and synchronization of coupled oscillators (Restrepo *et al.*, 2005) in networks. If the effective viral strength $\tau > \frac{1}{\lambda_1}$, the epidemic spreads over the network. If $\tau < \frac{1}{\lambda_1}$, then[6] the epidemic dies out. Sharp bounds or exact expressions for λ_1 are desirable to control these processes. Bounds for λ_2 and λ_N in connected graphs follow from the general bounds (A.175)

---

[5] Since the graph G is assumed to be undirected, a link between node n_m and n_{m+1} is denoted by $n_m \sim n_{m+1}$, which reflects both possible directions $n_m \leftrightarrows n_{m+1}$.

[6] The basic reproduction number $R_0 = \tau \lambda_1$ separates the two epidemic phases: if $R_0 < 1$, an epidemic dies out, else it spreads. We assume here a mean-field approximation. The more precise analysis can be found in Van Mieghem and van de Bovenkamp (2013) and Prasse *et al.* (2021).

and (A.176), respectively, on eigenvalues of non-negative, irreducible, symmetric matrices in **art. 273**.

Due the importance of the spectral radius in complex networks, Stevanović (2015) recently collected a large variety of results, mostly bounds on λ_1, into a book.

We remark that the largest eigenvalue of a non-negative matrix, that is not necessarily symmetric, also obeys the Rayleigh principle (A.130) as can be verified from **art. 251** by incorporating the Perron-Frobenius Theorem 75. Hence, most of the deduced bounds in this Section 3.7 also apply to directed graphs, whose adjacency matrix is generally non-symmetric.

3.7.1 Lower bounds for the spectral radius λ_1

72. *Classical lower bound.* The Rayleigh inequalities in **art. 251** indicate that

$$\lambda_1 = \sup_{x \neq 0} \frac{x^T A x}{x^T x}$$

and that the maximum is attained if and only if x is the eigenvector of A belonging to λ_1, while for any other vector $y \neq x$, it holds that $\lambda_1 \geq \frac{y^T A y}{y^T y}$. By choosing the vector $y = u$, we obtain, with (2.5), the classical bound

$$\lambda_1 \geq \frac{u^T A u}{u^T u} = \frac{2L}{N} \tag{3.63}$$

Equality is reached in a regular graph, because the average degree is $E[D] = \frac{2L}{N} = r$ since $d_j = r$ for each node j, and because r is the largest eigenvalue of A belonging to the eigenvector u (Theorem 8). The differences $\lambda_1 - E[D]$ and $d_{\max} - \lambda_1$ can be considered as measures for the irregularity of a graph.

Combining the relation $2L = \sum_{k=1}^{N} \lambda_k^2$ in (3.7) with the classical lower bound (3.63) indicates, for any graph, that

$$N\lambda_1 \geq \sum_{k=1}^{N} \lambda_k^2$$

with equality only for a regular graph. Hence, we can determine from the spectrum of the adjacency matrix that the graph is regular if it holds that $N\lambda_1 = \sum_{k=1}^{N} \lambda_k^2$.

73. *Variations on the Rayleigh inequality.* A series of other bounds can be deduced from the Rayleigh inequality $\lambda_1 \geq \frac{y^T A y}{y^T y}$. Applying the Rayleigh inequalities to A^k, invoking **art. 243**, and choosing $y = A^m u$ in Rayleigh's inequality $\lambda_1^k \geq \frac{u^T A^m A^k A^m u}{u^T A^{2m} u}$ leads, for non-negative integers m and k, with the definition $N_j = u^T A^j u$ in (3.26) of the total number of walks to

$$\lambda_1^k \geq \frac{N_{2m+k}}{N_{2m}} \tag{3.64}$$

For $m = 0$ in (3.64), we find

$$\lambda_1 \geq \left(\frac{N_k}{N}\right)^{\frac{1}{k}} \tag{3.65}$$

The particular case of $k = 2$ in (3.65) becomes with $N_2 = d^T d$ in **art. 59**

$$\lambda_1 \geq \sqrt{\frac{1}{N}\sum_{k=1}^{N} d_k^2} = \sqrt{\text{Var}\,[D] + (E\,[D])^2} = \frac{2L}{N}\sqrt{1 + \frac{\text{Var}\,[D]}{(E\,[D])^2}} \tag{3.66}$$

Since the variance $\text{Var}[D] \geq 0$ and $\text{Var}[D] = 0$ only for regular graphs, the lower bound (3.66) is thus *always better* than the classical bound (3.63) for non-regular graphs. Beside regular graphs, equality in (3.66) also occurs in complete bipartite graphs $K_{m,N-m}$ (Section 6.7).

74. *Exact expression for the spectral radius of irregular graphs.* From the inequality (3.30) in **art. 59**, we deduce for irregular graphs that

$$\left(\frac{N_k}{N}\right)^{\frac{1}{k}} < \left(\frac{N_{2k}}{N}\right)^{\frac{1}{2k}}$$

Thus, the sequence $\frac{N_1}{N}, \left(\frac{N_2}{N}\right)^{\frac{1}{2}}, \left(\frac{N_4}{N}\right)^{\frac{1}{4}}, \ldots$ is increasing[7], while each term is bounded by the spectral radius λ_1 and we find

$$\lim_{k\to\infty} \left(\frac{N_k}{N}\right)^{\frac{1}{k}} = \lambda_1 \tag{3.67}$$

which complements **art. 58**. The Fiol-Garriga inequality $N_k < \sum_{j=1}^{N} d_j^k$ in (3.41) for $k > 2$ indicates that

$$\left(\frac{N_k}{N}\right)^{\frac{1}{k}} < \left(\frac{1}{N}\sum_{j=1}^{N} d_j^k\right)^{\frac{1}{k}} = \left(E\,[D^k]\right)^{\frac{1}{k}} = \frac{\|d\|_k}{N^{\frac{1}{k}}} \tag{3.68}$$

illustrating that $\left(E\,[D^k]\right)^{\frac{1}{k}}$ is *increasing* in k, while the Hölder norm $\|d\|_k$ by (A.14) is *non-increasing* in k. Since $\lim_{k\to\infty} \left(\frac{1}{N}\sum_{j=1}^{N} d_j^k\right)^{\frac{1}{k}} = d_{\max}$, we find with (3.67) and (3.68) again the upper bound $\lambda_1 \leq d_{\max}$ in **art. 42** with equality only for regular graphs. Combined with (3.66), it holds that $\left(E\,[D^2]\right)^{\frac{1}{2}} \leq \lambda_1 < \lim_{k\to\infty}$ $\left(E\,[D^k]\right)^{\frac{1}{k}}$. Since $\left(E\,[D^q]\right)^{\frac{1}{q}}$ is continuous and increasing in real $q > 0$, there exists one value of $q \geq 2$ for irregular graphs that satisfies

$$\lambda_1 = \left(\frac{1}{N}\sum_{j=1}^{N} d_j^q\right)^{\frac{1}{q}} \tag{3.69}$$

---

[7] Regular graphs with degree r have $\lambda_1 = \left(\frac{N_k}{N}\right)^{\frac{1}{k}} = r$ by (3.29), independent of k.

Formula (3.69) was proved differently by Hofmeister (1988). Computations of (3.69) on Erdős-Rényi random graphs show, for most instances, that $q \in (2, 5)$. Hofmeister (1988) has proved for $n \in \mathbb{N}$ that if $q = 2n+1$ in (3.69), then λ_1 is an integer and if $q = 2n$, then λ_1^2 is an integer. Indeed, if $q = k \in \mathbb{N}$, then (3.69) shows that $\lambda_1^k \in \mathbb{Q}^+$. But $\lambda_1 \notin \mathbb{Q}\backslash\mathbb{Z}$ (**art.** 45), so that $\lambda_1^k \in \mathbb{N}$, implying that λ_1 satisfies $x^k - m = 0$ for $m = \lambda_1^k \in \mathbb{N}$. If k is odd, $x^k - m = 0$ has only one real zero, λ_1; if k is even, then $x^k - m = 0$ has two real zeros, $\pm\lambda_1$, and thus $\lambda_1^2 \in \mathbb{N}$.

75. Spectral radius in subgraphs. The Interlacing Theorem 71 states that λ_1 is larger than or equal to the largest eigenvalue of any subgraph G_s of G:

$$\lambda_1 \geq \max_{\text{all } G_s \subset G} (\lambda_1 (A_{G_s})) \tag{3.70}$$

The lower bounds deduced in this Section 3.7, such as (3.63) and (3.66), also apply to each individual subgraph G_s. It is a matter of ingenuity to find that subgraph G_s with highest largest eigenvalue $\lambda_1 (A_{G_s})$. The lower bound (3.70) can also be deduced from the Rayleigh inequality by choosing zero components in the vector y such that $y^T A y = w^T A_{G_s} w$, where the vector w contains the non-zero components of y and A_{G_s} is the subgraph obtained by deleting those rows and columns that correspond to the zero components in y.

Examples The spectral radius of the star $K_{1,N}$ is computed in Section 6.7 as $\lambda_1 (A_{K_{1,N}}) = \sqrt{N}$. Since any node j in a connected graph is locally a star K_{1,d_j} with $d_j + 1$ nodes and $\lambda_1 \left(A_{K_{1,d_j}} \right) = \sqrt{d_j}$, we find that the lower bound in (3.70)

$$\lambda_1 \geq \max_{\text{all } j \in \mathcal{N}} \left(\lambda_1 \left(A_{K_{1,d_j}} \right) \right) = \sqrt{d_{\max}}$$

Another derivation follows from the bound $\left(A^k \right)_{jj} \leq \lambda_1^k$ in (3.24), which holds for any node j and any integer $k \geq 1$,

$$\lambda_1 \geq \sqrt[k]{\max_{1 \leq j \leq N} \left(A^k \right)_{jj}} \tag{3.71}$$

The largest eigenvalue λ_1 of the adjacency matrix is at least as large as the m-th root of the largest number of m-hop cycles around a node in the graph. For $k = 2$ and with $\left(A^2 \right)_{jj} = d_j$ in (2.20), we again find

$$\lambda_1 \geq \sqrt{d_{\max}}$$

with equality in the star $K_{1,N-1}$.

If G_s is the largest clique in G containing ω nodes, then $\lambda_1 (A_{K_\omega}) = \omega - 1$ and (3.70) leads to

$$\lambda_1 \geq \omega - 1$$

where ω is clique number (see **art.** 92).

76. Rayleigh's inequality and the walk generating function. Continuing as in

art. 73, we propose to choose in the Rayleigh inequality $\lambda_1^k \geq \frac{y^T A^k y}{y^T y}$, the vector

$$y = \sum_{j=0}^{\infty} A^j u \, z^j = (I - Az)^{-1} u = u + zd + z^2 Ad + \dots$$

which converges for $|z| < \lambda_1^{-1}$ (**art. 62**). The quadratic form

$$y^T A^k y = u^T (I - Az)^{-1} A^k (I - Az)^{-1} u$$

is only a norm for even $k = 2m$, $y^T A^{2m} y = \left\| A^m (I - Az)^{-1} u \right\|_2^2 \geq 0$ and nonnegative for all $|z| < \lambda_1^{-1}$. Further, using the matrix norm inequality (A.25),

$$\frac{y^T A^{2m} y}{y^T y} = \frac{\left\| A^m (I - Az)^{-1} u \right\|_2^2}{\left\| (I - Az)^{-1} u \right\|_2^2} \leq \|A^m\|_2^2 = \lambda_1^{2m}$$

where the last inequality follows from (A.23). The Cauchy product of power series yields

$$y^T A^k y = \sum_{m=0}^{\infty} u^T A^m \, z^m \sum_{j=0}^{\infty} A^{j+k} u \, z^j = \sum_{m=0}^{\infty} \left(\sum_{j=0}^{m} u^T A^{m-j} A^{j+k} u \right) z^m$$

$$= \sum_{m=0}^{\infty} (m+1) \, u^T A^{m+k} u \, z^m = \sum_{m=0}^{\infty} (m+1) \, N_{k+m} \, z^m$$

We write the sum $y^T A^k y = \sum_{m=0}^{\infty} (m+1) N_{k+m} \, z^m$ in terms of the generating function $N_G(z) = \sum_{k=0}^{\infty} N_k z^k$ in (3.35) of the total number of walks in a graph G

$$N_G(z) = \sum_{m=0}^{\infty} N_m z^m = \sum_{m=-k}^{\infty} N_{k+m} z^{k+m} = z^k \left(\sum_{m=-k}^{-1} N_{k+m} z^m + \sum_{m=0}^{\infty} N_{k+m} z^m \right)$$

$$= \sum_{m=1}^{k} N_{k-m} z^{k-m} + z^k \sum_{m=0}^{\infty} N_{k+m} z^m$$

and

$$y^T A^k y = \sum_{m=0}^{\infty} (m+1) N_{k+m} z^m = \frac{d}{dz} \left(z^{1-k} N_G(z) - \sum_{m=1}^{k} N_{k-m} z^{1-m} \right)$$

$$= \frac{d \left(z^{1-k} N_G(z) \right)}{dz} + \sum_{m=2}^{k} (m-1) N_{k-m} z^{-m}$$

Rayleigh's inequality becomes, for $|z| < \lambda_1^{-1}$,

$$\lambda_1^k \geq z^{-k} \frac{z \frac{dN_G(z)}{dz} + (1-k) N_G(z) + \sum_{j=0}^{k-2} (k-1-j) N_j z^j}{z \frac{dN_G(z)}{dz} + N_G(z)}$$

which simplifies most for $k = 1$,

$$\lambda_1 \geq \frac{\frac{dN_G(z)}{dz}}{\frac{d}{dz}(zN_G(z))} = \frac{\frac{dN_G(z)}{dz}}{z\frac{dN_G(z)}{dz} + N_G(z)} \tag{3.72}$$

This general lower bound (3.72) can be written in terms of the logarithmic derivative of the generating function $N_G(z)$ as $\frac{1}{\lambda_1} \leq z + \frac{1}{\frac{d\log N_G(z)}{dz}}$.

77. Deductions from the walk generating function lower bound (3.72). Differentiating the right-hand side the Rayleigh quotient $r_1(z) = \frac{y^T A y}{y^T y} = \frac{\frac{dN_G(z)}{dz}}{\frac{d}{dz}(zN_G(z))}$ in (3.72) with respect to z gives

$$\frac{dr_1(z)}{dz} = \frac{N_G(z)\frac{d^2 N_G(z)}{dz^2} - 2\left(\frac{dN_G(z)}{dz}\right)^2}{\left(\frac{d}{dz}(zN_G(z))\right)^2}$$

illustrating that if $N_G(z)\frac{d^2 N_G(z)}{dz^2} - 2\left(\frac{dN_G(z)}{dz}\right)^2 > 0$, then $r_1(z)$ is increasing and until its maximum at $\frac{dr_1(z)}{dz} = 0$. The solution of the differential equation $\frac{dr_1(z)}{dz} = 0$ as well as (3.72) with equality sign is precisely the generating function (3.38) of regular graphs.

(a) When $z = 0$, (3.72) reduces to $\lambda_1 \geq \frac{N_1}{N}$, which is the classical lower bound (3.63).

(b) For small z, we substitute the power series of $N_G(z)$ up to order three in z in the right-hand side of (3.72),

$$\lambda_1 \geq \frac{N_1 + 2N_2 z + 3N_3 z^2 + O(z^3)}{N_0 + 2N_1 z + 3N_2 z^2 + O(z^3)}$$

Maximizing the lower bound for z yields, after a tedious calculation,

$$\lambda_1 \geq \frac{N_0 N_3 - N_1 N_2 + \sqrt{N_0^2 N_3^2 - 6N_0 N_1 N_2 N_3 - 3N_1^2 N_2^2 + 4(N_1^3 N_3 + N_0 N_2^3)}}{2(N_0 N_2 - N_1^2)} \tag{3.73}$$

Numerical results in Table 3.1 and Fig. 8.6 show that the bound (3.73) is better than (3.66), which is not surprising because (3.73) includes via N_3 additional information about the graph.

(c) For large, positive and real z, the Rayleigh quotient is

$$r_k(z) = \frac{y^T A^k y}{y^T y} = \frac{\sum_{m=0}^{\infty}(m+1)N_{k+m} z^m}{\sum_{m=0}^{\infty}(m+1)N_m z^m} = \lim_{n\to\infty}\frac{\sum_{m=0}^{n}(m+1)N_{k+m} z^m}{\sum_{m=0}^{n}(m+1)N_m z^m}$$

$$= \lim_{n\to\infty}\frac{N_{k+n}}{N_n} \lim_{n\to\infty}\frac{1 + \sum_{m=0}^{n-1}\frac{(m+1)N_{k+m}}{(n+1)N_{k+n}} z^{m-n}}{1 + \sum_{m=0}^{n-1}\frac{(m+1)N_m}{(n+1)N_n} z^{m-n}}$$

and

$$r_k(z) = \lim_{n\to\infty}\frac{N_{k+n}}{N_n} \lim_{n\to\infty}\frac{1 + \frac{nN_{k+n-1}}{(n+1)N_{k+n}} z^{-1} + O\left(\frac{1}{z^2}\right)}{1 + \frac{nN_{n-1}}{(n+1)N_n} z^{-1} + O\left(\frac{1}{z^2}\right)}$$

	Air transport	Random ER	Complete bipartite
λ_1 exact	80.9576	19.3405	105.5557
bound (3.63) $= \frac{2L}{N}$	18.3079	18.3304	17.8701
bound (3.66)	42.7942	18.8005	105.5557
bound (3.73)	75.9029	19.2867	105.5557

Table 3.1. *Comparison of a few lower bounds for λ_1. All networks have $N = 1247$ nodes. The European direct airport-to-airport traffic network is obtained from Eurostat, while the Erdős-Rénji graph is defined in Section 1.5 and the complete bipartite graph $K_{n,m}$ in Section 6.7.*

which illustrates that $r_k(z) = \lim_{n\to\infty} \frac{N_{k+n}}{N_n} = \lambda_1^k$ for sufficiently large, positive z, because then the last limit tends to 1. Although the series $y = \sum_{j=0}^{\infty} A^j u\, z^j$ only converges for $|z| < \lambda_1^{-1}$, the Rayleigh quotient $r_k(z) = \frac{y^T A^k y}{y^T y}$ exists for all non-negative real z.

78. Another improvement of the classical bound (3.63) in terms of the total number (3.26) of walks N_k is derived in Van Mieghem (2007) and improved in Walker and Van Mieghem (2008),

$$\lambda_1 \geq \frac{N_1}{N} + 2\left(\frac{N_3}{2N} - \frac{N_1 N_2}{N^2} + \frac{N_1^3}{2N^3}\right)\lambda_0^{-2} + O(t^{-4}) \tag{3.74}$$

where $t \geq T$, $\lambda_0 = t\sqrt{N}$,

$$T = \frac{1}{\sqrt{N}} \max_{1\leq j\leq m}\left(a_{jj} + \sum_{i\neq j}|a_{ij}|\right) \tag{3.75}$$

Since $N_1 = u^T A u = 2L$, the first term in (3.74) is the classical bound (3.63). The Lagrange series (3.74) with terms containing powers of λ_0^{-2j} for $j > 0$ measures the irregularity $\lambda_1 - E[D]$ of the graph.

The basic idea in Walker and Van Mieghem (2008) starts from the matrix function $f(M)$ of any symmetric matrix M, where $f(t)$ is an arbitrary increasing function in the real number t, whose inverse function $f^{-1}(t)$ around $\frac{N_1}{N}$ exists. A function of a matrix (see **art.** 232) can be written in terms of a Taylor series, $f(Mt) = \sum_{k=0}^{\infty} f_k M^k t^k$, where t is a scalar, properly chosen to guarantee convergence of the Taylor series. The classical bound (3.63) applied to the matrix $f(Mt)$ with largest eigenvalue $f(\lambda_1(Mt))$ (see **art.** 257) yields

$$\frac{u^T f(Mt) u}{N} \leq f(t\lambda_1(M))$$

Since $f^{-1}(t)$ is also increasing in t, it holds that

$$\lambda_1(M) \geq \frac{1}{t} f^{-1} \left(\frac{u^T f(Mt) u}{N} \right) = \frac{1}{t} f^{-1} \left(\sum_{k=0}^{\infty} f_k \frac{u^T M^k u}{N} t^k \right) \tag{3.76}$$

where the inverse function f^{-1} can be expanded in a Langrange series. The function $f(t) = t$ returns the classical bound (3.63). An interesting property of (3.76) for positive semidefinite matrices is that any real increasing function $f(t)$, different than $f(t) = t$, provides a tighter lower bound than the classical bound (3.63). Indeed, using $\frac{u^T M^k u}{N} \geq \left(\frac{u^T M u}{N} \right)^k$ for any integer k as demonstrated in **art. 61**, we find that $\sum_{k=0}^{\infty} f_k \frac{u^T M^k u}{N} t^k \geq \sum_{k=0}^{\infty} f_k \left(\frac{u^T M u}{N} t \right)^k = f\left(\frac{u^T M u}{N} t \right)$ and $\frac{1}{t} f^{-1} \left(\sum_{k=0}^{\infty} f_k \frac{u^T M^k u}{N} t^k \right) \geq \frac{u^T M u}{N}$, which demonstrates the property.

3.7.2 Upper bounds for the spectral radius λ_1

79. A sharper upper bound than $\lambda_1 \leq \sqrt{2L \left(1 - \frac{1}{N}\right)}$ in **art. 42**

$$\lambda_1 \leq \sqrt{2L \left(1 - \frac{1}{\omega}\right)} \tag{3.77}$$

where $\omega \leq N$ is the clique number (**art. 92**), is deduced by Nikiforov (2002), using the Motzkin-Straus Theorem 17 in **art. 93**. Squaring (3.3) and applying the Cauchy-Schwarz inequality (A.72) yields

$$\lambda_1^2 = 4 \left(\sum_{l=1}^{L} (x_1)_{l+} (x_1)_{l-} \right)^2 \leq 4L \sum_{l=1}^{L} (x_1)_{l+}^2 (x_1)_{l-}^2$$

After substituting the vector component $x_{l+} \rightarrow (x_1)_{l+}^2$ into the Motzkin-Straus Theorem 17, so that the requirement $1 = \sum_{j=1}^{N} x_j = \sum_{j=1}^{N} (x_1)_j^2$ is satisfied, we find (3.77).

More generally, combining Theorem 109, **art. 42** and **art. 73** gives for any integer $k \geq 1$,

$$\left(\frac{N_{2k}}{N} \right)^{\frac{1}{2k}} \leq \lambda_1 \leq \min \left\{ \left(\frac{W_{2k}}{1 + (N-1)^{1-2k}} \right)^{\frac{1}{2k}}, d_{\max} \right\} \tag{3.78}$$

where W_k is the number of closed walks with k hops. For $k = 1$, (3.78) reduces to

$$\frac{2L}{N} \sqrt{1 + \frac{\text{Var}[D]}{(E[D])^2}} \leq \lambda_1 \leq \min \left\{ \sqrt{\frac{2L(N-1)}{N}}, d_{\max} \right\} \tag{3.79}$$

When we assume that $\left| \frac{\lambda_n}{\lambda_1} \right| < 1$ for all $2 \leq n \leq N$, which, as mentioned in

art. 73, excludes bipartite graphs, the definition of W_k in **art.** 65 indicates that

$$W_k = \lambda_1^k \left\{ 1 + \sum_{j=2}^{N} \left(\frac{\lambda_j}{\lambda_1} \right)^k \right\} < \lambda_1^k \left\{ 1 + (N-1) \left| \frac{\max(\lambda_2, \lambda_N)}{\lambda_1} \right|^k \right\}$$

This implies that $W_k^{1/k}$ is decreasing in k, because $(1+x)^{1/k}$ is for $x > 0$ and, in addition, $\left(\frac{\max(\lambda_2,\lambda_N)}{\lambda_1} \right)^k$ is exponentially decaying in k. Hence, $\lim_{k\to\infty} W_k^{1/k} = \lambda_1$. While the left-hand side of (3.78) is increasing in k (**art.** 73), the right-hand side is decreasing in k. Together, they provide increasingly sharp bounds for λ_1 when k increases.

An upper bound, related to the lower bound (3.71),

$$\lambda_1 \le \max_{m \ge 1} \sqrt[m]{\max_{1 \le i \le N} \sum_{j=1}^{N} (A^m)_{ij}}$$

follows from (A.26) and (A.21). Since $W_m = \sum_{j=1}^{N} (A^m)_{jj}$, the above upper bound is different from (3.78).

80. *Bounds for connected graphs.* A connected graph has an adjacency matrix that is irreducible (Section 10.6). We apply the bounds (A.171) in **art.** 270 to A^2 by choosing $y = u$,

$$\min_{1 \le i \le N} \left(A^2 u \right)_i \le \lambda_1^2 \le \max_{1 \le i \le N} \left(A^2 u \right)_i$$

where $\left(A^2 u \right)_i = (Ad)_i = \sum_{j=1}^{N} a_{ij} d_j = \sum_{j \in \text{neighbors}(i)} d_j$. Thus,

$$\min_{1 \le i \le N} \sqrt{\sum_{j \in \text{neighbors}(i)} d_j} \le \lambda_1 \le \max_{1 \le i \le N} \sqrt{\sum_{j \in \text{neighbors}(i)} d_j} \tag{3.80}$$

Invoking the basic law of the degree (2.3), we have

$$\left(A^2 u \right)_i = 2L - d_i - \sum_{j \notin \text{neighbors}(i)} d_j \le 2L - d_i - (N - 1 - d_i)$$

where the inequality arises from the connectivity of the graph, which implies that the degree d_j of each node j is at least one. Thus, $\max_{1 \le i \le N} \left(A^2 u \right)_i = 2L - N + 1$ and this maximum is reached in the complete graph K_N and in the star $K_{1,N-1}$. Hence, for any connected graph, we obtain the bound

$$\lambda_1 \le \sqrt{2L - N + 1} \tag{3.81}$$

which is sharper than $\lambda_1 \le \sqrt{2L - \frac{2L}{N}}$ in **art.** 42, but the latter bound did not assume connectivity of the graph. In particular, for any tree where $L = N - 1$, the upper bound (3.81) shows that $\lambda_1 \le \sqrt{N-1}$. When the maximum degree in a tree is known, this bound is complemented by Theorem 19 in **art.** 95.

When choosing $y = d$ in (A.171) in **art.** 270, we obtain a companion of (3.80) for connected graphs:

$$\min_{1 \leq i \leq N} \frac{1}{d_i} \sum_{j \in \text{neighbors}(i)} d_j \leq \lambda_1 \leq \max_{1 \leq i \leq N} \frac{1}{d_i} \sum_{j \in \text{neighbors}(i)} d_j \qquad (3.82)$$

81. *Upper bounds for λ_1 in irregular graphs.* Since the largest eigenvalue in a regular graph with degree r equals $\lambda_1 = d_{\max} = r$, we omit regular graphs here. We consider the Laplacian matrix $Q = \Delta - A$ with eigenvalues $\mu_1 \geq \mu_2 \geq \cdots \geq \mu_N = 0$ and their corresponding normalized eigenvectors $y_1, y_2, \ldots y_N = \frac{u}{\sqrt{N}}$ such that $y_k^T y_k = 1$. The definition of an eigenvalue in terms of its corresponding, normalized eigenvector shows that

$$\mu_1 = y_1^T Q y_1 = \sum_{j=1}^{N} d_j (y_1)_j^2 - 2 \sum_{l=1}^{L} (y_1)_{l+} (y_1)_{l-} = \sum_{l=1}^{L} ((y_1)_{l+} - (y_1)_{l-})^2$$

where the last equality follows from **art.** 102. Similarly, using Rayleigh's principle (**art.** 251), we have the bound

$$\mu_1 \geq x_1^T Q x = \sum_{j=1}^{N} d_j (x_1)_j^2 - \lambda_1 (A) = \sum_{l=1}^{L} ((x_1)_{l+} - (x_1)_{l-})^2$$

from which we deduce that

$$d_{\max} - \lambda_1 (A) = \sum_{j=1}^{N} (d_{\max} - d_j) (x_1)_j^2 + \sum_{l=1}^{L} ((x_1)_{l+} - (x_1)_{l-})^2$$

The last sum can be lower bounded in a similar vein as in **art.** 138 by considering a path as a subgraph of a connected graph. After bounding skillfully the right-hand side of the last equality, Stevanović (2004) derived the upper bound for irregular graphs

$$\lambda_1 < d_{\max} - \frac{1}{2N (N d_{\max} - 1) d_{\max}^2}$$

Stevanović's upper bound has been improved several times. A discussion of several improvements is given in Stevanović (2015, p. 54-62). Using the diameter ρ of the graph G, the two best improvements so far are

$$d_{\max} - \lambda_1 > \frac{1}{N\rho} \qquad (3.83)$$

due to Cioabă (2007) and

$$d_{\max} - \lambda_1 > \frac{1}{(N - d_{\min}) \rho + \frac{1}{d_{\max} - E[D]} - \binom{\rho}{2}} \qquad (3.84)$$

due to Shi (2009). For more regular graphs, Cioabă's (3.83) bound is better, while Shi's (3.84) expression is sharper for more irregular graphs. Both proofs use similar ingredients as in Section 4.3, in particular **art.** 138, and are omitted.

3.8 Eigenvalue spacings

The difference $\lambda_k - \lambda_{k+1}$, for $1 \leq k \leq N - 1$, between two consecutive eigenvalues λ_k and λ_{k+1} of the adjacency matrix A is called the k-th eigenvalue spacing of A. Only basic and simple, but general relations are deduced. Higher order differences (see **art. 306**) are not considered, nor the combination with the powerful Interlacing Theorem 71 in **art. 263**. Recently, Kollár and Sarnak (2021) study eigenvalue gap intervals in cubic graphs, connected regular graphs with degree $r = 3$, and list gap intervals $\left(2\sqrt{2}, 3\right)$ in cubic Ramanujan graphs, $[-3, -2)$ in line graphs and $(-1, 1)$ in planar graphs.

82. *Spectral gap.* The difference between the largest eigenvalue λ_1 and second largest λ_2, called the spectral gap, is never larger than N:

$$\lambda_1 - \lambda_2 \leq N \tag{3.85}$$

Indeed, since $\lambda_1 > 0$ as indicated by the bounds (3.79), it follows from (3.5) that

$$0 = \sum_{k=1}^{N} \lambda_k = \lambda_1 + \sum_{k=2}^{N} \lambda_k \leq \lambda_1 + (N-1)\lambda_2$$

such that $\lambda_2 \geq -\frac{\lambda_1}{N-1}$. Hence,

$$\lambda_1 - \lambda_2 \leq \lambda_1 + \frac{\lambda_1}{N-1} = \frac{N\lambda_1}{N-1}$$

Art. 42 states that the largest possible eigenvalue is $\lambda_1 = N - 1$, attained in the complete graph, which proves (3.85). The equality sign in (3.85) occurs in case of the complete graph (see Section 6.1). When a link is removed in the complete graph, the spectral gap drops by at least 1 (see Section 6.10). The spectral gap plays an important role in the dynamics of processes on graphs (**art. 99**) and it characterizes the robustness of a graph due to its relation with the algebraic connectivity (**art. 110** and Section 4.3).

83. *Eigenvalue spacings.* The sum over all spacings between two consecutive eigenvalues equals

$$\sum_{k=1}^{N-1} (\lambda_k - \lambda_{k+1}) = \lambda_1 - \lambda_N \tag{3.86}$$

Since each spacing $\lambda_k - \lambda_{k+1} \geq 0$, the largest possible spacing occurs when all but one spacing is zero, in which case $\max_{1 \leq k \leq N-1} \lambda_k - \lambda_{k+1}$ is equal to $\lambda_1 - \lambda_N$. However, each spacing consists of two consecutive eigenvalues, which implies that $\lambda_N = \lambda_2$ or $\lambda_{N-1} = \lambda_1$. **Art.** 82 shows that the largest possible spacing is attained in the complete graph and is equal to N, the largest possible spectral gap.

Let $\Delta\lambda$ denote an arbitrary spacing between two consecutive eigenvalues, then the telescoping series (3.86) shows that its average equals

$$E\left[\Delta\lambda\right] = \frac{\lambda_1 - \lambda_N}{N-1} \leq \frac{2d_{\max}}{N-1}$$

Abel's partial summation

$$\sum_{k=1}^{n} a_k b_k = \sum_{k=1}^{n-1} \left(\sum_{l=1}^{k} a_l \right) (b_k - b_{k+1}) + b_n \left(\sum_{l=1}^{n} a_l \right) \tag{3.87}$$

applied to $\sum_{k=1}^{N} \lambda_k = 0$ in (3.5) shows that

$$\sum_{k=1}^{N-1} k \left(\lambda_k - \lambda_{k+1} \right) = -N \lambda_N$$

The inequality (Hardy *et al.*, 1999)

$$\min_{1 \leq k \leq n} \frac{r_k}{q_k} \leq \frac{r_1 + r_2 + \cdots + r_n}{q_1 + q_2 + \cdots + q_n} \leq \max_{1 \leq k \leq n} \frac{r_k}{q_k} \tag{3.88}$$

where $q_1, q_2, \ldots, q_n$ are positive real numbers and $r_1, r_2, \ldots, r_n$ are real numbers, yields $r_k = k \left(\lambda_k - \lambda_{k+1} \right)$ and $q_k = k$ bounds for the minimum and maximum spacing between consecutive eigenvalues of the adjacency matrix A:

$$0 \leq \min_{1 \leq k \leq N-1} \left(\lambda_k - \lambda_{k+1} \right) \leq \frac{-2\lambda_N}{N-1} \leq \max_{1 \leq k \leq N-1} \left(\lambda_k - \lambda_{k+1} \right) \tag{3.89}$$

Relation (A.176) in **art. 273** implies that

$$-\lambda_N \leq \left\lceil \frac{N}{2} \right\rceil$$

such that the minimum spacing is never larger than

$$\min_{1 \leq k \leq N-1} \left(\lambda_k - \lambda_{k+1} \right) \leq \frac{N}{N-1}$$

With $\mathrm{Var}[\lambda] = \frac{2L}{N}$ in **art. 49**, Lupas' upper bound (B.72) in **art. 345** is

$$\min_{1 \leq k < j \leq N} |\lambda_k - \lambda_j| \leq 2 \sqrt{\frac{L}{\binom{N-1}{3}}} \leq \frac{2\sqrt{3}}{N-1} \sqrt{E[D]}$$

84. *Inequalities for λ_N.* Besides the general bounds in **art. 273**, new bounds for the smallest eigenvalue λ_N of the adjacency matrix A can be deduced, when known relationships are rewritten in terms of the spacings.

Partial summation (3.87) of the total number of closed walks (3.52) yields, for any integer $0 \leq m \leq k$,

$$W_k = \sum_{j=1}^{N} \lambda_l^k = \sum_{j=1}^{N-1} \left(\sum_{l=1}^{j} \lambda_l^m \right) (\lambda_j^{k-m} - \lambda_{j+1}^{k-m}) + \lambda_N^{k-m} W_m \tag{3.90}$$

while the generalization of the telescoping series (3.86) is, for any n,

$$\sum_{j=1}^{N-1} \left(\lambda_j^n - \lambda_{j+1}^n \right) = \lambda_1^n - \lambda_N^n \tag{3.91}$$

The difference $\lambda_j^n - \lambda_{j+1}^n$ can be negative when eigenvalues are negative and n is even. The sum $\sum_{l=1}^{j} \lambda_l^m$ is always positive for $j < N$, which is immediate for even m. For odd m and denoting by q the index such that $\lambda_q \geq 0$ and $\lambda_{q+1} < 0$, we can write for $j > q$

$$\sum_{l=1}^{j} \lambda_l^m = \sum_{l=1}^{q} \lambda_l^m + \sum_{l=q+1}^{j} \lambda_l^m$$

where the first sum is strictly positive and maximal $\sum_{l=1}^{q} \lambda_l^m \geq \sum_{l=1}^{j} \lambda_l^m$ for any $1 \leq j \leq N$ and the second is strictly negative. The second sum decreases with increasing j and is thus larger than or equal to $\sum_{l=q+1}^{N-1} \lambda_l^m$. However, in that extreme case where $j = N - 1$, the sum $\sum_{l=1}^{N-1} \lambda_l^m = W_m - \lambda_N^m > 0$. The minimum value of the sum $\sum_{l=1}^{j} \lambda_l^m$ is attained for even m at $j = 1$ and for odd m at either $j = 1$, if $\lambda_1^m < W_m - \lambda_N^m$, or at $j = N - 1$, if $\lambda_1^m > W_m - \lambda_N^m$. If $m = 1$, the minimum occurs at $j = N - 1$ provided $\lambda_1 > |\lambda_N|$, which excludes, as in **art. 73**, bipartite graphs.

With this preparation, the inequality (3.88), with $r_j = q_j \sum_{l=1}^{j} \lambda_l^m$, $q_j = \lambda_j^{k-m} - \lambda_{j+1}^{k-m} > 0$ and $k - m$ is odd, becomes, using (3.90) and (3.91),

$$\frac{W_k - \lambda_N^{k-m} W_m}{\lambda_1^{k-m} - \lambda_N^{k-m}} \geq \min_{1 \leq j \leq N-1} \sum_{l=1}^{j} \lambda_l^m = \min\left(W_m - \lambda_N^m, \lambda_1^m\right) 1_{\{m \text{ is odd}\}} + \lambda_1^m 1_{\{m \text{ is even}\}}$$

from which we arrive at the bound, for even m and for odd m provided $W_m - \lambda_N^m > \lambda_1^m$,

$$\frac{W_k - \lambda_1^k}{W_m - \lambda_1^m} \geq \lambda_N^{k-m} \tag{3.92}$$

and, for odd m provided $W_m - \lambda_N^m < \lambda_1^m$,

$$\frac{W_k - \lambda_N^k}{W_m - \lambda_N^m} \geq \lambda_1^{k-m} \tag{3.93}$$

For example, for $k - m = 1$ and excluding bipartite graphs, (3.93) reduces for $k = 2$ and $m = 1$ and using (3.7) to $\frac{2L - \lambda_N^2}{-\lambda_N} \geq \lambda_1$, from which the lower bound follows

$$\lambda_N \geq \frac{1}{2}\left(\lambda_1 - \sqrt{\lambda_1^2 + 8L}\right)$$

For $k = 3$ and $m = 2$, and using (3.8), (3.92) generates the upper bound

$$\frac{6 \blacktriangle_G - \lambda_1^3}{2L - \lambda_1^2} \geq \lambda_N$$

Since we can only compute the sum $\sum_{j=1}^{N-1}\left(\sum_{l=1}^{j} \lambda_l^m\right)$ for $m = 0$, the inequality (3.88), with $r_j = q_j(\lambda_j^k - \lambda_{j+1}^k)$ and $q_j = j$, yields for all integer $k \geq 0$, using (3.90),

$$\min_{1 \leq j \leq N-1}\left(\lambda_j^k - \lambda_{j+1}^k\right) \leq \frac{2\left(W_k - N\lambda_N^k\right)}{N(N-1)} \leq \max_{1 \leq j \leq N-1}\left(\lambda_j^k - \lambda_{j+1}^k\right) \tag{3.94}$$

which is the generalization of (3.89).

3.9 Adding or removing nodes or links

This section relates the adjacency eigenvalues of the original graph to those in the resulting graph after topology changes such as node and link additions and removals.

85. *Addition of a node to a graph.* When node $N + 1$ is added to a graph G_N to form the graph G_{N+1}, the adjacency matrix of the latter is expressed as

$$A_{N+1} = \begin{bmatrix} A_N & v_{N \times 1} \\ (v^T)_{1 \times N} & 0 \end{bmatrix} \tag{3.95}$$

where $v_{N \times 1}$ is the zero-one connection vector of the new node $N + 1$ to any other node in G_N. The degree of node $N + 1$ is $d_{N+1} = v^T v$. The matrix (3.95) is a special case of (A.154) in **art. 259**. The analysis in **art. 259-261** and **art. 264** readily applies to relate the spectrum of A_N and A_{N+1}.

Suppose that v is an eigenvector of the adjacency matrix A_N with eigenvalue λ_v, then $\lambda_v = \lambda_{\max}(A_N) = \lambda_1$ by the Perron-Frobenius Theorem 75 because v has non-negative components[8]. Hence, if v is the eigenvector belonging to the largest eigenvalue, then $A_N v = \lambda_1 v$ and $(A_N - \lambda I)^{-1} v = (\lambda_1 - \lambda)^{-1} v$ for $\lambda \neq \lambda_1$ such that

$$v^T (A_N - \lambda I)^{-1} v = \frac{d_{N+1}}{\lambda_1 - \lambda}$$

The general determinant equation (A.157) becomes

$$\det (A_{N+1} - \lambda I) = (\lambda^2 - \lambda_1 \lambda - d_{N+1}) \frac{\det (A_N - \lambda I)}{\lambda_1 - \lambda}$$

Hence, if v is the (unscaled) eigenvector of A_N belonging to the largest eigenvalue whose norm is $\|v\|_2^2 = v^T v = d_{N+1}$, then the spectrum of A_{N+1} consists of all eigenvalues of A_N, except for $\lambda = \lambda_1$ and two new eigenvalues,

$$\frac{\lambda_1}{2} \left(1 \pm \sqrt{1 + 4 \frac{d_{N+1}}{\lambda_1^2}} \right)$$

In other words, the largest eigenvalue λ_1 of A_N is split up into a slightly larger one and a smaller one with strength related to the degree d_{N+1}. Such a vector v exists, for example, when $v = u$ and A_N is the adjacency matrix of a regular graph (**art. 55**). The node $N + 1$ is then the cone of the regular graph with degree $d_{N+1} = N$ (see **art. 86** and **art. 166** for the Laplacian spectrum of the cone). Moreover, the analysis also holds for weighted, undirected graphs.

---

[8] If v has zero components, then A is reducible, which implies that the graph G is disconnected.

86. *Cone of a graph.* Invoking the Schur complement (A.59), we obtain the alternative expression for the determinant

$$\det \begin{bmatrix} A_N - \lambda I & v \\ v^T & -\lambda \end{bmatrix} = -\lambda \det \left(A_N - \lambda I + \frac{v v^T}{\lambda} \right) \qquad (3.96)$$

For any complex number z, the determinant $\det(A_{N+1} - \lambda I)$ in (3.95) can be split into two others by (A.32):

$$\det \begin{bmatrix} A_N - \lambda I & v \\ v^T & -\lambda \end{bmatrix} = \det \begin{bmatrix} A_N - \lambda I & v \\ v^T - w^T & -\lambda - z \end{bmatrix} + \det \begin{bmatrix} A_N - \lambda I & v \\ w^T & z \end{bmatrix}$$

When choosing $w = v$, then

$$\det \begin{bmatrix} A_N - \lambda I & v \\ v^T & -\lambda \end{bmatrix} = -\det \begin{bmatrix} A_N - \lambda I & v \\ 0_{1 \times N} & \lambda + z \end{bmatrix} + \det \begin{bmatrix} A_N - \lambda I & v \\ v^T & z \end{bmatrix}$$

$$= -(\lambda + z) \det(A_N - \lambda I) - z \det \left(A_N - \lambda I + \frac{v v^T}{z} \right)$$

where the Schur complement (A.59) is used. The particular case of the cone, where $v = u$, then reduces, with $J = u.u^T$, to

$$\det(A_{N+1} - \lambda I) = z \det \left(A_N - \frac{1}{z} J - \lambda I \right) - (\lambda + z) \det(A_N - \lambda I) \qquad (3.97)$$

Since the adjacency matrix of the complement G^c equals $A^c = J - I - A$, choosing $z = 1$ in (3.97) results in

$$\det(A_{N+1} - \lambda I) = (-1)^N \det(A_N^c + (\lambda + 1) I) - (\lambda + 1) \det(A_N - \lambda I)$$

When a node that connects to all other nodes in G_N is added such that $v = u$, the resulting graph G_{N+1} is called the *cone* of G_N. The cone is always a connected graph. The cone construction is useful to convert a reducible matrix into an irreducible one, or to connect a graph with several disconnected clusters of components. An interesting application occurs in Google's PageRank as discussed in Van Mieghem (2014, pp. 251-255).

87. *Removing m nodes from a graph.* Let $\mathcal{N}_m$ denote the set of the m nodes that are removed from G, and $G_m(\mathcal{N}) = G \backslash \mathcal{N}_m$ is the resulting graph after the removal of m nodes from G. We can always relabel the nodes, without affecting the eigenvalues (**art.** 239), in such a way that the adjacency matrix A of G has the form

$$A = \begin{bmatrix} A_1 & B \\ B^T & A_2 \end{bmatrix}$$

where A_1 is the adjacency matrix of $G_m(\mathcal{N})$ and A_2 is the adjacency matrix of the removed subgraph on m nodes. Lemma 10 shows that $\lambda_1(A) > \lambda_1(A_1)$ and $\lambda_1(A) > \lambda_1(A_2)$ for $m > 0$, provided the graph is connected, else the upper bound is not strict. We assume here that G is connected. Invoking (A.152) in **art.** 258

where x_1 is the eigenvector of A belonging to $\lambda_1(A)$ and writing out the quadratic form leads to

$$\lambda_1(A) > \lambda_1(A_1) \geq \frac{\lambda_1(A)\left(1 - 2\sum_{n \in \mathcal{N}_m}(x_1)_n^2\right) + \sum_{j \in \mathcal{N}_m}\sum_{i \in \mathcal{N}_m} a_{ij}(x_1)_i(x_1)_j}{1 - \sum_{n \in \mathcal{N}_m}(x_1)_n^2}$$

(3.98)

which sharpens[9] the inequality in Li *et al.* (2012), where the denominator is absent. The upper bound in (3.98) of $\lambda_1(A_1)$ states that the spectral radius λ_1 of a graph G is always larger than or equal to the largest eigenvalue of any subgraph G_s of G,

$$\lambda_1 \geq \max_{\text{all } G_s \subset G}(\lambda_1(A_{G_s}))$$

which is another proof for (3.70) in **art. 72**. If only $m = 1$ node is removed, then (3.98) simplifies, because $a_{ii} = 0$, to

$$\lambda_1(A) > \lambda_1\left(A_{G \backslash \{n\}}\right) \geq \lambda_1(A)\frac{1 - 2(x_1)_n^2}{1 - (x_1)_n^2}$$

(3.99)

The highest lower bound in (3.99) occurs for the removal of node n with smallest principal eigenvector component, which is positive in a connected graph (**art. 269**).

The addition of a node to a graph G_N was discussed in **art. 85**. In particular, when G_{N+1} is the cone of a regular graph G_N, the spectral radius $\lambda_1(A_{N+1})$ of G_{N+1} equals $\frac{\lambda_1(A_N)}{2}\left(1 + \sqrt{1 + 4\frac{d_n}{\lambda_1(A_N)^2}}\right)$, where $\lambda_1(A_N)$ is the spectral radius of G_N and $d_n = N$ is the degree of the added cone node. Hence, the increase of the spectral radius is related to the degree d_n. The lower bound in (3.99) underlines the interpretation of a principal eigenvector component as an importance or centrality measure (see Section 8.7.1). For, the more important the node n is, the higher the value of $(x_1)_n$, and the larger the possible decrease in spectral radius when this node n is removed.

Applying (A.153) and interlacing (**art. 263**) leads to

$$\lambda_N(A) \leq \lambda_N(A_1) \leq \frac{\left(1 - 2\sum_{n \in \mathcal{N}_m}(x_N)_n^2\right)\lambda_N(A) + \sum_{j \in \mathcal{N}_m}\sum_{i \in \mathcal{N}_m} a_{ij}(x_N)_i(x_N)_j}{1 - \sum_{n \in \mathcal{N}_m}(x_N)_n^2}$$

(3.100)

which sharpens the upper bound in Xing and Zhou (2013).

88. *Removing m links from a graph.* After removing the set $\mathcal{L}_m$ of the m links from G, the resulting graph is $G_m(\mathcal{L}) = G \backslash \mathcal{L}_m$. The adjacency matrix $A_m(\mathcal{L})$ of $G_m(\mathcal{L})$ is still a symmetric matrix. The eigenvector z_1 of $A_m(\mathcal{L})$, corresponding to the largest eigenvalue $\lambda_1(A_m(\mathcal{L}))$ in the graph $G_m(\mathcal{L})$, is normalized such that $z_1^T z_1 = 1$. Let e_j be a base vector in the N-dimensional space, where the i-th component equals $(e_j)_i = \delta_{ij}$ and δ_{ij} is the Kronecker delta. Then, the adjacency

---

[9] A similar observation was made in Stevanović (2015, p. 42).

matrix that represents the single link between nodes i and j equals

$$\hat{A}_{ij} = e_i e_j^T + e_j e_i^T \qquad (3.101)$$

Thus, $\hat{A}_{ij}$ equals the zero matrix, except that $\left(\hat{A}_{ij}\right)_{ij} = \left(\hat{A}_{ij}\right)_{ji} = 1$. Clearly, $\det\left(\hat{A}_{ij} - \lambda I\right) = (-1)^N \lambda^{N-2} \left(\lambda^2 - 1\right)$, such that the largest eigenvalue of $\hat{A}_{ij}$ is 1. For any vector y, we have

$$y^T \hat{A}_{ij} y = y^T \left(e_i e_j^T + e_j e_i^T\right) y = y^T e_i e_j^T y + y^T e_j e_i^T y = 2 y_i y_j \qquad (3.102)$$

Art. 44 shows that $2 y_i y_j \leq 1$.

After these preliminaries, we now provide a general bound on the difference between the largest eigenvalues in G and $G_m(\mathcal{L})$, where m links are removed.

Lemma 4 *For any graph G and $G_m(\mathcal{L}) = G \backslash \mathcal{L}_m$, it holds that*

$$2 \sum_{l \in \mathcal{L}_m} (z_1)_{l^+} (z_1)_{l^-} \leq \lambda_1 (A) - \lambda_1 (A_m(\mathcal{L})) \leq 2 \sum_{l \in \mathcal{L}_m} (x_1)_{l^+} (x_1)_{l^-} \qquad (3.103)$$

where x_1 and z_1 are the eigenvectors of A and A_m corresponding to the largest eigenvalues $\lambda_1 (A)$ and $\lambda_1 (A_m)$, respectively, and where a link l joins the nodes l^+ and l^-.

Proof: Since $A_m = A - \sum_{l \in \mathcal{L}_m} \hat{A}_{l^+ l^-}$ where the left-hand side (or start) of the link l is the node l^+ and the right-hand side (or end) of the link l is the node l^- and with the normalization $x_1^T x_1 = 1$, **art.** 44 shows that

$$\lambda_1 (A) = x_1^T A x_1 = x_1^T \left(A_m + \sum_{l \in \mathcal{L}_m} \hat{A}_{l^+ l^-}\right) x_1 = x_1^T A_m x_1 + \sum_{l \in \mathcal{L}_m} x_1^T \hat{A}_{l^+ l^-} x_1$$

Using (3.102) yields $x_1^T \hat{A}_{l^+ l^-} x_1 = 2 (x_1)_{l^+} (x_1)_{l^-}$ and we arrive at

$$\lambda_1 (A) = x_1^T A_m x_1 + 2 \sum_{l \in \mathcal{L}_m} (x_1)_{l^+} (x_1)_{l^-}$$

The Rayleigh principle (**art.** 251) states that, for any normalized vector w with $w^T w = 1$, it holds that $w^T A w \leq \lambda_1 (A)$, where equality is only attained if w equals the eigenvector of A belonging to $\lambda_1 (A)$. Hence, using $x_1^T A_m x_1 \leq \lambda_1 (A_m)$ leads to

$$\lambda_1 (A) = x_1^T A_m x_1 + 2 \sum_{l \in \mathcal{L}_m} (x_1)_{l^+} (x_1)_{l^-} \leq \lambda_1 (A_m) + 2 \sum_{l \in \mathcal{L}_m} (x_1)_{l^+} (x_1)_{l^-}$$

from which the upper bound in (3.103) is immediate. When repeating the analysis from the point of view of A_m rather than from A, then

$$\lambda_1 (A_m) = z_1^T A_m z_1 = z_1^T \left(A - \sum_{l \in \mathcal{L}_m} \hat{A}_{l^+ l^-}\right) z_1 = z_1^T A z_1 - 2 \sum_{l \in \mathcal{L}_m} (z_1)_{l^+} (z_1)_{l^-}$$

By invoking the Rayleigh principle again, we arrive at the lower bound.   □

For connected graphs G and G_m, it is known that $\lambda_1(A) - \lambda_1(A_m) > 0$ (see Lemma 10). The same conclusion also follows from Lemma 4 because the Perron-Frobenius Theorem 75 states that all vector components of z_1 (and x_1) are positive in a connected graph G_m. Lemma 4 indicates that, when those m links are removed that maximize $2\sum_{l \in \mathcal{M}_m} (x_1)_{l+}(x_1)_{l-}$, then the upper bound in (3.103) is maximal, which may lead to the largest possible difference $\lambda_1(A) - \lambda_1(A_m)$. However, those removed links do not necessarily also maximize the lower bound $2\sum_{l \in \mathcal{M}_m} (z_1)_{l+}(z_1)_{l-}$. Hence, the greedy strategy of removing consecutively the link l with the highest product $(x_1)_{l+}(x_1)_{l-}$ is not necessarily guaranteed to lead to the overall optimum. The fact that the problem to find m links in G, whose removal minimizes $\lambda_1(A_m)$, is NP-hard as proved in Van Mieghem *et al.* (2011), underlines this remark.

89. *Graphs that optimize the spectral radius.* Given the class $\mathcal{G}$ of all graphs with N nodes and L links, which graph in this class has the lowest, respectively, highest spectral radius? The first question is answered in Theorem 14. Surprisingly, the second problem of finding the *connected* graph with the largest spectral radius in that class $\mathcal{G}$ turns out to be difficult.

Theorem 14 *Among all graphs G with N nodes and L links, the regular graph has the lowest spectral radius.*

Proof: We give two proofs. (a) Let A be the adjacency matrix of a regular graph and consider, with the definition (3.101),

$$\widetilde{A} = A + \hat{A}_{kj} - \hat{A}_{il}$$

which is the adjacency matrix of the graph, constructed from the regular graph by adding the link (k, j) and removing the link (i, l). After applying the Rayleigh inequality (**art. 251**), we obtain

$$\lambda_1\left(\widetilde{A}\right) \geq \frac{u^T \widetilde{A} u}{u^T u} = \lambda_1(A)$$

where we have used $u^T\left(\hat{A}_{kj} - \hat{A}_{il}\right) u = 0$, which follows from (3.102) and the fact that u is the eigenvector belonging to the largest eigenvalue of the adjacency matrix A of a regular graph (Theorem 8). The argument also shows that any construction leading from A to $\widetilde{A}$ by adding and removing m links from a regular graph maintains the inequality $\lambda_1\left(\widetilde{A}\right) \geq \lambda_1(A)$, because $u^T\left(\sum_m \left\{\hat{A}_{kj} - \hat{A}_{il}\right\}\right) u = 0$.
(b) Theorem 14 follows directly from the inequality (3.66), because $\text{Var}[D] = 0$ for a regular graph and equality in (3.66) is only reached for a regular graph.   □

We now discuss the second problem. Among all graphs G with N nodes and L links, Rowlinson (1988) proved that the graph with largest spectral radius consists of a clique and a node adjacent to at least one node of the clique, possibly all, and a certain number of isolated nodes. Thus, when N and L is fixed, there is only one such graph that maximizes the spectral radius. However, when we require, in addition, that the graph must be connected, the problem becomes more difficult as outlined by Simić et al. (2010). The general subclass of connected graphs in $\mathcal{G}$ that maximizes the spectral radius are nested split graphs. The set of nodes in a split graph can be divided into a coclique and a clique (**art. 92**) with some cross links joining a node from the coclique to a node in the clique. Simić et al. (2010) draw the structure of a connected nested split graph and they provide a set of lower and upper bounds for the spectral radius of nested split graphs.

90. *A link joining two disjoint graphs.* The graphs G_1 and G_2 are disjoint graphs implying that the nodal set $\mathcal{N}_1$ of G_1 and $\mathcal{N}_2$ of G_2 are disjoint sets. Let $n = |\mathcal{N}_1|$ and $m = |\mathcal{N}_2|$. Consider the graph G that is created after connecting the disjoint graphs G_1 and G_2 by one link. The link $l = i \sim j$ that connects the separate graphs G_1 and G_2 is the link between nodes $i \in \mathcal{N}_1$ and $j \in \mathcal{N}_2$. The corresponding adjacency matrix of the graph G is

$$A_G = \begin{bmatrix} (A_{G_1})_{n \times n} & \left(e_i e_j^T\right)_{n \times m} \\ \left(e_i e_j^T\right)^T_{m \times n} & (A_{G_2})_{m \times m} \end{bmatrix}$$

where the nodal set $\mathcal{N}_1$ is numbered from 1 to n and the set $\mathcal{N}_2$ from $n + 1$ to $n + m$ and where e_i is an $n \times 1$ basic vector and e_j is an $m \times 1$ basic vector. The $n \times m$ matrix $e_i e_j^T$ has zero elements, except for the element on row i and column j that equals 1. This matrix can be written as a Kronecker product (**art. 286**) as $e_i e_j^T = e_i \otimes e_j^T$.

Theorem 15 (Heilbronner) *The characteristic polynomial $c_{A_G}(\lambda)$ of the adjacency matrix A_G of the graph G consisting of two disjoint graphs G_1 and G_2 connected by a link between the nodes $i \in G_1$ and $j \in G_2$ is*

$$c_{A_G}(\lambda) = \det(A_G - \lambda I)$$
$$= \det(A_{G_1} - \lambda I) \det(A_{G_2} - \lambda I) - \det\left(A_{G_1 \setminus \{i\}} - \lambda I\right) \det\left(A_{G_2 \setminus \{j\}} - \lambda I\right)$$
$$(3.104)$$

Theorem 15 appears in Cvetković et al. (1995, Section 2.3) and is attributed to Heilbronner (1953). We give our own proof and show below that generalizations to graphs that connect two disjoint graphs by two and more links are not obvious to derive.

Proof: The characteristic polynomial of G is

$$\det(A_G - \lambda I) = \begin{vmatrix} (A_{G_1} - \lambda I)_{n \times n} & \left(e_i e_j^T\right)_{n \times m} \\ \left(e_j e_i^T\right)_{m \times n} & (A_{G_2} - \lambda I)_{m \times m} \end{vmatrix}$$

Invoking the Schur complement (A.57) yields

$$\det(A_G - \lambda I) = \det(A_{G_1} - \lambda I) \det\left(A_{G_2} - \lambda I - e_j e_i^T (A_{G_1} - \lambda I)^{-1} e_i e_j^T\right)$$

For any $n \times n$ matrix Y, it holds that $e_j e_i^T Y e_i e_j^T = y_{ii} e_j e_j^T$, which equals the zero matrix, except that the j-th diagonal element is y_{ii}. Thus,

$$e_j e_i^T \left(A_{G_1} - \lambda I\right)^{-1} e_i e_j^T = \left(A_{G_1} - \lambda I\right)_{ii}^{-1} e_j e_j^T = \frac{\det \left(A_{G_1 \setminus \{i\}} - \lambda I\right)}{\det \left(A_{G_1} - \lambda I\right)} e_j e_j^T$$

where we have used (A.52) and where $G \setminus \{i\}$ represents the graph G from which node i and all incident links are removed. Hence,

$$\det \left(A_G - \lambda I\right) = \det \left(A_{G_1} - \lambda I\right) \det \left(A_{G_2} - \lambda I - \frac{\det \left(A_{G_1 \setminus \{i\}} - \lambda I\right)}{\det \left(A_{G_1} - \lambda I\right)} e_j e_j^T\right)$$

(3.105)

and

$$\lambda I + \frac{\det \left(A_{G_1 \setminus \{i\}} - \lambda I\right)}{\det \left(A_{G_1} - \lambda I\right)} e_j e_j^T = \text{diag} \left(\lambda, \lambda, \ldots, \lambda + \frac{\det \left(A_{G_1 \setminus \{i\}} - \lambda I\right)}{\det \left(A_{G_1} - \lambda I\right)}, \lambda, \ldots, \lambda\right)$$

Using the column addition property of the determinant (A.32) in **art. 209**, we can write the last determinant in (3.105) as

$$R = \det \left(A_{G_2} - \lambda I - \frac{\det \left(A_{G_1 \setminus \{i\}} - \lambda I\right)}{\det \left(A_{G_1} - \lambda I\right)} e_j e_j^T\right)$$

$$= \det \left(A_{G_2} - \lambda I\right) + \det \left(A_{G_2 : \text{col } j = 0} - \text{diag} \left(\lambda, \ldots, \frac{\det \left(A_{G_1 \setminus \{i\}} - \lambda I\right)}{\det \left(A_{G_1} - \lambda I\right)}, \ldots, \lambda\right)\right)$$

$$= \det \left(A_{G_2} - \lambda I\right) - \frac{\det \left(A_{G_1 \setminus \{i\}} - \lambda I\right)}{\det \left(A_{G_1} - \lambda I\right)} \det \left(A_{G_2 \setminus \{j\}} - \lambda I\right)$$

so that (3.105) reduces to (3.104). □

For the special case where the graph $G_2 = \{j\}$ is a single node connected to node i in G_1, (3.104) reduces to

$$c_{A_G} (\lambda) = -\lambda \det \left(A_{G \setminus \{j\}} - \lambda I\right) - \det \left(A_{G \setminus \{i,j\}} - \lambda I\right) \qquad (3.106)$$

which can be computed directly by expanding $\det \left(A_G - \lambda I\right)$ in cofactors along row i using Theorem 59. Formula (3.106) is useful for graphs with degree 1 nodes (like node j here), in particular, in trees as shown in Section 6.4 for the path graph P_N.

If the link (i, j) is absent, then the last sum in (3.104) is absent as well and (3.104) reduces to the well-known case of the characteristic polynomial of two disjoint graphs (**art. 116**). The largest eigenvalue of two disjoint graphs is $\lambda_1 (G_1 + G_2) = \max \left(\lambda_1 (G_1), \lambda_1 (G_2)\right)$, where $G = G_1 + G_2$ is the direct sum of two graphs $G_1 (\mathcal{N}_1, \mathcal{L}_1)$ and $G_2 (\mathcal{N}_2, \mathcal{L}_2)$ where $G (\mathcal{N}, \mathcal{L})$ satisfies $\mathcal{N} = \mathcal{N}_1 \cup \mathcal{N}_2$ and $\mathcal{L} = \mathcal{L}_1 \cup \mathcal{L}_2$. By the Interlacing Theorem 71, we know that $\lambda_1 (G_1) \geq \lambda_1 \left(G_{1 \setminus \{i\}}\right)$ and $\lambda_1 (G_2) \geq \lambda_1 \left(G_{2 \setminus \{j\}}\right)$, so that the largest zero of the second polynomial in (3.104) obeys

$$\lambda_1 \left(G_{1 \setminus \{i\}} + G_{2 \setminus \{j\}}\right) = \max \left(\lambda_1 \left(G_{1 \setminus \{i\}}\right), \lambda_1 \left(G_{2 \setminus \{j\}}\right)\right) \leq \max \left(\lambda_1 (G_1), \lambda_1 (G_2)\right)$$

The zeros of the characteristic polynomial $c_{A_G} (\lambda)$ lie at the intersections of

$\det \left(A_{G_1} - \lambda I \right) \det \left(A_{G_2} - \lambda I \right)$ and $\det \left(A_{G_1 \setminus \{i\}} - \lambda I \right) \det \left(A_{G_2 \setminus \{j\}} - \lambda I \right)$. Since the eigenvalues of $A_{G_1 \setminus \{i\}}$ interlace those of A_{G_1}, the zeros of A_G are, in general, different from either of the zeros of the polynomials in (3.104). The sign of $c_{A_G}(\lambda)$ evaluated at $\lambda = \lambda_1 (G_1 + G_2)$ equals $(-1)^{n+m-1}$, which is minus the sign of the second polynomial for $\lambda \to \infty$. The sign of the first polynomial for $\lambda \to \infty$ is $(-1)^{n+m}$, which shows that the largest eigenvalue $\lambda_1(G)$ is larger than $\lambda_1(G_1 + G_2)$. The adjacency matrix that represents the single link between node i and j equals $e_i e_j^T + e_j e_i^T$, whose largest eigenvalue is $\lambda_1 \left(e_i e_j^T + e_j e_i^T \right) = 1$. Lemma 7 shows that $\lambda_1(G) \le \lambda_1(G_1 + G_2) + 1$. In conclusion, the largest eigenvalue of G is bounded by

$$\lambda_1(G_1 + G_2) < \lambda_1(G) \le \lambda_1(G_1 + G_2) + 1 \tag{3.107}$$

91. *Two links joining two disjoint graphs.* Only the addition of one link between the disjoint graphs G_1 and G_2 leads to a simple expression as (3.104). Indeed, consider the addition of an additional link (k, l) between G_1 and G_2. Then,

$$A_G = \begin{bmatrix} (A_{G_1})_{n \times n} & \left(e_i e_j^T + e_k e_l^T \right)_{n \times m} \\ \left(e_i e_j^T + e_k e_l^T \right)^T_{m \times n} & (A_{G_2})_{m \times m} \end{bmatrix}$$

and the Schur complement (A.57) indicates that

$$\frac{\det (A_G - \lambda I)}{\det (A_{G_1} - \lambda I)} = \det \left(A_{G_2} - \lambda I - \left(e_j e_i^T + e_l e_k^T \right) (A_{G_1} - \lambda I)^{-1} \left(e_i e_j^T + e_k e_l^T \right) \right)$$

Since

$$\left(e_j e_i^T + e_l e_k^T \right) Y \left(e_i e_j^T + e_k e_l^T \right) = y_{ii} e_j e_j^T + y_{kk} e_l e_l^T + y_{ik} e_j e_l^T + y_{ki} e_l e_j^T$$

and invoking the structure of the inverse of a matrix (**art. 262**), we need to interpret

$$(A_{G_1} - \lambda I)^{-1}_{ij} = (-1)^{i+j} \frac{\det \left(A_{G_1 \setminus \text{row}_i \setminus \text{col}_j} - \lambda I \right)}{\det (A_{G_1} - \lambda I)}$$

The adjacency matrix $A_{G_1 \setminus \text{row}_i \setminus \text{col}_j}$ is not symmetric anymore and represents a graph where the out-degree links of node i and the in-degree links of node j are removed. A node is only removed if all its in- and out-degree links are removed. Thus,

$$\begin{aligned} R^* &= - \left(e_j e_i^T + e_l e_k^T \right) (A_{G_1} - \lambda I)^{-1} \left(e_i e_j^T + e_k e_l^T \right) \\ &= - \frac{\det \left(A_{G_1 \setminus \{i\}} - \lambda I \right)}{\det (A_{G_1} - \lambda I)} e_j e_j^T - \frac{\det \left(A_{G_1 \setminus \{k\}} - \lambda I \right)}{\det (A_{G_1} - \lambda I)} e_l e_l^T \\ &\quad - (-1)^{i+k} \frac{\det \left(A_{G_1 \setminus \text{row}_i \setminus \text{col}_k} - \lambda I \right)}{\det (A_{G_1} - \lambda I)} e_j e_l^T \\ &\quad - (-1)^{i+k} \frac{\det \left(A_{G_1 \setminus \text{row}_k \setminus \text{col}_i} - \lambda I \right)}{\det (A_{G_1} - \lambda I)} e_l e_j^T \end{aligned}$$

Since the matrices $e_j e_l^T$ and $e_l e_j^T$ contain off-diagonal elements, the matrix R^*

has four non-zero elements on rows and columns with the same indices j and l. Moreover, as $A_{\setminus \text{row}_k \setminus \text{col}_i} = \left(A_{\setminus \text{row}_i \setminus \text{col}_k} \right)^T$ for a symmetric matrix A and $\det A = \det \left(A^T \right)$, we find that $\det \left(A_{G_1 \setminus \text{row}_i \setminus \text{col}_k} - \lambda I \right) = \det \left(A_{G_1 \setminus \text{row}_k \setminus \text{col}_i} - \lambda I \right)$ and that R^* is a symmetric matrix, with non-zero elements

$$R^*_{jj} = -\frac{\det \left(A_{G_1 \setminus \{i\}} - \lambda I \right)}{\det \left(A_{G_1} - \lambda I \right)}$$

$$R^*_{ll} = -\frac{\det \left(A_{G_1 \setminus \{k\}} - \lambda I \right)}{\det \left(A_{G_1} - \lambda I \right)}$$

$$R^*_{jl} = R^*_{lj} = -(-1)^{i+k} \frac{\det \left(A_{G_1 \setminus \text{row}_i \setminus \text{col}_k} - \lambda I \right)}{\det \left(A_{G_1} - \lambda I \right)}$$

When writing the matrix R^* as a row of column vectors,

$$R = \begin{bmatrix} r_1 & \cdots & r_j & \cdots & r_l & \cdots & r_m \end{bmatrix}$$

all vectors r_i are zero, except for r_j and r_l that both contain two non-zero elements on row j and row l.

Using the column addition property of the determinant (A.32) in **art.** 209, we obtain

$$\det \left(A_{G_2} - \lambda I + R^* \right) = \det \left(A_{G_2} - \lambda I \right) + \det \left(A_{G_2} - \lambda I |_{r_j} \right)$$
$$+ \det \left(A_{G_2} - \lambda I |_{r_l} \right) + \det \left(A_{G_2} - \lambda I |_{r_j, r_l} \right)$$

where in the matrix $A_{G_2} - \lambda I |_{r_l}$, the l-th column is replaced by the vector r_l and $A_{G_2} - \lambda I |_{r_j, r_l}$ has column j and l replaced by the vector r_j and the vector r_l, respectively. Expanding the determinant $\det \left(A_{G_2} - \lambda I |_{r_j} \right)$ in cofactors of the j-th column yields

$$\det \left(A_{G_2} - \lambda I |_{r_j} \right) = R^*_{jj} \det \left(A_{G_2 \setminus \{j\}} - \lambda I \right) + (-1)^{j+l} R^*_{lj} \det \left(A_{G_2 \setminus \text{row}_l \setminus \text{col}_j} - \lambda I \right)$$

and, similarly,

$$\det \left(A_{G_2} - \lambda I |_{r_l} \right) = R^*_{ll} \det \left(A_{G_2 \setminus \{l\}} - \lambda I \right) + (-1)^{j+l} R^*_{jl} \det \left(A_{G_2 \setminus \text{row}_j \setminus \text{col}_l} - \lambda I \right)$$

Expanding $\det \left(A_{G_2} - \lambda I |_{r_j, r_l} \right)$ first in cofactors of the j-th column gives

$$\det \left(A_{G_2} - \lambda I |_{r_j, r_l} \right) = R^*_{jj} \det \left(A_{G_2 \setminus \{j\}} - \lambda I |_{r_l} \right)$$
$$+ (-1)^{j+l} R^*_{lj} \det \left(A_{G_2 \setminus \text{row}_l \setminus \text{col}_j} - \lambda I |_{r_l} \right)$$

The determinant $\det \left(A_{G_2 \setminus \{j\}} - \lambda I |_{r_l} \right)$ contains a single element on the l-th row and column so that $\det \left(A_{G_2 \setminus \{j\}} - \lambda I |_{r_l} \right) = R^*_{ll} \det \left(A_{G_2 \setminus \{j,l\}} - \lambda I \right)$ and, similarly, taking into account the sign due to size reduction,

$$\det \left(A_{G_2 \setminus \text{row}_l \setminus \text{col}_j} - \lambda I |_{r_l} \right) = (-1)^{j+l-1} R^*_{jl} \det \left(A_{G_2 \setminus \{j,l\}} - \lambda I \right)$$

Collecting all pieces in $\det\left(A_G - \lambda I\right) = \det\left(A_{G_1} - \lambda I\right)\det\left(A_{G_2} - \lambda I + R^*\right)$ and using $\det\left(A_{G_2\setminus\,\mathrm{row}_l\setminus\mathrm{col}_j} - \lambda I\right) = \det\left(A_{G_2\setminus\,\mathrm{row}_j\setminus\mathrm{col}_l} - \lambda I\right)$ yields

$$
\begin{aligned}
\det\left(A_G - \lambda I\right) = {}& \det\left(A_{G_1} - \lambda I\right)\det\left(A_{G_2} - \lambda I\right)\\
& - \det\left(A_{G_1\setminus\{i\}} - \lambda I\right)\det\left(A_{G_2\setminus\{j\}} - \lambda I\right)\\
& - \det\left(A_{G_1\setminus\{k\}} - \lambda I\right)\det\left(A_{G_2\setminus\{l\}} - \lambda I\right)\\
& - 2\left(-1\right)^{i+j+k+l}\det\left(A_{G_1\setminus\,\mathrm{row}_i\setminus\mathrm{col}_k} - \lambda I\right)\det\left(A_{G_2\setminus\,\mathrm{row}_j\setminus\mathrm{col}_l} - \lambda I\right)\\
& + \frac{\det\left(A_{G_1\setminus\{k\}} - \lambda I\right)\det\left(A_{G_1\setminus\{i\}} - \lambda I\right) - \det^2\left(A_{G_1\setminus\,\mathrm{row}_i\setminus\mathrm{col}_k} - \lambda I\right)}{\det\left(A_{G_1} - \lambda I\right)}\\
& \times \det\left(A_{G_2\setminus\{j,l\}} - \lambda I\right)
\end{aligned}
$$

Symmetry in A_G – we can repeat the analysis with G_1 and G_2 interchanged by (A.59) – requires that

$$
\begin{aligned}
s ={}& \det\left(A_{G_1\setminus\{i,k\}} - \lambda I\right)\\
={}& \frac{\det\left(A_{G_1\setminus\{i\}} - \lambda I\right)\det\left(A_{G_1\setminus\{k\}} - \lambda I\right) - \left(\det\left(A_{G_1\setminus\,\mathrm{row}_i\setminus\mathrm{col}_k} - \lambda I\right)\right)^2}{\det\left(A_{G_1} - \lambda I\right)}
\end{aligned}
\tag{3.108}
$$

The identity (3.108) can be used to compute the characteristic polynomial $c_{A_G}\left(\lambda\right)$ of a graph when the characteristic polynomials $c_{A_{G\setminus\{i\}}}\left(\lambda\right)$, $c_{A_{G\setminus\{k\}}}\left(\lambda\right)$, $c_{A_{G\setminus\{i,k\}}}\left(\lambda\right)$ and $\det\left(A_{G_1\setminus\,\mathrm{row}_i\setminus\mathrm{col}_k} - \lambda I\right)$ are easier to determine, due to the flexibility to choose an arbitrary pair of different nodes i and k of G. In fact, (3.108) is a special case of (A.51) for symmetric matrices, which in turn is a special case of Jacobi's famous Theorem 61 in **art.** 214. Invoking identity (3.108), we finally arrive at

Theorem 16 *The characteristic polynomial $c_{A_G}\left(\lambda\right)$ of the adjacency matrix A_G of the graph G consisting of two disjoint graphs G_1 and G_2 connected by two different links (i,j) and (k,l), where the nodes $i,k \in G_1$ and $j,l \in G_2$, is given by*

$$
\begin{aligned}
\det\left(A_G - \lambda I\right) = {}& \det\left(A_{G_1} - \lambda I\right)\det\left(A_{G_2} - \lambda I\right)\\
& - \det\left(A_{G_1\setminus\{i\}} - \lambda I\right)\det\left(A_{G_2\setminus\{j\}} - \lambda I\right)\\
& - \det\left(A_{G_1\setminus\{k\}} - \lambda I\right)\det\left(A_{G_2\setminus\{l\}} - \lambda I\right)\\
& - 2\left(-1\right)^{i+j+k+l}\det\left(A_{G_1\setminus\,\mathrm{row}_i\setminus\mathrm{col}_k} - \lambda I\right)\det\left(A_{G_2\setminus\,\mathrm{row}_j\setminus\mathrm{col}_l} - \lambda I\right)\\
& + \det\left(A_{G_1\setminus\{i,k\}} - \lambda I\right)\det\left(A_{G_2\setminus\{j,l\}} - \lambda I\right)
\end{aligned}
\tag{3.109}
$$

The method can be generalized to any number of links between two disjoint graphs, although the resulting expression will be prohibitively complex. If $i = k$, then (3.108) shows that $\det \left(A_{G_1 \setminus \{i,k\}} - \lambda I \right) = 0$ and (3.109) reduces to

$$\det \left(A_G - \lambda I \right) = \det \left(A_{G_1} - \lambda I \right) \det \left(A_{G_2} - \lambda I \right) - \det \left(A_{G_1 \setminus \{i\}} - \lambda I \right)$$
$$\times \left\{ \det \left(A_{G_2 \setminus \{j\}} - \lambda I \right) + \det \left(A_{G_2 \setminus \{l\}} - \lambda I \right) \right.$$
$$\left. + 2 \left(-1 \right)^{j+l} \det \left(A_{G_2 \setminus \text{row}_j \setminus \text{col}_l} - \lambda I \right) \right\}$$

If $i = k$ and $j = l$, then (3.109) reduces to Heilbronner's formula (3.104).

3.10 Additional properties

92. *Cliques and cocliques.* A clique of size m in a graph G with $N \geq m$ nodes is a set of m pairwise adjacent nodes. Only when $m = N$ or the clique is a disjoint subgraph of G, the clique is a complete graph and each node has degree $m - 1$. A coclique, the complement of a clique, is a set of pairwise non-adjacent nodes. The clique number ω is the size of the largest clique in G, while the independence number is the size of the largest coclique.

Suppose that G has a coclique of size c. We can always relabel the nodes such that the nodes belonging to that coclique possess the first c labels. The corresponding adjacency matrix A has the form

$$A = \begin{bmatrix} O_{c \times c} & F_{c \times (N-c)} \\ F^T_{(N-c) \times c} & \widetilde{F}_{(N-c) \times (N-c)} \end{bmatrix} \tag{3.110}$$

Since the principal matrix $O_{c \times c}$ has c eigenvalues equal to zero, the Interlacing Theorem 71 shows that, for $1 \leq j \leq c$,

$$\lambda_{N-c+j} \left(A \right) \leq 0 \leq \lambda_j \left(A \right)$$

Hence, the adjacency matrix A has at least c non-negative and $N - c + 1$ non-positive eigenvalues. The converse is that the number $n_+ = \{ j : \lambda_j \left(A \right) \geq 0 \}$ of non-negative eigenvalues of A provides an upper bound for the independence number. Also, the number $n_- = \{ j : \lambda_j \left(A \right) \leq 0 \}$ of non-positive eigenvalues of A bounds the independence number by $N - n_-$.

Only for the complete graph K_N, where $c = 1$ in (3.110), there is only one positive eigenvalue. If one link (e.g. between node 1 and 2) in the complete graph is removed, the coclique has size $c = 2$, and two eigenvalues are non-negative. Consequently, the second largest eigenvalue λ_2 in any graph apart from K_N is at least equal to zero. Another argument is that, apart from the complete graph, any graph possesses the star $K_{1,2}$ as a subgraph, whose adjacency eigenvalues are computed in Section 6.7. It follows then again from the Interlacing Theorem 71 that $\lambda_2 \geq 0$.

Similarly, if G has a clique of size c, then, after relabeling, the adjacency matrix has the form

$$A = \begin{bmatrix} (J - I)_{c \times c} & \tilde{G}_{c \times (N-c)} \\ G^T_{(N-c) \times c} & \tilde{G}_{(N-c) \times (N-c)} \end{bmatrix}$$

Since the principal matrix $(J - I)_{c \times c}$ has an eigenvalue $c - 1$ and $(-1)^{[c-1]}$ eigenvalues by (6.1), the Interlacing Theorem 71 shows that,

$$\lambda_{N-c+1}(A) \leq c - 1 \leq \lambda_1(A)$$

and, for $2 \leq j \leq c$,

$$\lambda_{N-c+j}(A) \leq -1 \leq \lambda_j(A)$$

The bounds for the clique are less elegant than those for the coclique.

93. *The clique number.* The determination of the clique number in a given graph G is an NP-complete problem. Motzkin and Straus (1965) found a remarkable result that specifies the clique number ω in a graph G:

Theorem 17 (Motzkin-Straus) *For a given graph G, the maximum value of $F(x) = \sum_{l \in \mathcal{L}} x_{l-} x_{l+}$ subject to $u^T x = \sum_{j=1}^N x_j = 1$ and $x_j \geq 0$ for $1 \leq j \leq N$ equals $\frac{1}{2}\left(1 - \frac{1}{\omega}\right)$.*

The Motzkin-Straus Theorem 17 can be reformulated (**art. 44**) as

$$\left(1 - \frac{1}{\omega}\right) = \max_{x \in \mathcal{S}} x^T A x \tag{3.111}$$

where the simplex $\mathcal{S}$ contains all vectors x that lie in the hyperplane $u^T x = 1$ and possess non-negative components.

Before concentrating on the proof of Motzkin-Straus Theorem 17, we consider the Lagrangian

$$L(x_1, x_2, \ldots, x_N) = F(x_1, x_2, \ldots, x_N) - \xi \left(\sum_{j=1}^N x_j - 1\right)$$

where

$$F(x_1, x_2, \ldots, x_N) = \frac{1}{2} \sum_{i=1}^N \sum_{j=1}^N a_{ij} x_i x_j = \frac{1}{2} x^T A x = \sum_{l \in \mathcal{L}} x_{l-} x_{l+}$$

The partial derivative with respect to x_k obeys, since $a_{jj} = 0$ and $A = A^T$,

$$\frac{\partial L}{\partial x_k} = \frac{\partial F}{\partial x_k} - \xi = \sum_{i=1}^N a_{ki} x_i - \xi$$

A necessary condition for the extremal vector x^* is that $\left.\frac{\partial L}{\partial x_k}\right|_{x=x^*} = 0$ or $\left.\frac{\partial F}{\partial x_k}\right|_{x=x^*} =$

ξ for all $1 \leq k \leq N$. When the constraint is the usual normalization $x^T x = 1$, the corresponding Lagrangian

$$\widetilde{L}(x_1, x_2, \ldots, x_N) = F(x_1, x_2, \ldots, x_N) - \xi \left(\sum_{j=1}^{N} x_j^2 - 1 \right)$$

has the partial derivatives $\frac{\partial \widetilde{L}}{\partial x_k} = \frac{\partial F}{\partial x_k} - 2\xi x_k = \sum_{i=1}^{N} a_{ki} x_i - 2\xi x_k$, and the extremal vector x^* needs to obey the eigenvalue equation $Ax^* = 2\xi x^*$. The Lagrangian method thus provides another demonstration that equality in the Rayleigh inequalities (**art. 251**) is achieved for the eigenvectors.

Proof of the Motzkin-Straus Theorem 17: We denote the maximal vector by x^*, so that $\max_{x \in \mathcal{S}} F(x) = F(x^*) = F(x_1^*, x_2^*, \ldots, x_N^*)$.

a) *Lower bound.* We can always relabel nodes in G, so that $1, 2, \ldots, c = \omega$ are the nodes in the largest clique of G (**art. 92**). After choosing $x_j = \frac{1}{c}$ for $1 \leq j \leq c$ and $x_l = 0$ for $c + 1 \leq l \leq N$, we obtain

$$F = \frac{1}{2} \begin{bmatrix} \frac{1}{c} \left(u^T \right)_{1 \times c} & 0 \end{bmatrix} \begin{bmatrix} (J - I)_{c \times c} & G_{c \times (N-c)} \\ G_{(N-c) \times c}^T & \widetilde{G}_{(N-c) \times (N-c)} \end{bmatrix} \begin{bmatrix} \frac{1}{c} u_{c \times 1} \\ 0 \end{bmatrix}$$

$$= \frac{1}{2c^2} \left(u^T \right)_{1 \times c} (J - I)_{c \times c} u_{c \times 1}$$

Further, using $J = u.u^T$,

$$F = \frac{1}{2c^2} \left(\left(u^T u \right)^2 - u^T u \right) = \frac{c^2 - c}{2c^2} = \frac{1}{2} \left(1 - \frac{1}{c} \right)$$

With this choice of the vector x, we arrive with $c = \omega$ at the lower bound

$$F(x^*) \geq \frac{1}{2} \left(1 - \frac{1}{\omega} \right)$$

b) *Upper bound.* The remainder of the proof consists of demonstrating $F(x) \leq \frac{1}{2} \left(1 - \frac{1}{\omega} \right)$ for any vector $x \in \mathcal{S}$. First, if G is the complete graph K_N, then

$$F(x) = \frac{1}{2} x^T (J - I) x = \frac{1}{2} \left(\left(x^T u \right)^2 - x^T x \right)$$

The constraint shows that $x^T u = 1$, while the Cauchy-Schwarz inequality (A.12) indicates that $\frac{1}{N} \leq x^T x$ so that

$$F(x) = \frac{1}{2} \left(1 - x^T x \right) \leq \frac{1}{2} \left(1 - \frac{1}{N} \right) = \frac{1}{2} \left(1 - \frac{1}{\omega} \right)$$

and the theorem holds for the complete graph K_N. Second, for a graph with $N = 1$ node, $F(x) = 0$ as well as $\frac{1}{2} \left(1 - \frac{1}{\omega} \right)$, because $\omega = 1$. Suppose now that for a graph G' with $N - 1$ nodes, it holds that

$$F(x_{G'}^*) = \frac{1}{2} \left(1 - \frac{1}{\omega_{G'}} \right)$$

which is the induction hypothesis. There are now two cases to consider for a graph G with N nodes. If x_G^* lies on the boundary of the simplex $\mathcal{S}$, then one of the coordinates $x_j^* = 0$ and $F(x_G^*) = F(x_{G'}^*)$, where G' is obtained from G by deleting node j. By the induction hypothesis, the theorem holds for G' so that

$$F(x_G^*) = F(x_{G'}^*) = \frac{1}{2}\left(1 - \frac{1}{w_{G'}}\right) \le \frac{1}{2}\left(1 - \frac{1}{w_G}\right)$$

which illustrates that the theorem holds in general when x^* lies on the boundary of $\mathcal{S}$. It remains to focus on the case where x_G^* lies in the interior of $\mathcal{S}$ and G is different from the complete graph K_N. After evaluating the Taylor series (A.8) in **art.** 200 at the vector $w = x^* + h$ for $h = (-y, y, 0, \dots, 0)$, which obeys the constraint $u^T w = 1$ and $w_j \ge 0$ provided $y \le x_1^*$,

$$F(x_1^* - y, x_2^* + y, \dots, x_N^*) = F(x_1^*, x_2^*, \dots, x_N^*) - y\left.\frac{\partial F}{\partial x_1}\right|_{x=x^*} + y\left.\frac{\partial F}{\partial x_2}\right|_{x=x^*}$$
$$-\frac{1}{2}\sum_{i=1}^{2}\sum_{j=1}^{2} a_{ij}y^2$$

and taking into account the Lagrangian condition $\left.\frac{\partial F}{\partial x_j}\right|_{x=x^*} = \xi$, for all $1 \le j \le N$, we find, for any $0 \le y \le x_1^*$,

$$F(x_1^* - y, x_2^* + y, \dots, x_N^*) = F(x_1^*, x_2^*, \dots, x_N^*) - a_{12}y^2$$

Since $G \ne K_N$, there is always a link absent, which we label to be between node 1 and 2 such that $a_{12} = 0$ and, for any $0 \le y \le x_1^*$,

$$F(x_1^* - y, x_2^* + y, \dots, x_N^*) = F(x_1^*, x_2^*, \dots, x_N^*) \equiv F(x_G^*)$$

which illustrates that there is a continuum of optimal vectors w^* as long as $0 \le y \le x_1^*$, where $y = 0$ is the trivial case. Finally, if $y = x_1^*$, then

$$F(0, x_1^* + x_2^*, \dots, x_N^*) = F(x_G^*)$$

which means that the maximum is attained for the subgraph G' by deleting node 1 from G. By the induction hypothesis, the theorem holds for G' and the induction principle then states that the theorem holds for G as well. $\square$

For vectors x normalized as $x^T x = 1$, the Rayleigh inequalities (**art.** 251) demonstrate that $x^T A x \le \lambda_1$, with equality only if $x = x_1$ is the (normalized) eigenvector of A belonging to the spectral radius λ_1. When choosing $x = \frac{x_1}{u^T x_1}$ in (3.111), Wilf (1986) found that

$$\left(1 - \frac{1}{w}\right) = \max_{x \in \mathcal{S}} x^T A x \ge \frac{x_1^T A x_1}{(u^T x_1)^2} = \frac{\lambda_1}{w_1^2} \qquad (3.112)$$

where $w_1 = u^T x_1$ is the fundamental weight (3.2) in **art.** 43. Hence, the clique number is lower bounded by

$$\omega \geq \frac{w_1^2}{w_1^2 - \lambda_1} \geq \frac{N}{N - \lambda_1} \tag{3.113}$$

where the last inequality stems from $w_1 = u^T x_1 \leq \sqrt{N}$ (**art. 68**). Alternatively, Wilf's bound leads to a lower bound for the fundamental weight w_1, besides $w_1 \geq 1$ (**art. 43** and **art. 68**),

$$\max\left(1, \sqrt{\frac{\lambda_1}{1 - \frac{1}{\omega}}}\right) \leq w_1 \leq \sqrt{N} \tag{3.114}$$

The Motzkin-Straus Theorem 17 for $x = \frac{u}{N}$ yields

$$\left(1 - \frac{1}{\omega}\right) \geq \frac{u^T A u}{N^2} = \frac{2L}{N^2}$$

If a connected graph G does not possess triangles, then $\omega = 2$, so that $L \leq \frac{N^2}{4}$, which provides another proof of Mantel's Theorem 7.

94. *Equitable partitions.* If π is an equitable partition of the connected graph G, then the adjacency matrix A and the corresponding quotient matrix A^π have the same spectral radius.

Indeed, **art. 37** shows that the eigenvalues of the quotient matrix A^π corresponding to an equitable partition are a subset of the eigenvalues of the symmetric matrix A. Moreover, any eigenvector v of A^π belonging to eigenvalue λ is transformed to an eigenvector Sv with the same eigenvalue λ. The Perron-Frobenius Theorem 75 states that the eigenvector belonging to λ_1 is the only one with non-negative components. Both A and A^π are non-negative matrices. Since the characteristic matrix S of the partition (**art. 36**) has non-negative elements, both the eigenvector v and Sv have non-negative vector components and, thus, must belong to the spectral radius.

In the terminology of **art. 263**, the eigenvalues of the quotient matrix A^π corresponding to an equitable partition interlace tightly the eigenvalues of the symmetric matrix A and of any permuted matrix $_\pi A$. An interesting consequence is Hoffman's coclique bound for regular graphs:

Theorem 18 (Hoffman) *Consider a regular graph G with degree r and smallest adjacency eigenvalue λ_N, then the size c of the coclique obeys $c \leq \frac{|\lambda_N|}{r + |\lambda_N|} N$.*

Proof: The quotient matrix A^π in **art. 37** of the adjacency matrix $_\pi A$ in (3.110) is

$$A^\pi = \begin{bmatrix} 0 & r \\ \frac{cr}{N-c} & r - \frac{cr}{N-c} \end{bmatrix}$$

and A^π has eigenvalues r and $-\frac{cr}{N-c}$. The Interlacing Theorem 71 indicates that $\lambda_N \leq -\frac{cr}{N-c}$, from which the bound $c \leq \frac{\lambda_N N}{\lambda_N - r}$ follows. $\square$

Any node in G outside a Hoffman coclique is adjacent to $\frac{cr}{N-c}$ nodes of the Hoffman coclique.

95. *Spectral radius of a tree.*

Theorem 19 *The spectral radius λ_1 of any tree with maximum degree $d_{\max} > 1$ is smaller than $2\sqrt{d_{\max} - 1}$.*

There are several proofs of Theorem 19 for which we refer to Stevanović (2015, Sec. 3.3.1). The upper bound (3.81) for any connected graph (**art.** 80) shows that, in any tree, $\lambda_1 \leq \sqrt{N-1}$ with equality for the star $K_{1,N-1}$.

96. *Eigenvalue equation of $\Xi = X \circ X$.* **Art.** 274 relates the diagonal elements of a symmetric matrix to its eigenvalues. Since $a_{jj} = 0$, the matrix equation (A.179) becomes $\Xi\lambda = 0$, where the eigenvalue vector $\lambda = (\lambda_1, \lambda_2, \ldots, \lambda_N)$ and where the non-negative, asymmetric matrix Ξ in (A.178) consists of column vectors $\xi_j = \left((x_k)_1^2, (x_k)_2^2, \ldots, (x_k)_N^2 \right)$, where $(x_k)_j$ is the j-th component of the k-th eigenvector of A belonging to λ_k. Geometrically, $\Xi\lambda = 0$ means that the vector λ is orthogonal to all N vectors ξ_j and, in order to have a non-zero solution for λ, it must hold that $\det\Xi = 0$. This means that the matrix Ξ corresponding to the adjacency matrix A has a zero eigenvalue, while all other eigenvalues of Ξ lie, as shown in **art.** 274, within the unit circle and the largest eigenvalue is precisely equal to 1. The eigenvector u of the *asymmetric* Ξ belonging to eigenvalue 1 and the eigenvector λ of Ξ belonging to eigenvalue 0 are orthogonal, i.e. $u^T\lambda = 0$, agreeing with $\mathrm{trace}(A) = \sum_{k=1}^N \lambda_k = 0$ in (3.5). In addition, $\det\Xi = 0$ implies that the set of vectors $\xi_1, \xi_2, \ldots, \xi_N$ is linearly dependent and $\mathrm{rank}(\Xi) < N$. Since the k-th row of Ξ equals the vector $z_j = \left((x_1)_j^2, (x_2)_j^2, \ldots, (x_N)_j^2 \right)$, the property $\det\Xi = \det\Xi^T = 0$ also implies that the vectors $z_1, z_2, \ldots, z_N$ are linearly dependent.

Since $\left(A^2 \right)_{jj} = d_j$, another instance of (A.180) gives

$$\Xi\lambda^2 = d$$

where the vector $\lambda^2 = \left(\lambda_1^2, \lambda_2^2, \ldots, \lambda_N^2 \right)$ and $d = (d_1, d_2, \ldots, d_N)$ is the degree vector.

97. *Co-eigenvector graphs.* If the orthogonal matrix X of the adjacency matrix A is known and if $\mathrm{rank}(\Xi) = N - 1$, then **art.** 96 shows that the eigenvalue equation $\Xi\lambda = 0$ has a unique eigenvector λ. In that case, the orthogonal matrix X specifies the symmetric adjacency matrix $A = X\Lambda X^T$ of the graph G uniquely, where $\Lambda = \mathrm{diag}(\lambda)$.

If $\mathrm{rank}(\Xi) = N - m < N - 1$, then the kernel space of Ξ has dimension m and contains, apart from the eigenvalue vector λ, precisely $m - 1$ other linearly independent vectors. Those $m - 1$ other independent vectors *may* generate one or more eigenvalue vectors λ_ν for which $A_\nu = X\Lambda_\nu X^T$ is an adjacency matrix of a graph G_ν. All such graphs G_ν are called co-eigenvector graphs of the graph G.

98. *The adjacency matrix and Hadamard products.* Since A is a symmetric 0-1 matrix, the Hadamard product $A \circ A = A$ and, more general, the k-fold Hadamard product $A^{k\circ} = A$. If there would exist a relation between the eigenvalues of the matrix $M_1 \circ M_2$ in terms of the eigenvalues of the matrix M_1 and M_2, then the eigenvalues of the adjacency matrix are invariant under the Hadamard product in the sense that $\lambda_j \left(A^{k\circ} \right) = \lambda_j \left(A \right)$ for any integer $k \geq 1$ and $1 \leq j \leq N$.

An alternative derivation of $\Xi \lambda^2 = d$ uses the Hadamard product $A = A \circ A$ and its spectral decomposition (A.140) in **art. 256**,

$$A = \sum_{k=1}^{N} \sum_{m=1}^{N} \lambda_k \lambda_m \left(x_k \circ x_m \right) \left(x_k \circ x_m \right)^T$$

Since $\left(x_k \circ x_m \right)^T u = x_k^T x_m = \delta_{km}$ due to orthogonality (A.124) of the eigenvectors of a symmetric matrix (**art. 248**) and with $d = Au$ in (2.4), it holds that

$$d = \sum_{k=1}^{N} \sum_{m=1}^{N} \lambda_k \lambda_m \left(x_k \circ x_m \right) \delta_{km} = \sum_{k=1}^{N} \lambda_k^2 \left(x_k \circ x_k \right)$$

which is written in matrix form as $\Xi \lambda^2 = d$.

We generalize the above method and compute $A = A^{3\circ}$ as

$$a_{ij} = \left(A^{3\circ} \right)_{ij} = \sum_{k=1}^{N} \sum_{l=1}^{N} \sum_{m=1}^{N} \lambda_k \lambda_l \lambda_m \left(\left(x_k \circ x_m \right) \left(x_k \circ x_m \right)^T \right)_{ij} \left(x_l x_l^T \right)_{ij}$$

Using $\left(x_k w_l^T \right)_{ij} \left(v_m y_q^T \right)_{ij} = \left(\left(x_k \circ v_m \right) \left(w_l \circ y_q \right)^T \right)_{ij}$ in **art. 256** yields

$$a_{ij} = \left(A^{3\circ} \right)_{ij} = \sum_{k=1}^{N} \sum_{l=1}^{N} \sum_{m=1}^{N} \lambda_k \lambda_l \lambda_m \left(\left(x_k \circ x_l \circ x_m \right) \left(x_k \circ x_l \circ x_m \right)^T \right)_{ij}$$

The procedure is readily generalized to $A = A^{k\circ}$ resulting in the k-fold Hadamard product decomposition

$$A = \sum_{m_1=1}^{N} \sum_{m_2=1}^{N} \cdots \sum_{m_k=1}^{N} \left(\prod_{j=1}^{k} \lambda_{m_j} \right) \left(x_{m_1} \circ x_{m_2} \circ \cdots \circ x_{m_k} \right) \left(x_{m_1} \circ x_{m_2} \circ \cdots \circ x_{m_k} \right)^T$$

$$\tag{3.115}$$

With $\left(\left(x_{m_1} \circ x_{m_2} \circ \cdots \circ x_{m_k} \right) \left(x_{m_1} \circ x_{m_2} \circ \cdots \circ x_{m_k} \right)^T \right)_{ij} = \prod_{r=1}^{k} \left(x_{m_r} \right)_i \left(x_{m_r} \right)_j$, we verify that the corresponding (i,j)-th element is

$$a_{ij} = \sum_{m_1=1}^{N} \sum_{m_2=1}^{N} \cdots \sum_{m_k=1}^{N} \left(\prod_{r=1}^{k} \lambda_{m_r} \left(x_{m_r} \right)_i \left(x_{m_r} \right)_j \right) = \left(\sum_{m=1}^{N} \lambda_m \left(x_m \right)_i \left(x_m \right)_j \right)^k = a_{ij}^k$$

Right-multiplication of both sides in (3.115) by the all-one vector u presents, for any integer $k \geq 1$, the degree vector

$$d = \sum_{m_1=1}^{N} \sum_{m_2=1}^{N} \cdots \sum_{m_k=1}^{N} \left(\sum_{i=1}^{N} \prod_{j=1}^{k} \lambda_{m_j} \left(x_{m_j} \right)_i \right) \left(x_{m_1} \circ x_{m_2} \circ \cdots \circ x_{m_k} \right)$$

as a linear combination of the vectors $x_{m_1} \circ x_{m_2} \circ \cdots \circ x_{m_k}$ and generalizes $d = \Xi \lambda^2$, corresponding to $k = 2$, in **art.** 96. However, only for $k = 2$, the orthogonality of the eigenvectors applies and leads to an elegant result.

3.11 The stochastic matrix $P = \Delta^{-1} A$

99. The stochastic matrix $P = \Delta^{-1} A$, introduced in **art.** 8, characterizes a random walk on a graph. A discrete-time random walk is a stochastic process that starts at a node i at discrete time $k = 0$, moves in the next step $k = 1$ to node j with probability $p_{ij} = \frac{1}{d_i} a_{ij}$, then at $k = 2$ to node l with probability p_{jl} and so continues, at each discrete time k, to jump to nodes in the graph. A random walk is described by a finite Markov chain that is time-reversible[10]. If $s[k]$ denotes the $1 \times N$ state vector at discrete time k with component $s_i[k] = \Pr[X_k = i]$, where $X_k \in \mathcal{N}$ is the random variable of the random walk at discrete time k, then the Markov governing equation is $s[k+1] = s[k]P$ as derived in Van Mieghem (2014, Section 9.2), where the transition probability is $p_{ij} = \Pr[X_{k+1} = j | X_k = i]$. Random walks on graphs have many applications in different fields (see, e.g., the survey by Lovász (1993) and the relation with electric networks by Doyle and Snell (1984)); perhaps the most important application is randomly searching or sampling.

The combination of Markov theory and algebra leads to interesting properties of $P = \Delta^{-1} A$. In a connected graph, the left-eigenvector of P belonging to eigenvalue $\lambda = 1$ is the steady-state vector π (which is a $1 \times N$ row vector, see Van Mieghem (2014)). The corresponding right-eigenvector is the all-one vector u. These eigenvectors obey the eigenvalue equations $P^T \pi^T = \pi^T$ and $Pu = u$ and the orthogonality relation $\pi u = 1$ (**art.** 237). If $d = (d_1, d_2, \ldots, d_N)$ is the degree vector, then the basic law for the degree (2.5) is rewritten as $\left(\frac{d}{2L}\right)^T u = 1$. The steady-state eigenvector π of an aperiodic, irreducible Markov chain is unique (Van Mieghem, 2014, Chapter 9) such that the equations $\pi u = 1$ and $\left(\frac{d}{2L}\right)^T u = 1$ imply that the steady-state vector is $\pi = \left(\frac{d}{2L}\right)^T$ or

$$\pi_j = \frac{d_j}{2L} \tag{3.116}$$

In general, the transition probability matrix P is not symmetric, but, after a similarity transform $H = \Delta^{1/2}$, a symmetric matrix $R = \Delta^{1/2} P \Delta^{-1/2} = \Delta^{-1/2} A \Delta^{-1/2}$ is obtained whose eigenvalues are the same as those of P (**art.** 239). The powerful property (**art.** 247) of symmetric matrices shows that all eigenvalues are real and that $R = U \text{diag}(\lambda_R) U^T$, where the columns of the orthogonal matrix U consist of the normalized eigenvectors v_k that obey $v_j^T v_k = \delta_{jk}$. Explicitly written in terms

---

[10] Alternatively, a time-reversible Markov chain can be viewed as a random walk on an undirected graph.

of these eigenvectors gives (**art. 254**)

$$R = \sum_{k=1}^{N} \lambda_k \left(P \right) v_k v_k^T$$

where, with the Perron-Frobenius Theorem 75, the real eigenvalues are ordered as $1 = \lambda_1 \left(P \right) \geq \lambda_2 \left(P \right) \geq \cdots \geq \lambda_N \left(P \right) \geq -1$. If we exclude bipartite graphs, where the set of nodes is $\mathcal{N} = \mathcal{N}_1 \cup \mathcal{N}_2$ with $\mathcal{N}_1 \cap \mathcal{N}_2 = \varnothing$ and where each link connects a node in $\mathcal{N}_1$ and in $\mathcal{N}_2$, and reducible or periodic Markov chains (**art. 268**), then $|\lambda_k \left(P \right)| < 1$, for $k > 1$. **Art.** 239 shows that the similarity transform $H = \Delta^{1/2}$ maps the steady state vector π into $v_1 = H^{-1}\pi^T$ and, with (3.116),

$$v_1 = \frac{\Delta^{-1/2}\pi^T}{\left\| \Delta^{-1/2}\pi^T \right\|_2}$$

or

$$v_{1j} = \frac{\frac{\sqrt{d_j}}{2L}}{\sqrt{\sum_{j=1}^{N} \left(\frac{\sqrt{d_j}}{2L} \right)^2}} = \sqrt{\frac{d_j}{2L}} = \sqrt{\pi_j}$$

Finally, since $P = \Delta^{-1/2}R\Delta^{1/2}$, the spectral decomposition of the transition probability matrix of a random walk on a graph with adjacency matrix A is

$$P = \sum_{k=1}^{N} \lambda_k \left(P \right) \Delta^{-1/2} v_k v_k^T \Delta^{1/2} = u\pi + \sum_{k=2}^{N} \lambda_k \left(P \right) \Delta^{-1/2} v_k v_k^T \Delta^{1/2}$$

The n-step transition probability is, with $\left(v_k v_k^T \right)_{ij} = v_{ki} v_{kj}$ and (3.116),

$$P_{ij}^n = \frac{d_j}{2L} + \sqrt{\frac{d_j}{d_i}} \sum_{k=2}^{N} \lambda_k^n \left(P \right) v_{ki} v_{kj}$$

The convergence rate towards the unique steady state π_j in a connected graph, also coined the "mixing rate", can be estimated from

$$\left| P_{ij}^n - \pi_j \right| \leq \sqrt{\frac{d_j}{d_i}} \sum_{k=2}^{N} |\lambda_k^n \left(P \right)| \, |v_{ki}| \, |v_{kj}| < \sqrt{\frac{d_j}{d_i}} \sum_{k=2}^{N} |\lambda_k^n \left(P \right)|$$

Denoting by $\xi = \max \left(|\lambda_2 \left(P \right)|, |\lambda_N \left(P \right)| \right) < 1$ and by ξ_0 the largest element of the reduced set $\{ |\lambda_k \left(P \right)| \} \setminus \{ \xi \}$ with $2 \leq k \leq N$, we obtain

$$\left| P_{ij}^n - \pi_j \right| < \sqrt{\frac{d_j}{d_i}} \xi^n + O \left(\xi_0^n \right)$$

Hence, the smaller ξ or, equivalently, the larger the spectral gap $|\lambda_1 \left(P \right)| - |\lambda_2 \left(P \right)| \geq 1 - \xi$, the faster the random walk converges to its steady-state.

100. The stochastic matrix $P = \Delta^{-1}A$ can also be expressed in terms of the Laplacian $Q = \Delta - A$ as $P = I - \Delta^{-1}Q$. This shows that the eigenvector x

of P with corresponding eigenvalue $\lambda(P)$ is the same as that of the *normalized* Laplacian $\Delta^{-1}Q$ belonging to $\tilde{\mu} = 1 - \lambda(P)$ and $0 \leq \tilde{\mu} \leq 2$. Hence, the spectral gap of a stochastic matrix P also equals the second smallest eigenvalue of *normalized* Laplacian $\Delta^{-1}Q$. Moreover, $\text{trace}(P) = \text{trace}(A) = 0$ and $\text{trace}(P^2) = \text{trace}(R^2)$ implies, with $(R)_{ij} = \frac{a_{ij}}{\sqrt{d_i d_j}}$, that

$$\sum_{k=1}^{N} \lambda_k^2(P) = \sum_{i=1}^{N} \sum_{j=1}^{N} \frac{a_{ij}}{\sqrt{d_i d_j}} \frac{a_{ji}}{\sqrt{d_i d_j}} = \sum_{i=1}^{N} \sum_{j=1}^{N} \frac{a_{ij}}{d_i d_j}$$

With $\frac{1}{d_i d_j} = \frac{1}{2} \left\{ \frac{1}{d_i^2} + \frac{1}{d_j^2} - \left(\frac{1}{d_i} - \frac{1}{d_j} \right)^2 \right\}$, we obtain that

$$\sum_{i=1}^{N} \sum_{j=1}^{N} \frac{a_{ij}}{d_i d_j} = \frac{1}{2} \sum_{i=1}^{N} \frac{1}{d_i^2} \sum_{j=1}^{N} a_{ij} + \frac{1}{2} \sum_{j=1}^{N} \frac{1}{d_j^2} \sum_{i=1}^{N} a_{ji} - \frac{1}{2} \sum_{i=1}^{N} \sum_{j=1}^{N} a_{ij} \left(\frac{1}{d_i} - \frac{1}{d_j} \right)^2$$

$$= \sum_{i=1}^{N} \frac{1}{d_i} - \sum_{i=1}^{N} \sum_{j=1}^{i-1} a_{ij} \left(\frac{1}{d_i} - \frac{1}{d_j} \right)^2$$

Thus,

$$\sum_{k=1}^{N} \lambda_k^2(P) = \sum_{i=1}^{N} \frac{1}{d_i} - \sum_{i=1}^{N} \sum_{j=1}^{i-1} a_{ij} \left(\frac{1}{d_i} - \frac{1}{d_j} \right)^2$$

which shows that $\sum_{k=1}^{N} \lambda_k^2(P) \leq \sum_{i=1}^{N} \frac{1}{d_i}$, where $\frac{1}{N} \sum_{i=1}^{N} \frac{1}{d_i} = E\left[\frac{1}{D}\right]$ is the harmonic mean of the degree set $\{d_i\}_{1 \leq i \leq N}$. Only for regular graphs where $d_i = r$, the double sum disappears and $\sum_{k=1}^{N} \lambda_k^2(P) = \frac{N}{r}$. Since

$$\sum_{k=1}^{N} \lambda_k^2(P) = \sum_{k=1}^{N} \left(1 - \lambda_k\left(\Delta^{-1}Q\right)\right)^2 = 1 + \sum_{k=1}^{N-1} (1 - \tilde{\mu}_k)^2 \leq 1 + (N-1)(1 - \tilde{\mu}_2)^2$$

we find, for regular graphs, an upper bound for the spectral gap $\tilde{\mu}_2 \leq 1 - \sqrt{\frac{N-r}{r(N-1)}}$. A tight upper bound

$$\tilde{\mu}_2 \leq 1 - 2\frac{\sqrt{d_{\max} - 1}}{d_{\max}} \left(1 - \frac{2}{\rho}\right) + \frac{2}{\rho}$$

for a graph with diameter $\rho \geq 4$ is derived by Nilli (1991) using Rayleigh's equation (4.21) and some ingenuity.

4

Eigenvalues of the Laplacian Q

In the sequel, we denote the eigenvalues and eigenvectors of the $N \times N$ Laplacian matrix Q by μ and z, respectively, to distinguish them from the eigenvalues λ and eigenvectors x of the adjacency matrix A. The Laplacian eigenvalue equation is $Qz_k = \mu_k z_k$, where the eigenvalue μ_k belongs to the eigenvector z_k.

4.1 General properties

101. The Laplacian matrix is defined by $Q = BB^T = \Delta - A$ in (2.6) in **art.** 4, from which symmetry $Q = Q^T$ follows. Eigenvalues and eigenvectors of a symmetric matrix are real (**art.** 247). The spectral decomposition of any symmetric matrix in **art.** 254 shows that

$$Q = Z \text{diag}(\mu) Z^T = \sum_{k=1}^{N} \mu_k z_k z_k^T \tag{4.1}$$

where Z is the orthogonal matrix with the Laplacian eigenvectors $z_1, z_2, \ldots z_N$ in its columns, obeying $ZZ^T = Z^T Z = I$. We order the N real eigenvalues of the Laplacian Q as $\mu_N \le \mu_{N-1} \le \cdots \le \mu_1$. Similarly as for the adjacency matrix (**art.** 45), none of the eigenvalues of the Laplacian Q is a fraction of the form $\frac{a}{b}$, where a and b are coprime and $b > 1$. A Laplacian eigenvalue can only be an integer or an irrational number.

102. The quadratic form in **art.** 199,

$$x^T Q x = x^T BB^T x = \left\| B^T x \right\|_2^2 \ge 0 \tag{4.2}$$

is positive semidefinite, which implies that all eigenvalues of the Laplacian Q are non-negative and at least one is zero because $\det Q = 0$ as shown in **art.** 4. Thus, the zero eigenvalue is the smallest eigenvalue of Q. Since $Qu = 0$, because the row sum $\sum_{k=1}^{N} q_{ik} = 0$ for each row $1 \le i \le N$, is an instance of the eigenvalue equation (1.3), the eigenvector belonging to the zero eigenvalue is the all-one vector u.

The l-th component of $\left(B^T x\right)_l = x_i - x_j$, where the link $l = i \to j$ connects node

i and j, starting at node $i = l^+$ and ending at node $j = l^-$, allows us to write

$$x^T Q x = \left\| B^T x \right\|_2^2 = \sum_{l \in \mathcal{L}} (x_{l+} - x_{l-})^2 \tag{4.3}$$

If a link l contains a weight w_l, then $\widetilde{Q} = B \operatorname{diag}(w_l) B^T$, as shown in **art. 14** for an electrical resistor network where $w_l = \frac{1}{r_l}$, and the quadratic form

$$x^T \widetilde{Q} x = \sum_{l \in \mathcal{L}} w_l (x_{l+} - x_{l-})^2 \tag{4.4}$$

generalizes (4.3). In terms of the basic vectors $\{e_k\}_{1 \le k \le N}$, the l-th component is $(B^T x)_l = x_{l+} - x_{l-} = (e_{l+} - e_{l-})^T x$ and $(x_{l+} - x_{l-})^2 = x^T (e_{l+} - e_{l-})(e_{l+} - e_{l-})^T x$. Substitution into (4.4) produces the link decomposition of the weighted Laplacian

$$\widetilde{Q} = \sum_{l \in \mathcal{L}} w_l (e_{l+} - e_{l-})(e_{l+} - e_{l-})^T \tag{4.5}$$

which complements the eigenvalue decomposition (4.1) as a sum over the nodes.

Since (4.3) holds for any real vector x, we may consider the component x_n as a real function $f(n)$ acting on a node n. With $x_{l+} = f(l^+)$ and $x_{l-} = f(l^-)$, we alternatively have

$$(Qf, f) = \sum_{l \in \mathcal{L}} \left(f(l^+) - f(l^-) \right)^2$$

where $(g, f) = \sum_{x \in \mathcal{N}} f(x) g(x)$ denotes the scalar product in **art. 350** of two real functions f and g belonging to $L^2(\mathcal{N})$, the space of all real functions on the set of nodes $\mathcal{N}$ for which the norm $\|f\|^2 = (f, f)$ exists.

103. Since Q is a symmetric matrix, all eigenvectors $z_1, z_2, \ldots, z_N$ are orthogonal (**art. 247**). **Art. 102** shows that the eigenvector $z_N = \frac{u}{\sqrt{N}}$ belonging to the smallest eigenvalue $\mu_N = 0$, such that, for all $1 \le j \le N - 1$,

$$u^T z_j = \sum_{k=1}^N (z_j)_k = 0$$

Thus, the sum of all vector components of a Laplacian eigenvector, different from $z_N = \frac{u}{\sqrt{N}}$, is zero. When these eigenvector components are ranked in increasing order, then the smallest and largest eigenvector component of $z_j \ne \frac{u}{\sqrt{N}}$, with $1 \le j \le N - 1$, have a different sign.

If x in (4.3) is the normalized eigenvector z_k belonging to eigenvalue μ_k, satisfying $z_k^T z_m = \delta_{km}$, then the k-th largest Laplacian eigenvalue $\mu_k = z_k^T Q z_k$,

$$\mu_k = \sum_{l \in \mathcal{L}} \left((z_k)_{l+} - (z_k)_{l-} \right)^2 \tag{4.6}$$

equals the sum over all links in the graph of the square differences of the eigenvector components over the end points of a link l. The weighted analogue $\widetilde{\mu}_k = \widetilde{z}_k^T \widetilde{Q} \widetilde{z}_k$ in (4.4) suggests a physical interpretation of a Laplacian eigenvalue as an energy, e.g. the energy dissipated in a resistor network (**art.** 14), and a similar energy interpretation also follows from the Schrödinger equation in Section 1.3. Since the eigenvector z_k is normalized, any component lies within the interval $(z_k)_j \in (-1, 1)$ and the square $\left((z_k)_{l+} - (z_k)_{l-} \right)^2 \in (0, 2)$. The eigenvalue μ_k in (4.6) increases from zero at $k = N$, because all components $(z_N)_j = \frac{1}{\sqrt{N}}$ are the same, to μ_1 at $k = 1$, where the eigenvalue components at both sides of a link have largest probability to be of different sign. This observation means that $(z_k)_n$ as a function $f_k(n)$ of the node n oscillates, on average over all links of the graph, increasingly heavily with decreasing index k. Thus, as well-known in Fourier analysis of functions and illustrated by the spectrum of the circulant matrix in Section 6.2.1, the higher the eigenfrequency (eigenvalue), the more the corresponding eigenfunction (eigenvector) oscillates and (4.6) is the discrete analogue of that spectral property.

104. Gerschgorin's Theorem 65 states that each eigenvalue μ of the Laplacian $Q = \Delta - A$ lies in an interval $|\mu - d_j| \le d_j$ around a degree d_j-value. Hence,

$$0 \le \mu \le 2d_j$$

which shows that Gerschgorin's Theorem 65, alternatively to **art.** 101, demonstrates that Q is positive semidefinite. Moreover, $\mu_1 \le 2d_{\max}$. This same bound (4.20) is also found by considering the non-negative matrix $d_{\max} I - Q$ whose largest eigenvalue is $d_{\max}$ and smallest eigenvalue is $d_{\max} - \mu_1$. The Perron-Frobenius Theorem 75 states that the positive largest eigenvalue is larger than the absolute value of any other one eigenvalue, whence $d_{\max} \ge |d_{\max} - \mu_1|$. This inequality is essentially the same as Gerschgorin's.

A tighter bound than $\mu_1 \le 2d_{\max}$ for the largest Laplacian eigenvalue μ_1 follows from Gerschgorin's Theorem 65 applied to the matrix $B^T B$, that possesses the same non-zero eigenvalues as Q by Lemma 11 in **art.** 284. Indeed, it follows from **art.** 25 that $\left(B^T B \right)_{jj} = \sum_{n=1}^N b_{lj}^2 = 2$, while the radius r_l in Gerschgorin's Theorem 65 equals

$$r_l = \sum_{k=1; k \neq l}^L \left| B^T B \right|_{lk} = \sum_{k=1}^L \left| B^T B \right|_{lk} - 2 \le \sum_{k=1}^L \left(R^T R \right)_{lk} - 2 = \left(R^T R u \right)_l - 2$$

where the matrix R is the unsigned incidence matrix. **Art.** 26 shows for a link l with end nodes l^+ and l^- that $(R^T R u)_l = d_{l+} + d_{l-}$ so that

$$\mu_1 \le \max_{l \in \mathcal{L}} (d_{l+} + d_{l-}) \le 2d_{\max}$$

Hence, the spectral radius μ_1 of the Laplacian Q is smaller than or equal to the largest sum of the nodal degrees on both sides of a link in the graph G. This inequality appears in Anderson and Morley (1985), but is proved differently, based on the Perron-Frobenius Theorem 75.

105. The definition of $Q = \Delta - A$ shows that $\text{trace}(Q) = \text{trace}(\Delta) = \sum_{j=1}^N d_j$.

The basic law of the degree (2.3) and the general trace formula (A.99) combine to

$$\sum_{k=1}^{N} \mu_k = 2L \tag{4.7}$$

Hence, the average value of a Laplacian eigenvalue equals the average degree, $E[\mu] = E[D]$.

Corollary 4 or (A.181) shows that any partial sum with $1 \leq j \leq N$ ordered eigenvalues satisfies

$$\sum_{k=1}^{j} d_{(k)} \leq \sum_{k=1}^{j} \mu_k \tag{4.8}$$

where $d_{(k)}$ denotes the k-th largest degree in the graph, i.e., $d_{(N)} \leq d_{(N-1)} \leq \cdots \leq d_{(1)}$.

106. Applying the general trace relation (A.118) to the Laplacian Q yields

$$\sum_{k=1}^{N} \mu_k^2 = \text{trace}\left(Q^2\right)$$

The square equals $Q^2 = (\Delta - A)^2 = \Delta^2 + A^2 - \left(\Delta A + (\Delta A)^T\right)$ and $\text{trace}\left(Q^2\right) = \sum_{k=1}^{N} d_k^2 + \text{trace}\left(A^2\right)$. Using (A.118) and (3.7) leads to

$$\sum_{k=1}^{N} \mu_k^2 = \sum_{k=1}^{N} d_k^2 + 2L \tag{4.9}$$

Stochastically[1], when considering the eigenvalue μ and the degree D in a graph as a random variable, (4.9) translates with $E[\mu] = E[D]$ in **art.** 105 to

$$\text{Var}[\mu] = \text{Var}[D] + E[D]$$

where the variance $\text{Var}[X] = E\left[X^2\right] - (E[X])^2$ for any random variable X. Since $E[D] > 0$ (excluding graphs without links), the variability of the Laplacian eigenvalues is larger than that of the degree D in the graph, i.e. $\text{Var}[\mu] \geq \text{Var}[D]$. Furthermore, since $\text{Var}[D] \geq 0$ and $E[\mu] = E[D]$ from (4.7), we find for any graph the inequality $E[\mu] \leq \text{Var}[\mu]$, which is written in terms of the Laplacian eigenvalues as

$$\left(\sum_{k=1}^{N} \mu_k\right)^2 \leq N \sum_{k=1}^{N} \left(\mu_k^2 - \mu_k\right) \tag{4.10}$$

Equality in (4.10) and $E[\mu] = \text{Var}[\mu]$ only holds for a regular graph with $\text{Var}[D] = 0$. Hence, similarly as for the adjacency eigenvalues in **art.** 72, we conclude that the

---

[1] Each of the values $\mu_1, \mu_2, \ldots, \mu_N$ is interpreted as a realization (outcome) of the random variable μ and the mean of the m-th powers is computed as $E[\mu^m] = \frac{1}{N} \sum_{k=1}^{N} \mu_k^m$.

Laplacian spectrum can determine whether a graph is regular if $\left(\sum_{k=1}^{N} \mu_k\right)^2 = N \sum_{k=1}^{N} \left(\mu_k^2 - \mu_k\right)$ holds.

Applying Corollary 4 yields, for $1 \le j \le N$,

$$\sum_{k=1}^{j} d_{(k)}^2 + \sum_{k=1}^{j} d_{(k)} \le \sum_{k=1}^{j} \mu_k^2 \qquad (4.11)$$

where $d_{(k)}$ denotes the k-th largest degree in the graph. Hence, for $j = 1$, we find the bound

$$\sqrt{d_{\max}\left(d_{\max} + 1\right)} \le \mu_1 \qquad (4.12)$$

107. The case for the third powers in (A.118) needs the computation of the trace of

$$\begin{aligned} Q^3 &= (\Delta - A)^3 \\ &= \Delta^3 - \Delta^2 A - \Delta A \Delta + \Delta A^2 - A \Delta^2 + A \Delta A + A^2 \Delta - A^3 \end{aligned}$$

Since $a_{jj} = 0$, all matrices to first power in A have a vanishing trace. By computing the product of the matrices, we find that

$$\text{trace}\left(\Delta A^2\right) = \text{trace}\left(A \Delta A\right) = \text{trace}\left(A^2 \Delta\right) = \sum_{k=1}^{N} d_k^2$$

Hence,

$$\text{trace}\left(Q^3\right) = \sum_{k=1}^{N} d_k^3 + 3 \sum_{k=1}^{N} d_k^2 - \text{trace}\left(A^3\right)$$

where $\text{trace}\left(A^3\right) = \sum_{j=1}^{N} \sum_{k=1}^{N} \sum_{l=1}^{N} a_{jk} a_{kl} a_{lj} = \sum_{k=1}^{N} \lambda_k^3$ and (3.8) shows that $\text{trace}\left(A^3\right)$ equals six times the number of triangles in the graph, which we denote by $\blacktriangle_G$. Combining all yields

$$\sum_{k=1}^{N} \mu_k^3 = \sum_{k=1}^{N} d_k^3 + 3 \sum_{k=1}^{N} d_k^2 - 6 \blacktriangle_G \qquad (4.13)$$

For the complete graph, we have that $\text{trace}\left(A^3\right) = N\left(N - 1\right)\left(N - 2\right)$ and $\sum_{k=1}^{N} d_k^2 = N\left(N - 1\right)^2$ such that, for $N > 3$,

$$3 \sum_{k=1}^{N} d_k^2 - 6 \blacktriangle_G = 3N\left(N - 1\right)\left(3 - N\right) < 0$$

while for a tree, where $\blacktriangle_G = 0$, the last two sums are $3 \sum_{k=1}^{N} d_k^2 - 6 \blacktriangle_G > 0$. Thus, the sum of the third powers of the Laplacian eigenvalues can be lower and higher than the corresponding sum of the degrees.

Stochastically, the third centered moment, which quantifies the skewness of the distribution, follows from (4.13), (4.9) and $E[\mu] = E[D]$ as

$$E\left[(\mu - E[\mu])^3\right] = E\left[(D - E[D])^3\right] + 3\mathrm{Var}[D] - \frac{6\blacktriangle_G}{N}$$

The third centered moment of the Laplacian eigenvalue μ differs from that of the degree D by an amount $3\left(\mathrm{Var}[D] - \frac{2\blacktriangle_G}{N}\right)$.

108. Due to the non-commutativity of the matrices A and Δ, it is difficult to extend the computation

$$\mathrm{trace}\,(Q^m) = \mathrm{trace}\,((\Delta - A)^m)$$

to the case $m > 3$, because the trace operator only preserves cyclic permutations

$$\mathrm{trace}\,(ABC) = \mathrm{trace}\,(BCA) = \mathrm{trace}\,(CAB) \tag{4.14}$$

but not arbitrary permutations,

$$\mathrm{trace}\,(ABC) \neq \mathrm{trace}\,(ACB) \neq \mathrm{trace}\,(BAC)$$

In general, we can expand the matrix product as

$$(A + B)^m = \sum_{j=0}^{2^m-1} \prod_{k=0}^{m-1} \{c_k(j)A + (1 - c_k(j))B\} \tag{4.15}$$

where $c_k(n) = \frac{1}{2}\left(1 - (-1)^{\lfloor \frac{n}{2^k} \rfloor}\right)$ is the k-binary digit of the number of $n = \sum_{k=0}^{\lfloor \log_2 n \rfloor} c_k(n)2^k$ and the matrix product operator on the right-hand side is non-commutative. If A and B commute, we readily verify that (4.15) reduces to the binomial formula $(A + B)^m = \sum_{k=0}^{m} \binom{m}{k} A^{m-k} B^k$. For $m = 4$, formula (4.15) yields

$$(A + B)^4 = B^4 + AB^3 + BAB^2 + A^2B^2 + B^2AB + ABAB + BA^2B + A^3B$$
$$+ B^3A + AB^2A + BABA + A^2BA + B^2A^2 + ABA^2 + BA^3 + A^4$$

from which we find

$$\mathrm{trace}\,(A + B)^4 = \mathrm{trace}\,(B^4) + 4\,\mathrm{trace}\,(B^3A) + 4\,\mathrm{trace}\,(B^2A^2) + 4\,\mathrm{trace}\,(BA^3)$$
$$+ 2\,\mathrm{trace}\,(BABA) + \mathrm{trace}\,(A^4)$$

where $\mathrm{trace}(BABA) \neq \mathrm{trace}(B^2A^2)$ causes a deviation from the binomial formula.

Only in regular graphs, Δ and A commute and the binomial expansion yields, for any integer m,

$$\mathrm{trace}\,((\Delta - A)^m) = \sum_{k=0}^{m} \binom{m}{k} (-1)^{m-k}\,\mathrm{trace}\,(\Delta^k A^{m-k})$$

Since $\Delta^k = \mathrm{diag}(d_1^k, d_2^k, \ldots, d_N^k)$, we then have

$$\mathrm{trace}\,(\Delta^k A^{m-k}) = \sum_{l=1}^{N} (\Delta^k A^{m-k})_{ll} = \sum_{l=1}^{N} d_l^k (A^{m-k})_{ll}$$

and

$$\text{trace}\left(Q^m\right) = \sum_{k=0}^{m} \binom{m}{k} (-1)^{m-k} \sum_{l=1}^{N} d_l^k \left(A^{m-k}\right)_{ll}$$

Taking into account that $A_{ll} = 0$ and using (A.118), the m-th moment of the Laplacian eigenvalues of a regular graph with degree r are expressed in terms of the number of closed walks W_j in (3.52) in **art. 65** of length j starting and returning at node l (**art. 6**), as

$$\sum_{k=1}^{N} \mu_k^m = Nr^m + \binom{m}{2} Nr^{m-1} + (-1)^m \sum_{k=0}^{m-2} \binom{m}{k} (-r)^k W_{m-k} \qquad (4.16)$$

109. Art. 274 relates the diagonal elements of a symmetric matrix to its eigenvalues and so provides another relation between the degree d_j of node j and the set of Laplacian eigenvalues $0 = \mu_N \le \mu_{N-1} \le \cdots \le \mu_1$. The matrix equation (A.179) applied to the degree vector $d = (d_1, d_2, \ldots, d_N)$ becomes

$$d = \Xi_Q \mu \qquad (4.17)$$

where the eigenvalue vector $\mu = (\mu_1, \mu_2, \ldots, \mu_N)$ and where the stochastic matrix Ξ_Q in (A.178) consists of column vectors $\xi_j = \left((z_1)_j^2, (z_2)_j^2, \ldots, (z_N)_j^2 \right)$, where $(z_k)_j$ is the j-th component of the k-th eigenvector of Q belonging to μ_k.

Analogously to the adjacency matrix in **art. 96**, also for the Laplacian the determinant is singular, $\det \Xi_Q = 0$. This follows from orthogonality (A.124) of eigenvectors and the fact that $z_N = \frac{1}{\sqrt{N}} u$, because the sum of the first $N-1$ columns in Ξ_Q is a multiple of the last column. Hence, beside the largest eigenvalue at 1, Ξ_Q and Ξ_Q^T have also a zero eigenvalue. The obvious consequence is that $\Xi_Q \mu = d$ in (4.17) cannot be inverted. However, when deleting the last column, corresponding to $\mu_N = 0$, and the last row, the resulting matrix $\widetilde{\Xi}_Q$ can be inverted and the eigenvalues $\mu_1, \mu_2, \ldots, \mu_{N-1}$ can be determined if the degree vector d is known.

110. If G is regular, where all nodes have the same degree, $d_j = r$ for all $1 \le j \le N$, then the eigenvalues of the Laplacian Q and the adjacency matrix A are directly connected because $\det\left(Q - \mu I\right) = \det\left((r - \mu) I - A\right)$. Thus, for all $1 \le j \le N$,

$$\mu_j\left(Q\right) = r - \lambda_{N+1-j}\left(A\right) \qquad (4.18)$$

Since $\mu_N\left(Q\right) = 0$, we find again as in **art. 55** that the largest eigenvalue of the adjacency matrix in a regular graph equals $\lambda_1\left(A\right) = r$.

From (4.18), the difference for all $1 \le j \le N$,

$$\mu_{j-1}\left(Q\right) - \mu_j\left(Q\right) = \lambda_{N+1-j}\left(A\right) - \lambda_{N+2-j}\left(A\right)$$

shows that the spectral gap (**art. 82**) in a regular graph equals $\lambda_1\left(A\right) - \lambda_2\left(A\right) = \mu_{N-1}\left(Q\right)$. This relation might suggest that the spectral gap in any graph is related to the second smallest eigenvalue μ_{N-1} of the Laplacian, whose properties are further explored in Section 4.3. However, Section 8.5.2 exhibits a graph with large spectral gap and small μ_{N-1}.

111. A direct application of Lemma 7 to $A = \Delta - Q$ yields, for any eigenvalue $1 \leq k \leq N$,

$$d_{\min} - \mu_k(Q) \leq \lambda_k(A) \leq d_{\max} - \mu_k(Q)$$

and

$$d_{(k)} - \lambda_1(A) \leq \mu_k(Q) \leq d_{(k)} - \lambda_N(A)$$

Equality is only reached when $d_{\min} = d_{\max} = r$ as in a regular graph (**art.** 110).

112. *The Laplacian spectrum of the complement G^c of G.* From the adjacency matrix $A^c = J - I - A$ of the complement G^c of a graph G (**art.** 1), the Laplacian of the complement G^c is immediate as

$$Q^c = \Delta^c - A^c = (N-1)I - \Delta - J + I + A$$
$$= NI - J - Q$$

Let $z_1, z_2, \ldots, z_N = u$ denote the eigenvectors of Q belonging to the eigenvalues $\mu_1, \mu_2, \ldots$, and $\mu_N = 0$, respectively. The eigenvalues of J are N and $[0]^{N-1}$ as shown in (6.1). Since $Ju = Nu$ and $Jz_j = 0$ for $1 \leq j \leq N-1$ as demonstrated in **art.** 125, we observe that $Q^c u = 0$ and

$$Q^c z_j = (N - \mu_j) z_j$$

Hence, the set of eigenvectors of Q and of the complement Q^c are the same, while the ordered eigenvalues, for $1 \leq j \leq N-1$, are

$$\mu_j(Q^c) = N - \mu_{N-j}(Q) \tag{4.19}$$

Art. 101 and alternatively **art.** 104 indicate that all eigenvalues of a Laplacian matrix are non-negative, hence $\mu_j(Q^c) \geq 0$ for all $1 \leq j \leq N$ such that (4.19) implies that $N - \mu_{N-j}(Q) \geq 0$. Thus, all Laplacian eigenvalues must lie in the interval $[0, N]$. Hence, the upper bound for μ_1 in **art.** 104 needs to be refined to

$$\mu_1 \leq \min\left(N, \max_{l \in \mathcal{L}}(d_{l+} + d_{l-})\right) \tag{4.20}$$

Several other upper bounds for μ_1 are discussed in Brankov *et al.* (2006).

113. Art. 103 shows that the eigenvector z of Q belonging to μ_{N-1} must satisfy $z^T u = 0$. By requiring this additional constraint and choosing the scaling of the eigenvector such that $z^T z = 1$, Rayleigh's principle (**art.** 251) applied to the second smallest eigenvalue of the Laplacian results in

$$\mu_{N-1} = \min_{\|x\|_2^2 = 1 \text{ and } x^T u = 0} x^T Q x \tag{4.21}$$

Applied to the complement Q^c and with (4.19), we obtain

$$\mu_{N-1}\left(Q^c\right) = N - \mu_1\left(Q\right) = \min_{\|x\|_2^2 = 1 \text{ and } x^T u = 0} x^T Q^c x$$

Since $x^T Q^c x = x^T \left(NI - J - Q\right) x = N - x^T Q x$ as follows from **art. 112**, we have

$$N - \mu_1\left(Q\right) = \min_{\|x\|_2^2 = 1 \text{ and } x^T u = 0} \left(N - x^T Q x\right) = N - \max_{\|x\|_2^2 = 1 \text{ and } x^T u = 0} x^T Q x$$

Hence, the largest eigenvalue of Q obeys

$$\mu_1\left(Q\right) = \max_{\|x\|_2^2 = 1 \text{ and } x^T u = 0} x^T Q x = N - \mu_{N-1}\left(Q^c\right)$$

114. *Threshold graphs.* A weighted threshold graph on N nodes, coded by the vector $W_T = (w_2, w_3, \ldots, w_N)$, is constructed, starting from node 1, by sequentially adding a node $n \in \{2, 3, \ldots, N\}$, which is connected to all previous nodes $1, 2, \ldots, n-1$ with link weight w_n. Hence, the weighted degree (**art. 5**) of node n is $\tilde{d}_n = (n-1) w_n + \sum_{j=n+1}^{N} w_j$. An example of a threshold graph is the uniform degree graph in Section 6.11. If all link weights $w_n \in \{0, 1\}$, then Hammer and Kelmans (1996) prove that the Laplacian eigenvalues of a threshold graph are integers (which follows from iterates of the cone of a graph in **art. 166**) and that the Laplacian eigenvalue vector μ is almost the same as the ordered degree vector d of the threshold graph.

4.1.1 Eigenvalues and connectivity

115. Disconnectivity is a special case of the reducibility of a matrix (**art. 268**) and expresses that there is no walk nor path between two nodes in a different component or cluster. A component of a graph G is a largest or maximally connected subgraph of G.

Theorem 20 *The graph G is connected if and only if $\mu_{N-1} > 0$.*

Proof: The theorem is a consequence of the Perron-Frobenius Theorem 75 for a non-negative, irreducible matrix. Indeed, consider the non-negative matrix $\alpha I - Q$, where $\alpha \geq d_{\max}$. If G is connected, then $\alpha I - Q$ is irreducible and the Perron-Frobenius Theorem 75 states that the largest eigenvalue r of $\alpha I - Q$ is positive and simple, the corresponding eigenvector x_r has positive components and satisfies $Q x_r = (\alpha - r) x_r$. Since eigenvectors of a symmetric matrix are orthogonal (**art. 247**) while $u^T x_r > 0$, the eigenvector x_r must be proportional to the all-one vector u, and thus $\mu_N = \alpha - r = 0$. Since there is only one such eigenvector x_r and since the eigenvalue r exceeds all others, all other eigenvalues of Q must exceed zero, otherwise $rI - Q$ would have a larger eigenvalue than r. $\square$

116. A graph G has k components or clusters, if there exists a relabeling of the nodes such that the adjacency matrix has the structure

$$A = \begin{bmatrix} A_1 & O & \cdots & O \\ O & A_2 & & \vdots \\ \vdots & & \ddots & \\ O & & \cdots & A_k \end{bmatrix}$$

where the square submatrix A_m is the adjacency matrix of the connected component m. The corresponding Laplacian is

$$Q = \begin{bmatrix} Q_1 & O & \cdots & O \\ O & Q_2 & & \vdots \\ \vdots & & \ddots & \\ O & & \cdots & Q_k \end{bmatrix}$$

Using (A.57) indicates that

$$\det(Q - \mu I) = \prod_{m=1}^{k} \det(Q_m - \mu_m I)$$

Since each block matrix Q_m is a Laplacian, whose row sum is zero and $\det Q_m = 0$, the characteristic polynomial $\det(Q - \mu I)$ has at least a k-fold zero eigenvalue. If each block matrix Q_m is irreducible, i.e., the m-th cluster is connected, Theorem 20 shows that Q_m has only one zero eigenvalue. Hence, we have proved:

Theorem 21 *The multiplicity of the smallest eigenvalue $\mu = 0$ of the Laplacian Q is equal to the number of components in the graph G.*

If Q has only one zero eigenvalue with corresponding eigenvector u (**art.** 101), then the graph is connected; it has only one component. Theorem 21 as well as Theorem 20 also imply that, if the second smallest eigenvalue μ_{N-1} of Q is zero, the graph G is disconnected.

The following Corollary for the maximum possible eigenvalue $\mu_1 = N$ appeared in Anderson and Morley (1985):

Corollary 1 *If $\mu_1 = N$ in a graph G on N nodes, then G is connected.*

Proof: If $\mu_1 = N$, then the Laplacian complement formula (4.19) indicates that $\mu_{N-1}(Q^c) = N - \mu_1(Q) = 0$. Theorem 20 applied to the complement graph G^c then states that G^c is disconnected. Theorem 3 then implies that $(G^c)^c = G$ is connected. $\qquad\square$

The converse of Corollary 1 is not true, as can be verified for the path graph, whose Laplacian spectra is given in (6.15).

4.1.2 The number of spanning trees and the Laplacian Q

117. *Matrix Tree Theorem.* The coefficients $\{c_k(Q)\}_{0 \le k \le N}$ of the characteristic polynomial of the Laplacian

$$c_Q(x) = \det(Q - xI) = \sum_{k=0}^{N} c_k(Q) x^k \qquad (4.22)$$

can be expressed in terms of sums over minors (see **art.** 235). Apart from $c_N = (-1)^N$, we apply (A.96) for $0 \le m < N$ to the Laplacian $Q = BB^T$

$$(-1)^{N-m} c_{N-m}(Q) = \sum_{all} \text{minor}_m(BB^T) = \sum_{all} \det\left((BB^T)_m\right)$$

where $(Q)_m = (BB^T)_m$ denotes an $m \times m$ submatrix of Q obtained by deleting the same set of $N - m$ rows and columns and where the sum is of over all $\binom{N}{m} = \binom{N}{N-m}$ ways in which $N - m$ rows can be deleted among the N rows. Since $Q = BB^T$ and $q_{ij} = \sum_{k=1}^{L} b_{ik} b_{jk}$, deleting a row i in Q translates to deleting row i in B. Thus, $(B)_m$ is an $m \times L$ submatrix of B in which the same $N - m$ rows in B as in Q are deleted.

We apply (A.70) in the Binet-Cauchy Theorem 62 to $\det(BB^T)_m$,

$$\det(BB^T)_m = \sum_{k_1=1}^{L} \sum_{k_2=k_1+1}^{L} \cdots \sum_{k_m=k_{m-1}+1}^{L} \begin{vmatrix} b_{1k_1} & \cdots & b_{1k_m} \\ \vdots & \cdots & \vdots \\ b_{mk_1} & \cdots & b_{mk_m} \end{vmatrix}^2$$

which illustrates that $\det(BB^T)_m$ is non-zero and $(-1)^k c_k(Q)$ is non-negative integer. Hence, the characteristic polynomial $c_Q(-x) = \sum_{k=0}^{N} (-1)^k c_k(Q) x^k$ has all non-zero integer coefficients and $c_Q(-x) > 0$ for real $x > 0$. Descartes' rule of signs in Theorem 87 shows that $c_Q(-x)$ has no positive real zeros, i.e. $c_Q(x)$ has only non-negative zeros, in agreement with the positive semidefinite nature of the Laplacian (**art.** 101).

Poincaré's Theorem 2 in **art.** 12 tells us that the square of the above determinant in the multiple sum is either zero or one. It remains to investigate for which set $(k_1, k_2, \ldots, k_m)$ the determinant is non-zero, hence, of rank m. **Art.** 12 shows that, only if the subgraph formed by the m links (columns in the matrix of the above determinant) is a spanning tree, the determinant is non-zero.

To conclude, $\det(BB^T)_m$ equals the total number of trees with m links that can be formed in the graph on $m + 1$ given nodes. The coefficient $(-1)^{N-m} c_{N-m}(Q)$ then counts all these spanning trees with m links over all possible ways of deleting $N - m$ nodes in the graph. In summary, we have demonstrated the famous Matrix Tree Theorem:

Theorem 22 (Matrix Tree Theorem) *In a graph G with N nodes, the coefficient*

$(-1)^{N-m}c_{N-m}(Q)$ *of the characteristic polynomial of the Laplacian Q equals the number of all spanning trees with m links in all subgraphs of G.*

Clearly, $c_0(Q) = \det Q = 0$ because there does not exist a tree with N links that spans the N nodes in a graph. The other extreme is, by convention, $(-1)^N c_N(Q) = 1$. Further, $(-1)^{N-1} c_{N-1}(Q) = 2L$, equals the number of spanning trees in G each consisting of one link, which equals twice the number of links in G. Indeed, $\det(BB^T)_1 = \sum_{k_1=1}^{L}|b_{1k_1}|$ is the number of neighbors of node 1; taking the sum over all possible ways to delete one row results in $(-1)^{N-1}c_{N-1}(Q) = \sum_{i=1}^{N}\sum_{k_i=1}^{L}|b_{ik_i}|$, which is the sum of the absolute value of all elements in B. This result also follows from the general relation (A.99) for the second highest degree coefficient in any polynomial and **art.** 105. When $m = N-1$, **art.** 12 shows that $\det(BB^T)_{N-1}$ equals the number of all spanning trees with $N-1$ links in the graph G. Since there are precisely N ways to remove one node (i.e. one row in B), the coefficient $-c_1(Q)$ counts N times all trees spanning all N nodes in G.

The characteristic polynomial of the Laplacian of the example graph in Fig. 2.1 with $N = 6$ nodes and $L = 9$ links is

$$c_Q(x) = x^6 - 18x^5 + 125x^4 - 416x^3 + 659x^2 - 396x$$

$$= x\left(x - \frac{7-\sqrt{13}}{2}\right)\left(x - \frac{7-\sqrt{5}}{2}\right)(x-4)\left(x - \frac{7+\sqrt{5}}{2}\right)\left(x - \frac{7+\sqrt{14}}{2}\right)$$

The example graph has 18 spanning trees with one link, 125 consisting of two links, ..., and 66 spanning trees with five links ($396 = 6 \times 66$) spanning all $N = 6$ nodes.

118. *Matrix Tree Theorem for a weighed Laplacian $\widetilde{Q}$.* **Art.** 14 has introduced the weighted Laplacian $\widetilde{Q} = \widetilde{B}\widetilde{B}^T$, with $N \times L$ weighted incidence matrix $\widetilde{B} = B\mathrm{diag}\left(\frac{1}{\sqrt{r_{ij}}}\right)$, in the context of electrical resistor networks, where the link weight $w_{ij} = \frac{1}{r_{ij}}$ is the inverse of the resistance $r_{ij} = r_l$ of the link l between node i and j. The weighted incidence matrix $\widetilde{B}$ has the same zero elements as the incidence matrix B, but the non-zero elements are different.

Similar to **art.** 117 for $\det(BB^T)_m$, the Binet-Cauchy Theorem 62 becomes

$$\det\left(\widetilde{B}\widetilde{B}^T\right)_m = \sum_{k_1=1}^{L}\sum_{k_2=k_1+1}^{L}\cdots\sum_{k_m=k_{m-1}+1}^{L}\prod_{j=1}^{m}\frac{1}{r_{k_j}}\left|\begin{matrix} b_{1k_1} & \cdots & b_{1k_m} \\ \vdots & \cdots & \vdots \\ b_{mk_1} & \cdots & b_{mk_m} \end{matrix}\right|^2$$

Poincaré's Theorem 2 in **art.** 12 then shows that the remaining determinant is only non-zero when the corresponding $m \times m$ submatrix corresponds to a tree T with m links spanning $m+1$ nodes in the graph (**art.** 11). If the set of all spanning trees on k nodes is denoted by $\mathcal{T}(k)$ with cardinality $|\mathcal{T}(k)|$, then

$$\det\left(\widetilde{B}\widetilde{B}^T\right)_m = \sum_{T\in\mathcal{T}(m+1)}\prod_{l\in T}\frac{1}{r_l}$$

In contrast to the common, additive definition $\sum_{l \in T} w_l$ of the weight in **art.** 7, we define here the weight of a tree T as the *product* of the weight $w_l = \frac{1}{r_l}$ of each link l in the tree T,

$$w(T) = \prod_{l \in T} w_l$$

and define the "weighted" complexity as

$$\widetilde{\xi}(G) = \sum_{T \in \mathcal{T}(N)} w(T) \tag{4.23}$$

If all link weights $w_l = 1$, then the complexity $\xi(G) = |\mathcal{T}(N)|$ of an unweighted graph G equals the total number of spanning trees on all N nodes of G.

119. There is another Matrix Tree Theorem variant for the coefficients of the characteristic polynomial of Q due to Kelmans and Chelnokov (1974) based on the notion of a forest. A forest is a collection of trees. A k-forest, denoted by F_k, is a forest consisting of k components and a 1-forest is a tree. A component j is a set $\mathcal{N}_j$ of nodes of G and two different components possess different nodes such that $\mathcal{N}_j \cap \mathcal{N}_l = \varnothing$ for each component j and l of a k-forest. A k-spanning forest of G is a k-forest whose union of components consists of all nodes of G, thus $\cup_{l=1}^{k} \mathcal{N}_l = \mathcal{N}$, and a k-spanning forest of G has $N - k$ links. Two k-spanning forests are different if they have different sets of links.

Theorem 23 (Matrix Tree Theorem according to Kelmans) *In a graph G with N nodes, the coefficient $(-1)^m c_m(Q)$ of the characteristic polynomial of the Laplacian Q equals $c_0(Q) = 0$ for $m = 0$ and, for $1 \leq m \leq N$,*

$$(-1)^m c_m(Q) = \sum_{\text{all } F_m} \gamma(F_m)$$

where the sum is over all possible m-spanning forests of the graph G with precisely m components and where $\gamma(F_k) = \prod_{l=1}^{k} n_l$ with $n_l = |\mathcal{N}_l|$.

Kelmans' Theorem 23 is used in **art.** 127. Besides $-c_1(Q) = N\xi(G)$ and $(-1)^{N-1} c_{N-1}(Q) = 2L$, Kelmans and Chelnokov (1974) also give[2]

$$(-1)^{N-2} c_{N-2}(Q) = 2L^2 - L - \frac{1}{2}\sum_{k=1}^{N} d_k^2$$

$$(-1)^{N-3} c_{N-3}(Q) = \frac{4}{3}L^3 - 2L^2 - (L-1)\sum_{k=1}^{N} d_k^2 + \frac{1}{3}\sum_{k=1}^{N} d_k^3 - 2\blacktriangle_G$$

where $\blacktriangle_G$ is the number of triangles in G. Invoking the Newton identities in **art.** 294, we may verify that these expressions for the coefficients $c_k(Q)$ are consistent with (4.9) and (4.13).

---

[2] The first result is presented without proof, but a reference to the Russian PhD thesis of Kelmans is given, while the second result is obtained by using special types of graphs.

120. *The sequence* $(-1)^k c_k(Q) \geq 0$ *is unimodal.* The Matrix Tree Theorem 22 indicates that the characteristic polynomial $c_Q(-x)$ has positive coefficients $(-1)^k c_k(Q) \geq 0$. The symmetric Laplacian matrix Q has real eigenvalues (**art.** 247). Thus, the zeros of $c_Q(x)$ are real. **Art.** 328 shows that the sequence of the coefficients $(-1)^k c_k(Q) \geq 0$ is unimodal with a plateau of two points or a peak. Newton's Theorem 97 in **art.** 327 provides the inequality, for $1 \leq k \leq N - 1$,

$$c_k^2(Q) \geq c_{k+1}(Q)\,c_{k-1}(Q)\,\frac{k+1}{k}\,\frac{N-k+1}{N-k} \tag{4.24}$$

For example, for $k = N - 1$, we find with $c_N(Q) = (-1)^N$ and $c_{N-1}(Q) = (-1)^{N-1}\,2L$ that

$$2L^2\frac{N-1}{N} \geq (-1)^{N-2}\,c_{N-2}(Q)$$

121. *Spacing between Laplacian eigenvalues.* If all the eigenvalues of the Laplacian Q are distinct, then Mahler's lower bound (B.71) in **art.** 344 for their spacing is

$$\min_{1\leq k < j \leq N} |\mu_k - \mu_j| > \frac{\sqrt{3}}{N^{\frac{N}{2}+1}\,(c_Q(-1))^{N-1}}$$

where $c_Q(-1) = \det(Q + I)$ in (4.22), while Lupas' upper bound (B.72) in **art.** 345 is

$$\min_{1\leq k < j \leq N} |\mu_k - \mu_j| \leq 2\sqrt{\frac{3\,\mathrm{Var}\,[\mu]}{N^2 - 1}}$$

where $\mathrm{Var}[\mu] = \mathrm{Var}[D] + E[D]$ in **art.** 106.

4.1.3 The complexity

122. As defined in **art.** 118, the *complexity* $\xi(G)$ of the graph G equals the number of all possible spanning trees in the graph. Let J denote the all-one matrix with $(J)_{ij} = 1$ and $J = u.u^T$, then

$$\mathrm{adj}Q = \xi(G)\,J \tag{4.25}$$

where $X^{-1} = \frac{\mathrm{adj}X}{\det X}$. Indeed, if $\mathrm{rank}(Q) < N - 1$, then every cofactor of Q is zero, thus $\mathrm{adj}Q = 0$ and (4.25) shows that $\xi(G) = 0$ implying that the graph is disconnected. If $\mathrm{rank}(Q) = N - 1$, then $Q\,\mathrm{adj}Q = I \det Q = 0$ which means that each column vector of $\mathrm{adj}Q$ is orthogonal to the $N - 1$ dimensional space spanned by the row vectors of Q. Thus, each column vector of $\mathrm{adj}Q$ belongs to the null-space or kernel of Q, which is one-dimensional and spanned by u, since $Qu = 0$. Hence, each column vector of $\mathrm{adj}Q$ is a multiple of the vector u. Since Q is symmetric, so is $\mathrm{adj}Q$ and all the multipliers must be equal such that $\mathrm{adj}Q = \alpha J$. Since $\mathrm{adj}Q = \det\left(\left(BB^T\right)_{N-1}\right)$, the Matrix Tree Theorem 22 in **art.** 117 shows that $\xi(G)$ equals the total number of trees that span N nodes. Equation (4.25) demonstrates that all elements of adj$Q$ are equal to $\xi(G)$.

Example We apply (4.25) to the complete graph K_N with Laplacian $Q_{K_N} = NI - J$. It suffices to compute one suitable element of adj$Q$, for example, $(\text{adj}Q)_{11}$, which is equal to the determinant of the $(N-1) \times (N-1)$ principal submatrix of Q obtained by deleting the first row and column in Q,

$$
(\text{adj}Q)_{11} = \det \begin{bmatrix} N-1 & -1 & \cdots & -1 \\ -1 & N-1 & \cdots & -1 \\ \vdots & & \ddots & \vdots \\ -1 & -1 & \cdots & N-1 \end{bmatrix}
$$

Adding all rows to the first and subsequently adding this new first row to all other rows gives

$$
(\text{adj}Q)_{11} = \det \begin{bmatrix} 1 & 1 & \cdots & 1 \\ -1 & N-1 & \cdots & -1 \\ \vdots & & \ddots & \vdots \\ -1 & -1 & \cdots & N-1 \end{bmatrix} = \det \begin{bmatrix} 1 & 1 & \cdots & 1 \\ 0 & N & \cdots & 0 \\ \vdots & & \ddots & \vdots \\ 0 & 0 & \cdots & N \end{bmatrix} = N^{N-2}
$$

Hence, the total number of spanning trees in the complete graph K_N, which is the largest number of possible spanning trees in any graph with N nodes, equals

$$
\xi(K_N) = N^{N-2} \tag{4.26}
$$

which is a famous result of Cayley of which many proofs exist, see, e.g., Lovász (2003), van Lint and Wilson (1996, Chapter 2) and Van Mieghem (2014, p. 631-633).

123. Equation (4.25) shows that all N minors M_{N-1} of Q are equal to $\xi(G)$. Application of the general relation (A.96) for the coefficients of the characteristic polynomial then gives $c_1 = -N\xi(G)$, as earlier established in **art. 117**. Using (A.100) and the fact that $\mu_N = 0$ (see **art. 101**) yields $c_1 = -\prod_{j=1}^{N-1} \mu_j$. By combining both, the total number of spanning trees $\xi(G)$ in a connected graph is expressed in terms of the eigenvalues of the Laplacian Q as

$$
\xi(G) = \frac{1}{N} \prod_{j=1}^{N-1} \mu_j \tag{4.27}
$$

124. The complexity of G is also given by

$$
\xi(G) = \frac{\det(\alpha J + Q)}{\alpha N^2} \tag{4.28}
$$

for any number $\alpha \neq 0$. Indeed, observe that $JQ = (JB)B^T = 0$ since $JB = 0$ as follows from $J = u.u^T$ and from (2.1) in **art. 1**. Hence, taking into account that $JQ = 0$ and $J^2 = NJ$, we have

$$
(NI - J)(\alpha J + Q) = \alpha NJ + NQ - \alpha J^2 - JQ = NQ
$$

and

$$\text{adj}\left((NI - J)(\alpha J + Q)\right) = \text{adj}\left(\alpha J + Q\right)\text{adj}\left(NI - J\right) = \text{adj}\left(NQ\right)$$

Since $Q_{K_N} = NI - J$ and as shown in **art. 122**, $\text{adj}(NI - J) = N^{N-2}J$ and since $\text{adj}(NQ) = N^{N-1}\text{adj}Q = N^{N-1}\xi\left(G\right)J$, where we have used (4.25),

$$\text{adj}\left(\alpha J + Q\right)J = N\xi\left(G\right)J$$

Left-multiplication with $\alpha J + Q$ finally gives

$$\left(\alpha J + Q\right)\text{adj}\left(\alpha J + Q\right)J = \alpha N^2\xi\left(G\right)J$$

which proves (4.28) for $\alpha \neq 0$, after invoking the definition (A.43) of the inverse $X^{-1} = \frac{\text{adj}X}{\det X}$, written as $X\text{adj}X = \det X$.

125. Since $Qu = 0$, we also have that $QJ = O$ and, after taking the transpose, $J^TQ^T = JQ = O$. Hence, the Laplacian $Q = \Delta - A$ commutes $QJ = JQ$ with the all-one matrix J. **Art.** 55 shows that the adjacency matrix A and the all-one matrix J only commute if the graph is regular. Since commuting matrices have a common, not necessarily complete set of eigenvectors on Lemma 13, Q and J have a common basis of eigenvectors. The all-one vector u is also an eigenvector of J with eigenvalue $\lambda\left(J\right) = N$. The eigenvalues (6.1) of the $N \times N$ rank 1 symmetric matrix $J = u.u^T$ are N and zero with multiplicity $N - 1$. If X is the matrix containing as columns the eigenvectors $j_1 = u, j_2, \ldots, j_N$ of J and $X^TX = I$, then $\text{diag}(\lambda_k\left(Q\right)) = X^TQX$. However, there are infinitely many sets of basis vectors that are also eigenvectors of J, but not necessarily of Q. Hence, the difficulty lies in finding X_Q among all those of X_J.

   Art. 193 indicates that the matrix $Y = I - \frac{1}{N}J$ projects any vector onto the space orthogonal to the vector u. Hence, a set of eigenvectors of J consists of $N - 1$ columns of Y and the vector u. Moreover, the Laplacian of the complete graph K_N is $Q_{K_N} = NI - J$ and the projector matrix $Y = I - \frac{1}{N}J = \frac{1}{N}Q_{K_N}$ will reappear in pseudoinverse $Q^\dagger$ of the Laplacian in **art. 128**.

126. Vice versa, if z_k is an eigenvector of Q belonging to $\mu_k > 0$, then it is also an eigenvector of J, because $Jz_k = 0$ for any z_k orthogonal to u. This means that the eigenvalues of $\alpha J + Q$ with $\alpha \neq 0$ consist of the eigenvalue αN with eigenvector u and the set $0 < \mu_j, \mu_{j-1}, \ldots, \mu_1$ where $j \leq N - 1$. Since $\mu_{N-1} > 0$, Theorem 21 shows that the graph is connected. Invoking (A.98), a connected graph satisfies $\det\left(\alpha J + Q\right) = \alpha N \prod_{j=1}^{N-1} \mu_j$ and the complexity via (4.28) leads again to (4.27).

   If G_r is a regular graph where all nodes have degree r, then **art.** 110 shows that $\mu_j = r - \lambda_{N+1-j}$. Substituted in (4.27) yields

$$\xi\left(G_r\right) = N^{-1}\prod_{j=1}^{N-1}\left(r - \lambda_{N+1-j}\right) = N^{-1}\prod_{m=2}^{N}\left(r - \lambda_m\right)$$

The characteristic polynomial $c_{A_r}(x)$ of the adjacency matrix of G_r equals

$$c_{A_r}(x) = (x - r) \prod_{m=2}^{N} (x - \lambda_m)$$

from which we deduce that

$$\left. \frac{dc_{A_r}(x)}{dx} \right|_{x=r} = \prod_{m=2}^{N} (r - \lambda_m) = N\xi(G_r)$$

127. Since $c_0(Q) = \det Q = 0$, the characteristic polynomial of the Laplacian is

$$c_Q(x) = x \sum_{k=0}^{N-1} c_{k+1}(Q) x^k$$

Applying the Newton equations (**art.** 294) to $\frac{c_Q(x)}{x}$ gives

$$\sum_{k=1}^{N-1} \frac{1}{\mu_k} = -\frac{c_2(Q)}{c_1(Q)}$$

Since all zeros of $\sum_{k=0}^{N-1} c_{k+1}(Q) x^k$ for a connected graph are positive and $\frac{c_{N-1}}{c_N} = -2L$ (**art.** 117), **art.** 294 provides the bound

$$-\frac{c_2(Q)}{c_1(Q)} \geq \frac{(N-1)^2}{2L}$$

Art. 123 shows that $c_1(Q) = -N\xi(G)$, while the Matrix Tree Theorem 22 in **art.** 117 indicates that $c_2(Q)$ equals the number of all spanning trees with $N-2$ links in all subgraphs of G that are obtained after deleting any pair of two nodes in G.

For a tree $G = T$, we have that $\xi(G) = 1$ and $c_1(Q) = -N$, while Kelmans' Theorem 23 states that

$$c_2(Q) = \sum_{\text{all } F_2} \gamma(F_2)$$

where the sum is over all possible 2-spanning forests of the graph G with precisely two components. A 2-spanning forest F_2 is constructed from a spanning tree of G in which one link is deleted such that two disjoint trees T_1 on $n_1 = |T_1|$ nodes and T_2 with $n_2 = |T_2|$ nodes are obtained. Now, $\gamma(F_2) = n_1 n_2$ is also equal to the number of ways of choosing a node v_1 in tree T_1 (component 1) and a node v_2 in T_2 (component 2). Since G is a tree, the number of pairs (T_1, v_1) and (T_2, v_2) equals the distance $h(v_1, v_2)$ in hops between node v_1 and node v_2, because (T_1, v_1) and (T_2, v_2) can only be obtained by deleting one of the links in G on the single path from v_1 to v_2. Thus,

$$c_2(Q) = \sum_{v_1 \in \mathcal{N}} \sum_{v_2 \neq v_1 \in \mathcal{N}} h(v_1, v_2) = \frac{N(N-1)}{2} E[H_T]$$

where H_T is the hopcount in the tree T. Hence, the average hopcount in any tree satisfies

$$E\left[H_T\right] = \frac{2}{N-1} \sum_{k=1}^{N-1} \frac{1}{\mu_k} \tag{4.29}$$

Mohar (1991) has attributed formula (4.29) to Brendan McKay, who provided me with the above derivation. Section 5.2 demonstrates, via inequality (5.46), that the right-hand side of (4.29) is a lower bound for the average hopcount in any graph.

4.2 The pseudoinverse matrix $Q^\dagger$ of the weighted Laplacian $\widetilde{Q}$

We study the inversion problem of the fundamental relation $x = \widetilde{Q}v$ in (2.15) of **art. 14** between the $N \times 1$ injected current flow vector x into nodes of the network and the $N \times 1$ voltage vector v at the nodes, where $\widetilde{Q}$ is a weighted Laplacian matrix. To simplify the notation, we omit the tildes in the eigenvalues and eigenvectors of $\widetilde{Q}$. In fact, the subsequent algebraic manipulations equally hold for the Laplacian Q and the weighted Laplacian $\widetilde{Q}$.

128. *The pseudoinverse $Q^\dagger$.* Due to the zero eigenvalue $\mu_N = 0$ leading to $\det \widetilde{Q} = 0$ in **art. 101**, the matrix equation $x = \widetilde{Q}v$ cannot be inverted. We write the spectral decomposition (4.1) as

$$\widetilde{Q} = \sum_{k=1}^{N-1} \mu_k z_k z_k^T + \mu_N \frac{u}{\sqrt{N}} \frac{u^T}{\sqrt{N}} = \sum_{k=1}^{N-1} \mu_k z_k z_k^T$$

If the graph G is connected, all eigenvalues $\mu_k > 0$ of $\widetilde{Q}$ for $1 \le k < N$ so that the $N \times N$ symmetric matrix

$$Q^\dagger = \sum_{k=1}^{N-1} \mu_k^{-1} z_k z_k^T \tag{4.30}$$

exists. Furthermore, we verify that $\widetilde{Q}Q^\dagger = Q^\dagger \widetilde{Q}$ and invoking the orthogonality of the eigenvectors $z_k^T z_m = \delta_{km}$ yields

$$\widetilde{Q}Q^\dagger = \sum_{k=1}^{N-1}\sum_{m=1}^{N-1} \frac{\mu_k}{\mu_m} z_k \left(z_k^T z_m\right) z_m^T = \sum_{k=1}^{N-1} z_k z_k^T = \sum_{k=1}^{N} z_k z_k^T - \frac{u}{\sqrt{N}} \frac{u^T}{\sqrt{N}}$$

With $\sum_{k=1}^{N} z_k z_k^T = ZZ^T = I$ and using the all-one matrix $J = u.u^T$, we arrive at

$$\widetilde{Q}Q^\dagger = Q^\dagger \widetilde{Q} = I - \frac{1}{N}J \tag{4.31}$$

Relation (4.31) illustrates that the matrix $Q^\dagger$ commutes with the weighted Laplacian $\widetilde{Q}$ and that the product $Q^\dagger \widetilde{Q}$ equals the orthogonal projector $Y = I - \frac{1}{N}J$ onto the hyperplane through the origin that is orthogonal to the vector u in **art. 125**. The matrix $Q^\dagger$ is also called the Moore-Penrose pseudoinverse. Since the vector u is orthogonal to any other eigenvector, it follows from (4.30) that

$Q^\dagger u = u^T Q^\dagger = 0$. For any positive real number $\beta > 0$, the above argument and (4.30) show that $\widetilde{Q}^\beta = \sum_{k=1}^{N-1} \mu_k^\beta z_k z_k^T$, $\left(\widetilde{Q}^\beta\right)^\dagger = \sum_{k=1}^{N-1} \mu_k^{-\beta} z_k z_k^T = \left(Q^\dagger\right)^\beta$ and $\widetilde{Q}^\beta \left(Q^\dagger\right)^\beta = \left(Q^\dagger\right)^\beta \widetilde{Q}^\beta = I - \frac{1}{N}J$.

Multiplying both sides of the injected current-voltage relation $x = \widetilde{Q}v$ by the pseudoinverse $Q^\dagger$ of the weighted Laplacian $\widetilde{Q}$ and using (4.31) yields

$$v = Q^\dagger x + \frac{u^T v}{N} u \qquad (4.32)$$

where the average voltage in the network equals $v_{\mathrm{av}} = \frac{u^T v}{N}$. The solution (4.32), indeed, coincides physically with the fact that only the potential difference matters and that a voltage is only determined with respect to a reference. In other words, we can always choose the voltage reference at will and by choosing $v_{\mathrm{av}} = 0$, the solution (4.32) is most close to the standard inversion.

129. We present an alternative expression to (4.30) for the pseudoinverse $Q^\dagger$ of the (possibly weighted) Laplacian $\widetilde{Q}$. Since $Q^\dagger J = Q^\dagger u . u^T = O$ because $Q^\dagger u = 0$, we observe for any number α that

$$\left(\widetilde{Q} + \alpha J\right) Q^\dagger = I - \frac{1}{N}J \qquad (4.33)$$

The $N \times N$ matrix $\widetilde{Q} + \alpha J = \sum_{k=1}^{N-1} \mu_k z_k z_k^T + \alpha u . u^T = \sum_{k=1}^{N} \mu_k z_k z_k^T$, with μ_N here meaning $\widetilde{\mu}_N = \alpha N$ and $z_N = \frac{u}{\sqrt{N}}$, has an inverse, provided $\widetilde{\mu}_N \neq 0$ and thus $\alpha \neq 0$, as for any connected graph (**art.** 126). The general determinant formula (A.98) indicates that $\det\left(\widetilde{Q} + \alpha J\right) = \alpha N \prod_{k=1}^{N-1} \mu_k$. Since the inverse $\left(\widetilde{Q} + \alpha J\right)^{-1}$ exists for $\alpha \neq 0$, the general formula (A.88) shows that

$$\left(\widetilde{Q} + \alpha J\right)^{-1} = \sum_{k=1}^{N} \mu_k^{-1} z_k z_k^T = \sum_{k=1}^{N-1} \mu_k^{-1} z_k z_k^T + \frac{1}{\alpha N^2} J \qquad (4.34)$$

Thus, for any non-zero real number α, we obtain from (4.33) an alternative expression to (4.30) for the pseudoinverse of the weighted Laplacian $\widetilde{Q}$,

$$Q^\dagger = \left(\widetilde{Q} + \alpha J\right)^{-1} \left(I - \frac{1}{N}J\right) \qquad (4.35)$$

We can also write (4.35) in terms of the Laplacian $Q_{K_N} = NI - J$ of the complete graph K_N as $\widetilde{Q}^\dagger = N\left(\widetilde{Q} + \alpha J\right)^{-1} Q_{K_N}$. Additionally, comparing (4.30) and (4.34) leads to

$$Q^\dagger = \left(\widetilde{Q} + \alpha J\right)^{-1} - \frac{1}{\alpha N^2} J \qquad (4.36)$$

which illustrates that the right-hand side of (4.36) is independent of $\alpha \neq 0$, because (4.30) is.

The definition (A.44) of the inverse matrix in **art**. 212, applied to (4.36), shows that

$$Q_{ij}^{\dagger} = (-1)^{i+j} \frac{\det\left(\widetilde{Q}_{\backslash \text{row } i \backslash \text{col } j} + \alpha u_{N-1}.u_{N-1}^T\right)}{\det\left(\widetilde{Q} + \alpha J\right)} - \frac{1}{\alpha N^2}$$

Using the "rank one update" formula (A.65) in **art**. 219

$$\det\left(\widetilde{Q}_{\backslash \text{row } i \backslash \text{col } j} + \alpha u_{N-1}.u_{N-1}^T\right) = \det\begin{bmatrix} \widetilde{Q}_{\backslash \text{row } i \backslash \text{col } j} & -\alpha u_{N-1} \\ u_{N-1}^T & 1 \end{bmatrix}$$

$$= \det\left(\widetilde{Q}_{\backslash \text{row } i \backslash \text{col } j}\right)\left(1 + \alpha u^T \left(\widetilde{Q}_{\backslash \text{row } i \backslash \text{col } j}\right)^{-1} u\right)$$

and $\det\left(\widetilde{Q} + \alpha J\right) = \alpha N \prod_{k=1}^{N-1} \mu_k$ yields

$$Q_{ij}^{\dagger} = \frac{1}{\alpha N}\left(\frac{\det\left(\widetilde{Q}_{\backslash \text{row } i \backslash \text{col } j}\right)}{(-1)^{i+j} \prod_{k=1}^{N-1} \mu_k} - \frac{1}{N}\right) + \frac{\det\left(\widetilde{Q}_{\backslash \text{row } i \backslash \text{col } j}\right) u^T \left(\widetilde{Q}_{\backslash \text{row } i \backslash \text{col } j}\right)^{-1} u}{(-1)^{i+j} N \prod_{k=1}^{N-1} \mu_k}$$

Since the right-hand side must hold for any $\alpha \neq 0$, we conclude that

$$(-1)^{i+j} \det\left(\widetilde{Q}_{\backslash \text{row } i \backslash \text{col } j}\right) = \frac{1}{N} \prod_{k=1}^{N-1} \mu_k \tag{4.37}$$

which generalizes the complexity $\xi(G)$ in (4.27) and **art**. 124 to *weighted* Laplacians. Section 8.8 shows that any principal submatrix of a (weighted) Laplacian is positive definite. Moreover, (4.37) indicates that the removal of any row and column in the (weighted) Laplacian leads, apart from the factor $(-1)^{i+j}$, to the same result, similar as (4.25). With (4.37), we arrive at

$$Q_{ij}^{\dagger} = \frac{u^T \left(\widetilde{Q}_{\backslash \text{row } i \backslash \text{col } j}\right)^{-1} u}{N^2} \tag{4.38}$$

Again the definition (A.43) of the inverse matrix shows that

$$Q_{ij}^{\dagger} = \frac{u^T \left(\text{adj}\widetilde{Q}_{\backslash \text{row } i \backslash \text{col } j}\right) u}{N^2 \det\left(\widetilde{Q}_{\backslash \text{row } i \backslash \text{col } j}\right)} = \frac{\sum_{k=1}^{N-1}\sum_{l=1}^{N-1} (-1)^{k+l} \det \widetilde{Q}_{\backslash \text{row}(i,k) \backslash \text{col}(j,l)}}{N \prod_{k=1}^{N-1} \mu_k} \tag{4.39}$$

Yet another representation

$$Q_{ij}^{\dagger} = (-1)^{i+j} \frac{\det\begin{bmatrix} \widetilde{Q}_{\backslash \text{row } i \backslash \text{col } j} & -\alpha u_{N-1} \\ u_{N-1}^T & 1 \end{bmatrix}}{\alpha N \prod_{k=1}^{N-1} \mu_k} - \frac{1}{\alpha N^2}$$

simplifies most for $\alpha = -1$, which bears resemblance with Fiedler's block inverse in (5.17). We proceed further in Section 5.1, where the effective resistance of a (weighted) graph is introduced. The element $Q^\dagger_{ij}$ can be expressed in terms of the effective resistance as shown in (Van Mieghem *et al.*, 2017, Appendix B), which demonstrates that $Q^\dagger_{ii} \geq Q^\dagger_{ij}$ for each i and j.

Example Consider the weighted adjacency matrix $W = b(J - I)$ of the complete graph K_N, where $b \neq 0$. The corresponding weighted Laplacian equals $\widetilde{Q}_b = (N-1)bI - b(J-I) = b(NI - J)$ and $\widetilde{Q}_b + \alpha J = bNI + (\alpha - b)J$. Since (4.35) holds for any $\alpha \neq 0$, the easiest choice is $\alpha = b$ leading to the inverse $\left(\widetilde{Q} + bJ\right)^{-1} = \frac{1}{Nb}I$. Hence, a pseudoinverse of the weighted Laplacian $\widetilde{Q}_b = b(NI - J)$ for the complete graph K_N follows from (4.35) as

$$Q^\dagger = \frac{1}{Nb}\left(I - \frac{1}{N}J\right) = \frac{1}{N^2 b}(NI - J) = \frac{1}{N^2 b}Q_{K_N} \qquad (4.40)$$

which is again a weighted Laplacian $\widetilde{Q}_{b^\dagger} = b^\dagger(NI - J)$ of the complete graph K_N with $b^\dagger = \frac{1}{N^2 b}$.

130. *Cramer's method.* We solve $x = \widetilde{Q}v$ by Cramer's method in **art. 220**. We assume that the graph is connected, i.e. $\mu_{N-1} > 0$ by Theorem 21. The rank of the $N \times N$ weighted Laplacian $\widetilde{Q}$ is $N - 1$, because of a vanishing smallest eigenvalue, $\mu_N = 0$, belonging to the eigenvector u. We recall from **art. 14** that $u^T x = u^T \widetilde{Q}v = 0$, physically meaning that the sum of injected currents in the graph is zero. We ignore the trivial case $x = 0$ in which the potential vector $v = \alpha u$ and thus assume that at least two components of x are non-zero. There are basically two approaches[3] to determine the N unknowns $v_1, v_2, \ldots, v_N$: (i) one of the N equations/rows in $x = \widetilde{Q}v$ can be replaced by an additional equation as explored below and (ii) the set is rewritten in $N - 1$ unknowns in terms of one of them, say v_N. The analysis of (ii) is omitted, because the resulting expressions for v_N are less general as those in (i).

We replace an arbitrary equation or row in the set $\widetilde{Q}v = x$ by a new linear equation $c^T v = \sum_{j=1}^N c_j v_j$, where c is a real vector. Following **art. 128**, we choose $c = \frac{u}{N}$ so that $v_{av} = c^T v$, which we can choose at will. Without loss of generality, we first replace the N-th equation in $\widetilde{Q}v = x$ by $u^T v = N v_{av}$ and the resulting set of linear equations becomes

$$\begin{bmatrix} \widetilde{Q}_{\backslash \text{row } N} \\ u \end{bmatrix} v = \begin{bmatrix} x_{\backslash \text{row } N} \\ N v_{av} \end{bmatrix}$$

where $\widetilde{Q}_{\backslash \text{row } N}$ is the $(N-1) \times N$ matrix obtained from $\widetilde{Q}$ by removing row N. Clearly, the potential $v_N = N v_{av} - \sum_{j=1}^{N-1} v_j$. Cramer's solution (A.68) in **art. 220**

---

[3] These two approaches are similar to computing the adjoint matrix $Q(\lambda) = c_A(\lambda)(\lambda I - A)^{-1}$, whose columns are eigenvectors (see **art. 230**).

yields, for $1 \leq j < N$,

$$v_j = \frac{\begin{vmatrix} \widetilde{Q}_{\backslash \text{ row } N} & & \begin{bmatrix} x_{\backslash \text{ row } N} \\ N v_{\text{av}} \end{bmatrix} \end{vmatrix}_{\text{col } j=}}{\begin{vmatrix} \widetilde{Q}_{\backslash \text{ row } N} \\ u \end{vmatrix}} = \frac{\eta_j}{\det\left(\widetilde{Q}_{\text{row } N=u}\right)}$$

and[4] $\det\left(\widetilde{Q}_{\text{row } N=u}\right) = N \det\left(\widetilde{Q}_{\backslash \text{ row } N \backslash \text{ col } N}\right)$ is the same factor, appearing in each v_j for $1 \leq j < N$ and equal to $\prod_{k=1}^{N-1} \mu_k = N\xi(G)$ by (4.37), where $\xi(G)$ in (4.27) is the complexity in a weighted graph. The numerator η_j in v_j is

$$\eta_j = \begin{vmatrix} \widetilde{q}_{11} & \cdots & \widetilde{q}_{1;j-1} & x_1 & \widetilde{q}_{1;j+1} & \cdots & \widetilde{q}_{1N} \\ \widetilde{q}_{21} & \cdots & \widetilde{q}_{2;j-1} & x_2 & \widetilde{q}_{2;j+1} & \cdots & \widetilde{q}_{2N} \\ \vdots & & \vdots & \vdots & \vdots & & \vdots \\ \widetilde{q}_{N-1;1} & & \widetilde{q}_{N-1;j-1} & x_{N-1} & \widetilde{q}_{N-1;j+1} & & \widetilde{q}_{N-1;N} \\ 1 & \cdots & 1 & N v_{\text{av}} & 1 & \cdots & 1 \end{vmatrix}$$

We use the basic rules for determinants in **art.** 209. After[5] adding all columns in η_j, except for column j, to the last one and using $\sum_{l=1}^{N} \widetilde{q}_{il} = 0$, the last column becomes $-\widetilde{q}_{ij}$, except for the last row, which is $N - 1$. After changing the sign in the last row and interchanging column j and N, we obtain

$$v_j = \frac{1}{N\xi(G)} \begin{vmatrix} \widetilde{q}_{11} & \cdots & \widetilde{q}_{1;j} & \cdots & \widetilde{q}_{1;N-1} & x_1 \\ \widetilde{q}_{21} & \cdots & \widetilde{q}_{2;j} & \cdots & \widetilde{q}_{2;N-1} & x_2 \\ \vdots & & \vdots & & \vdots & \vdots \\ \widetilde{q}_{N-1;1} & & \widetilde{q}_{N-1;j} & & \widetilde{q}_{N-1;N-1} & x_{N-1} \\ 1 & \cdots & 1-N & \cdots & 1 & N v_{\text{av}} \end{vmatrix} \quad (4.41)$$

$$= \frac{1}{N\xi(G)} \det\begin{pmatrix} \widetilde{Q}_{\backslash \text{ row}(N)\backslash \text{ col}(N)} & x_{\backslash \text{ row } N} \\ s_j^T & N v_{\text{av}} \end{pmatrix}$$

where the $(N-1) \times 1$ vector s_j has all ones, except for component $(s_j)_j = 1 - N$. Only the element $(s_j)_j$ for v_j differs from $(s_m)_m$ for each v_m when both $j < N$ and $m < N$.

---

[4] Add all columns in $\widetilde{Q}_{\text{row } N=u}$ to the last column, which becomes zero due to $Qu = 0$, except the last row element equals N. Expand the determinant to the last row (Theorem 59).

[5] After adding all *rows* to the last one, using $\sum_{k=1}^{N-1} \widetilde{q}_{kl} = -\widetilde{q}_{kN}$ and $\sum_{k=1}^{N-1} x_k = -x_N$ because $u^T x = 0$, the (N, j)-th element is $N v_{\text{av}} - x_N$, while (N, l)-th element, with $l \neq j$, is $1 - \widetilde{q}_{N;l}$. Splitting the resulting determinant along the last row into two determinants leads to a determinant equal to η_j and another $\det\left(\widetilde{Q}_{\text{col } j=x}\right)$, that must be zero. Expanding $\det\left(\widetilde{Q}_{\text{col } j=x}\right) = 0$ along column j gives $\sum_{k=1}^{N} (-1)^{k+j} x_k \det\left(\widetilde{Q}_{\backslash \text{ row}(k)\backslash \text{ col}(j)}\right) = 0$, which is only possible for any vector x obeying $x^T u = 0$, if $(-1)^{k+j} \det\left(\widetilde{Q}_{\backslash \text{ row}(k)\backslash \text{ col}(j)}\right)$ is a constant. This demonstrates (4.37) again.

We apply twice Theorem 59: first, we expand the determinant in the numerator of v_j in (4.41) with respect to column N and then to row N. We arrive, for $1 \le j < N$, at

$$v_j = v_{\text{av}} - \frac{1}{N\xi(G)} \sum_{k=1}^{N-1} (-1)^k x_k \sum_{l=1}^{N-1} (s_j)_l (-1)^l \det\left(\widetilde{Q}_{\backslash \text{row}(k,N)\backslash \text{col}(l,N)}\right) \quad (4.42)$$

which gives the solution of $x = \widetilde{Q}v$ and equals $v_j = \left(\widetilde{Q}^\dagger x\right)_j$ if $v_{\text{av}} = 0$. Since

$$(s_i)_l - (s_j)_l = \begin{cases} N & \text{for } l = j \\ -N & \text{for } l = i \\ 0 & \text{if } l \ne \{i, j\} \end{cases}$$

the potential difference between node i and j is

$$v_i - v_j = \frac{\sum_{k=1}^{N-1}(-1)^k x_k \left\{ (-1)^i \det\left(\widetilde{Q}_{\backslash \text{row}(k,N)\backslash \text{col}(i,N)}\right) - (-1)^j \det\left(\widetilde{Q}_{\backslash \text{row}(k,N)\backslash \text{col}(j,N)}\right) \right\}}{\xi(G)}$$

$$(4.43)$$

We will compute the effective resistance ω_{ij} from (4.43) in Section 5.6, from which a triangle closure equation (5.37) for effective resistances is deduced.

131. *The pseudoinverse $Q^\dagger$ is not always a Laplacian.* The pseudoinverse $Q^\dagger$ of the weighted Laplacian $\widetilde{Q}$ is not necessary a Laplacian, because off-diagonal elements of $Q^\dagger$ can be positive, in contrast to the Laplacian $\widetilde{Q}$. We demonstrate the observation by an example of the path graph[6], whose explicit pseudoinverse $Q^\dagger$ is computed in Section 6.4.3. From (6.19), $q_N(m) = q_N(-m)$ and (6.18), it follows that

$$(Q_{\text{path}})^\dagger_{i,i+1} = \frac{1}{Nb} \left\{ i^2 - iN + (N-1)\frac{2N-1}{6} \right\}$$

Thus, $(Q_{\text{path}})^\dagger_{i,i+1} = (Q_{\text{path}})^\dagger_{i+1,i} > 0$ if

$$i^2 - iN + (N-1)\frac{2N-1}{6} > 0$$

The discriminant of this quadratic equation is $-\frac{1}{3}\left(N^2 - 6N + 2\right)$, which is negative if $N > 5$, implying that there is no intersection with the real axis and all solutions in i are positive. Hence, from $N = 6$ on, the pseudoinverse $Q^\dagger$ of the path graph has positive elements in the band one below until one above the diagonal.

We give a physical argument. Consider an electrical resistor network (**art.** 14) where a current I_c is injected in node j, while all other nodes are sinks, which leads to a current vector $x = I_c\left(e_j - \frac{1}{N}u\right)$. The potential vector $v = Q^\dagger x$ in (4.32) then equals for a unit current $I_c = 1$ ampere

$$v = Q^\dagger\left(e_j - \frac{u}{N}\right) = Q^\dagger e_j = \text{col}_j\, Q^\dagger$$

---

[6] A similar verification holds for the cycle graph.

and

$$v_j = Q_{jj}^\dagger > 0 \tag{4.44}$$

is the largest positive potential (**art.** 129) in $\mathrm{col}_j\, Q^\dagger$, because $Q_{jj}^\dagger \geq Q_{kj}^\dagger$ for any $1 \leq k \leq N$ as shown in Corollary 2. It is possible for a node $i \neq j$ that its potential $v_i = Q_{ij}^\dagger > 0$ if the resistance r_{ij} is small.

132. *The diagonal elements of the pseudoinverse $Q^\dagger$.* Similarly as for the Laplacian Q and the weighted Laplacian $\widetilde{Q}$, the positive diagonal elements $Q_{mm}^\dagger$ of the pseudoinverse $Q^\dagger$ play an important role, as shown in Section 5.1. From (4.30), a diagonal element of the pseudoinverse $Q^\dagger$ equals

$$Q_{mm}^\dagger = \sum_{k=1}^{N-1} \frac{1}{\mu_k} (z_k)_m^2$$

Using the doubly stochastic matrix Ξ_Q for the Laplacian, defined in (A.178) and in **art.** 109, the vector with the diagonal elements $\zeta = \left(Q_{11}^\dagger, Q_{22}^\dagger, \ldots, Q_{NN}^\dagger \right)$ is

$$\zeta = \Xi_Q \frac{1}{\mu} \tag{4.45}$$

where the vector $\frac{1}{\mu} = \left(\frac{1}{\mu_1}, \frac{1}{\mu_2}, \ldots, \frac{1}{\mu_{N-1}}, 0 \right)$. Relation (4.45), which corresponds to that of the degree vector $d = \Xi_Q \mu$ in (4.17), implies (see **art.** 275) that the vector $\frac{1}{\mu}$ *majorizes* the vector ζ, while the vector μ majorizes the degree vector d.

4.3 Second smallest eigenvalue of the Laplacian Q

The second smallest eigenvalue μ_{N-1} of the Laplacian has many interesting properties and was coined by Fiedler (1973), the *algebraic connectivity* of a graph. After Fiedler's seminal paper of 1973, results on the algebraic connectivity up to 2006 are reviewed by de Abreu (2007). In this section, mainly general bounds are presented, whereas **art.** 144 provides the major motivation to focus in depth on the algebraic connectivity μ_{N-1}. Bounds on μ_{N-1} in trees are given in Section 6.8.4.

4.3.1 Upper bounds for μ_{N-1}

133. The all-one eigenvector u of the Laplacian Q belongs (**art.** 350) to the smallest eigenvalue $\mu_N = 0$. In the terminology of **art.** 101 and **art.** 350, any constant function $f(x) = c$ is an eigenfunction of μ_N. Rayleigh's theorem (**art.** 251) states that $\mu_{N-1}(f, f) \leq (Qf, f)$ for any function f orthogonal to a constant function c and that the minimizer, for which equality holds, is the eigenfunction belonging to the second smallest eigenvalue μ_{N-1}. With **art.** 101, we obtain

$$\mu_{N-1} \leq \frac{\sum_{l \in \mathcal{L}} \left(f(l^+) - f(l^-) \right)^2}{\sum_{n \in \mathcal{N}} f^2(n)} \tag{4.46}$$

for any f that satisfies $(f, c) = c \sum_{n \in \mathcal{N}} f(n) = 0$. The latter condition is always fulfilled if we choose $f(x) = g(x) - \frac{1}{N} \sum_{n \in \mathcal{N}} g(n)$, where the last term can be interpreted as an average of g over all nodes of the graph. In addition, for such a choice, $(f(l^+) - f(l^-)) = (g(l^+) - g(l^-))$ such that

$$\mu_{N-1} \leq \frac{\sum_{l \in \mathcal{L}} (g(l^+) - g(l^-))^2}{\sum_{n \in \mathcal{N}} g^2(n) - \frac{1}{N} \left(\sum_{u \in \mathcal{N}} g(u) \right)^2}$$

for any non-constant function g.

For example, choose the vector or eigenfunction f as $f(u) = 1$, $f(v) = -1$ and $f(n) = 0$ for any node $n \neq v \neq u$. This vector is orthogonal to the constant, $(f, c) = 0$. Inequality (4.46) then gives

$$\mu_{N-1} \leq \frac{d_u + d_v}{2}$$

A sharper bound using the same method is obtained in (4.51).

Invoking the Koebe-Andreev-Thurston Theorem[7] of planar graphs and (4.46), Spielman and Teng (2007) shows that the algebraic connectivity of a planar graph with maximum degree $d_{\max}$ is bounded by $\mu_{N-1} \leq \frac{8 d_{\max}}{N}$.

134. *Fiedler's expressions for μ_{N-1}.* There is an alternative representation for (f, f) or for $z^T z = \|z\|_2^2$ due to Fiedler. From the special case of Cauchy's identity (A.71) as explored in **art. 68**,

$$\sum_{i=1}^{N} \sum_{j=1}^{N} (z_i - z_j)^2 = \sum_{i=1}^{N} \sum_{j=1}^{N} z_i^2 - 2 \sum_{i=1}^{N} z_i \sum_{j=1}^{N} z_j + \sum_{i=1}^{N} \sum_{j=1}^{N} z_j^2 = 2N z^T z - 2 \left(u^T z \right)^2$$

we find that

$$z^T z = \frac{1}{2N} \sum_{i=1}^{N} \sum_{j=1}^{N} (z_i - z_j)^2$$

because any Laplacian eigenvector z that does not belong to $\mu_N = 0$ is orthogonal to u. If $f = z_{N-1}$ is the eigenfunction of Q belonging to μ_{N-1}, equality holds in (4.46) and introducing the above, we arrive at Fiedler's expression for the algebraic connectivity

$$\mu_{N-1} = \frac{2N \sum_{l \in \mathcal{L}} (f(l^+) - f(l^-))^2}{\sum_{u \in \mathcal{N}} \sum_{v \in \mathcal{N}} (f(u) - f(v))^2} \tag{4.47}$$

Fiedler's formula (4.47) can also be written in terms of the adjacency matrix elements as

$$\mu_{N-1} = N \frac{\sum_{u \in \mathcal{N}} \sum_{v \in \mathcal{N}} a_{uv} (f(u) - f(v))^2}{\sum_{u \in \mathcal{N}} \sum_{v \in \mathcal{N}} (f(u) - f(v))^2} \tag{4.48}$$

---

[7] For a planar graph with N nodes, there exists a set of disks $\{D_1, D_2, \ldots, D_N\}$ in the plane with disjoint interiors such that D_i touches D_j if and only if $(i, j) \in \mathcal{L}$.

from which

$$\mu_{N-1} = N \left(1 - \frac{\sum_{u \in \mathcal{N}} \sum_{v \in \mathcal{N}} (1 - a_{uv}) (f(u) - f(v))^2}{\sum_{u \in \mathcal{N}} \sum_{v \in \mathcal{N}} (f(u) - f(v))^2} \right) = N \left(1 - \frac{\mu_1 (Q^c)}{N} \right)$$

agreeing with the general relation (**art. 112**) between Laplacian eigenvalues of the graph G and of its complement G^c. Further, we have

$$\mu_{N-1} = \frac{N}{1 + \frac{\sum_{u \in \mathcal{N}} \sum_{v \in \mathcal{N}} (1 - a_{uv})(f(u) - f(v))^2}{\sum_{u \in \mathcal{N}} \sum_{v \in \mathcal{N}} a_{uv}(f(u) - f(v))^2}} = \frac{N}{1 + \frac{\sum_{l \in \mathcal{L}^c} (f(l^+) - f(l^-))^2}{\sum_{l \in \mathcal{L}} (f(l^+) - f(l^-))^2}} \tag{4.49}$$

where the sum over links in the graph G generally contains all different non-negative terms than the sum over links in its complement G^c. In a dense graph G, where $L = |\mathcal{L}|$ is large and thus $L^c = |\mathcal{L}^c|$ is small, we expect a large algebraic connectivity μ_{N-1}, but not larger than N. That maximum $\mu_{N-1} = N$ can only occur if $L^c = 0$, thus only if the complement G^c is the empty graph and G is the complete graph K_N.

The numerator and denominator in (4.47) are invariant to the addition of a constant. If f is orthogonal to the constant eigenfunction c of Q belonging to μ_N, Rayleigh's principle in **art. 251** states that

$$\mu_{N-1} \leq \frac{2N \sum_{l \in \mathcal{L}} (f(l^+) - f(l^-))^2}{\sum_{u \in \mathcal{N}} \sum_{v \in \mathcal{N}} (f(u) - f(v))^2} \tag{4.50}$$

The advantage of Fiedler's inequality (4.50) is, that explicit orthogonality $(c, f) = 0$ for f to the constant function c, is not required anymore since it is implicitly incorporated into the denominator. For example, choosing now the eigenfunction f as $f(x) = 1_{\{x=w\}}$ leads, with $(f(u) - f(v))^2 = 1_{\{u=w\}} + 1_{\{v=w\}}$ provided $u \neq v$ and

$$\sum_{u \in \mathcal{N}} \sum_{v \in \mathcal{N}} (f(u) - f(v))^2 = \sum_{u \in \mathcal{N}} \sum_{v \in \mathcal{N} \setminus \{u\}} 1_{\{u=w\}} + \sum_{v \in \mathcal{N}} \sum_{u \in \mathcal{N} \setminus \{v\}} 1_{\{v=w\}}$$

$$= 2 \sum_{u \in \mathcal{N}} 1_{\{u=w\}} \sum_{v \in \mathcal{N} \setminus \{u\}} 1 = 2(N-1)$$

to $\mu_{N-1} \leq \frac{N}{N-1} d_w$. Since the inequality holds for any node w, the sharpest bound is reached when $d_w = d_{\min}$ and we find Fiedler's inequality for the second smallest eigenvalue of the Laplacian

$$\mu_{N-1} \leq \frac{N}{N-1} d_{\min} \tag{4.51}$$

Since equality is attained for the complete graph K_N as shown in Section 6.1, the bound (4.51) is generally the best possible. This inequality also follows from (A.189) in Fiedler's Theorem 80 for symmetric, positive semidefinite matrices. The bound (4.51) is also derived from the Alon-Milman inequality (4.73) as shown in **art. 144**.

135. *Fiedler eigenvector.* The eigenvector z_{N-1} of the Laplacian Q belonging to the algebraic connectivity μ_{N-1} is called the Fiedler eigenvector, with component $(z_{N-1})_v = f(v)$ for node v. We know already that the Fiedler eigenvector must be orthogonal $(c, f) = 0$ to the all-one vector u, which is equivalent to $\sum_{v \in \mathcal{N}} f(v) = 0$. **Art.** 112 shows that the Fiedler eigenvector $z_{N-1}(Q)$ is also the eigenvector $z_1(Q^c)$ of the Laplacian Q^c of the complementary graph G^c belonging to the largest eigenvalue $\mu_1(Q^c)$.

Suppose that $(f(u) - f(v))^2 = c$, where c is a positive constant, holds for all nodes u and v, then (4.49) shows that

$$\mu_{N-1} = \frac{N}{1 + \frac{L^c}{L}} = \frac{N}{1 + \frac{1}{L}\left(\binom{N}{2} - L\right)} = \frac{2L}{N-1} = \frac{N}{N-1} E[D]$$

which is contradicted by Fiedler's upper bound (4.51) in non-regular graphs, because then $E[D] > d_{\min}$. Hence, in non-regular graphs, the absolute value of the difference of the Fiedler eigenvector components cannot be the same for all node pairs (u, v). Thus, $f(u) - f(v) = \pm\sqrt{c}$ cannot hold!

136. *Lower bound for μ_1.* We apply (4.51) to the complement G^c of a graph G,

$$\mu_{N-1}(Q^c) \leq \frac{N}{N-1} d_{\min}(G^c) = \frac{N}{N-1}(N - 1 - d_{\max}(G)) = N - \frac{N}{N-1} d_{\max}(G)$$

Using (4.19) yields a lower bound for the largest eigenvalue of the Laplacian

$$\frac{N}{N-1} d_{\max} \leq \mu_1 \leq \min(N, 2d_{\max}) \tag{4.52}$$

where the upper bound follows from (4.20). The weaker lower bound, $\mu_1 \geq d_{\max}$, is immediate from (4.8), but the lower $d_{\max}\sqrt{1 + \frac{1}{d_{\max}}} \leq \mu_1$ in (4.12) can be better than (4.52) for small $d_{\max}$.

Grone and Merris (1994) succeeded in improving Fiedler's lower bound (4.52):

$$\mu_1 \geq d_{\max} + 1 \tag{4.53}$$

which is a strict inequality when $d_{\max} < N - 1$ and excludes[8] the complete graph K_N. Rayleigh's principle (**art.** 251) applied to the largest eigenvalue μ_1 of the Laplacian Q yields for any vector $x \neq 0$, using (4.3),

$$\mu_1 \geq \frac{x^T Q x}{x^T x} = \frac{\sum_{l \in \mathcal{L}}(x_{l+} - x_{l-})^2}{x^T x}$$

The particular choice of the vector x, where $x_j = d_j$ and $x_k = -1$ for each node $k \in \text{neighbor}(j)$ else $x_k = 0$, obeys $u^T x = 0$ and we find that

$$\mu_1 \geq \frac{x^T Q x}{x^T x} = \frac{d_j(d_j + 1)^2}{d_j^2 + d_j} = d_j + 1$$

Since the inequality $\mu_1 \geq d_j + 1$ holds for any node j, we arrive at (4.53).

---

[8] In the complete graph K_N, all non-zero Laplacian eigenvalues are equal (see p. 193) to $\mu_k = N$, for $k > 0$.

Applying (4.53) to the complement G^c then shows that

$$\mu_1(Q^c) \geq d_{\max}(G^c) + 1 = N - d_{\min}(G)$$

and, with (4.19), that

$$\mu_{N-1}(Q) \leq d_{\min}(G) \tag{4.54}$$

valid for any graph G, except for K_N. Equality in (4.54) is reached, for example, in the star $K_{1,N}$ (see Section 6.7). Clearly, the Grone-Merris upper bound (4.54) is sharper than Fiedler's upper bound (4.51).

4.3.2 Lower bounds for μ_{N-1}

137. *Lower bounds for any Laplacian eigenvalue.* Brouwer and Haemers (2008) have impressively extended the Grone-Merris lower bound (4.53):

Theorem 24 (Brouwer and Haemers) *For any graph but $K_m + (N - m) K_1$, the disjoint union of the complete graph K_m and $N - m$ isolated nodes, the j-th largest Laplacian eigenvalue is lower bounded, for $1 \leq j \leq N$, by*

$$\mu_j \geq d_{(j)} - j + 2 \tag{4.55}$$

where $d_{(j)}$ is the j-th largest nodal degree.

Proof: The proof of Brouwer and Haemers (2008) cleverly combines the generalized interlacing Theorem 72 applied to a specific quotient matrix K_π, defined in **art.** 37. The proof is rather complex and omitted. □

Brouwer and Haemers (2008) also discuss graphs for which equality in (4.55) is reached. Also, unweighted threshold graphs (**art.** 114) possess integer Laplacian eigenvalues close to the degrees (Hammer and Kelmans, 1996, Theorem 5.3).

Since $\mu_j \geq 0$, the bound (4.55) becomes useless when $d_{(j)} < j - 2$. In fact, we may introduce slack variables $\epsilon_j \geq 0$ in (4.55) to obtain the equality

$$\mu_j = d_{(j)} - j + 2 + \epsilon_j$$

Substitution into the m-th moment formula (4.16) specifies the moments $\sum_{j=1}^{N} \epsilon_j^m$. For example, for $m = 1$, we find from (4.7), using $\sum_{j=1}^{N} d_{(j)} = \sum_{j=1}^{N} d_j$,

$$\sum_{j=1}^{N} \epsilon_j = \frac{N(N-3)}{2}$$

which shows that the average of the ϵ_j's increases linearly with N. The cases for higher values of m are more involved, as illustrated for $m = 2$, which is derived from (4.9),

$$\sum_{j=1}^{N} \epsilon_j^2 = \frac{N(2N^2 - 9N + 13)}{6} + 2L + 2\sum_{j=1}^{N} d_j^2 + 2\sum_{j=1}^{N} j(\mu_j - d_j) - 2\sum_{j=1}^{N} d_j \mu_j$$

The last sum is related to the covariance $E[D\mu] - E[D]E[\mu]$ and, in general, difficult to assess. The above method of equating moments (see also **art. 354**) suggests to consider $\mu_j = d_{(j)} + \tilde{\epsilon}_j$, where the difference $\tilde{\epsilon}_j$ can be negative as well as positive, but the average difference is zero.

In their book, Brouwer and Haemers (2012) conjecture, for any integer $1 \le k \le N$, the upper bound

$$\sum_{j=1}^{k} \mu_j \le L + \binom{k+1}{2} \tag{4.56}$$

The Brouwer conjecture (4.56) on the partial sum of the largest Laplacian eigenvalues of a graph seems hard to prove in general, but it has been verified for many graph types, like regular graphs. Moreover, the inequality (4.56) seems to hold for weighted Laplacians as well.

138. *Lower bounds for μ_{N-1}.* We apply the functional framework of **art. 102** to derive a lower bound for the second smallest eigenvalue of the Laplacian. Assume that f is the eigenfunction of Q belonging to μ_{N-1} for which the equality sign holds in (4.46),

$$\mu_{N-1} = \frac{\sum_{l \in \mathcal{L}} (f(l^+) - f(l^-))^2}{\sum_{n \in \mathcal{N}} f^2(n)}$$

Let node u for which $|f(u)| = \max_{n \in \mathcal{N}} |f(n)| > 0$. Clearly, $\sum_{n \in \mathcal{N}} f^2(n) \le N f^2(u)$. Since $\sum_{n \in \mathcal{N}} f(n) = 0$ due to $u^T z_{N-1} = 0$ as shown in **art. 133**, there exists a node v for which $f(u)f(v) < 0$. Since $\mu_{N-1} > 0$ for a connected graph, it means that there exists a path $\mathcal{P}(v, u)$ from v to u with hopcount $h(\mathcal{P}(v, u))$. The minimum number of links to connect a graph occurs in a minimum spanning tree (MST) consisting of $N - 1$ links. Only if the diameter $\rho \ge h(\mathcal{P}(v, u))$ of G is smaller than $N - 1$, we have a strict inequality in

$$\sum_{l \in \mathcal{L}} (f(l^+) - f(l^-))^2 \ge \sum_{l \in \text{MST}} (f(l^+) - f(l^-))^2 \ge \sum_{l \in \mathcal{P}(v,u)} (f(l^+) - f(l^-))^2$$

By the Cauchy-Schwarz inequality (A.12), we have

$$h(\mathcal{P}(v, u)) \sum_{l \in \mathcal{P}(v,u)} (f(l^+) - f(l^-))^2 \ge \left(\sum_{l \in \mathcal{P}(v,u)} (f(l^+) - f(l^-)) \right)^2$$

and with $\sum_{l \in \mathcal{P}(v,u)} (f(l^+) - f(l^-)) = f(v) - f(u)$, we find

$$h(\mathcal{P}(v, u)) \sum_{l \in \mathcal{P}(v,u)} (f(l^+) - f(l^-))^2 \ge (f(v) - f(u))^2 \tag{4.57}$$

Using $(f(v) - f(u))^2 \ge f^2(u)$ because $f(u)f(v) < 0$, we obtain

$$\sum_{l \in \mathcal{P}(v,u)} (f(l^+) - f(l^-))^2 \ge \frac{f^2(u)}{h(\mathcal{P})} \ge \frac{f^2(u)}{\rho}$$

Combining all inequalities leads to $\mu_{N-1} \geq \frac{1}{\rho N}$.

139. *Betweenness and weighted betweenness.* The bound $\mu_{N-1} \geq \frac{1}{\rho N}$ can be improved by summing the Cauchy-Schwarz bound (4.57) over all node pairs,

$$\sum_{u \in \mathcal{N}} \sum_{v \in \mathcal{N}} (f(u) - f(v))^2 \leq \sum_{u \in \mathcal{N}} \sum_{v \in \mathcal{N}} h(\mathcal{P}(v,u)) \sum_{l \in \mathcal{P}(v,u)} (f(l^+) - f(l^-))^2$$

$$= \sum_{u \in \mathcal{N}} \sum_{v \in \mathcal{N}} h(\mathcal{P}(v,u)) \sum_{l \in \mathcal{L}} (f(l^+) - f(l^-))^2 1_{\{l \in \mathcal{P}(v,u)\}}$$

$$= \sum_{l \in \mathcal{L}} (f(l^+) - f(l^-))^2 \sum_{u \in \mathcal{N}} \sum_{v \in \mathcal{N}} h(\mathcal{P}(v,u)) 1_{\{l \in \mathcal{P}(v,u)\}}$$

and

$$\sum_{u \in \mathcal{N}} \sum_{v \in \mathcal{N}} (f(u) - f(v))^2 \leq 2 \sum_{l \in \mathcal{L}} (f(l^+) - f(l^-))^2 r_l \tag{4.58}$$

where we define

$$r_l = \frac{1}{2} \sum_{u \in \mathcal{N}} \sum_{v \in \mathcal{N}} h(\mathcal{P}(v,u)) 1_{\{l \in \mathcal{P}(v,u)\}} \tag{4.59}$$

which is an integer, measuring the importance of a link l. Hence, we deduce from Fiedler's definition (4.47) that

$$\mu_{N-1} \geq \frac{N \sum_{l \in \mathcal{L}} (f(l^+) - f(l^-))^2}{\sum_{l \in \mathcal{L}} (f(l^+) - f(l^-))^2 r_l} \tag{4.60}$$

Since the definition (4.59) holds for any path $\mathcal{P}(v,u)$ between a node pair (v,u), there will exist a set of particular paths that minimizes or maximizes r_l. That minimum or maximum of r_l can be regarded as a "centrality" metric for a link l defined as the minimum or maximum of the sum of all hopcounts of paths $\mathcal{P}(v,u)$ between all pairs of nodes (v,u) that contain that link l. Also, it does not necessarily hold that the shortest hopcount paths[9] between a node pair (v,u) will lead to the minimum of r_l. It also follows from the definition (4.59) that

$$\min_{\mathcal{P}} h(\mathcal{P}) B_l \leq \sum_{u \in \mathcal{N}} \sum_{v \in \mathcal{N}} h(\mathcal{P}(v,u)) 1_{\{l \in \mathcal{P}(v,u)\}} \leq \max_{\mathcal{P}} h(\mathcal{P}) B_l$$

where the betweenness of a link l, defined as

$$B_l = \frac{1}{2} \sum_{u \in \mathcal{N}} \sum_{v \in \mathcal{N}} 1_{\{l \in \mathcal{P}(v,u)\}} \tag{4.61}$$

equals the total number of shortest paths in the graph that traverse or contain link l. In a connected graph, the minimum hopcount is $\min_{\mathcal{P}} h(\mathcal{P}) = 1$, namely the direct link between any two nodes, because $h(\mathcal{P}(u,u)) = 0$ as $\mathcal{P}(u,u)$ cannot be a

---

[9] Generally, there are several shortest hopcount paths between a node pair in a graph with unit link weights.

path in absence of self-loops, and the maximum hopcount $\max_{\mathcal{P}} h(\mathcal{P}) = \rho$ equals the diameter of the graph, so that

$$B_l \le r_l \le \rho B_l \qquad (4.62)$$

In view of the correspondence with the betweenness B_l, defined in (4.61) below, we may call r_l a *weighted betweenness* of the link l. By summing (4.59) over all links and using $\sum_{l=1}^{L} 1_{\{l \in \mathcal{P}(v,u)\}} = h(\mathcal{P}(v,u))$, we find

$$\sum_{l=1}^{L} B_l = \frac{1}{2} \sum_{u \in \mathcal{N}} \sum_{v \in \mathcal{N}} h(\mathcal{P}(v,u)) = \binom{N}{2} E[H]$$

and

$$\sum_{l=1}^{L} r_l = \frac{1}{2} \sum_{u \in \mathcal{N}} \sum_{v \in \mathcal{N}} h^2(\mathcal{P}(v,u))$$

from which the average weighted betweenness $r_{\mathrm{av}} = \frac{1}{L}\sum_{l=1}^{L} r_l$ follows as

$$r_{\mathrm{av}} = \frac{1}{2L} \sum_{u \in \mathcal{N}} \sum_{v \in \mathcal{N}} h^2(\mathcal{P}(v,u)) = \frac{N(N-1)}{2L} E[H^2] = \frac{(N-1)E[H^2]}{E[D]}$$

With $E[H^2] = (E[H])^2 + \mathrm{Var}H$ and the average betweenness $B_{\mathrm{av}} = \frac{1}{L}\sum_{l=1}^{L} B_l$, we find that

$$r_{\mathrm{av}} = \frac{E[D]}{(N-1)} B_{\mathrm{av}}^2 + \frac{(N-1)}{E[D]} \mathrm{Var}H$$

In general, the right-hand side in (4.58) can be bounded as

$$\min_{l \in \mathcal{L}} r_l \sum_{l \in \mathcal{L}} (f(l^+) - f(l^-))^2 \le \sum_{l \in \mathcal{L}} (f(l^+) - f(l^-))^2 r_l \le \max_{l \in \mathcal{L}} r_l \sum_{l \in \mathcal{L}} (f(l^+) - f(l^-))^2$$

While the smallest value m_r of $\sum_{l \in \mathcal{L}} (f(l^+) - f(l^-))^2 r_l$ would yield in (4.60) the largest lower bound for μ_{N-1}, we cannot guarantee that m_r can attain the lower bound $\min_{l \in \mathcal{L}} r_l \sum_{l \in \mathcal{L}} (f(l^+) - f(l^-))^2$ nor that the ensuing substitution in (4.60) leading to $\frac{N}{\min_{l \in \mathcal{L}} r_l}$ will still lower bound μ_{N-1}. Simulations in Martin-Hernandez *et al.* (2014) have led us to consider the average weighted betweenness $r_{\mathrm{av}} = \frac{1}{L}\sum_{l=1}^{L} r_l$ to approximate the algebraic connectivity as $\mu_{N-1} \approx \frac{N}{r_{\mathrm{av}}}$ and

$$\mu_{N-1} \approx \frac{NE[D]}{(N-1)E[H^2]} \qquad (4.63)$$

where H is the hopcount of an arbitrary path in G. For the complete graph K_N, equality holds in (4.63). For a path graph on N nodes, $E[H^2] = \frac{N^2-1}{6}$ (as follows from Van Mieghem (2014, p. 629)) and the algebraic connectivity follows from (6.15) as $\mu_{N-1} = 2\left(1 - \cos\left(\frac{\pi}{N}\right)\right) = \frac{\pi^2}{N^2} + O\left(\frac{1}{N^4}\right)$ so that, for large N, $\mu_{N-1} \sim$

$\frac{\pi^2}{N^2} < \frac{NE[D]}{(N-1)E[H^2]} = \frac{12}{N^2-1} \sim \frac{12}{N^2}$. Since $E[H^2] \le \rho^2$ and $\frac{NE[D]}{(N-1)E[H^2]} > \frac{E[D]}{E[H^2]}$, we deduce from the estimate (4.63) the "approximate" inequality

$$\mu_{N-1} \gtrsim \frac{E[D]}{\rho^2}$$

and the suggestion that the diameter may be "close" to $\rho \approx \frac{\sqrt{E[D]}}{\sqrt{\mu_{N-1}}}$. For sparse graphs, $\mu_{N-1} \approx \frac{E[D]}{\rho^2}$ seems accurate. However, the right-hand side in (4.63) can be lower and larger than the algebraic connectivity μ_{N-1} and, even for large N, the example of the path graph disproves asymptotic equality. These considerations force us to continue with the worst alternative, $\max_{l \in \mathcal{L}} r_l$, so that

$$\mu_{N-1} \ge \frac{N}{\max_{l \in \mathcal{L}} r_l} \tag{4.64}$$

Furthermore, using the inequality $r_l \le \rho B_l$ from (4.62) into (4.58) yields

$$\sum_{l \in \mathcal{L}} \left(f\left(l^+\right) - f\left(l^-\right) \right)^2 r_l \le \rho \sum_{l \in \mathcal{L}} \left(f\left(l^+\right) - f\left(l^-\right) \right)^2 B_l$$

As shown by Wang *et al.* (2008), the maximum betweenness in any graph is

$$B_l = \frac{1}{2} \sum_{u \in \mathcal{N}} \sum_{v \in \mathcal{N}} 1_{\{l \in \mathcal{P}(v,u)\}} \le \left[\frac{N^2}{4} \right] \tag{4.65}$$

and

$$\sum_{l \in \mathcal{L}} \left(f\left(l^+\right) - f\left(l^-\right) \right)^2 r_l \le \left[\frac{N^2}{4} \right] \rho \sum_{l \in \mathcal{L}} \left(f\left(l^+\right) - f\left(l^-\right) \right)^2$$

Since $\left[\frac{N^2}{4} \right] \le \frac{N^2}{4}$, introduction in (4.60) leads to

$$\mu_{N-1} \ge \frac{4}{\rho N} \tag{4.66}$$

which is clearly inferior to (4.64). Nevertheless, equality in this lower bound (4.66) can be reached. As mentioned by Mohar (1991), McKay has shown that in a tree of diameter $\rho = t + 2$, obtained from a t-hop path, where k nodes are connected to each of its end-nodes such that $N = t + 1 + 2k$, (4.66) is sharp if $\frac{k}{t} \to \infty$.

140. We present another interpretation, deduced from **art. 139**, by rewriting the definition (4.59) as

$$2r_l = \sum_{u \in \mathcal{N}} \sum_{v \in \mathcal{N}} h\left(\mathcal{P}\left(v,u\right)\right) 1_{\{l \in \mathcal{P}(v,u)\}}$$

replacing $1_{\{l \in \mathcal{P}(v,u)\}} = \left(1_{\{l \in \mathcal{P}(v,u)\}} - 1 + 1 \right)$ and splitting the sums

$$2r_l = \sum_{u \in \mathcal{N}} \sum_{v \in \mathcal{N}} h\left(\mathcal{P}\left(v,u\right)\right) - \sum_{u \in \mathcal{N}} \sum_{v \in \mathcal{N} \setminus \{u\}} h\left(\mathcal{P}\left(v,u\right)\right) \left(1 - 1_{\{l \in \mathcal{P}(v,u)\}} \right)$$

because $h\left(\mathcal{P}\left(u,u\right)\right) = 0$. In all other cases where nodes u and v are different, $h\left(\mathcal{P}\left(v,u\right)\right) \geq 1$, such that the last sum is lower bounded:

$$\tilde{r} = \sum_{u\in\mathcal{N}}\sum_{v\in\mathcal{N}\setminus\{u\}} h\left(\mathcal{P}\left(v,u\right)\right)\left(1 - 1_{\{l\in\mathcal{P}(v,u)\}}\right) \geq \sum_{u\in\mathcal{N}}\sum_{v\in\mathcal{N}\setminus\{u\}}\left(1 - 1_{\{l\in\mathcal{P}(v,u)\}}\right)$$

$$= 2\binom{N}{2} - \sum_{u\in\mathcal{N}}\sum_{v\in\mathcal{N}\setminus\{u\}} 1_{\{l\in\mathcal{P}(v,u)\}} \geq 2\binom{N}{2} - 2\frac{N^2}{4} = \frac{N(N-2)}{2}$$

where in the last line (4.65) has been used. With the definition of the average hopcount, $\sum_{u\in\mathcal{N}}\sum_{v\in\mathcal{N}} h\left(\mathcal{P}\left(v,u\right)\right) = N\left(N-1\right)E\left[H\right]$, we find

$$2r_l = \sum_{u\in\mathcal{N}}\sum_{v\in\mathcal{N}} h\left(\mathcal{P}\left(v,u\right)\right) 1_{\{l\in\mathcal{P}(v,u)\}} \leq N\left(N-1\right)E\left[H\right] - \frac{N(N-2)}{2}$$

The lower bound $\mu_{N-1} \geq \frac{N}{\max_{l\in\mathcal{L}} r_l}$ in (4.64) shows that

$$\mu_{N-1} \geq \frac{2}{(N-1)E\left[H\right] - \frac{(N-2)}{2}}$$

or

$$E\left[H\right] \geq \frac{2}{(N-1)\mu_{N-1}} + \frac{1}{2} - \frac{1}{2\left(N-1\right)}$$

141. *Another type of lower bound for μ_{N-1}.* Let f be the eigenfunction of the Laplacian Q belonging to μ_{N-1}, then the eigenvalue equation in (1.3) is $Qf\left(u\right) = \mu_{N-1}f\left(u\right)$ for each nodal component $u \in \mathcal{N}$. Since f is non-zero and orthogonal to the constant function,

$$0 = \sum_{n\in\mathcal{N}} f\left(n\right) = \sum_{n^+\in\mathcal{N}} f\left(n^+\right) - \sum_{n^-\in\mathcal{N}} \left|f\left(n^-\right)\right|$$

where, for $n^+ \in \mathcal{N}$, $f\left(n^+\right) > 0$ and $n^- \in \mathcal{N}$, $f\left(n^-\right) \leq 0$. Let us define the set of positive nodes $\mathcal{N}^+ = \{n \in \mathcal{N} : f\left(n\right) > 0\}$ and $\mathcal{N}^- = \mathcal{N}\setminus\mathcal{N}^+$. Similarly, let $\mathcal{L}^+ = \{u^+v^+ \in \mathcal{N}: u^+, v^+ \in \mathcal{N}^+\}$ denote the set of all links between positive nodes and $\mathcal{L}^- = \{u^+v^- \in \mathcal{N}: u^+ \in \mathcal{N}^+, v^- \in \mathcal{N}^-\}$ denote the set of all links between positive nodes and negative nodes. Multiplying both sides of the eigenvalue equation by $f\left(u\right)$ and summing over positive nodes yields

$$\mu_{N-1} = \frac{\sum_{v\in\mathcal{N}^+} Qf\left(v\right) f\left(v\right)}{\sum_{v\in\mathcal{N}^+} f^2\left(v\right)}$$

Using the definition in **art.** 4 of the Laplacian $Q = \Delta - A$,

$$\sum_{v \in \mathcal{N}+} Qf(v) f(v) = \sum_{v \in \mathcal{N}+} f(v)(\Delta f - Af)(v)$$

$$= \sum_{v \in \mathcal{N}+} f(v) \left(d(v) f(v) - \sum_{u \in \text{neighbors}(v)} f(u) \right)$$

$$= \sum_{v \in \mathcal{N}+} \sum_{u \in \text{neighbors}(v)} f(v)(f(v) - f(u))$$

Further, after splitting the neighbors into positive and negative nodes,

$$\sum_{v \in \mathcal{N}+} Qf(v) f(v) = \sum_{v^+ u^+ \in \mathcal{L}^+} f(v)(f(v) - f(u)) + \sum_{v^+ u^- \in \mathcal{L}^-} f(v)(f(v) - f(u))$$

Since the graph is bidirectional, i.e., $A^T = A$, the link $v^+ u^+ = u^+ v^+$ appears twice in the sum such that

$$\sum_{v^+ u^+ \in \mathcal{L}^+} f(v)(f(v) - f(u)) = \sum_{u^+ v^+ \in \mathcal{L}^+} \{ f(v)(f(v) - f(u)) + f(u)(f(u) - f(v)) \}$$

$$= \sum_{u^+ v^+ \in \mathcal{L}^+} (f(v) - f(u))^2$$

where a link $u^+ v^+ \in \mathcal{L}^+$ is only counted once. Similarly as before, we denote the link $l = u^+ v^+$ by the head of link as $l^+ = u^+$ and by the tail as $l^- = v^+$. Thus, we arrive at

$$\sum_{v \in \mathcal{N}+} Qf(v) f(v) = \sum_{l \in \mathcal{L}^+} (f(l^+) - f(l^-))^2 + \sum_{v^+ u^- \in \mathcal{L}^-} f(v)(f(v) - f(u))$$

and

$$\mu_{N-1} = \frac{\sum_{l \in \mathcal{L}^+} (f(l^+) - f(l^-))^2 + \sum_{v^+ u^- \in \mathcal{L}^-} f(v)(f(v) - f(u))}{\sum_{v \in \mathcal{N}+} f^2(v)}$$

Since $f(v^+)(f(v^+) - f(u^-)) > 0$, the last sum in the numerator is non-negative, which leads to a lower bound

$$\mu_{N-1} \geq \frac{\sum_{l \in \mathcal{L}^+} (f(l^+) - f(l^-))^2}{\sum_{n \in \mathcal{N}+} f^2(n)} \tag{4.67}$$

The lower bound (4.67) resembles the upper bound (4.46), except that only positive nodes and links are considered and that f is not arbitrary, but the eigenfunction of Q belonging to the eigenvalue μ_{N-1}.

We can improve this lower bound (4.67) by incorporating positive terms in $\sum_{v^+ u^- \in \mathcal{L}^-} f(v)(f(v) - f(u))$, that we have neglected. This means that also links outside the positive cluster are taken into account. Following Alon (1986), we can define $g(v) = f(v) 1_{\{v \in \mathcal{N}+\}}$ such that

$$\sum_{v^+ u^- \in \mathcal{L}^-} f(v)(f(v) - f(u)) \geq \sum_{v^+ u^- \in \mathcal{L}^-} (g(v) - g(u))^2$$

With this function, the first sum remains unaltered,

$$\sum_{l \in \mathcal{L}^+} \left(f\left(l^+\right) - f\left(l^-\right) \right)^2 = \sum_{l \in \mathcal{L}^+} \left(g\left(l^+\right) - g\left(l^-\right) \right)^2 = \sum_{l \in \mathcal{L}} \left(g\left(l^+\right) - g\left(l^-\right) \right)^2$$

and, also $\sum_{n \in \mathcal{N}^+} f^2(n) = \sum_{n \in \mathcal{N}^+} g^2(n) = \sum_{n \in \mathcal{N}} g^2(n)$. Thus, the improved lower bound is

$$\mu_{N-1} \geq \frac{\sum_{l \in \mathcal{L}} \left(g\left(l^+\right) - g\left(l^-\right) \right)^2}{\sum_{n \in \mathcal{N}} g^2(n)} \tag{4.68}$$

142. Let $G + \{e\}$ denote a graph obtained from G by adding a link e between two nodes of G. For any f orthogonal to a constant function, we have that

$$\frac{\sum_{l \in \mathcal{L}(G+\{e\})} \left(f\left(l^+\right) - f\left(l^-\right) \right)^2}{\sum_{n \in \mathcal{N}} f^2(n)} = \frac{\sum_{l \in \mathcal{L}(G)} \left(f\left(l^+\right) - f\left(l^-\right) \right)^2}{\sum_{n \in \mathcal{N}} f^2(n)} + \frac{\left(f\left(e^+\right) - f\left(e^-\right) \right)^2}{\sum_{n \in \mathcal{N}} f^2(n)}$$

If $f = f_{G+\{e\}}$ is an eigenfunction of $G + \{e\}$ corresponding to $\mu_{N-1}(G + \{e\})$, then (4.46) shows that

$$\mu_{N-1}(G + \{e\}) \leq \mu_{N-1}(G) + \frac{\left(f_{G+\{e\}}\left(e^+\right) - f_{G+\{e\}}\left(e^-\right) \right)^2}{\sum_{n \in \mathcal{N}} f_{G+\{e\}}^2(n)}$$

On the other hand, if $f = f_G$ is an eigenfunction of G corresponding to $\mu_{N-1}(G)$, then

$$\mu_{N-1}(G + \{e\}) \geq \mu_{N-1}(G) + \frac{\left(f_G\left(e^+\right) - f_G\left(e^-\right) \right)^2}{\sum_{n \in \mathcal{N}} f_G^2(n)}$$

In the first bound,

$$
\begin{aligned}
b &= \frac{\left(f_{G+\{e\}}\left(e^+\right) - f_{G+\{e\}}\left(e^-\right) \right)^2}{\sum_{n \in \mathcal{N}} f_{G+\{e\}}^2(n)} \\
&= \frac{f_{G+\{e\}}^2\left(e^+\right) + f_{G+\{e\}}^2\left(e^-\right) - 2 f_{G+\{e\}}\left(e^+\right) f_{G+\{e\}}\left(e^-\right)}{f_{G+\{e\}}^2\left(e^+\right) + f_{G+\{e\}}^2\left(e^-\right) + \sum_{n \in \mathcal{N} \setminus \{e^+, e^-\}} f_{G+\{e\}}^2(n)} \\
&\leq \frac{f_{G+\{e\}}^2\left(e^+\right) + f_{G+\{e\}}^2\left(e^-\right) + 2 \left| f_{G+\{e\}}\left(e^+\right) \right| \left| f_{G+\{e\}}\left(e^-\right) \right|}{f_{G+\{e\}}^2\left(e^+\right) + f_{G+\{e\}}^2\left(e^-\right)} \leq 2
\end{aligned}
$$

because $\max_{x \geq 0, y \geq 0} \frac{2xy}{x^2 + y^2} = \max_{r = \frac{y}{x} \geq 0} \frac{2r}{1 + r^2} = 1$. With $\frac{\left(f_G\left(e^+\right) - f_G\left(e^-\right) \right)^2}{\sum_{n \in \mathcal{N}} f_G^2(n)} \geq 0$ in the second bound, we arrive at

$$\mu_{N-1}(G) \leq \mu_{N-1}(G + \{e\}) \leq \mu_{N-1}(G) + 2 \tag{4.69}$$

The same bounds (4.69) are elegantly proved by invoking interlacing (**art. 267**) on $Q_{G+\{e\}} = Q_G + Q_{\{e\}}$. Indeed, the Laplacian $Q_{\{e_{ij}\}}$ of a link e_{ij} between node i and j has precisely four non-zero elements: $q_{ij} = q_{ji} = -1$ and $q_{ii} = q_{jj} = 1$.

The eigenvalues of $Q_{\{e_{ij}\}}$ are obtained from $\det\left(Q_{\{e\}} - \mu I\right)$ after expanding the determinant in cofactors over row i (or j),

$$\det\left(Q_{\{e\}} - \mu I\right) = (-1)^{2i}(-\mu)^{N-2}(1-\mu)^2 - (-1)^{i+j}(-1)^{j+i}(-\mu)^{N-2}$$
$$= (-\mu)^{N-1}(\mu - 2)$$

The eigenvalues of $Q_{\{e_{ij}\}}$ are thus $[0]^{N-1}$ and 2; the interlacing inequality (A.165) leads to (4.69).

Art. 105 shows that, by adding one link, the sum of all eigenvalues increases by 2. Hence, when the upper bound in (4.69) is achieved, all other eigenvalues of $Q_{G+\{e\}}$ are precisely equal to those of Q_G.

4.4 Partitioning of a graph

The problem of graph partitioning consists of dividing the nodes of a graph into a number of disjoint groups, also called partitions (see **art. 36**), such that a certain criterion is met. The most popular criterion is that the number of links between these disjoint groups is minimized. Sometimes, the number of those partitions and their individual size is prescribed. Most, but not all (see **art. 143**) variants of the graph partitioning problem are NP-hard. We refer to Spielman and Teng (2007) for the history of spectral methods for graph partitioning.

143. *Graph partitioning into two disjoint subsets.* When confining to a graph partitioning into two disjoint subsets (subgroups, clusters, partitions,...), an index vector y can be defined with vector component $y_j = 1$ if the node j belongs to one partition and $y_j = -1$ if node j belongs to the other partition. The number of links R between the two disjoint subsets, also called the cut size or size of the separator, elegantly follows from the characteristic property (4.3) of the Laplacian,

$$R = \frac{1}{4}\sum_{l\in\mathcal{L}}(y_{l+} - y_{l-})^2 = \frac{1}{4}y^T Q y \tag{4.70}$$

because, only if the starting node l^+ and the ending node l^- of a link l belong to a different partition, $(y_{l+} - y_{l-})^2 = 4$, else $y_{l+} = y_{l-}$. The minimum cut size is

$$R_{\min} = \min_{y\in\mathbb{Y}}\frac{1}{4}y^T Q y$$

where $\mathbb{Y}$ is the set of all possible index vectors of the N-dimensional space with either -1 or 1 components.

Since all eigenvectors $\{z_k\}_{1\le k\le N}$ of the Laplacian Q are orthogonal (**art. 247**), any vector can be written as a linear combination. Let $y = \sum_{j=1}^N \alpha_j z_j$, then

$$R = \frac{1}{4}\sum_{j=1}^N \alpha_j \sum_{k=1}^N \alpha_k z_j^T Q z_k$$

and using the orthogonality property (A.121) in **art. 247**, we obtain

$$R = \frac{1}{4} \sum_{j=1}^{N} \alpha_j^2 \mu_j \qquad (4.71)$$

Since $\mu_N = 0$ and all other eigenvalues are larger than zero for a connected graph (Theorem 20), the alternative eigenvalue expression (4.71) shows that R is a sum of positive real numbers.

Although Stoer and Wagner (1997) have presented a highly efficient, non-spectral min-cut algorithm with a computational complexity of $O\left(NL + N^2 \log N\right)$, which demonstrates that the min-cut problem is not NP-hard, the minimization of (4.71) is generally difficult. However, if one chooses in (4.70) $y = \alpha_{N-1} z_{N-1}$, then $R = \frac{1}{4} \alpha_{N-1}^2 \mu_{N-1}$, which is, in view of (4.71), obviously the best possible to minimize R. Unfortunately, choosing the index vector y parallel to the Fiedler vector z_{N-1} is generally not possible, because $z_{N-1} \notin \mathbb{Y}$. A good strategy is to choose the sign of the components in y according to the sign of the corresponding component in the Fiedler vector. A slightly better approach is the choice $y = \alpha_N u + z_{N-1}$, since the eigenvector u belonging to $\mu_N = 0$ does not affect the value of R in (4.71) and it provides a higher degree of freedom to choose the size of each partition. This strategy agrees with Fiedler's graph partitioning explained in **art. 150**.

144. *The Alon-Milman inequality.* Another approach to the separator problem is to establish useful bounds. As we will demonstrate here, it turns out that the algebraic connectivity μ_{N-1} plays an important role in such bounds. Our starting point is the upper bound in (4.46) for μ_{N-1}. The ingenuity lies in finding a function f, introduced in **art. 133**, satisfying $(f, c) = 0$ that has both a graph interpretation and that provides a tight bound for μ_{N-1} in (4.46). Alon and Milman (1985) have proposed the function

$$g(u) = \frac{1}{a} - \left(\frac{1}{a} + \frac{1}{b}\right) \frac{\min(h, h(u, A))}{h}$$

where $(g, c) \neq 0$ such that $f = g - \bar{g}$ in **art. 133**, with $\bar{g} = \frac{1}{N} \sum_{n \in \mathcal{N}} g(n)$. Further, h is the distance (in hops) between two disjoint subsets A and B of $\mathcal{N}$, $h(u, A)$ is the shortest distance of node $u \in \mathcal{N}$ to a node of the set A and $a = \frac{N_A}{N}$ and $b = \frac{N_B}{N}$, where $N_k = |\mathcal{N}_k|$ is the number of nodes of set $\mathcal{N}_k$. If $u \in A$, then $g(u) = \frac{1}{a}$, while, if $u \in B$, then $h(u, A) = h$ and $g(u) = -\frac{1}{b}$. Moreover, if u and v are adjacent, i.e., they are either head $(u = l^+)$ or tail $(u = l^-)$ of a link l, then

$$|g(u) - g(v)| \leq \frac{1}{h}\left(\frac{1}{a} + \frac{1}{b}\right) \qquad (4.72)$$

Indeed, if u and v belong to the same set, then $g(u) - g(v) = 0$. If $u \in A$ and $v \notin A$, then $h(v, A) = 1$, because u and v are adjacent and $g(u) - g(v) = \left(\frac{1}{a} + \frac{1}{b}\right)\frac{1}{h}$. If both u and v do not belong to A, then $|h(v, A) - h(u, A)| \leq 1$ and

$$|g(u) - g(v)| = \frac{1}{h}\left(\frac{1}{a} + \frac{1}{b}\right)|-\min(h, h(u, A)) + \min(h, h(v, A))|$$

where the difference of the min-operator is largest and equal to 1 if not both $h(u, A)$ and $h(v, A)$ are larger than h. This proves (4.72). Using this bound (4.72), the numerator in (4.46) is

$$\sum_{l \in \mathcal{L}} \left(f(l^+) - f(l^-)\right)^2 = \sum_{l \in \mathcal{L}} \left(g(l^+) - g(l^-)\right)^2 = \sum_{l \in \mathcal{L} \setminus \{A \cup B\}} \left(g(l^+) - g(l^-)\right)^2$$

$$\leq \frac{1}{h^2}\left(\frac{1}{a} + \frac{1}{b}\right)^2 (L - L_A - L_B)$$

where L_A and L_B are the number of links in the sets A and B, respectively. The denominator of (4.46) is

$$\sum_{n \in \mathcal{N}} f^2(n) \geq \sum_{n \in (A \cup B)} f^2(n) = \sum_{n \in A} (g(n) - \bar{g})^2 + \sum_{n \in B} (g(n) - \bar{g})^2$$

$$= N_A \left(\frac{1}{a} - \bar{g}\right)^2 + N_B \left(\frac{1}{b} + \bar{g}\right)^2$$

$$= N\left(\frac{1}{a} + \frac{1}{b} + (a + b)\bar{g}^2\right) \geq N\left(\frac{1}{a} + \frac{1}{b}\right)$$

Finally, with (4.46), Alon and Milman (1985) arrive at

$$\mu_{N-1} \leq \frac{1}{h^2}\left(\frac{1}{N_A} + \frac{1}{N_B}\right)(L - L_A - L_B) \tag{4.73}$$

The Alon-Milman inequality (4.73) shows that a large algebraic connectivity μ_{N-1} leads to a high number of links between the two clusters A and B. Indeed, consider all subsets A and B in a graph G with a fixed number of nodes N_A and N_B and same separation h, then a large μ_{N-1} implies a large number of links $L - L_A - L_B$ between any pair – thus also minimal pairs – of subsets A and B. Hence, a large μ_{N-1} means a higher inter-twined subgraph structure and, consequently, it is more difficult to cut away a subgraph from G. A graph with large second smallest Laplacian eigenvalue μ_{N-1} is thus more "robust", in the sense of being better connected or interlinked. Just this property of μ_{N-1} has made the second smallest Laplacian eigenvalue a fundamental characterizer of the robustness of a graph.

However, the algebraic connectivity μ_{N-1} should not be viewed as a strict disconnectivity or robustness metric. Fig. 4.1 depicts two graphs G_1 and G_2, each with $N = 7$ nodes, $L = 10$ links and diameter $\rho = 4$, but with different algebraic connectivity $\mu_{N-1}(G_1) = 0.6338$ and $\mu_{N-1}(G_2) = 0.5858$. Although $\mu_{N-1}(G_1) > \mu_{N-1}(G_2)$, it is easier to disconnect G_1 than G_2, because one link removal disconnects G_1, while two links need to be deleted in G_2.

145. *Bounds for the separator.* The Alon-Milman method of **art.** 144 can be

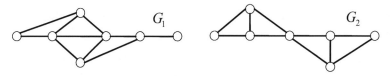

Fig. 4.1. Two graphs G_1 and G_2, each with $N = 7$ nodes, $L = 10$ links and diameter $\rho = 4$, but with different algebraic connectivity.

extended to deduce bounds for the separator S of two disjoint subsets A and B, that are at a distance h from each other. The separator S is the set of nodes at a distance less than h hops from A and not belonging to A nor B,

$$S = \{u \in \mathcal{N} \backslash A : h(u, A) < h\}$$

and $A \cup B \cup S = \mathcal{N}$. Sometimes, when $h = 1$, the separator is called the cut size, since there is a cut that splits the graph into two partitions. As in **art. 144**, we define $a = \frac{N_A}{N}$, $b = \frac{N_B}{N}$ and $s = \frac{N_S}{N}$, where N_C is the number of nodes in the set C. Instead of using the inequality (4.46), Pothen *et al.* (1990) start from the Fiedler inequality (4.50) in which they use

$$f(u) = 1 - \frac{2}{h}\min(h, h(u, A))$$

which is recognized as the Alon-Milman function $g(u)$ with $a = b = 1$. If $u \in S$, then $f(u) = 1 - \frac{2}{h}h(u, A)$ and $1 - \frac{2}{h} \geq f(u) \geq 1 - \frac{2(h-1)}{h} = -\left(1 - \frac{2}{h}\right)$. The numerator in (4.50) is computed precisely as in **art. 144** with $a = b = 1$,

$$\sum_{l \in \mathcal{L}} \left(f(l^+) - f(l^-)\right)^2 \leq \left(\frac{2}{h}\right)^2 (L - L_A - L_B) \leq \left(\frac{2}{h}\right)^2 sN d_{\max}$$

The denominator $n = \frac{1}{2}\sum_{u \in \mathcal{N}}\sum_{v \in \mathcal{N}}\left(f(u) - f(v)\right)^2$ in (4.50) is

$$n = \left(\sum_{u \in A}\sum_{v \in S} + \sum_{u \in A}\sum_{v \in B} + \sum_{u \in B}\sum_{v \in S} + \sum_{u \in S}\sum_{v \in S : v > u}\right)(f(u) - f(v))^2$$

$$\geq \left(\sum_{u \in A}\sum_{v \in S} + \sum_{u \in A}\sum_{v \in B} + \sum_{u \in B}\sum_{v \in S}\right)(f(u) - f(v))^2$$

$$\geq \left(1 - \left(1 - \frac{2}{h}\right)\right)^2 N^2 as + (1 - (-1))^2 N^2 ab + \left(-1 + \left(1 - \frac{2}{h}\right)\right)^2 N^2 bs$$

$$= \left(\frac{2}{h}\right)^2 N^2 \left\{s(a + b) + h^2 ab\right\}$$

With $b = 1 - a - s$, we arrive at

$$\mu_{N-1} \leq \frac{s d_{\max}}{s(1 - s) + a(1 - a - s)h^2} \tag{4.74}$$

which provides a quadratic inequality in s, from which a lower bound for s can be derived.

146. Pothen *et al.* (1990) present another inequality for the normalized size s of the separator, that is a direct application of the Wielandt-Hoffman inequality (A.168) for symmetric matrices. We can always relabel the nodes in the graph G corresponding to the sets A, B and S such that the Laplacian becomes

$$Q = \begin{bmatrix} Q_{Na \times Na} & O_{Na \times Nb} & Q_{Na \times Ns} \\ O_{Nb \times Na} & Q_{Nb \times Nb} & Q_{Nb \times Ns} \\ (Q_{Na \times Ns})^T & (Q_{Nb \times Ns})^T & Q_{Ns \times Ns} \end{bmatrix}$$

The idea, then, is to consider another matrix, whose eigenvalues are all known, such as $M = \mathrm{diag}(J_{Na \times Na}, J_{Nb \times Nb}, J_{Ns \times Ns})$, where J is the all-one matrix. The eigenvalues of M are those of the separate block matrices, that follow from (6.1) as Na, Nb, Nc and all the others are zero. Let us assume that $a \geq b \geq c$. We apply the Wielandt-Hoffman inequality (A.168) to M and $-Q$ (to have consistent ordering in the eigenvalues) such that

$$\sum_{k=1}^{n} \lambda_k (-Q) \lambda_k (M) = -\sum_{k=1}^{n} \mu_{N+1-k} \lambda_k (M) = -(0.Na + \mu_{N-1} Nb + \mu_{N-2} Ns)$$

while $\mathrm{trace}(-QM) = -\mathrm{trace}(QM)$ and, with the shorter notation for the square matrix $R_{Nl \times Nl} = R_{Nl}$,

$$\mathrm{trace}\,(QM) = \mathrm{trace}\,(Q_{Na} J_{Na}) + \mathrm{trace}\,(Q_{Nb} J_{Nb}) + \mathrm{trace}\,(Q_{Ns} J_{Ns})$$

$$= \left(\sum_{u \in A} + \sum_{u \in B} + \sum_{u \in S} \right) (d_u - d_u^*)$$

$$= 2 (L - L_A - L_B - L_S)$$

where d_u^* is the number of links incident to the node u and with end node in the same set as u. Substituting both in (A.168) yields

$$\mu_{N-1} Nb + \mu_{N-2} Ns \leq 2 (L - L_A - L_B - L_S)$$
$$\leq 2 (L - L_A - L_B) \leq 2Nsd_{\max}$$

from which, using $b = 1 - a - s$, a lower bound for the size of the separator follows as

$$s \geq \frac{(1-a)\,\mu_{N-1}}{2d_{\max} - (\mu_{N-2} - \mu_{N-1})}$$

This inequality, that contains beside the algebraic connectivity μ_{N-1} also the gap $\mu_{N-2} - \mu_{N-1}$, complements the inequality (4.74).

147. *Applications of the Alon-Milman bound (4.73).* Alon and Milman (1985) mention the following applications of the bound (4.73).

First, let $A = \{u\}$ and $B = \mathcal{N} \backslash \{u\}$, then $h = 1$ and $L - L_A - L_B = d_u$, the degree of node u. Since this inequality holds for any node u, the tightest bound is obtained by choosing a node u with minimum degree $d_{\min} = \min_{u \in \mathcal{N}} d_u$, which leads again to (4.51).

Second, let $a = b = \frac{1}{2}$, then the set of all links connecting a node in A to a node in B is called the bisector of G. The minimum number of the bisector is related to min-cut, max-flow problems. The Alon-Milman bound (4.73) shows a lower bound for the bisector,

$$\frac{N}{4}\mu_{N-1} \leq \text{ bisector } (G)$$

Third, if $h > 1$, then every link in the set $\mathcal{L} \backslash (\mathcal{L}_A \cup \mathcal{L}_B)$ is incident with at least one of the $N - N_A - N_B$ nodes of the set $S = \mathcal{N} \backslash (\mathcal{N}_A \cup \mathcal{N}_B)$, such that $L - L_A - L_B \leq (N - N_A - N_B)d_{\max}$. The Alon-Milman bound (4.73) becomes, using $a + b < 1$,

$$\mu_{N-1} \leq \frac{1}{h^2}\left(\frac{1}{a} + \frac{1}{b}\right)(1 - a - b)d_{\max} \leq \frac{1}{abh^2}(1 - a - b)d_{\max}$$

$$= \frac{sd_{\max}}{a(1 - a - s)h^2}$$

which is clearly weaker than (4.74) because $0 < s < 1$. It provides a lower bound for the fraction $b = \frac{N_B}{N}$ as

$$b \leq \frac{1 - a}{1 + \frac{ah^2\mu_{N-1}}{d_{\max}}} \tag{4.75}$$

where $h > 1$.

Based on (4.75), Alon and Milman (1985) also derive a second bound

$$b \leq (1 - a)\exp\left(-\ln(1 + 2a)\left\lfloor h\sqrt{\frac{\mu_{N-1}}{2d_{\max}}}\right\rfloor\right) \tag{4.76}$$

where $\lfloor x \rfloor$ denotes[10] the largest integer smaller than or equal to x.

Proof: The idea is to construct subsets A_r of $\mathcal{N}$ that include, beside the original subset A, additional nodes of $\mathcal{N}$ within distance $r \in \mathbb{R}$ hops from A, i.e., $A_r = \{v \in \mathcal{N} : d(v, A) \leq r\}$. We construct a sequence on distance $r = j\beta$ for $j = 0, 1, \ldots, k$ of those subsets such that $A_{j\beta}$ and $\mathcal{N} \backslash A_{(j+1)\beta}$ are more than $h > 1$ hops separated, which requires that $h > \beta > 1$. For those subsets $A \subset A_\beta \subset A_{2\beta} \subset \cdots A_{k\beta} \subseteq \mathcal{N}$, application of (4.75) yields

$$1 - a_{(j+1)\beta} = \frac{1 - a_{j\beta}}{1 + \frac{a_{j\beta}h^2\mu_{N-1}}{d_{\max}}} \leq \frac{1 - a_{j\beta}}{1 + a\frac{\beta^2\mu_{N-1}}{d_{\max}}}$$

The largest possible k is such that $k\beta \leq \rho$, where ρ is the diameter of the graph. With (4.51), we observe that, for $N \geq 2$,

$$\frac{1}{\mu_{N-1}} \geq \frac{N - 1}{N}\frac{1}{d_{\min}} \geq \frac{1}{2d_{\max}}$$

---

[10] Likewise, $\lceil x \rceil$ denotes the smallest integer larger than or equal to x.

such that $\frac{2d_{\max}}{\mu_{N-1}} > 1$. Let $\beta^2 = \frac{2d_{\max}}{\mu_{N-1}} > 1$, then $1 - a_{(j+1)\beta} \le (1 - a_{j\beta})\frac{1}{1+2a}$ for $0 \le j < k$. Multiplying those inequalities yields

$$1 - a_{k\beta} \le (1 - a)\frac{1}{(1 + 2a)^k} = (1 - a)e^{-k\ln(1+2a)}$$

and by construction $B \subseteq \mathcal{N}\backslash A_{k\beta}$ or $b < 1 - a_{k\beta}$ and $k < \frac{\rho}{\beta} = \rho\sqrt{\frac{\mu_{N-1}}{2d_{\max}}}$. This proves (4.76) for any $h < \rho$. $\qquad\square$

148. *Isoperimetric constant η.* If we choose the set B equal to $\mathcal{N}\backslash\mathcal{N}_A$, then $\mathcal{L} - \mathcal{L}_A - \mathcal{L}_B$ is the set of links with one end in A and the other in B. Thus, $\partial A = L - L_A - L_B$ is the number of links between A and its complement $\mathcal{N}\backslash\mathcal{N}_A$ and ∂A is called the cut size . The isoperimetric constant of the graph G is defined[11] as

$$\eta = \min_{\mathcal{N}_A}\frac{\partial A}{N_A} \qquad (4.77)$$

where the minimum is over all non-empty subsets $\mathcal{N}_A$ of $\mathcal{N}$ satisfying $N_A \le \lfloor\frac{N}{2}\rfloor$. The isoperimetric constant is also called the *Cheeger constant*.

The Alon-Milman bound (4.73) reduces (with $h = 1$) to

$$\mu_{N-1} \le \partial A\left(\frac{1}{N_A} + \frac{1}{N - N_A}\right)$$

If we denote $\eta_k = \min_{\mathcal{N}_A}\left\{\frac{\partial A}{N_A}\,\Big|\,N_A = k\right\}$, then $\mu_{N-1} \le \frac{\partial A}{k}\frac{N}{N-k}$ and this inequality holds for any set N_A, also for the minimizer of the right-hand side. Thus, $\mu_{N-1} \le \eta_k\frac{N}{N-k}$ and

$$\frac{N - k}{N}\mu_{N-1} \le \eta_k$$

We may further minimize both sides over all $k = 1, 2, \ldots, \lfloor\frac{N}{2}\rfloor$. Observe that $\eta = \min_{1\le k\le\lfloor\frac{N}{2}\rfloor}\eta_k$. Hence, the Alon-Milman bound (4.73) leads to a lower bound for the isoperimetric constant

$$\frac{\mu_{N-1}}{2} \le \eta$$

Using Alon's machinery of **art.** 141 that led to the lower bound (4.68), Mohar (1989) showed that, for $N > 3$,

$$\eta \le \sqrt{\mu_{N-1}(2d_{\max} - \mu_{N-1})}$$

Tighter, though more complex, bounds for the cut size as well as for the isoperimetric constant are derived in Devriendt and Van Mieghem (2019b).

149. *Expanders.* A graph G with N nodes is a c-expander if every subset $\mathcal{N}_A$ with $N_A \le \lfloor\frac{N}{2}\rfloor$ nodes is connected to its complement $\mathcal{N}\backslash\mathcal{N}_A$ by at least cN_A links.

---

[11] The computation of the isoperimetric constant is an NP-complete problem as shown by Mohar (1989).

Since $\partial A \geq c N_A$, **art.** 148 indicates that $c = \eta$. Expanders are thus difficult to disconnect because every set of nodes in G is well connected to its complement. This "robustness" property makes expanders highly desirable in the design of fault tolerant networks such as man-made infrastructures like communications networks and electric power transmission networks. A part of the network can only be cut off by destroying a large number of individual connections. In particular, sparse expanders, graphs with few links, have great interest, because the cost of a network usually increases with the number of links.

A well-studied subclass of expanders are regular graphs. In Govers *et al.* (2008), Wigderson mentions that almost every regular graph with degree $r \geq 3$ is an expander. The proof is probabilistic and does not provide insight how to construct a regular c-expander. Although nearly any regular graph is an expander, it turns out that there are only few methods to construct them explicitly. It follows from the bounds in **art.** 148 and $c = \eta$ that

$$\frac{1}{2}\mu_{N-1} \leq c \leq \sqrt{\mu_{N-1}(2r - \mu_{N-1})}$$

where $\mu_{N-1} = r - \lambda_2(A)$ also equals the spectral gap (**art.** 82 and **art.** 110). The larger the spectral gap or the smaller $\lambda_2(A)$, the larger c and the stronger or the more robust the expander is. A remarkable achievement is the discovery that, for all r-regular graphs, $\lambda_2(A) \geq 2\sqrt{r - 1}$ and that equality is only attained in Ramanujan graphs, where $r - 1$ is a prime power, as shown by Lubotzky *et al.* (1988).

150. *Graph partitioning.* Since $rI - Q$ is a non-negative matrix for $r > d_{\max}$, a direct application of Fiedler's Theorem 77 in **art.** 272 for $k = 2$ shows that a connected graph G can be partitioned into two distinct, connected components G_1 and G_2, where the nodes of $G_1 = G \backslash G_2$ are elements of the set $\mathcal{M} = \left\{ j \in \mathcal{N} : (x_{N-1})_j \geq \alpha \right\}$, where x_{N-1} is the eigenvector belonging to the second smallest eigenvalue μ_{N-1} of the Laplacian Q and α is some threshold value that specifies different disjoint partitions. If $\alpha > \max_{1 \leq j \leq N} (x_{N-1})_j$ or if $\alpha < \min_{1 \leq j \leq N} (x_{N-1})_j$, there is only the "trivial" partition consisting of the original graph G itself. Fiedler (1975) demonstrates that, by varying the threshold $\alpha \geq 0$, all possible cuts that separate the graph $G = G_1 \cup G_2$ into two distinct ($G_1 \cap G_2 = \varnothing$) connected components G_1 and G_2 can be obtained in this way.

Art. 103 indicates that the sum over all positive vector components equals the sum over all negative ones. This means that the value $\alpha = 0$ in Fiedler's partitioning algorithm divides the graph into two "equivalent" partitions, where "equivalent" is measured with respect to the second smallest Laplacian eigenvector. It does not imply, however, that both partitions have the same number of nodes.

4.5 The modularity and the modularity matrix M

151. *Modularity.* The modularity, proposed by Newman and Girvan (2004), is a measure of the quality of a particular division of the network. The modularity is proportional to the number of links falling within clusters or groups minus the expected number in an equivalent network with links placed at random. Thus, if the number of links within a group is no better than random, the modularity is zero. A modularity approaching one reflects networks with strong community structure: a dense intra-group and a sparse inter-group connection pattern.

If links are placed at random, then the expected number of links between node i and node j equals $\frac{d_i d_j}{2L}$. The modularity m is defined by Newman (2006) as

$$m = \frac{1}{2L} \sum_{i=1}^{N} \sum_{j=1}^{N} \left(a_{ij} - \frac{d_i d_j}{2L} \right) 1_{\{i \text{ and } j \text{ belong to the same cluster}\}} \tag{4.78}$$

We consider first a network partitioning into two clusters or subgraphs as in **art. 143**. The indicator function is rewritten in terms of the y vector, defined in **art. 143**, as

$$1_{\{i \text{ and } j \text{ belong to the same cluster}\}} = \frac{1}{2} (y_i y_j + 1)$$

so that

$$m = \frac{1}{4L} \sum_{i=1}^{N} \sum_{j=1}^{N} \left(a_{ij} - \frac{d_i d_j}{2L} \right) y_i y_j$$

because, by the basic law for the degree (2.3) and by (2.2),

$$\sum_{j=1}^{N} \left(a_{ij} - \frac{d_i d_j}{2L} \right) = \sum_{j=1}^{N} a_{ij} - \frac{d_i}{2L} \sum_{j=1}^{N} d_j = 0 \tag{4.79}$$

If there is only one partition to which all nodes belong, then $y = u$ and the modularity is $m = 0$ as follows from (4.79).

After defining the symmetric modularity matrix

$$M = A - \frac{1}{2L} d.d^T \tag{4.80}$$

with elements $m_{ij} = a_{ij} - \frac{d_i d_j}{2L}$, we rewrite the modularity m, with respect to a partitioning into two clusters specified by the vector y, as a quadratic form

$$m = \frac{1}{4L} y^T M y$$

which is analogous to the number of links R in (4.70) between the two disjoint partitions.

152. *A graph with c communities.* For a partitioning of the network into c clusters, instead of the vector y, the $N \times c$ community matrix S, defined in **art. 36**, can be used to rephrase the condition as

$$1_{\{i \text{ and } j \text{ belong to the same cluster}\}} = \sum_{k=1}^{c} S_{ik} S_{jk}$$

which leads to the matrix representation of the modularity (Van Mieghem *et al.*, 2010)

$$m = \frac{1}{2L} \sum_{k=1}^{c} \sum_{i=1}^{N} \sum_{j=1}^{N} S_{ik} m_{ij} S_{jk} = \frac{\text{trace}\left(S^T M S\right)}{2L} \quad (4.81)$$

We define the community vector s_k as the k-th column of the community matrix S, which specifies the k-th cluster: all components of s_k, corresponding to nodes belonging to cluster C_k, are equal to one, otherwise they are zero. For $c = 2$ clusters, the vector $y = s_1 - s_2$ and only one vector suffices for the partitioning, instead of s_1 and s_2.

Using the eigenvalue decomposition (**art. 254**) of the symmetric modularity matrix $M = W \text{diag}(\lambda_j (M)) W^T$, where W is the orthogonal $N \times N$ matrix with the j-th eigenvector w_j belonging to $\lambda_j (M)$ in column j, the general spectral expression for the modularity m for any number of clusters c follows from (4.81) as

$$m = \frac{\text{trace}\left(\left(W^T S\right)^T \text{diag}(\lambda_j (M)) W^T S\right)}{2L}$$

$$= \frac{1}{2L} \sum_{j=1}^{N} \left(\sum_{k=1}^{c} \left(w_j^T s_k\right)^2 \right) \lambda_j (M) \quad (4.82)$$

because $\left(W^T S\right)_{jk} = \sum_{q=1}^{N} W_{qj} S_{qk} = w_j^T s_k$. The scalar product $w_j^T s_k = \sum_{q \in C_k} (w_j)_q$ is the sum of those eigenvector components of w_j that belong to cluster C_k. If we write the community vector $s_k = \sum_{j=1}^{N} \beta_{kj} w_j$ as a linear combination of the eigenvectors of M, then the orthogonality of eigenvectors indicates that the coefficients equal $\beta_{kj} = w_j^T s_k$. Moreover, **art. 36** shows that the vectors $s_1, s_2, \ldots, s_c$ are orthogonal vectors, and, by definition, that $\sum_{k=1}^{c} s_k = u$. Since u is an eigenvector of M belonging to the zero eigenvalue as follows from (4.79), we observe that

$$\sum_{k=1}^{c} w_j^T s_k = 0$$

provided the eigenvector $w_j \neq u$. Using the Cauchy identity (A.71)

$$c \sum_{k=1}^{c} \left(w_j^T s_k\right)^2 - \left(\sum_{k=1}^{c} w_j^T s_k\right)^2 = \sum_{m=2}^{c} \sum_{k=1}^{m-1} \left(w_j^T (s_m - s_k)\right)^2$$

we find that

$$m = \frac{1}{2Lc} \sum_{j=1}^{N} \left(\sum_{m=2}^{c} \sum_{k=1}^{m-1} \left(w_j^T (s_m - s_k)\right)^2 \right) \lambda_j (M)$$

which reduces for $c = 2$ and $y = s_1 - s_2$ to (4.99) below.

Since $WW^T = I$ (**art. 247**), we have that $\text{trace}\left(\left(W^T S\right)^T W^T S\right) = \text{trace}(S^T S) =$

N (**art. 36**), such that we obtain a companion of (4.82)

$$\sum_{j=1}^{N}\sum_{k=1}^{c}\left(w_j^T s_k\right)^2 = N \tag{4.83}$$

Let $w_q = \frac{u}{\sqrt{N}}$ denote the eigenvector of M belonging to the eigenvalue $\lambda_q\left(M\right) = 0$, then

$$\sum_{k=1}^{c}\left(w_q^T s_k\right)^2 = \frac{1}{N}\sum_{k=1}^{c}\left(u^T s_k\right)^2 = \frac{1}{N}\sum_{k=1}^{c}n_k^2$$

where n_k is the number of nodes in cluster C_k. Invoking the inequality (3.88) to (4.82) subject to (4.83) yields

$$\frac{\sum_{j=1;j\neq q}^{N}\left(\sum_{k=1}^{c}\left(w_j^T s_k\right)^2\right)\lambda_j\left(M\right)}{\sum_{j=1;j\neq q}^{N}\sum_{k=1}^{c}\left(w_j^T s_k\right)^2} \leq \max_{1\leq j\leq N}\frac{\left(\sum_{k=1}^{c}\left(w_j^T s_k\right)^2\right)\lambda_j\left(M\right)}{\sum_{k=1}^{c}\left(w_j^T s_k\right)^2} = \lambda_1\left(M\right)$$

from which we find, with $E\left[D\right] = \frac{2L}{N}$, a spectral upper bound for the modularity

$$m \leq \frac{\lambda_1\left(M\right)}{E\left[D\right]}\left(1 - \frac{1}{N^2}\sum_{k=1}^{c}n_k^2\right) \tag{4.84}$$

This bound can also be written as

$$m \leq \frac{\lambda_1\left(M\right)}{E\left[D\right]}\left(1 - \frac{1}{c} - \frac{c}{N^2}\mathrm{Var}\left[n_C\right]\right)$$

where n_C is the number of nodes in an arbitrary cluster, because $E\left[n_C\right] = \frac{1}{c}\sum_{k=1}^{c}n_k = \frac{N}{c}$. The spectrum of the non-back tracking matrix can accurately determine the number c of clusters in a graph as shown in Budel and Van Mieghem (2021).

153. *Upper bound for the modularity.* Newman's definition (4.78) is first rewritten as follows. We transform the nodal representation to a counting over links $l = i \sim j$ such that

$$\sum_{i=1}^{N}\sum_{j=1}^{N}a_{ij}1_{\{i \text{ and } j \text{ belong to the same cluster}\}} = 2\sum_{k=1}^{c}L_k$$

where L_k is the number of links of cluster C_k, and the factor 2 arises from the fact that all links are counted twice, due the symmetry $A = A^T$ of the adjacency matrix. If we denote by L_{inter} the number of inter-community links, i.e. the number of links that are cut by partitioning the network into c communities or clusters, then $L = \sum_{k=1}^{c}L_k + L_{\mathrm{inter}}$. Similarly,

$$\sum_{i=1}^{N}\sum_{j=1}^{N}d_i d_j 1_{\{i \text{ and } j \text{ belong to the same cluster}\}} = \sum_{k=1}^{c}\left(\sum_{i\in C_k}d_i\right)\left(\sum_{j\in C_k}d_j\right) = \sum_{k=1}^{c}D_{C_k}^2$$

where $D_{C_k} = \sum_{i\in C_k}d_i$ is the sum of the degrees of all nodes that belong to cluster

C_k. Clearly, $D_{C_k} \geq 2L_k$, because some nodes in cluster C_k may possess links connected to nodes in other clusters. The basic law of the degree (2.3) then shows that $\sum_{k=1}^{c} D_{C_k} = 2L$. Substituting these expressions in the definition (4.78) leads to an alternative expression[12] for the modularity

$$m = \sum_{k=1}^{c} \left(\frac{L_k}{L} - \left(\frac{D_{C_k}}{2L} \right)^2 \right) \tag{4.85}$$

Subject to the basic law of the degree, $\sum_{k=1}^{c} D_{C_k} = 2L$, the sum $\sum_{k=1}^{c} D_{C_k}^2$ is maximized when $D_{C_k} = \frac{2L}{c}$ for all $1 \leq k \leq c$. Indeed, the corresponding Lagrangian

$$\mathcal{L} = \sum_{k=1}^{c} D_{C_k}^2 + \xi \left(\sum_{k=1}^{c} D_{C_k} - 2L \right)$$

where ξ is a Lagrange multiplier, supplies the set of equations for the optimal solution, $\frac{\partial \mathcal{L}}{\partial D_{C_j}} = 2D_{C_j} + \xi = 0$ for $1 \leq j \leq c$ and $\frac{\partial \mathcal{L}}{\partial \xi} = \sum_{k=1}^{c} D_{C_k} - 2L = 0$, which is satisfied for $\xi = -\frac{4L}{c}$ and $D_{C_j} = \frac{2L}{c}$ for all $1 \leq j \leq c$. Hence, $\sum_{k=1}^{c} D_{C_k}^2 \leq \frac{(2L)^2}{c}$. The modularity in (4.85) is minimized, for $c > 1$, if $L_k = 0$ for $1 \leq k \leq c$ and $\sum_{k=1}^{c} D_{C_k}^2$ is maximized such that $m \geq -\frac{1}{c}$. In conclusion, the modularity of any graph is never smaller than $-\frac{1}{2}$, and this minimum is obtained for the complete bipartite graph.

Invoking the Cauchy identity (A.71) and $\sum_{k=1}^{c} D_{C_k} = 2L$,

$$\sum_{k=1}^{c} D_{C_k}^2 = \frac{(2L)^2}{c} + \frac{1}{c} \sum_{j=2}^{c} \sum_{k=1}^{j-1} \left(D_{C_j} - D_{C_k} \right)^2$$

results in yet another expression for the modularity

$$m = 1 - \frac{L_{\text{inter}}}{L} - \frac{1}{c} - \frac{1}{c} \sum_{j=2}^{c} \sum_{k=1}^{j-1} \left(\frac{D_{C_j} - D_{C_k}}{2L} \right)^2 \tag{4.86}$$

Since the double sum is always positive, (4.86) provides us with an upper bound for the modularity,

$$m \leq 1 - \frac{1}{c} - \frac{L_{\text{inter}}}{L} \tag{4.87}$$

The upper bound (4.87) is only attained if the degree sum of all clusters is the same. In addition, the upper bound (4.87) shows that $m \leq 1$ and that a modularity of 1 is only reached asymptotically, when the number of clusters $c \to \infty$ and $L_{\text{inter}} = o(L)$, implying that the fraction of inter-community links over the total number of links L is vanishingly small for large graphs ($N \to \infty$ and $L \to \infty$).

154. *Lower bound for the modularity.* Let $D_{\Delta C} = \max_{\{C_j, C_k\}} \left| D_{C_j} - D_{C_k} \right|$, then

---

[12] Newman (2010) presents still another expression for the modularity.

a lower bound of the modularity, deduced from (4.86), is

$$m \geq 1 - \frac{L_{\text{inter}}}{L} - \frac{1}{c} - \frac{(c-1)}{2} \left(\frac{D_{\Delta C}}{2L} \right)^2 \tag{4.88}$$

Only if $D_{\Delta C} = 0$, the lower bound (4.88) equals the upper bound (4.87) and the equality sign can occur. Excluding the case that $D_{\Delta C} = 0$, then not all D_{C_j} are equal, and we may assume an ordering $D_{C_1} \geq D_{C_2} \geq \ldots \geq D_{C_c}$, with at least one strict inequality. We demonstrate that, for $c > 2$, not all differences $D_{C_j} - D_{C_k} = D_{\Delta C} > 0$ for any pair (j, k). For, assume the contrary so that $D_{C_1} - D_{C_2} = D_{C_2} - D_{C_3} = D_{C_1} - D_{C_3} = D_{\Delta C} > 0$, then $D_{\Delta C} = D_{C_1} - D_{C_3} = (D_{C_1} - D_{C_2}) + (D_{C_2} - D_{C_3}) = 2D_{\Delta C}$, which cannot hold for $D_{\Delta C} > 0$. Hence, if $D_{\Delta C} > 0$, the inequality in (4.88) is strict; alternatively, the lower bound (4.88) is not attainable in that case.

In order for a network to have modular structure, the modularity must be positive. The requirement that the lower bound (4.88) is non-negative, supplies us with an upper bound for the maximum difference $D_{\Delta C}$ in the nodal degree sum between two clusters in a "modular" graph

$$D_{\Delta C} \leq 2L \sqrt{\frac{2}{c-1} \left(1 - \frac{L_{\text{inter}}}{L} - \frac{1}{c} \right)} \tag{4.89}$$

For $c > 1$, (4.89) demonstrates that $D_{\Delta C} < 2L$. Ignoring the integer nature of c, the lower bound (4.88) is maximized with respect to the number of clusters c when

$$c^* = \frac{2\sqrt{2}L}{D_{\Delta C}} > \sqrt{2} \tag{4.90}$$

resulting in

$$m \geq 1 - \frac{L_{\text{inter}}}{L} - \sqrt{2} \left(\frac{D_{\Delta C}}{2L} \right) + \frac{1}{2} \left(\frac{D_{\Delta C}}{2L} \right)^2$$

The right-hand side in this lower bound is positive provided that $1 > \frac{D_{\Delta C}}{2L} > \sqrt{2} \left(1 - \sqrt{\frac{L_{\text{inter}}}{L}} \right)$. When this lower bound for $\frac{D_{\Delta C}}{2L}$ is satisfied, the modularity m is certainly positive, implying that the graph exhibits modular structure.

155. *Spectrum of the modularity matrix M.* Since the row sum (4.79) of the modularity matrix M is zero, which translates to $Mu = 0$, the modularity matrix has a zero eigenvalue corresponding to the eigenvector u, similar to the Laplacian matrix (**art.** 4). Unlike the Laplacian Q, the modularity matrix M always has negative eigenvalues. Indeed, from (A.99) and **art.** 46, the sum of the eigenvalues of M equals

$$\sum_{j=1}^{N} \lambda_j (M) = -\frac{1}{2L} \sum_{j=1}^{N} d_j^2 = -\frac{N_2}{N_1} \tag{4.91}$$

where N_k is the total number of walks of length k (**art. 59**). The second order moment of the modularity eigenvalues are $\sum_{j=1}^{N} \lambda_j^2(M) = \text{trace}(A^2) - \frac{1}{L}\text{trace}(Add^T) + \frac{1}{(2L)^2}\sum_{j=1}^{N} d_j^2\text{trace}(d.d^T)$. Using (A.99) and **art. 59**, we have

$$\sum_{j=1}^{N} \lambda_j^2(M) = 2L - \frac{1}{L}d^T A d + \left(\frac{1}{2L}\sum_{j=1}^{N} d_j^2\right)^2 = N_1 - \frac{2N_3}{N_1} + \left(\frac{N_2}{N_1}\right)^2 \quad (4.92)$$

In general, M and A do not commute. Hence, **art. 284** shows that the set of eigenvectors $\{w_k\}_{1\leq k\leq N}$ of M is different from the set of eigenvectors $\{x_k\}_{1\leq k\leq N}$ of A.

The eigenvalues of the modularity matrix $M = A - \frac{1}{2L}d.d^T$ are zeros of the characteristic polynomial

$$\det(M - \lambda I) = \det\left(A - \lambda I - \frac{d.d^T}{2L}\right) = \det(A - \lambda I)\det\left(I - (A - \lambda I)^{-1}\frac{d.d^T}{2L}\right)$$

Using the "rank one update" formula (A.66), we have

$$\det(M - \lambda I) = \det(A - \lambda I)\left(1 - \frac{1}{2L}d^T(A - \lambda I)^{-1}d\right) \quad (4.93)$$

We invoke the resolvent $d^T(A - \lambda I)^{-1}d = \sum_{m=1}^{N}\frac{(d^T x_m)^2}{\lambda_m - \lambda}$ in (A.162) in **art. 262**, where x_m is the eigenvector of A belonging to eigenvalue λ_m. Using $d^T x_m = u^T A x_m = \lambda_m u^T x_m$, $N_1 = 2L$ and (3.28) produces

$$1 - \frac{1}{2L}d^T(A - \lambda I)^{-1}d = \frac{1}{2L}\left\{\sum_{m=1}^{N}\lambda_m(u^T x_m)^2 - \sum_{m=1}^{N}\frac{\lambda_m^2(u^T x_m)^2}{\lambda_m - \lambda}\right\}$$

$$= \frac{-\lambda}{2L}\sum_{m=1}^{N}\frac{(u^T x_m)^2\lambda_m}{\lambda_m - \lambda}$$

which can, in view of (3.37), be written in terms of the generating function $N_G(z)$ of the total number of walks (**art. 62**). Thus, we arrive at[13]

$$\det(M - \lambda I) = \frac{\lambda}{2L}\det(A - \lambda I)\left(N_G\left(\frac{1}{\lambda}\right) - N\right) \quad (4.94)$$

Since $\lim_{\lambda\to 0} N_G\left(\frac{1}{\lambda}\right) = 0$, the characteristic polynomial (4.94) of M illustrates that $\lambda = 0$ is an eigenvalue of M, corresponding to the eigenvalue u as shown above. By a same argument as in **art. 263**, the function $N_G\left(\frac{1}{\lambda}\right) - N$ has simple zeros that lie in between two consecutive eigenvalues of the adjacency matrix A.

In summary, the eigenvalues of the modularity matrix M interlace with the

---

[13] Invoking (3.40) and $c_A(\lambda) = \det(A - \lambda I)$, another expression is

$$\det(M - \lambda I) = \frac{\lambda}{2L}\left((-1)^N \lambda c_{A^c}(-\lambda - 1) - (\lambda + N)c_A(\lambda)\right)$$

eigenvalues of adjacency matrix A: $\lambda_1(A) \geq \lambda_1(M) \geq \lambda_2(A) \geq \lambda_2(M) \geq \ldots \geq \lambda_N(A) \geq \lambda_N(M)$.

156. *Spectrum of the modularity matrix M for regular graphs.* For regular graphs, where each node has degree r and $Au = ru$ (**art. 55**), we have that $(A - \lambda I)u = (r - \lambda)u$ from which $(A - \lambda I)^{-1} u = (r - \lambda)^{-1} u$. Substituted in (4.93) yields, with the degree vector $d = r.u$,

$$\det(M - \lambda I) = \det(A - \lambda I)\left(1 - \frac{r}{N}u^T(A - \lambda I)^{-1}u\right)$$

$$= \det(A - \lambda I)\left(1 - \frac{r}{N(r - \lambda)}u^T u\right) = \frac{\lambda}{\lambda - r}\det(A - \lambda I)$$

After invoking the basic relation (A.97), we arrive at

$$\det(M - \lambda I) = \frac{\lambda}{\lambda - r}\prod_{k=1}^{N}(\lambda_k(A) - \lambda) = -\lambda\prod_{k=2}^{N}(\lambda_k(A) - \lambda)$$

In summary, the eigenvalues of the modularity matrix M of a regular graph are precisely equal to the eigenvalues of the corresponding adjacency matrix A, except that the largest eigenvalue $\lambda_1(A) = r$ is replaced by the eigenvalue at zero.

157. *The largest eigenvalue of the modularity matrix.* Since $N_G\left(\frac{1}{\lambda}\right) - N > 0$ in (4.94) for $\lambda \geq \lambda_1$ as follows from (3.35) in **art. 62**, $\lambda_1(M) \leq \lambda_1(A)$. This inequality is also found from the interlacing property of M and A derived in **art. 155**. We will show here that $\lambda_1(M) < \lambda_1(A)$.

Since $\lambda = 0$ is always an eigenvalue of M (**art. 155**), there cannot be a smaller largest eigenvalue than zero. The interlacing property bounds the largest eigenvalue from below, $\lambda_1(M) \geq \lambda_2(A)$, and **art. 92** demonstrates that all graphs have a non-negative second largest eigenvalue $\lambda_2(A) \geq 0$, except for the complete graph. The modularity matrix of the complete graph K_N is $M_{K_N} = \frac{1}{N}J - I$, whose characteristic polynomial is $\det(M - \lambda I) = (-1)^N \lambda(1 + \lambda)^{N-1}$ as follows from (6.1). This illustrates that the largest eigenvalue of the complete graph is $\lambda_1(M_{K_N}) = 0$, which is also the smallest possible largest modularity eigenvalue of all graphs.

The eigenvector w_1 of M belonging to $\lambda_1(M)$ has negative components (in contrast to the largest eigenvector x_1 of A), because $u^T w_1 = 0$, which is similar to the eigenvectors of the Laplacian Q (**art. 103**). The Rayleigh equation (A.130) and the Rayleigh inequalities in **art. 251** demonstrate that

$$\lambda_1(M) = \frac{w_1^T M w_1}{w_1^T w_1} = \frac{w_1^T A w_1}{w_1^T w_1} - \frac{1}{2L}\frac{(w_1^T d)^2}{w_1^T w_1} \leq \lambda_1(A) - \frac{1}{2L}(w_1^T d)^2 \qquad (4.95)$$

because $w_1^T w_1 = 1$ as the orthogonal eigenvectors are normalized (**art. 247**). The scalar product $w_1^T d$ is only zero for regular graphs, where each node has degree r, because the degree vector is $d = r.u$ and $w_1^T u = 0$, provided $w_1 \neq \frac{u}{\sqrt{N}}$ (as in the complete graph). However, **art. 156** shows that the largest eigenvalue for regular graphs equals $\lambda_1(M_r) = \max(0, \lambda_2(A_r)) < \lambda_1(A_r)$, where the subscript

r explicitly refers to regular graphs. Due to interlacing (**art. 155**), of all graphs, the regular graph has the smallest largest eigenvalue of the modularity matrix. Because the last term in the above upper bound is always strictly positive for non-regular graphs, we obtain the range of $\lambda_1(M)$ for any graph: $0 \leq \lambda_1(M) < \lambda_1(A)$. In summary, the largest eigenvalue of the modularity matrix M is always strictly smaller than the largest eigenvalue of the corresponding adjacency matrix A.

We apply the Rayleigh principle to the adjacency matrix A,

$$\lambda_1(A) = \frac{x_1^T A x_1}{x_1^T x_1} = \frac{x_1^T M x_1}{x_1^T x_1} + \frac{1}{2L}\left(x_1^T d\right)^2 \leq \lambda_1(M) + \frac{1}{2L}\left(x_1^T d\right)^2 \qquad (4.96)$$

Combining both Rayleigh inequalities (4.96) and (4.95), we obtain bounds for the difference $\lambda_1(A) - \lambda_1(M) > 0$,

$$\frac{1}{2L}\left(w_1^T d\right)^2 \leq \lambda_1(A) - \lambda_1(M) \leq \frac{1}{2L}\left(x_1^T d\right)^2$$

Since $x_1^T d = x_1^T A^T u = (A x_1)^T u = \lambda_1(A) x_1^T u$ and invoking interlacing, we arrive from (4.96) at the lower bound

$$\max\left(\lambda_2(A), \lambda_1(A)\left\{1 - \frac{\left(x_1^T u\right)^2}{2L}\lambda_1(A)\right\}\right) \leq \lambda_1(M)$$

which is only useful when the fundamental weight $\left(x_1^T u\right)^2$ can be determined accurately. On the other hand, the scalar product $x_1^T d$ is maximal if $x_1 = \frac{d}{\sqrt{d^T d}}$, such that, using (4.91),

$$\frac{1}{2L}\left(x_1^T d\right)^2 \leq \frac{d^T d}{2L} = \frac{N_2}{N_1} = -\sum_{j=1}^N \lambda_j(M)$$

from which we obtain, together with (4.96), the upper bound

$$\lambda_1(A) \leq -\sum_{j=2}^N \lambda_j(M)$$

158. *Bounds for the largest eigenvalue of the modularity matrix.* Applying $d^T M d = d^T A d - \frac{1}{2L}\left(d^T d\right)^2 = N_3 - \frac{N_2^2}{N_1}$, we obtain with $d = \sum_{k=1}^N \gamma_k w_k$, where $\gamma_k = d^T w_k$, the decomposition

$$N_3 - \frac{N_2^2}{N_1} = \sum_{k=1}^N \gamma_k^2 \lambda_k(M) \qquad (4.97)$$

As shown in Section 8.5, the sign of (4.97) determines whether a graph is assortative (positive sign) or disassortative (negative sign). Similarly, from $d^T M^2 d$, we deduce that

$$N_4 - 2\frac{N_3 N_2}{N_1} + \frac{N_2^3}{N_1^2} = \sum_{k=1}^N \gamma_k^2 \lambda_k^2(M)$$

By applying the inequality (3.88), we obtain

$$\frac{N_3}{N_2} - \frac{N_2}{N_1} = \frac{\sum_{k=1}^{N} \gamma_k^2 \lambda_k(M)}{\sum_{k=1}^{N} \gamma_k^2} \leq \max_{1 \leq k \leq n} \frac{\gamma_k^2 \lambda_k(M)}{\gamma_k^2} = \lambda_1(M)$$

and

$$\frac{N_4}{N_2} - 2\frac{N_3}{N_1} + \left(\frac{N_2}{N_1}\right)^2 \leq \lambda_1^2(M)$$

Application of Laguerre's Theorem 110, combined with **art.** 294 and trace relations (4.91) and (4.92), yields the rather complicated upper bound

$$\lambda_1(M) \leq -\frac{1}{N}\left(\frac{N_2}{N_1}\right) + \frac{N-1}{N}\sqrt{\left(\frac{N_2}{N_1}\right)^2 - \frac{N}{N-1}\left(\frac{2N_3}{N_1} - N_1\right)} \qquad (4.98)$$

For regular graphs where $N_k = Nr^k$ and $0 < \lambda_1(M_r) = \lambda_2(A_r)$, the bound (4.98) provides an upper bound for the second largest eigenvalue of the adjacency matrix,

$$\lambda_2(A_r) \leq -\frac{r}{N} + \frac{1}{N}\sqrt{r(N-1)}\sqrt{N^2 - (N+1)r}$$

For the complete graph K_N, where $r = N-1$ and $\lambda_1(M_{K_N}) = 0$, the bound (4.98) is exact. In view of the upper bound (4.84) for the modularity, the bound (4.98) is only useful when the right-hand side is smaller than the average degree $E[D]$. Numerical evaluations indicate that the bound (4.98) is seldom sharp.

159. *Maximizing the modularity.* Maximizing the modularity m consists of finding the best $N \times c$ community matrix S in either definition (4.81) or (4.82). Numerous algorithms exist, that approximate the best community matrix S, for which we refer to Newman (2010). Here, we concentrate on a spectral method.

Starting from the quadratic form $m = \frac{1}{4L}y^T M y$ for the modularity, where the number of clusters $c = 2$, Newman (2006) mimics the method in **art.** 143 by writing the vector $y = \sum_{j=1}^{N} \beta_j w_j$ with $\beta_j = y^T w_j$ as a linear combination of the orthogonal eigenvectors $w_1, w_2, \ldots, w_N$ of M,

$$m = \frac{1}{4L}\sum_{j=1}^{N} \beta_j^2 \lambda_j(M) \qquad (4.99)$$

Maximizing the modularity m is thus equal to choosing the vector y as a linear combination of the few largest eigenvectors, such that components of y are either -1 and $+1$, which is difficult as mentioned above in **art.** 143. Newman (2006) proposes to maximize $\beta_1 = y^T w_1$ and the maximum $\beta_1 = \sum_{j=1}^{N} \left|(w_1)_j\right|$ is reached when each component $y_j = -1$ if $(w_1)_j < 0$ or $y_j = 1$ if $(w_1)_j \geq 0$. Moreover, using properties of norms (**art.** 203), we find that $\beta_1 = \|w_1\|_1 \geq \|w_1\|_2 = 1$, and by construction and the orthogonality of the eigenvectors, $\beta_j < \|w_j\|_1$.

This separation of nodes into two partitions according to the sign of the vector components in the largest eigenvector w_1 of M is similar in spirit to Fiedler's algorithm (**art. 150**). Apart from the sign considered so far, a large eigenvector component contributes more to the modularity m in (4.99) than a small (in absolute value) component. Thus, the magnitude (in absolute value) of the components in w_1 measures how firmly the corresponding node in the graph belongs to its assigned group, which is a general characteristic of a class of spectral measures called "eigenvalue centralities", defined in Section 8.7.1.

Since u is the eigenvector belonging to $\lambda(M) = 0$, the trivial partition of the network in one group is excluded from modularity, because $\lambda(M) = 0$ does not contribute to the sum in (4.99) and that any other eigenvector, due to the orthogonality (**art. 247**), must have at least one negative component. In contrast to the Fiedler partitioning based on the Laplacian, the situation where all non-zero eigenvalues of M are negative might occur (as in the complete graph, for example; **art. 157**), which indicates that there is no partition, except for the trivial one, and that the modularity m in (4.99) is negative. This observation is important: Newman (2006) exploits the fact that $m < 0$ to *not* partition a (sub)network.

4.6 Bounds for the diameter

160. *Exponential growth of a graph.* Mohar (1991) has derived a beautiful formula that relates the algebraic connectivity μ_{N-1} with the "growth" of a graph G. Let $B_k(v)$ be the set of nodes of G lying at a distance of at most k hops from an arbitrary node $v \in \mathcal{N}$ and denote the cardinality of $B_k(v)$ by $b_k = |B_k(v)|$. Mohar (1991) defines the growth of the graph G by the increase of the numbers b_k with the number of hops k from v.

Mohar (1991) starts by applying Fiedler's inequality (4.50) for the algebraic connectivity in **art. 134** to the eigenfunction

$$f(u) = \begin{cases} 1 & \text{if } u \in B_{k-1} \\ 0 & \text{if } u \in B_k \backslash B_{k-1} \\ -1 & \text{if } u \notin B_k \end{cases}$$

Executing the sums in (4.50) yields

$$\sum_{u \in \mathcal{N}} \sum_{w \in \mathcal{N}} (f(u) - f(v))^2 = \sum_{u \in B_{k-1}} \sum_{w \in B_k \backslash B_{k-1}} 1 + 4 \sum_{u \in B_{k-1}} \sum_{w \notin B_k} 1$$
$$+ \sum_{u \in B_k \backslash B_{k-1}} \sum_{w \in B_{k-1}} 1 + \sum_{u \in B_k \backslash B_{k-1}} \sum_{w \notin B_k} 1$$
$$+ 4 \sum_{u \notin B_k} \sum_{w \in B_{k-1}} + \sum_{u \notin B_k} \sum_{w \in B_k \backslash B_{k-1}} 1$$
$$= 2b_{k-1} m_k + 2(N - b_k) m_k + 8b_{k-1}(N - b_k)$$

where $m_k = \sum_{w \in B_k \backslash B_{k-1}} 1 = b_k - b_{k-1}$, the number of nodes at k hops from an arbitrary node v, and

$$\sum_{l \in \mathcal{L}} \left(f\left(l^+\right) - f\left(l^-\right)\right)^2 = \gamma_{k-1} + \gamma_k$$

where γ_k is the number of links $l = (l^+, l^-)$ with one end node $l^+ \in B_k$ and the other end node $l^- \in B_{k+1} \backslash B_k$. Introducing the above summations into Fiedler's inequality (4.50) for the algebraic connectivity μ_{N-1} results in the inequality

$$N(\gamma_{k-1} + \gamma_k) \geq \mu_{N-1} \left(b_{k-1}m_k + (N - b_k)m_k + 4b_{k-1}(N - b_k)\right)$$
$$= \mu_{N-1} \left(2b_k(N - b_k) + 2b_{k-1}(N - b_{k-1}) - Nm_k + m_k^2\right)$$

where the last equality is readily verified by working out the products. After rewriting the inequality as

$$N\left(\frac{\gamma_{k-1} + \gamma_k}{\mu_{N-1}}\right) + Nm_k - m_k^2 \geq 2\left\{b_k(N - b_k) + b_{k-1}(N - b_{k-1})\right\}$$

Mohar (1991) bounds $\gamma_{k-1} + \gamma_k \leq d_{\max}m_k$, so that

$$N\left(\frac{d_{\max}}{\mu_{N-1}} + 1\right)m_k - m_k^2 \geq 2\left[b_k(N - b_k) + b_{k-1}(N - b_{k-1})\right]$$

He further omits the quadrate m_k^2 because $N\left(\frac{d_{\max}}{\mu_{N-1}} + 1\right)m_k \geq N\left(\frac{d_{\max}}{\mu_{N-1}} + 1\right)m_k - m_k^2$ and, finally, arrives at a lower bound for the number m_k of nodes at k hops from an arbitrary node v,

$$N\left(\frac{d_{\max}}{\mu_{N-1}} + 1\right)m_k \geq 2\left[b_k(N - b_k) + b_{k-1}(N - b_{k-1})\right] \qquad (4.100)$$

Mohar (1991) proposes the function $y(t)$ with the property at integer values $t = k$ that $y(k) = b_k$. By a remarkable insight[14], he further relates the inequality (4.100) of $m_k = b_k - b_{k-1}$ to the logistic differential equation

$$y'(t) = \frac{\beta}{N}y(t)(N - y(t))$$

where $\beta = \frac{4\mu_{N-1}}{d_{\max} + \mu_{N-1}}$. The solution for the number $y(t)$ of nodes at distance t with the initial condition $y(0) = 1$ is

$$y(t) = \frac{N}{1 + (N - 1)e^{-\beta t}}$$

---

[14] Mohar (1991) expands $b_{k-1} = y(k - h) = y(k) - y'(k)h + o(h)$ to first order in h, thus treating h as arbitrarily small and ignoring terms of the order $o(h)$, while actually $h = 1$. The replacement in (4.100) results in a differential inequality

$$N\left(\frac{d_{\max}}{\mu_{N-1}} + 1\right)y'(k)h \geq 2\left[y(k)(N - y(k)) + (y(k) - y'(k)h)(N - y(k) + y'(k)h)\right]$$
$$= 4y(k)(N - y(k)) - 2(N - 2y(k))y'(k)h + o(h)$$

which led Mohar to the logistic differential equation.

Mohar (1991) proves for a connected graph ($\mu_{N-1} > 0$) that,

$$\text{if } k > 1 \text{ and } b_k \leq \frac{N}{2}, \text{then } b_k > y(k). \qquad (4.101)$$

The inequality (4.101) implies that the graph G has exponential growth until about half of the nodes in G are reached. From then on, i.e. when $b_k > \frac{N}{2}$, the finite size N of G limits the exponential law. When $y(l) = \frac{N}{2}$ so that $l = \frac{1}{\beta} \ln(N-1)$, it holds[15] that $b_l > \frac{N}{2}$ and more than half of the nodes are reached from an arbitrary node v, which leads to the bounds for the diameter: $\lfloor l \rfloor \leq \rho \leq 2 \lceil l \rceil$. Explicitly, Mohar bounds for the diameter as

$$\left\lfloor \frac{\mu_{N-1} + d_{\max}}{4\mu_{N-1}} \ln(N-1) \right\rfloor \leq \rho \leq 2 \left\lceil \frac{\mu_{N-1} + d_{\max}}{4\mu_{N-1}} \ln(N-1) \right\rceil \qquad (4.102)$$

161. *Distance between non-overlapping subgaphs.* van Dam and Haemers (1995) ingeniously apply interlacing (**art. 266**) and the definition (**art. 21**) of the hopcount or distance between nodes in terms of powers of the adjacency matrix A. They start by defining two sets of nodes, $\mathcal{V}$ with $v = |\mathcal{V}|$ nodes and $\mathcal{W}$ containing $w = |\mathcal{W}|$ nodes, whose nodes are separated by at least $m+1$ hops (see **art. 23**). The union of both sets, $\mathcal{V} \cup \mathcal{W} = \mathcal{N}$, comprises all nodes in the graph G. **Art. 21** shows for node $k \in \mathcal{V}$ and $l \in \mathcal{W}$ that $(p_m(A))_{kl} = 0$ for any polynomial $p_m(z)$ of degree m. By a suitable node relabeling $\mathcal{V} = \{1, 2, \ldots, v\}$ and $\mathcal{W} = \{N - w + 1, N - w + 2, \ldots, N\}$, the matrix $p_m(A)$ can be written as block matrix

$$p_m(A) = \left[\begin{array}{cc} P_{v \times (N-w)} & O_{v \times w} \\ R_{(N-v) \times (N-w)} & S_{(N-v) \times w} \end{array} \right]$$

Next, van Dam and Haemers (1995) concentrate on a regular graph (**art. 55**) with degree r and construct the polynomial such that $p_m(r) = 1$. Since $Au = ru$ and $p_m(A)u = p_m(r)u = u$, working-out the block matrix $p_m(A)$ with corresponding block vector $u = \left[\begin{array}{cc} u_{1 \times (N-w)} & u_{1 \times w} \end{array} \right]^T$ yields $Pu_{(N-w) \times 1} = u_{v \times 1}$ and $Ru_{(N-w) \times 1} + Su_{w \times 1} = u_{(N-v) \times 1}$, but also from $(p_m(A))^T u = u$, we find that $P^T u_{v \times 1} + R^T u_{(N-v) \times 1} = u_{(N-w) \times 1}$ and $S^T u_{(N-v) \times 1} = u_{w \times 1}$. Inspired by the quotient matrix (**art. 37**) of a graph, the average row sums of the block matrices are determined as $\frac{u_{1 \times v} P u_{(N-w) \times 1}}{v} = 1$ and $\frac{u_{1 \times (N-v)} R u_{(N-w) \times 1}}{N-v} + \frac{u_{1 \times (N-v)} S u_{w \times 1}}{N-v} = 1$. With $u_{1 \times (N-v)} S u_{w \times 1} = \left(u_{1 \times (N-v)} S u_{w \times 1} \right)^T = w$, we find that $\frac{u_{1 \times (N-v)} S u_{w \times 1}}{N-v} = \frac{w}{N-v}$ and $\frac{u_{1 \times (N-v)} R u_{(N-w) \times 1}}{N-v} = 1 - \frac{w}{N-v}$. Rather than continuing with the matrix $p_m(A)$, van Dam and Haemers (1995) continue with the larger, symmetric, general bipartite matrix

$$M = \left[\begin{array}{cc} O & p_m(A) \\ p_m(A) & O \end{array} \right]$$

---

[15] Indeed, assume the opposite, namely that $b_l \leq \frac{N}{2}$, then (4.101) shows that $b_l > y(l) = \frac{N}{2}$, which leads to a contradiction.

whose eigenvalues are $\pm p_m\left(\lambda_k\right)$ for $1 \leq k \leq N$ and whose corresponding average row sum matrix is

$$
C = \begin{bmatrix}
0 & 0 & 1 & 0 \\
0 & 0 & 1 - \frac{w}{N-v} & \frac{w}{N-v} \\
\frac{v}{N-w} & 1 - \frac{v}{N-w} & 0 & 0 \\
0 & 1 & 0 & 0
\end{bmatrix}
$$

The eigenvalues of the matrix C (see Section 6.8) are $\lambda_1\left(C\right) = -\lambda_4\left(C\right)$ and $\lambda_2\left(C\right) = -\lambda_3\left(C\right) = \sqrt{\frac{vw}{(N-v)(N-w)}}$. The general interlacing Theorem 72 in **art. 266** states that the eigenvalues of the matrices C and M interlace, thus

$$
\lambda_2\left(C\right) \leq \lambda_2\left(M\right) = \max_{k>1}\left|p_m\left(\lambda_k\right)\right|
$$

In summary, if the $v = |\mathcal{V}|$ nodes in the set $\mathcal{V}$ and the $w = |\mathcal{W}|$ nodes in the set $\mathcal{W}$ in a regular graph G with degree r are separated by at least $m+1$ hops, then the van Dam-Haemers inequality states that

$$
\frac{vw}{(N-v)(N-w)} \leq \max_{k>1} p_m^2\left(\lambda_k\right) \tag{4.103}
$$

for each polynomial $p_m\left(z\right)$ of degree m obeying $p_m\left(r\right) = 1$.

162. *Diameter ρ.* Another consequence of the Alon-Milman bound (4.73) is:

Theorem 25 (Alon-Milman) *The diameter ρ of a connected graph is at most*

$$
\rho \leq \left\lfloor \sqrt{\frac{2d_{\max}}{\mu_{N-1}}} \log_2 N \right\rfloor + 1 \tag{4.104}
$$

Proof: If B is the set of all nodes of G at a larger distance than h from A and A contains at least half of the nodes $(a \geq \frac{1}{2})$, then (4.76) gives

$$
b \leq \frac{1}{2} \exp\left(-\ln(2) \left\lfloor h\sqrt{\frac{\mu_{N-1}}{2d_{\max}}} \right\rfloor \right)
$$

If we require that $\exp\left(-\ln(2) \left\lfloor h\sqrt{\frac{\mu_{N-1}}{2d_{\max}}} \right\rfloor \right) \leq \frac{1}{N}$, then

$$
h \leq \sqrt{\frac{2d_{\max}}{\mu_{N-1}}} \log_2 N < \left\lfloor \sqrt{\frac{2d_{\max}}{\mu_{N-1}}} \log_2 N \right\rfloor + 1
$$

By construction, for such h, it holds that $b < \frac{1}{N}$ or $B = \varnothing$, which implies that $A = \mathcal{N}$. Next, if $v \in \mathcal{N}$, then the subset $\{v_h\}$ of nodes that is reached within h hops of node v contains more than $N/2$ nodes. Indeed, suppose the converse and define $A = \mathcal{N} \backslash \{v_h\}$. Then $a = A/N > \frac{1}{2}$. But, we have shown that, if $h = \rho$, then $A = \mathcal{N}$. This contradicts the hypothesis. Hence, all nodes in G are reached from an arbitrary node within $h = \rho$ hops, where ρ is specified in (4.104). $\qquad\square$

For $\alpha > 1$, Mohar (1991) has derived the bound

$$\rho \leq 2 \left\lceil \sqrt{\frac{\mu_1}{\mu_{N-1}}} \sqrt{\frac{\alpha^2 - 1}{4\alpha}} + 1 \right\rceil \left\lceil \log_\alpha \frac{N}{2} \right\rceil \tag{4.105}$$

which can be minimized for α. Mohar (1991) showed that (4.105) is sharper than Alon-Milman's bound (4.104), because $\mu_1 \leq 2d_{\max}$ as shown in **art. 104**.

Theorem 26 (Mohar) *The diameter ρ of a connected graph is at most*

$$\rho \leq 2 \left\lceil \frac{\log \frac{N}{2}}{\log \left(\frac{d_{\max} + \eta}{d_{\max} - \eta} \right)} \right\rceil \tag{4.106}$$

where η is the isoperimetric constant.

Proof: Mohar (1989) considers the subsets $A_u(r) = \{v \in \mathcal{N} : d(v, u) \leq r\}$ at distance r of node u. The definition (4.77) shows that, for $|A_u(r)| \leq \left[\frac{N}{2}\right]$,

$$\eta \left(|A_u(r)| + |A_u(r - 1)| \right) \leq \partial A_u(r) + \partial A_u(r - 1)$$

where $\partial A_u(r)$ contains all the links between the set $A_u(r) \setminus A_u(r-1)$ and the set $A_u(r+1) \setminus A_u(r)$. Hence, $\partial A_u(r) + \partial A_u(r-1)$ contains all links in two-hop shortest paths between the set $A_u(r-1) \setminus A_u(r-2)$ and the set $A_u(r+1) \setminus A_u(r)$, which equals

$$\partial A_u(r) + \partial A_u(r - 1) = \sum_{v \in A_u(r) \setminus A_u(r-1)} d_v \leq d_{\max} \left(|A_u(r)| - |A_u(r - 1)| \right)$$

Thus, $\eta \left(|A_u(r)| + |A_u(r-1)| \right) \leq d_{\max} \left(|A_u(r)| - |A_u(r-1)| \right)$ from which, for $|A_u(r)| \leq \left[\frac{N}{2}\right]$,

$$|A_u(r)| \geq \frac{d_{\max} + \eta}{d_{\max} - \eta} |A_u(r - 1)|$$

Since $|A_u(0)| = 1$ and $|A_u(1)| = d_u$, iterating the inequality yields

$$|A_u(r)| \geq \left(\frac{d_{\max} + \eta}{d_{\max} - \eta} \right)^r$$

provided $|A_u(r)| \leq \left[\frac{N}{2}\right]$, which restricts $r_{\max} \leq \left\lceil \frac{\log \frac{N}{2}}{\log \left(\frac{d_{\max} + \eta}{d_{\max} - \eta} \right)} \right\rceil$. This maximum hopcount reaches half of the nodes. To reach also the other half of nodes in the complement, at most $2r_{\max}$ hops are needed, which proves (4.106). □

Theorem 27 (Chung, Faber and Manteuffel) *The diameter ρ of a connected graph is at most*

$$\rho \leq \left\lfloor \frac{\text{arccosh}\,(N - 1)}{\text{arccosh} \left(\frac{\mu_1 + \mu_{N-1}}{\mu_1 - \mu_{N-1}} \right)} \right\rfloor + 1 \tag{4.107}$$

Proof: Chung *et al.* (1994) start from the characterization of the diameter ρ in **art. 22**: the diameter ρ is the smallest integer such that, for any given node pair (i,j) in the graph G, there exists a polynomial $p_m(z) = \sum_{k=0}^m a_k z^k$ of degree $m \le \rho$ in z such that the element (i,j) in the corresponding matrix polynomial $p_m(Q)$ is non-zero. Alternatively, if $(p_m(Q))_{ij} > 0$ for each node pair (i,j), then the diameter $\rho \le m$.

First, Chung *et al.* (1994) derive an upper bound for any element of an $N \times N$ matrix R with zero row sum and zero column sum, thus $Ru = 0$ and $u^T R = 0$, where u is the all-one vector. Chung *et al.* (1994) propose the vector $v_i = e_i - \frac{u}{N}$, which satisfies $v_i^T \frac{u}{\sqrt{N}} = 0$ and $\|v_i\|_2^2 = v_i^T v_i = 1 - \frac{1}{N}$. Then, invoking norms (**art. 205**),

$$R_{ij} = e_i^T R e_j = v_i^T R v_j \le \|v_i\|_2 \|R\|_2 \|v_j\|_2 = \|R\|_2 \left(1 - \frac{1}{N}\right) \qquad (4.108)$$

The Laplacian Q is a special case of the matrix R, but $p_m(Q) u = \sum_{k=0}^m a_k Q^k u = a_0 u$ is not zero, unless $a_0 = 0$. Because properties of the Chebyshev polynomials $T_n(z) = \cos(n \arccos z)$ in Section 12.7 will be used later, Chung *et al.* (1994) consider $R = p_m(Q) - \frac{1}{N} J$ and $a_0 = p_m(0) = 1$, which satisfies (4.108) and thus possesses a zero eigenvalue, while all other eigenvalues (**art. 243**) are $p_m(\mu_j)$ for $N - 1 \le j \le 1$. Using the inequality $|a| - |b| \le |a + b|$ and (4.108) shows that

$$(p_m(Q))_{ij} = R_{ij} + \frac{1}{N} \ge \frac{1}{N} - |R_{ij}| = \frac{1}{N} - \|R\|_2 \left(1 - \frac{1}{N}\right)$$

In order for $(p_m(Q))_{ij} > 0$ for any node pair (i,j) such that $\rho \le m$, we must require that $\|R\|_2 = \left\|p_m(Q) - \frac{1}{N} J\right\|_2 < \frac{1}{N-1}$. Applying (A.23) yields $\left\|p_m(Q) - \frac{1}{N} J\right\|_2 = \max_{N-1 \le j \le 1} |p_m(\mu_j)| < \frac{1}{N-1}$. It remains to find a polynomial $p_m(z)$ with $p_m(0) = 1$ that is bounded on the interval $[\mu_{N-1}, \mu_1]$ by $\frac{1}{N-1}$. The Chebyshev polynomials $T_n(z)$ possess optimality properties (**art. 343**) on an interval $[a,b]$ such that the polynomial

$$p_m(z) = \frac{T_m\left(\frac{a+b-2z}{b-a}\right)}{T_m\left(\frac{a+b}{b-a}\right)} \qquad (4.109)$$

satisfies $p_m(0) = 1$ and $\max_{x \in [a,b]} |p_m(x)| = \frac{1}{T_m\left(\frac{a+b}{b-a}\right)}$. Applying the latter expression to the requirement $\max_{N-1 \le j \le 1} |p_m(\mu_j)| < \frac{1}{N-1}$ gives the bound

$$T_m\left(\frac{\mu_1 + \mu_{N-1}}{\mu_1 - \mu_{N-1}}\right) > N - 1$$

Finally, using the definition $T_m(z) = \cosh(m \operatorname{arccosh} z)$ in **art. 375** leads to

$$m > \frac{\operatorname{arccosh}(N-1)}{\operatorname{arccosh}\left(\frac{\mu_1 + \mu_{N-1}}{\mu_1 - \mu_{N-1}}\right)}$$

Since m is an integer and $\rho \le m$, we arrive at (4.107). $\qquad \square$

Chung *et al.* (1994) remark that the more "aesthetically pleasing" inequality

$$\rho \leq \left\lceil \frac{\operatorname{arccosh}(N-1)}{\operatorname{arccosh}\left(\frac{\mu_1+\mu_{N-1}}{\mu_1-\mu_{N-1}}\right)} \right\rceil$$

fails for graphs with diameter $\rho = 1 + m$, where $m = \frac{\operatorname{arccosh}(N-1)}{\operatorname{arccosh}\left(\frac{\mu_1+\mu_{N-1}}{\mu_1-\mu_{N-1}}\right)}$, but questioned whether such graphs exist. The affirmative answer was published by Merris (1999), who found that the cocktail party graph is a member of such graphs that do not obey the above inequality. At about the same time of the work by Chung *et al.* (1994), van Dam and Haemers (1995) found almost the same bound (4.110) for the diameter as (4.107).

Theorem 28 (van Dam-Haemers) *The diameter ρ of a connected graph is at most*

$$\rho \leq \left\lfloor \frac{\log\left(2\left(N-1\right)\right)}{\log\left(\sqrt{\mu_1}+\sqrt{\mu_{N-1}}\right) - \log\left(\sqrt{\mu_1}-\sqrt{\mu_{N-1}}\right)} \right\rfloor + 1 \qquad (4.110)$$

Proof: The van Dam-Haemers inequality (4.103) in **art. 161** is sharpest for the polynomial that minimizes $\max_{k>1} p_m^2\left(\lambda_k\right)$. By "relaxing" this criterion to the minimization of $\max_{x\in[\lambda_N,\lambda_2]}\left|p_m\left(x\right)\right|$, we arrive at the Chebyshev polynomials $T_n\left(z\right)$ (see **art. 343** and Section 12.7). In particular, van Dam and Haemers (1995) find a similar polynomial (4.109) as in Chung *et al.* (1994), namely $p_m\left(z\right) = \frac{T_m\left(\frac{a+b-2z}{b-a}\right)}{T_m\left(\frac{a+b-2r}{b-a}\right)}$ with $a = \lambda_2$ and $b = \lambda_N$. The diameter ρ corresponds to $m = \rho - 1$ in the van Dam-Haemers inequality (4.103) for at least two nodes so that $v = w = 1$, from which, similarly as in the proof of Theorem 27, the upper bound for the diameter follows as

$$\rho \leq \left\lfloor \frac{\log\left(2\left(N-1\right)\right)}{\log\left(\frac{\sqrt{k-\lambda_N}+\sqrt{k-\lambda_2}}{\sqrt{k-\lambda_N}-\sqrt{k-\lambda_2}}\right)} \right\rfloor + 1$$

van Dam and Haemers (1995) remark that any non-regular graph can be made regular with degree r by the addition of self-loops, that (a) do not alter the Laplacian (**art. 4**) and (b) allow to substitute $k - \lambda_j$ by μ_j based on the Laplacian $Q = rI - A$ of a regular graph. After substitution, (4.110) is found. □

4.7 Eigenvalues of graphs and subgraphs

163. *Laplacian eigenvalues of a subgraph.* If G_1 and G_2 are link-disjoint graphs on the same set of nodes, then the union $G = G_1 \cup G_2$ possesses the adjacency matrix $A_G = A_{G_1} + A_{G_2}$ and the Laplacian $Q_G = Q_{G_1} + Q_{G_2}$. Interlacing in **art. 267** then states that, for each eigenvalue $1 \le k \le N$,

$$\lambda_N (G_1) + \lambda_k (G_2) \le \lambda_k (G) \le \lambda_k (G_2) + \lambda_1 (G_1)$$
$$\mu_k (G_2) \le \mu_k (G) \le \mu_k (G_2) + \mu_1 (G_1)$$

This shows that the Laplacian eigenvalues $\mu_k (G)$ are non-decreasing if links are added in a graph, or, more generally, if $G_2 \subseteq G$ and both have the same number of nodes, $N_{G_2} = N_G$, then $\mu_k (G_2) \le \mu_k (G)$.

164. *Addition of a link.* The general result in **art.** 163 can be sharpened for the specific case of adding one link to a graph. If $G+\{l\}$ is the graph obtained from G by adding a link l, then the incidence matrix $B_{G+\{l\}}$ consists of the incidence matrix B_G with one added column containing the vector z, that has only two non-zero elements, 1 at row $e^+ = i$ and -1 at row $e^- = j$,

$$Q_{G+\{l\}} = B_G B_G^T + zz^T = Q_G + zz^T \tag{4.111}$$

In the terminology of **art.** 90, $zz^T = -e_i e_j^T - e_j e_i^T + e_i e_i^T + e_j e_j^T$, where e_k is a $N \times 1$ base vector. Further, an application of Schur's complement (A.65) leads to

$$\det \left(Q_{G+\{l\}} - \mu I \right) = \det \left(Q_G + zz^T - \mu I \right)$$
$$= \det \left(Q_G - \mu I \right) \det \left(I + z^T \left(Q_G - \mu I \right)^{-1} z \right)$$

Applying the "rank one update" formula (A.66) yields

$$\det \left(I + z^T \left(Q_G - \mu I \right)^{-1} z \right) = 1 + z \left(Q_G - \mu I \right)^{-1} z^T$$

The same argument as in **art.** 263 shows that the strictly increasing rational function $\frac{\det \left(Q_{G+\{l\}} - \mu I \right)}{\det \left(Q_G - \mu I \right)}$ only possesses simple poles and zeros that lie in between the poles. From the common zero $\mu_N (G) = \mu_N (G + \{l\}) = 0$ on, the function $\frac{\det \left(Q_{G+\{l\}} - \mu I \right)}{\det \left(Q_G - \mu I \right)}$ increases implying that first the pole at $\mu_{N-1} (G)$ is reached before the zero at $\mu_{N-1} (G + \{l\})$. Hence, interlacing results in

$$\mu_j (G) \le \mu_j (G + \{l\}) \le \mu_{j-1} (G) \tag{4.112}$$

for all $1 < i \le N$ and $\mu_1 (G) \le \mu_1 (G + \{l\})$ for $i = 1$. Comparing this bound for $j = N - 1$ with (4.69) in **art.** 142 yields

$$\mu_{N-1} (G) \le \mu_{N-1} (G + \{l\}) \le \min \left(\mu_{N-2} (G), \mu_{N-1} (G) + 2 \right)$$

165. *Addition of a link without changing μ_{N-1}.* Let y denote the Fiedler eigenvector of Q_G belonging to the algebraic connectivity $\mu_{N-1} (G)$, normalized such that $y^T y = 1$. For any vector w, there holds that $w^T zz^T w = \left(w^T z \right)^2 = (w_i - w_j)^2$ and (4.111) indicates that $y^T Q_{G+\{l\}} y = y^T \left(Q_G + zz^T \right) y = \mu_{N-1} (G) + (y_i - y_j)^2$. **Art.** 113 shows that $y^T Q_{G+\{l\}} y \ge \mu_{N-1} (G + \{l\})$ and using (4.112), we obtain

$$(y_i - y_j)^2 \ge \mu_{N-1} (G + \{l\}) - \mu_{N-1} (G) \ge 0$$

which demonstrates that the algebraic connectivity does not change when a link $l = i \sim j$ is added between two nodes i and j with equal Fiedler eigenvector components, i.e. $y_i = y_j$.

166. *Laplacian spectrum of the cone of a graph.* When a node $N + 1$ is added to a graph G_N with N nodes, a similar analysis as in **art.** 85 applies. The Laplacian, corresponding to the adjacency matrix (3.95), is

$$Q_{(N+1)\times(N+1)} = \begin{bmatrix} Q_{N\times N} + \text{diag}\,(v) & -v_{N\times 1} \\ -\left(v^T\right)_{1\times N} & d_{N+1} \end{bmatrix}$$

The special case $v = u$, where the new node with label $N + 1$ is connected to all nodes in graph G_N, forms the cone of the graph G_N. Let w_k be the eigenvector of $Q_{N\times N}$ belonging to μ_k for $1 \le k < N$, then, for the vector $z_k^T = \begin{bmatrix} w_k^T & 0 \end{bmatrix}$,

$$Q_{(N+1)\times(N+1)}z_k = \begin{bmatrix} Q_{N\times N} + I & -u_{N\times 1} \\ -\left(u^T\right)_{1\times N} & N \end{bmatrix}\begin{bmatrix} w_k \\ 0 \end{bmatrix} = \begin{bmatrix} (\mu_k + 1)\,w_k \\ -u^T w_k \end{bmatrix}$$

Any eigenvector w of $Q_{N\times N}$, orthogonal to u so that $u^T w_k = 0$, results in an eigenvector z_k of $Q_{(N+1)\times(N+1)}$ belonging to $\mu_k + 1$. Hence, in addition to the zero Laplacian eigenvalue, $N - 1$ eigenvalues of the Laplacian of the cone of G are $\{\mu_k + 1\}_{1 \le k < N}$. The largest eigenvalue $N + 1$ follows from (4.7) or is determined by Corollary 1.

Alternatively, as shown by Das (2004), the entire spectrum can be deduced by considering the complement G_{N+1}^c of the cone of G_N. Since the cone node has degree N, the complement G_{N+1}^c is disconnected. Theorem 20 states that the Laplacian of G_{N+1}^c has at least two eigenvalues $\mu_N^c = \mu_{N-1}^c = 0$, while **art.** 116 tells us that the remaining Laplacian eigenvalues of G_{N+1}^c are those of G_N^c. Using (4.19) then shows that the eigenvalues $\left\{\mu_j\left(Q_{(N+1)\times(N+1)}\right)\right\}_{1 \le j \le N+1}$ of the cone of a graph are $N + 1$, $\mu_j\left(Q_{N\times N}\right) + 1$ for $1 \le j \le N - 1$, and zero.

167. *Removal of a node.* Let us consider the graph $G\backslash\{j\}$ obtained by removing an arbitrary node j and its incident links from G. **Art.** 166 shows that the Laplacian eigenvalues of the cone of $G\backslash\{j\}$ equal N, $\mu_1\left(G\backslash\{j\}\right) + 1$, $\mu_2\left(G\backslash\{j\}\right) + 1$, ..., $\mu_{N-2}\left(G\backslash\{j\}\right) + 1$ and 0. The original graph G is a subgraph of the cone of $G\backslash\{j\}$. Since the Laplacian eigenvalues are non-decreasing if links are added to the graph (**art.** 163), we conclude that, for all $1 \le k \le N - 2$,

$$\mu_k\left(G\backslash\{j\}\right) + 1 \ge \mu_{k+1}\left(G\right)$$

168. *Vertex connectivity $\kappa_N\,(G)$.* The vertex connectivity of a graph, $\kappa_N\,(G)$, is the minimum number of nodes whose removal (together with adjacent links) disconnects the graph G. The Rayleigh principle (**art.** 251) shows, for any other connection vector $v \ne u$ in **art.** 166, that $z^T Q_{(N+1)\times(N+1)}z \ge \mu_{N-1}\left(Q_{(N+1)\times(N+1)}\right)$ such that

$$\mu_{N-1}\left(Q_{(N+1)\times(N+1)}\right) \le \mu_{N-1}\left(Q_{N\times N}\right) + 1$$

Repeating the argument gives $\mu_{N-1}\left(Q_{(N+k)\times(N+k)}\right) \leq \mu_{N-1}\left(Q_{N\times N}\right) + k$. If $\kappa_{\mathcal{N}}(G) = k$, the above relation shows that (Fiedler, 1973)

$$\mu_{N-1} \leq \kappa_{\mathcal{N}}(G) \tag{4.113}$$

Indeed, for a disconnected graph $\mu_{N-1}(Q_{N\times N}) = 0$ and the addition of minimum $\kappa(G) = k$ nodes connects the graph, i.e., $\mu_{N-1}\left(Q_{(N+k)\times(N+k)}\right) > 0$.

169. *Edge connectivity $\kappa_{\mathcal{L}}(G)$.* The edge connectivity of a graph, $\kappa_{\mathcal{L}}(G)$, is the minimum number of links whose removal disconnects the graph G. For any connected graph G, it holds that

$$\kappa_{\mathcal{N}}(G) \leq \kappa_{\mathcal{L}}(G) \leq d_{\min}(G) \tag{4.114}$$

Indeed, let us concentrate on a connected graph G that is not a complete graph. Since $d_{\min}(G)$ is the minimum degree of a node, say n, in G, by removing all links of node n, G becomes disconnected. By definition, since $\kappa_{\mathcal{L}}(G)$ is the minimum number of links that leads to disconnectivity, it follows that $\kappa_{\mathcal{L}}(G) \leq d_{\min}(G)$ and $\kappa_{\mathcal{L}}(G) \leq N - 2$ because G is not a complete graph and consequently the minimum nodal degree is at most $N - 2$. Furthermore, the definition of $\kappa_{\mathcal{L}}(G)$ implies that there exists a set S of $\kappa_{\mathcal{L}}(G)$ links whose removal splits the graph G into two connected subgraphs G_1 and G_2, as illustrated in Fig. 4.2. Any link of that set S connects a node in G_1 to a node in G_2. Indeed, adding an arbitrary link of that set makes G again connected. But G can be disconnected into the same two connected subgraphs by removing nodes in G_1 and/or G_2. Since possible disconnectivity inside either G_1 or G_2 can occur before $\kappa_{\mathcal{L}}(G)$ nodes are removed, it follows that $\kappa_{\mathcal{N}}(G)$ cannot exceed $\kappa_{\mathcal{L}}(G)$, which establishes the inequality (4.114).

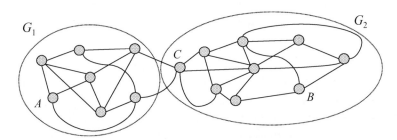

Fig. 4.2. A graph G with $N = 16$ nodes and $L = 32$ links. Two connected subgraphs G_1 and G_2 are shown. The graph's connectivity parameters are $\kappa_{\mathcal{N}}(G) = 1$ (removal of node C), $\kappa_{\mathcal{L}}(G) = 2$ (removal of links from C to G_1), $d_{\min}(G) = 3$ and $E[D] = \frac{2L}{N} = 4$.

Let us proceed to find the number of link-disjoint paths between A and B in a connected graph G. Suppose that H is a set of links whose removal separates A from B. Thus, the removal of all links in the set H destroys all paths from A to B. The maximum number of link-disjoint paths between A and B cannot exceed the number of links in H. However, this property holds for any set H, and thus also for the set with the smallest possible number of links. A similar argument applies to node-disjoint paths. Hence, we end up with Theorem 29:

Theorem 29 (Menger's Theorem) *The maximum number of link- (node)-disjoint paths between A and B is equal to the minimum number of links (nodes) separating or disconnecting A and B.*

The edge connectivity $\kappa_{\mathcal{L}}(G)$ (analogously the vertex connectivity $\kappa_{\mathcal{N}}(G)$) is the minimum number of links (nodes) whose removal disconnects G. By Menger's Theorem, it follows that there are at least $\kappa_{\mathcal{L}}(G)$ link-disjoint paths and at least $\kappa_{\mathcal{N}}(G)$ node-disjoint paths between any pair of nodes in G.

170. *Edge connectivity $\kappa_{\mathcal{L}}(G)$ and the algebraic connectivity μ_{N-1}.* Fiedler (1973) has proved a lower bound for μ_{N-1} in terms of the edge connectivity $\kappa_{\mathcal{L}}(G)$.

Theorem 30 (Fiedler) *For any graph G with L links and N nodes,*

$$\mu_{N-1} \geq 2\kappa_{\mathcal{L}}(G)\left(1 - \cos\frac{\pi}{N}\right) \tag{4.115}$$

Proof: Consider the symmetric, stochastic matrix $P = I - \frac{1}{d_{\max}}Q$ in Theorem 79. The spectral gap of P equals $1 - \lambda_2(P) = \frac{\mu_{N-1}}{d_{\max}}$, and is lower bounded in (A.184) by $\psi_n(r(P)) \leq 1 - \lambda_2(P)$, where $\psi_n(x) = 2x\left(1 - \cos\frac{\pi}{n}\right)$ for $x \leq \frac{1}{2}$ and the measure of irreducibility $r(P)$, defined in (A.169), equals $r(P) = \frac{\kappa_{\mathcal{L}}(G)}{d_{\max}}$. Indeed, by Merger's Theorem 29, the maximum number of link-disjoint paths between node A and B equals the minimum number of links that separates A from B. Hence, there are at least $\kappa_{\mathcal{L}}(G)$ link-disjoint paths between any pair of nodes in G. □

The function $\psi_n(x)$ in Theorem 79 on p. 387 provides a second bound

$$\mu_{N-1} \geq 2\kappa_{\mathcal{L}}(G)\left(\cos\frac{\pi}{N} - \cos\frac{2\pi}{N}\right) - 2d_{\max}\left(1 - \cos\frac{\pi}{N}\right)\cos\frac{\pi}{N}$$

which is only better than (4.115) if and only if $2\kappa_{\mathcal{L}}(G) > d_{\max}$. The algebraic connectivity $\mu_{N-1}(C_N) = 2\left(1 - \cos\frac{2\pi}{N}\right)$ of a circuit C_N follows from (6.6), while $\mu_{N-1}(P_N) = 2\left(1 - \cos\frac{\pi}{N}\right)$ for a path P_N follows from (6.10). Also, $\kappa_{\mathcal{L}}(C_N) = \kappa_{\mathcal{N}}(C_N) = 2$ and $\kappa_{\mathcal{L}}(P_N) = \kappa_{\mathcal{N}}(P_N) = 1$ show that equality is achieved in the bound (4.115) for the path P_N. However, in most cases as verified for example from Fig. 4.1, the lower bound (4.115) is rather weak.

171. *Pendants in a graph.* A node with degree one is called a pendant. Many complex networks possess pendants. If a connected graph G has a pendant, then the second smallest eigenvalue $\mu_{N-1} \leq 1$ as follows from (4.53) in **art. 136**.

Theorem 31 *If pendants are not adjacent to the highest degree node, then $\mu_{N-1} < 1$.*

Proof: Let node i be a pendant, connected to node j in G. The complement G^c of G has at least one node of degree $N - 2$, namely the pendant i in G^c, while the degree $d_j^c \geq 1$, because node j is not the highest degree node. We can construct a spanning tree T from the star $K_{1,N-2}$ with center at node i by precisely adding one link at a leaf node of the star to node j in G^c. **Art. 164** shows for $T = K_{1,N-2} + \{l\}$ that $\mu_1(T) > \mu_1(K_{1,N-2})$ and $\mu_1(K_{1,N-2}) = N-1$ is computed in Section 6.7. Since the spanning tree T is a subtree graph of G^c, **art. 163** implies that $\mu_1(G^c) \geq \mu_1(T) > N-1$, such that we arrive, with the complement formula (4.19), at $\mu_{N-1}(G) < 1$. $\qquad\square$

The following theorem is due to Das (2004):

Theorem 32 (Das) *If G is a connected graph with a Laplacian eigenvalue $0 < \mu < 1$, then the diameter of G is at least 3.*

Proof: Let z denote the eigenvector of the Laplacian Q belonging to μ, then the eigenvalue equation (1.3) for the j-th component is

$$\mu z_j = \sum_{k=1}^{N} q_{jk} z_k = d_j z_j - \sum_{k=1}^{N} a_{jk} z_k = d_j z_j - \sum_{k \in \text{neighbors}(j)} z_k$$

which we rewrite for a particular neighbor k_j of node j as

$$z_{k_j} = (d_j - \mu) z_j - \sum_{k \in \text{neighbors}(j)\backslash k_j} z_k$$

If $z_{k_j} = \min_{k \in \text{neighbors}(j)} z_k = z_{\min(j)}$, then $\sum_{k \in \text{neighbors}(j)\backslash k_j} z_k \leq (d_j - 1) z_{\max(j)} \geq (d_j - 1) z_{\max}$, where $z_{\max} = \max_{1 \leq k \leq N} z_k$, and

$$z_{\min(j)} \geq (d_j - \mu) z_j - (d_j - 1) z_{\max}$$

If we now choose node j such that $z_j = z_{\max}$, then we obtain the bound $z_{\min(j)} \geq (1 - \mu) z_{\max}$. **Art. 103** shows that the sign of $z_{\max}$ is positive. Hence, if $0 < \mu < 1$, all eigenvector components corresponding to the neighbors of the node j with largest eigenvector component, have the same sign as $z_j = z_{\max}$.

Similarly, if $z_{k_j} = \max_{k \in \text{neighbors}(j)} = z_{\max(j)}$, then $\sum_{k \in \text{neighbors}(j)\backslash k_j} z_k \geq (d_j - 1) z_{\min(j)} \geq (d_j - 1) z_{\min}$, where $z_{\min} = \min_{1 \leq k \leq N} z_k$, and

$$z_{\max(j)} \leq (d_j - \mu) z_j - (d_j - 1) z_{\min}$$

Choosing node j such that $z_j = z_{\min}$, then yields $z_{\max(j)} \leq (1 - \mu) z_{\min}$. **Art. 103** shows, for $\mu \neq \mu_N = 0$, that the sign of $z_{\min}$ is negative. Hence, if $0 < \mu < 1$, all eigenvector components corresponding to the neighbors of the node j with smallest eigenvector component, have the same sign as $z_j = z_{\min}$, opposite to $z_{\max}$. This implies that the nodes with largest and smallest eigenvector component are not neighbors (not directly connected), nor have neighbors in common. Since G is connected, this means that the diameter in G is at least 3. $\qquad\square$

5

Effective resistance matrix

After the adjacency matrix and the Laplacian matrix of a graph, we believe that the effective resistance matrix deserves a third position of importance. The effective resistance matrix is a distance matrix and intimately related to the pseudoinverse $Q^\dagger$ of the Laplacian matrix introduced in Section 4.2. Geometrically, as shown in Devriendt and Van Mieghem (2019a), the elements in the effective resistance matrix equal the squared distances between vertices in the simplex of the graph. For more details on the effective resistance matrix, we refer to Fiedler (2009) and Devriendt (2022b).

5.1 Effective resistance matrix Ω

We confine ourselves to a connected resistor network (**art.** 14) in which the injected nodal current vector is specified by $x = \widetilde{Q}v$ in (2.15) in terms of the nodal potential vector v. The weight of link $l = (i,j)$ between node i and j is $w_l = w_{ij} = \frac{1}{r_l}$, where r_l is the resistance of link l. The inverse relation $v = Q^\dagger x$ in (4.32) in Section 4.2 assumes that the reference potential is chosen equal to the average voltage in the network $v_{av} = \frac{u^T v}{N} = 0$. We aim to determine the effective resistance matrix Ω with elements ω_{ab} that, for a constant current $I_c > 0$, satisfy $v_a - v_b = \omega_{ab} I_c$, which is Ohm's law (2.12). The effective resistance ω_{ab} is a generalization of the classical series and parallel formulas for the resistance to any graph configuration.

Due to the linearity of the flow dynamics in **art.** 14, any nodal current vector x can be decomposed in several, elementary current injections with some magnitude $I_c > 0$ at some node a and leaving the network at some node b. Such elementary current injection is represented by $x = I_c (e_a - e_b)$, where e_k is the basic vector with components $(e_k)_m = 1_{\{k=m\}}$ and generates a potential at each node in the network, specified by the inverse relation $v = Q^\dagger x$ in (4.32) as

$$v = I_c Q^\dagger (e_a - e_b) \tag{5.1}$$

Ohm's law (2.12) states that the resistance is the proportionality constant or ratio between the potential difference $v_a - v_b$ at the nodes a and b in a graph and the current I_c injected at node a and leaving at node b. The ratio $\frac{v_a - v_b}{I_c}$ thus measures the resistance of a subgraph over which the injected current I_c in node a spreads towards node b and ω_{ab} is called the "effective" resistance between nodes a and b. The pseudoinverse $Q^\dagger$ can be computed in different ways, by (4.30) in **art. 128** and by (4.36) and related formulas in **art. 129**. The definition $v_a - v_b = \omega_{ab}I_c$ of effective resistance and $v_a - v_b = (e_a - e_b)^T v$ combined with (5.1) then leads to the quadratic form

$$\omega_{ab} = (e_a - e_b)^T Q^\dagger (e_a - e_b) \tag{5.2}$$

Multiplying (5.2) out yields

$$\omega_{ab} = Q^\dagger_{aa} + Q^\dagger_{bb} - 2Q^\dagger_{ab} \tag{5.3}$$

from which the symmetric effective resistance matrix Ω is obtained as

$$\Omega = \zeta u^T + u\zeta^T - 2Q^\dagger \tag{5.4}$$

where the vector $\zeta = \left(Q^\dagger_{11}, Q^\dagger_{22}, \ldots, Q^\dagger_{NN}\right)$. All diagonal elements of Ω are zero, as follows from the definition $v_a - v_b = \omega_{ab}I_c$ or from (5.2). The explicit form of the $N \times N$ matrix $\zeta u^T + u\zeta^T$ is

$$\zeta u^T + u\zeta^T = \begin{bmatrix} 2\zeta_1 & \zeta_1 + \zeta_2 & \zeta_1 + \zeta_3 & \cdots & \zeta_1 + \zeta_N \\ \zeta_1 + \zeta_2 & 2\zeta_2 & \zeta_2 + \zeta_3 & \cdots & \zeta_2 + \zeta_N \\ \zeta_1 + \zeta_3 & \zeta_2 + \zeta_3 & 2\zeta_3 & \cdots & \zeta_3 + \zeta_N \\ \vdots & \vdots & \vdots & \ddots & \vdots \\ \zeta_1 + \zeta_N & \zeta_2 + \zeta_N & \zeta_3 + \zeta_N & \cdots & 2\zeta_N \end{bmatrix}$$

Example The Laplacian pseudoinverse of the complete graph with $r_l = r_{K_N}$ for each link l is $Q^\dagger_{K_N} = \frac{1}{Nr_{K_N}}\left(I - \frac{1}{N}J\right)$ in (4.40), with $\zeta_{K_N} = \frac{\left(1 - \frac{1}{N}\right)}{Nr_{K_N}}u$. Formula (5.4) provides the effective resistance matrix $\Omega_{K_N} = \frac{2}{Nr_{K_N}}(J - I) = \frac{2}{N}\widetilde{A}_{K_N}$, where each link has link weight $w_{ij} = \frac{1}{r_{K_N}}$ in the weighted adjacency matrix $\widetilde{A}_{K_N}$.

Substituting the spectral decomposition $Q^\dagger = \sum_{k=1}^{N-1} \mu_k^{-1} z_k z_k^T$ in (4.30) in the definition (5.2) yields, with $(e_a - e_b)^T z_k = (z_k)_a - (z_k)_b$,

$$\omega_{ab} = \sum_{k=1}^{N-1} \frac{1}{\mu_k}\left((z_k)_a - (z_k)_b\right)^2$$

illustrating that $\omega_{ab} \geq 0$ and that the effective resistance ω_{ab} between node a and b increases with increasing difference between the vector components of Laplacian eigenvectors. For a small algebraic connectivity μ_{N-1}, the Fiedler vector z_{N-1} contributes significantly. Since $Q^\dagger \widetilde{Q} Q^\dagger = Q^\dagger \left(I - \frac{1}{N}J\right) = Q^\dagger$ and substituting $\widetilde{Q} = B\text{diag}\left(\frac{1}{r_{ij}}\right)B^T$ in (2.14), the quadratic form (5.2) is transformed to a Euclidean norm,

$$\omega_{ab} = (e_a - e_b)^T Q^\dagger \widetilde{Q} Q^\dagger (e_a - e_b) = \left\| \text{diag}\left(\frac{1}{\sqrt{r_{ij}}}\right) B^T Q^\dagger (e_a - e_b) \right\|_2^2 \tag{5.5}$$

The effective resistance ω_l of a link $l = (i,j)$ between node i and j in graph G equals the "parallel resistor formula"

$$\frac{1}{\omega_l} = \frac{1}{r_l} + \frac{1}{\left(\omega_{G\backslash l}\right)_{ij}} \tag{5.6}$$

where $\left(\omega_{G\backslash l}\right)_{ij}$ is the effective resistance between node i and j in the graph $G\backslash l$ obtained from the graph G after deletion of the link $l = (i,j)$. Indeed, the current $x = I_c\left(e_i - e_j\right)$ injected in node i and leaving at node j flows through the resistor r_l of direct link l and through the remaining part of the network. Applying the law of Ohm and the definition $v_a - v_b = \omega_{ab}I_c$ of effective resistance leads to (5.6), from which $\left(\omega_{G\backslash l}\right)_{ij} \leq \left(\omega_G\right)_{ij} = \omega_l$. If the direct link is absent, then collapse all intermediate nodes between node i and $j \neq i$ to a single node v, resulting in the graph $\widehat{G}$ with $\left(\omega_{\widehat{G}}\right)_{ij} \leq \left(\omega_G\right)_{ij}$, because all resistances among the intermediate nodes are put to zero in the graph $\widehat{G}$. The parallel resistor formula indicates that $\left(\omega_{\widehat{G}}\right)_{iv}^{-1} = \sum_{k=1}^{d_i} \frac{1}{w_{ik}}$ and $\left(\omega_{\widehat{G}}\right)_{vj}^{-1} = \sum_{k=1}^{d_j} \frac{1}{w_{jk}}$, while the series connection gives $\left(\omega_{\widehat{G}}\right)_{ij} = \left(\omega_{\widehat{G}}\right)_{iv} + \left(\omega_{\widehat{G}}\right)_{vj}$. If all links have a unit resistance $w_{kl} = 1$ and $a_{ij} = 0$, then we arrive at

$$\omega_{ij} \geq \frac{1}{d_i} + \frac{1}{d_j} \qquad \text{no direct link between } i \text{ and } j \tag{5.7}$$

Coppersmith *et al.* (1996) elegantly show that

$$\omega_{ij} \geq \frac{1}{1+d_i} + \frac{1}{1+d_j}$$

between any two nodes i and j in a graph G.

The parallel resistor formula (5.6) shows that the *relative resistance* $\frac{\omega_l}{r_l} \leq 1$, with equality only if $\left(\omega_{G\backslash l}\right)_{ij} \to \infty$, implying that the removal of link l in G disconnects the graph. The relative resistance $\omega_{ij}w_{ij}$, coined by Devriendt and Lambiotte (2022), appears in Foster's theorem in (5.20), in spanning trees (Section 5.6), in sparsification (Section 8.9) and in the resistance curvature $p_i = 1 - \frac{1}{2}\sum_{k=1}^N a_{ki}\omega_{ik}w_{ik}$ at node i where $w_{ik} = \frac{1}{r_{ik}}$, defined and studied by Devriendt and Lambiotte (2022).

5.2 Effective graph resistance

The effective graph resistance, defined as

$$R_G = \frac{1}{2}\sum_{a=1}^N \sum_{b=1}^N \omega_{ab} = \frac{1}{2}u^T \Omega u \tag{5.8}$$

can be regarded as a graph metric that measures the difficulty of transport in a graph G. The smaller R_G, the better transport is facilitated in the graph. In a nicely written book, Doyle and Snell (1984) extensively treat the connection between electric resistor networks and random walks, for which we also refer to

Chandra *et al.* (1997), Ellens *et al.* (2011) and Ghosh *et al.* (2008). In chemical graph theory, the effective graph resistance R_G is called the Kirchhoff index.

The effective graph resistance $R_G = \frac{1}{2}u^T \Omega u$ in (5.8) becomes with (5.4)

$$R_G = \frac{1}{2}u^T \zeta.u^T u + \frac{1}{2}u^T u.\zeta^T u - u^T Q^\dagger u = Nu^T \zeta = N\text{trace}\left(Q^\dagger\right) \tag{5.9}$$

because $u^T Q^\dagger u = 0$ as the vector u is orthogonal to each other eigenvector of Q. The trace-formula (A.99) leads to

$$R_G = N \sum_{k=1}^{N-1} \frac{1}{\mu_k} \tag{5.10}$$

The vector $\zeta = \left(Q_{11}^\dagger, Q_{22}^\dagger, \ldots, Q_{NN}^\dagger\right)$ is a graph metric vector and as important as the degree vector $d = (Q_{11}, Q_{22}, \ldots, Q_{NN})$. As shown in Van Mieghem *et al.* (2017), the component k of the vector ζ, that satisfies $Q_{kk}^\dagger \leq Q_{jj}^\dagger$ for $1 \leq j \leq N$, can be regarded as the best spreader node in the graph or as the node lying in the center of gravity of the graph.

For the undirected version of the graph in Fig. 2.1, the effective resistance matrix Ω, computed from (5.4), is

$$\Omega = \frac{1}{66} \begin{bmatrix} 0 & 31 & 37 & 64 & 49 & 36 \\ 31 & 0 & 34 & 55 & 34 & 31 \\ 37 & 34 & 0 & 45 & 48 & 49 \\ 64 & 55 & 45 & 0 & 45 & 64 \\ 49 & 34 & 48 & 45 & 0 & 37 \\ 36 & 31 & 49 & 64 & 37 & 0 \end{bmatrix}$$

and the corresponding effective graph resistance (5.10) is $R_G = \frac{659}{66}$. A more interesting example is the effective resistance of the chain of cliques $G_D^*(n_1, n_2, \ldots, n_{D+1})$, defined in Section 6.13. By using Theorem 45, **art.** 127 and the explicit relations for the coefficients $c_2(D)$ and $c_1(D)$ of the characteristic polynomial $p_D(\mu)$ in Van Mieghem and Wang (2009), the effective resistance of the chain of cliques $G_D^*(n_1, n_2, \ldots, n_{D+1})$ is

$$R_{G_D^*} = \sum_{q=2}^{D+1} \frac{\left(N - \sum_{k=1}^{q-1} n_k\right)}{n_{q-1}n_q} \sum_{k=1}^{q-1} n_k + N \sum_{k=1}^{D+1} \frac{n_k - 1}{n_{k-1} + n_k + n_{k+1}} \tag{5.11}$$

where $n_0 = n_{D+2} = 0$, the number of nodes $N = \sum_{j=1}^{D+1} n_j$ in (6.50) and the number of links $L = \sum_{j=1}^{D+1} \binom{n_j}{2} + \sum_{j=1}^{D} n_j n_{j+1}$ in (6.51). Theorem 41 shows that the minimum effective resistance in any graph with N nodes and diameter D is achieved in the class $G_D^*(n_1, n_2, \ldots, n_{D+1})$. Hence, minimizing (5.11) with respect to $n_1, n_2, \ldots, n_{D+1}$ subject to (6.50) yields the smallest possible effective resistance in any graph with N nodes and diameter D. An extreme case is the path P_{D+1}

with D-hops (see the end of Section 6.13), for which all $n_j = 1$ such that $N = D+1$ and the effective graph resistance, computed via (5.11) and (5.10) with (6.15), is

$$R_{P_{D+1}} = \frac{D(D+1)(D+2)}{6} = \frac{D+1}{2}\sum_{k=1}^{D}\frac{1}{1-\cos\frac{k\pi}{D+1}}$$

5.3 Properties of the effective resistance

5.3.1 The effective resistance ω_{ab} and $\sqrt{\omega_{ab}}$ are both a metric

The effective resistance matrix Ω is a distance matrix (**art.** 8) obeying (a) non-negativity, $\omega_{ab} \geq 0$, $\omega_{aa} = 0$, (b) symmetry $\omega_{ab} = \omega_{ba}$ and (c) the triangle inequality, $\omega_{ab} \leq \omega_{ac} + \omega_{cb}$, which follows from the simplex representation of an undirected graph (Fiedler, 2009; Devriendt, 2022a). This metric property of Ω has been discovered by Klein and Randić (1993) and by Gvishiani and Gurvich (1987) (in Russian).

Indeed, injecting a current I_c in node a, which leaves the network at node b, translates to $v_a - v_b = \omega_{ab}I_c$. The potential of any other node m lies in between $v_a \geq v_m \geq v_b$, else node m would be the source node if $v_m > v_a$ or a sink node if $v_m < v_b$. This property is known as the maximum principle of harmonic functions (**art.** 15): the potential v is a harmonic function with boundary conditions at node a and b. If y_{am} is the current flowing from node a to node m, then $v_a - v_m = \omega_{am}y_{am}$ and, similarly, $v_m - v_b = \omega_{mb}y_{mb}$, so that $v_a - v_b = \omega_{am}y_{am} + \omega_{mb}y_{mb}$ and $\omega_{ab} = \omega_{am}\frac{y_{am}}{I_c} + \omega_{mb}\frac{y_{mb}}{I_c}$. The law of current conservation tells that both the current $y_{am} \leq I_c$ and $y_{mb} \leq I_c$ cannot exceed I_c, resulting in

$$\omega_{ab} \leq \omega_{am} + \omega_{mb}$$

which proves the triangle inequality. Equality for $m \neq a$ nor $m \neq b$ only holds if $y_{am} = I_c$ and $y_{mb} = I_c$, meaning that the effective resistances ω_{am} and ω_{mb} are in series and m is a "cut" node. In Section 5.6, we deduce the triangle closure $\omega_{am} + \omega_{mb} - \omega_{ab}$ in (5.37).

The spectral decomposition of the Laplacian $Q = ZMZ^T$, where $M = \text{diag}(\mu)$, and its positive semidefiniteness allow us to write $Q = Z\text{diag}(\sqrt{\mu})\text{diag}(\sqrt{\mu})Z^T = S^T S$ with $S = \text{diag}(\sqrt{\mu})Z^T$ and similarly for the pseudoinverse. From the Gram decomposition $Q^\dagger = (S^\dagger)^T S^\dagger$, the square of the Euclidian distance between two vertices a and b with coordinates $p_a = S^\dagger e_a$ and $p_b = S^\dagger e_b$ in the inverse simplex (Devriendt and Van Mieghem, 2019a) equals $\|p_a - p_b\|^2 = (p_a - p_b)^T(p_a - p_b) = (e_a - e_b)^T(S^\dagger)^T S^\dagger(e_a - e_b)$. The definition (5.2) indicates that $\|p_a - p_b\|^2 = \omega_{ab}$ and, by the triangle inequality $\|p_a - p_b\| \leq \|p_a - p_m\| + \|p_m - p_b\|$, we arrive[1] at

$$\sqrt{\omega_{ab}} \leq \sqrt{\omega_{am}} + \sqrt{\omega_{mb}}$$

---

[1] A direct demonstration (**art.** 203) follows from $\sqrt{\omega_{ab}} = \left\|\text{diag}\left(\frac{1}{\sqrt{r_{ij}}}\right)B^T Q^\dagger(e_a - e_b)\right\|_2$ in (5.5).

In summary, the effective resistances obey two triangle inequalities. Thus, both ω_{ab} and $\sqrt{\omega_{ab}}$ are a metric.

5.3.2 The effective resistance ω_{ab} as minimization of electrical power

The power, the energy per unit time (in watts), dissipated in a resistor network is the sum of the power dissipated in each resistor, which equals $\mathcal{P} = v^T x$. The voltage vector $v = Q^\dagger x$ in (4.32) gives $\mathcal{P} = x^T Q^\dagger x$, while the injected current vector $x = \tilde{Q} v$ in (2.15) leads to $\mathcal{P} = v^T \tilde{Q} v$. The effective resistance matrix $\Omega = \zeta . u^T + u . \zeta^T - 2Q^\dagger$ in (5.4) and the conservation law $u^T x = 0$ of current in **art. 14** expresses the power in $\mathcal{P} = x^T Q^\dagger x$ as

$$\mathcal{P} = -\frac{1}{2} x^T \Omega x = -\frac{1}{2} \sum_{i=1}^N \sum_{j=1}^N \omega_{ij} x_i x_j = -\sum_{i=1}^N \sum_{j=1}^{i-1} \omega_{ij} x_i x_j \qquad (5.12)$$

where at least one component x_j of the injected current x is negative. Since the voltage is specified with respect to a voltage reference, the power equals $\mathcal{P} = \beta^2 (v^*)^T \tilde{Q} v^*$ for $v = \beta (v^* + \alpha u)$ for any real α and β.

If we inject a current I_c in node a, which leaves the resistor network at node b, then the potential difference at the nodes a and b is $v_a - v_b = \omega_{ab} I_c = \beta (v_a^* - v_b^*)$. If we choose $\alpha = \frac{1}{\beta} v_b$, then $v_a^* = \frac{1}{\beta} (v_a - v_b) = \frac{1}{\beta} \omega_{ab} I_c$ and $v_b^* = 0$. Finally, we choose $\beta = \omega_{ab} I_c$ so that $v_a^* = 1$ and $v_b^* = 0$ and the normalized vector $v^* = \frac{1}{\omega_{ab} I_c} (v - v_b u)$ is dimensionless. Using (4.4) and $w_l = \frac{1}{r_l}$, the dissipated power is

$$\mathcal{P} = v^T \tilde{Q} v = \sum_{l \in \mathcal{L}} \frac{1}{r_l} (v_{l+} - v_{l-})^2 = \sum_{i=1}^N \sum_{j=i+1}^N \frac{a_{ij}}{r_{ij}} (v_i - v_j)^2$$

The power $\mathcal{P}$ is minimized (**art. 200**) with respect to the nodal voltages if the corresponding gradient $\nabla \mathcal{P} = 0$, i.e. each partial derivative $\frac{\partial \mathcal{P}}{\partial v_k} = 0$ for $k \in \mathcal{N}$. Now,

$$\mathcal{P} = \sum_{i=1; i \neq k}^N \sum_{j=i+1; j \neq k}^N \frac{a_{ij}}{r_{ij}} (v_i - v_j)^2 + \sum_{j=k+1}^N \frac{a_{kj}}{r_{kj}} (v_k - v_j)^2 + \sum_{i=1}^{k-1} \frac{a_{ik}}{r_{ik}} (v_i - v_k)^2$$

and

$$\frac{\partial \mathcal{P}}{\partial v_k} = 2 \sum_{j=k+1}^N \frac{a_{kj}}{r_{kj}} (v_k - v_j) - 2 \sum_{i=1}^{k-1} \frac{a_{ik}}{r_{ik}} (v_i - v_k) = 2 \sum_{j=1}^N \frac{a_{jk}}{r_{jk}} (v_k - v_j) = 2 x_k$$

where the last equality follows from the flow conservation law in (2.11). By construction, the injected current is $x_k = 0$ in all nodes $k \in \mathcal{N}$, except for node a and b, where $x_a = -x_b = I_c$. But, in those nodes a and b, the voltage v_a and v_b is given and is not variable, i.e. $\mathcal{P}$ cannot be varied over those voltages. This means that the vector $v^* = \frac{1}{\omega_{ab} I_c} (v - v_b u)$ minimizes the power $\mathcal{P} = (\omega_{ab} I_c)^2 (v^*)^T \tilde{Q} v^*$, given the normalized voltage $v_a^* = 1$ and $v_b^* = 0$, and consistent with the maximum principle of harmonic functions (**art. 15**). Invoking (5.1) in $v^* = \frac{1}{w_{ab}I_c}(v - v_b u)$, the minimized power is

$$\mathcal{P}_{min} = (w_{ab}I_c)^2 (v^*)^T \widetilde{Q}v^* = I_c^2 (e_a - e_b)^T Q^\dagger \widetilde{Q} Q^\dagger (e_a - e_b) = w_{ab}I_c^2$$

where the last equality follows from (5.5). Hence, the effective resistance w_{ab} is proportional to the minimized power $\mathcal{P}_{min}$, given the injected current vector $x = I_c(e_a - e_b)$. On the other hand, since $\frac{v_a - v_b}{w_{ab}} = I_c$ and $\mathcal{P}_{min} = \frac{(v_a - v_b)^2}{w_{ab}}$, the effective resistance w_{ab} is inversely proportional to the minimized power $\mathcal{P}_{min}$, given the potential v_a and v_b. Explicitly as in Batson *et al.* (2013), we find that $w_{ab}^{-1} = \min\limits_{v^*:v_a^*=1;v_b^*=0}(v^*)^T \widetilde{Q}v^*$.

5.3.3 *Eigenvalue equation of the matrix $\widetilde{Q}\Omega$*

From the definition (5.4) of the effective resistance matrix Ω, we obtain $\widetilde{Q}\Omega = \widetilde{Q}\zeta u^T + \widetilde{Q}u\zeta^T - 2\widetilde{Q}Q^\dagger$ and using the basic inversion product $\widetilde{Q}Q^\dagger = I - \frac{1}{N}J$ in (4.31) and Laplacian characterizing eigenvalue equation $\widetilde{Q}u = 0$,

$$\widetilde{Q}\Omega = \widetilde{Q}\zeta u^T - 2\left(I - \frac{1}{N}J\right) \tag{5.13}$$

Right-multiplication of (5.13) with any vector x orthogonal to the all-one vector u, thus satisfying $u^T x = 0$, leads for a weighted Laplacian matrix $\widetilde{Q}$ with corresponding effective resistance matrix Ω to the eigenvalue equation

$$\widetilde{Q}\Omega x = -2x \tag{5.14}$$

Since each column of a weighted Laplacian matrix $\widetilde{Q}$ sums to zero, (5.14) leads to

$$\widetilde{Q}\Omega\widetilde{Q} = -2\widetilde{Q} \tag{5.15}$$

The eigenvalues of the $N \times N$ asymmetric matrix $\widetilde{Q}\Omega$ in (5.13) are the zeros (**art. 235**) in λ of the characteristic polynomial $c_{\widetilde{Q}\Omega}(\lambda) = \det\left(\widetilde{Q}\Omega - \lambda I\right)$, which is, with $J = uu^T$ and (5.13), $c_{\widetilde{Q}\Omega}(\lambda) = \det\left(\left(\widetilde{Q}\zeta + \frac{2}{N}u\right)u^T - (\lambda + 2)I\right)$. Invoking the "rank one update" formula (A.66), $\det\left(I + cd^T\right) = 1 + d^T c$, yields

$$c_{\widetilde{Q}\Omega}(\lambda) = (-1)^N \lambda(\lambda + 2)^{N-1}$$

Hence, the matrix $\widetilde{Q}\Omega$ has $N - 1$ eigenvalues equal to $\lambda = -2$, belonging to each possible external current x orthogonal to u, and one zero eigenvalue whose eigenvector must be a linear combination[2] of the eigenvector u and x. Hence, $\widetilde{Q}\Omega(au + bx) = aN\widetilde{Q}\zeta - 2bx = 0$, so that $x = \frac{aN}{2b}\widetilde{Q}\zeta$ and the eigenvector belonging

---

[2] Since $\widetilde{Q}\Omega$ is not symmetric, the eigenvectors are not necessarily orthogonal, but independent.

to $\lambda = 0$ equals $u + \frac{N}{2}\widetilde{Q}\zeta$. Since $u^T\left(u + \frac{N}{2}\widetilde{Q}\zeta\right) = N$, it is convenient to scale that eigenvector of $\widetilde{Q}\Omega$ belonging to the single zero eigenvalue as

$$p = \frac{1}{2}\widetilde{Q}\zeta + \frac{u}{N} \tag{5.16}$$

which satisfies $u^T p = 1$ and

$$\Omega p = 2\left(\frac{\zeta^T\widetilde{Q}\zeta}{4} + R_G\right)u$$

Alternatively, a solution for p in $\widetilde{Q}\Omega p = \widetilde{Q}(\Omega p) = 0$ is immediate from $\widetilde{Q}u = 0$ as $\Omega p = cu$ for some constant $c \neq 0$. Devriendt and Lambiotte (2022) demonstrate that the vector p has several fundamental properties and that $\frac{\zeta^T\widetilde{Q}\zeta}{4} + R_G = \sigma^2$ can be interpreted as a variance of a distribution on a graph. The above properties of p are recast into the matrix equation

$$\begin{pmatrix} 0 & u^T \\ u & \Omega \end{pmatrix}\begin{pmatrix} -2\sigma^2 \\ p \end{pmatrix} = \begin{pmatrix} 1 \\ 0 \end{pmatrix}$$

which has been generalized by Fiedler (2009) to Fiedler's block matrix identity,

$$\begin{pmatrix} 0 & u^T \\ u & \Omega \end{pmatrix}\begin{pmatrix} -2\sigma^2 & p^T \\ p & -\frac{1}{2}\widetilde{Q} \end{pmatrix} = I$$

which is verified taking into account the definition (5.16) of p and $\Omega\widetilde{Q}$ in (5.13). Since $\left(\Omega\widetilde{Q}\right)^T = \widetilde{Q}\Omega$, Fiedler's block matrix identity shows that

$$\widetilde{Q}\Omega = 2pu^T - 2I$$

Fiedler's block matrix identity, equivalent[3] to

$$\begin{pmatrix} 0 & u^T \\ u & \Omega \end{pmatrix}^{-1} = \begin{pmatrix} -2\sigma^2 & p^T \\ p & -\frac{1}{2}\widetilde{Q} \end{pmatrix} \qquad \text{with } \Omega p = 2\sigma^2 u \tag{5.17}$$

relates the effective resistance matrix Ω to the (weighted) Laplacian $\widetilde{Q}$ of a graph and possesses many deep, geometric properties of the simplex geometry of a graph, for which we refer to Fiedler (2009); Van Mieghem *et al.* (2017); Devriendt and Van Mieghem (2019a); Devriendt (2022a). Applying the block inverse (A.61) to Fiedler's block matrix identity (5.17) indicates that $2\sigma^2 = \frac{1}{u^T\Omega^{-1}u}$ and the vector $p = \frac{1}{u^T\Omega^{-1}u}\Omega^{-1}u$, while the inverse of the effective resistance matrix is

$$\Omega^{-1} = \frac{1}{2\sigma^2}p.p^T - \frac{1}{2}\widetilde{Q} \tag{5.18}$$

which was earlier found by Bapat (2004)[Theorem 3] after a longer computation without resorting to Fiedler's block matrix identity (5.17).

---

[3] The nice identity (5.17) has independently been derived and studied by Subak-Sharpe (1990).

5.3.3.1 Foster's Theorem

Applying the trace-formula (A.99) to the eigenvalue equation (5.14) yields

$$\text{trace}\left(\left(\widetilde{Q}\Omega \right)^k \right) = (-1)^k \, 2^k \, (N-1) \tag{5.19}$$

for any non-negative integer k. For any two $n \times n$ matrices A and B, it holds that $\text{trace}\left(A^T B \right) = \sum_{i=1}^{n} \sum_{j=1}^{n} a_{ij} b_{ij} = u^T (A \circ B) u$, where an element of the Hadamard product $A \circ B$ equals $(A \circ B)_{ij} = a_{ij} b_{ij}$. Hence[4], for $k = 1$ in (5.19), $\text{trace}(\widetilde{Q}\Omega) = \sum_{i=1}^{N} \sum_{j=1}^{N} \widetilde{q}_{ij} \omega_{ij}$ and since $\omega_{ii} = 0$, we find with $\widetilde{Q} = \widetilde{\Delta} - \widetilde{A}$ that $\text{trace}(\widetilde{Q}\Omega) = -\sum_{i=1}^{N} \sum_{j=1}^{N} \widetilde{a}_{ij} \omega_{ij}$, which maps (5.19) to

$$\frac{1}{2} \sum_{i=1}^{N} \sum_{j=1}^{N} a_{ij} \frac{\omega_{ij}}{r_{ij}} = N - 1 \tag{5.20}$$

In summary, the sum (5.20) over all weighted links of the *relative resistances* equals $\sum_{i \sim j} \frac{\omega_{ij}}{r_{ij}} = \sum_{l \in \mathcal{L}} \frac{\omega_l}{r_l} = N - 1$ or, in terms of the Hadamard product, $u^T \left(\widetilde{A} \circ \Omega \right) u = 2 (N - 1)$. Klein (2002)[Corollary C] mentions that (5.20) was first discovered by Foster (1949) and (5.20) is known as Foster's Theorem. The (weighted) effective graph resistance (5.8) is written in terms of the Hadamard product as

$$\widetilde{R}_G = \frac{1}{2} u^T \Omega u = \frac{1}{2} u^T (J \circ \Omega) u = \frac{1}{2} u^T \left(\left(J - \widetilde{A} + \widetilde{A} \right) \circ \Omega \right) u$$
$$= \frac{1}{2} u^T \left(\left(J - \widetilde{A} \right) \circ \Omega \right) u + \frac{1}{2} u^T \left(\widetilde{A} \circ \Omega \right) u$$

Foster's formula (5.20) expresses the (weighted) effective graph resistance as

$$\widetilde{R}_G = N - 1 + \frac{1}{2} u^T \left(\left(J - \widetilde{A} \right) \circ \Omega \right) u \tag{5.21}$$

5.3.3.2 Beyond Foster's Theorem

Klein and Randić (1993) have considered the matrix $\widetilde{Q}\Phi\widetilde{Q}\Omega$, where Φ is an arbitrary $N \times N$ symmetric matrix, which equals with (5.13)

$$\widetilde{Q}\Phi\widetilde{Q}\Omega = -2\widetilde{Q}\Phi + \widetilde{Q} \left(\Phi\widetilde{Q}\zeta + \frac{2}{N}\Phi u \right) u^T$$

After taking the trace,

$$\text{trace}\left(\widetilde{Q}\Phi\widetilde{Q}\Omega \right) = -2\text{trace}\left(\widetilde{Q}\Phi \right) + \text{trace}\left(\widetilde{Q} \left(\Phi\widetilde{Q}\zeta + \frac{2}{N}\Phi u \right) u^T \right)$$

---

[4] For $k = 2$ in (5.19), we find after tedious computations the less physically interpretable relation

$$4 (N - 1) = \sum_{i=1}^{N} \sum_{r=1}^{N} \omega_{ir}^2 \widetilde{q}_{ii} \widetilde{q}_{rr} - 2 \sum_{r=1}^{N} \sum_{l=1}^{N} \widetilde{a}_{rl} \sum_{i=1}^{N} \widetilde{q}_{ii} \omega_{il} \omega_{ir} + \sum_{i=1}^{N} \sum_{m=1}^{N} \sum_{r=1}^{N} \sum_{l=1}^{N} \widetilde{a}_{im} \widetilde{a}_{rl} \omega_{mr} \omega_{li}$$

where the quadratic form $\sum_{i=1}^{N} \sum_{r=1}^{N} \omega_{ir}^2 \widetilde{q}_{ii} \widetilde{q}_{rr} = \widetilde{q}^T (\Omega \circ \Omega) \widetilde{q}$ and the vector $\widetilde{q} = (\widetilde{q}_{11}, \widetilde{q}_{22}, \ldots, \widetilde{q}_{NN})$ with $\widetilde{q}_{ii} = \sum_{k=1}^{N} \frac{a_{ik}}{r_{ik}}$. The quadratic form $\widetilde{q}^T \Omega \widetilde{q}$ is related to the Kemeny constant $\frac{d^T \Omega d}{4L}$, as shown in Wang et al. (2017, Corollary 1).

and using the cyclic permutation property (4.14) of the trace, we obtain

$$\text{trace}\left(\widetilde{Q}\Phi\widetilde{Q}\Omega\right) = -2\text{trace}\left(\widetilde{Q}\Phi\right) \tag{5.22}$$

because $\text{trace}\left(\widetilde{Q}\left(\Phi\widetilde{Q}\zeta + \frac{2}{N}\Phi u\right)u^T\right) = \text{trace}\left(\left(\Phi\widetilde{Q}\zeta + \frac{2}{N}\Phi u\right)u^T\widetilde{Q}\right) = 0$. If $\Phi = \Omega\left(\widetilde{Q}\Omega\right)^{k-2}$, then (5.22) shows that $\text{trace}\left(\left(\widetilde{Q}\Omega\right)^k\right) = -2\text{trace}\left(\left(\widetilde{Q}\Omega\right)^{k-1}\right)$, which leads, after iteration in k, to (5.19), illustrating that (5.22) is a generalization of (5.19).

5.4 The pseudoinverse $Q^\dagger$ and the effective resistance matrix Ω

We rewrite $Q^\dagger_{ij} = e_i^T Q^\dagger e_j$ with the property that $Q^\dagger u = 0$ as

$$Q^\dagger_{ij} = e_i^T Q^\dagger e_j = \left(e_i - \frac{u}{N}\right)^T Q^\dagger \left(e_j - \frac{u}{N}\right)$$

Using $x^T Q^\dagger x = -\frac{1}{2}x^T \Omega x$ in (5.12) for any vector x obeying $x^T u = 0$, we obtain

$$Q^\dagger_{ij} = -\frac{1}{2}\left(e_i - \frac{u}{N}\right)^T \Omega \left(e_j - \frac{u}{N}\right) = \frac{u^T\Omega}{2N}\left(e_i + e_j\right) - \frac{1}{2}\omega_{ij} - \frac{u^T\Omega u}{2N^2}$$

With the definition (5.8) of the effective graph resistance $\widetilde{R}_G = \frac{u^T\Omega u}{2}$, the elements of the Laplacian pseudoinverse are expressed in terms of those of the effective resistance matrix

$$Q^\dagger_{ij} = \frac{1}{2}\left(\frac{1}{N}\sum_{k=1}^{N}\omega_{ik} + \frac{1}{N}\sum_{k=1}^{N}\omega_{jk}\right) - \frac{1}{2}\omega_{ij} - \frac{R_G}{N^2} \tag{5.23}$$

Corollary 2 *In each row (or column) of the pseudoinverse $Q^\dagger$, the diagonal element is the largest: $Q^\dagger_{ii} \geq Q^\dagger_{ij}$ for each row $1 \leq i \leq N$.*

Proof: The difference $Q^\dagger_{ii} - Q^\dagger_{ij}$ in (5.23) with $\omega_{ii} = 0$ gives

$$Q^\dagger_{ii} - Q^\dagger_{ij} = \frac{1}{2}\omega_{ij} + \frac{u^T\Omega}{2N}(e_i - e_j) = \frac{1}{2N}\sum_{k=1}^{N}\{\omega_{ki} + \omega_{ij} - \omega_{kj}\}$$

Section 5.3.1 shows that each element in the effective resistance matrix Ω satisfies the triangle inequality $\omega_{ki} + \omega_{ij} \geq \omega_{kj}$, so that $Q^\dagger_{ii} - Q^\dagger_{ij} \geq 0$, for any j. □

5.5 The spectrum of the effective resistance matrix Ω

Let us denote the eigenvalue equation of the $N \times N$ symmetric, non-negative effective resistance matrix Ω by

$$\Omega v_k = \rho_k v_k \tag{5.24}$$

where ρ_k is the k-th eigenvalue belonging to the normalized eigenvector v_k, i.e. $v_k^T v_k = 1$. The real eigenvalues are ordered as usual: $\rho_1 \geq \rho_2 \geq \cdots \geq \rho_N$. The eigenvalue decomposition in matrix form is

$$\Omega = VRV^T \tag{5.25}$$

where V is an orthogonal matrix, the $N \times 1$ vector $\rho = (\rho_1, \rho_2, \cdots, \rho_N)^T$ with eigenvalues of Ω and $R = \mathrm{diag}\,(\rho)$. Invoking the definition (5.4) of Ω in the eigenvalue equation (5.24) leads to

$$\zeta.u^T v_k + u.\zeta^T v_k - 2Q^\dagger v_k = \rho_k v_k \tag{5.26}$$

Taking into account that $u^T Q^\dagger = 0$, we obtain $u^T \zeta.u^T v_k + u^T u.\zeta^T v_k = \rho_k u^T v_k$. The definition (5.8) of the effective graph resistance $R_G = \frac{1}{2} u^T \Omega u$, complemented by $u^T \zeta = \frac{R_G}{N}$, shows that

$$\rho_k = \frac{R_G}{N} + N \frac{\zeta^T v_k}{u^T v_k} \tag{5.27}$$

In a connected graph, the effective resistance matrix Ω has full rank, i.e. $\det \Omega \neq 0$, because the inverse Ω^{-1} exists as shown in (5.18). Hence, we conclude that the effective resistance matrix Ω does not possess a zero eigenvalue in a connected graph and $\rho_k \neq 0$ for $1 \leq k \leq N$. A powerful theorem, that has appeared already in Fiedler (2009, Corollary 6.2.9), is

Theorem 33 *In a connected graph, the eigenvalues* $\rho_2, \rho_3, \ldots, \rho_N$ *of the effective resistance matrix* Ω *interlace with those of the Laplacian matrix as*

$$0 > -\frac{2}{\mu_1} \geq \rho_2 \geq -\frac{2}{\mu_2} \geq \cdots \geq -\frac{2}{\mu_{N-2}} \geq \rho_{N-1} \geq -\frac{2}{\mu_{N-1}} \geq \rho_N \tag{5.28}$$

Proof: We include the proof of Sun *et al.* (2015). Let the $(N-1) \times 1$ vector $\mu = (\mu_1, \mu_2, \ldots, \mu_{N-1})$ denote the positive eigenvalues of the weighted Laplacian of a connected graph (Section 4.1.1). The spectral decomposition (4.1) is then $\widetilde{Q} = Z \begin{bmatrix} \mathrm{diag}\,(\mu) & 0 \\ 0 & 0 \end{bmatrix} Z^T$. Let the matrix $\widehat{S} = Z^T \Omega Z$, then $\Omega = Z \widehat{S} Z^T$, which we write as $\Omega = Z \begin{bmatrix} S & s \\ s^T & s_{NN} \end{bmatrix} Z^T$, where S is a symmetric $(N-1) \times (N-1)$ matrix. Combining $\widetilde{Q}\Omega\widetilde{Q} = -2\widetilde{Q}$ in (5.15) shows that $\mathrm{diag}(\mu)\,S\,\mathrm{diag}(\mu) = -2\mathrm{diag}(\mu)$. Hence, $S = -2\,(\mathrm{diag}\,(\mu))^{-1}$ is the principal submatrix of Ω and the Interlacing Theorem 71 then leads to (5.28). $\qquad\square$

Theorem 34 *In a connected graph, the effective resistance matrix* Ω *has one positive and* $N-1$ *negative eigenvalues.*

Proof: Theorem 34 follows from Theorem 33 and the Perron-Frobenius Theorem

75 for non-negative matrices as Ω. We add a second proof. For any vector z, definition (5.4) of Ω indicates that

$$z^T \Omega z = 2 \left(z^T \zeta \right) \left(z^T u \right) - 2 z^T Q^\dagger z \tag{5.29}$$

If $z^T u = 0$ or $z^T \zeta = 0$, then $z^T \Omega z \leq 0$, because $Q^\dagger$ is positive semidefinite. In other words, for any vector z orthogonal to the all-one vector u or to the vector ζ, the quadratic form $z^T \Omega z$ is negative, but $\frac{1}{2} u^T \Omega u = 2 R_N > 0$ and $\zeta^T \Omega \zeta > 0$. Since Ω is of full rank N and has no zero eigenvalue, there are $(N-1)$ negative eigenvalues and one positive eigenvalue. $\qquad\square$

We now concentrate on the largest eigenvalue ρ_1. A consequence of Theorem 34 and (A.99) with the zero diagonal in Ω lead to

$$\rho_1 = - \sum_{k=2}^{N} \rho_k \text{ and } \rho_1 = \sum_{k=2}^{N} |\rho_k|$$

Gerschgorin's Theorem 65 states that each eigenvalue ρ_k lies in a circle around the origin – each diagonal element of Ω is zero – with radius $(\Omega u)_i$ for $1 \leq i \leq N$. It follows from the definition (5.4) and $\zeta^T u = \frac{R_G}{N}$ in (5.9) that

$$\Omega u = \zeta N + u \frac{R_G}{N} \tag{5.30}$$

and Gerschgorin's Theorem 65 becomes, for a certain k and i,

$$|\rho_k| \leq \frac{R_G}{N} + N \zeta_i \tag{5.31}$$

In particular, Gerschgorin's Theorem 65 provides the upper bound

$$\rho_1 \leq \frac{R_G}{N} + N \max_i \zeta_i \tag{5.32}$$

Theorem 33 or the Rayleigh inequality $\rho_1 \geq \frac{z^T \Omega z}{z^T z}$ in (A.129) for $z = u$ give the lower bound

$$\rho_1 \geq \frac{u^T \Omega u}{u^T u} = \frac{2 R_G}{N} \tag{5.33}$$

Combining $\rho_k = \frac{R_G}{N} + N \frac{\zeta^T v_k}{u^T v_k}$ in (5.27) and Gerschgorin's bound (5.31) implies that there exists a component i in ζ so that $\frac{|\zeta^T v_k|}{|u^T v_k|} \leq \frac{2 R_G}{N^2} + \zeta_i$ for $k > 1$. Since the components of the principal eigenvector v_1 are non-negative by the Perron-Frobenius Theorem 75, it holds that

$$\frac{\zeta^T v_1}{u^T v_1} = \frac{\sum_{j=1}^{N} \zeta_j v_j}{\sum_{j=1}^{N} v_j} \leq \max_i \zeta_i$$

where equality only holds if $\zeta = \alpha u$ and relation $\zeta^T u = \frac{R_G}{N}$ in (5.9) implies that $\alpha = \frac{R_G}{N^2}$ and $\zeta = \frac{R_G}{N^2} u$.

Theorem 35 *If $\zeta = \frac{R_G}{N^2}u$, then the eigenvalues of effective resistance matrix Ω are $\rho_1 = \frac{2R_G}{N}$ and $\rho_k = -\frac{2}{\mu_k}$ for $k > 1$.*

Proof: If $\zeta = \frac{R_G}{N^2}u$, then $\Omega u = \zeta N + u\frac{R_G}{N}$ in (5.30) reduces to the eigenvalue equation $\Omega u = \frac{2R_G}{N}u$ illustrating that $\rho_1 = \frac{2R_G}{N}$ and $v_1 = \frac{u}{\sqrt{N}}$. For $\zeta = \frac{R_G}{N^2}u$, the eigenvalue equation (5.26) becomes

$$2\left(\frac{R_G}{N^2}J - Q^\dagger\right)v_k = \rho_k v_k$$

which we rewrite with $Q^\dagger = \left(\widetilde{Q} + \alpha J\right)^{-1} - \frac{1}{\alpha N^2}J$ in (4.36) after choosing $\alpha = -R_G$ as $-2\left(\widetilde{Q} - R_G J\right)^{-1}v_k = \rho_k v_k$, equivalent to $-\frac{2}{\rho_k}v_k = \left(\widetilde{Q} - R_G J\right)v_k$. For $k > 1$, the orthogonality of eigenvectors of the symmetric matrix Ω implies that $v_k^T u = 0$ and $R_G J v_k = 0$. Thus, we arrive at the eigenvalue equation $-\frac{2}{\rho_k}v_k = \widetilde{Q}v_k$, which indicates, for $k > 1$, that $\mu_k = -\frac{2}{\rho_k}$ if $\zeta = \frac{R_G}{N^2}u$. $\qquad\square$

Earlier, Zhou *et al.* (2016, Theorem 12) have demonstrated that equality in (5.28) only holds if Ω is "resistance regular", i.e. $\Omega u = c.u$, equivalent to $\zeta = c'.u$, for some positive real numbers c and c'.

With $-\frac{2}{\mu_1} \geq \rho_2 \geq -\frac{2}{\mu_2}$ in (5.28), the spectral gap $\rho_1 - \rho_2$ of the effective resistance matrix Ω is bounded by $\rho_1 + \frac{2}{\mu_1} \leq \rho_1 - \rho_2 \leq \rho_1 + \frac{2}{\mu_2}$. With the lower (5.33) and upper (5.32) bound, the spectral gap lies between

$$\frac{2R_G}{N} + \frac{2}{\mu_1} \leq \rho_1 - \rho_2 \leq \frac{R_G}{N} + N\max_i \zeta_i + \frac{2}{\mu_2}$$

Since $\mu_1 \leq N$ by (4.20), the spectral gap of Ω is always larger than

$$\rho_1 - \rho_2 \geq \frac{2(R_G + 1)}{N} \qquad (5.34)$$

Theorem 35 shows that $(\rho_1 - \rho_2)_{K_N} = 2$ and equality in (5.34) occurs in the complete graph K_N. The spectral gap of Ω in K_N is smallest among all graphs.

5.6 The effective resistance and spanning trees

As explained in **art.** 16, Kirchhoff (1847) has proposed an explicit solution of a variant of the inversion problem of the fundamental relation $x = \widetilde{Q}v$ in (2.15). Here, we present another method and deduce two forms (5.35) and (5.36) for the effective resistance ω_{ij}.

Fiedler's inverse block matrix (5.17) elegantly[5] leads to

$$\omega_{ij} = \frac{\det \widetilde{Q}_{\setminus\{i,j\}}}{\det \widetilde{Q}_{\setminus\{i\}}} = \frac{\det\left(\widetilde{Q}_{\setminus \text{row}(i,j)\setminus \text{col}(i,j)}\right)}{\xi(G)} \qquad (5.35)$$

---

[5] The derivation is based on computations of Karel Devriendt.

where $M_{\setminus\{i_1,\ldots,i_m\}}$ is the $(N-m)\times(N-m)$ submatrix obtained from the $N\times N$ matrix M by removing the rows and same columns $i_1,\ldots,i_m$. Indeed, we rewrite (5.17) as

$$\begin{pmatrix} 0 & 1 & u^T \\ 1 & 0 & \omega_j^T \\ u & \omega_j & \Omega_{\setminus\{j\}} \end{pmatrix}^{-1} = \begin{pmatrix} -2\sigma^2 & p_j & p^T \\ p_j & -\frac{1}{2}\tilde{Q}_{jj} & \tilde{Q}_j \\ p & \tilde{Q}_j & -\frac{1}{2}\tilde{Q}_{\setminus\{j\}} \end{pmatrix}$$

The Schur block matrix inverse formula (A.60) in **art. 217** shows that

$$-\frac{1}{2}\tilde{Q}_{\setminus\{j\}} = \left(\Omega_{\setminus\{j\}} - \begin{bmatrix} u & \omega_j \end{bmatrix}\begin{bmatrix} 0 & 1 \\ 1 & 0 \end{bmatrix}^{-1}\begin{bmatrix} u^T \\ \omega_j^T \end{bmatrix}\right)^{-1} = \left(\Omega_{\setminus\{j\}} - \omega_j u^T - u\omega_j^T\right)^{-1}$$

from which, after inversion, the element $-2\omega_{ij} = -2\left(\tilde{Q}_{\setminus\{j\}}\right)^{-1}_{ii}$ is found and the definition (A.43) of the inverse of matrix and $\det\tilde{Q}_{\setminus\{i\}} = \xi(G)$ by (4.37) then demonstrates (5.35).

If $x_i = 1$ and $x_j = -1$ in the potential difference (4.43) in **art. 130**, then the definition of the effective resistance between node i and j shows that $v_i - v_j = \omega_{ij}$ and we find

$$\omega_{ij} = \frac{\det\left(\tilde{Q}_{\setminus\text{row}(j,N)\setminus\text{col}(j,N)}\right) + \det\left(\tilde{Q}_{\setminus\text{row}(i,N)\setminus\text{col}(i,N)}\right)}{\xi(G)}$$
$$- \frac{(-1)^{j+i}\left(\det\left(\tilde{Q}_{\setminus\text{row}(i,N)\setminus\text{col}(j,N)}\right) + \det\left(\tilde{Q}_{\setminus\text{row}(j,N)\setminus\text{col}(i,N)}\right)\right)}{\xi(G)}$$

Since $\tilde{Q} = \tilde{Q}^T$, it holds that $\det\left(\tilde{Q}_{\setminus\text{row}(i,N)\setminus\text{col}(j,N)}\right) = \det\left(\tilde{Q}_{\setminus\text{row}(j,N)\setminus\text{col}(i,N)}\right)$. Cramer's method in **art. 130** holds for an arbitrarily removed row in $\tilde{Q}v = x$ and we arrive, after replacing row N by a row $k \neq \{i,j\}$, at

$$\omega_{ij} = \frac{\det\left(\tilde{Q}_{\setminus\text{row}(i,k)\setminus\text{col}(i,k)}\right) + \det\left(\tilde{Q}_{\setminus\text{row}(j,k)\setminus\text{col}(j,k)}\right) - 2(-1)^{i+j}\det\left(\tilde{Q}_{\setminus\text{row}(i,k)\setminus\text{col}(j,k)}\right)}{\xi(G)}$$
(5.36)

We rewrite (5.36) with (5.35) as a *triangle closure equation* for the distance matrix Ω,

$$(\omega_{ik} + \omega_{kj}) - \omega_{ij} = 2(-1)^{i+j}\frac{\det\left(\tilde{Q}_{\setminus\text{row}(i,k)\setminus\text{col}(j,k)}\right)}{\xi(G)} \geq 0 \qquad (5.37)$$

where[6] the last inequality is due to the triangle inequality (see Section 5.3.1).

---

[6] Comparing the definition (5.3) of effective resistance $\omega_{ij} = Q_{ii}^\dagger + Q_{jj}^\dagger - 2Q_{ij}^\dagger$ and (5.36) would hint to $Q_{ij}^\dagger = \frac{(-1)^{i+j}\det(\tilde{Q}_{\setminus\text{row}(i,k)\setminus\text{col}(j,k)})}{\det(\tilde{Q}_{\setminus\{k\}})}$, which corresponds to the definition of the inverse (A.43), apart from an arbitrary row and column deletion (due to $\det\tilde{Q} = 0$). But, $Q_{ij}^\dagger \geq 0$ is not true in general. The correct expression for $Q_{ij}^\dagger$ is given in (4.39).

The Matrix Tree Theorem for the weighted Laplacian in **art.** 118 additionally tells us that

$$\det\left(\widetilde{Q}_{\backslash \text{ row } i \backslash \text{ col } j}\right) = \sum_{T \in \mathcal{T}(N)} w\left(T\right) \qquad \text{with } w\left(T\right) = \prod_{l \in T} \frac{1}{r_l}$$

where $\mathcal{T}\left(N\right)$ is the set of trees spanning all N nodes in weighted graph with weight $w_l = \frac{1}{r_l}$ of link l and the total number of such spanning trees equals the complexity $\xi\left(G\right) = |\mathcal{T}\left(N\right)|$ in **art.** 122.

The numerator $\det\left(\widetilde{Q}_{\backslash \text{row}(i,j)\backslash \text{col}(i,j)}\right)$ in (5.35) can also be expressed in terms of spanning trees. The idea is to merge a pair of linked nodes (i, j) of the graph G into a new node k by letting $r_{ij} \to 0$ or $w_{ij} = \frac{1}{r_{ij}} \to \infty$ so that the voltages $v_i = v_j$. The merging transforms the graph G into a new graph G' with weighted Laplacian Q', where row i and j and column i and j in $\widetilde{Q}$ are replaced by a row and column for node k, containing the links from node i and j to the other nodes in G. The link weights from node k to node m in the graph G' is $w'_{km} = w_{im} + w_{jm}$ for all nodes $m \in \mathcal{N}\backslash\{i, j\}$. Then, we obtain that $Q'_{\backslash\{k\}} = \widetilde{Q}_{\backslash\{i,j\}}$, which does not contain row k; thus, neither w' link weights nor the link weight $w_{ij} = \frac{1}{r_{ij}}$. The Matrix Tree Theorem for the weighted Laplacian in **art.** 118 then states that

$$\det\left(\widetilde{Q}_{\backslash \text{row}(i,j)\backslash \text{col}(i,j)}\right) = \det\left(Q'_{\backslash\{k\}}\right) = \sum_{T \in \mathcal{T}'(N-1)} w\left(T\right)$$

where $\mathcal{T}'\left(N-1\right)$ is the set of spanning trees on all $N-1$ nodes in the weighted graph G', thus also node k. Since node k is the merger of the linked nodes (i, j), this means that each spanning tree of $\mathcal{T}'\left(N-1\right)$ in G' contains the link (i, j) in the original graph G, but the link (i, j) is not weighted. In other words, $\mathcal{T}'\left(N-1\right) = \mathcal{T}_{(i,j)}\left(N\right)$ is the set of spanning trees in G that contain the link (i, j) and the weight of a tree $T' \in \mathcal{T}'\left(N-1\right)$ and its corresponding tree $T \in \mathcal{T}\left(N\right)$ are related by $w\left(T'\right) = \frac{w(T)}{w_{ij}}$. In summary, we have proved

Theorem 36 *The effective resistance between node i and j in (5.35) equals*

$$\omega_{ij} = \frac{1}{w_{ij}} \frac{\sum_{T \in \mathcal{T}_{(i,j)}(N)} w\left(T\right)}{\sum_{T \in \mathcal{T}(N)} w\left(T\right)} = \frac{\sum_{T \in \mathcal{T}_{(i,j)}(N)} \prod_{l \in T; l \neq (i,j)} \frac{1}{r_l}}{\sum_{T \in \mathcal{T}(N)} \prod_{l \in T} \frac{1}{r_l}} \tag{5.38}$$

where $\mathcal{T}\left(N\right)$ is the set of trees spanning all N nodes in weighted graph with weight $w_l = \frac{1}{r_l}$ and $\mathcal{T}_{(i,j)}\left(N\right)$ is the set of spanning trees in G that contain the link (i, j).

The relative resistance $w_{ij}\omega_{ij}$ in (5.38), the effective resistance $\omega_l = \omega_{ij}$ of the link $l = (i, j)$ multiplied by its weight $w_{ij} = w_l = \frac{1}{r_l}$, equals the probability that the link l appears in a random spanning tree of G. The *relative resistance* $w_{ij}\omega_{ij}$ reflects the importance of a link for connectivity of the graph. For unit resistances $r_l = 1$, the effective resistance in (5.38) simplifies to $\omega_{ij} = \frac{|\mathcal{T}_{(i,j)}(N)|}{|\mathcal{T}(N)|} = \frac{|\mathcal{T}_{(i,j)}(N)|}{\xi(G)}$, which is a rational number, and equals the fraction of all spanning trees of the graph G on N nodes that contain the link (i, j).

5.7 Bounds for the effective resistance matrix Ω

A constrained system can never reach a lower global minimum of the system dynamics than a system without constraints. Hence, transport restricted to a single path is never more efficient than unrestricted transports over all possible paths, which implies (**art. 7**) that the weight $w\left(\mathcal{P}_{ij}^*\right)$ of the shortest path $\mathcal{P}_{ij}^*$ between nodes i and j is lower bounded by the effective resistance:

$$\omega_{ij} \leq s_{ij} = w\left(\mathcal{P}_{ij}^*\right) \tag{5.39}$$

In other words, the $N \times N$ difference matrix $S - \Omega$ is a non-negative matrix. Equality in (5.39) occurs for trees, where there is only a single path between nodes.

Besides the upper bound $\omega_{ij} \leq s_{ij}$ in (5.39), Theorem 37 presents a lower bound, given in Lyons and Peres (2016, ex 2.129; p. 602):

Theorem 37 *The effective resistance ω_{ij} can be lower-bounded by the hopcount h_{ij} of shortest path $\mathcal{P}_{ij}^*$ as*

$$\frac{1}{m} h_{ij}^2 \leq \omega_{ij} \tag{5.40}$$

where $m = \sum_{l \in \mathcal{L}} w_l^{-1}$ is the sum over all links of the inverse link weights.

Equality in (5.40) occurs in an unweighted graph, where two nodes i and j are connected by k paths of h hops.

Proof: For a pair of nodes i and j, an $i-j$ cut consists of a set of links such that removing these links from the graph disconnects i from j. If $\mathcal{C}_{ij}$ is a collection of $i-j$ cuts which are independent, i.e. no two cuts share a link, then the inequality of Nash-Williams (1959) states that

$$\sum_{C \in \mathcal{C}_{ij}} w(C) \leq \omega_{ij} \tag{5.41}$$

where $w(C) = \left(\sum_{(a,b) \in C} w_{ab}^{-1}\right)^{-1}$ is the weight of a cut $C \in \mathcal{C}_{ij}$. Nash-Williams' inequality (5.41) follows from Rayleigh's monotonicity law by identifying the start and end nodes of all links in each of the cuts. For two nodes i and j which are h_{ij} hops removed from each other, we consider the following collection of $i-j$ cuts $\mathcal{C}_{ij} = \{C_k\}_{k=0}^{h_{ij}-1}$, where the cut $C_k = \{(a,b) \in \mathcal{L} : h_{ia} = k, h_{ib} = k+1\}$ contains all links between one node at shortest path hop distance k from i and the other node at distance $k+1$. Nash-Williams' inequality (5.41) shows that $\omega_{ij} \geq \sum_{k=0}^{h_{ij}-1} w(C_k)$. After multiplying both sides with $m = \sum_{l \in \mathcal{L}} w_l^{-1}$ and noticing that $m \geq \sum_{k=0}^{h_{ij}-1} w^{-1}(C_k)$, because the right-hand side only counts the weights of links in the cuts, we obtain

$$m\omega_{ij} \geq \sum_{k=0}^{h_{ij}-1} w(C_k) \sum_{k=0}^{h_{ij}-1} w^{-1}(C_k) \geq h_{ij}^2$$

where the last step follows from the Cauchy-Schwarz inequality (A.72) in **art**. 222 that erases the link weights. This proves the lower bound in (5.40).   □

5.8 Lower and upper bound for the effective graph resistance R_G

We end this chapter by establishing a lower and upper bound for the effective graph resistance R_G. **Art**. 127 demonstrates that

$$R_G = \frac{c_2(Q)}{\xi(G)} \geq \frac{(N-1)^2}{E[D]} \tag{5.42}$$

where the complexity $\xi(G)$ of the graph G equals the number of all possible spanning trees in the graph and where $c_2(Q)$ equals the number of all spanning trees with $N-2$ links in all subgraphs of G (see the Matrix Tree Theorem 22). The lower bound in (5.42) for the effective graph resistance R_G is attained by the complete graph K_N, for which $R_{K_N} = N - 1$. This lowest value of the effective graph resistance R_G follows from (5.21) in the unweighted case, where $\left(J - \widetilde{A}\right) \circ \Omega = A^c \circ \Omega$,

$$R_G = N - 1 + \frac{1}{2}u^T\left(A^c \circ \Omega\right)u = N - 1 + \sum_{i \sim j \in G^c} \omega_{ij} \tag{5.43}$$

because $A^c \circ \Omega$ is a non-negative matrix, which reduces to the zero matrix only for the complete graph K_N.

Applying Schur's argument in **art**. 276 for a convex function g to the degree vector $d = \Xi_Q \mu$ in (4.17) in **art**. 109 and the diagonal element vector $\zeta = \Xi_Q \frac{1}{\mu}$ in (4.45) in **art**. 132 yields

$$\begin{cases} \sum_{m=1}^{N} g\left(\frac{N}{N-1}d_m\right) \leq \frac{N}{N-1}\sum_{k=1}^{N-1} g\left(\mu_k\right) \\ \sum_{m=1}^{N} g\left(\frac{N}{N-1}Q_{mm}^\dagger\right) \leq \frac{N}{N-1}\sum_{k=1}^{N-1} g\left(\frac{1}{\mu_k}\right) \end{cases}$$

For example, $g(x) = \frac{1}{x}$ is convex for $x > 0$ and, hence,

$$\begin{cases} \frac{N-1}{N}\sum_{m=1}^{N}\frac{1}{d_m} \leq \frac{N}{N-1}\sum_{k=1}^{N-1}\frac{1}{\mu_k} = \frac{R_G}{N-1} \\ \frac{N-1}{N}\sum_{m=1}^{N}\frac{1}{Q_{mm}^\dagger} \leq \frac{N}{N-1}\sum_{k=1}^{N-1}\mu_k = \frac{2LN}{N-1} \end{cases}$$

The harmonic, geometric and arithmetic mean inequality (6.38) indicates that $\frac{N^2}{2L} = \frac{N}{E[D]} \leq \sum_{m=1}^{N}\frac{1}{d_m} = NE\left[\frac{1}{D}\right]$. Finally, we obtain the inequality, differently found in Van Mieghem *et al.* (2017),

$$\frac{(N-1)^2}{E[D]} \leq (N-1)^2 E\left[\frac{1}{D}\right] \leq R_G \tag{5.44}$$

which is slightly sharper than (5.42). Combining (5.43) and (5.7) shows that

$$R_G \geq N - 1 + \sum_{i \sim j \in G^c}\left(\frac{1}{d_i} + \frac{1}{d_j}\right)$$

Now, $\sum_{i\sim j\in G^c}\left(\frac{1}{d_i}+\frac{1}{d_j}\right)=\frac{1}{2}\sum_{i=1}^N\sum_{j=1;i\neq j}^N(1-a_{ij})\left(\frac{1}{d_i}+\frac{1}{d_j}\right)$ and

$$\sum_{i=1}^N\sum_{j=1;;i\neq j}^N(1-a_{ij})\left(\frac{1}{d_i}+\frac{1}{d_j}\right)=\sum_{i=1}^N\frac{1}{d_i}\sum_{j=1;;i\neq j}^N(1-a_{ij})+\sum_{j=1}^N\frac{1}{d_j}\sum_{i=1;i\neq j}^N(1-a_{ij})$$

$$=2\sum_{i=1}^N\frac{N-1-d_i}{d_i}=2\left((N-1)\sum_{i=1}^N\frac{1}{d_i}-N\right)$$

lead to a lower bound, tighter than (5.44),

$$R_G\geq(N-1)\sum_{i=1}^N\frac{1}{d_i}-1 \tag{5.45}$$

that appeared in Zhou and Trinasjtić (2008).

An upper bound $\omega_{ij}\leq h_{ij}$ follows from (5.39), where the distance matrix H is defined in **art.** 8. This bound in (5.8) yields

$$R_G\leq\binom{N}{2}E[H]$$

where $E[H]$ is the average hopcount in the graph. With $R_G=N\sum_{k=1}^{N-1}\frac{1}{\mu_k}$ in (5.10), this bound is equivalent to

$$E[H]\geq\frac{2}{N-1}\sum_{k=1}^{N-1}\frac{1}{\mu_k} \tag{5.46}$$

where equality is obtained for a tree as shown in (4.29) of **art.** 127.

6

Spectra of special types of graphs

This chapter presents spectra of graphs that are known in closed form.

6.1 The complete graph

The eigenvalues of the adjacency matrix of the complete graph K_N are $\lambda_1 = N - 1$ and $\lambda_2 = \ldots = \lambda_N = -1$. Since K_N is a regular graph (**art.** 110), the eigenvalues of the Laplacian are, apart from $\mu_N = 0$, all equal to $\mu_j = N$ for $1 \leq j \leq N - 1$.

The adjacency matrix of the complete graph is $A_{K_N} = J - I$ and $J = u.u^T$. A direct computation of the determinant $\det(A_{K_N} - \lambda I)$ in (A.94) is

$$\det(J - I - \lambda I) = \det\left(u.u^T - (\lambda + 1)I\right) = (-(\lambda + 1))^N \det\left(I - \frac{u.u^T}{\lambda + 1}\right)$$

Using the "rank 1 update" formula (A.66) and $u^T u = N$, we obtain

$$\det(J - I - \lambda I) = (-1)^N (\lambda + 1)^{N-1} (\lambda + 1 - N)$$

from which the eigenvalues of the adjacency matrix of the complete graph K_N are immediate. In summary,

$$\det(J - xI)_{n \times n} = (-1)^n x^{n-1} (x - n) \tag{6.1}$$

A computation of $\det(Q_{K_N} - \mu I)$ follows the same determinant manipulations as in the example in **art.** 122, after replacing $N - 1$ by $N - 1 - \mu$, to obtain $\det(Q_{K_N} - \mu I) = -\mu(N - \mu)^{N-1}$.

6.2 A small-world graph

In a small-world graph $G_{SWk;N}$, each node is placed on a ring as illustrated in Fig. 6.1 and has links to precisely k subsequent neighbors and, by the cyclic structure of the ring, also to k previous neighbors. The small-world graph has been proposed by Watts and Strogatz (1998) – and is further discussed in Watts (1999) – to study the effect of adding random links to a regular network or of rewiring links randomly. The thus modified small-world graphs are found to be highly clustered, like regular graphs. As mentioned in Section 1.5, depending on the rewiring process of links, typical paths may have a large hopcount, unlike in random graphs.

The adjacency matrix $A_{SW\,k;N}$ is of the type of a symmetric circulant, Toeplitz matrix whose eigenvalue structure (eigenvalues and eigenvectors) can be exactly determined by the Fourier matrix.

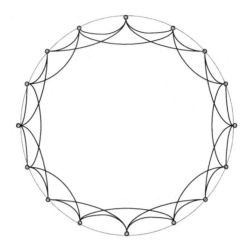

Fig. 6.1. A Watts-Strogatz small-world graph $G_{SW\,k;N}$ with $k = 2$ is a regular graph on $N = 16$ nodes with degree $d = 4$.

6.2.1 The eigenvalue structure of a circulant matrix

A circulant matrix C is an $n \times n$ matrix with the form

$$C = \begin{bmatrix} c_0 & c_{n-1} & c_{n-2} & \cdots & c_1 \\ c_1 & c_0 & c_{n-1} & \cdots & c_2 \\ c_2 & c_1 & c_0 & \ddots & c_3 \\ \vdots & \vdots & \ddots & \ddots & \vdots \\ c_{n-1} & c_{n-2} & c_{n-3} & \cdots & c_0 \end{bmatrix}$$

Each column is precisely the same as the previous one, but the elements are shifted one position down and wrapped around at the bottom. In fact, $c_{jk} = c_{(j-k)\,\mathrm{mod}\,n}$, which shows that diagonals parallel to the main diagonal contain the same elements. The elementary circulant matrix E has all zero elements except for $c_1 = 1$,

$$E = \begin{bmatrix} 0 & 0 & 0 & \cdots & 1 \\ 1 & 0 & 0 & \cdots & 0 \\ 0 & 1 & 0 & \ddots & 0 \\ \vdots & \vdots & \ddots & \ddots & \vdots \\ 0 & 0 & 0 & \cdots & 0 \end{bmatrix}$$

and represents a unit-shift relabeling transformation of nodes: $1 \to 2, 2 \to 3, \ldots, n \to 1$. Thus, the unit-shift relabeling transformation, which is a particular example of a permutation (**art. 31**), maps the vector $x = (x_1, x_2, \ldots, x_n)$ into $Ex = (x_n, x_1, x_2, \ldots, x_{n-1})$. Again applying the unit-shift relabeling transformation maps Ex into $E^2 x = (x_{n-1}, x_n, x_1, \ldots, x_{n-2})$, which is a two-shift relabeling transformation, and

$$
E^2 = \begin{bmatrix}
0 & 0 & 0 & \cdots & 1 & 0 \\
0 & 0 & 0 & \cdots & 0 & 1 \\
1 & 0 & 0 & \ddots & \ddots & 0 \\
0 & 1 & \ddots & \ddots & \ddots & \vdots \\
\vdots & \ddots & \ddots & \ddots & \ddots & 0 \\
0 & \cdots & 0 & 1 & 0 & 0
\end{bmatrix}
$$

Hence, we observe that E^2 equals the circulant matrix C with all $c_j = 0$, except for $c_2 = 1$. In general, E^k represents a k-shift relabeling transformation, where each node label $n_j \to n_{(j+k) \bmod n}$ and E^k equals C with all $c_j = 0$, except for $c_k = 1$. Alternatively, a general circulant matrix C can be decomposed into elementary k-shift relabeling matrices E^k, with $E^0 = I$ and $E^n = I$, as

$$
C = c_0 I + c_1 E + c_2 E^2 + \cdots + c_{n-1} E^{n-1} = \sum_{k=0}^{n-1} c_k E^k
$$

Denoting the polynomial $p(x) = \sum_{k=0}^{n-1} c_k x^k$, we can write that $C = p(E)$.

The eigenstructure of E can be found quite elegantly. Indeed, the eigenvalue equation $Ex = \lambda x$ is equivalent to solving the set, for both λ and each component x_j of the $n \times 1$ eigenvector x,

$$
x_n = \lambda x_1
$$
$$
x_1 = \lambda x_2
$$
$$
x_2 = \lambda x_3
$$
$$
\vdots
$$
$$
x_{n-1} = \lambda x_n
$$

After multiplying all equations, we find $\prod_{j=1}^{n} x_j = \lambda^n \prod_{j=1}^{n} x_j$, from which $\lambda^n = 1$ and $\lambda_k = e^{\frac{2\pi i k}{n}}$, for $k = 0, 1, \ldots, n-1$. The roots of unity $\lambda_k = e^{\frac{2\pi i k}{n}}$ obey $\lambda_k^* = e^{-\frac{2\pi i k}{n}}$, $\lambda_k^* \lambda_k = |\lambda_k|^2 = 1$ and, thus with (A.98) in **art. 235**, we obtain $\det E = \prod_{k=0}^{n-1} \lambda_k = (-1)^{n-1}$. Since any eigenvector is only determined apart from a scaling factor, we may choose the first vector component $x_1 = \alpha$ and, after backsubstitution in the above set, we find that $x_k = \lambda^{1-k} \alpha$ for all $k = 1, \ldots, n-1$ and $x_n = \lambda = \lambda^{1-n}$, because $\lambda^n = 1$. Thus, the eigenvector x of E belonging to the eigenvalue λ_k equals $x = \alpha \left(1, \lambda_k^{-1}, \lambda_k^{-2}, \ldots, \lambda_k^{-n+1}\right)$ and the matrix X containing the eigenvectors of E as column vectors is, with $\xi = e^{-\frac{2\pi i}{n}}$,

$$
X = \alpha \begin{bmatrix}
1 & 1 & 1 & \cdots & 1 \\
1 & \xi & \xi^2 & \cdots & \xi^{n-1} \\
1 & \xi^2 & \xi^4 & \cdots & \xi^{2(n-1)} \\
\vdots & \vdots & \vdots & & \vdots \\
1 & \xi^{n-1} & \xi^{2(n-1)} & \cdots & \xi^{(n-1)(n-1)}
\end{bmatrix}
\tag{6.2}
$$

where $(X)_{kj} = \xi^{(k-1)(j-1)}$. We observe that $X^T = X$. If x and y are the eigenvectors belonging to eigenvalue λ_k and λ_j, respectively, then the inner product $x^H y$ (**art. 246**) is

$$
x^H y = \sum_{l=1}^{n} x_l^* y_l = \alpha^2 \sum_{l=1}^{n} \left(\lambda_k^{1-l}\right)^* \lambda_j^{1-l} = \alpha^2 e^{\frac{2\pi i(j-k)}{n}} \sum_{l=1}^{n} \left(e^{-\frac{2\pi i(j-k)}{n}}\right)^l
$$

$$
= \alpha^2 \sum_{l=0}^{n-1} \left(e^{-\frac{2\pi i(j-k)}{n}}\right)^l = \alpha^2 \frac{1 - e^{-2\pi i(j-k)}}{1 - e^{-\frac{2\pi i(j-k)}{n}}}
$$

Since $e^{2\pi i m} = 1$ for any integer m, we find that $x^H y = 0$ if $k \neq j$, and $x^H y = n\alpha^2$ if $k = j$, which suggests the normalization $\alpha^2 n = 1$. Hence, with $\alpha = \frac{1}{\sqrt{n}}$, we have shown that X in (6.2) is a unitary matrix (**art. 247**), that obeys $X^H X = X X^H = I$. The matrix X is also called the Fourier matrix. The eigenvalue equation, written in terms of the matrix X, is $EX = X\Lambda$, where

$$
\Lambda = \operatorname{diag}\left(1, e^{\frac{2\pi i}{n}}, e^{\frac{4\pi i}{n}}, \ldots, e^{\frac{2\pi i k}{n}}, \ldots, e^{\frac{2\pi i(n-1)}{n}}\right)
$$

$$
= \operatorname{diag}\left(1, \xi^{-1}, \xi^{-2}, \ldots, \xi^{-(n-1)}\right)
$$

Clearly, $\Lambda^n = I$ confirming the general property $P^n = I$ of an $n \times n$ permutation matrix P in **art. 31**. Using the unitary property results, after left multiplication of both sides in $EX = X\Lambda$ by X^H, in

$$
X^H E X = \operatorname{diag}\left(1, \xi^{-1}, \xi^{-2}, \ldots, \xi^{-(n-1)}\right)
$$

and

$$
X^H E^k X = \operatorname{diag}\left(1, \xi^{-k}, \xi^{-2k}, \ldots, \xi^{-(n-1)k}\right)
$$

Since $\det E \neq 0$, the inverse E^{-1} exists and is found as

$$
E^{-1} = X \operatorname{diag}\left(1, \xi, \xi^2, \ldots, \xi^{(n-1)}\right) X^H
$$

Explicitly,

$$
\left(E^{-1}\right)_{kj} = \frac{1}{n} \sum_{m=1}^{n} (X)_{km} \, \xi^{m-1} \left(X^H\right)_{mj} = \frac{1}{n} \sum_{m=1}^{n} \xi^{(k-1)(m-1)+(m-1)-(m-1)(j-1)}
$$

$$
= \frac{1}{n} \sum_{m=0}^{n-1} e^{-\frac{2\pi i(k-j+1)}{n}m} = \frac{1}{n} \frac{1 - e^{-2\pi i(k-j+1)}}{1 - e^{-\frac{2\pi i(k-j+1)}{n}}} = 1_{\{k=j-1\}}
$$

from which we find $E^{-1} = E^{n-1}$, which corresponds to a unit-shift relabeling transformation in the other direction: $1 \to n$, $2 \to 1,\ldots, n \to n-1$.

Finally, the eigenvalue structure of a general circulant matrix $C = p(E) = \sum_{k=0}^{n-1} c_k E^k$ is

$$X^H C X = X^H \left(\sum_{k=0}^{n-1} c_k E^k \right) X = \sum_{k=0}^{n-1} c_k X^H E^k X$$

$$= \sum_{k=0}^{n-1} c_k \mathrm{diag} \left(1, \xi^{-k}, \xi^{-2k}, \ldots, \xi^{-(n-1)k} \right)$$

$$= \mathrm{diag} \left(\sum_{k=0}^{n-1} c_k, \sum_{k=0}^{n-1} c_k \xi^{-k}, \sum_{k=0}^{n-1} c_k \xi^{-2k}, \ldots, \sum_{k=0}^{n-1} c_k \xi^{-(n-1)k} \right)$$

In terms of the polynomial $p(x) = \sum_{k=0}^{n-1} c_k x^k$, we arrive at the eigenvalue decomposition of a general circulant matrix C,

$$X^H C X = \mathrm{diag} \left(p(1), p\left(\xi^{-1}\right), p\left(\xi^{-2}\right), \ldots, p\left(\xi^{-(n-1)}\right) \right) \tag{6.3}$$

Remark: Let $w_k = \frac{1}{\sqrt{n}} \left(1, \lambda_k^{-1}, \lambda_k^{-2}, \ldots, \lambda_k^{-n+1} \right)$ with $\lambda_k = e^{\frac{2\pi i k}{n}}$ be the eigenvector of C belonging to eigenvalue $p\left(\xi^{-k}\right) = p\left(e^{\frac{2\pi i k}{n}}\right)$, so that $C w_k = p\left(e^{\frac{2\pi i k}{n}}\right) w_k$. If C is real and $p\left(e^{\frac{2\pi i k}{n}}\right)$ is real, then separating both real and imaginary parts of w_k in the eigenvalue equation gives us two equations

$$\begin{cases} C\left(\mathrm{Re}\, w_k\right) = p\left(e^{\frac{2\pi i k}{n}}\right) \left(\mathrm{Re}\, w_k\right) \\ C\left(\mathrm{Im}\, w_k\right) = p\left(e^{\frac{2\pi i k}{n}}\right) \left(\mathrm{Im}\, w_k\right) \end{cases}$$

Hence, the real eigenvalue $p\left(e^{\frac{2\pi i k}{n}}\right)$ of the real matrix C possesses two real, orthogonal eigenvectors $\mathrm{Re}\, w_k$ and $\mathrm{Im}\, w_k$, satisfying $\left(\mathrm{Re}\, w_k\right)^T \left(\mathrm{Im}\, w_k\right) = 0$, and must have multiplicity at least two.

6.2.2 The spectrum of a small-world graph

The adjacency matrix $A_{\mathrm{SW}k;N}$ of a small-world graph where each node, placed on a ring, has links to precisely k subsequent and k previous neighbors (see Fig. 6.1), is a symmetric circulant matrix where $c_{N-j} = c_j$ and $c_0 = 0$, $c_j = 1_{\{j \in [1,k]\}}$, where 1_y is the indicator function. Since the degree of each node is $2k$ and the maximum possible degree is $N-1$, the value of k is limited to $k \le \frac{N-1}{2}$. The corresponding polynomial is denoted as $p_{\mathrm{SW}k;N}(z) = \sum_{j=0}^{N-1} c_j z^j$. Since $c_0 = 0$, we have that

$$p_{\mathrm{SW}k;N}(z) = \sum_{j=1}^{N-1} c_j z^j = \sum_{j=1}^{a} c_j z^j + \sum_{j=a+1}^{N-1} c_j z^j$$

Changing the j-index in the last summation to $m = N - j$ yields $\sum_{j=a+1}^{N-1} c_j z^j = \sum_{m=1}^{N-a-1} c_{N-m} z^{N-m}$ and invoking symmetry $c_{N-m} = c_m$, we find, for any integer $a \in [1, N-1]$, that $p_{\mathrm{SW}k;N}(z) = \sum_{j=1}^{a} c_j z^j + \sum_{j=1}^{N-a-1} c_j z^{N-j}$. When choosing $a = k$, the bound $2k + 1 \leq N$ implies that $N - a - 1 = N - k - 1 \geq k$. Introducing $c_j = 1_{\{j \in [1,k]\}}$ and $a = k$, we obtain

$$p_{\mathrm{SW}k;N}(z) = \sum_{j=1}^{k} z^j + z^N \sum_{j=1}^{k} z^{-j} = z\frac{1 - z^k}{1 - z} + z^{N-1}\frac{1 - z^{-k}}{1 - z^{-1}}$$

An eigenvalue $(\lambda_{\mathrm{SW}k;N})_m$ of $A_{\mathrm{SW}k;N}$, belonging to eigenvector

$$w_{m-1} = \frac{1}{\sqrt{n}}\left(1, \lambda_{m-1}^{-1}, \lambda_{m-1}^{-2}, \ldots, \lambda_{m-1}^{-n+1}\right)$$

with $\lambda_k = e^{\frac{2\pi i k}{n}}$, follows for $m = 1, \ldots, N$ from (6.3) as

$$(\lambda_{\mathrm{SW}k;N})_m = p_{\mathrm{SW}k;N}\left(\xi^{1-m}\right) = \xi^{1-m}\frac{1 - \xi^{(1-m)k}}{1 - \xi^{1-m}} + \xi^{(1-m)(N-1)}\frac{1 - \xi^{(m-1)k}}{1 - \xi^{m-1}}$$

$$= 2\,\mathrm{Re}\left(e^{\frac{2\pi i}{N}(m-1)}\frac{1 - e^{\frac{2\pi i}{N}(m-1)k}}{1 - e^{\frac{2\pi i}{N}(m-1)}}\right)$$

After rewriting $\frac{1 - e^{\frac{2\pi i}{N}(m-1)k}}{1 - e^{\frac{2\pi i}{N}(m-1)}} = e^{\frac{\pi i}{N}(m-1)(k-1)}\frac{\sin\left(\frac{\pi(m-1)k}{N}\right)}{\sin\left(\frac{\pi(m-1)}{N}\right)}$, the eigenvalue with index m of $A_{\mathrm{SW}k;N}$ is

$$(\lambda_{\mathrm{SW}k;N})_m = 2\frac{\sin\left(\frac{\pi(m-1)k}{N}\right)}{\sin\left(\frac{\pi(m-1)}{N}\right)}\cos\left(\frac{\pi(m-1)(k+1)}{N}\right)$$

Finally, using $2\sin(\alpha k)\cos(\alpha(k+1)) = \sin(\alpha(2k+1)) - \sin(\alpha)$, the unordered[1] eigenvalues of $A_{\mathrm{SW}k;N}$ are, for $1 \leq m \leq N$,

$$(\lambda_{\mathrm{SW}k;N})_m = \frac{\sin\left(\frac{\pi(m-1)(2k+1)}{N}\right)}{\sin\left(\frac{\pi(m-1)}{N}\right)} - 1 \tag{6.4}$$

We find from (6.4) that the spectral radius $(\lambda_{\mathrm{SW}k;N})_1 = 2k$, which is equal to the degree in a regular graph (**art. 55**). The complete spectrum for $N = 101$ is drawn in Fig. 6.2, which is representative for values of N roughly above 50. Fig. 6.2 illustrates the spectral evolution (as function of k in the abscissa) from a circuit ($k = 1$) towards the complete graph ($k = \left[\frac{N-1}{2}\right]$). If $2k + 1 = N$ in which case $A_{\mathrm{SW}k;N} = J - I$, then we obtain from (6.4) that $\lim_{2k+1 \to N}(\lambda_{\mathrm{SW}k;N})_j = -1$ for all $m \neq 1$, while $(\lambda_{\mathrm{SW}k;N})_1 = N - 1$, agreeing with the computations in Section 6.1.

---

[1] The m-th ordered eigenvalue $(\lambda_{\mathrm{SW}k})_{(m)}$, satisfying $(\lambda_{\mathrm{SW}k})_{(m)} \geq (\lambda_{\mathrm{SW}k})_{(m+1)}$, for $1 \leq m < N$, is not easy to determine as Fig. 6.2 suggests.

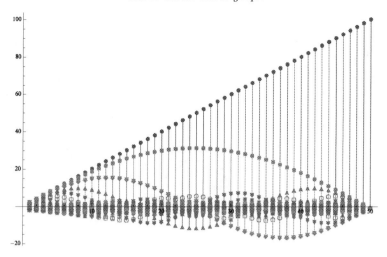

Fig. 6.2. The complete spectrum (6.4) for $N = 101$. The x-axis plots all values of k from 1 to $\left[\frac{n-1}{2}\right] = 50$, and for each k, all $1 \leq m \leq N$ values of $(\lambda_{\text{SW}\,k;N})_m$ are shown.

In terms of the Chebyshev polynomial of the second kind (B.135), we can write (6.4) as

$$(\lambda_{\text{SW}\,k;N})_m = U_{2k}\left(\cos\left(\frac{\pi\,(m-1)}{N}\right)\right) - 1$$

Applying $\sin(x + n\pi) = (-1)^n \sin(x)$, valid for any integer n, we find additional symmetry in the eigenvalue spectrum,

$$(\lambda_{\text{SW}\,k;N})_m = \frac{\sin\left(\frac{\pi(2k+1)\{N-(m-1)\}}{N}\right)}{\sin\left(\frac{\pi\{N-(m-1)\}}{N}\right)} - 1 = (\lambda_{\text{SW}\,k;N})_{N+2-m}$$

for $2 \leq m \leq N$. In general, deducing more symmetry is difficult because, if N is a prime, precisely $\left[\frac{N}{2} + 1\right]$ eigenvalues are distinct for any $k < \frac{N-1}{2}$. Theorem 11 on p. 75 states that the diameter of $G_{\text{SW}\,k;N}$ is at most $\left[\frac{N}{2}\right]$ when N is prime. Fig. 6.3 reflects the irregular dependence of the number of different eigenvalues, which reminds us of the irregular structure of quantities in number theory, such as the number of divisors and the prime number factorization.

Since a small-world graph is a regular graph, the Laplacian $Q_{\text{SW}\,k;N} = 2kI - A_{\text{SW}\,k;N}$ and the corresponding unordered spectrum is (**art. 110**)

$$(\mu_{\text{SW}\,k;N})_{N+1-m} = 2k - (\lambda_{\text{SW}\,k;N})_m = 2k + 1 - \frac{\sin\left(\frac{\pi(m-1)(2k+1)}{N}\right)}{\sin\left(\frac{\pi(m-1)}{N}\right)}$$

As Theorem 20 on p. 119 prescribes for a connected graph, there is precisely one zero eigenvalue $(\mu_{\text{SW}\,k;N})_N = 0$. **Art. 110** demonstrates that $(\mu_{\text{SW}\,k;N})_{N-1}$ equals

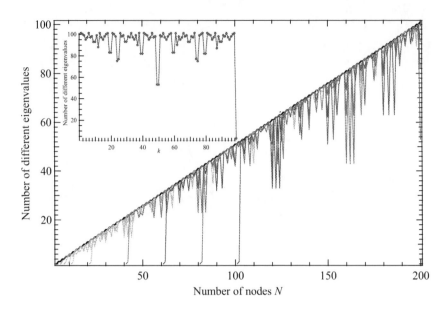

Fig. 6.3. The number of different eigenvalues in $A_{\text{SW}\,k;N}$ as a function of N for different values of $k = 1, 2, 3, 4, 5$ and $k = 10, 20, 30, 40, 50$. The insert shows the number of different eigenvalues for $N = 200$ versus k.

the spectral gap, the difference between the largest and second largest eigenvalue at each k as illustrated in Fig. 6.2.

The largest negative eigenvalue of $(\lambda_{\text{SW}\,k;N})_m$ lies between $\frac{N}{2k+1} < m - 1 < \frac{2N}{2k+1}$ and, by symmetry $m \to N + 2 - m$, $\frac{2kN}{2k+1} > m - 1 > \frac{(2k-1)N}{2k+1}$. Indeed, if we let $x = \frac{\pi(m-1)}{N}$ and $x \in [0, \pi)$, then the function $f(x) = \frac{\sin(2k+1)x}{\sin x} - 1$ has the same derivative as $\tilde{f}(x) = \frac{\sin(2k+1)x}{\sin x}$, which has zeros at $x = \frac{l\pi}{2k+1} \in [0, \pi)$ for $l = 1, 2, \ldots, 2k$. By Rolle's Theorem, $f'(x)$ has always a zero in an interval between two zeros of $f(x)$, because $f(x)$ is continuous. Since $\sin x$ has the same sign in $x \in (0, \pi)$, the largest absolute values of $f(x)$ will occur near $x \to 0$ and $x \to \pi$, where $\sin x$ has zeros. A good estimate for the value at which the largest negative eigenvalue occurs is half of the interval, hence, $m_{\min} = \left[\frac{3N}{2(2k+1)} + 1\right]$. The corresponding eigenvalue is, approximately,

$$(\lambda_{\text{SW}\,k;N})_{m_{\min}} \approx -\frac{1}{\sin\left(\frac{3\pi}{2(2k+1)}\right)} - 1 < -2$$

Numerical values indicate that $m_{\min} = \left[\frac{3N}{2(2k+1)} + 1\right]$ is, in many cases, exact. Hence, the eigenvalues $\lambda_{\text{SW}\,k;N}$ of the adjacency matrix $A_{\text{SW}\,k;N}$ lie in the interval $\left[(\lambda_{\text{SW}\,k;N})_{m_{\min}}, 2k\right]$, and most of them lie in the interval $\left[(\lambda_{\text{SW}\,k;N})_{m_{\min}}, 0\right]$. This interval is, to a good approximation, independent of the size of the graph N, but only a function of the degree of each node, which is $2k$. The approximation fails for the complete graph K_N when $2k + 1 = N$ and $(\lambda_{\text{SW}\,k;N})_{m_{\min}} = -1$.

6.3 A cycle on N nodes

A circuit or cyle C_N is a ring topology on N nodes, where each node on the ring is connected to its previous and subsequent neighbor on the ring. Hence, the circuit is a special case of the small-world graph for $k = 1$. The adjacency matrix of the circuit is $A_C = E + E^{-1} = E + E^T$,

$$
A_C = \begin{bmatrix}
0 & 1 & 0 & \cdots & 0 & 1 \\
1 & 0 & 1 & \cdots & 0 & 0 \\
0 & 1 & 0 & \ddots & 0 & 0 \\
\vdots & \vdots & \ddots & \ddots & \ddots & \vdots \\
0 & 0 & 0 & \cdots & 0 & 1 \\
1 & 0 & 0 & \cdots & 1 & 0
\end{bmatrix}
$$

The eigenvalues of the adjacency matrix A_C of the circuit follow directly from (6.4) as

$$
(\lambda_C)_m = \frac{\sin\left(3\frac{\pi(m-1)}{N}\right)}{\sin\left(\frac{\pi(m-1)}{N}\right)} - 1
$$

Using the identities $\sin 3x = 3\sin x - 4\sin^3 x$ and $1 - 2\sin^2 x = \cos 2x$ yields, for $m = 1, \ldots, N$,

$$
(\lambda_C)_m = 2\cos\left(\frac{2\pi(m-1)}{N}\right) \tag{6.5}
$$

which shows that $(\lambda_C)_m = (\lambda_C)_{N-m+2}$ and that $-2 \le (\lambda_C)_m \le 2$. The lower bound of -2 is only attained if N is even. **Art.** 29 shows that the line graph of the circuit is the circuit itself: $l(C_N) = C_N$. Since the number of links $L = N$ in the circuit C_N, the prefactor $(\lambda + 2)^{L-N}$ in the general expression (2.33) of the characteristic polynomial of a line graph vanishes. Nevertheless, only if N is even, this line graph $l(C_N)$ still has an eigenvalue equal to -2, while all other eigenvalues are larger (**art.** 27).

The real eigenvectors x_m of the adjacency matrix A_C belonging to eigenvalue $(\lambda_C)_m = e^{\frac{2\pi i}{N}(m-1)} + e^{-\frac{2\pi i}{N}(m-1)}$ follow from the eigenvector matrix X of the circulant matrix in (6.2) as mentioned in the remark of Section 6.2.1. Indeed, let us denote the eigenvector x_m belonging to the eigenvalue $\lambda_m = e^{\frac{2\pi i}{N}(m-1)}$ with vector components $(x_m)_k = X_{km} = \xi^{(k-1)(m-1)}$ for $1 \le m, k \le N$ with $\xi = e^{-\frac{2\pi i}{N}}$. Then, the unscaled eigenvectors

$$
x_m = \left(1, e^{-\frac{2\pi i(m-1)}{N}}, e^{-\frac{4\pi i(m-1)}{N}}, e^{-\frac{6\pi i(m-1)}{N}}, \ldots, e^{-\frac{2(N-1)\pi i(m-1)}{N}}\right)
$$

and

$$x_{-m} = x_m^* = \left(1, e^{\frac{2\pi i(m-1)}{N}}, e^{\frac{4\pi i(m-1)}{N}}, e^{\frac{6\pi i(m-1)}{N}}, \ldots, e^{\frac{2(N-1)\pi i(m-1)}{N}}\right)$$

obey the eigenvalue equations (see Section 6.2.1)

$$Ex_m = e^{\frac{2\pi i}{N}(m-1)}x_m$$

$$E^{-1}x_m^* = e^{-\frac{2\pi i}{N}(m-1)}x_m^*$$

We obtain after addition and after addition of the conjugates

$$Ex_m + E^{-1}x_m^* = e^{\frac{2\pi i}{N}(m-1)}x_m + e^{-\frac{2\pi i}{N}(m-1)}x_m^*$$

$$Ex_m^* + E^{-1}x_m = e^{-\frac{2\pi i}{N}(m-1)}x_m^* + e^{\frac{2\pi i}{N}(m-1)}x_m$$

Again, adding and subtracting yields

$$\left(E + E^{-1}\right)(x_m + x_m^*) = \left(e^{\frac{2\pi i}{N}(m-1)} + e^{-\frac{2\pi i}{N}(m-1)}\right)(x_m + x_m^*)$$

$$\left(E + E^{-1}\right)(x_m - x_m^*) = \left(e^{\frac{2\pi i}{N}(m-1)} + e^{-\frac{2\pi i}{N}(m-1)}\right)(x_m - x_m^*)$$

Since $A_C = E + E^{-1}$ and $(\lambda_C)_m = e^{\frac{2\pi i}{N}(m-1)} + e^{-\frac{2\pi i}{N}(m-1)}$, we find that both

$$\frac{\mathrm{Re}(x_m^*)}{\sqrt{N}} = \frac{1}{\sqrt{N}}\left(1, \cos\frac{2\pi(m-1)}{N}, \cos\frac{4\pi(m-1)}{N}, \ldots, \cos\frac{2(N-1)\pi(m-1)}{N}\right)$$

and

$$\frac{\mathrm{Im}(x_m^*)}{\sqrt{N}} = \frac{1}{\sqrt{N}}\left(0, \sin\frac{2\pi(m-1)}{N}, \sin\frac{4\pi(m-1)}{N}, \ldots, \sin\frac{2(N-1)\pi(m-1)}{N}\right)$$

are two real, orthogonal eigenvectors belonging to the same real eigenvalue $(\lambda_C)_m = (\lambda_C)_{N+2-m}$ in (6.5). If $N = 2n+1$ is odd, then with $1 \le m \le n+1$, there are $2n+1$ eigenvectors, because $\frac{\mathrm{Im}(x_0^*)}{\sqrt{N}} = 0$ is never an eigenvector. If $N = 2n$, then the range $1 \le m < n+1$ contains $2n-1$ eigenvectors and $\frac{\mathrm{Re}(x_{n+1}^*)}{\sqrt{N}}$ is the eigenvector belonging to $(\lambda_C)_{n+1} = -2$.

The corresponding Laplacian Q_C possesses the spectrum (**art. 110**),

$$(\mu_C)_{N+1-m} = 2 - 2\cos\left(\frac{2\pi(m-1)}{N}\right) \qquad m = 1, \ldots, N \tag{6.6}$$

and contains the same eigenvectors as the adjacency matrix A_C.

The characteristic polynomial of the circuit C is

$$c_C(\lambda) = \prod_{m=1}^{N}\left(2\cos\left(\frac{2\pi(m-1)}{N}\right) - \lambda\right) = \prod_{m=0}^{N-1}\left(2\cos\left(\frac{2\pi m}{N}\right) - \lambda\right)$$

Since $\prod_{m=0}^{N-1}\left(2\cos\left(\frac{2\pi m}{N}\right) - \lambda\right) = (2-\lambda)\prod_{m=1}^{N-1}\left(2\cos\left(\frac{2\pi m}{N}\right) - \lambda\right)$, we have

$$c_C(\lambda) = \prod_{m=1}^{N}\left(2\cos\left(\frac{2\pi m}{N}\right) - \lambda\right) = (-1)^N 2^N \prod_{m=1}^{N}\left(\frac{\lambda}{2} - \cos\left(\frac{2\pi m}{N}\right)\right)$$

Using the product form $T_n (x) - 1 = 2^{n-1} \prod_{m=1}^{n} \left(x - \cos \left(\frac{2m\pi}{n} \right) \right)$ in (B.128) of the Chebyshev polynomial in (B.127) shows that

$$c_C (\lambda) = 2(-1)^N \left(T_N \left(\frac{\lambda}{2} \right) - 1 \right)$$

6.4 A path of N − 1 hops

6.4.1 The adjacency matrix A_P of the path graph

A path P_N on N nodes, consisting of $N-1$ hops, has an adjacency matrix A_P, where each row has precisely one non-zero element in the upper triangular part. There exists a relabeling transformation that transforms the $N \times N$ adjacency matrix A_P of the path graph P_N on N nodes in a tri-diagonal Toeplitz matrix, where each non-zero element appears on the line parallel and just above the main diagonal. The eigenstructure of the general $N \times N$ tri-diagonal Toeplitz matrix,

$$T_N (a, b, c) = \begin{bmatrix} b & a & & & \\ c & b & a & & \\ & \ddots & \ddots & \ddots & \\ & & c & b & a \\ & & & c & b \end{bmatrix} \tag{6.7}$$

is computed in Van Mieghem (2014, Section A.6.2.1). The matrix $T_N (a, b, c)$ has N distinct eigenvalues λ_m, for $1 \leq m \leq N$,

$$\lambda_m = b + 2\sqrt{ac} \cos \left(\frac{\pi m}{N+1} \right) \tag{6.8}$$

The components $(x_m)_k$ of the eigenvector x_m belonging to λ_m are, for $1 \leq k \leq N$,

$$(x_m)_k = 2\alpha \left(\frac{c}{a} \right)^{\frac{k}{2}} \sin \left(\frac{\pi m k}{N+1} \right) \tag{6.9}$$

Since the eigenvalues are invariant under a similarity transform such as a relabeling transformation (**art.** 239), the complete eigenvalue and eigenvector system of A_{P_N} follows, for $a = c = 1$ and $b = 0$, from the eigenstructure of the general $N \times N$ tri-diagonal Toeplitz matrix for $m = 1, \ldots N$, as

$$\lambda_m (P_N) = 2 \cos \left(\frac{\pi m}{N+1} \right) \tag{6.10}$$

Formula (6.10) shows that $\lambda_m (P_N) = -\lambda_{N+1-m} (P_N)$ and that all eigenvalues of the $N-1$ hops path P_N are strictly smaller than 2, in particular, $-2 < \lambda_m (P_N) < 2$. The largest eigenvalue of the path P_N is the smallest largest adjacency eigenvalue among any connected graph as proved by Lovász and Pelikán (1973). We provide another reasoning: Lemma 10 shows that a tree has the smallest λ_1, because it is the connected graph with the minimum number of links. Further, the tree with minimum maximum degree ($d_{\max} = 2$) and minimum degree variance is the path. According to the bound (3.79) and $L = N - 1$ in any tree, the bounds for the largest eigenvalue of the path satisfies

$$2\left(1 - \frac{1}{N}\right) \leq 2\sqrt{1 - \frac{6}{N}} \leq \lambda_1\left(P_N\right) \leq 2$$

and the lower bound even tends to the upper bound for large N. Any other tree has a larger variance, thus a larger lower bound in (3.79), while also the upper bound $\lambda_1 \leq d_{\max}$ in **art.** 42 is larger, because $d_{\max} > 2$.

The characteristic polynomial of the path is

$$c_{P_N}\left(\lambda\right) = \prod_{m=1}^{N}\left(2\cos\left(\frac{\pi m}{N+1}\right) - \lambda\right) = (-1)^N U_N\left(\frac{\lambda}{2}\right) \tag{6.11}$$

where the Chebyshev polynomial $U_N\left(x\right)$ of the second kind (B.135) has been used.

The characteristic polynomial $c_{P_N}\left(\lambda\right)$ is elegantly derived by Harary *et al.* (1971) from the corresponding generating function $\varphi_P\left(z\right) = \sum_{N=0}^{\infty} c_{P_N}\left(\lambda\right) z^N$. The recursion

$$c_{P_N}\left(\lambda\right) = -\lambda c_{P_{N-1}}\left(\lambda\right) - c_{P_{N-2}}\left(\lambda\right) \tag{6.12}$$

follows directly from (3.106). Since $c_{P_1}\left(\lambda\right) = -\lambda$ and $c_{P_2}\left(\lambda\right) = \lambda^2 - 1$ and defining $c_{P_0}\left(\lambda\right) = 1$, we multiply both sides of the recursion (6.12) by z^N and sum over $N \geq 2$,

$$\sum_{N=2}^{\infty} c_{P_N}\left(\lambda\right) z^N = -\lambda \sum_{N=2}^{\infty} c_{P_{N-1}}\left(\lambda\right) z^N - \sum_{N=2}^{\infty} c_{P_{N-2}}\left(\lambda\right) z^N$$

In terms of $\varphi_P\left(z\right) = \sum_{N=0}^{\infty} c_{P_N}\left(\lambda\right) z^N$, we obtain the generating function

$$\varphi_P\left(z\right) = \frac{1}{1 + \lambda z + z^2}$$

The generating function $\sum_{k=0}^{\infty} U_k\left(x\right) z^k = \frac{1}{1-2zx+z^2}$ in (B.138) again leads to (6.11).

6.4.2 The Laplacian matrix Q_P of the path graph

After a suitable relabeling, the Laplacian Q_P of an $N - 1$ hops path is, except for the first and last row, a Toeplitz matrix,

$$Q_P = \begin{bmatrix} 1 & -1 & & & & \\ -1 & 2 & -1 & & & \\ & \ddots & \ddots & \ddots & & \\ & & -1 & 2 & -1 \\ & & & -1 & 1 \end{bmatrix} \tag{6.13}$$

We compute here the eigenstructure of Q_P analogous to the derivation of the eigenstructure of the general $N \times N$ tri-diagonal Toeplitz matrix in Van Mieghem (2014, Section A.6.2.1) by considering a pseudo tri-diagonal Toeplitz matrix

$$
\tilde{T} = \begin{bmatrix} d & f & & & \\ c & b & a & & \\ & \ddots & \ddots & \ddots & \\ & & c & b & a \\ & & & g & h \end{bmatrix}
\tag{6.14}
$$

6.4.2.1 Laplacian eigenvalues of the path graph

An eigenvector x corresponding to eigenvalue ξ satisfies $\left(\tilde{T} - \xi I\right) x = 0$ or, written per component,

$$
(d - \xi)x_1 + fx_2 = 0
$$
$$
cx_{k-1} + (b - \xi)x_k + ax_{k+1} = 0 \qquad 2 \le k \le N - 1
$$
$$
gx_{N-1} + (h - \xi)x_N = 0
$$

We consider the generating function $G(z) = \sum_{k=1}^{N} x_k z^k$, where all x_k are real because all eigenvalues ξ of $\tilde{T}$ are real (**art. 370**). After multiplying the k-th vector component equation by z^k and summing over all $k \in [2, N-1]$, the above difference equation is transformed into

$$
cz \sum_{k=1}^{N-2} x_k z^k + (b - \xi) \sum_{k=2}^{N-1} x_k z^k + az^{-1} \sum_{k=3}^{N} x_k z^k = 0
$$

and, in terms of $G(z)$,

$$
cz\left(G(z) - x_{N-1}z^{N-1} - x_N z^N\right) + (b-\xi)\left(G(z) - x_1 z - x_N z^N\right) + a\frac{G(z) - x_2 z^2 - x_1 z}{z} = 0
$$

Thus,

$$
\left(cz^2 + (b - \xi)z + a\right) G(z) = cx_{N-1}z^{N+1} + cx_N z^{N+2} + (b - \xi)x_1 z^2
$$
$$
+ (b - \xi)x_N z^{N+1} + ax_2 z^2 + ax_1 z
$$
$$
= x_N z^{N+1} \left\{ c\frac{\xi - h}{g} + (b - \xi) + cz \right\}
$$
$$
+ x_1 z \left\{ a + \left(a\frac{\xi - d}{f} + (b - \xi) \right) z \right\}
$$

where in the last step $x_2 = \frac{\xi - d}{f} x_1$ and $x_{N-1} = \frac{\xi - h}{g} x_N$ have been substituted from the first and last vector component equation. Solving for $G(z)$ yields

$$
G(z) = z\frac{x_N z^N \left\{ z + \left(\frac{1}{g} - \frac{1}{c}\right)\xi + \frac{b}{c} - \frac{h}{g} \right\} + \frac{1}{c}x_1 \left\{ \left(\left(\frac{a}{f} - 1\right)\xi + b - \frac{ad}{f}\right)z + a \right\}}{c(z - r_1)(z - r_2)}
$$

where r_1 and r_2 are the roots of the polynomial $z^2 + \frac{b-\xi}{c}z + \frac{a}{c} = 0$, thus obeying $r_1 + r_2 = \frac{b-\xi}{c}$ and $r_1 r_2 = \frac{a}{c} \neq 0$. These roots r_1 and r_2 cannot be zero and depend on the yet unknown eigenvalue ξ. Since $G(z)$ is a polynomial of order N, the zeros r_1 and r_2 must also be zeros of the numerator

$$t(z) = x_N z^N \left\{ z + \left(\frac{1}{g} - \frac{1}{c} \right) \xi + \frac{b}{c} - \frac{h}{g} \right\} + \frac{1}{c} x_1 \left\{ \left(\left(\frac{a}{f} - 1 \right) \xi + b - \frac{ad}{f} \right) z + a \right\}$$

Proceeding is only possible if the zeros of $t(z)$ are known. With $\beta \neq 0$, we can factor $t(z)$ as

$$t(z) = \left(\beta x_N z^N + \frac{1}{c} x_1 \right) (z + a)$$

whose zeros are $-a$, and $\left(-\frac{x_1}{c\beta x_N} \right)^{1/N} e^{\frac{2\pi i m}{N}}$ for $m = 0, 1, \ldots, N-1$, provided that

$$\begin{cases} \beta = \left(\frac{a}{f} - 1 \right) \xi + b - \frac{ad}{f} \\ \beta \left(\left(\frac{1}{g} - \frac{1}{c} \right) \xi + \frac{b}{c} - \frac{h}{g} \right) = a \end{cases}$$

which is only possible for any eigenvalue ξ if $a = f$ and $g = c$. Then, the above requirement on elements of the pseudo tri-diagonal Toeplitz matrix $\widetilde{T}$ in (6.14) simplifies to

$$\begin{cases} b - d = \beta \\ b - h = \frac{ac}{\beta} \end{cases}$$

or $(b-h)(b-d) = ac$, which is fulfilled for Q_P, where $a = c = -1$, $b = 2$ and $d = h = 1$.

We confine ourselves in the sequel to the Laplacian Q_P, in which case, r_1 and r_2 must be either 1 or $\left(\frac{x_1}{x_N} \right)^{1/N} e^{\frac{2\pi i m}{N}}$ for $m = 0, 1, \ldots, N-1$. If $r_1 = 1$, then also $r_2 = 1$ in which case $\mu = 0$ and $x_1 = x_N$ as follows by raising $\left(\frac{x_1}{x_N} \right)^{1/N} e^{\frac{2\pi i m}{N}} = 1$ to the power N. In that case, $G(z) = \frac{x_N z (z^N - 1)}{(z-1)} = x_N \sum_{k=1}^{N} z^k$ such that the corresponding eigenvector is, indeed, the scaled all-one vector αu with $\alpha = x_N$ (**art. 101**). All positive eigenvalues $\mu > 0$ correspond to distinct zeros $r_1 = \left(\frac{x_1}{x_N} \right)^{1/N} e^{\frac{2\pi i m}{N}}$ for $m = 0, 1, 2, \ldots, N-1$. But, since $r_2 = r_1^{-1}$, the zero r_2 also must be of this form, $r_2 = \left(\frac{x_1}{x_N} \right)^{1/N} e^{\frac{2\pi i n}{N}}$ for some $0 \leq n \neq m \leq N-1$. Thus, the product

$$r_1 r_2 = \left(\frac{x_1}{x_N} \right)^{2/N} e^{\frac{2\pi i (m+n)}{N}} = 1$$

raised to the power N, shows that $\frac{x_1}{x_N} = \pm 1 = e^{i\pi k}$ such that $r_1 = e^{\frac{\pi i (2m+k)}{N}}$ and $r_2 = e^{\frac{\pi i (2n+k)}{N}}$. Requiring that $r_2 = r_1^{-1}$ results in $r_1 = e^{\frac{\pi i (m-n)}{N}}$ and $r_2 = e^{\frac{\pi i (n-m)}{N}} = e^{-\frac{\pi i (m-n)}{N}}$. Now, r_1 changes with m, while r_2 with n. For each $m =$

$1, 2, \ldots, N - 1$, there must correspond to $l = m - n$ in the exponent of r_1, a value $-l = -(m - n)$ in r_2, only by changing $n \neq m$, thus $n = m - l$. The extent over which the integer l can range is $-(N - 1) \leq l \leq N - 1$ and to each l there must correspond a $-l$. Hence, for $l = 1, 2, \ldots, N - 1$, we finally find that $\mu_l = 2 - (r_1 + r_2)$ and

$$\mu_l = 2 - \left(e^{\frac{\pi i l}{N}} + e^{-\frac{\pi i l}{N}}\right) = 2\left(1 - \cos\left(\frac{\pi l}{N}\right)\right) = 4\sin^2\frac{\pi l}{2N}$$

and to $l = 0$, the case $r_1 = r_2 = 1$ corresponds with $\mu_0 = 0$. In summary, the *ordered* Laplacian eigenvalues of the $N - 1$ hops path are

$$(\mu_P)_{N-m} = 2\left(1 - \cos\left(\frac{\pi m}{N}\right)\right) \qquad m = 0, 1, \ldots, N - 1 \qquad (6.15)$$

All Laplacian eigenvalues of the path are simple, while most of the cycle Laplacian eigenvalues in (6.6) have double multiplicity. Moreover, the Laplacian eigenvalues of the path P_N are the same as the Laplacian eigenvalues of the cycle C_{2N}.

6.4.2.2 Laplacian eigenvectors of the path graph

We now determine the Laplacian eigenvectors $z_1, z_2, \ldots, z_{N-1}$ and use the notation $(z_m)_j$ for the j-th component of the eigenvector z_m, where $1 \leq j \leq N$ points towards node j in the path graph P_N. The eigenvector z_m corresponding to $\mu_m = (\mu_P)_{N-m} > 0$ has the generating function $G_m(w) = \sum_{k=1}^{N}(z_m)_k w^k$,

$$G_m(w) = \frac{(z_m)_N (w - 1) w\left(w^N - \frac{(z_m)_1}{(z_m)_N}\right)}{\left(w - e^{\frac{\pi i m}{N}}\right)\left(w - e^{-\frac{\pi i m}{N}}\right)}$$

Invoking **art. 313** to the polynomial

$$p_0(w) = (z_m)_N (w - 1) w\left(w^N - \frac{(z_m)_1}{(z_m)_N}\right)$$
$$= (z_m)_N w^{N+2} - (z_m)_N w^{N+1} - (z_m)_1 w^2 + (z_m)_1 w$$

yields, with $p_0(w) = \sum_{j=0}^{N+2} a_j w^j$ where all coefficients $a_j = 0$ are zero, except for $a_{N+2} = -a_{N+1} = (z_m)_N$ and $a_1 = -a_2 = (z_m)_1$,

$$G_m(w) = \frac{1}{e^{\frac{\pi i m}{N}} - e^{-\frac{\pi i m}{N}}} \sum_{k=0}^{N}\left\{\sum_{j=k+1}^{N+2} a_j \left(e^{(j-k-1)\frac{\pi i m}{N}} - e^{-(j-k-1)\frac{\pi i m}{N}}\right)\right\} w^k$$

$$= \frac{1}{\sin\frac{\pi m}{N}} \sum_{k=0}^{N}\left\{\sum_{j=k+1}^{N+2} a_j \sin\left(\frac{\pi m}{N}(j - k - 1)\right)\right\} w^k$$

Equating corresponding powers in w in $G_m(w) = \sum_{k=1}^{N}(z_m)_k w^k$ yields $(z_m)_k = \frac{1}{\sin\frac{\pi m}{N}} \sum_{j=k+1}^{N+2} a_j \sin\left(\frac{\pi m}{N}(j - k - 1)\right)$ for $0 \leq k \leq N$. Since $(z_m)_0 = 0$, introducing the coefficients a_j reduces to

$$0 = \sum_{j=1}^{N+2} a_j \sin\left(\frac{\pi m}{N}(j-1)\right) = (z_m)_N \sin\left(\pi m + \frac{\pi m}{N}\right) - (z_m)_1 \sin\left(\frac{\pi m}{N}\right)$$

such that $(z_m)_N = (-1)^m (z_m)_1$. For $k > 0$ and $a_{N+2} = -a_{N+1} = (z_m)_N$, we have that

$$(z_m)_k = \frac{1}{\sin\frac{\pi m}{N}} \sum_{j=k+2}^{N+2} a_j \sin\left(\frac{\pi m}{N}(j-k-1)\right)$$

$$= \frac{(z_m)_N}{\sin\frac{\pi m}{N}} \left\{\sin\left(\frac{\pi m}{N}(N-k+1)\right) - \sin\left(\frac{\pi m}{N}(N-k)\right)\right\}$$

$$= \frac{2(z_m)_N}{\sin\frac{\pi m}{N}} \sin\left(\frac{\pi m}{2N}\right) \cos\left(\frac{\pi m}{2N}(2N-2k+1)\right)$$

Using $(z_m)_N = (-1)^m (z_m)_1$, we finally find the k-th component of the eigenvector z_m belonging to the eigenvalue $\mu_m = (\mu_P)_{N-m} > 0$,

$$(z_m)_k = \frac{(z_m)_1}{\cos\frac{\pi m}{2N}} \cos\left(\frac{\pi m}{2N}(2k-1)\right)$$

A proper normalization of the eigenvectors, obeying $z_k^T z_m = \delta_{km}$ as in **art. 247**, is readily obtained for $1 \le m \le N-1$ as

$$(z_m)_k = \sqrt{\frac{2}{N}} \cos\frac{\pi m}{2N}(2k-1) \text{ for } 1 \le k \le N \qquad (6.16)$$

which illustrates the Laplacian property in **art. 103** of more oscillations in the eigenvector z_m with higher frequencies m. For $m = 0$ in (6.15), the eigenvector belonging to $\mu_N = 0$ is $z_N = \frac{u}{\sqrt{N}}$.

6.4.3 The pseudoinverse matrix $Q^\dagger$ of the Laplacian of the path graph

For a path graph with equal link weights b, the weighted Laplacian in (6.13) equals $\widetilde{Q} = bQ_P$. The positive eigenvalues μ_k of the weighted Laplacian $\widetilde{Q}$ of the path graph follow from (6.15) as $(\widetilde{\mu}_P)_{N-k} = 2b\left(1 - \cos\left(\frac{\pi k}{N}\right)\right) = 4b\sin^2\left(\frac{\pi k}{2N}\right)$ where $1 \le k \le N-1$ (and $\mu_0 = (\widetilde{\mu}_P)_N = 0$). The normalized eigenvector elements of the Laplacian Q_P of the path graph, corresponding to $\mu_k = (\widetilde{\mu}_P)_{N-k}$, are specified in (6.16). The elements of the pseudoinverse of the path graph Laplacian follow from (4.30) as

$$Q_{ij}^\dagger = \sum_{k=1}^{N-1} \frac{(z_k)_i (z_k)_j}{\mu_k} = \frac{1}{Nb} \sum_{k=1}^{N-1} \frac{\cos\left(\frac{\pi ki}{N} - \frac{\pi k}{2N}\right)\cos\left(\frac{\pi kj}{N} - \frac{\pi k}{2N}\right)}{1 - \cos\left(\frac{\pi k}{N}\right)}$$

$$= \frac{1}{2Nb} \sum_{k=1}^{N-1} \frac{\cos\left(\frac{\pi k(i+j-1)}{N}\right) + \cos\left(\frac{\pi k(i-j)}{N}\right)}{1 - \cos\left(\frac{\pi k}{N}\right)}$$

which is rewritten as (6.19) by

Theorem 38 *If we define the trigonometric sum*

$$q_N(m) = \sum_{k=1}^{N-1} \frac{\cos\left(\frac{\pi km}{N}\right)}{1 - \cos\left(\frac{\pi k}{N}\right)} = \frac{1}{2}\sum_{k=1}^{N-1} \frac{\cos\left(\frac{\pi km}{N}\right)}{\sin^2\left(\frac{\pi k}{2N}\right)} \qquad (6.17)$$

which is an even function in m, i.e. $q_N(m) = q_N(-m)$ and equal, for $0 \le m \le 2N$, to

$$q_N(m) = \frac{m^2}{2} - m\left(N + \frac{1}{2}\right) + \frac{(2m+1) + (-1)^{m+1}}{4} + \frac{N^2 - 1}{3} \qquad (6.18)$$

then we can compactly express each element (i, j) of the symmetric pseudoinverse matrix $Q^\dagger_{path}$ of the path graph as

$$(Q_{path})^\dagger_{ij} = \frac{1}{2Nb}\{q_N(i+j-1) + q_N(i-j)\} \qquad (6.19)$$

In fact, (6.19) shows that the symmetric pseudoinverse matrix $Q^\dagger$ is the sum of two symmetric matrices Q_1 and Q_2, where all elements in Q_1 along parallels of the anti-diagonal are the same, whereas all elements in Q_2 along parallels of the diagonal are the same. Since $\cos\left(\frac{\pi k}{N}(m + 2jN)\right) = \cos\left(\frac{\pi k}{N}m + 2\pi kj\right) = \cos\left(\frac{\pi k}{N}m\right)$ for any integer j, we find periodicity $q_N(m) = q_N(m + 2jN)$ in N. Invoking an identity, proved in (Van Mieghem *et al.*, 2017, Appendix), we have

$$q_N(0) = \frac{1}{2}\sum_{k=1}^{N-1} \frac{1}{\sin^2\left(\frac{\pi k}{2N}\right)} = \frac{N^2 - 1}{3} \qquad (6.20)$$

which is the maximum value of $q_N(m)$, because $|q_N(m)| \le \frac{1}{2}\sum_{k=1}^{N-1} \frac{\left|\cos\left(\frac{\pi km}{N}\right)\right|}{\sin^2\left(\frac{\pi k}{2N}\right)} \le \frac{1}{2}\sum_{k=1}^{N-1} \frac{1}{\sin^2\left(\frac{\pi k}{2N}\right)} = q_N(0)$.

Proof of Theorem 38: The trigonometric sum $q_N(m)$ in (6.17) is evaluated by first deriving a difference equation for $q_N(m)$, which is then solved.

(a) Difference equation for $q_N(m)$. Using

$$\cos\left(\frac{\pi km}{N}\right) - \cos\left(\frac{\pi k(m-1)}{N}\right) = -2\sin\left(\frac{\pi km}{N} - \frac{\pi k}{2N}\right)\sin\left(\frac{\pi k}{2N}\right)$$

the difference $\Delta q_N(m) = q_N(m) - q_N(m-1)$ is

$$\Delta q_N(m) = \frac{1}{2}\sum_{k=1}^{N-1} \frac{\cos\left(\frac{\pi km}{N}\right) - \cos\left(\frac{\pi k(m-1)}{N}\right)}{\sin^2\left(\frac{\pi k}{2N}\right)} = -\sum_{k=1}^{N-1} \frac{\sin\left(\frac{\pi km}{N} - \frac{\pi k}{2N}\right)}{\sin\left(\frac{\pi k}{2N}\right)}$$

from which we find that $\Delta q_N(m)|_{m=0} = q_N(0) - q_N(-1) = q_N(0) - q_N(1) = N-1$. Observing that

$$\sin\left(\frac{\pi k(m+1)}{N} - \frac{\pi k}{2N}\right) - \sin\left(\frac{\pi km}{N} - \frac{\pi k}{2N}\right) = 2\sin\left(\frac{\pi k}{2N}\right)\cos\left(\frac{\pi km}{N}\right)$$

the second order difference $\Delta^2 q_N\,(m+1) = \Delta q_N\,(m+1) - \Delta q_N\,(m) = q_N\,(m+1) - 2q_N\,(m) + q_N\,(m-1)$ is

$$\Delta^2 q_N\,(m+1) = -\sum_{k=1}^{N-1} \frac{\sin\left(\frac{\pi k(m+1)}{N} - \frac{\pi k}{2N}\right) - \sin\left(\frac{\pi k m}{N} - \frac{\pi k}{2N}\right)}{\sin\left(\frac{\pi k}{2N}\right)} = -2\sum_{k=1}^{N-1} \cos\left(\frac{\pi k m}{N}\right)$$

Taking the real part of the geometric sum $\sum_{k=0}^{N-1} e^{ikx} = \frac{e^{iNx}-1}{e^{ix}-1} = e^{\frac{i(N-1)x}{2}}\frac{\sin\left(\frac{Nx}{2}\right)}{\sin\left(\frac{x}{2}\right)}$,

$$\sum_{k=0}^{N-1} \cos kx = \frac{\sin\left(x\left(N-\frac{1}{2}\right)\right)}{2\sin\left(\frac{x}{2}\right)} + \frac{1}{2} \tag{6.21}$$

and evaluating at $x = \frac{\pi m}{N} > 0$ yields $\sum_{k=1}^{N-1} \cos k\left(\frac{\pi m}{N}\right) = \frac{1}{2}\left((-1)^{m-1} - 1\right)$, while $\sum_{k=1}^{N-1} \cos k\left(\frac{\pi m}{N}\right) = N-1$ for $m=0$. The second order difference $\Delta^2 q_N\,(m+1) = q_N\,(m+1) - 2q_N\,(m) + q_N\,(m-1)$ becomes for $m \neq 0$,

$$q_N\,(m+1) - 2q_N\,(m) + q_N\,(m-1) = 1 + (-1)^m \tag{6.22}$$

while, for $m = 0$,

$$q_N\,(1) - 2q_N\,(0) + q_N\,(-1) = 2\,(q_N\,(1) - q_N\,(0)) = -2\,(N-1) \tag{6.23}$$

(b) Solving the difference equation (6.22) for $q_N\,(m)$. The general solution of the difference equation $q_N\,(m+1) - 2q_N\,(m) + q_N\,(m-1) = f\,(m)$ for integers $m \neq 0$ and an arbitrary function of $f\,(m)$ can be deduced with generating functions,

$$T\,(z) = \sum_{m=0}^{\infty} q_N\,(m)\, z^m \tag{6.24}$$

After multiplying both sides by z^m and summing over all $m > 0$, the difference equation becomes

$$\sum_{m=1}^{\infty} q_N\,(m+1)\, z^m - 2\sum_{m=1}^{\infty} q_N\,(m)\, z^m + \sum_{m=1}^{\infty} q_N\,(m-1)\, z^m = \sum_{m=1}^{\infty} f\,(m)\, z^m$$

Written in terms of $F\,(z) = \sum_{m=0}^{\infty} f\,(m)\, z^m$ and $T\,(z)$,

$$\frac{1}{z}\,(T\,(z) - q_N\,(1)\, z - q_N\,(0)) - 2\,(T\,(z) - q_N\,(0)) + zT\,(z) = F\,(z) - f\,(0)$$

Rearranged,

$$\left(\frac{1}{z} - 2 + z\right) T\,(z) = F\,(z) + \{q_N\,(1) - f\,(0) - 2q_N\,(0)\} + \frac{q_N\,(0)}{z}$$

invoking the conditions (6.23) and $q_N\,(0) - (N-1) = q_N\,(1)$ yields

$$T\,(z) = \frac{z}{(1-z)^2} F\,(z) - \{(N-1) + q_N\,(0) + f\,(0)\} \frac{z}{(1-z)^2} + \frac{q_N\,(0)}{(1-z)^2}$$

After expanding the Taylor series around $z = 0$ and using the Cauchy product $\sum_{m=0}^{\infty} f_m z^m \sum_{m=0}^{\infty} g_m z^m = \sum_{m=0}^{\infty} \left(\sum_{k=0}^{m} f_{m-k} g_k \right) z^m$, we obtain

$$T(z) = \sum_{m=0}^{\infty} \left\{ \sum_{k=0}^{m} k f(m-k) - m \{(N-1) + q_N(0) + f(0)\} + q_N(0)(m+1) \right\} z^m$$

where we have used the derivative of the geometric series, $\frac{d}{dz} \frac{1}{1-z} = \frac{1}{(1-z)^2} = \sum_{m=1}^{\infty} m z^{m-1}$. Equating corresponding powers in z yields the general solution as a function of $f(m)$ and the initial conditions $q_N(0)$,

$$q_N(m) = \sum_{k=0}^{m} k f(m-k) - m \{(N-1) + f(0)\} + q_N(0) \qquad (6.25)$$

For $f(m) = 1 + (-1)^m$, the sum in the general solution (6.25) becomes

$$\sum_{k=0}^{m} k f(m-k) = \sum_{k=0}^{m} k \left(1 + (-1)^{k-m}\right) = \frac{m(m+1)}{2} + \frac{(2m+1) + (-1)^{m+1}}{4}$$

and (6.25) reduces to

$$q_N(m) = \frac{m(m+1)}{2} - m(N+1) + \frac{(2m+1) + (-1)^{m+1}}{4} + q_N(0)$$

Finally, with (6.20), we arrive at (6.18), for $0 \leq m \leq 2N$ due to periodicity and $q_N(-m) = q_N(m)$. $\qquad \square$

6.5 A path of h hops

A path of $h > 0$ hops/links in a graph with N nodes has h non-zero rows with one non-zero element in the upper triangular part. After a similarity transform, the corresponding adjacency matrix can be transformed into

$$A_{h\text{-hop path}} = \begin{bmatrix} (A_P)_{(h+1) \times (h+1)} & O_{(h+1) \times (N-h-1)} \\ O_{(N-h-1) \times (h+1)} & O_{(N-h-1) \times (N-h-1)} \end{bmatrix}$$

where A_P is the tri-diagonal Toeplitz adjacency matrix of an h hops path in a graph with $h+1$ nodes. Invoking (6.10), the spectrum of an h hops path possesses a zero eigenvalue of multiplicity $N - h - 1$ and $h + 1$ eigenvalues for $k = 1, \ldots, h+1$,

$$(\lambda_{h\text{-hop path}})_k = 2 \cos \left(\frac{\pi k}{h+2} \right)$$

6.6 The wheel W_{N+1}

The wheel graph W_{N+1} is the graph obtained by adding to the circuit graph one central node n with links or "spokes" to each node of the circuit. Thus, the wheel graph is the cone of the circuit graph. The adjacency matrix is a special case of **art. 85**,

$$A_W = \begin{bmatrix} A_C & u_{N \times 1} \\ u_{1 \times N} & 0 \end{bmatrix}$$

Since u is an eigenvector of A_C belonging to $\lambda_C = 2$ because the circuit is a regular graph, all eigenvalues of A_C are the same as those of $A_{W_{N+1}}$, except for the largest eigenvalue $\lambda_C = 2$, which is replaced by two new ones, $1 \pm \sqrt{1+N}$, as derived in **art. 85**. Hence, the spectrum of the wheel with $N + 1$ nodes is $-\sqrt{1+N}+1$, $\left\{ 2\cos\left(\frac{2\pi(m-1)}{N}\right) \right\}_{2 \le m \le N}$ and $1 + \sqrt{1+N}$.

The Laplacian spectrum follows from **art. 166** and (6.6) as, $(\mu_W)_{N+1} = 0$, $(\mu_W)_1 = N + 1$ and $(\mu_W)_{N+2-m} = 3 - 2\cos\left(\frac{2\pi(m-1)}{N}\right)$ for $m = 2, \ldots, N$.

6.7 The complete bipartite graph $K_{m,n}$

The complete bipartite graph $K_{m,n}$ consists of two sets $\mathcal{M}$ and $\mathcal{N}$ with $m = |\mathcal{M}|$ and $n = |\mathcal{N}|$ nodes respectively, where each node of one set is connected to all other nodes of the other set. There are no links between nodes of a same set. The adjacency matrix of $K_{m,n}$ is, with $N = m + n$,

$$A_{K_{m,n}} = \begin{bmatrix} O_{m \times m} & J_{m \times n} \\ J_{n \times m} & O_{n \times n} \end{bmatrix} \tag{6.26}$$

and the characteristic polynomial is

$$\det\left(A_{K_{m,n}} - \lambda I\right) = \begin{vmatrix} -\lambda I_{m \times m} & J_{m \times n} \\ J_{n \times m} & -\lambda I_{n \times n} \end{vmatrix}$$

Invoking (A.57) and $J_{k \times n}J_{n \times l} = nJ_{k \times l}$ gives

$$\det\left(A_{K_{m,n}} - \lambda I\right) = (-\lambda)^m \det\left(-\lambda I_{n \times n} + \frac{1}{\lambda}J_{n \times m}J_{m \times n}\right)$$

$$= (-\lambda)^m \det\left(\frac{m}{\lambda}J - \lambda I\right)_{n \times n} = (-\lambda)^m \left(\frac{m}{\lambda}\right)^n \det\left(J - \frac{\lambda^2}{m}I\right)_{n \times n}$$

Using $\det(J - xI)_{n \times n} = (-1)^n x^{n-1}(x - n)$ in (6.1), the characteristic polynomial of $K_{m,n}$ is

$$\det\left(A_{K_{m,n}} - \lambda I\right) = (-1)^{m+n-1}\lambda^{m+n-2}\left(\lambda^2 - mn\right) \tag{6.27}$$

from which the eigenvalues[2] follow as $-\lambda_{\max}$, $[0]^{N-2}$ and $\lambda_{\max} = \sqrt{mn}$. With $N = m + n$, the spectrum of a star topology $K_{1,n}$ for $m = 1$ is $-\sqrt{N-1}$, $[0]^{N-2}$ and $\lambda_{\max} = \sqrt{N-1}$.

---

[2] We denote the multiplicity m of an eigenvalue λ by $[\lambda]^m$

The Laplacian of the complete bipartite graph $K_{m,n}$ is

$$Q_{K_{m,n}} = \begin{bmatrix} nI_{m \times m} & -J_{m \times n} \\ -J_{n \times m} & mI_{n \times n} \end{bmatrix}$$

and the characteristic polynomial is

$$\det \left(Q_{K_{m,n}} - \mu I \right) = \begin{vmatrix} (n - \mu) I_{m \times m} & -J_{m \times n} \\ -J_{n \times m} & (m - \mu) I_{n \times n} \end{vmatrix}$$

A derivation similar to the above results in

$$\det \left(Q_{K_{m,n}} - \mu I \right) = (m - \mu)^{n-1} (n - \mu)^{m-1} ((m - \mu)(n - \mu) - nm)$$

The eigenvalues of $Q_{K_{m,n}}$ are $0, [m]^{n-1}, [n]^{m-1}$ and $m + n = N$. In the case of the star $K_{1,n}$, the eigenvalues of $Q_{K_{1,n}}$ are $0, [1]^{n-1}$ and $n + 1$. The complexity $\xi(N)$, the number of trees in $K_{m,n}$, is found from (4.27) as

$$\xi_{K_{m,n}}(N) = \frac{1}{N} \prod_{k=1}^{N-1} \mu_k = m^{n-1} n^{m-1}$$

and clearly, for the star where $m = 1$, $\xi_{K_{1,n}}(N) = 1$.

The eigenvector $x_1^T = \begin{bmatrix} x_{C_{m \times 1}} & x_{B_{n \times 1}} \end{bmatrix}^T$ belonging to the largest eigenvalue $\lambda_{\max} = \sqrt{mn}$ obeys

$$\begin{bmatrix} O_{m \times m} & J_{m \times n} \\ J_{n \times m} & O_{n \times n} \end{bmatrix} \begin{bmatrix} x_{C_{m \times 1}} \\ x_{B_{n \times 1}} \end{bmatrix} = \begin{bmatrix} J_{m \times n} x_B \\ J_{n \times m} x_C \end{bmatrix} = \lambda_{\max} \begin{bmatrix} x_C \\ x_B \end{bmatrix}$$

Thus, the k-th component of x_C must satisfy $\sum_{j=1}^{n} (x_B)_j = \sqrt{mn} \, (x_C)_k$ for $1 \le k \le m$ and similarly, the j-th component of x_B must satisfy $\sum_{k=1}^{m} (x_C)_k = \sqrt{mn} \, (x_B)_j$ for $1 \le j \le n$. This implies that all vector components of x_C and x_B are the same, i.e. $(x_C)_k = x_c$ for $1 \le k \le m$ and $(x_B)_j = x_b$ for $1 \le j \le n$. The eigenvalue equation simplifies to

$$\begin{cases} nx_b = \sqrt{mn} x_c \\ mx_c = \sqrt{mn} x_b \end{cases}$$

whose unscaled solution is $x_b = 1$ and $x_c = \sqrt{\frac{n}{m}}$. Finally, normalizing the eigenvector so that $x_1^T x_1 = 1$ yields

$$x_1^T = \begin{bmatrix} \frac{1}{\sqrt{2m}} u_{m \times 1} & \frac{1}{\sqrt{2n}} u_{n \times 1} \end{bmatrix}^T \tag{6.28}$$

Alternatively, we may solve the block eigenvalue equations by left-multiplying $J_{m \times n} x_B = \lambda x_C$ by $J_{n \times m}$. Using $J_{k \times n} J_{n \times l} = n J_{k \times l}$ yields $m J_{n \times n} x_B = \lambda J_{n \times m} x_C$. With the second equation $J_{n \times m} x_C = \lambda x_B$, we find that $J_{n \times n} x_B = \frac{\lambda^2}{m} x_B$. The eigenvector of $J_{n \times n}$ belonging to the only non-zero eigenvalue n is u. Hence, $x_B = u_{n \times 1}$ and $\frac{\lambda^2}{m} = n$ or $\lambda = \pm\sqrt{mn}$. Substituted into the first equation gives $x_C = \frac{1}{\sqrt{mn}} J_{m \times n} u_{n \times 1} = \sqrt{\frac{n}{m}} u_{m \times 1}$. After normalization, we arrive again at (6.28) for $\lambda = \sqrt{mn}$. This approach also leads to the normalized eigenvector belonging to $\lambda = -\sqrt{mn}$,

$$x_N^T = \left[\begin{array}{cc} \frac{1}{\sqrt{2m}} u_{m \times 1} & -\frac{1}{\sqrt{2n}} u_{n \times 1} \end{array} \right]^T \tag{6.29}$$

It follows from $A^k = \sum_{m=1}^{N} \lambda_m^k x_m x_m^T$ in (3.19) that

$$A_{K_{m,n}}^k = \lambda_1^k \left\{ x_1 x_1^T + (-1)^k x_N x_N^T \right\}$$

Introducing the eigenvectors in (6.28) and (6.29),

$$A_{K_{m,n}}^k = (mn)^{\frac{k}{2}} \left\{ \left[\begin{array}{cc} \frac{J_{m \times m}}{2m} & \frac{J_{m \times n}}{2\sqrt{mn}} \\ \frac{J_{n \times m}}{2\sqrt{mn}} & \frac{J_{n \times n}}{2n} \end{array} \right] + (-1)^k \left[\begin{array}{cc} \frac{J_{m \times m}}{2m} & -\frac{J_{m \times n}}{2\sqrt{mn}} \\ -\frac{J_{n \times m}}{2\sqrt{mn}} & \frac{J_{n \times n}}{2n} \end{array} \right] \right\}$$

leads to an explicit form for the integer powers k of the adjacency matrix of the complete bipartite graph $K_{m,n}$

$$A_{K_{m,n}}^k = (mn)^{\frac{k-1}{2}} \left[\begin{array}{cc} \frac{1+(-1)^k}{2} \sqrt{\frac{n}{m}} J_{m \times m} & \frac{1-(-1)^k}{2} J_{m \times n} \\ \frac{1-(-1)^k}{2} J_{n \times m} & \frac{1+(-1)^k}{2} \sqrt{\frac{m}{n}} J_{n \times n} \end{array} \right] \tag{6.30}$$

illustrating the alternating bipartite structure for even k, which is a general property of bipartite graphs as shown in Section 6.8.

6.8 A general bipartite graph

6.8.1 Undirected bipartite graph

Instead of connecting each of the $m \leq n$ nodes in the set $\mathcal{M}$ to each of the n nodes in the other set $\mathcal{N}$, we may consider an arbitrary linking between the two sets represented by a matrix $R_{m \times n}$, resulting in a general bipartite graph $B_{m,n}$ with adjacency matrix

$$A_{B_{m,n}} = \left[\begin{array}{cc} O_{m \times m} & R_{m \times n} \\ R_{n \times m}^T & O_{n \times n} \end{array} \right]$$

Using (A.59) when $m \leq n$, the characteristic polynomial is

$$\det \left(A_{B_{m,n}} - \lambda I \right) = (-\lambda)^n \det \left(-\lambda I_{m \times m} + \frac{1}{\lambda} R_{m \times n} R_{n \times m}^T \right)$$

$$= (-\lambda)^{n-m} \det \left(R_{m \times n} R_{n \times m}^T - \lambda^2 I_{m \times m} \right)$$

while, using (A.57) when $m > n$, we obtain

$$\det \left(A_{B_{m,n}} - \lambda I \right) = (-\lambda)^{m-n} \det \left(R_{n \times m}^T R_{m \times n} - \lambda^2 I_{n \times n} \right)$$

These two forms for $m \leq n$ and $m > n$ are an illustration of Lemma 11. In the sequel, we confine to the case where $m \leq n$ without loss of generality.

The singular value decomposition of R is $R = U_{m \times m} (\Sigma_R)_{m \times n} V_{n \times n}^T$, where $\Sigma_R = \text{diag}(\sigma_1, \ldots, \sigma_m, 0, \ldots, 0)$, because the rank of R cannot be larger than $m \leq n$

and where U and V are orthonormal matrices (**art.** 247). From $UU^T = I$ and $RR^T = U_{m \times m} \Sigma_{m \times m}^2 U_{m \times m}^T$, we see that

$$\det \left(R_{m \times n} R_{n \times m}^T - \lambda^2 I_{m \times m} \right) = \det \left(U_{m \times m} \left(\Sigma_{m \times m}^2 - \lambda^2 I \right) U_{m \times m}^T \right) = \prod_{j=1}^{m} \left(\sigma_j^2 - \lambda^2 \right)$$

Hence, the spectrum of the general bipartite graph is

$$\det \left(A_{B_{m,n}} - \lambda I \right) = (-1)^{m-n} \lambda^{n-m} \prod_{j=1}^{m} \left(\sigma_j^2 - \lambda^2 \right)$$

which show that, apart from the zero eigenvalues, it is completely determined by the singular values of R, because $\lambda = \pm \sigma_j$ for $j = 1, \ldots, m$.

Since

$$A_{B_{m,n}}^2 = \begin{bmatrix} R_{m \times n} R_{n \times m}^T & O_{m \times n} \\ O_{n \times m} & R_{n \times m}^T R_{m \times n} \end{bmatrix}$$

and, further for any integer $k \geq 1$,

$$A_{B_{m,n}}^{2k} = \begin{bmatrix} \left(R_{m \times n} R_{n \times m}^T \right)^k & O_{m \times n} \\ O_{n \times m} & \left(R_{n \times m}^T R_{m \times n} \right)^k \end{bmatrix}$$

and

$$A_{B_{m,n}}^{2k+1} = \begin{bmatrix} O_{m \times m} & \left(R_{m \times n} R_{n \times m}^T \right)^k R_{m \times n} \\ \left(R_{n \times m}^T R_{m \times n} \right)^k R_{n \times m}^T & O_{n \times n} \end{bmatrix}$$

the even powers $A_{B_{m,n}}^{2k}$ are reducible non-negative matrices (**art.** 268), while the odd powers again represent a "bipartite" matrix structure.

6.8.2 Directed bipartite graph

A general *directed* bipartite graph BG has an adjacency matrix,

$$A_{BG} = \begin{bmatrix} O_{m \times m} & B_{m \times n} \\ C_{n \times m} & O_{n \times n} \end{bmatrix} \tag{6.31}$$

Any tree T on $N = n + m$ nodes can be represented in the form of a levelset. Denote by $\left\{ X_N^{(k)} \right\}$ the k-th levelset of a tree T, which is the set of nodes in the tree T at hopcount k from the root in a graph with N nodes (**art.** 23), and by $X_N^{(k)}$ the number of elements in the set $\left\{ X_N^{(k)} \right\}$. Then, we have $X_N^{(0)} = 1$ because the zeroth level can only contain the root node itself. For all $k > 0$, it holds that $0 \leq X_N^{(k)} \leq N - 1$ and that

$$\sum_{k=0}^{N-1} X_N^{(k)} = N \tag{6.32}$$

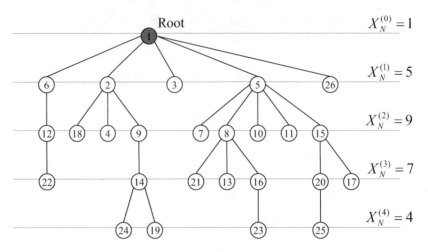

Fig. 6.4. An instance of a tree with $N = 26$ nodes organized per level $0 \leq k \leq 4$. The nodes in the tree are arbitrarily labeled.

Nodes $X_N^{(k)}$ at a same level k are not interconnected. Fig. 6.4 draws a tree organized per level.

The levelset can be folded level by level to form a general bipartite graph. Indeed, the root connects to the nodes $X_N^{(1)}$ at hop 1; those $X_N^{(1)}$ are the ancestors of all the nodes on levelset $X_N^{(2)}$. We may arrange these $X_N^{(2)}$ nodes at the side of the root. Next, these $X_N^{(2)}$ are the ancestors of all $X_N^{(3)}$ nodes, which we move to the other side of the $X_N^{(1)}$ node. In this way, all even levels are placed at the side of the root and all odd levels at the other side, thus creating a general directed bipartite graph. Hence, *the adjacency matrix of any tree can be recast in the form of (6.31)*, where $m = \sum_{k=0}^{\frac{N-1}{2}} X_N^{(2k)}$ and $n = \sum_{k=0}^{\frac{N-2}{2}} X_N^{(2k+1)} = N - m$. In a stochastic setting, where $E\left[X_N^{(k)}\right] = N \Pr\left[H_N = k\right]$, we observe that the average multiplicity of the zero eigenvalue of A_{BG} in (6.33) equals (assuming $n > m$)

$$E\left[n - m\right] = N \sum_{k=0}^{N-1} \Pr\left[H_N = k\right] (-1)^k = N\varphi_{H_N}(-1)$$

where the probability generating function of the hopcount in a random tree is $\varphi_{H_N}(z) = E\left[z^{H_N}\right] = \sum_{k=0}^{N-1} \Pr\left[H_N = k\right] z^k$.

If $x^T = \begin{bmatrix} x_C & x_B \end{bmatrix}^T$ is an eigenvector belonging to eigenvalue λ, which means that

$$\begin{bmatrix} O_{m \times m} & B_{m \times n} \\ C_{n \times m} & O_{n \times n} \end{bmatrix} \begin{bmatrix} x_{C_{m \times 1}} \\ x_{B_{n \times 1}} \end{bmatrix} = \begin{bmatrix} B x_B \\ C x_C \end{bmatrix} = \begin{bmatrix} \lambda x_C \\ \lambda x_B \end{bmatrix}$$

then also $x^T = \begin{bmatrix} x_C & -x_B \end{bmatrix}^T$ is an eigenvector belonging to the eigenvalue $-\lambda$, which shows that the spectrum is symmetric around $\lambda = 0$. The same result can be derived from (A.57) analogously to the spectrum of $A_{B_{m,n}}$ above as

$$\det \left(A_{BG} - \lambda I \right) = (-1)^{m-n} \lambda^{n-m} \det \left(B_{m \times n} C_{n \times m} - \lambda^2 I_{m \times m} \right) \qquad (6.33)$$

Hence, A_{BG} has, at least, $n - m$ zero eigenvalues. Consequently, we have demonstrated:

Theorem 39 *The spectrum of the adjacency matrix of any tree is symmetric around* $\lambda = 0$ *with, at least,* $n - m$ *zero eigenvalues.*

6.8.3 Symmetry in the spectrum of an adjacency matrix A

Theorem 39 can also be proven as follows. If the spectrum of A is symmetric, then the characteristic polynomial $c_A(x) = c_A(-x)$ is even, which implies that the odd coefficients c_{2k+1} of the characteristic polynomial $c_A(x)$ are all zero. Thus, for each eigenvalue $\lambda = r > 0$ of A, there is an eigenvalue $\lambda = -r$ of A with the same multiplicity. **Art.** 51 shows that the product $a_{1p_1} a_{2p_2} \ldots a_{kp_k}$ of a permutation $p = (p_1, p_2, \ldots, p_k)$ of $(1, 2, \ldots, k)$ in a tree is always zero for odd k. Hence, (3.9) indicates that the spectrum of the adjacency matrix of any tree is symmetric. Indeed, the only non-zero product $a_{1p_1} a_{2p_2} \ldots a_{kp_k}$ in a tree is obtained when a link is traversed in both directions. Loops longer than two hops are not possible due to the tree structure and the permutation requirement for the determinant. The latter only admits paths of k hops as subgraphs G_k in **art.** 51 because all first as well as second indices need to differ in $a_{1p_1} a_{2p_2} \ldots a_{kp_k}$, because $a_{jj} = 0$. A longer even loop containing more than two hops will visit the intermediate nodes of the k-hop path twice, which the determinant structure does not allow.

The skewness s_λ, defined in **art.** 65, is zero for a tree, which again agrees with Theorem 39. The reverse of Theorem 39 is:

Theorem 40 *If the spectrum of an adjacency matrix A is symmetric around $\lambda = 0$, then the corresponding graph is a bipartite graph.*

Proof: Consider the adjacency matrix

$$A = \left[\begin{array}{cc} O_{m \times m} & B_{m \times n} \\ C_{n \times m} & D_{n \times n} \end{array} \right]$$

where $C = B^T$ if the graph is undirected. Any adjacency matrix can be written in this form for $m \geq 1$, because $a_{jj} = 0$. The characteristic polynomial $\det(A - \lambda I)$ follows from (A.57) as

$$\det(A - \lambda I) = (-\lambda)^{m-n} \det \left(\lambda D_{n \times n} - \lambda^2 I_{n \times n} - C_{n \times m} B_{m \times n} \right)$$

The determinant on the right-hand side is only a symmetric polynomial in λ if $D = O$. In that case, A equals the adjacency matrix (6.31) of a bipartite graph. $\square$

6.8.4 Laplacian spectrum of a tree

We have shown in Section 6.8.2 that any tree can be represented by a general bipartite graph after properly folding the levelsets. **Art.** 25 indicates that the unsigned incidence matrix R and the incidence matrix B satisfy $B^T B = R^T R$, provided the links are directed from a node at an even levelset $X_N^{(2k)}$ to a node at an odd levelset $X_N^{(2k+1)}$ (or all in the opposite direction), for all levels $0 \leq k < N-1$. Under this condition, the relation (2.34) in **art.** 27 applies such that, with $L = N-1$ in any tree T,

$$\mu_j(T) = \lambda_j\left(A_{l(T)}\right) + 2 \tag{6.34}$$

for $1 \leq j < N$ and $\mu_N(T) = 0$, as for any graph. Hence, the j-th Laplacian eigenvalue of a tree T equals the j-th eigenvalue of the $(N-1) \times (N-1)$ adjacency matrix of its corresponding line graph $l(T)$.

   Petrović and Gutman (2002) have elegantly proven that the path with $N-1$ hops is the tree with smallest largest Laplacian eigenvalue. Their arguments are as follows. When adding a link to a graph, the Laplacian eigenvalues are non-decreasing as shown in **art.** 164 such that, among all connected graphs, some tree will have the smallest largest Laplacian, because a tree has the minimum number of links of any connected graph. Now, (6.34) shows that the smallest largest Laplacian in any tree is attained in the tree whose line graph has the smallest largest adjacency eigenvalue. This line graph has $N-1$ nodes. The line graph of any tree on $N > 4$ nodes possesses cycles, except for the path P_N with $N-1$ hops. Any connected cycle-containing graph G on $N-1$ nodes has a spanning tree that contains the minimum number of links and whose largest adjacency eigenvalue is smaller than $\lambda_1(G)$ by Lemma 10. As shown in Section 6.4, the path P_{N-1} has the smallest largest adjacency eigenvalue among all trees on $N-1$ nodes. Since the line graph of the path P_N is the path P_{N-1}, we conclude that the path P_N has the smallest largest Laplacian eigenvalue among connected graphs. Combining (6.10) for P_{N-1} and (6.34) yields $\mu_1(P_N) = 2\left(1 + \cos\left(\frac{\pi}{N}\right)\right) < 4$, which agrees, indeed, with (6.15).

   Kolokolnikov (2015) has proved that, in any tree T on N nodes and with maximum degree $d_{\max}$ fixed and not a function of N, the algebraic connectivity μ_{N-1} obeys, for large N,

$$\mu_{N-1} \leq 2\frac{(d_{\max} - 1)}{N} + O\left(\frac{\ln N}{N^2}\right)$$

Moreover, Kolokolnikov (2015) conjectures that the upper bound can be likely sharpened by replacing the factor 2 by $\frac{d_{\max}}{d_{\max}-1}$, in which case equality is achieved for a maximally balanced tree[3]. For large N, the algebraic connectivity μ_{N-1} of a tree with *fixed* $d_{\max}$ (because, in the star, $\mu_{N-1}(K_{1,N}) = 1$) tends to zero as $O(N^{-1})$.

---

[3] A maximally balanced tree is a tree whose leaves are all at the same distance from the root node and whose non-leaves have the same degree. Examples are well-balanced Bethe trees and Cayley trees.

6.9 Complete multipartite graph

Instead of two partitions, we now consider the complete m-partite graph, where each partition of k_j nodes with $1 \leq j \leq m$ is internally not connected, but fully connected to any other partition. The corresponding adjacency matrix is

$$
A_{m\text{-partite}} = \begin{bmatrix}
O_{k_1} & J_{k_1 \times k_2} & \cdots & J_{k_1 \times k_m} \\
J_{k_2 \times k_1} & O_{k_2} & \cdots & J_{k_2 \times k_m} \\
\vdots & & \ddots & \vdots \\
J_{k_m \times k_1} & \cdots & \cdots & O_{k_m}
\end{bmatrix}
$$

This complete m-partite graph is denoted as $K_{\{k_1, k_2, \ldots, k_m\}}$ and possesses $N = \sum_{j=1}^{m} k_j$ nodes and $L = \sum_{i=1}^{m} \sum_{j=i+1}^{m} k_i k_j$ links. The complement of the m-partite graph is the union of m cliques $K_{k_1}, K_{k_2}, \ldots, K_{k_m}$, whose spectrum is the union of the eigenvalues of each clique, given by (6.1). Thus, the eigenvalues of $A_{m\text{-partite}}^c$ are $\{k_j - 1\}_{1 \leq j \leq m}$ and $[-1]^{N-m}$, where $N = \sum_{j=1}^{m} k_j$. As we will see below, the eigenvalues of $A_{m\text{-partite}}^c$ via (3.39) are not quite helpful to derive those of $A_{m\text{-partite}}$.

If all $k_j = k$, then $A_{m\text{-partite}} = A_{K_m} \otimes J_{k \times k}$, whose corresponding spectrum follows from **art.** 286 as $\{\lambda_j (A_{K_m}) \lambda_l (J_{k \times k})\}_{1 \leq j \leq m, 1 \leq l \leq k}$, where, according to (6.1), $\lambda_j (A_{K_m}) \in \left\{ m - 1, [-1]^{m-1} \right\}$ and $\lambda_l (J_{k \times k}) \in \left\{ k, [0]^{k-1} \right\}$. Thus, when all m partitions are equal, $k_j = k$ for $1 \leq j \leq m$, the regular, complete m-partite graph has $N = km$ nodes, $L = k^2 \binom{m}{2} = \left(1 - \frac{1}{m}\right) \frac{N^2}{2}$ links and degree $r = \left(1 - \frac{1}{m}\right) N = N - k$. Moreover, the eigenvalues of $A_{m\text{-partite}}$ are $(m-1) k$, $[0]^{N-m}$ and $[-k]^{m-1}$.

When the number N of nodes and the number m of partitions is given, then the most regular complete m-partite graph is called *the Turán graph* $T_m (N)$. Specifically, if the number of nodes is $N = mk + l$, where $0 \leq l < m$ so that $l = N \bmod m$ and $k = \left\lfloor \frac{N}{m} \right\rfloor$, then the Turán graph $T_m (N)$ contains l partitions with $k + 1$ nodes, each of degree $N - (k+1)$, and $m - l$ partitions with k nodes, each of degree $N - k$. The number of links in $T_m (N)$ equals $L = \left\lfloor \left(1 - \frac{1}{m}\right) \frac{N^2}{2} \right\rfloor = l (k+1) (N - (k+1)) + (m - l) k (N - k)$. The Turán graph $T_m (N)$ is the graph on N nodes with the highest numbers of links that does *not* contain a clique K_{m+1}. This property of $T_m (N)$ was proved[4] by Pál Turán in 1941 and has marked the beginning of extremal graph theory (see e.g. Bollobás (2004)). An interesting corollary is: Any graph G on N nodes with more than $\left(1 - \frac{1}{m}\right) \frac{N^2}{2}$ links or average degree larger than $\left(1 - \frac{1}{m}\right) N$ contains a clique K_{m+1}. The special case for triangles ($m = 2$) was encountered in Mantel's Theorem 7. A more general result, due to Nikiforov *et al.* (2018), states that if a graph G on N nodes does not contain a graph H and $\lambda_1 (G) \geq \kappa N^{1-1/s}$, then G contains an induced copy of $K_{s,t}$ for $t \geq s \geq 3$, where $\kappa \geq (R(H, K_t))^{2/s} R(H, K_s)$ and $R(H, G)$ is the Ramsey number of H versus G.

The eigenvalues of the general complete multipartite graph $K_{\{k_1, k_2, \ldots, k_m\}}$ can be

---

[4] A proof in English, close to the original of Pál Turán, is given in Diestel (2010). Five different proofs are given in Aigner and Ziegler (2003, Chapter 32).

obtained using the quotient matrix (**art. 37**). Since the row sum of each block matrix in $A_{m\text{-partite}}$ is the same, the partition is equitable, with corresponding quotient matrix

$$
\left(A^{\pi}\right)_{m\text{-partite}} =
\begin{bmatrix}
0 & k_2 & \cdots & k_m \\
k_1 & 0 & \cdots & k_m \\
\vdots & & \ddots & \vdots \\
k_1 & k_2 & \cdots & 0
\end{bmatrix}
= (J - I)_{m \times m} \operatorname{diag}(k_j)
$$

The eigenvalues of $\left(A^{\pi}\right)_{m\text{-partite}}$ are the non-trivial eigenvalues of $A_{m\text{-partite}}$. The remaining $N - m$ eigenvalues of $A_{m\text{-partite}}$ are zero, because, only when $\lambda = 0$, the matrix $A_{m\text{-partite}} - \lambda I$ has in each block k_j identical rows. The eigenvalues of $\left(A^{\pi}\right)_{m\text{-partite}}$ are obtained by subtracting the first row from all the others, which results in

$$
\det\left(\left(A^{\pi}\right)_{m\text{-partite}} - \lambda I\right) =
\begin{vmatrix}
-\lambda & k_2 & k_3 & \cdots & k_m \\
k_1 + \lambda & -(\lambda + k_2) & 0 & \cdots & 0 \\
k_1 + \lambda & 0 & -(\lambda + k_3) & \cdots & 0 \\
\vdots & \vdots & \vdots & & \vdots \\
k_1 + \lambda & 0 & 0 & \cdots & -(\lambda + k_m)
\end{vmatrix}
$$

$$
=
\begin{vmatrix}
-\lambda & y^T \\
(k_1 + \lambda)\, u & \operatorname{diag}\left(-(\lambda + k_j)\right)
\end{vmatrix}
$$

where the vector $y = (k_2, k_3, \ldots, k_m)$. Using the Schur-complement (A.59),

$$
\det\left(\left(A^{\pi}\right)_{m\text{-partite}} - \lambda I\right) = (-1)^m \prod_{j=2}^{m} (\lambda + k_j) \det\left(\lambda + y^T \operatorname{diag}\left(\frac{-1}{\lambda + k_j}\right)(k_1 + \lambda)\, u\right)
$$

$$
= (-1)^m \prod_{j=2}^{m} (\lambda + k_j) \left(\lambda - (k_1 + \lambda) \sum_{j=2}^{m} \frac{k_j}{\lambda + k_j}\right)
$$

replacing λ by $(k_1 + \lambda) - k_1$ leads to

$$
\det\left(\left(A^{\pi}\right)_{m\text{-partite}} - \lambda I\right) = (-1)^m \prod_{j=1}^{m} (\lambda + k_j) \left(1 - \sum_{j=1}^{m} \frac{k_j}{\lambda + k_j}\right) \tag{6.35}
$$

and the polynomial of degree m in λ

$$
\det\left(\left(A^{\pi}\right)_{m\text{-partite}} - \lambda I\right) = (-1)^m \prod_{j=1}^{m} (\lambda + k_j) - (-1)^m \sum_{j=1}^{m} k_j \prod_{l=1; l \neq j}^{m} (\lambda + k_l)
$$

When multiplying out, we find that all coefficients c_j in

$$
\det\left(\left(A^{\pi}\right)_{m\text{-partite}} - \lambda I\right) = (-1)^{m-1} \sum_{j=0}^{m} c_j \lambda^j
$$

are positive, except for $c_m = -1$ and $c_{m-1} = 0$. Explicitly, we have

$$c_0 = (-1)^{m-1} \det \left((A^\pi)_{m\text{-partite}} \right) = (m-1) \prod_{j=1}^{m} k_j$$

and

$$c_j = (m - j - 1) \, e_{m-j} \, (k_1, k_2, \ldots, k_n)$$

for $0 \le j \le m - 2$, where e_k is the elementary symmetric polynomial (**art.** 297). **Art.** 320 demonstrates that $\det \left((A^\pi)_{m\text{-partite}} - \lambda I \right)$ has only one positive zero, while all others are negative. Equation (6.35) shows that all eigenvalues of $(A^\pi)_{m\text{-partite}}$ satisfy

$$\sum_{j=1}^{m} \frac{1}{1 + \frac{\lambda}{k_j}} = 1 \tag{6.36}$$

The partial fraction $g(x) = \sum_{j=1}^{m} \frac{k_j}{x+k_j}$ in (6.36) has simple poles at $x = -k_j$ with $g(-k_j - \varepsilon) < 0$ and $g(-k_j + \varepsilon) > 0$ and, $g(0) = m$ and $\lim_{x \to \pm\infty} g(x) = 0$, and $g(x)$ is strictly decreasing for all, finite real x, because $g'(x) < 0$. These properties indicate that, if all k_j are different and ranked as $k_{(1)} > k_{(2)} > \cdots > k_{(m)}$, then the eigenvalues as a solution of $g(x) = 1$ lie between $k_{(j)}$ and $k_{(j-1)}$ for $1 < j < m$ and only one eigenvalue is strict positive. That largest eigenvalue of $A_{m\text{-partite}}$ and $(A^\pi)_{m\text{-partite}}$ is the unique, positive solution of (6.36) which shows that $(m-1) \min_{1 \le j \le m} k_j \le \lambda \le (m-1) \max_{1 \le j \le m} k_j$.

The spectral gap of $A_{m\text{-partite}}$ is equal to the largest eigenvalue of $(A^\pi)_{m\text{-partite}}$, because $\lambda_2(A_{m\text{-partite}}) = 0$. Therefore, an explicit expression for the largest eigenvalue $\lambda_1(A_{m\text{-partite}})$ is desirable to estimate the influence of the partitions k_j on the spectral gap. Below, we devote some effort and present two different expansions for $\lambda_1(A_{m\text{-partite}})$. If all $k_j = k$, then, as found above,

$$\det \left((A^\pi)_{m\text{-partite}} - \lambda I \right) = (-1)^m (\lambda + k)^{m-1} (\lambda - (m-1)k)$$

which reduces to the characteristic polynomial (6.1) of the complete graph K_m if $k = 1$. The spectral gap is $km - k = N - k$, which equals that of the complete graph K_N minus k. When not all k_j are equal, the spectral gap is smaller than $N - k$ as verified from Lagrange optimization of (6.36) for all k_j subject to $N = \sum_{j=1}^{m} k_j$. This underlines that regularity in a graph's structure scores highest in terms of robustness.

We can rewrite (6.36) as

$$\lambda = \frac{m-1}{\sum_{j=1}^{m} \frac{1}{\lambda + k_j}} \tag{6.37}$$

from which the positive $\lambda > \frac{m-1}{\sum_{j=1}^{m} \frac{1}{k_j}}$. Iterating (6.37) once gives a sharper lower bound

$$\lambda > \frac{m-1}{\sum_{j_1=1}^{m}\frac{1}{k_{j_1}+\frac{m-1}{\sum_{j_2=1}^{m}\frac{1}{k_{j_2}}}}} = \frac{m-1}{\left(\sum_{j_2=1}^{m}\frac{1}{k_{j_2}}\right)\left(\sum_{j_1=1}^{m}\frac{1}{m-1+\sum_{j_2=1}^{m}\frac{k_{j_1}}{k_{j_2}}}\right)}$$

After q-times iterating the equation, we obtain a finite continued fraction expansion

$$\lambda_1\left(A_{m\text{-partite}}\right) > \cfrac{m-1}{\sum_{j_1=1}^{m}\cfrac{1}{k_{j_1}+\cfrac{m-1}{\sum_{j_2=1}^{m}\cfrac{1}{k_{j_2}+\cfrac{m-1}{\ddots\cfrac{m-1}{\sum_{j_q=1}^{m}\frac{1}{k_{j_q}}}}}}}$$

that approaches $\lambda_1\left(A_{m\text{-partite}}\right)$ arbitrarily close from below for sufficiently large q.

Finally, for real positive numbers $a_1, a_2, \ldots, a_n$, the harmonic, geometric and arithmetic mean inequality (Hardy *et al.*, 1999) is

$$\frac{n}{\sum_{j=1}^{n}\frac{1}{a_j}} \leq \sqrt[n]{\prod_{j=1}^{n}a_j} \leq \frac{1}{n}\sum_{j=1}^{n}a_j \tag{6.38}$$

with equality only if all a_j are equal. Applied to (6.37) yields

$$\lambda = \frac{m-1}{m}\frac{m}{\sum_{j=1}^{m}\frac{1}{\lambda+k_j}} \leq \frac{m-1}{m}\frac{1}{m}\left(\sum_{j=1}^{m}\lambda+\sum_{j=1}^{m}k_j\right) = \frac{m-1}{m}\left(\lambda+\frac{N}{m}\right)$$

from which an upper bound is deduced

$$\lambda_1\left(A_{m\text{-partite}}\right) \leq \left(1-\frac{1}{m}\right)N$$

where the right-hand side equals the degree of the *regular* complete m-partite graph when all $k_j = k$. Only when $m = N$ in case all $k_j = 1$, the largest eigenvalue $\lambda_1\left(A_{m\text{-partite}}\right)$ equals that of the complete graph K_N.

6.10 An m-fully meshed star topology

In the complete bipartite graph $K_{m,n}$, the m star nodes are not interconnected among themselves. The opposite variant, which we now consider, is essentially $K_{m,n}$ where all nodes in the m set are fully connected. We denote this topology by $G_{m\text{star}}$. The adjacency matrix of a graph of m stars, in which node 1 up to node m has degree $N-1$ while all other nodes have degree m, is

$$A_{m\text{star}} = \left[\begin{array}{cc} (J-I)_{m\times m} & J_{m\times(N-m)} \\ J_{(N-m)\times m} & O_{(N-m)\times(N-m)} \end{array}\right]$$

Observe that $A_{m\text{star}} = A_{K_{m,n}} + \breve{A}_{K_m}$, where

$$\breve{A}_{K_m} = \left[\begin{array}{cc} (J-I)_{m\times m} & O_{m\times(N-m)} \\ O_{(N-m)\times m} & O_{(N-m)\times(N-m)} \end{array}\right]$$

The characteristic polynomial is

$$\det\left(A_{mstar} - \lambda I\right) = \det \begin{bmatrix} (J - (\lambda + 1)\,I)_{m \times m} & J_{m \times (N-m)} \\ J_{(N-m) \times m} & -\lambda I_{(N-m) \times (N-m)} \end{bmatrix}$$

which will be solved in two ways by applying (A.57) first and then (A.59).

Applying (A.57) and using

$$X_{m \times m} = (J - (\lambda + 1)\,I)_{m \times m} \tag{6.39}$$

gives $\det\left(A_{mstar} - \lambda I\right) = \det\left(X\right)\det\left(-\lambda I - J_{(N-m) \times m} X_{m \times m}^{-1} J_{m \times (N-m)}\right)$. We first need to compute the inverse $X^{-1} = \frac{\mathrm{adj}\,X}{\det X}$ of $X_{m \times m} = (J - (\lambda + 1)\,I)_{m \times m}$, where the adjoint matrix $\mathrm{adj}(A)$ is the transpose of the matrix of the cofactors of A in **art**. 212. An inspection of the matrix X shows that there are precisely two types of cofactors. The cofactor $\mathring{X}_{jj}$ of a diagonal element of X equals

$$\mathring{X}_{jj} = \det \begin{bmatrix} -\lambda & 1 & \dots & 1 \\ 1 & -\lambda & \dots & 1 \\ \vdots & & \ddots & \vdots \\ 1 & 1 & \dots & -\lambda \end{bmatrix} = \det\left(J - (\lambda + 1)I\right)_{(m-1) \times (m-1)}$$

The off-diagonal cofactor $\mathring{X}_{ij}$ (with $i \neq j$) is

$$\mathring{X}_{ij} = (-1)^{i+j} \det \begin{bmatrix} & & & {}^{i\text{-th col}} & & \\ -\lambda & 1 & \dots & 1 & \dots & 1 \\ 1 & -\lambda & \dots & 1 & \dots & 1 \\ \vdots & \vdots & \ddots & \vdots & & \vdots \\ 1 & 1 & \dots & 1 & \dots & 1 \\ \vdots & \vdots & & \vdots & \ddots & \vdots \\ 1 & 1 & \dots & 1 & \dots & -\lambda \end{bmatrix}$$

where the j-th row and the i-th column consist of all ones. Subtracting row j from all other rows yields

$$\mathring{X}_{ij} = (-1)^{i+j} \det \begin{bmatrix} & & & {}^{i\text{-th col}} & & \\ -\lambda - 1 & 0 & \dots & 0 & \dots & 0 \\ 0 & -\lambda - 1 & \dots & 0 & \dots & 0 \\ \vdots & \vdots & \ddots & \vdots & & \vdots \\ 1 & 1 & \dots & 1 & \dots & 1 \\ \vdots & \vdots & & \vdots & \ddots & \vdots \\ 0 & 0 & \dots & 0 & \dots & -\lambda - 1 \end{bmatrix}$$

The i-th column now has only one non-zero element at row j, such that the determinant is equal to $(-1)^{i+j}$ times the minor of element (j, i), which is $(-1)^{m-2}(\lambda +$

$1)^{m-2}$. Hence, the adjoint matrix has all elements equal to $(-1)^{m-2}(\lambda+1)^{m-2}$ except for the diagonal elements that are equal to $(-1)^{m-2}(\lambda+1)^{m-2}(\lambda+2-m)$,

$$\mathrm{adj}\,(X) = (-1)^{m-2}(\lambda+1)^{m-2}J$$
$$+ \left(-(-1)^{m-2}(\lambda+1)^{m-2} + (-1)^{m-2}(\lambda+1)^{m-2}(\lambda+2-m)\right)I$$
$$= (-1)^{m-2}(\lambda+1)^{m-2}(J + (\lambda+1-m)I)$$

and, since $\det X = (-1)^{m-1}(\lambda+1)^{m-1}(\lambda+1-m)$, the inverse matrix of $X = J - (\lambda+1)I$ is

$$X^{-1} = (J - (\lambda+1)I)^{-1} = \frac{-1}{(\lambda+1)(\lambda+1-m)}(J + (\lambda+1-m)I)_{m\times m} \qquad (6.40)$$

We now compute $Y = J_{(N-m)\times m}X^{-1}_{m\times m}J_{m\times(N-m)}$,

$$Y = -\frac{1}{(\lambda+1)(\lambda+1-m)}J_{(N-m)\times m}(J_{m\times m} + (\lambda+1-m)I_{m\times m})J_{m\times(N-m)}$$

Using $J_{k\times n}J_{n\times l} = nJ_{k\times l}$ gives

$$Y = -\frac{1}{(\lambda+1)(\lambda+1-m)}\left(m^2 J_{(N-m)\times(N-m)} + m(\lambda+1-m)J_{(N-m)\times(N-m)}\right)$$

whence

$$Y = -\frac{m}{(\lambda+1-m)}J_{(N-m)\times(N-m)} \qquad (6.41)$$

Combining all in $\det(A_{mstar} - \lambda I) = \det(X)\det(-\lambda I - Y)$ yields

$$\det(A_{mstar} - \lambda I) = \det(J - (\lambda+1)I)_{m\times m}\det\left(\frac{m}{\lambda+1-m}J - \lambda I\right)_{(N-m)\times(N-m)}$$

Finally, using (6.1) leads to

$$\det(A_{mstar} - \lambda I) = (-1)^N(\lambda+1)^{m-1}\lambda^{N-m-1}(\lambda(\lambda+1-m) - m(N-m))$$
$$= (-1)^N(\lambda+1)^{m-1}\lambda^{N-m-1}(\lambda-\alpha_-)(\lambda-\alpha_+) \qquad (6.42)$$

where

$$\alpha_\pm = \frac{m-1}{2} \pm \sqrt{m(N-m) + \left(\frac{m-1}{2}\right)^2}$$

The eigenvalues of A_{mstar} are $\alpha_-, [-1]^{m-1}, [0]^{N-1-m}$, and $(\lambda_{max})_{mstar} = \alpha_+$, which is larger than $(\lambda_{max})_{K_{m,n}} = \sqrt{m(N-m)}$ as was expected from Gerschgorin's Theorem 65 on p. 355. When viewing the complete spectrum, we observe that the spectrum is not symmetric in λ anymore for $m > 1$.

If $m = N - 1$, the $mstar$ topology equals K_N. It is readily verified that, indeed, for $m = N - 1$, the spectrum reduces to that of K_N. If $m = N - 2$, then the $mstar$

topology equals K_N minus one link, for which the eigenvalues are $\alpha_-, [-1]^{N-3}, 0,$ and

$$
(\lambda_{\max})_{mstar} = \alpha_+ = \frac{N-3}{2}\left\{ 1 + \sqrt{1 + \frac{8(N-2)}{(N-3)^2}} \right\}
$$

$$
= N - 1 - 2\frac{(N^2 - 2N - 1)}{(N-3)^3} + O\left(N^{-2}\right)
$$

Hence, by deleting one link in the complete graph K_N, the spectral gap (**art. 82**) reduces from N to $\alpha_+ < N - 1$. The spectral gap of the complete multipartite graph (Section 6.9) equals $N - 2$, when $k = 2$ and $N = 2m$. In that case, m links are removed from the complete graph K_N in such a way that each node has still degree $N - 2$.

The second, considerably more efficient way of computing $\det(A_{mstar} - \lambda I)$ is based on (A.59),

$$
\det(A_{mstar} - \lambda I) = (-\lambda)^{N-m}\det\left((J - (\lambda+1)I)_{m \times m} + \frac{1}{\lambda}J_{m \times (N-m)}J_{(N-m) \times m} \right)
$$

Using $J_{k \times n}J_{n \times l} = nJ_{k \times l}$ and (6.1) leads, after some manipulations, to (6.42). The first, elaborate computation supplies us with the matrices X^{-1} in (6.40) and Y in (6.41), that will be of use later in Sections 6.10.2 and 6.10.3.

The spectrum of A_{mstar} can be determined in yet another way[5]. Since A_{mstar} has $N - m$ identical rows, it has an eigenvalue 0 with multiplicity at least $N - m - 1$. Further, since $A_{mstar} + I$ has m identical rows, it follows that A_{mstar} has an eigenvalue -1 with multiplicity at least equal to $m - 1$. The remaining two other eigenvalues are obtained after determining the eigenvector that is orthogonal to the eigenvector (with constant components) belonging to $\lambda = 0$ and that belonging to $\lambda = -1$.

The remainder of this section computes the spectra of several subgraphs of G_{mstar}.

6.10.1 Fully-interconnected stars linked to two separate groups

In stead of the $J_{m \times (N-m)}$ matrix in A_{mstar} of Section 6.10, a next step is to consider some matrix B. Thus, instead of connecting each of the m fully interconnected stars to all other non-star nodes, each such star does not necessarily need to connect to all other nodes, but to a few.

Let us consider

$$
A_{lmnstar} = \begin{bmatrix} A_{m \times m} & B_{m \times (N-m)} \\ B^T_{(N-m) \times m} & O_{(N-m) \times (N-m)} \end{bmatrix}
$$

---

[5] This method was pointed out to me by E. van Dam.

where

$$B = \left[\begin{array}{cc} J_{n \times l} & O_{n \times (N-m-l)} \\ O_{(m-n) \times l} & J_{(m-n) \times (N-m-l)} \end{array} \right]$$

which means that n stars all reach the same l nodes and $m - n$ stars all reach $N - m - l$ other nodes. The eigenvalue analysis is simplified if we consider $A = O$. Then, using (A.57) gives

$$\det \left(A_{lmnstar} - \lambda I \right) = (-\lambda)^m \det \left(-\lambda I + \frac{1}{\lambda} B^T B \right)$$

where

$$B^T B = \left[\begin{array}{cc} J^T_{n \times l} & O^T_{(m-n) \times l} \\ O^T_{n \times (N-m-l)} & J^T_{(m-n) \times (N-m-l)} \end{array} \right] \left[\begin{array}{cc} J_{n \times l} & O_{n \times (N-m-l)} \\ O_{(m-n) \times l} & J_{(m-n) \times (N-m-l)} \end{array} \right]$$

$$= \left[\begin{array}{cc} n J_{l \times l} & O_{l \times (N-m-l)} \\ O_{(N-m-l) \times l} & (m-n) J_{(N-m-l) \times (N-m-l)} \end{array} \right]$$

With the dimensions of $B_{m \times (N-m)}$ and $\left(B^T B \right)_{(N-m) \times (N-m)}$, we have

$$b = \det \left(-\lambda I + \frac{1}{\lambda} B^T B \right) = \lambda^{m-N} \det \left(B^T B - \lambda^2 I \right)$$

$$= \lambda^{m-N} \det \left[\begin{array}{cc} n J_{l \times l} - \lambda^2 I & O_{l \times (N-m-l)} \\ O_{(N-m-l) \times l} & (m-n) J_{(N-m-l) \times (N-m-l)} - \lambda^2 I \end{array} \right]$$

$$= \lambda^{m-N} \det \left(n J_{l \times l} - \lambda^2 I \right) \det \left((m-n) J_{(N-m-l) \times (N-m-l)} - \lambda^2 I \right)$$

$$= \lambda^{m-N} n^l (m-n)^{N-m-l} \det \left(J_{l \times l} - \frac{\lambda^2}{n} I \right) \det \left(J_{(N-m-l) \times (N-m-l)} - \frac{\lambda^2}{m-n} I \right)$$

With (6.1), we arrive at

$$\det \left(A_{lmnstar} - \lambda I \right) = (-1)^N \lambda^{N-4} \left(\lambda^2 - nl \right) \left(\lambda^2 - (N-m-l)(m-n) \right)$$

and the eigenvalues of $A_{lmnstarG}$ are $\pm \sqrt{nl}, \pm \sqrt{(N-m-l)(m-n)}$ and $[0]^{N-4}$. For $l = n = 0$, the spectrum reduces to that of $K_{m,N-m}$.

6.10.2 Star-like, two-hierarchical structure

We compute the spectrum of a classical star-like, two-hierarchical telephony network where

$$A_{mdoublestar} = \left[\begin{array}{cc} (J-I)_{m \times m} & B_{m \times (N-m)} \\ B^T_{(N-m) \times m} & O_{(N-m) \times (N-m)} \end{array} \right]$$

where

$$B = \left[\begin{array}{ccccccccc} 1 & \cdots & 1 & 0 & \cdots & 0 & 0 & \cdots & 0 \\ 0 & \cdots & 0 & 1 & \cdots & 1 & 0 & \cdots & 0 \\ \vdots & & & & \vdots & & & \vdots \\ 0 & \cdots & & & & \cdots & 1 & \cdots & 1 \end{array} \right] = I_{m \times m} \otimes u_{1 \times l}$$

with $u_{1 \times l}$ is the l component long all-one vector and the Kronecker product is defined in **art. 286**. Thus, the dimension of B is $m \times lm$ and $N - m = lm$, and the number of nodes in $A_{mdoublestar}$ is $N = (l+1)\,m$. All m fully interconnected nodes $(A_{m \times m} = (J - I)_{m \times m})$ may represent the highest level core in a telephony network. Each of these m nodes connects to l different lower level nodes, the local exchanges, in the telephony network.

Applying (A.57) and denoting $X_{m \times m} = (J - (\lambda + 1)\,I)_{m \times m}$, the characteristic polynomial is

$$\det\left(A_{mdoublestar} - \lambda I\right) = \det\left(X\right) \det\left(-\lambda I - B^T_{(N-m) \times m} X^{-1}_{m \times m} B_{m \times (N-m)}\right)$$

In Section 6.10, the inverse of $X_{m \times m} = (J - (\lambda + 1)\,I)_{m \times m}$ is computed in (6.40),

$$B^T_{(N-m) \times m} X^{-1}_{m \times m} B_{m \times (N-m)} = -\frac{B^T_{(N-m) \times m}\left(J + (\lambda + 1 - m)\,I\right)_{m \times m} B_{m \times (N-m)}}{(\lambda + 1)(\lambda + 1 - m)}$$

Using properties of the Kronecker product (Meyer, 2000, p. 598),

$$J_{m \times m} B_{m \times (N-m)} = J_{m \times m}\left(I_{m \times m} \otimes u_{1 \times l}\right) = \left(J_{m \times m} \otimes u_{1 \times 1}\right)\left(I_{m \times m} \otimes u_{1 \times l}\right)$$

$$= \left(J_{m \times m} I_{m \times m} \otimes u_{1 \times 1} u_{1 \times l}\right) = J_{m \times m} \otimes u_{1 \times l} = J_{m \times ml}$$

and, similarly,

$$B^T_{(N-m) \times m} J_{m \times ml} = \left(I_{m \times m} \otimes u_{l \times 1}\right)\left(J_{m \times m} \otimes u_{1 \times l}\right) = J_{m \times m} \otimes u_{l \times 1} u_{1 \times l}$$

$$= J_{m \times m} \otimes J_{l \times l} = J_{ml \times ml}$$

the matrix $V = B^T_{(N-m) \times m}\left(J + (\lambda + 1 - m)\,I\right)_{m \times m} B_{m \times (N-m)}$ is

$$V = J_{ml \times ml} + (\lambda + 1 - m)\,B^T_{(N-m) \times m} B_{m \times (N-m)}$$

Further, $B^T_{(N-m) \times m} B_{m \times (N-m)} = \left(I_{m \times m} \otimes u_{l \times 1}\right)\left(I_{m \times m} \otimes u_{1 \times l}\right) = I_{m \times m} \otimes J_{l \times l}$ and, with $J_{ml \times ml} = J_{m \times m} \otimes J_{l \times l}$, we have $V = \{J_{m \times m} + (\lambda + 1 - m)\,I_{m \times m}\} \otimes J_{l \times l}$. Hence,

$$C = \det\left(-\lambda I - B^T_{(N-m) \times m} X^{-1}_{m \times m} B_{m \times (N-m)}\right)$$

$$= \det\left(-\lambda I + \frac{1}{(\lambda + 1)(\lambda + 1 - m)}\left\{J_{m \times m} + (\lambda + 1 - m)\,I_{m \times m}\right\} \otimes J_{l \times l}\right)$$

$$= \frac{\det\left(-\lambda(\lambda + 1)(\lambda + 1 - m)\,I + \left\{J_{m \times m} + (\lambda + 1 - m)\,I_{m \times m}\right\} \otimes J_{l \times l}\right)}{(\lambda + 1)^{ml}(\lambda + 1 - m)^{ml}}$$

The eigenvalues of $D_{m \times m} \otimes E_{l \times l}$ are the ml numbers $\{\lambda_j(D)\,\lambda_k(E)\}_{1 \le j \le m, 1 \le k \le l}$ (**art. 286**). The eigenvalues of $D = J_{m \times m} + (\lambda + 1 - m)\,I_{m \times m}$ follow from (6.1) as $\lambda(D) = \left\{[\lambda + 1 - m]^{m-1}, \lambda + 1\right\}$, while the eigenvalues of $E = J_{l \times l}$ are $\lambda(E) = \left\{[0]^{l-1}, l\right\}$. With $z = \lambda(\lambda + 1)(\lambda + 1 - m)$, $\det\left(D_{m \times m} \otimes E_{l \times l} - zI\right) = 0$ has the zeros $[0]^{ml-m}, l(\lambda + 1)$ and $[l(\lambda + 1 - m)]^{m-1}$ and the same as the polynomial

$$z^{ml-m}\left(z - l(\lambda + 1)\right)\left(z - l(\lambda + 1 - m)\right)^{m-1}$$

such that the polynomial C in λ is

$$C = \left. \frac{z^{ml-m}\,(z - l\,(\lambda + 1))\,(z - l\,(\lambda + 1 - m))^{m-1}}{(\lambda + 1)^{ml}\,(\lambda + 1 - m)^{ml}} \right|_{z = \lambda(\lambda+1)(\lambda+1-m)}$$

Combining all and using (6.1), yields

$$\det\left(A_{m\text{doublestar}} - \lambda I\right) = \alpha(-1)^m\,(\lambda + 1)^{m-1}\,(\lambda + 1 - m)$$
$$\times \frac{z^{ml-m}\,(z - l\,(\lambda + 1))\,(z - l\,(\lambda + 1 - m))^{m-1}}{(\lambda + 1)^{ml}\,(\lambda + 1 - m)^{ml}}$$

which simplifies with $z = \lambda\,(\lambda + 1)\,(\lambda + 1 - m)$ to

$$\det\left(A_{m\text{doublestar}} - \lambda I\right) = \alpha(-1)^m \lambda^{ml-m}\,(\lambda\,(\lambda + 1 - m) - l)\,(\lambda\,(\lambda + 1) - l)^{m-1}$$

The eigenvalues of $A_{m\text{doublestar}}$ with $N = (l+1)\,m$ nodes are, beside a high-multiplicity root at zero $[0]^{ml-m}$, $\frac{m-1}{2} \pm \frac{1}{2}\sqrt{(m-1)^2 + 4l}$ and $\left[\frac{-1}{2} \pm \frac{1}{2}\sqrt{1 + 4l}\right]^{m-1}$. The number of different eigenvalues equals four, which implies that the diameter is three (**art. 69**). The largest eigenvalue of the double star with $m = 2$ and $N = 2\,(l + 1)$ was given earlier by Das and Kumar (2004),

$$\lambda_{\max}\left(A_{2\text{doublestar}}\right) = \sqrt{\frac{(N-1) + \sqrt{2N - 3}}{2}} = \frac{1}{2} + \frac{1}{2}\sqrt{2N - 3}$$

6.10.3 *Complementary double cone*

We consider a complete graph K_N to which two nodes, labeled by $N + 1$ and $N + 2$, are connected. Node $N + 1$ is connected to m nodes in K_N and node $N + 2$ to the $N - m$ other nodes. The corresponding adjacency matrix of this "complementary double cone" (CDC) on K_N is

$$A_{CDC} = \begin{bmatrix} (J - I)_{N \times N} & B_{N \times 2} \\ B^T_{2 \times N} & O_{2 \times 2} \end{bmatrix}_{(N+2) \times (N+2)}$$

where

$$B_{N \times 2} = \begin{bmatrix} u_{m \times 1} & 0_{m \times 1} \\ 0_{(N-m) \times 1} & u_{(N-m) \times 1} \end{bmatrix}$$

The CDC graph has diameter 3 and each other graph with diameter 3 is a subgraph of CDC (see also **art. 56** on strongly regular graphs). The corresponding Laplacian is

$$Q_{CDC} = \begin{bmatrix} NI - (J - I)_{N \times N} & -B_{N \times 2} \\ -B^T_{2 \times N} & \operatorname{diag}(m, N - m) \end{bmatrix}$$

whose eigenvalues follow from

$$\det\left(Q_{CDC} - \mu I\right) = \det \begin{bmatrix} (N + 1 - \mu)\,I - J_{N \times N} & -B_{N \times 2} \\ -B^T_{2 \times N} & \operatorname{diag}(m - \mu, N - m - \mu) \end{bmatrix}$$

We apply the Schur complement (A.59) with $D = \text{diag}(m - \mu, N - m - \mu)$ and

$$BD^{-1}C = B_{N\times 2}\text{diag}\left((m-\mu)^{-1}, (N-m-\mu)^{-1}\right) B_{2\times N}^T$$

$$= \begin{bmatrix} u_{m\times 1} & 0_{m\times 1} \\ 0_{(N-m)\times 1} & u_{(N-m)\times 1} \end{bmatrix} \begin{bmatrix} \frac{1}{m-\mu} & 0 \\ 0 & \frac{1}{N-m-\mu} \end{bmatrix} \begin{bmatrix} u_{1\times m} & 0_{1\times(N-m)} \\ 0_{1\times m} & u_{1\times(N-m)} \end{bmatrix}$$

$$= \begin{bmatrix} \frac{1}{m-\mu}J_{m\times m} & O_{m\times(N-m)} \\ O_{(N-m)\times m} & \frac{1}{N-m-\mu}J_{(N-m)\times(N-m)} \end{bmatrix}$$

such that

$$T = (N + 1 - \mu)I - J_{N\times N} - BD^{-1}C$$

$$= \begin{bmatrix} (N+1-\mu)I_{m\times m} - \left(\frac{1}{m-\mu}+1\right)J & -J_{m\times(N-m)} \\ -J_{(N-m)\times m} & (N+1-\mu)I - \left(\frac{1}{N-m-\mu}+1\right)J \end{bmatrix}$$

Hence,

$$\det(Q_{CDC} - \mu I) = \det D \det T = (m - \mu)(N - m - \mu) \det T$$

The determinant of T is computed with (A.57). The computation is similar to those of m fully connected stars in Section 6.10. Using (6.39), we express the matrix as

$$(N+1-\mu)I_{m\times m} - \frac{1+m-\mu}{m-\mu}J_{m\times m} = -\frac{1+m-\mu}{m-\mu}X_{m\times m}$$

where $\lambda + 1 = \frac{m-\mu}{1+m-\mu}(N+1-\mu)$. With (6.41), we have

$$-\frac{m-\mu}{1+m-\mu}J_{(N-m)\times m}X^{-1}J_{m\times(N-m)} = \frac{(m-\mu)}{(1+m-\mu)}\frac{m}{(\lambda+1-m)}J_{(N-m)\times(N-m)}$$

and with $\theta = \frac{(m-\mu)}{(1+m-\mu)}\frac{m}{(\lambda+1-m)} + \frac{N-m-\mu+1}{N-m-\mu}$,

$$\det T = \det\left(-\frac{1+m-\mu}{m-\mu}X_{m\times m}\right)\det\left((N+1-\mu)I_{(N-m)\times(N-m)} - \theta J\right)$$

$$= \left(\frac{1+m-\mu}{\mu-m}\right)^m (-\theta)^{N-m}\det(J-(\lambda+1)I)\det\left(J - \frac{N+1-\mu}{\theta}I\right)$$

Using (6.1) yields

$$\det T = \left(\frac{1+m-\mu}{m-\mu}\right)(\lambda+1-m)(N+1-\mu)^{N-2}(N+1-\mu-(N-m)\theta)$$

After simplification, we find that

$$\theta = \frac{m(N-m)(N+1) - \mu\left\{(N+1)^2 - m + m(N-m)\right\} + 2(N+1)\mu^2 - \mu^3}{(m(N-m) - \mu(N+1) + \mu^2)(N-m-\mu)}$$

We now compute $N + 1 - \mu - (N-m)\theta = \frac{s(\mu)}{r}$, where

$$r = \left(m(N-m) - \mu(N+1) + \mu^2\right)(N-m-\mu)$$

The result is

$$s(\mu) = \mu\left(\mu^3 - 2\mu^2(N+1) + \mu\left\{(N+1)^2 + m(N-m)\right\} - m(N-m)(N+2)\right)$$

The polynomial $\frac{s(\mu)}{\mu}$ has degree 3 in μ and the sum of its zeros is $2(N+1)$, while the product is $m(N-m)(N+2)$. Combining all factors yields

$$\det T = \frac{1}{(m-\mu)(N-m-\mu)}(N+1-\mu)^{N-2}s(\mu)$$

and

$$\det(Q_{CDC} - \mu I) = \det D \det T = (N+1-\mu)^{N-2}s(\mu)$$

In summary, the eigenvalues of $(Q_{CDC})_{(N+2)\times(N+2)}$ are 0, $[N+1]^{N-2}$, and the three real positive roots of $s(\mu)$.

6.11 Uniform degree graph

We define the uniform degree graph Υ_N by the adjacency matrix

$$A_{\Upsilon_N} = \begin{bmatrix} (J-I)_{[\frac{N}{2}]\times[\frac{N}{2}]} & \nabla_{[\frac{N}{2}]\times[\frac{N+1}{2}]} \\ \nabla^T_{[\frac{N+1}{2}]\times[\frac{N}{2}]} & O_{[\frac{N+1}{2}]\times[\frac{N+1}{2}]} \end{bmatrix} \tag{6.43}$$

where $\nabla_{n\times n}$ is square and symmetric Hankel matrix for even $N = 2n$

$$\nabla_{n\times n} = \begin{bmatrix} 1 & 1 & 1 & \cdots & & 1 \\ \vdots & \vdots & \vdots & & 0 & \\ 1 & 1 & 1 & 0 & & \\ 1 & 1 & 0 & & & \\ 1 & 0 & & & & \end{bmatrix}$$

but $\nabla_{n\times(n+1)}$ is non-square and asymmetric for odd $N = 2n+1$,

$$\nabla_{n\times(n+1)} = \begin{bmatrix} 1 & 1 & 1 & \cdots & & 1 \\ \vdots & \vdots & \vdots & & 0 & \\ 1 & 1 & 1 & 0 & & \\ 1 & 1 & 0 & & & \\ 1 & 1 & & & & \end{bmatrix}$$

The uniform degree graph Υ_N consists of a union of $\left[\frac{N}{2}\right]$ stars with different size, $\Upsilon_N = \cup_{j=\left[\frac{N+1}{2}\right]}^{N} K_{1,j-1}$. Indeed, the star $K_{1,N-1}$ with center at node 1 spans all nodes, the star $K_{1,N-2}$ with center at node 2 spans all nodes but node N, the star $K_{1,N-3}$ with center at node 3 spans all nodes but node N and $N-1$ and so on. Except for two nodes (**art. 3**), each node in a uniform degree graph Υ_N has a different degree and the degree vector is

$$d = \left(1, 2, \ldots, \left[\frac{N}{2}\right], \left[\frac{N}{2}\right], \left[\frac{N}{2}\right] + 1, \ldots, N-1\right)$$

Behzad and Chartrand (1967) prove that there exist only two graphs with two nodes of equal degree (also called *antiregular* graphs): G_N with equal degree $\left[\frac{N}{2}\right]$ and its complement G_N^c with equal degree $\left[\frac{N-1}{2}\right]$. Bapat (2013) states that antiregular graphs are threshold graphs (**art. 114**). Hence, the uniform degree graph Υ_N is a special case of an unweighted threshold graph. The number of links in Υ_N equals $L_{\Upsilon_N} = \left[\frac{N^2}{4}\right]$, while the number of triangles (**art. 50**) is $\blacktriangle_{\Upsilon_{2n}} = \frac{(n-1)n(2n-1)}{6}$ and $\blacktriangle_{\Upsilon_{2n+1}} = 2\binom{n+1}{3}$. The complement of the uniform degree graph Υ_N is a uniform degree graph Υ_{N-1} and one isolated node (with degree 0). Finally, the bipartite graph $\widetilde{\Upsilon}_N$ derived from the uniform degree graph Υ_N, without the major clique of size $\left[\frac{N}{2}\right]$ and with adjacency matrix

$$
A_{\widetilde{\Upsilon}_N} = \begin{bmatrix} O_{\left[\frac{N}{2}\right]\times\left[\frac{N}{2}\right]} & \nabla_{\left[\frac{N}{2}\right]\times\left[\frac{N+1}{2}\right]} \\ \nabla^T_{\left[\frac{N+1}{2}\right]\times\left[\frac{N}{2}\right]} & O_{\left[\frac{N+1}{2}\right]\times\left[\frac{N+1}{2}\right]} \end{bmatrix} \tag{6.44}
$$

has the property, for even[6] N, that each degree from $1, 2, \ldots, \left[\frac{N}{2}\right]$ occurs precisely twice.

6.11.1 The characteristic polynomial of Υ_{2n} and $\widetilde{\Upsilon}_{2n}$

We confine ourselves to the even case with $N = 2n$ and let $q = 1$ for Υ_{2n} and $q = 0$ for $\widetilde{\Upsilon}_{2n}$. The computation of the eigenvalues of $A_{\Upsilon_{2n}}$ and $A_{\widetilde{\Upsilon}_{2n}}$ is based on (A.59),

$$
\det\left(A_{\left(q\Upsilon_{2n}+(1-q)\widetilde{\Upsilon}_{2n}\right)} - \lambda I\right) = \det \begin{bmatrix} qA_{K_n} - \lambda I & \nabla_{n\times n} \\ \nabla^T_{n\times n} & -\lambda I_n \end{bmatrix}
$$

$$
= (-\lambda)^n \det\left(qA_{K_n} - \lambda I_n + \frac{1}{\lambda}\nabla^2_{n\times n}\right)
$$

$$
= \det\left(\lambda^2 I_n - \lambda q A_{K_n} - \nabla^2_{n\times n}\right)
$$

since $\nabla^T_{n\times n} = \nabla_{n\times n}$. The matrix $\nabla^2_{n\times n}$ has the particular form

$$
\nabla^2_{n\times n} = \begin{bmatrix} n & n-1 & n-2 & \cdots & 2 & 1 \\ n-1 & n-1 & n-2 & \cdots & 2 & 1 \\ n-2 & n-2 & n-2 & \cdots & 2 & 1 \\ \vdots & \vdots & \vdots & \ddots & \vdots & \vdots \\ 2 & 2 & 2 & \cdots & 2 & 1 \\ 1 & 1 & 1 & \cdots & 1 & 1 \end{bmatrix}
$$

---

[6] When N is odd, $\widetilde{\Upsilon}_{2m+1}$ is a disconnected graph with one isolated node, and apart from the isolated nodes, each degree occurs still twice.

Denoting $Y(q) = (-1)^n \det\left(A_{(q\Upsilon_{2n} + (1-q)\tilde{\Upsilon}_{2n})} - \lambda I\right)$ yields

$$Y(q) = \begin{vmatrix} n - \lambda^2 & n - 1 + q\lambda & n - 2 + q\lambda & \cdots & 2 + q\lambda & 1 + q\lambda \\ n - 1 + q\lambda & n - 1 - \lambda^2 & n - 2 + q\lambda & \cdots & 2 + q\lambda & 1 + q\lambda \\ n - 2 + q\lambda & n - 2 + q\lambda & n - 2 - \lambda^2 & \cdots & 2 + q\lambda & 1 + q\lambda \\ \vdots & \vdots & \vdots & \ddots & \vdots & \vdots \\ 2 + q\lambda & 2 + q\lambda & 2 + q\lambda & \cdots & 2 - \lambda^2 & 1 + q\lambda \\ 1 + q\lambda & 1 + q\lambda & 1 + q\lambda & \cdots & 1 + q\lambda & 1 - \lambda^2 \end{vmatrix}$$

We first subtract in the determinant (**art.** 209) column j from $j - 1$, starting from $j = 2$ to $j = n$, and obtain

$$Y(q) = \begin{vmatrix} 1 - q\lambda - \lambda^2 & 1 & 1 & \cdots & 1 & 1 + q\lambda \\ \lambda^2 + q\lambda & 1 - q\lambda - \lambda^2 & 1 & \cdots & 1 & 1 + q\lambda \\ 0 & \lambda^2 + q\lambda & 1 - q\lambda - \lambda^2 & \cdots & 1 & 1 + q\lambda \\ \vdots & \vdots & \vdots & \ddots & \vdots & \vdots \\ 0 & 0 & 0 & \cdots & 1 - q\lambda - \lambda^2 & 1 + q\lambda \\ 0 & 0 & 0 & \cdots & \lambda^2 + q\lambda & 1 - \lambda^2 \end{vmatrix}$$

Next, we repeat the same action on the rows and subtract row i from row $i - 1$, starting from $i = 2$ to $i = n$, and find, with $y = \lambda^2 + q\lambda$, that the determinant

$$Y(q) = \begin{vmatrix} 1 - 2y & y & 0 & \cdots & 0 & 0 \\ y & 1 - 2y & y & \cdots & 0 & 0 \\ 0 & y & 1 - 2y & \cdots & 0 & 0 \\ \vdots & \vdots & \ddots & \ddots & \ddots & \vdots \\ 0 & 0 & 0 & \cdots & 1 - 2y & y \\ 0 & 0 & 0 & \cdots & y & 1 - \lambda^2 \end{vmatrix}$$

has a tri-diagonal Toeplitz form (6.7), except for the element $y_{nn} = 1 - \lambda^2$. The eigenvalues of this pseudo tri-diagonal Toeplitz matrix of the form (6.14) cannot straightforwardly be solved with the generating function method of Section 6.4. The determinant $Y(q)$ can be expanded in a continued fraction as in **art.** 373.

6.11.2 The characteristic polynomial of ∇^{-2}

As shown in Section 6.8, the eigenvalues of the bipartite matrix $A_{\tilde{\Upsilon}_{2n}}$ are plus and minus those of $\nabla_{n \times n}$. Let us concentrate on the eigenvalues of $\nabla_{n \times n}$. The j-th row of the eigenvalue equation $\nabla x = \delta x$ yields

$$\sum_{m=1}^{n+1-j} x_m = \delta x_j$$

Subtracting row $j + 1$ from row j gives us

$$x_{n+1-j} = \delta\left(x_j - x_{j+1}\right) \tag{6.45}$$

which is valid for $1 \le j \le n$ with the assumption that $x_{n+1} = 0$, because $x_1 = \delta x_n$. After the index transformation $j \to n + 1 - j$, we obtain

$$x_j = \delta \left(x_{n+1-j} - x_{n+2-j} \right) \tag{6.46}$$

Assuming the existence of the inverse ∇^{-1}, the eigenvalue equation becomes $x = \delta \nabla^{-1} x$, and comparison with (6.46) illustrates that the inverse ∇^{-1} of the Hankel matrix ∇,

$$
\nabla^{-1}_{n \times n} =
\begin{bmatrix}
0 & 0 & 0 & \cdots & 1 \\
\vdots & \vdots & \vdots & \diagup & -1 \\
0 & 0 & 1 & \diagup & 0 \\
0 & 1 & -1 & 0 & \vdots \\
1 & -1 & 0 & \cdots & 0
\end{bmatrix}
$$

is again a Hankel matrix. Moreover, ∇^{-1} exists so that the eigenvalue $\delta \ne 0$. Iteratively applying the cofactor expansion (**art.** 212) with respect to the last element of the first row, the determinant is evaluated as $\det \nabla^{-1}_{n \times n} = (-1)^{\left[\frac{n}{2} \right]} = \det \nabla_{n \times n}$, while $\operatorname{trace}\left(\nabla^{-1}_{n \times n} \right) = (-1)^{n-1}$ and $\operatorname{trace}(\nabla_{n \times n}) = \left[\frac{n}{2} \right]$. **Art.** 235 then indicates that

$$\prod_{k=1}^{n} \lambda_k (\nabla) = (-1)^{\left[\frac{n}{2} \right]} \quad \sum_{k=1}^{n} \lambda_k (\nabla) = \left[\frac{n}{2} \right] \quad \sum_{k=1}^{n} \frac{1}{\lambda_k (\nabla)} = (-1)^{n-1}$$

More interestingly, ∇^{-2} is a pseudo tri-diagonal Toeplitz matrix (6.14), equal to the Laplacian matrix Q_P of the path on p. 204, except for the last diagonal element that is $\left(\nabla^{-2} \right)_{nn} = 2$ instead of $(Q_P)_{nn} = 1$, thus $\nabla^{-2} = Q_P + e_n e_n^T$. This difference is also manifested in $\det \nabla^{-2}_{n \times n} = 1$ and $\operatorname{trace}\left(\nabla^{-2}_{n \times n} \right) = 2n - 1$, while $\det Q_P = 0$ and $\operatorname{trace}(Q_P) = 2n - 2$, and is in agreement with the analysis of $Y(q)$ above. Alternatively with the Toeplitz matrix (6.7), $\nabla^{-2} = T_n (-1, 2, -1) - e_1 e_1^T$. By interlacing (Lemma 7) and $\lambda_1 \left(e_1 e_1^T \right) = 1$, but $\lambda_k \left(e_1 e_1^T \right) = 0$ for $k > 1$, the eigenvalues are upper bounded by those of $T_n (-1, 2, -1)$,

$$1 + 2 \cos \left(\frac{\pi m}{n+1} \right) \le \delta_m^{-2} \le 2 + 2 \cos \left(\frac{\pi m}{n+1} \right)$$

We now evaluate the characteristic polynomial $c_n (\lambda) = \det \left(\nabla^{-2}_{n \times n} - \lambda I \right)$ in closed form. Expanding the determinant $c_n (\lambda) = \det \left(\nabla^{-2}_{n \times n} - \lambda I \right)$ towards the first row yields

$$c_n (\lambda) = (1 - \lambda) \left| T_{n-1} (-1, 2 - \lambda, -1) \right| - \left| T_{n-2} (-1, 2 - \lambda, -1) \right|$$

Invoking (A.98) with (6.8),

$$c_n (\lambda) = (1 - \lambda) \prod_{k=1}^{n-1} \left(2 - \lambda + 2 \cos \frac{k\pi}{n} \right) - \prod_{k=1}^{n-2} \left(2 - \lambda + 2 \cos \frac{k\pi}{n-1} \right)$$

$$= (1 - \lambda) (-1)^n U_{n-1} \left(\frac{\lambda - 2}{2} \right) - (-1)^{n-2} U_{n-2} \left(\frac{\lambda - 2}{2} \right)$$

After introducing the series representation (B.135) of the Chebyshev polynomial $U_n(z)$ of the second kind and some manipulations, we find that the characteristic polynomial $c_n(\lambda)$ of $\nabla_{n \times n}^{-2}$ is

$$c_n(\lambda) = \sum_{k=0}^{n} \binom{n+k}{2k} (-\lambda)^k \tag{6.47}$$

Since the binomial coefficients are non-negative, Descartes' rule of signs (Theorem 87) states that $c_n(\lambda)$ has only positive real zeros. The same result also follows from the eigenvalue equation $\nabla^{-2} x = \delta^{-2} x$. The interval of eigenvalues $\lambda = \delta^{-2}$ of ∇^{-2} is $[0, 4]$, where the maximum eigenvalue follows from Gerschgorin's Theorem 65.

Introducing $\binom{\alpha}{k} = \frac{1}{2\pi i} \int_{C(0)} \frac{(1+z)^\alpha}{z^{k+1}} dz$ into (6.47) yields

$$c_n(\lambda) = \frac{1}{2\pi i} \int_{C(0)} \sum_{k=0}^{n} \frac{(1+z)^{n+k}}{z^{2k+1}} (-\lambda)^k \, dz$$

$$= -\frac{(-\lambda)^{n+1}}{2\pi i} \int_{C(0)} \frac{(1+z)^{2n+1}}{(\lambda(1+z) + z^2) z^{2n+1}} dz + \frac{1}{2\pi i} \int_{C(0)} \frac{(1+z)^n z}{\lambda(1+z) + z^2} dz$$

If $\lambda = 0$, the first term vanishes and the second term equals one. If $\lambda \neq 0$, the second term is zero, because the integrand is analytic at $z = 0$. The first integral can be closed over the entire complex plane, except for the origin, thereby enclosing the simple poles of $\lambda(1+z) + z^2 = (z - z_1)(z - z_2)$ at $z_1 = \frac{-\lambda + \sqrt{\lambda^2 - 4\lambda}}{2}$ and $z_2 = \frac{-\lambda - \sqrt{\lambda^2 - 4\lambda}}{2}$ in clockwise sense. By Cauchy's residue theorem, it holds that

$$\frac{1}{2\pi i} \int_{C(0)} \frac{(1+z)^{2n+1}}{(z - z_1)(z - z_2) z^{2n+1}} dz = \frac{1}{(z_2 - z_1)} \left\{ \left(\frac{1 + z_1}{z_1} \right)^{2n+1} - \left(\frac{1 + z_2}{z_2} \right)^{2n+1} \right\}$$

By using properties of the zeros of a quadratic equation, we arrive, after some manipulations, to the closed form of the characteristic polynomial $c_n(\lambda)$ of $\nabla_{n \times n}^{-2}$ with $c_n(0) = 1$ and for $\lambda \neq 0$,

$$c_n(\lambda) = \frac{\left(\lambda + \sqrt{\lambda^2 - 4\lambda}\right) \left(2 - \lambda - \sqrt{\lambda^2 - 4\lambda}\right)^n - \left(\lambda - \sqrt{\lambda^2 - 4\lambda}\right) \left(2 - \lambda + \sqrt{\lambda^2 - 4\lambda}\right)^n}{2^{n+1} \sqrt{\lambda^2 - 4\lambda}} \tag{6.48}$$

which bears resemblance to Chebyshev polynomials (**art. 377**). After invoking the polar representation of complex numbers, the alternative form of characteristic polynomial $c_n(\lambda) = \det\left(\nabla_{n \times n}^{-2} - \lambda I\right)$ is

$$c_n(\lambda) = \frac{1}{\sqrt{1 - \frac{\lambda}{4}}} \sin\left(\arccos \frac{\sqrt{\lambda}}{2} - n \arccos \left(1 - \frac{\lambda}{2}\right) \right) \tag{6.49}$$

6.12 A link joining two disconnected graphs

We consider two applications of Theorem 15 in **art.** 90. Another application on the kite graph $P_n K_m$, that consists of a complete graph K_m and a path graph P_n attached to one of the nodes of K_m, is presented in Van Mieghem (2015b).

Example 1 We consider the complete graphs $G_1 = K_n$ and $G_2 = K_m$, where $n \geq m$, that are connected by one link. Then $G_1 \backslash \{i\} = K_{n-1}$ and $G_2 \backslash \{j\} = K_{m-1}$, because a removal of node in the complete graph is a complete graph with size minus 1. By using (6.1) into (3.104), the graph G with $N = n + m$ nodes has the characteristic polynomial in $y = \lambda + 1$

$$\det(A_G - \lambda I) = (-1)^{n+m} y^{n+m-4} \left\{ y^2 (y - n)(y - m) - (y - n + 1)(y - m + 1) \right\}$$
$$= (-1)^{n+m} y^{n+m-4} p(y)$$

where

$$p(y) = y^4 - (n + m) y^3 + (nm - 1) y^2 + (n + m - 2) y - (n - 1)(m - 1)$$

The zeros $z_1 \geq z_2 \geq z_3 \geq z_4$ of the fourth degree polynomial $p(y)$ can be computed exactly. Unfortunately, the algebraic expressions are cumbersome and fail to provide insight.

The sum of zeros equals $z_1 + z_2 + z_3 + z_4 = n + m = N$, while the product equals $z_1 z_2 z_3 z_4 = -(n-1)(m-1)$. Since $\lambda_1(K_n) = n - 1$, we know from (3.107) that the largest zero lies[7] between $n < z_1 < n + 1$. We note that

$$p(n) = -1 - (n - m) \quad p(m) = -1 + (n - m) \quad p(m - 1) = -(m - 1)^2 p(n)$$

and

$$p(1) = N - 3 \quad p(0) = -(n - 1)(m - 1) \quad p(-1) = N + 1$$

If $n - m = 1$, then $p(m) = 0$ and $z_2 = m$ is the second highest zero. If $n - m > 1$, then $p(m) > 0$ and $p(n) < 0$ indicate that there is at least one zero between m and n. If $n = m$, the second largest zero z_2 lies between $n - 1$ and n, because $p(n-1) = -(n-1)^2 p(n)$. The fact that $p(0) < 0$ and $p(1) > 0$ illustrates that z_3 must lie in between 0 and 1 (for $m \geq 2$). The fact that $p(0) < 0$ and $p(-1) > 0$ illustrates that z_4 must lie in between -1 and 0 and z_4 is the only negative zero. Moreover, in case $n - m \geq 1$, $z_1 + z_2 > n + m$ so that $z_3 + z_4 < 0$, which shows that $|z_4| > z_3$. In summary for $n > m$, the zeros of $p(y)$ obey $n < z_1 < n + 1$, $m < z_2 < n$, $0 < z_3 < 1$ and $-1 < z_4 < 0$.

Next, we consider the Lagrange expansion (**art.** 342) around $y = n$. We expand the polynomial $p(y)$ in a Taylor series around $y = n$ and obtain, with $x = y - n$,

$$p(x + n) = x^4 + \{3n - m\} x^3 + \{3n^2 - 2nm - 1\} x^2$$
$$+ \{n^3 - mn^2 - (n - m) - 2\} x - \{1 + (n - m)\}$$

---

[7] Numerical computations show that $z_1 < n + \frac{1}{2}$ for $n \geq m > 2$.

Application of the general Lagrange series (B.68) and recalling that $y = \lambda + 1$, yields, up to third order in $\xi = \frac{1+(n-m)}{n^2(n-m)-(n-m)-2}$,

$$\lambda_1(G) = n - 1 + \xi - \frac{3n^2 - 2nm - 1}{n^3 - mn^2 - (n-m) - 2}\xi^2$$
$$+ \left[\frac{(3n-m)(n^3 - mn^2 - (n-m) - 2) - 2(3n^2 - 2nm - 1)^2}{n^3 - mn^2 - (n-m) - 2}\right]\xi^3 + O(\xi^4)$$

This Lagrange series converges rapidly for large n and large $n - m$, since ξ is then small, but diverges for $n = m$. Since the second order term is positive, we find for $n - m \geq 1$ that

$$\Delta\lambda_1 = \lambda_1(G) - \lambda_1(G\backslash\{i \sim j\}) \leq \xi = \frac{1 + (n-m)}{n^2(n-m) - (n-m) - 2}$$

which is also satisfied for $n = m \geq 2$ when the absolute value is taken. This bound is increasingly sharp in $n - m$ and shows that $\Delta\lambda_1$ is decreasing with $n - m$. In other words, the maximum $\Delta\lambda_1$ is obtained for the symmetrical case where $n = m$.

Example 2 The case where both $G_1 = K_{1,n}$ and $G_2 = K_{1,m}$ are stars can also be evaluated exactly with (6.27). We have that $G_1\backslash\{i\} = K_{1,n-1}$ and $G_2\backslash\{j\} = K_{1,m-1}$ when i and j are not the center node. (a) Application of (3.104) yields

$$\det(A_G - \lambda I) = (-1)^{n+m}\lambda^{n+m-4}\{\lambda^2(\lambda^2 - n)(\lambda^2 - m) - (\lambda^2 - n + 1)(\lambda^2 - m + 1)\}$$
$$= (-1)^{n+m}\lambda^{n+m-4}q(\lambda^2)$$

where

$$q(\lambda) = \lambda^3 - (n + m + 1)\lambda^2 + (nm + n + m - 2)\lambda - (n - 1)(m - 1)$$

The zeros of the third degree polynomial $q(\lambda)$ in λ^2 again can be computed exactly. Cardano's explicit expressions are unfortunately still unattractively complex.

(b) If i is a center node, then $G_1\backslash\{i\}$ consists of $n - 1$ disconnected nodes and $A_{G_1\backslash\{i\}} = O$; then

$$\det(A_G - \lambda I) = (-1)^{n+m}\lambda^{n+m-2}\{(\lambda^2 - n)(\lambda^2 - m) - (\lambda^2 - m + 1)\}$$
$$= (-1)^{n+m}\lambda^{n+m-2}\{\lambda^4 - (n + m + 1)\lambda^2 + (n + 1)m - 1\}$$

The zeros of the quadratic polynomial in λ^2 are

$$\lambda_\pm^2 = \frac{1}{2}\left\{(n + m + 1) \pm \sqrt{(n + 1 - m)^2 + 4}\right\}$$

from which

$$\lambda_1(G) = \sqrt{\frac{1}{2}\left\{(n + 1 + m) + \sqrt{(n + 1 - m)^2 + 4}\right\}}$$
$$= \sqrt{n + 1 + \frac{1}{n + 1 - m} + O\left(\frac{1}{(n + 1 - m)^3}\right)}$$

This eigenvalue is largest when $n = m$ and equal to $\lambda_1(G) = \sqrt{n + \frac{1+\sqrt{5}}{2}}$.

(c) In case both i and j are center nodes of the star, we have

$$\det(A_G - \lambda I) = (-1)^{n+m} \lambda^{n+m-2} (\lambda^2 - n)(\lambda^2 - m) - (-1)^{n+m} \lambda^{n+m}$$

$$= (-1)^{n+m} \lambda^{n+m-2} \{\lambda^4 - (n+m+1)\lambda^2 + nm\}$$

and with roots $\lambda_{\pm}^2 = \frac{1}{2}\left\{(n+m+1) \pm \sqrt{(n-m)^2 + 2(n+m) + 1}\right\}$. In case $n = m$, we find $\lambda_1(G) = \sqrt{n + \frac{1+\sqrt{4n+1}}{2}}$.

6.13 A chain of cliques

A chain of $D+1$ cliques is a graph $G_D^*(n_1, n_2, ..., n_{D+1})$ consisting of $D+1$ complete graphs K_{n_j} or cliques with $1 \leq j \leq D+1$, where each clique K_{n_j} is fully interconnected with its neighboring cliques $K_{n_{j-1}}$ and $K_{n_{j+1}}$. Two graphs G_1 and G_2 are fully interconnected if each node in G_1 is connected to each node in G_2. An example of a member of the class $G_D^*(n_1, n_2, ..., n_{D+1})$ is drawn in Fig. 6.5. The

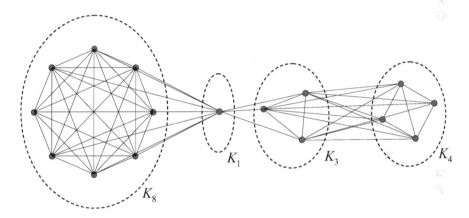

Fig. 6.5. A chain of cliques $G_4^*(8, 1, 3, 4)$.

total number of nodes in $G_D^*(n_1, n_2, ..., n_{D+1})$ is

$$N = \sum_{j=1}^{D+1} n_j \tag{6.50}$$

The total number of links in G_D^* is

$$L = \sum_{j=1}^{D+1} \binom{n_j}{2} + \sum_{j=1}^{D} n_j n_{j+1} \tag{6.51}$$

where the first sum equals the number of intra-cluster links and the second the number of inter-cluster links. The main motivation to study the class of graphs

$G_D^*(n_1, n_2, ..., n_{D+1})$ with $n_j \geq 1$ is its extremal properties, which are proved in Wang *et al.* (2010):

Theorem 41 *Any graph $G(N, D)$ with N nodes and diameter D is a subgraph of at least one graph in the class $G_D^*(n_1 = 1, n_2, ..., n_D, n_{D+1} = 1)$.*

Theorem 42 *The maximum of any Laplacian eigenvalue $\mu_i(G_D^*)$ for $i \in [1, N]$ achieved in the class $G_D^*(n_1 = 1, n_2, ..., n_D, n_{D+1} = 1)$ is also the maximum among all the graphs with N nodes and diameter D.*

Theorem 43 *The maximum number of links in a graph with given size N and diameter D is $L_{\max}(N, D) = \binom{N-D+2}{2} + D - 3$, which can only be obtained by either $G_D^*(1, ..., 1, n_j = N - D, 1, ..., 1)$ with $j \in [2, D]$, where only one clique has size larger than one, or by $G_D^*(1, ..., 1, n_j > 1, n_{j+1} > 1, 1, ..., 1)$ with $j \in [2, D-1]$ where only two cliques have size larger than one and they are next to each other.*

Another theorem, due to van Dam (2007) and related to Theorem 43, is:

Theorem 44 *The graph $G_D^*(n_1, n_2, ..., n_{D+1})$ with $n_{\lceil \frac{D+1}{2} \rceil} = N - D$ and all other $n_j = 1$ is the graph with the largest eigenvalue of the adjacency matrix among all graphs with a same diameter D and number of nodes N.*

Here, we will compute the Laplacian spectrum of $G_D^*(n_1, n_2, ..., n_{D-1}, n_D, n_{D+1})$: we will show that $N - D$ eigenvalues are exactly known, while the remaining D eigenvalues are the positive zeros of an orthogonal polynomial. The adjacency matrix $A_{G_D^*}$ of $G_D^*(n_1, n_2, ..., n_{D-1}, n_D, n_{D+1})$ is

$$
\begin{bmatrix}
\tilde{J}_{n_1 \times n_1} & J_{n_1 \times n_2} & & & & \\
J_{n_2 \times n_1} & \tilde{J}_{n_2 \times n_2} & J_{n_2 \times n_3} & & & \\
& & \ddots & & & \\
& & J_{n_i \times n_{i-1}} & \tilde{J}_{n_i \times n_i} & J_{n_i \times n_{i+1}} & \\
& & & & \ddots & \\
& & & & J_{n_{D+1} \times n_{D+1}} & \tilde{J}_{n_{D+1} \times n_{D+1}}
\end{bmatrix}
$$

where $\tilde{J} = J - I$.

Theorem 45 *The characteristic polynomial of the Laplacian $Q_{G_D^*}$ of the class of graphs $G_D^*(n_1, n_2, ..., n_{D+1})$ equals*

$$\det\left(Q_{G_D^*} - \mu I\right) = p_D(\mu) \prod_{j=1}^{D+1} (d_j + 1 - \mu)^{n_j - 1} \tag{6.52}$$

where $d_j = n_{j-1} + n_j + n_{j+1} - 1$ denotes the degree of a node in clique j. The polynomial $p_D(\mu) = \prod_{j=1}^{D+1} \theta_j$ is of degree $D+1$ in μ and the function $\theta_j = \theta_j(D; \mu)$

obeys the recursion

$$\theta_j = (d_j + 1 - \mu) - \left(\frac{n_{j-1}}{\theta_{j-1}} + 1\right) n_j \qquad (6.53)$$

with initial condition $\theta_0 = 1$ and with the convention that $n_0 = n_{D+2} = 0$.

The proof below elegantly uses the concept of a quotient matrix, defined in Section 2.5. An elementary, though more elaborated proof, which is basically an extension of the derivation in Section 6.10.3, is found in Van Mieghem and Wang (2009). Consider the k-partition of a graph G that separates the node set $\mathcal{N}$ of G into $k \in [1, N]$ disjoint, non-empty subsets $\{\mathcal{N}_1, \mathcal{N}_2, ..., \mathcal{N}_k\}$. Correspondingly, the quotient matrix A^π of the adjacency matrix of G is a $k \times k$ matrix, where $A^\pi_{i,j}$ is the average number of neighbors in $\mathcal{N}_j$ of nodes in $\mathcal{N}_i$. Similarly, the quotient matrix Q^π of the Laplacian matrix Q of G is a $k \times k$ matrix, where

$$Q^\pi_{i,j} = \begin{cases} -A^\pi_{i,j}, & \text{if } i \neq j \\ \sum_{i \neq k} A^\pi_{i,k}, & \text{if } i = j \end{cases}$$

As defined in **art. 37**, a partition is called regular or equitable if for all $1 \leq i, j \leq k$ the number of neighbors in $\mathcal{N}_j$ is the same for all the nodes in $\mathcal{N}_i$. The eigenvalues derived from the quotient matrix A^π (Q^π) of the adjacency A (Laplacian Q) matrix are also eigenvalues of A (Laplacian Q) given the partition is equitable (see **art. 37**).

Proof: The partition that separates the graph $G_D^*(n_1, n_2, ..., n_{D+1})$ into the $D + 1$ cliques $K_{n_1}, K_{n_2}, ..., K_{n_{D+1}}$ is equitable. The quotient matrix Q^π of the Laplacian matrix Q of G is

$$Q^\pi = \begin{bmatrix} n_2 & -n_2 & & & & & \\ -n_1 & n_1 + n_3 & -n_3 & & & & \\ & -n_2 & n_2 + n_4 & -n_4 & & & \\ & & & \ddots & & & \\ & & & & -n_{D-1} & n_{D-1} + n_{D+1} & -n_{D+1} \\ & & & & & -n_D & n_D \end{bmatrix} \qquad (6.54)$$

We use (A.57) to det $(Q^\pi - \mu I)$

$$= \begin{vmatrix} n_2 - \mu & -n_2 \\ -n_1 & n_1 + n_3 - \mu & -n_3 \\ & -n_2 & n_2 + n_4 - \mu & -n_4 \\ & & & \ddots \\ & & & & -n_{D-1} & n_{D-1} + n_{D+1} - \mu & -n_{D+1} \\ & & & & & -n_D & n_D - \mu \end{vmatrix}$$

$$= (n_2 - \mu) \begin{vmatrix} n_1 + n_3 - \mu - \frac{n_1 n_2}{n_2 - \mu} & -n_3 \\ -n_2 & n_2 + n_4 - \mu & -n_4 \\ & & \ddots \\ & & & n_D - \mu \end{vmatrix}$$

We repeat the method and obtain

$$\det(Q^\pi - \mu I) = (n_2 - \mu)\left(n_1 + n_3 - \mu - \frac{n_1 n_2}{n_2 - \mu}\right) \times$$

$$\det \begin{bmatrix} n_2 + n_4 - \mu - \dfrac{n_2 n_3}{\left(n_1 + n_3 - \mu - \frac{n_1 n_2}{n_2 - \mu}\right)} & -n_4 \\ & \ddots \\ & & -n_{D-1} & n_{D-1} + n_{D+1} - \mu & -n_{D+1} \\ & & & -n_D & n_D - \mu \end{bmatrix}$$

Eventually, after subsequent expansions using (A.57), we find

$$\det(Q^\pi - \mu I) = \prod_{j=1}^{D+1} \theta_j = p_D(\mu)$$

where θ_j follows the recursion

$$\theta_j = (n_{j-1} + n_{j+1} - \lambda) - \frac{n_{j-1} n_j}{\theta_{j-1}}$$

with initial condition $\theta_0 = 1$ and with the convention that $n_0 = n_{D+2} = 0$. When written in terms of the degree $d_j = n_{j-1} + n_j + n_{j+1} - 1$, we obtain (6.53).

Any two nodes s and t in a same clique K_{n_i} of G_D^* are connected to each other and they are connected to the same set of neighbors. The two rows in det $(Q_{G_D^*} - \mu I)$ corresponding to node s and t are the same when $\mu = d_i + 1$, where d_i is the degree of all nodes in clique K_{n_i}. In this case, det $(Q_{G_D^*} - \mu I) = 0$ since the rank of $Q_{G_D^*} - \mu I$ is reduced by 1. Hence, $\mu = d_i + 1$ is an eigenvalue of the Laplacian matrix $Q_{G_D^*}$. The corresponding eigenvector x has only two non-zero components, $x_s = -x_t \neq 0$. Since the $D + 1$ partitions of $G_D^*(n_1, n_2, ..., n_{D+1})$ are equitable, the $D + 1$ eigenvalues of Q^π, which are the roots of det $(Q^\pi - \mu I) = 0$, are also the eigenvalues of the Laplacian matrix $Q_{G_D^*}$. Each eigenvector of $Q_{G_D^*}$, belonging to the $D + 1$ eigenvalues, has the same elements $x_s = x_t$ if the nodes s and t belong to the same clique. Hence, the Laplacian matrix $Q_{G_D^*}$ has $D + 1$ non-trivial eigenvalues, which are the roots of $\det(Q^\pi - \mu I) = 0$ and trivial eigenvalues $d_j + 1$ with multiplicity $n_j - 1$ for $1 \leq j \leq D + 1$.   □

6.13.1 Orthogonal polynomials

In the sequel, we will show that the polynomial $p_D(\mu)$ in Theorem 45 belongs to a set of orthogonal polynomials (see Chapter 12). The dependence of θ_j on the diameter D and on μ is further on explicitly written.

Lemma 5 *For all $j \geq 0$, the functions $\theta_j(D; x)$ are rational functions*

$$\theta_j(D; x) = \frac{t_j(D; x)}{t_{j-1}(D; x)} \tag{6.55}$$

where $t_j(x)$ is a polynomial of degree j in $x = -\mu$ and $t_0(D; x) = 1$.

Proof: It holds for $j = 1$ as verified from (6.53) because $\theta_0(D; x) = 1$. Let us assume that (6.53) holds for $j - 1$ (induction argument). Substitution of (6.55) into the right-hand side of (6.53),

$$\theta_j(D; x) = \begin{cases} \frac{(x + n_{j-1} + n_{j+1})t_{j-1}(D;x) - n_{j-1}n_j t_{j-2}(D;x)}{t_{j-1}(D;x)} & 1 \leq j \leq D \\ \frac{(x + n_D)t_D(D;x) - n_D n_{D+1}t_{D-1}(D;x)}{t_D(D;x)} & j = D + 1 \end{cases}$$

indeed shows that the left-hand side is of the form (6.55) for j. This demonstrates the induction argument and proves the lemma.   □

The polynomial of interest,

$$p_D(\mu) = \prod_{j=1}^{D+1}\theta_j(D; \mu) = \sum_{k=0}^{D+1} c_k(D)\mu^k = \prod_{k=1}^{D+1}(z_k - \mu) \tag{6.56}$$

(where the product with the zeros $z_{D+1} \leq z_D \leq \cdots \leq z_1$ follows from the definition of the eigenvalue equation (A.97)) equals with (6.55)

$$p_D(-x) = \frac{\prod_{j=1}^{D+1} t_j(D; x)}{\prod_{j=1}^{D+1} t_{j-1}(D; x)} = t_{D+1}(D; x)$$

We rewrite (6.55) as $t_j(D; x) = \theta_j(D; x)t_{j-1}(D; x)$ and with (6.53), we obtain the set of polynomials

$$\begin{cases} t_{D+1}(D; x) = (x + n_D)t_D(D; x) - n_D n_{D+1}t_{D-1}(D; x) \\ t_j(D; x) = (x + n_{j-1} + n_{j+1})t_{j-1}(D; x) - n_{j-1}n_j t_{j-2}(D; x) & \text{for } 1 \leq j \leq D \\ t_1(D; x) = (x + n_2)t_0(D; x) \end{cases}$$

$$\tag{6.57}$$

where $t_0(D; x) = 1$. **Art. 357** demonstrates that, for a fixed D, the sequence $\{t_j(D; x)\}_{0 \leq j \leq D+1}$ is a set of orthogonal polynomials because (6.57) obeys Favard's three-term recurrence relation. By Theorem 112, the zeros of any set of orthogonal polynomials are all simple, real and lie in the orthogonality interval $[a, b]$, which is here for the Laplacian equal to $[0, N]$.

By iterating the equation upwards, we find that

$$
t_j (D; 0) = \begin{cases} \displaystyle\prod_{m=2}^{j+1} n_m & 1 \le j \le D \\ 0 & j = D + 1 \end{cases}
\tag{6.58}
$$

Thus, $t_{D+1} (D; 0) = 0$ and thus $\theta_{D+1} (D; 0) = 0$ implies that $p_D (\mu)$ must have a zero at $\mu = 0$, which is, indeed, a general property of any Laplacian (**art. 101**). From (6.55), it then follows that $\theta_j (D; 0) = n_{j+1} > 0$. The eigenvalues of the Jacobi matrix (**art. 370**),

$$
M = \begin{bmatrix}
-n_2 & 1 & & & \\
n_1 n_2 & -(n_1 + n_3) & 1 & & \\
& \ddots & \ddots & \ddots & \\
& & n_{D-1} n_D & -(n_{D-1} + n_{D+1}) & 1 \\
& & & n_D n_{D+1} & -n_D
\end{bmatrix}
\tag{6.59}
$$

are equal to the zeros of $p_D (-x)$. Moreover, we observe that also the quotient matrix Q^π in (6.54) possesses the same eigenvalues as the Jacobi matrix M. Since the eigenvalues of M are simple, **art. 239** shows that there exists a similarity transform that maps the Jacobi matrix M into the quotient matrix Q^π (and vice versa). Moreover, the matrix M can be symmetrized by a similarity transform,

$$
H = \mathrm{diag}\left(1, \frac{1}{\sqrt{n_1 n_2}}, \dots, \frac{1}{\sqrt{n_1 n_j} \displaystyle\prod_{k=2}^{j-1} n_k}, \dots, \frac{1}{\sqrt{n_1 n_{D+1}} \displaystyle\prod_{k=2}^{D} n_k} \right)
$$

and the eigenvector belonging to zero equals

$$
\widetilde{\tau} (D; 0) = H\tau (D; 0) = \begin{bmatrix} 1 & \sqrt{\frac{n_2}{n_1}} & \cdots & \sqrt{\frac{n_{D-1}}{n_1}} & \sqrt{\frac{n_D}{n_1}} \end{bmatrix}^T
$$

After the similarity transform H, the result is $\widetilde{M} = HMH^{-1}$,

$$
\widetilde{M} = \begin{bmatrix}
-n_2 & \sqrt{n_1 n_2} & & & \\
\sqrt{n_1 n_2} & -(n_1 + n_3) & \sqrt{n_2 n_3} & & \\
& \ddots & \ddots & \ddots & \\
& & \sqrt{n_{D-1} n_D} & -(n_{D-1} + n_{D+1}) & \sqrt{n_D n_{D+1}} \\
& & & \sqrt{n_D n_{D+1}} & -n_D
\end{bmatrix}
$$

The corresponding square root matrix A of the Gram matrix $-M = A^T A$ can be computed explicitly as

$$
A = \begin{bmatrix}
\sqrt{n_2} & -\sqrt{n_1} & & & & \\
0 & \sqrt{n_3} & -\sqrt{n_2} & & & \\
& \ddots & \ddots & \ddots & & \\
& & 0 & \sqrt{n_{D+1}} & -\sqrt{n_D} \\
& & & 0 & 0
\end{bmatrix}
$$

in contrast to the general theory in **art.** 374, where each element is a continued fraction.

In summary, all non-trivial eigenvalues of $Q_{G_D^*}$ are also eigenvalues of the simpler matrices Q^π, $-M$ or $-\widetilde{M}$. Properties and bounds on those non-trivial eigenvalues and zeros of $p_D(\mu)$ as well as the spectrum of the corresponding adjacency matrix are studied in Van Mieghem and Wang (2009). We mention the asymptotic scaling law:

Theorem 46 *For a constant diameter D and a large number N of nodes, all non-trivial eigenvalues of both the adjacency and Laplacian matrix of any graph in the class $G_D^*(n_1, n_2, ..., n_{D+1})$ scale linearly with N, the number of nodes.*

All coefficients $c_k(D)$ of $p_D(\mu)$ in (6.56) can be computed explicitly in terms of the clique sizes $n_1, n_2, \ldots, n_{D+1}$ for which we refer to Van Mieghem and Wang (2009). We merely list here the first few polynomials $q_D(\mu) = \frac{p_D(\mu)}{-\mu}$:

$q_1(\mu) = -(\mu - N)$

$q_2(\mu) = \mu^2 - (N + n_2)\mu + Nn_2 = (\mu - N)(\mu - n_2)$

$q_3(\mu) = -\mu^3 + (2N - n_1 - n_4)\mu^2 - (n_2^2 + n_3^2 + n_1n_2 + n_1n_3 + n_1n_4 + 3n_2n_3 + n_2n_4 + n_3n_4)\mu$
$\qquad + Nn_2n_3$

$q_4(\mu) = \mu^4 - (2N - n_1 - n_5)\mu^3 + (n_2^2 + n_3^2 + n_4^2 + n_4n_5 + n_3(3n_4 + n_5))\mu^2$
$\qquad + (n_2(3n_3 + 3n_4 + 2n_5) + n_1(n_2 + n_3 + 2n_4 + n_5))\mu^2$
$\qquad - (n_3n_4(n_3 + n_4 + n_5) + n_2\{n_3^2 + n_4^2 + 4n_3n_4 + (n_3 + n_4)n_5 + n_2(n_3 + n_4 + n_5)\}$
$\qquad + n_1(n_2 + n_4)(n_3 + n_4 + n_5))\mu + Nn_2n_3n_4$

For increasing D, the explicit expressions rapidly become involved without a simple structure. There is one exception: $G_D^*(n_1, n_2, ..., n_{D+1})$ with all unit size cliques, $n_j = 1$, is a D-hop line topology, whose spectrum is exactly given in (6.15), such that

$$
q_D\left(\mu; \{n_j = 1\}_{1 \le j \le D+1}\right) = \prod_{k=1}^{D} \left(2\left(1 - \cos\left(\frac{k\pi}{D+1}\right)\right) - \mu\right)
$$

Finally, we mention that $q_3(\mu)$ appears as the polynomial $s(\mu)$ in the Laplacian spectrum of the complementary double cone (CDC) in Section 6.10.3. The CDC, written as $G_3^*(1, m, N - m, 1)$, is clearly a member of the class $G_D^*(n_1, n_2, ..., n_{D+1})$ with $\sum_{j=1}^4 n_j = N + 2$.

6.14 The lattice

Consider a rectangular lattice with size z_1 and z_2 where at each lattice point with two integer coordinates (k, l) a node is placed. A node at (k, l) is connected to its direct neighbors at $(k + 1, l)$, $(k - 1, l)$, $(k, l + 1)$ and $(k, l - 1)$ where possible. At border points, nodes only have three neighbors and at the four corner points only two. The number of lattice points (nodes) equals $N = (z_1 + 1)(z_2 + 1)$ and the number of links is $L = 2z_1 z_2 + (z_1 + z_2)$. Meyer (2000) nicely relates the Laplacian of the lattice $G_{\text{La}(N)}$ to the discrete version of the Laplacian operator,

$$\frac{\partial^2 u}{\partial x^2} + \frac{\partial^2 u}{\partial y^2}$$

In a similar vein, Cvetković *et al.* (2009, Chapter 9) discuss the Laplacian operator and its discretization in the solution of the wave equation with rectangular boundary.

The adjacency matrix, following Meyer (2000), is

$$A_{\text{La}(N)} = \begin{bmatrix} T_{(z_1+1)\times(z_1+1)} & I_{(z_1+1)\times(z_1+1)} & & & & \\ I_{(z_1+1)\times(z_1+1)} & T_{(z_1+1)\times(z_1+1)} & I & & & \\ & I_{(z_1+1)\times(z_1+1)} & \ddots & & \ddots & \\ & & & \ddots & T_{(z_1+1)\times(z_1+1)} & I_{(z_1+1)\times(z_1+1)} \\ & & & & I_{(z_1+1)\times(z_1+1)} & T_{(z_1+1)\times(z_1+1)} \end{bmatrix}$$

where the Toeplitz matrix

$$T_{(z_1+1)\times(z_1+1)} = \begin{bmatrix} 0 & 1 & & & \\ 1 & 0 & 1 & & \\ & 1 & \ddots & \ddots & \\ & & \ddots & 0 & 1 \\ & & & 1 & 0 \end{bmatrix}$$

is the adjacency matrix of a z_1 hops path whose eigenvalues are given by (6.10) with $N \to z_1 + 1$. The Laplacian $Q_{\text{La}(N)}$ is not easily given in general form because the sum of the rows in $A_{\text{La}(N)}$ or the degree of a node is not constant. The adjacency matrix $A_{\text{La}(N)}$ is a block Toeplitz matrix whose structure is most elegantly written in terms of a Kronecker product. We may verify that[8]

$$A_{\text{La}(N)} = I_{(z_2+1)\times(z_2+1)} \otimes T_{(z_1+1)\times(z_1+1)} + T_{(z_2+1)\times(z_2+1)} \otimes I_{(z_1+1)\times(z_1+1)} \quad (6.60)$$

The eigenvalues of $A_{\text{La}(N)}$ are immediate from **art. 286**. For $1 \leq j \leq z_1 + 1, 1 \leq k \leq z_2 + 1$, it holds that $\lambda_{jk}\left(A_{\text{La}(N)}\right) = \lambda_j\left(T_{(z_1+1)\times(z_1+1)}\right) + \lambda_k\left(T_{(z_2+1)\times(z_2+1)}\right)$

---

[8] When applying the identity $(A_1 \otimes B_1)(A_2 \otimes B_2) = (A_1 A_2 \otimes B_1 B_2)$ in (Meyer, 2000, p. 597) to

and with (6.10), we arrive at

$$\lambda_{jk}\left(A_{\mathrm{La}(N)}\right) = 2\cos\left(\frac{\pi j}{z_1 + 2}\right) + 2\cos\left(\frac{\pi k}{z_2 + 2}\right)$$

where $1 \le j \le z_1 + 1, 1 \le k \le z_2 + 1$.

Several extensions are possible. For a cubic or three dimensional lattice, the adjacency matrix generalizes to

$$\begin{aligned}
A_{\mathrm{La}(N)} &= I_{(z_3+1)\times(z_3+1)} \otimes I_{(z_2+1)\times(z_2+1)} \otimes T_{(z_1+1)\times(z_1+1)} \\
&\quad + I_{(z_3+1)\times(z_3+1)} \otimes T_{(z_2+1)\times(z_2+1)} \otimes I_{(z_1+1)\times(z_1+1)} \\
&\quad + T_{(z_3+1)\times(z_3+1)} \otimes I_{(z_2+1)\times(z_2+1)} \otimes I_{(z_1+1)\times(z_1+1)}
\end{aligned}$$

with spectrum

$$\lambda_{jkl}\left(A_{\mathrm{La}(N)}\right) = 2\cos\left(\frac{\pi j}{z_1 + 2}\right) + 2\cos\left(\frac{\pi k}{z_2 + 2}\right) + 2\cos\left(\frac{\pi l}{z_3 + 2}\right)$$

where $1 \le j \le z_1 + 1, 1 \le k \le z_2 + 1, 1 \le l \le z_3 + 1$. The Kronecker product where the Toeplitz matrix T of the path is changed for the circulant Toeplitz matrix of the circuit represents a lattice on a torus (Cvetković *et al.*, 1995, p. 74).

We end this section by considering the m-dimensional lattice La_m with lengths $z_1, z_2, \ldots, z_m$ in each dimension, respectively, and where at each lattice point with integer coordinates a node is placed that is connected to its nearest neighbors whose coordinates only differ by one in only one component. The total number of nodes in La_m is $N = (z_1 + 1) \times (z_2 + 1) \times \ldots \times (z_m + 1)$. The lattice graph can be written as a Cartesian product (Cvetković *et al.*, 1995) of m path graphs, which we denote by $\mathrm{La}_m = P_{(z_1+1)}\square P_{(z_2+1)}\square \ldots \square P_{(z_m+1)}$. According to Cvetković *et al.* (1995), the eigenvalues of La_m can be written as a sum of one combination of eigenvalues of path graphs and the corresponding eigenvector is the Kronecker product of the corresponding eigenvectors of the same path graphs,

$$\begin{aligned}
\lambda_{i_1 i_2 \ldots i_N}(\mathrm{La}_m) &= \sum_{j=1}^{m} \lambda_{i_j}\left(P_{(z_j+1)}\right) \\
x_{i_1 i_2 \ldots i_m}(\mathrm{La}_m) &= x_{i_1}\left(P_{(z_1+1)}\right) \otimes x_{i_2}\left(P_{(z_2+1)}\right) \otimes \ldots \otimes x_{i_m}\left(P_{(z_m+1)}\right)
\end{aligned} \tag{6.61}$$

where $i_j \in \{1, 2, \ldots, z_j + 1\}$ for each $j \in \{1, 2, \ldots, m\}$. Since both the adjacency and the Laplacian spectrum of the path P_N graph are completely known (Section

---

compute the square of $A_{\mathrm{La}(N)}^2$ given by (6.60), powers of the Toeplitz matrix appear. However,

$$T_{(z_1+1)\times(z_1+1)}^2 = \begin{bmatrix} 1 & 0 & 1 & & & \\ 0 & 2 & 0 & \ddots & & \\ 1 & 0 & \ddots & \ddots & 1 \\ & \ddots & \ddots & 2 & 0 \\ & & 1 & 0 & 1 \end{bmatrix}$$

shows that the Toeplitz structure is destroyed.

6.4), the corresponding spectra of the m-dimensional lattice La_m can also analytically be computed from (6.61) by substituting $N = z_j + 1$ in the derivations in Section 6.4.

Lemma 6 *The number L of links in the m-dimensional lattice La_m is, for $m \geq 1$,*

$$L = \prod_{j=1}^{m}(z_j + 1) \sum_{j=1}^{m} \frac{z_j}{z_j + 1}$$

Proof: We will prove the lemma by induction. Let the number of links in the k-dimensional lattice La_k be $l(z_1, z_2, \ldots, z_k)$. For $k = 1$, we have a path graph P_{z_1+1} and its number of links is $L = l(z_1) = z_1 = (z_1 + 1)\frac{z_1}{z_1+1}$. Let us assume that the lemma holds for k-dimensional lattices. We consider the $(k + 1)$-dimensional lattice La_{k+1}, which is constructed from k different k-dimensional lattices

$$La_{(z_{i_1}+1)\times(z_{i_2}+1)\times\ldots\times(z_{i_k}+1)}, \text{ where } i_1, i_2, \ldots, i_k \in \{1, 2, \ldots, (k+1)\}$$

in the following way. We position a total of $\left(z_{i_{k+1}} + 1\right)$ such k-dimensional lattices $La_{(z_{i_1}+1)\times(z_{i_2}+1)\times\ldots\times(z_{i_k}+1)}$ next to each other in the direction of dimension i_{k+1}. In this way, every link is counted k-times in each dimension. Intuitively, this construction is easier to imagine in three dimensions, where the three dimensional lattice $La_{(z_1+1)\times(z_2+1)\times(z_3+1)}$ is constructed by (z_3+1) consecutive two dimensional $La_{(z_1+1)\times(z_2+1)}$ planes that are positioned next to each other in the direction of the third dimension, $(z_2 + 1)$ consecutive two dimensional $La_{(z_1+1)\times(z_3+1)}$ planes that are positioned next to each other in the direction of the second dimension and, finally, $(z_1 + 1)$ consecutive two dimensional La $L_{(z_2+1)\times(z_3+1)}$ planes that are positioned next to each other in the direction of the first dimension. All links in this process are counted twice. Returning to the k-dimensional case, we thus deduce that

$$l(z_1, z_2, \ldots, z_{k+1}) = \frac{1}{k} \sum_{i=1}^{k+1}(z_i + 1)l(z_{j_1}, z_{j_2}, \ldots, z_{j_k})$$

where $j_w \neq i$ for each $i = 1, 2, \ldots, k + 1$ and $w = 1, 2, \ldots, k$. Introducing the induction hypothesis for k-dimensional lattices, we obtain

$$l(z_1, z_2, \ldots, z_{k+1}) = \frac{1}{k} \sum_{i=1}^{k+1}(z_i + 1) \prod_{j=1,j\neq i}^{k+1}(z_j + 1) \sum_{j=1,j\neq i}^{k+1} \frac{z_j}{z_j + 1}$$

$$= \frac{1}{k}\prod_{j=1}^{k+1}(z_j + 1) \sum_{i=1}^{k+1}\sum_{j=1,j\neq i}^{k+1} \frac{z_j}{z_j + 1}$$

$$= \frac{1}{k}\prod_{j=1}^{k+1}(z_j + 1)k \sum_{i=1}^{k+1} \frac{z_j}{z_j + 1}$$

which illustrates that the induction hypothesis is true for $k + 1$, and consequently, the lemma is true for each dimension $m \geq 1$. $\qquad\square$

7

Density function of the eigenvalues

General properties of the density function of eigenvalues are studied. Most articles in this chapter implicitly consider the eigenvalues of the adjacency matrix A. Especially for large graphs and random graphs, a probabilistic setting in terms of the density function is more suitable than the list of eigenvalues.

7.1 Definitions

172. *The Dirac function.* The Dirac function, also called impulse or delta function, is the continuous counterpart of the Kronecker delta or indicator function. The Dirac function is a generalized function with characteristic property

$$g(x) = \int_L g(u)\, \delta(x - u)\, du \tag{7.1}$$

where L is a path in the complex plane containing x and g is a function that is defined and finite along L. For example, when L is the real axis and $g = 1$, we find the well-known property that $\int_{-\infty}^{\infty} \delta(t)\, dt = 1$. Since the Dirac function is a generalized function, there exist several representations.

A first class of representations is deduced from integral transform pairs. For example, from the double-sided Laplace transform pair

$$F(s) = \int_{-\infty}^{\infty} f(x)\, e^{-sx} dx \quad \Leftrightarrow \quad f(x) = \frac{1}{2\pi i} \int_{c-i\infty}^{c+i\infty} F(s)\, e^{sx} ds$$

where c is the smallest real part of s for which $\int_{-\infty}^{\infty} f(x)\, e^{-sx} dx$ exists, we find formally after a reversal in integration that

$$f(x) = \int_{-\infty}^{\infty} du\, f(u) \left(\frac{1}{2\pi i} \int_{c-i\infty}^{c+i\infty} ds\, e^{s(x-u)} \right)$$

Comparison with the property (7.1) leads to the representation of the Dirac function as a complex integral

$$\delta(t) = \frac{1}{2\pi i} \int_{c-i\infty}^{c+i\infty} e^{zt} dz \tag{7.2}$$

valid for any finite real number c. If $\mathrm{Re}\,(t) > 0$ (similarly $\mathrm{Re}\,(t) < 0$), the contour in (7.2) can be closed over the negative (positive) $\mathrm{Re}\,(z)$-plane. Since e^{zt} is analytic inside the contour, Cauchy's integral theorem (see e.g. Titchmarsh (1964)) states that the integral vanishes; except if $t = 0$, then the integral is unbounded.

A second class represents the Dirac function as the limit of a sequence of functions. For example, executing the integral (7.2) as $\delta\,(t) = \lim_{y\to\infty} \frac{1}{2\pi i} \int_{c-iy}^{c+iy} e^{zt} dz$ leads to

$$\delta\,(t) = \frac{1}{2\pi i} \lim_{y\to\infty} \frac{e^{(c+iy)t} - e^{(c-iy)t}}{t} = \frac{e^{ct}}{\pi t} \lim_{y\to\infty} \sin yt$$

but the latter limit with a value in $[-1,1]$ does not exist, although[1] $\int_{-\infty}^{\infty} \delta\,(t)\,dt = 1$, and illustrates difficulties, which has led to the development of a theory of generalized functions. A particularly interesting class of functions are probability density functions $f_X\,(t)$ of a continuous random variable X, because of their property $\int_{-\infty}^{\infty} f_X\,(t)\,dt = 1$. For example (see e.g. Van Mieghem (2014)), the limit of Gaussian probability density functions with variance tending to zero yields

$$\delta\,(t) = \lim_{\sigma\to 0} \frac{e^{-\frac{t^2}{2\sigma^2}}}{\sqrt{2\pi}\sigma}$$

while the limit of Cauchy probability density functions with width η tending to zero leads to

$$\delta\,(t) = \lim_{\eta\to 0} \frac{1}{\pi} \frac{\eta}{t^2 + \eta^2}$$

which is written in complex notation as

$$\delta\,(t) = -\frac{1}{\pi} \lim_{\eta\to 0} \mathrm{Im}\, \frac{1}{t + i\eta} \tag{7.3}$$

This representation (7.3) is related to the Cauchy transform (see **art. 361**).

173. The density function of the eigenvalues $\{\lambda_m\}_{1\le m\le N}$ is defined by

$$f_\lambda\,(t) = \frac{1}{N} \sum_{m=1}^{N} \delta\,(t - \lambda_m) \tag{7.4}$$

Using the representation (7.2) of the Dirac function, we have for $c > 0$,

$$f_\lambda\,(t) = \frac{1}{2\pi i} \int_{c-i\infty}^{c+i\infty} e^{zt} \varphi_\lambda\,(z)\,dz \tag{7.5}$$

---

[1] Indeed,

$$\int_{-\infty}^{\infty} \delta\,(t)\,dt = \frac{1}{2\pi i} \lim_{u\to\infty} \int_{c-i\infty}^{c+i\infty} \int_{-u}^{u} e^{zt} dt\,dz = \frac{1}{2\pi i} \lim_{u\to\infty} \int_{c-i\infty}^{c+i\infty} \frac{e^{zu} - e^{-zu}}{z} dz = 1$$

where the latter integral follows from Cauchy's integral theorem because, for $c > 0$, $\frac{1}{2\pi i} \int_{c-i\infty}^{c+i\infty} \frac{e^{zu}}{z} dz = 1$ if $\mathrm{Re}\,(u) > 0$, else it is zero.

where, analogous to the definition of a probability generating function (pgf),

$$\varphi_\lambda(z) = \frac{1}{N} \sum_{m=1}^{N} \exp(-z\lambda_m) \tag{7.6}$$

can be interpreted as the generating function of the density function of the eigenvalues $\{\lambda_m\}_{1 \leq m \leq N}$.

The representation (7.3) of the Dirac function leads to

$$f_\lambda(t) = -\frac{1}{\pi N} \lim_{\eta \to 0} \text{Im} \sum_{m=1}^{N} \frac{1}{t - \lambda_m + i\eta}$$

Invoking $\sum_{k=1}^{n} \frac{1}{z-\lambda_k} = \text{trace}(zI - A)^{-1}$ in (A.163) of the resolvent $(zI - A)^{-1}$ in **art. 262** yields

$$f_\lambda(t) = -\frac{1}{\pi N} \lim_{\eta \to 0} \text{Im trace}\left(((t + i\eta)I - A)^{-1}\right) \tag{7.7}$$

where $\frac{1}{N}\text{trace}\left((zI - A)^{-1}\right)$ is called the Stieltjes transform of the matrix A.

174. *Probabilistic setting.* The eigenvalues $\lambda_1, \lambda_2, \ldots, \lambda_N$, now ordered as $\lambda_1 \leq \lambda_2 \leq \cdots \leq \lambda_N$, can be regarded as a complete set of all realizations of the random variable λ, in which case the probability that λ is smaller than or equal to a real number u equals

$$\Pr[\lambda \leq u] = \frac{1}{N} \sum_{k=1}^{N} 1_{\{\lambda_k \leq u\}}$$

which shows, for $u = \lambda_j$, that[2]

$$\Pr[\lambda \leq \lambda_j] = \frac{j}{N}$$

Usually only a limited set of realizations of a random variable can be measured and the above representation for its probability is then approximate and called the empirical distribution (see e.g. (Van Mieghem, 2014, p. 580-581)).

By applying Abel summation (3.87) to the pgf in (7.6)

$$\varphi_\lambda(z) = \frac{1}{N} \sum_{k=1}^{N-1} k\left(\exp(-z\lambda_k) - \exp(-z\lambda_{k+1})\right) + \exp(-z\lambda_N)$$

replacing $\exp(-z\lambda_k) - \exp(-z\lambda_{k+1}) = -z \int_{\lambda_{k+1}}^{\lambda_k} e^{-zu} du$,

$$\varphi_\lambda(z) = z \sum_{k=1}^{N-1} \frac{k}{N} \int_{\lambda_k}^{\lambda_{k+1}} e^{-zu} du + \exp(-z\lambda_N)$$

---

[2] Our usual ordering $\lambda_N \leq \lambda_{N-1} \leq \cdots \leq \lambda_1$ results in $\Pr[\lambda \leq \lambda_j] = \frac{N+1-j}{N}$ and unnecessarily complicates the derivations below.

we introduce $\Pr\left[\lambda \leq \lambda_j\right] = \frac{j}{N}$ in the k-sum

$$\sum_{k=1}^{N-1} \frac{k}{N} \int_{\lambda_k}^{\lambda_{k+1}} e^{-zu} du = \sum_{k=1}^{N-1} \int_{\lambda_k}^{\lambda_{k+1}} \Pr\left[\lambda \leq u\right] e^{-zu} du$$

$$= \left\{ \int_{\lambda_1}^{\lambda_2} + \int_{\lambda_2}^{\lambda_3} + \cdots + \int_{\lambda_{N-1}}^{\lambda_N} \right\} \Pr\left[\lambda \leq u\right] e^{-zu} du$$

and arrive, for the ordering $\lambda_1 \leq \lambda_2 \leq \cdots \leq \lambda_N$, at

$$\varphi_\lambda(z) = \exp\left(-z\lambda_N\right) - z \int_{\lambda_N}^{\lambda_1} \Pr\left[\lambda \leq u\right] e^{-zu} du$$

After partial integration with the probability density function $f_\lambda(u) = \frac{d}{du} \Pr\left[\lambda \leq u\right]$, the usual expression for the probability generating function of the real random variable $-N < \lambda < N$ is found

$$\varphi_\lambda(z) = \frac{e^{-z\lambda_1}}{N} + \int_{\lambda_1}^{\lambda_N} e^{-zu} f_\lambda(u) du = \int_{-\infty}^{\infty} e^{-zu} f_\lambda(u) du$$

175. *Trace representation of $\varphi_\lambda(z)$.* **Art. 234** shows that $e^{-zA} = \sum_{k=0}^{\infty} \frac{A^k}{k!} (-z)^k$. Introducing $A^k = X \operatorname{diag}\left(\lambda_m^k\right) X^T$ in the Taylor series, where the orthogonal matrix X has the eigenvectors of A as columns (**art. 247**), yields $e^{-zA} = X \operatorname{diag}\left(e^{-z\lambda_m}\right) X^T$. Hence, if $\{\lambda_m\}_{1 \leq m \leq N}$ are the eigenvalues of a symmetric matrix $A_{N \times N}$, then $\left\{e^{-z\lambda_m}\right\}_{1 \leq m \leq N}$ are the eigenvalues of e^{-zA} and the eigenvector of A belonging to the eigenvalue λ_m is also the eigenvector of e^{-zA} belonging to the eigenvalue $e^{-z\lambda_m}$. After cyclic permutation (4.14) with $X^T X = I$, we arrive at the trace representation

$$\varphi_\lambda(z) = \frac{1}{N} \operatorname{trace}\left(e^{-zA}\right) \tag{7.8}$$

The relation with a probability generating function, $\varphi_\lambda(z) = E\left[e^{-z\lambda}\right]$, suggests that the moments $E\left[\lambda^n\right] = (-1)^n \varphi_\lambda^{(n)}(0)$, and with (7.8) that

$$E\left[\lambda^n\right] = \frac{1}{N} \operatorname{trace}\left(A^n\right) \tag{7.9}$$

Relation (7.9) lies at the basis of Wigner's moment approach in **art. 187** to computing the eigenvalues of random matrices.

176. *Lower bound $\varphi_\lambda(z) \geq 1$ for real z.* Since e^{-z} is convex for real z, the general convexity bound (Van Mieghem, 2014, eq. (5.5)), from which also Jensen's inequality $E\left[e^{-z\lambda}\right] \geq e^{-zE[\lambda]}$ is derived, gives

$$\varphi_\lambda(z) = \frac{1}{N} \sum_{m=1}^{N} \exp\left(-z\lambda_m\right) \geq \exp\left(-\frac{z}{N} \sum_{m=1}^{N} \lambda_m\right) = 1$$

because $E\left[\lambda\right] = \frac{1}{N} \sum_{m=1}^{N} \lambda_m = 0$ for the adjacency matrix A (**art. 46**).

7.2 The density when $N \to \infty$

Usually analysis simplifies in limit cases. For a few graphs, where the number of nodes N tends to infinity, the spectrum can be computed, which requires the evaluation of $\varphi_{\lambda;\infty}(z) = \lim_{N\to\infty} \frac{1}{N} \sum_{m=1}^{N} \exp(-z\lambda_m)$.

177. *Replacing a sum by an integral.* The basic summation formula (Titchmarsh and Heath-Brown, 1986, p. 13),

$$\sum_{a<k\leq b} f(k) = \int_a^b f(x)\,dx + \int_a^b \left(x - [x] - \frac{1}{2}\right) \frac{df(x)}{dx}\,dx + \left(a - [a] - \frac{1}{2}\right) f(a)$$

$$- \left(b - [b] - \frac{1}{2}\right) f(b) \tag{7.10}$$

is valid for any function $f(x)$ with continuous derivative in the interval $[a, b]$. We define the continuous eigenvalue function $\Lambda(x)$ on $[0, N]$ such that $\Lambda(m) = \lambda_m$. Since, for any integer $m \in [1, N]$, **art.** 42 shows that $-(N-1) < \lambda_m \leq N-1$ and since $\lambda_N \leq \lambda_{N-1} \leq \dots \leq \lambda_1$, the eigenvalue continuous function $|\Lambda(x)|$ is bounded on $[1, N]$ by $N-1$ and $\Lambda(m) \leq \Lambda(m-1)$ for any m. Thus, we assume that $\Lambda(x)$ is continuous and not increasing on $[0, N]$. The continuous eigenvalue function $\Lambda(x)$ can be obtained by Hermite or Bernstein interpolation (**art.** 304), but not by Lagrange interpolation (**art.** 303), because the Lagrange interpolating polynomial is not necessarily increasing at each real $x \in [1, N]$.

Application of the summation formula (7.10) to the pgf (7.6) yields

$$\varphi_\lambda(z) = \frac{1}{N} \int_0^N e^{-z\Lambda(x)}\,dx - y_N(z) + \frac{e^{-z\Lambda(N)} - e^{-z\Lambda(0)}}{2N}$$

where

$$y_N(z) = \frac{z}{N} \int_0^N \left(x - [x] - \frac{1}{2}\right) e^{-z\Lambda(x)} \frac{d\Lambda(x)}{dx}\,dx$$

Since $-\frac{1}{2} \leq x - [x] - \frac{1}{2} \leq \frac{1}{2}$ and $\frac{d\Lambda(x)}{dx} \leq 0$, we may bound $y_N(z)$ for real z as,

$$\frac{z}{2N} \int_0^N e^{-z\Lambda(x)} \frac{d\Lambda(x)}{dx}\,dx \leq y_N(z) \leq -\frac{z}{2N} \int_0^N e^{-z\Lambda(x)} \frac{d\Lambda(x)}{dx}\,dx$$

With $\int_0^N e^{-z\Lambda(x)} \frac{d\Lambda(x)}{dx}\,dx = \int_{\Lambda(0)}^{\Lambda(N)} e^{-z\Lambda(x)}\,d\Lambda(x) = \frac{e^{-z\Lambda(0)} - e^{-z\Lambda(N)}}{z}$, we have that

$$\frac{e^{-z\Lambda(0)} - e^{-z\Lambda(N)}}{2N} \leq y_N(z) \leq \frac{e^{-z\Lambda(N)} - e^{-z\Lambda(0)}}{2N}$$

Thus, for real z, we obtain the bounds for the pgf $\varphi_\lambda(z)$ in (7.6)

$$\frac{1}{N} \int_0^N e^{-z\Lambda(x)}\,dx \leq \varphi_\lambda(z) \leq \frac{1}{N} \int_0^N e^{-z\Lambda(x)}\,dx + \frac{e^{-z\Lambda(N)} - e^{-z\Lambda(0)}}{N}$$

The density function $f_\lambda(t)$ involves a line integration (7.5) over $\mathrm{Re}(z) = c > 0$.

If, for $\operatorname{Re}(z) > 0$,

$$\lim_{N \to \infty} \frac{e^{-z\Lambda(N)} - e^{-z\Lambda(0)}}{N} = \lim_{N \to \infty} \frac{e^{-z\Lambda(N)}}{N} = 0$$

then the limit

$$\lim_{N \to \infty} \varphi_\lambda(z) = \lim_{N \to \infty} \frac{1}{N} \int_0^N e^{-z\Lambda(x)} dx$$

exists and, hence, also $\lim_{N \to \infty} f_\lambda(t)$. The condition means that the absolute value of the smallest eigenvalue $\Lambda(N) = \lambda_N < 0$ grows as $|\lambda_N| = O(\log N)$ at most, for $\operatorname{Re}(z) = c > 0$, but arbitrarily small. This condition is quite restrictive and suggests to consider the spectrum of *normalized* eigenvalues.

178. *Moments.* We start from $E[\lambda^m] = \frac{1}{N}\operatorname{trace}(A^m)$ in (7.9) and the number of closed walks $W_m = \operatorname{trace}(A^m)$ in **art.** 65 and use the Stieltjes integral (**art.** 350),

$$E[\lambda^m] = \int_{-\infty}^{\infty} x^m dF_\lambda(G_N) = \frac{1}{N}\operatorname{trace}\left(A_{N \times N}^m\right)$$

If $\lim_{N \to \infty} \frac{1}{N}\operatorname{trace}\left(A_{N \times N}^m\right) = w_G(m)$ exists and the distribution $F_\lambda(G_N)$ tends to $F_{\lambda\infty}$, then

$$\int_{-\infty}^{\infty} x^m dF_{\lambda\infty} = w_G(m) \tag{7.11}$$

which implies that the limiting distribution $F_{\lambda\infty}$ of the eigenvalues of the infinitely large graph $G_\infty = \lim_{N \to \infty} G_N$ exists. Assuming that this distribution is also differentiable, then $\int_{-\infty}^{\infty} x^m dF_{\lambda\infty} = \int_{-\infty}^{\infty} x^m f_{\lambda\infty}(x)\, dx$. Since $\operatorname{trace}(A_{N \times N}) = 0$ and $\operatorname{trace}\left(A_{N \times N}^2\right) = 2L$, we find, beside $\int_{-\infty}^{\infty} f_{\lambda\infty}(x)\, dx = 1$, that

$$\int_{-\infty}^{\infty} x f_{\lambda\infty}(x)\, dx = 0$$

$$\int_{-\infty}^{\infty} x^2 f_{\lambda\infty}(x)\, dx = E[D]$$

Multiplying both sides in (7.11) by $\frac{(-z)^m}{m!}$ and summing over all integers m yields again the pgf

$$\varphi_{\lambda\infty}(z) = \int_{-\infty}^{\infty} e^{-zx} dF_{\lambda\infty} = \sum_{m=0}^{\infty} \frac{(-z)^m}{m!} w_G(m) \tag{7.12}$$

7.3 Examples of spectral density functions

Only for a few graphs, the spectral density functions can be computed analytically with the methods **art.** 175-178.

179. *Infinite line topology.* Applying (7.6) to a path P on N nodes with eigenvalue $\lambda_k(P) = 2\cos\left(\frac{\pi k}{N+1}\right)$ for $1 \le k \le N$ in (6.10) yields the pgf of the eigenvalues of the path P

$$\varphi_{\lambda_P}(z) = \frac{1}{N} \sum_{k=1}^{N} \exp\left(-2z\cos\left(\frac{\pi k}{N+1}\right)\right)$$

Since $\Lambda(x) = 2\cos\left(\frac{\pi x}{N+1}\right)$ and $\lim_{N\to\infty} \frac{e^{-z\Lambda(N)}}{N} = \lim_{N\to\infty} \frac{e^{-2z\cos\left(\frac{\pi N}{N+1}\right)}}{N} = 0$, **art. 177** shows that the limit generating function exists,

$$\lim_{N\to\infty} \varphi_{\lambda_P}(z) = \lim_{N\to\infty} \frac{1}{N}\int_0^N e^{-2z\cos\left(\frac{\pi x}{N+1}\right)} dx = \lim_{N\to\infty} \frac{N+1}{N\pi}\int_0^{\frac{\pi N}{N+1}} e^{-2z\cos\theta}d\theta$$

Hence,

$$\lim_{N\to\infty} \varphi_{\lambda_P}(z) = \frac{1}{\pi}\int_0^\pi \exp\left(-2z\cos\theta\right) d\theta = I_0(-2z) = I_0(2z)$$

where $I_n(z)$ is the modified Bessel function (Abramowitz and Stegun, 1968, Section 9.6.19); (Olver *et al.*, 2010, Chapter 10). The inverse Laplace transform is

$$\lim_{N\to\infty} f_\lambda(t) = \frac{1}{2\pi i}\int_{c-i\infty}^{c+i\infty} e^{zt} I_0(2z)\, dz = \frac{1}{2\pi^2 i}\int_{c-i\infty}^{c+i\infty} e^{z(t+2)}\left\{\pi e^{-2z} I_0(2z)\right\} dz$$

and with the Laplace pair in Abramowitz and Stegun (1968, Section 29.3.124), we arrive at the spectral density function of an infinitely long path:

$$\lim_{N\to\infty} f_\lambda(t) = \frac{1}{\pi}\frac{1}{\sqrt{4-t^2}} 1_{|t|<2} \tag{7.13}$$

180. The spectrum of an arbitrary path in a graph with N nodes can be computed if the distribution of the hopcount $H_N > 0$ of that path is known. Indeed, using the law of total probability (Van Mieghem, 2014, p. 23) yields

$$\Pr\left[\lambda_{\text{arbitrary path}} \le t\right] = \sum_{k=1}^{N-1} \Pr\left[\lambda_{\text{arbitrary path}} \le t \mid H_N = k\right] \Pr\left[H_N = k\right]$$

$$= \sum_{k=1}^{N-1} \Pr\left[\lambda_{k\text{-hop path}} \le t\right] \Pr\left[H_N = k\right]$$

Differentiation gives us the density,

$$f_{\lambda_{\text{arbitrary path}}}(t) = \sum_{k=1}^{N-1} f_{\lambda_{k\text{-hop path}}}(t) \Pr\left[H_N = k\right]$$

Introducing the definition (7.4) combined with the spectrum specified in Section 6.5,

$$f_{\lambda_{k\text{-hop path}}}(t) = \frac{N-k-1}{N}\delta(t) + \frac{1}{N}\sum_{m=1}^{k+1}\delta\left(t - 2\cos\left(\frac{\pi m}{k+2}\right)\right)$$

gives

$$f_{\lambda_{\text{arbitrary path}}}(t) = \frac{N - 1 - E[H_N]}{N}\delta(t)$$

$$+ \frac{1}{N}\sum_{k=1}^{N-1}\sum_{m=1}^{k+1}\delta\left(t - 2\cos\left(\frac{\pi m}{k+2}\right)\right)\Pr[H_N = k]$$

The spectral peak at $t = 0$ has a strength equal to $\frac{N-1-E[H_N]}{N}$. Just as for the $N-1$ hop path, the spectrum lies in the interval $(-2, 2)$ at discrete values $t = 2\cos\left(\frac{\pi m}{k+2}\right)$ that range over more possible values than a constant hop path. Moreover, the strength or amplitude of a peak is modulated by the hopcount distribution.

181. *Small-world graph $SW_{k;N}$.* Applying (7.6) to the small-world graph $SW_{k;N}$, with (6.4), gives

$$\varphi_{\lambda_{\text{SW }k;N}}(z) = \frac{e^z}{N}\sum_{m=1}^{N}\exp\left(-z\frac{\sin\left(\frac{\pi(m-1)(2k+1)}{N}\right)}{\sin\left(\frac{\pi(m-1)}{N}\right)}\right)$$

Since $\Lambda(x) = \frac{\sin\left(\frac{\pi(x-1)(2k+1)}{N}\right)}{\sin\left(\frac{\pi(x-1)}{N}\right)} - 1$ and $\Lambda(N) = \frac{\sin\left(\frac{\pi(N-1)(2k+1)}{N}\right)}{\sin\left(\frac{\pi(N-1)}{N}\right)} - 1 \geq (\lambda_{\text{SW }k;N})_{\min}$, which is independent of N, the limit generating function exists

$$\lim_{N\to\infty}\varphi_{\lambda_{\text{SW }k;N}}(z) = \lim_{N\to\infty}\frac{e^z}{N}\int_0^N\exp\left(-z\frac{\sin\left(\frac{\pi(x-1)(2k+1)}{N}\right)}{\sin\left(\frac{\pi(x-1)}{N}\right)}\right)dx$$

After the substitution $\theta = \frac{x-1}{N}$ and executing the limit, we find

$$\varphi_{\lambda_{\text{SW }k;\infty}}(z) = \frac{e^z}{\pi}\int_0^\pi\exp\left(-z\frac{\sin(2k+1)\theta}{\sin\theta}\right)d\theta \qquad (7.14)$$

In terms of the Chebyshev polynomial $U_n(x)$ of the second kind,

$$\varphi_{\lambda_{\text{SW }k;\infty}}(z) = \frac{e^z}{\pi}\int_0^\pi\exp\left(-zU_{2k}(\cos\theta)\right)d\theta$$

Since $\frac{\sin(2k+1)x}{\sin x} = \frac{\sin(2k+1)(\pi-x)}{\sin(\pi-x)}$, the definition (B.135) shows that $U_{2k}(\cos\theta) = U_{2k}(\cos(\pi-\theta))$ and

$$\varphi_{\lambda_{\text{SW }k;\infty}}(z) = \frac{2e^z}{\pi}\int_0^{\frac{\pi}{2}}\exp\left(-zU_{2k}(\cos\theta)\right)d\theta \qquad (7.15)$$

With $U_{2k}(\cos\theta) = \frac{\sin(2k+1)\theta}{\sin\theta} = 1 + 2\sum_{j=1}^k\cos 2j\theta$, we have

$$\varphi_{\lambda_{\text{SW }k;\infty}}(z) = \frac{1}{\pi}\int_0^\pi\exp\left(-2z\sum_{j=1}^k\cos 2j\theta\right)d\theta$$

Applying the generating function (Abramowitz and Stegun, 1968, Section 9.6.33) of the modified Bessel function $I_n(z) = I_{-n}(z)$ for integer n,

$$e^{z\cos x} = \sum_{n=-\infty}^{\infty} I_n(z)\, e^{inx}$$

and

$$\exp\left(-2z\sum_{j=1}^{k}\cos 2j\theta\right) = \prod_{m=1}^{k} e^{-2z\cos 2m\theta} = \prod_{m=1}^{k}\sum_{n=-\infty}^{\infty} I_n(-2z)\, e^{i2mn\theta}$$

$$= \sum_{n_1=-\infty}^{\infty} I_{n_1}(-2z)\cdots \sum_{n_k=-\infty}^{\infty} I_{n_k}(-2z)\, e^{i2\theta\sum_{m=1}^{k} mn_m}$$

yields

$$\varphi_{\lambda_{\mathrm{SW}\,k;\infty}}(z) = \sum_{n_1=-\infty}^{\infty}\sum_{n_2=-\infty}^{\infty}\cdots\sum_{n_k=-\infty}^{\infty}\prod_{m=1}^{k} I_{n_m}(-2z)\,\frac{1}{\pi}\int_0^{\pi} e^{i2\theta\sum_{m=1}^{k} mn_m}\, d\theta$$

The integral $\frac{1}{\pi}\int_0^{\pi} e^{i2\theta\sum_{m=1}^{k} mn_m}\, d\theta = 1_{\{\sum_{m=1}^{k} mn_m=0\}}$. Translating that condition as $n_1 = -\sum_{m=2}^{k} mn_m$ for $k > 1$, we arrive at

$$\varphi_{\lambda_{\mathrm{SW}\,k;\infty}}(z) = \sum_{n_2=-\infty}^{\infty}\cdots\sum_{n_k=-\infty}^{\infty} I_{\sum_{m=2}^{k} mn_m}(-2z)\prod_{m=2}^{k} I_{n_m}(-2z) \qquad (7.16)$$

while for $k = 1$,

$$\varphi_{\lambda_{\mathrm{SW}\,k=1;\infty}}(z) = I_0(-2z) = I_0(2z)$$

which shows that the limit density of the infinite cycle ($k = 1$) is the same as that of the infinite path in (7.13). If $k = 2$, then (7.16) becomes with $I_n(-z) = (-1)^n I_n(z)$,

$$\varphi_{\lambda_{\mathrm{SW}\,k=2;\infty}}(z) = \sum_{n=-\infty}^{\infty} I_n(-2z)\, I_{2n}(-2z) = I_0^2(2z) + 2\sum_{n=1}^{\infty}(-1)^n I_n(2z)\, I_{2n}(2z)$$

Unfortunately, we cannot evaluate the inverse Laplace transform of $\varphi_{\lambda_{\mathrm{SW}\,k;\infty}}(z)$ in (7.16) for $k > 1$, but **art. 182** computes the Taylor expansion of $\varphi_{\lambda_{\mathrm{SW}\,k;\infty}}(z)$ around $z_0 = 0$ exactly.

182. *Moments of the small-world graph* $SW_{k;\infty}$. Taylor expansion of the pgf $\varphi_\lambda(z) = E\left[e^{-\lambda z}\right]$ in (7.6) shows that $E[\lambda^m] = \frac{1}{N}\sum_{n=1}^{N}\lambda_n^m = (-1)^m \varphi_\lambda^{(m)}(0)$. Expanding the exponential in (7.14) in a Taylor series,

$$\varphi_{\lambda_{\mathrm{SW}\,k;\infty}}(z) = e^z \sum_{m=0}^{\infty}\frac{(-z)^m}{m!}\frac{1}{\pi}\int_0^{\pi}\left(\frac{\sin(2k+1)\theta}{\sin\theta}\right)^m d\theta$$

indicates that $E\left[e^{-(\lambda_{SW\,k;\infty}+1)z}\right] = \sum_{m=0}^{\infty} \frac{(-z)^m}{m!} \frac{1}{\pi} \int_0^\pi \left(\frac{\sin(2k+1)\theta}{\sin\theta}\right)^m d\theta$ and that, by equating corresponding powers in z, the centered moments around $c = -1$ are

$$E\left[(\lambda_{SW\,k;\infty}+1)^m\right] = \frac{1}{\pi}\int_0^\pi \left(\frac{\sin(2k+1)\theta}{\sin\theta}\right)^m d\theta$$

Since $\int_\pi^{2\pi}\left(\frac{\sin(2k+1)\theta}{\sin\theta}\right)^m d\theta = \int_0^\pi \left(\frac{\sin(2k+1)u}{\sin u}\right)^m du$, it holds that

$$\int_0^\pi \left(\frac{\sin(2k+1)\theta}{\sin\theta}\right)^m d\theta = \frac{1}{2}\int_0^{2\pi}\left(\frac{e^{i(2k+1)\theta} - e^{-i(2k+1)\theta}}{e^{i\theta} - e^{-i\theta}}\right)^m d\theta$$

The complex transformation $w = e^{i\theta}$ is

$$\int_0^{2\pi}\left(\frac{e^{i(2k+1)\theta} - e^{-i(2k+1)\theta}}{e^{i\theta} - e^{-i\theta}}\right)^m = \frac{1}{i}\int_{|w|=1}\left(\frac{w^{4k+2} - 1}{w^2 - 1}\right)^m w^{-2km-1}dw$$

The integrand is analytic inside the unit circle, except for a pole of order $2km+1$ at the origin $w = 0$. By Cauchy's integral theorem $\frac{1}{k!}\frac{d^k f(z)}{dz^k}\Big|_{z=z_0} = \frac{1}{2\pi i}\int_{C(z_0)}\frac{f(\omega)\,d\omega}{(\omega-z_0)^{k+1}}$, we find that

$$\frac{1}{i}\int_{|w|=1}\left(\frac{w^{4k+2} - 1}{w^2 - 1}\right)^m w^{-2km-1}dw = \frac{2\pi}{(2km)!}\lim_{w\to 0}\frac{d^{2km}}{dw^{2km}}\left(\frac{w^{4k+2} - 1}{w^2 - 1}\right)^m$$

Invoking the binomial series $(1+x)^\alpha = \sum_{j=0}^{\infty}\binom{\alpha}{j}x^j$ yields

$$\left(1 - w^{4k+2}\right)^m \left(1 - w^2\right)^{-m} = \sum_{j=0}^{\infty}\sum_{n=0}^{\infty}\binom{m}{j}\binom{-m}{n}(-1)^{j+n}w^{2n+4kj+2j}$$

Using $\frac{d^l}{dx^l}x^\alpha = \frac{\alpha!}{(\alpha-l)!}x^{\alpha-l}$ and $\binom{-z}{j} = (-1)^j\frac{\Gamma(z+j)}{j!\Gamma(z)} = (-1)^j\binom{z-1+j}{j}$ leads to

$$\lim_{w\to 0}\frac{d^{2km}}{dw^{2km}}\left(\frac{w^{4k+2} - 1}{w^2 - 1}\right)^m = \sum_{j=0}^{\infty}\sum_{n=0}^{\infty}\frac{(-1)^j\binom{m}{j}\binom{m-1+n}{n}(2n+4kj+2j)!}{(2n+4kj+2j-2km)!}1_X$$

The condition X is $\{n + 2kj + j - km = 0\}$ and specifies $n = km - (2k+1)j$. Since $n \geq 0$ or $km - (2k+1)j \geq 0$, it holds that $j \leq \frac{km}{2k+1}$ and we arrive at

$$\int_0^\pi \left(\frac{\sin(2k+1)\theta}{\sin\theta}\right)^m d\theta = \pi \sum_{j=0}^{\left\lfloor\frac{km}{2k+1}\right\rfloor}\binom{m}{j}\binom{(k+1)m - 1 - (2k+1)j}{m-1}(-1)^j$$

In summary, the pgf of the eigenvalues of an infinitely large small-world graph $SW_{k;\infty}$ with degree $r = 2k$ is

$$\varphi_{\lambda_{SW\,k;\infty}}(z) = e^z \sum_{m=0}^{\infty}\frac{(-z)^m}{m!}E\left[(\lambda_{SW\,k;\infty}+1)^m\right]$$

where the centered moments $E\left[(\lambda_{SW\,k;\infty} - c)^m\right]$ around $c = -1$,

$$E\left[(\lambda_{SW\,k;\infty} + 1)^m\right] = \sum_{j=0}^{\left\lfloor \frac{km}{2k+1} \right\rfloor} \binom{m}{j} \binom{(k+1)\,m - 1 - (2k+1)\,j}{m - 1} (-1)^j \qquad (7.17)$$

are positive integers. We find that $E\left[(\lambda_{SW\,k;\infty} + 1)\right] = 1$, $E\left[(\lambda_{SW\,k;\infty} + 1)^2\right] = 2k + 1$ and $E\left[(\lambda_{SW\,k;\infty} + 1)^3\right] = \binom{3k+2}{2} - 3\binom{k+1}{2}$. The moments

$$E\left[\lambda_{SW\,k;\infty}^m\right] = (-1)^m \sum_{l=0}^{m} \binom{m}{l} (-1)^l E\left[(\lambda_{SW\,k;\infty} + 1)^l\right]$$

being the average number of walks W_k, are also all positive integers and can be computed from (7.17): $E\left[\lambda_{SW\,k;\infty}\right] = 0$, $E\left[\lambda_{SW\,k;\infty}^2\right] = 2k$, $E\left[\lambda_{SW\,k;\infty}^3\right] = 6\binom{k}{2}$. For $k > 3$, we did not find simple expressions.

7.4 Density of a sparse regular graph

We present the ingenious method of McKay (1981), who succeeded in finding the asymptotic density of the eigenvalues of the adjacency matrix of a regular, sparse graph.

183. A large sparse, regular graph. Consider a regular graph $G(r; N)$, where each node has degree r. The sparseness of $G(r; N)$ is here understood in the sense that $G(r; N)$ has locally a tree-like structure. In other words, for small enough integers h, the graph induced by the nodes at hop distance $1, 2, \ldots, h$ from a certain node n is a tree, more specific a k-ary, regular tree with the out-degree $k = r - 1$. We will first determine the moments via trace(A^m), as explained in **art. 178**, where each element $(A^m)_{jj}$ equals the number of closed walks of m hops starting at node j and returning at j (**art. 6**). The regularity of $G(r; N)$ suggests for any node n that, for $N \to \infty$,

$$\frac{1}{N} \text{trace}\,(A^m) = \frac{1}{N} \sum_{j=1}^{N} (A^m)_{jj} \to (A^m)_{nn}$$

Hence, for large N and fixed m, the local structure around any node n is almost the same. In addition, as long as the contribution $c(N)$ of cycles to trace(A^m) is small, i.e., $c(N) = o(N)$, the above limit is unaltered, for $\frac{1}{N}\left(\sum_{j=1}^{N} (A^m)_{jj} \pm c(N)\right) \to (A^m)_{nn}$. The fact that the number of cycles in $G(r; N)$ grows less than proportionally with N is an alternative way to define the sparseness of $G(r; N)$.

184. Random walks and the reflection principle. McKay (1981) had the fortunate idea to relate the computation of the number of closed walks to the powerful reflection principle, primarily used in the theory of random walks.

The largest hop distance reached in a k-ary tree by a closed walk of m hops is

$\left[\frac{m}{2}\right]$. The length m of all closed walks in a k-ary tree is even. Moreover, all walks travel some hops down and return along the same path back to the root n. Due to the regular structure, the analogy with a path in a simple random walk is very effective.

In a *simple random walk*, an item moves along the (vertical) x-axis over the integers during n epochs, measured along the horizontal k-axis. At each epoch k, the item jumps either one step to the right ($x_k = x_{k-1} + 1$) or one step to the left ($x_k = x_{k-1} - 1$). Assuming that the item starts at the origin at epoch 0, then $x_0 = 0$, its position at $k = n$ equals $x_n = r - l$, where r and l are the total number of right and left steps, respectively. Geometrically, plotting the x distance versus discrete time k, the sequence $x_1, x_2, \ldots, x_n$ represents a path from the origin to the point (n, x_n). In general, x_k can be either negative, positive or zero. The number of such paths with r right steps is $\binom{n}{r} = \binom{n}{n-r} = \binom{n}{l}$, which is thus equal the number of paths with l left steps, because $l + r = n$. Writing in the sequel x for x_n and combining $x_n = r - l$ and $l + r = n$ leads to $r = \frac{x+n}{2}$. Hence, the number of paths from the origin to the point (n, x) is

$$T_{(n,x)} = \binom{n}{\frac{x+n}{2}} 1_{\{\frac{x+n}{2} \in \mathbb{N}\}} \tag{7.18}$$

The reflection principle states that:

Theorem 47 (Reflection principle) *The number of paths from the point $a = (m, |x|)$ to the point $b = (n, |y|)$ that cross or touch the k-axis is equal to the number of all paths from $-a = (m, -|x|)$ to that same point b.*

The reflection of a point (k, x) is the point $(k, -x)$.

Proof: The reflection principle is demonstrated by showing a one-to-one correspondence with the subpath from $a = (m, |x|)$ to $c = (\mu, 0)$ and the reflected subpath from $-a = (m, -|x|)$ to c. For each subpath from a to c, there corresponds precisely one subpath from $-a$ to c (and the sequel of c to b is the same in both cases). $\square$

A direct consequence of Theorem 47 is the so-called ballot theorem:

Theorem 48 (Ballot) *The number of paths from the origin to (n, x), where $n, x \in \mathbb{N}_0$, that never touch the discrete time k-axis equals $\frac{x}{n} T_{(n,x)}$.*

Never touching the x-axis implies that $x_1 > 0, x_2 > 0, \ldots, x_n = x > 0$. The proof is too nice to not include.

Proof: Since $x_1 > 0$, the first step in such a path is necessarily the point $(1, 1)$. Hence, the number of paths from the origin above the k-axis to the point (n, x) is equal to the number of paths from $(1, 1)$ to (n, x) lying above the k-axis. The total number of paths from $(1, 1)$ to (n, x) is $T_{(n-1,x-1)}$. All paths from $(1, -1)$ to (n, x) cross the k-axis and their number equals $T_{(n-1,x+1)}$. By the reflection principle, the number of paths that does not touch the k-axis equals $T_{(n-1,x-1)} - T_{(n-1,x+1)} = \frac{x}{n}T_{(n,x)}$, where the last equality follows from (7.18) and $\binom{n-1}{k-1} - \binom{n-1}{k} = \frac{2k-n}{n}\binom{n}{k}$. $\qquad\square$

With this preparation, we can determine $(A^m)_{nn}$. Each walk of m hops or m links, starting at epoch 0 at the origin where the root node n is placed, can be represented by the sequence of points $(0,0), (1,1), (2, h_2), \ldots (m, h_m)$, where $h_j \geq 0$ is the distance in hops from the root node n. A closed walk of m hops returns to the root, which means that $h_m = 0$. Each such walk of m hops may consist of smaller walks of l hops, each time when $h_j = 0$ for $0 < j < m$. In the language of a random walk, each time j, that the path starting from the origin and returning back to the origin, but only lying above the k-axis, it touches the k-axis at $(j, h_j = 0)$. We thus need to compute the number of such paths with l points touching the k-axis. Feller (1970, pp. 90-91) proves that this number of paths equals $\frac{l}{m-l}\binom{m-l}{\lfloor \frac{m}{2} \rfloor}1_{\{l \leq m, \frac{m}{2} \in \mathbb{N}\}}$. An elementary closed walk of $c \in [1, l]$ hops consists of one excursion to some maximum level H_c and back along the same track. The total number of such elementary closed walks of c hops is $r (r-1)^{H_c - 1}$, because the root has degree r, and from hop level 1 on, each node has outdegree $r - 1$. Only the upwards steps towards the local maxima at level H_c contribute to the determination of the total number of walks in a c-hop closed walk. Walk excursions that do not reach H_c are subwalks of elementary closed walks reaching the maximum level H_c. Since there are l such elementary closed walks, their total is $\prod_{c=1}^{l} r (1 - r)^{H_c - 1} = r^l (r-1)^{-l} (r-1)^{\sum_{c=1}^{l} H_c}$. Now, each closed walk has an even number of hops and precisely as many up as down in the k-ary tree. Hence, $\sum_{c=1}^{l} H_c = \lceil \frac{m}{2} \rceil$, the highest possible level to be reached. Thus, we end up with a total of $\frac{l}{m-l}\binom{m-l}{\lfloor \frac{m}{2} \rfloor}1_{\{\frac{m}{2} \in \mathbb{N}\}} r^l (r-1)^{\lceil \frac{m}{2} \rceil - l}$ walks with l touching points. Finally, summing over all possible l yields McKay's basic result

$$\left(A^{2m}\right)_{nn} = \sum_{l=1}^{m} \binom{2m-l}{m} \frac{l}{2m-l} r^l (r-1)^{m-l} \tag{7.19}$$

$$\left(A^{2m+1}\right)_{nn} = 0$$

185. *Asymptotic density $f_{\lambda\infty}(x)$.* The next hurdle is the inversion of (7.11) in **art.** 178. We assume that the limit density exists and is differentiable such that

$$\int_{-\infty}^{\infty} x^{2m} f_{\lambda\infty}(x)\, dx = \sum_{l=1}^{m} \binom{2m-l}{m} \frac{l}{2m-l} r^l (r-1)^{m-l}$$

and $f_{\lambda\infty}(x) = f_{\lambda\infty}(-x)$ is even to satisfy $\left(A^{2m+1}\right)_{nn} = 0$. Recall from Theorems 39 and 40 that symmetry in the spectrum of A is the unique fingerprint of a bipartite structure of which a tree is a special case. McKay succeeded in finding $f_{\lambda\infty}(x)$ by inverting this relation, using a rather complicated method.

He presents various alternative sums of (7.19) without derivation. Then he derives an asymptotic form of (7.19) for large m to conclude that the extent of $f_{\lambda\infty}(x)$

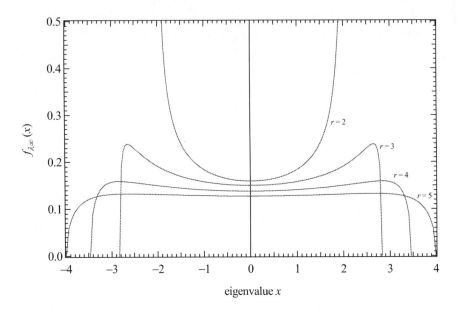

Fig. 7.1. The spectral density $f_{\lambda_\infty}(x)$ of a large sparse regular graph for various values of the degree r.

is bounded, i.e., $f_{\lambda_\infty}(x)$ exists only for $|x| \leq 2\sqrt{r-1}$. After normalizing the x-range to the interval $[-1,1]$, he employs Chebyshev polynomials (Section 12.7) and their orthogonality properties to execute the inversion, resulting in:

Theorem 49 (McKay's Law) *The asymptotic density $f_{\lambda_\infty}(x)$ of the eigenvalues of the adjacency matrix of a large, sparse regular graph with degree r equals*

$$f_{\lambda_\infty}(x) = \frac{r\sqrt{4(r-1)-x^2}}{2\pi(r^2-x^2)}1_{\{|x|\leq 2\sqrt{r-1}\}} \tag{7.20}$$

The corresponding distribution function $F_{\lambda_\infty}(x) = \Pr[\lambda_\infty \leq x]$ is, for $-2\sqrt{r-1} \leq x \leq 2\sqrt{r-1}$,

$$F_{\lambda_\infty}(x) = \frac{1}{2} + \frac{r}{2\pi}\left(\arcsin\frac{x}{2\sqrt{r-1}} - \frac{r-2}{r}\arctan\frac{\frac{r-2}{r}x}{\sqrt{4(r-1)-x^2}}\right)$$

The spectral density (7.20) is plotted in Fig. 7.1. For $r = 2$, we again find the spectral density (7.13) of an infinitely long path.

7.5 Random matrix theory

186. Random matrix theory investigates the eigenvalues of an $N \times N$ matrix A whose elements a_{ij} are random variables with a given joint distribution. Even if all elements a_{ij} are independent, there does not exist a general expression for the distribution of the eigenvalues. However, nice results exist in particular cases, e.g. when the elements a_{ij} are Gaussian random variables. Moreover, if the elements a_{ij} are properly scaled, in various cases the spectrum in the limit $N \to \infty$ seems to converge rapidly to a deterministic limit distribution. The fascinating results of random matrix theory and applications from nuclear physics to the distributions of the non-trivial zeros of the Riemann Zeta function are reviewed by Mehta (1991). Recent advances in random matrix theory, discussed by Edelman and Raj Rao (2005), present a general framework that relates, among others, the laws of Wigner (Theorem 50), McKay (Theorem 49) and Marčenko-Pastur (Theorem 54) to Hermite, Jacobi and Laguerre orthogonal polynomials (see Chapter 12), respectively. A rigorous mathematical treatment of random matrix theory has appeared in Anderson *et al.* (2010).

Random matrix theory immediately applies to the adjacency matrix of the Erdős-Rényi random graph $G_p(N)$, where each element a_{ij} is 1 with probability p and zero with probability $1 - p$.

7.5.1 Wigner's Semicircle Law

187. Wigner's Semicircle Law is the fundamental result in the spectral theory of large random matrices.

Theorem 50 (Wigner's Semicircle Law) *Let A be a random $N \times N$ real symmetric matrix with independent and identically distributed elements a_{ij} with $\sigma^2 = Var[a_{ij}]$ and denote by $\lambda(A_N)$ an eigenvalue of the set of the N real eigenvalues of the scaled matrix $A_N = \frac{A}{\sqrt{N}}$. The probability density function $f_{\lambda(A_N)}(x)$ tends for $N \to \infty$ to*

$$\lim_{N \to \infty} f_{\lambda(A_N)}(x) = \frac{1}{2\pi\sigma^2} \sqrt{4\sigma^2 - x^2} \, 1_{|x| \leq 2\sigma} \tag{7.21}$$

Since the first proof of Theorem 50 by Wigner (1955) and his subsequent generalizations (Wigner, 1957, 1958) many proofs have been published. However, none of them is short and easy enough to include here. Wigner's Semicircle Law illustrates that, for sufficiently large N, the distribution of the eigenvalues of $\frac{A}{\sqrt{N}}$ does not depend anymore on the probability distribution of the elements a_{ij}. Hence, Wigner's Semicircle Law exhibits a universal property of a class of large, real symmetric matrices with independent random elements. Mehta (1991) suspects that, for a much broader class of large random matrices, a mysterious yet unknown law of large numbers must be hidden. Generalizing Wigner's Semicircle Law to asymmetric complex matrices, Tao and Vu (2010) have proved:

Theorem 51 (Circular Law) *Let A_N be the $N \times N$ random matrix whose entries are i.i.d. complex random variables with mean 0 and variance 1. The empirical* spectral distribution of $\frac{A_N}{\sqrt{N}}$ converges (both in probability and in the almost sure sense) to the uniform distribution on the unit disk.

The adjacency matrix of the Erdős-Rényi random graph satisfies the conditions in Theorem 50 with $\sigma^2 = p(1 - p)$ and its eigenvalues grow as $O\left(\sqrt{N}\right)$, apart from the largest eigenvalue (see art. 190). In order to obtain the finite limit distribution (7.21) scaling by $\frac{1}{\sqrt{N}}$ is necessary.

188. The moment relation (7.9) for the eigenvalues suggests us to compute the moments of Wigner's Semicircle Law (7.21),

$$E\left[\lambda^n\right] = \int_{-\infty}^{\infty} x^n \lim_{N\to\infty} f_{\lambda(A_N)}(x)\, dx = \frac{1}{2\pi\sigma^2} \int_{-2\sigma}^{2\sigma} x^n \sqrt{4\sigma^2 - x^2}\, dx$$

Thus,

$$E\left[\lambda^n\right] = \sigma^n C_n$$

where

$$C_n = \frac{2^{n+1}}{\pi} \int_{-1}^{1} t^n \sqrt{1 - t^2}\, dt \tag{7.22}$$

shows that $C_n = 0$ for odd values of n, because of integration of an odd function over an even interval. Using the integral of the Beta-function (Abramowitz and Stegun, 1968, Section 6.2.1) for $\mathrm{Re}(z) > 0$ and $\mathrm{Re}(w) > 0$,

$$B(z, w) = \int_0^1 t^{z-1}(1 - t)^{w-1}\, dt = \frac{\Gamma(z)\Gamma(w)}{\Gamma(z + w)}$$

we execute the integral in (7.22) for $n = 2k$,

$$\int_0^1 t^{2k}\sqrt{1 - t^2}\, dt = \int_0^1 x^{k-\frac{1}{2}}(1 - x)^{\frac{1}{2}}\, dx = \frac{\Gamma\left(k + \frac{1}{2}\right)\Gamma\left(\frac{3}{2}\right)}{\Gamma(k + 2)}$$

Using the functional equation $\Gamma(z + 1) = z\Gamma(z)$, $\Gamma\left(\frac{1}{2}\right) = \sqrt{\pi}$ and the duplication formula $\Gamma(2z) = \frac{2^{2z-\frac{1}{2}}}{\sqrt{2\pi}}\Gamma(z)\Gamma\left(z + \frac{1}{2}\right)$ in (Abramowitz and Stegun, 1968, Section 6.1.18), finally gives

$$C_{2k} = \frac{(2k)!}{(k + 1)!k!} = \frac{\binom{2k}{k}}{k + 1} \tag{7.23}$$

The numbers C_{2k} are known as Catalan numbers (Comtet, 1974). Reversely, since all moments uniquely define a probability distribution, the only distribution, whose moments are Catalan numbers, is the semicircle distribution, with density function given by (7.21).

Another derivation, that avoids the theory of the Gamma function, rewrites the integral (7.22) as

$$C_n = \frac{2^{n+1}}{\pi} \int_{-1}^{1} t^n \sqrt{1 - t^2}\, dt = -\frac{2^{n+1}}{\pi} \int_{-1}^{1} \frac{-t}{\sqrt{1 - t^2}}\left\{t^{n-1}\left(1 - t^2\right)\right\}\, dt$$

Since $\frac{d}{dt}\sqrt{1-t^2} = \frac{-t}{\sqrt{1-t^2}}$, partial integration gives

$$C_n = \frac{2^{n+1}}{\pi} \int_{-1}^{1} \sqrt{1-t^2}\left\{(n-1)\,t^{n-2} - (n+1)\,t^n\right\} dt$$

$$= 4\,(n-1)\,C_{n-2} - (n+1)\,C_n$$

which leads, with $C_0 = 1$ and $C_1 = 0$, to the recursion $C_n = \frac{4(n-1)}{n+2}C_{n-2}$. Iteration gives

$$C_n = 2^{2p}\,\frac{n-1}{n+2}\frac{n-3}{n}\frac{n-5}{n-2}\cdots\frac{n-(2p-1)}{n-(2p-4)}C_{n-2p}$$

If n is odd, $C_n = 0$ as found above, while if $n = 2k$ and $p = k$, then

$$C_{2k} = 2^{2k}\frac{2k-1}{2k+2}\frac{2k-3}{2k}\frac{2k-5}{2k-2}\cdots\frac{1}{4}$$

$$= 2^{2k}\frac{2k}{2k}\frac{2k-1}{2k+2}\frac{2k-2}{2k-2}\frac{2k-3}{2k}\frac{2k-4}{2k-4}\frac{2k-5}{2k-2}\cdots\frac{1}{4} = \frac{(2k)!}{(k+1)!k!}$$

which again results in the Catalan numbers (7.23).

The Catalan numbers appear in many combinatorial problems (see e.g., Comtet (1974)). For example, the number of paths in the simple random walk that never cross (but may touch) the k-axis and that start from the origin and return to the origin at time $n = 2m$, is deduced from the reflection principle (Theorem 47) as

$$T_{(2m,0)} - T_{(2m,-2)} = \binom{2m}{m} - \binom{2m}{m-1} = C_{2m}$$

Indeed, the number of paths from the origin to $(2m, 0)$ that never cross the k-axis equals the total number of paths from the origin to $(2m, 0)$, which is $T_{(2m,0)}$, minus the number of paths from the origin to $(2m, 0)$ that cross the k-axis at some point. A path that crosses the k-axis, touches the line $x = -1$. Instead of considering the reflection principle with respect to the $x = 0$ line (i.e. the k-axis), it evidently applies for a reflection around a line at $x = j \in \mathbb{Z}$. Thus, the number of paths from $(2m, 0)$ to the origin that touch or cross the line at $x = -1$ is equal to the total number of paths from $(2m, -2)$ to the origin. That latter number is $T_{(2m,-2)}$, which demonstrates the claim.

189. *Extensions of Wigner's Semicircle Law.* A single eigenvalue has measure zero and does not contribute to the limit probability density function (7.21). By using Wigner's method, Füredi and Komlós (1981) have extended Wigner's Theorem 50.

Theorem 52 (Füredi-Komlós) *Let A be a random $N \times N$ real symmetric matrix where the elements $a_{ij} = a_{ji}$ are independent, not necessarily identically distributed, random variables bounded by a common bound K. Assume that, for $i \neq j$, these random variables possess a common mean $E\left[a_{ij}\right] = \mu$ and common $Var\left[a_{ij}\right] = \sigma^2$, while $E\left[a_{ii}\right] = \nu$.*

(a) *If $\mu > 0$, then the distribution of the largest eigenvalue $\lambda_1(A)$ can be approximated to within order $O\left(\frac{1}{\sqrt{N}}\right)$ by a Gaussian distribution with mean*

$$E[\lambda_1(A)] \simeq (N-1)\mu + \nu + \frac{\sigma^2}{\mu}$$

and bounded variance

$$Var[\lambda_1(A)] \simeq 2\sigma^2$$

In addition, with probability tending to 1,

$$\max_{j>1} |\lambda_j(A)| < 2\sigma\sqrt{N} + O\left(N^{1/3}\log N\right) \qquad (7.24)$$

(b) *If $\mu = 0$, then all eigenvalues of A, including the largest, obey the last bound (7.24).*

The Füredi-Komlós Theorem 52 has been sharpened to $\max_{j>1} |\lambda_j(A)| < 2\sigma\sqrt{N} + O\left(N^{1/4}\log N\right)$ by Vu (2007). The so-called Gaussian Unitary Ensemble (GUE) is defined by an $N \times N$ Hermitian Wigner matrix W, where the diagonal elements w_{ii} are i.i.d. real Gaussian random variables $N(0,1)$, while both the real and the imaginary part of the complex off-diagonal elements w_{ij} are i.i.d. Gaussian random variables $N\left(0, \frac{1}{2}\right)$. Among many results, Tao and Vu (2011) mention a theorem of Gustavsson, illustrating that, for large N, an eigenvalue $\lambda_i(W)$ of a random Hermitian Wigner matrix W has Gaussian fluctuation:

Theorem 53 (Gustavsson) *If i varies with N such that $i/N \to c$, as $N \to \infty$, for some $0 < c < 1$, then the scaled i-th eigenvalue of an $N \times N$ random Hermitian Wigner matrix W tends, for $N \to \infty$, in distribution to*

$$\sqrt{\frac{4 - \left(F_W^{-1}\left(\frac{i}{N}\right)\right)^2}{2}} \frac{\lambda_i\left(W\sqrt{N}\right) - NF_W^{-1}\left(\frac{i}{N}\right)}{\sqrt{\log N}} \xrightarrow{d} N(0,1)$$

where $F_W^{-1}(x)$ is the inverse function of the normalized ($\sigma = 1$ in (7.21)) Wigner semi-circle distribution function,

$$F_W(x) = \frac{1}{2\pi} \int_{-2}^{x} \sqrt{4 - t^2} dt$$

Loosely speaking, Gustavsson's Theorem 53 states that

$$\lambda_i(W) \approx F_W^{-1}\left(\frac{i}{N}\right)\sqrt{N} + N\left(0, \frac{2\log N}{\left(4 - \left(F_W^{-1}\left(\frac{i}{N}\right)\right)^2\right)N}\right)$$

190. *Spectrum of the Erdős-Rényi random graph.* We apply the powerful Füredi-Komlós Theorem 52 to the Erdős-Rényi random graph $G_p(N)$. Since $\mu = p$, $\nu = 0$ and $\sigma^2 = p(1-p)$, Theorem 52 states that the largest eigenvalue λ_1 is a Gaussian random variable with mean $E[\lambda_1] = (N-2)p + 1 + O\left(\frac{1}{\sqrt{N}}\right)$ and $\mathrm{Var}[\lambda_1(A)] \simeq 2p(1-p)$, while all other eigenvalues are smaller in absolute value than $2\sqrt{p(1-p)N} + O\left(N^{1/4}\log N\right)$, the latter due to Vu (2007).

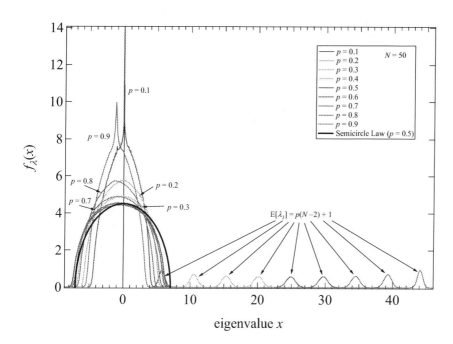

Fig. 7.2. The probability density function of an eigenvalue in $G_p(50)$ for various p. Wigner's Semicircle Law, rescaled and for $p = 0.5$ ($\sigma^2 = \frac{1}{4}$), is shown in bold. We observe that the spectrum for p and $1-p$ is similar, but slightly shifted. The high peak for $p = 0.1$ reflects disconnectivity, while the high peak at $p = 0.9$ shows the tendency to the spectrum of the complete graph where $N-1$ eigenvalues are precisely -1.

The spectrum of $G_p(50)$ together with the properly rescaled Wigner's Semicircle Law (7.21) is plotted in Fig. 7.2. Already for this small value of N, we observe that Wigner's Semicircle Law is a reasonable approximation for the intermediate p-region. The largest eigenvalue λ_1 for finite N, which is almost Gaussian distributed around $p(N-2)+1$ with variance $2p(1-p)$ by Theorem 52 and shown in Fig. 7.2, but which is not incorporated in Wigner's Semicircle Law, influences the average $E[\lambda] = \frac{1}{N}\sum_{k=1}^{N}\lambda_k = 0$ and causes the major bulk of the pdf around $x = 0$ to shift leftward compared to Wigner's Semicircle Law, which is perfectly centered around $x = 0$.

The finite size variant of the Wigner Semicircle Law for the eigenvalue distribution of the adjacency matrix of the Erdős-Rényi random graph $G_p(N)$ is

$$f_\lambda(x) \simeq \frac{\sqrt{4Np(1-p)-(x+p)^2}}{2\pi Np(1-p)}, \quad |x| \leq 2p(1-p)\sqrt{N} \qquad (7.25)$$

The expression (7.25) for the bulk density of eigenvalues, thus also ignoring the largest eigenvalue λ_1, agrees very well with simulations for finite N. Below, we sketch the derivation of (7.25). The probabilistic companion of (3.5) is

$$E\left[\lambda\right] = \sum_{k=-\infty}^{\infty} k \Pr\left[\lambda = k\right] = 0$$

while the discrete random variable λ needs to satisfy $\sum_{k=-\infty}^{\infty} \Pr\left[\lambda = k\right] = 1$. The Perron-Frobenius Theorem 75 states that any connected graph has one largest eigenvalue λ_1 with multiplicity one, such that $\Pr\left[\lambda = \lambda_1\right] = \frac{1}{N}$. Both the mean and the law of total probability can be written, for one realization of an Erdős-Rényi random graph, as

$$E\left[\lambda\right] = \lambda_1 \frac{1}{N} + \sum_{\text{All others}} k \Pr\left[\lambda = k\right] = 0 \qquad (7.26)$$

and $\sum_{\text{All others}} \Pr\left[\lambda = k\right] = 1 - \frac{1}{N}$. Fig. 7.2 suggests us to consider the Semicircle Law for finite N shifted over some value ε,

$$f_\lambda\left(x; \varepsilon\right) = \frac{\sqrt{4Np\left(1 - p\right) - \left(x + \varepsilon\right)^2}}{2\pi Np\left(1 - p\right)}, \quad |x| \le 2p\left(1 - p\right)\sqrt{N}$$

Denoting the radius $R = 2p\left(1 - p\right)\sqrt{N}$ and passing to the continuous random variable, relation (7.26) becomes

$$0 = \lambda_1 \frac{1}{N} + \int_{-R-\varepsilon}^{R-\varepsilon} x f_\lambda\left(x; \varepsilon\right) dx$$

$$= \lambda_1 \frac{1}{N} + \int_{-R-\varepsilon}^{R-\varepsilon} \left(x + \varepsilon\right) f_\lambda\left(x; \varepsilon\right) dx - \varepsilon \int_{-R-\varepsilon}^{R-\varepsilon} f_\lambda\left(x; \varepsilon\right) dx$$

Since $\int_{-R-\varepsilon}^{R-\varepsilon}\left(x + \varepsilon\right) f_\lambda\left(x; \varepsilon\right) dx = 0$ due to symmetry and $\int_{-R-\varepsilon}^{R-\varepsilon} f_\lambda\left(x; \varepsilon\right) dx = 1 - \frac{1}{N}$, we obtain $\lambda_1 \frac{1}{N} - \varepsilon\left(1 - \frac{1}{N}\right) = 0$. Finally, Theorem 52 states that $\lambda_1 = (N - 2) p + O\left(1\right)$ such that $\varepsilon = p + O\left(N^{-1}\right)$ leading to (7.25).

The complement of $G_p\left(N\right)$ is $\left(G_p(N)\right)^c = G_{1-p}\left(N\right)$, because a link in $G_p(N)$ is present with probability p and absent with probability $1 - p$ and $\left(G_p(N)\right)^c$ is also a random graph. For large N, there exists a large range of p values for which both $p \ge p_c$ and $1 - p \ge p_c$ such that both $G_p\left(N\right)$ and $\left(G_p(N)\right)^c$ are connected almost surely. Fig. 7.2 shows that the normalized spectra of $G_p\left(N\right)$ and $G_{1-p}\left(N\right)$ are, apart from a small shift and ignoring the largest eigenvalue, almost identical. Equation (3.39) and **art**. 62 indicate that the spectra of a graph and of its complement tend to each other if $u^T x_j \to 0$, except for the largest eigenvector x_1 which will tend to u. This seems to suggest that $G_p\left(N\right)$ and $G_{1-p}\left(N\right)$ are tending to a regular graph with degree $p\left(N - 1\right)$ and $\left(1 - p\right)\left(N - 1\right)$ and that these regular graphs, even for small N, have nearly the same spectrum, apart from the largest eigenvalue $p\left(N - 1\right)$

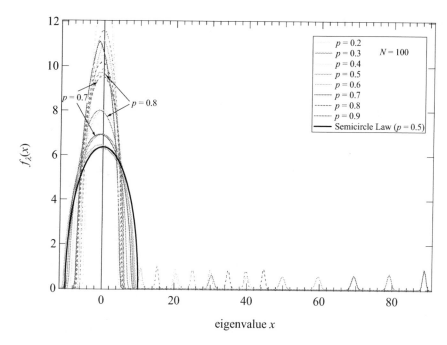

Fig. 7.3. The spectrum of the adjacency matrix of $G_p(100)$ (full lines) and of the corresponding matrix with i.i.d. uniform elements (dotted lines). The small peaks at higher values of x are due to λ_1.

and $(1-p)(N-1)$ respectively: $\frac{\lambda_{1-p}}{\sqrt{N}} \simeq -\frac{\lambda_p}{\sqrt{N}} - \frac{1}{\sqrt{N}}$ where λ_p is an eigenvalue of $G_p(N)$.

Fig. 7.3 shows the probability density function $f_\lambda(x)$ of the eigenvalues of the adjacency matrix A of $G_p(N)$ with $N = 100$ together with the eigenvalues of the corresponding matrix A_U where all one elements in the adjacency matrix of $G_p(100)$ are replaced by i.i.d. uniform random variables on $[0, 1]$. Since the elements of A_U are always smaller with probability 1 than those of A, the matrix norm $\|A_U\|_q < \|A\|_q$ and the inequality (A.26) imply that $\lambda_1(A_U) < \lambda_1(A)$. In addition, relation (3.7) shows that $\sum_{k=1}^{N} \lambda_k^2(A_U) < 2L$ such that $\mathrm{Var}[\lambda(A_U)] < \mathrm{Var}[\lambda(A)]$, which is manifested by a narrower and higher peaked pdf centered around $x = 0$.

7.5.2 The Marčenko-Pastur Law

The last of the classical laws in random matrix theory with an analytic density function for the eigenvalues is given in the next theorem without proof:

Theorem 54 (The Marčenko-Pastur law) *Let C be a random $m \times n$ matrix with independent and identically distributed complex elements c_{ij} with finite $\sigma^2 = Var[c_{ij}]$ and zero mean $E[c_{ij}] = 0$, or the complex elements c_{ij} are independently* *distributed with a finite fourth-order moment. Let $y = \frac{m}{n}$ as $n \to \infty$ and define* $a(y) = \sigma^2 \left(1 - \sqrt{y}\right)^2$ *and* $b(y) = \sigma^2 \left(1 + \sqrt{y}\right)^2$, *and denote by* $\lambda(S)$ *an eigenvalue of the set of the m real eigenvalues of the scaled Hermitian matrix* $S = \frac{1}{n}CC^*$. *The probability density function* $f_{\lambda(S)}(x)$ *tends for $n \to \infty$ to*

$$\lim_{n \to \infty} f_{\lambda(S)}(x) = \frac{1_{\{a(y) \leq x \leq b(y)\}}}{2\pi x y \sigma^2} \sqrt{(x - a(y))(b(y) - x)} + \left(1 - \frac{1}{y}\right)\delta(x) 1_{\{y > 1\}}$$

$$(7.27)$$

Marčenko and Pastur (1967) prove Theorem 54 by deriving a first-order partial differential equation, from whose solution the unique Stieltjes transform $m(\psi; z)$ of $\psi(x) = \lim_{n \to \infty} f_{\lambda(S)}(x)$ is found. The Cauchy or Stieltjes transform (**art. 362**) of a function $f(x)$, defined by

$$m(f, z) = \int_{-\infty}^{\infty} \frac{f(x)}{z - x} dx$$

is a special case of an integral of the Cauchy type, that is treated, together with its inverse, in **art. 361**. The method of Marčenko and Pastur (1967) is different from the moments method used by McKay, sketched in Section 7.4, and earlier by Wigner (1955).

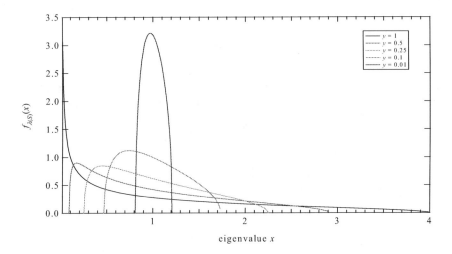

Fig. 7.4. The Marčenko-Pastur probability density function (7.27) for various values of y. Each curve starts at $x = a(y)$, which is increasing from 0 to 1 when y decreases from 1 to 0, and ends at $x = b(y)$, which decreases from 4 to 1 when y decreases from 1 to 0. When $y \to 0$, the Marčenko-Pastur probability density function tends to a delta function at $x = 1$.

The last term in (7.27), the point mass at $x = 0$, is a consequence of the non-square form of C. The rank$(CC^*) \leq \min(n, m)$ such that, for $y > 1$, the $m \times m$

matrix CC^* has $m - n = m\left(1 - \frac{1}{y}\right)$ zero eigenvalues, while all ny other eigenvalues are the same as those of C^*C, which follows from **art. 284**.

In the case $m = n$ and $y = 1$, and $C = C^T = \frac{A}{\sqrt{n}}$, the eigenvalues of S are the squares of those of $\frac{A}{\sqrt{n}}$. Since the latter eigenvalues obey Wigner's Semicircle Law (7.21) and since the density $f_{X^2}(x) = \frac{f_X(\sqrt{x}) + f_X(-\sqrt{x})}{2\sqrt{x}}$ for any random variable X as shown in Van Mieghem (2014, p. 50), we find, indeed for $y = 1$, that

$$f_{\lambda(S)} = \frac{f_{\lambda(A_n)}(\sqrt{x}) + f_{\lambda(A_n)}(-\sqrt{x})}{2\sqrt{x}} = \frac{f_{\lambda(A_n)}(\sqrt{x})}{\sqrt{x}}$$

Also, in that case, the matrix S represents a square covariance matrix. In general for m real random $n \times 1$ vectors, S represents the $m \times n$ covariance matrix, that appears in many applications of signal and information theory and physics. Fig. 7.4 illustrates the Marčenko-Pastur probability density function $f_{\lambda(S)}(x)$ for various values of $y \leq 1$.

7.5.3 Density of random graphs with arbitrary expected degrees

Raj Rao and Newman (2013) consider the configuration model that generates a random graph with a given degree distribution. The degree d_i of a node i is visualized as d_i stubs or half-links incident to node i. Given a degree sequence $d_1, d_2, \ldots d_N$, after a pairwise matching of stubs of different nodes, in which each matching appears with equal probability, the final configuration graph is obtained. Thus, each joining of two uniformly chosen and not yet paired stubs creates a link in the configuration graph and the process continues until all stubs have been joined. The expected number of links between nodes i and j equals $\frac{d_i d_j}{2L}$ for large N. This property of the configuration graphs relates naturally to the modularity matrix M defined by (4.80) in **art. 151**. However, the links in the configuration graph are not independent and to avoid this major complication, Raj Rao and Newman (2013) consider a modified graph in which the number of links between each pair (i, j) of nodes is an independent Poisson random variable with mean $\frac{d_i d_j}{2L}$. In particular, instead of the specific degree, they treat d_i as the *expected* degree of node i, which is, for large N, a good approximation because the actual degree is then narrowly peaked around the mean degree. These expected degrees, that are now real numbers x instead of integers, are drawn from the continuous probability density function $f_d(x)$. The corresponding adjacency matrix of the modified configuration graph is

$$A = \frac{1}{2L} d.d^T + M$$

where $\frac{1}{2L} d.d^T$ is the ensemble average of A and the modularity matrix M is the deviation from that ensemble average, whose elements are, by construction, independent but not identically distributed random variables with zero mean. Moreover, since the variance $\mathrm{Var}[a_{ij}] = \mathrm{Var}[m_{ij}]$ and since a_{ij} are Poisson random variables, we have that $\mathrm{Var}[m_{ij}] = E[a_{ij}] = \frac{d_i d_j}{2L}$ and $E\left[m_{ij}^2\right] = \frac{d_i d_j}{2L}$, because $E[m_{ij}] = 0$.

After using (7.7) for the density $f_{\lambda(M)}(x)$ of the eigenvalues of the modularity matrix M, repeatedly approximating $E[f(X)]$ by $f(E[X])$ justified by construction of the modified configuration graph and for large N and using the Cauchy or Stieltjes transform (see **art.** 361), Raj Rao and Newman (2013) end up with

$$f_{\lambda(M)}(x) = -\frac{d_{av}}{\pi x}\operatorname{Im}h^2(x)$$

where the average degree $d_{av} = \frac{2E[L]}{N}$ and where the function $h(x)$ satisfies the integral equation

$$h(z) = \frac{1}{d_{av}}\int_0^\infty \frac{xf_d(x)\,dx}{z - xh(z)} \tag{7.28}$$

Due to interlacing (**art.** 155) of eigenvalues of the adjacency and modularity matrix, the spectral density of the modularity matrix equals that of the adjacency matrix, $f_{\lambda(M)}(x) = f_{\lambda(A)}(x)$, except for the largest eigenvalue λ_1 of the adjacency matrix A. The largest eigenvalue $\lambda_1 = \lambda_1(A)$ is shown to satisfy $(\lambda_1 - 1)h(\lambda_1) = 1$, where h obeys (7.28).

Generally, for an arbitrary degree density function $f_d(k)$, the resulting spectral density $f_{\lambda(A)}(x)$ deviates from Wigner's Semicircle Law (7.21). The spectrum still consists of a main band, but nodes with exceptionally high degree, the so-called hubs, may give rise to eigenvalues that lie outside that band, akin to the energy of impurity states in solid state materials.

8

Spectra of complex networks

This chapter presents examples of the spectra of complex networks, which we have tried to interpret or to understand using the theory of previous chapters. In contrast to the mathematical rigor of the other chapters, this chapter is more intuitively oriented and it touches topics that are not yet understood or that lack maturity. Nevertheless, the examples may give a flavor of how real-world complex networks are analyzed as a sequence of small and partial steps towards, hopefully, complete understanding.

8.1 Simple observations

When we visualize the density function $f_\lambda(x)$ of the eigenvalues of the adjacency matrix of a graph, defined in **art. 173**, peaks at $x = 0$, $x = -1$ and $x = -2$ are often observed. The occurrence of adjacency eigenvalue at those integer values has a physical explanation. Integer eigenvalues are special (**art. 45**).

8.1.1 A graph with eigenvalue $\lambda(A) = 0$

A matrix has a zero eigenvalue if its determinant is zero (**art. 235**). A determinant is zero if two rows are identical or if some of the rows are linearly dependent (**art. 209**). For example, two rows are identical resulting in $\lambda(A) = 0$, if *two not mutually interconnected nodes are connected to a same set of nodes*. Since the elements a_{ij} of an adjacency matrix A are only 0 or 1, linear dependence of rows occurs every time the sum of a set of rows equals another row in the adjacency matrix. For example, consider the sum of two rows. If node n_1 is connected to the set S_1 of nodes and node n_2 is connected to the distinct set S_2, where $S_1 \cap S_2 = \varnothing$ and $n_1 \neq n_2$, then *the graph has a zero adjacency eigenvalue if another node $n_3 \neq n_2$ and $n_3 \neq n_1$ is connected to all nodes in the set $S_1 \cup S_2$*. These two types of zero eigenvalues occur when a graph possesses a "local bipartiteness". In real networks, this type of interconnection often occurs.

8.1.2 A graph with eigenvalue $\lambda(A) = -1$

An adjacency matrix A has an eigenvalue $\lambda(A) = -1$ every time *a node pair n_1 and n_2 in the graph is connected to a same set S of different nodes and n_1 and n_2 are mutually also interconnected.* Indeed, without loss of generality, we can relabel the nodes such that $n_1 = 1$ and $n_2 = 2$. In that case, the first two rows in A are of the form

$$\begin{matrix} 0 & 1 & a_{13} & a_{14} & \cdots & a_{1N} \\ 1 & 0 & a_{13} & a_{14} & \cdots & a_{1N} \end{matrix}$$

and the corresponding rows in $\det(A - \lambda I)$ of the characteristic polynomial are

$$\begin{matrix} -\lambda & 1 & a_{13} & a_{14} & \cdots & a_{1N} \\ 1 & -\lambda & a_{13} & a_{14} & \cdots & a_{1N} \end{matrix}$$

If two rows are identical, the determinant is zero. In order to make these rows identical, it suffices to take $\lambda = -1$ and $\det(A + I) = 0$, which shows that $\lambda = -1$ is an eigenvalue of A with this particular form. This observation generalizes to a graph where k nodes are fully meshed and, in addition, all k nodes are connected to the same set S of different nodes. Again, we may relabel nodes such that the first k rows describe these k nodes in a complete graph configuration, also called a clique. Let x denote a $(N - k) \times 1$ zero-one vector, then ux^T is a matrix with all rows identical and equal to x. The structure of $\det(A - \lambda I)$ is

$$\det(A - \lambda I) = \begin{vmatrix} (J - (\lambda + 1)I)_{k \times k} & u.x^T \\ B_{(N-k) \times k} & (C - \lambda I)_{(N-k) \times (N-k)} \end{vmatrix}$$

which shows that the first k rows are identical if $\lambda = -1$, implying that the multiplicity of this eigenvalue is $k - 1$. Observe that the spectrum in Section 6.1 of the complete graph K_N, where $k = N$, indeed contains an eigenvalue $\lambda = -1$ with multiplicity $N - 1$. We can also say that a peak in the density of the adjacency eigenvalues at $\lambda = -1$ reflects that a set of interconnected nodes all have the same neighbors, different from those in the interconnected set.

8.1.3 A graph with eigenvalue $\lambda(A) = -2$

If the graph is a line graph (**art. 25**), then **art. 27** demonstrates that the adjacency matrix has an eigenvalue equal to $\lambda(A) = -2$ with multiplicity $L - N$. However, it is in general rather difficult to conclude that a graph is a line graph. Each node with degree d – locally, a star $K_{1,d}$ – is transformed in the line graph into a clique with $\binom{d}{2}$ links. Thus, a line graph can be regarded as a collection of interconnected cliques K_{d_j}, where $1 \le j \le N$. The presence of an eigenvalue $\lambda(A) = -2$ is insufficient to deduce that a graph is a line graph. A more elaborate discussion on line graphs is found in Cvetković *et al.* (2004) and Cvetković *et al.* (2009, Section 3.4).

A peak in the density $f_\lambda(x)$ of the eigenvalues of the adjacency matrix at $\lambda(A) =$

-2 and $\lambda(A) = 2$ may correspond to a very long path (**art. 179**). As shown in Fig. 7.1, these peaks occur in large, sparse regular graphs with degree $r = 2$ by McKay's Theorem 49.

8.2 Distribution of the Laplacian eigenvalues and of the degree

Although the moments of the Laplacian eigenvalues (**art. 106-108**) can be expressed in terms of those of the degree, in most real-world networks the degree distribution and the Laplacian distribution are usually different. In this section, we present a curious example, where both distributions are remarkably alike.

Software is assembled from many interacting units and subsystems at several levels of granularity (subroutines, classes, source files, libraries, etc.) and the interactions and collaborations of those parts can be represented by a graph, which is called the software collaboration graph. Fig. 8.1 depicts the topology of the VTK network, which represents the collaborations in the VTK visualization C++ library that has been documented and studied by Myers (2003).

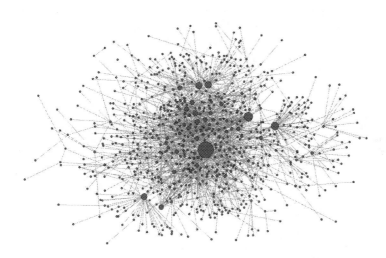

Fig. 8.1. The connected graph of the VTK network with $N = 771$ and $L = 1357$. The nodal size is drawn proportionally to its degree.

Fig. 8.2 shows the correspondence between the degree D and the Laplacian eigenvalue μ in the connected VTK graph with $N = 771$ nodes, $E[D] = E[\mu] = 3.5201$, $\text{Var}[D] = 33.0603$ and $\text{Var}[\mu] = 36.5804$, which agrees with the theory in **art. 106**. Both the degree D and the Laplacian eigenvalue μ of the VTK graph approximately follow a power law, a general characteristic of many complex networks, and each power law is specified by the fit in the legend in Fig. 8.2, where c_D and c_μ are normalization constants. The much more surprising fact is that the insert in Fig. 8.2 demonstrates how closely the ordered Laplacian eigenvalues μ_k follow the ordered degree $d_{(k)}$. Only in software collaboration networks, such as MySql studied in Myers (2003), have we observed such a close relationship between D and μ, which suggests that these graphs may be close to threshold graphs (**art. 114**).

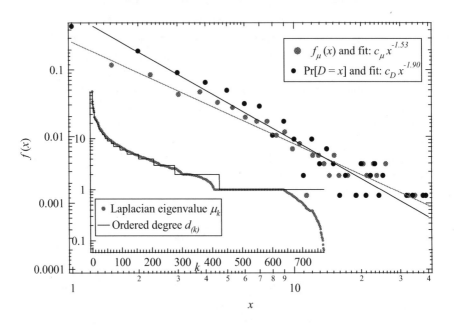

Fig. 8.2. The density function of the degree and of the Laplacian eigenvalues in the software dependence network VTK. The insert shows how close the ordered degree and Laplacian eigenvalues are.

The definition of the Laplacian $Q = \Delta - A$ hints that the influence of the adjacency matrix on the eigenvalues μ of the Laplacian is almost negligible. The bounds in **art. 106**, derived from the interlacing principle,

$$d_{(k)} - \lambda_1(A) \leq \mu_k(Q) \leq d_{(k)} - \lambda_N(A)$$

are too weak because $\lambda_1(A) = 11.46$ and $\lambda_N(A) = -9.13$. Our recent perturbation approximation (Van Mieghem, 2021) for a Laplacian eigenvalue μ_m expanded around the degree d_q of a node q in the graph

$$\mu_m \approx d_q + \frac{11}{16} \sum_{\substack{k=1 \\ k \neq q}}^{n} \frac{a_{kq}}{d_q - d_k} - \frac{5}{16} \sum_{\substack{r=1 \\ r \neq q}}^{n} \frac{a_{rq}}{d_q - d_r} \sum_{\substack{k=1 \\ k \neq q}}^{n} \frac{a_{qk}a_{kr}}{d_q - d_k} \tag{8.1}$$

$$+ \frac{1}{16} \left(\sum_{\substack{r=1 \\ r \neq q}}^{n} \frac{a_{rq}}{d_q - d_r} \sum_{\substack{l=1 \\ l \neq q}}^{n} \frac{a_{rl}}{d_q - d_l} \sum_{\substack{k=1 \\ k \neq q}}^{n} \frac{a_{kq}a_{kl}}{d_q - d_k} - \sum_{\substack{r=1 \\ r \neq q}}^{n} \frac{a_{rq}}{(d_q - d_r)^2} \sum_{\substack{k=1 \\ k \neq q}}^{n} \frac{a_{kq}}{d_q - d_k} \right)$$

is expected to be useful, e.g. when a few links are added or removed in threshold graphs (**art. 114**).

Fig. 8.3 presents the density function $f_\lambda(x)$ of the adjacency eigenvalues, which is typically tree-like: a high peak $f_\lambda(0) = 0.42$ at the origin $x = 0$ and the density function is almost symmetric around the origin, $f_\lambda(-x) \approx f_\lambda(x)$. If a graph is locally tree-like (**art. 183**), we would expect its density to approximately follow McKay's Theorem 49 drawn in Fig. 7.1. At first glance, the peaks in $f_\lambda(x)$ at roughly $x = -1$ and $x = 1$ may hint at such a locally tree-like structure, but McKay's Theorem 49 predicts singular behavior at $x = 2\sqrt{r-1} \approx 3.1$ for degree $r \simeq E[D] = 3.52$. The small variance $\mathrm{Var}[\lambda] = E[D] = 3.52$ (**art. 49**), which is much smaller than $\mathrm{Var}[D]$ and than $\mathrm{Var}[\mu] = \mathrm{Var}[D] + \mathrm{Var}[\lambda]$, supports the observation why the adjacency spectrum only marginally influences the Laplacian eigenvalues.

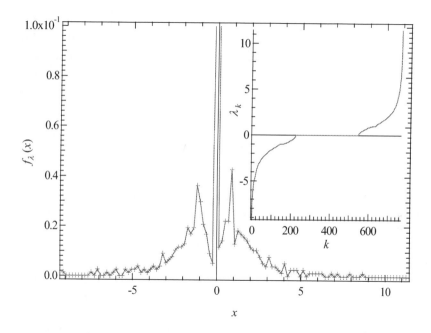

Fig. 8.3. The density of the eigenvalues of the adjacency matrix of the VTK graph. The insert shows the ordered eigenvalues λ_k versus their rank k, where $\lambda_1 = 11.46$ and $\lambda_N = -9.13$.

Finally, we mention the nice estimate of Dorogovtsev *et al.* (2003). Using an approximate analysis from statistical physics, but inspired by McKay's result (Section 7.4) based on random walks, Dorogovtsev *et al.* (2003) derived the asymptotic law for the tails of $f_\lambda(x)$ of locally tree-like graphs as

$$f_\lambda(x) \approx 2|x| \Pr\left[D = x^2\right]$$

for large x. For example, in a power law graph where $\Pr[D = k] = ck^{-\gamma}$, the asymptotic tail behavior of the density function of the adjacency eigenvalues is

$$f_\lambda(x) \approx 2c \, |x|^{1-2\gamma}$$

As shown in Fig. 8.2, the power law exponent for the VTK network is about $\gamma \simeq 1.9$ such that $2\gamma - 1 \simeq 2.8$, but fitting the tail region of $f_\lambda(x)$ in a log-log plot gives a slope of -1.7, which again seems to indicate that the VTK graph is not sufficiently close to a locally tree-like, power law graph.

8.3 Functional brain network

The interactions between brain areas can be represented by a functional brain network as shown by Stam and Reijneveld (2007) and Tewarie *et al.* (2021). The concept of functional connectivity refers to the statistical interdependencies between physiological time series recorded in various brain areas, and is thought to reflect communication between several brain areas. Magneto-encephalography (MEG), a recording of the brain's magnetic activity, is a method to assess functional connectivity within the brain. Each MEG channel is regarded as a node in the functional brain network, while the functional connectivity between each pair of channels is represented by a link, whose link weight reflects the strength of the connectivity, measured via the synchronization likelihood. It is based on the concept of general synchronization (Rulkov *et al.*, 1995), and takes linear as well as non-linear synchronization between two time series into account. The synchronization likelihood w_{ij} between time series i and j lies in the interval $[0,1]$, with $w_{ij} = 0$ indicating no synchronization, and $w_{ij} = 1$ meaning total synchronization. We adopt the convention that $w_{jj} = 0$, rather than $w_{jj} = 1$, because of the association with the adjacency matrix of the corresponding functional brain graph.

The weighted adjacency matrix W of the human functional brain network contains as elements w_{ij} the synchronization likelihood between the $N = 151$ different MEG channels, each probing a specific area in the human brain as detailed in Wang *et al.* (2010). Since all functional brain areas are correlated, the matrix W has the structure of the adjacency matrix A_{K_N} of the complete graph K_N, where the one-elements a_{ij} are substituted by the correlations $|w_{ij}| \le 1$. Since the matrix norm $\|W\|_q \le \|A_{K_N}\|_q$ because all elements $|w_{ij}| \le 1$, **art. 207** indicates that $\lambda_1(W) \le \|W\|_q$ and $\lambda_1(W) \le \lambda_1(A_{K_N}) = N - 1$. Fig. 8.4 shows the eigenvalues of the weighted adjacency matrix W of the functional brain network of a typical patient before and after surgery. The correlations w_{ij} before and after surgery are almost the same. The spectrum in Fig. 8.4 is closely related to that of the complete graph K_N: the $\lambda_N = -1$ eigenvalue with multiplicity $N - 1$ in K_N is here spread over the interval $[-1, \lambda_1)$. All eigenvalues are simple and the largest eigenvalue in $[14, 15]$ is clearly most sensitive to the changes in the weighted adjacency matrix W, as the insert in Fig. 8.4 shows. Hence, the changes in the few largest eigenvalues seem to be good indicators to evaluate the effect of the brain surgery.

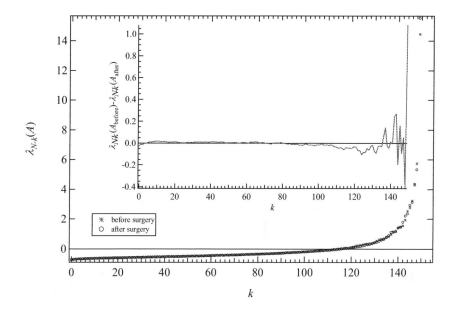

Fig. 8.4. The eigenvalues of the weighted adjacency matrix of the functional brain network before and after surgery in increasing order. The insert shows the differences between the eigenvalues before and after surgery.

8.4 Rewiring Watts-Strogatz small-world graphs

The spectrum of the Watts-Strogatz small-world graph $G_{SWk;N}$ without link rewiring is computed in Section 6.2. Recall that $G_{SWk;N}$ is a regular graph (**art.** 55) where each node has degree $r = 2k$. When links in $G_{SWk;N}$ are rewired, independently and with probability p_r, the graph's topology and properties change with p_r. Fig. 1.3 presents a rewired Watts-Strogatz small-world graph, while the original regular small-world graph $G_{SWk;N}$ is shown in Fig. 6.1. Here, we investigate the influence of the link rewiring probability p_r on the eigenvalues of the adjacency matrix of Watts-Strogatz small-world graphs.

Fig. 8.5 shows the pdf $f_\lambda(x)$ of an eigenvalue λ of the adjacency matrix of a Watts-Strogatz small-world graph. In absence of randomness $p_r = 0$, the spectrum is discrete, reflected by the peaks in Fig. 8.5 and drawn differently for all k in Fig. 6.2. When randomness is introduced by increasing $p_r > 0$, the peaks smooth out and Fig. 8.5 indicates that the pdf $f_\lambda(x)$ tends rapidly to that of the Erdős-Rényi random graph shown in Fig. 7.2.

Fig. 8.5 thus suggests that a bell-shape of the spectrum around the origin is a fingerprint of "randomness" in a graph, while peaks reflect "regularity" or "structure"[1]. We also observe that "irregularity" can be measured, as mentioned in

---

[1] The quotes here refer to an intuitive meaning. A commonly agreed and precise definition of "randomness" and "structure" of a graph is lacking.

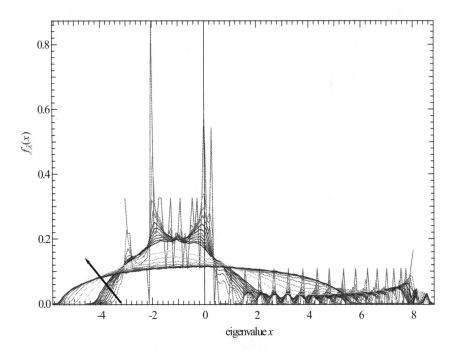

Fig. 8.5. The probability density function $f_\lambda(x)$ of an eigenvalue in Watts-Strogatz small-world graphs with $N = 200$ and $k = 4$ for various rewiring probabilities p_r ranging from 0 to 1, first in steps of 0.01 until $p_r = 0.1$, followed by an increase in steps of 0.1 up to $p_r = 1$. The arrow shows the direction of increasing p_r.

art. 72, by the amount that the largest eigenvalue deviates from the mean degree $E[D] = 2k$. Rewiring does not change the mean degree, because the number of links and nodes is kept constant and $E[D] = \frac{2L}{N}$, but the x-axis in Fig. 8.5 shows an increase of the largest eigenvalue from $\lambda_1 = 8 = 2k$ when $p_r = 0$ to about 9 for $p_r = 1$.

Fig. 6.3 has shown how irregular the number of different eigenvalues of $G_{SW\,k;N}$ without rewiring behaves as a function of N and k. Simulations indicate that, even for a small rewiring probability of $p_r = 0.01$, the spectrum only contains simple eigenvalues with high probability. When rewiring only one link in $G_{SW\,k;N}$ with $N = 200$ and $k = 4$, the number of distinct eigenvalues dramatically increases from 95 to about 190. Hence, destroying the regular adjacency matrix structure by even one element has a profound impact on the multiplicity of the eigenvalues. This very high sensitivity is a known fact in the study of zeros of polynomials (Wilkinson, 1965, Chapter 2): small perturbations of the coefficients of a polynomial may heavily impact the multiplicity of the real zeros and whether the perturbed zeros are still real. Another consequence is that the upper bound in Theorem 11 on p. 75 for the diameter ρ in terms of the number of different eigenvalues is almost useless in real-world graphs, where most of the eigenvalues are different, such that the bound in Theorem 11 reduces to $\rho \leq N - 1 = \rho_{\max}$. By rewiring links in $G_{\mathrm{SW}k;N}$, we observe even contrasting effects: the regular structure of $G_{\mathrm{SW}k;N}$ is destroyed, which causes the diameter ρ, in most cases, to shrink, while the number of different eigenvalues jumps to almost the maximum N.

8.5 Assortativity

8.5.1 Theory

"Mixing" in complex networks refers to the tendency of network nodes to connect preferentially to other nodes with either similar or opposite properties. Mixing is computed via the correlations between the properties, such as the degree, of nodes in a network. Here, we study the degree mixing in undirected graphs. Generally, the linear correlation coefficient between two random variables of X and Y is defined (Van Mieghem, 2014, p. 27) as

$$\rho(X, Y) = \frac{E[XY] - \mu_X \mu_Y}{\sigma_X \sigma_Y} \tag{8.2}$$

where $\mu_X = E[X]$ and $\sigma_X = \sqrt{\mathrm{Var}[X]}$ are the mean and standard deviation of the random variable X, respectively. Newman (2003a, eq. (21)) has expressed the linear degree correlation coefficient of a graph as

$$\rho_D = \frac{\sum\limits_{xy} xy \left(e_{xy} - a_x b_y\right)}{\sigma_a \sigma_b} \tag{8.3}$$

where e_{xy} is the fraction of all links that connect the nodes with degree x and y and where a_x and b_y are the fraction of links that start and end at nodes with degree x and y, satisfying the following three conditions: $\sum\limits_{xy} e_{xy} = 1$, $a_x = \sum\limits_y e_{xy}$ and $b_y = \sum\limits_x e_{xy}$. When $\rho_D > 0$, the graph possesses *assortative* mixing, a preference of high-degree nodes to connect to other high-degree nodes and, when $\rho_D < 0$, the graph features *disassortative* mixing, where high-degree nodes are connected to low-degree nodes. We refer to Noldus and Van Mieghem (2015) for review on assortativity.

The translation of (8.3) into the notation of random variables is presented as follows. Denote by D_i and D_j the node degree of two randomly chosen nodes i and j in an undirected graph with N nodes that are *connected*, thus with element $a_{ij} = 1$ in (1.1) of the symmetric adjacency matrix A. We are interested in the degree of nodes at both sides of a link, without taking the link, that we are looking at, into consideration. As Newman (2003a) points out, we need to consider the number of excess links at both sides and, thus, the degree $D_{l+} = D_i - 1$ and $D_{l-} = D_j - 1$, where the link l starts at $l^+ = i$ and ends at $l^- = j$. The linear correlation coefficient of those excess degrees is

$$\rho(D_{l+}, D_{l-}) = \frac{E[D_{l+}D_{l-}] - E[D_{l+}]E[D_{l-}]}{\sigma_{D_{l+}}\sigma_{D_{l-}}} = \frac{E[(D_{l+} - E[D_{l+}])(D_{l-} - E[D_{l-}])]}{\sigma_{D_{l+}}\sigma_{D_{l-}}}$$

where $\sigma_{D_i}^2 = \text{Var}[D_i] = E\left[(D_i - E\,[D_i])^2\right]$. Since $D_{l+} - E\,[D_{l+}] = D_i - E\,[D_i]$, subtracting one link everywhere does not change the linear correlation coefficient, provided $D_i > 0$ and $D_j > 0$, which is the case if there are no isolated nodes. Removing isolated nodes from the graph does not alter the linear degree correlation coefficient (8.3). Hence, we can assume that the graph has no zero-degree nodes. Since $E\,[D_i] = E\,[D_j]$, the linear degree correlation coefficient is

$$\rho\left(D_{l+}, D_{l-}\right) = \rho\left(D_i, D_j\right)\big|_{a_{ij}=1} = \frac{E\,[D_i D_j] - (E\,[D_i])^2}{\sigma_{D_i}^2} \qquad (8.4)$$

We express $E\,[D_i D_j]$, the mean $\mu_{D_i} = E\,[D_i]$ and variance $\sigma_{D_i}^2 = \text{Var}[D_i] = E\,[D_i^2] - \mu_{D_i}^2$ in the definition of $\rho\left(D_{l+}, D_{l-}\right)$ for undirected graphs in terms of the total number $N_k = u^T A^k u$ of walks with k hops (**art. 59**). First, we have that

$$E\,[D_i D_j] = \frac{1}{2L} \sum_{i=1}^{N} \sum_{j=1}^{N} d_i d_j a_{ij} = \frac{d^T A d}{2L} = \frac{N_3}{N_1}$$

and

$$E\,[D_i^2] = \frac{1}{2L} \sum_{i=1}^{N} \sum_{j=1}^{N} d_i^2 a_{ij} = \frac{1}{2L} \sum_{i=1}^{N} d_i^3$$

The average μ_{D_i} and μ_{D_j} are the mean node degree of the two connected nodes i and j, respectively, and not the mean of the degree D of a random node, which equals $E\,[D] = \frac{2L}{N}$. Thus,

$$\mu_{D_i} = \frac{1}{2L} \sum_{i=1}^{N} \sum_{j=1}^{N} d_i a_{ij} = \frac{1}{2L} \sum_{i=1}^{N} d_i^2 = \frac{d^T d}{2L} = \frac{N_2}{N_1}$$

illustrating that $\mu_{D_j} = \mu_{D_i}$ and $\sigma_{D_i}^2 = \sigma_{D_j}^2$. After substituting all terms into the linear degree correlation coefficient in (8.4), our reformulation of Newman's definition (8.3) in terms of N_k is

$$\rho_D = \rho\left(D_i, D_j\right) = \frac{N_1 N_3 - N_2^2}{N_1 \sum\limits_{i=1}^{N} d_i^3 - N_2^2} \qquad (8.5)$$

The crucial understanding of (dis)assortativity lies in the total number N_3 of walks with three hops, studied in Li *et al.* (2006), compared to $\sum\limits_{i=1}^{N} d_i^3 = N E\,[D^3]$.

8.5.1.1 Discussion of (8.5)

As shown in **art. 63**, the total number $N_k = u^T A^k u$ of walks of length k is upper bounded by

$$N_k \leq \sum_{j=1}^{N} d_j^k$$

with equality only if $k \leq 2$ and, for all k, only if the graph is regular. Hence, (8.5) shows that only if the graph is regular, $\rho_D = 1$, implying that maximum assortativity is only possible in regular graphs[2]. Since the variance of the degrees at one side of an arbitrary link

$$\sigma_{D_i}^2 = \frac{1}{N_1} \sum_{i=1}^{N} d_i^3 - \left(\frac{N_2}{N_1}\right)^2 \geq 0 \tag{8.6}$$

the sign of $N_1 N_3 - N_2^2$ in (8.5) distinguishes between assortativity ($\rho_D > 0$) and disassortativity ($\rho_D < 0$). The sign of $N_1 N_3 - N_2^2$ can also be determined from (4.97). Using $\sum_{j=1}^{N} d_j^3 - N_3 = \sum_{l \in \mathcal{L}} (d_{l+} - d_{l-})^2$ in (3.43) and denoting a link $l = i \sim j$, the degree correlation (8.5) can be rewritten as

$$\rho_D = 1 - \frac{\sum_{i \sim j} (d_i - d_j)^2}{\sum_{i=1}^{N} d_i^3 - \frac{1}{2L} \left(\sum_{i=1}^{N} d_i^2\right)^2} \tag{8.7}$$

The graph is zero assortative ($\rho_D = 0$) if

$$N_2^2 = N_1 N_3 \tag{8.8}$$

We can show that the connected Erdős-Rényi random graph $G_p(N)$ is zero-assortative for all N and link density $p = L / \binom{N}{2} > p_c$, where p_c is the disconnectivity threshold. Asymptotically for large N, the Barabási-Albert power law graph is zero-assortative as shown in Nikoloski *et al.* (2005).

Perfect disassortativity ($\rho_D = -1$ in (8.5)) implies that

$$N_2^2 = \frac{N_1}{2} \left(N_3 + \sum_{i=1}^{N} d_i^3\right) \tag{8.9}$$

For a complete bipartite graph $K_{m,n}$ (Section 6.7), we have that

$$\sum_{i \sim j} (d_i - d_j)^2 = mn (n - m)^2, \ \sum_{i=1}^{N} d_i^3 = nm (n^2 + m^2) \text{ and } \sum_{i=1}^{N} d_i^2 = nm (n + m)$$

such that (8.7) becomes $\rho_D = -1$, provided $m \neq n$. Hence, any complete bipartite graph $K_{m,n}$ is perfectly disassortative, irrespective of its size and structure (m, n), except for the regular graph variant where $m = n$. The perfect disassortativity of complete bipartite graphs is in line with the definition of disassortativity, because each node has only links to nodes of a different set with different properties. Nevertheless, the fact that all complete bipartite graphs $K_{m,n}$ with $m \neq n$ have

---

[2] The definition (8.5) is inadequate, due to a zero denominator and numerator, for a regular graph with degree r because N_k regular graph $= N r^k$ (**art. 59**). For regular graphs where $\sum_{i=1}^{N} d_i^3 = N_3$, the perfect disassortativity condition (8.9) becomes $N_2^2 = N_1 N_3$, which is equal to the zero assortativity condition (8.8). One may argue that $\rho_{D;\text{regular graph}} = 1$, since all degrees are equal and thus perfectly correlated. On the other hand, the complete graph K_N minus one link l has $\rho_D (K_N \backslash \{l\}) = \frac{-2}{N-1}$, which suggests that $\rho_D (K_N) = 0$ instead of 1.

$\rho_D = -1$, even those with nearly the same degrees $m = n \pm 1$ and thus close to regular graphs typified by $\rho_D = 1$, shows that assortativity and disassortativity of a graph is not easy to predict. It remains to be shown that the complete bipartite graph $K_{m,n}$ with $m \neq n$ is the only perfect disassortative graph.

There is an interesting relation between the linear degree correlation coefficient ρ_D of the graph G and the variance of the degree of a node in the corresponding line graph $l(G)$ in **art.** 25. The l-th component of the $L \times 1$ degree vector in the line graph $l(G)$ in **art.** 26 is $\left(d_{l(G)}\right)_l = d_i + d_j - 2$, where node i and node j are connected by link $l = i \sim j$. The variance of the degree $D_{l(G)}$ of a random node in the line graph $l(G)$ equals $\mathrm{Var}\left[D_{l(G)}\right] = E\left[(D_i + D_j)^2\right] - (E[D_i + D_j])^2$, which we rewrite as $\mathrm{Var}\left[D_{l(G)}\right] = 2\left(E\left[D_i^2\right] - \mu_{D_i}^2 + E[D_iD_j] - \mu_{D_i}^2\right)$. Using the definition of ρ_D in (8.4) leads to

$$\mathrm{Var}\left[D_{l(G)}\right] = 2\left(1 + \rho_D\right)\left(E\left[D_i^2\right] - \mu_{D_i}^2\right) = 2\left(1 + \rho_D\right)\mathrm{Var}\left[D_i\right]$$

$$= 2\left(1 + \rho_D\right)\left(\frac{1}{N_1}\sum_{i=1}^{N}d_i^3 - \left(\frac{N_2}{N_1}\right)^2\right) \tag{8.10}$$

The expression (8.10) shows for perfect disassortative graphs ($\rho_D = -1$) that $\mathrm{Var}\left[D_{l(G)}\right] = 0$. The latter means that $l(G)$ is then a regular graph, but this does not imply that the original graph G is regular. Indeed, if G is regular, then $l(G)$ is also regular as follows from the l-th component of the degree vector, $\left(d_{l(G)}\right)_l = d_i + d_j - 2$. However, the reverse is not necessarily true: it is possible that $l(G)$ is regular, while G is not, as shown above, for complete bipartite graphs $K_{m,n}$ with $m \neq n$ that are not regular. In summary, in both extreme cases $\rho_D = -1$ and $\rho_D = 1$, the corresponding line graph $l(G)$ is a regular graph.

8.5.1.2 Relation between λ_1 and ρ_D

We present a new lower bound for λ_1 in terms of the linear degree correlation coefficient ρ_D. For $k = 3$ in $\lambda_1^k \geq \frac{N_k}{N}$ in (3.65) and using (8.5), we obtain

$$\lambda_1^3 \geq \frac{N_3}{N} = \frac{1}{N}\left(\rho_D\left(\sum_{i=1}^{N}d_i^3 - \frac{N_2^2}{N_1}\right) + \frac{N_2^2}{N_1}\right) \tag{8.11}$$

This last inequality (8.11) with (8.6) shows that the lower bound for the largest eigenvalue λ_1 of the adjacency matrix A is strictly increasing in the linear degree correlation coefficient ρ_D, except for regular graphs. Given a constant degree vector d, inequality (8.11) shows that the largest eigenvalue λ_1 is obtained, when we succeed in increasing the assortativity of the graph by degree-preserving rewiring discussed in Section 8.5.2.

Fig. 8.6 illustrates how the largest eigenvalue λ_1 of the Barabási-Albert power law graph evolves as a function of the linear degree correlation coefficient ρ_D, which

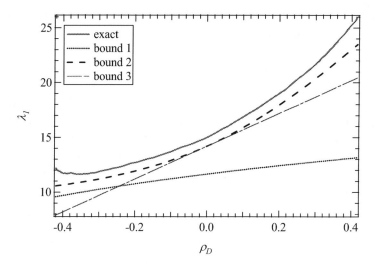

Fig. 8.6. The largest eigenvalue λ_1 of the Barabási-Albert power law graph with $N = 500$ nodes and $L = 1960$ links versus the linear degree correlation coefficient ρ_D. Various lower bounds are plotted: bound 1 is (8.11), bound 2 is (3.73) and bound 3 is $\lambda_1 \geq N_3/N_2$ in (3.64). The corresponding classical lower bound (3.63) is 7.84, while the lower bound (3.66) is 10.548. The latter two lower bounds are independent of ρ_D.

can be changed by degree-preserving rewiring. The optimized lower bound (3.73) outperforms the lower bound[3] (8.11).

8.5.1.3 Relation between μ_{N-1} and ρ_D

The Rayleigh principle in **art.** 133 provides an upper bound for the second smallest eigenvalue μ_{N-1} of the Laplacian Q for the choice $g(n) = d_n$, the degree of a node n,

$$\mu_{N-1} \leq \frac{\sum_{l \in \mathcal{L}} (d_{l+} - d_{l-})^2}{\sum_{j=1}^{N} d_j^2 - \frac{1}{N} \left(\sum_{j=1}^{N} d_j \right)^2}$$

After introducing (8.7), we find for any non-regular graph

$$\mu_{N-1} \leq (1 - \rho_D) \frac{\sum_{i=1}^{N} d_i^3 - \frac{1}{2L} \left(\sum_{i=1}^{N} d_i^2 \right)^2}{\sum_{j=1}^{N} d_j^2 - \frac{1}{N} \left(\sum_{j=1}^{N} d_j \right)^2} = (1 - \rho_D) \frac{E[D] E[D^3] - (E[D^2])^2}{E[D] \operatorname{Var}[D]}$$

$$(8.12)$$

which is an upper bound for the algebraic connectivity μ_{N-1} in terms of the linear correlation coefficient of the degree ρ_D. In degree-preserving rewiring, the fraction

---

[3] Especially for strong negative ρ_D, we found – very rarely though – that (3.73) can be slightly worse than (3.66).

in (8.12), which is always positive, is unchanged and we observe that the upper bound decreases linearly in ρ_D.

8.5.2 Degree-preserving rewiring

Degree-preserving rewiring changes links in a graph, while maintaining the degree distribution unchanged. If the degree vector d is constant and, consequently, that $N_1 = \sum_{i=1}^{N} d_i, N_2 = \sum_{i=1}^{N} d_i^2$ and $\sum_{i=1}^{N} d_i^3$ do not change during degree-preserving rewiring, only N_3 does, and, by (8.5), also the (dis)assortativity ρ_D.

Degree-preserving rewiring changes only the term $\sum_{i \sim j} (d_i - d_j)^2$ in (8.7), which allows us to understand how a degree-preserving rewiring operation changes the linear degree correlation ρ_D. Each step in degree-preserving random rewiring consists of first randomly selecting two links $i \sim j$ and $k \sim l$ associated with the four nodes i, j, k, l. Next, the links can be rewired either into $i \sim k$ and $j \sim l$ or into $i \sim l$ and $j \sim k$.

Theorem 55 *Given a graph in which two links are degree-preservingly rewired and the degree of the four involved nodes is ranked as $d_{(1)} \geq d_{(2)} \geq d_{(3)} \geq d_{(4)}$. The two links are associated with the four nodes $n_{d_{(1)}}, n_{d_{(2)}}, n_{d_{(3)}}$ and $n_{d_{(4)}}$ only in one of the following three ways: (a) $n_{d_{(1)}} \sim n_{d_{(2)}}, n_{d_{(3)}} \sim n_{d_{(4)}}$, (b) $n_{d_{(1)}} \sim n_{d_{(3)}}, n_{d_{(2)}} \sim n_{d_{(4)}}$ and (c) $n_{d_{(1)}} \sim n_{d_{(4)}}, n_{d_{(2)}} \sim n_{d_{(3)}}$. The corresponding linear degree correlation introduced by these three possibilities obeys $\rho_a \geq \rho_b \geq \rho_c$.*

Proof: In these three ways of placing the two links, the degree of each node remains the same. According to the definition (8.7), the linear degree correlation changes only via $\varepsilon = -\sum_{i \sim j} (d_i - d_j)^2$. Thus, the relative degree correlation difference between (a) and (b) is

$$\varepsilon_a - \varepsilon_b = -\left(d_{(1)} - d_{(2)}\right)^2 - \left(d_{(3)} - d_{(4)}\right)^2 + \left(d_{(1)} - d_{(3)}\right)^2 + \left(d_{(2)} - d_{(4)}\right)^2$$
$$= 2(d_{(2)} - d_{(3)})(d_{(1)} - d_{(4)}) \geq 0$$

since the rest of the graph remains the same in all three cases. Similarly,

$$\varepsilon_a - \varepsilon_c = 2(d_{(2)} - d_{(4)})(d_{(1)} - d_{(3)}) \geq 0$$
$$\varepsilon_b - \varepsilon_c = 2(d_{(1)} - d_{(2)})(d_{(3)} - d_{(4)}) \geq 0$$

These three inequalities prove the theorem. $\qquad\square$

A direct consequence of Theorem 55 is that we can now design a rewiring rule that increases or decreases the linear degree correlation ρ_D of a graph. We define *degree-preserving assortative rewiring* as follows: randomly select two links associated with four nodes and then rewire the two links such that, as in (a), the two nodes with the highest degree and the two lowest degree nodes are connected. If any of the new links exists before rewiring, discard this step and a new pair of links is randomly selected. Similarly, the procedure for *degree-preserving disassortative rewiring* is:

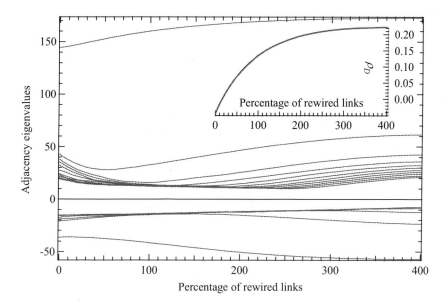

Fig. 8.7. The ten largest and five smallest eigenvalues of the adjacency matrix of the USA airport transport network versus the percentage of rewired links. The insert shows the linear degree correlation coefficient ρ_D as a function of the assortative degree-preserving rewiring.

randomly select two links associated with four nodes and then rewire the two links such that, as in (c), the highest degree node and the minimum degree node are connected, while also the remaining two nodes are linked provided the new links do not exist before rewiring. Theorem 55 shows that the degree-preserving assortative (disassortative) rewiring operations increase (decrease) the degree correlation of a graph.

The assortativity range, defined as difference $\max\rho_D - \min\rho_D$, may be regarded as a metric of a given degree vector d, which reflects its adaptivity in (dis)assortativity under degree-preserving rewiring. As shown earlier, for some graphs such as regular graphs, that difference $\max\rho_D - \min\rho_D = 0$, while $\max\rho_D - \min\rho_D \leq 2$ because $-1 \leq \rho_D \leq 1$.

Degree-preserving rewiring is an interesting tool to modify a graph in which the resources of the nodes are constrained. For instance, the number of outgoing links in a router as well as the number of daily flights at many airports are almost fixed.

We exemplify degree-preserving rewiring with the US air transportation network[4], where each node is an American airport and each link is a flight connection between two airports. We are interested in an infection process, where viruses are

---

[4] The number of nodes is $N = 2179$ and the number of links is $L = 31326$.

spread via airplanes from one city to another. From a topological point of view, the infection threshold $\tau_c = \frac{1}{\lambda_1}$ is the critical design parameter, which should be as high as possible, because an effective infection rate $\tau > \tau_c$ translates into a certain percentage of people that remain infected after sufficiently long time (for details see Pastor-Satorras *et al.* (2015)). Since most airports operate near to full capacity, the number of flights per airport should hardly change during the re-engineering to modify the largest eigenvalue λ_1. Fig. 8.7 shows how the adjacency eigenvalues of the US air transportation network change with degree-preserving assortative rewiring, while the disassortative companion figure is also shown in Van Mieghem *et al.* (2010). In each step of the rewiring process, only four elements 1 (i.e., two links) in the adjacency matrix change position. If we relabel the nodes in such a way that the link between 1 and 2 and between 3 and 4 (case (a) in Theorem 55) is rewired to either case (b) or (c), then only a 4×4 submatrix A_4 of the adjacency matrix A in

$$A = \begin{bmatrix} A_4 & C \\ C^T & A_c \end{bmatrix}$$

is altered. The Interlacing Theorem 71 states that $\lambda_{j+4}(A) \leq \lambda_j(A_c) \leq \lambda_j(A)$ for $1 \leq j \leq N-4$, which holds as well for A_r after just one degree-preserving rewiring step. Thus, most of the eigenvalues of A and A_r are interlaced, as observed from Fig. 8.7. The large bulk of the 2179 eigenvalues (not shown in Fig. 8.7) remains centered around zero and confined to the almost constant white strip between λ_{10} and λ_{N-5}. As shown in Section 8.5.1.2, assortative rewiring increases λ_1. Fig. 8.7 illustrates, in addition, that the spectral width or range $\lambda_1 - \lambda_N$ increases as well, while the spectral gap $\lambda_1 - \lambda_2$ remains high, in spite of the fact that the algebraic connectivity μ_{N-1} is small. In fact, Fig. 8.8 shows that μ_{N-1} decreases, in agreement with (8.12), and vanishes after about 10% of the link rewiring, which indicates (**art. 116**) that the graph is then disconnected. Fig. 8.8 further shows that by rewiring all links on average once (100%), assortative degree-preserved rewiring has dissected the US air transportation network into 20 disconnected clusters. Increasing assortativity implies that high-degree and low-degree nodes are linked increasingly more to each other, which, intuitively, explains why disconnectivity in more and more clusters starts occurring during the rewiring process.

The opposite occurs in disassortative rewiring: the algebraic connectivity μ_{N-1} was found to increase during degree-preserving rewiring from about 0.25 to almost 1, which is the maximum possible due to (4.54) and $d_{\min} = 1$ as follows from the insert in Fig. 8.8. Hence, in order to suppress virus propagation via air transport while guaranteeing connectivity, disassortative degree-preserving rewiring is advocated, which, in return, enhances the topological robustness as explained in **art. 144**.

Finally, we mention that highly disassortative graphs possess a zero eigenvalue of the adjacency matrix with large multiplicity, which can be understood from Section 8.1.1: high degree nodes are preferentially connected to a large set of low degree nodes, that are not interconnected among themselves.

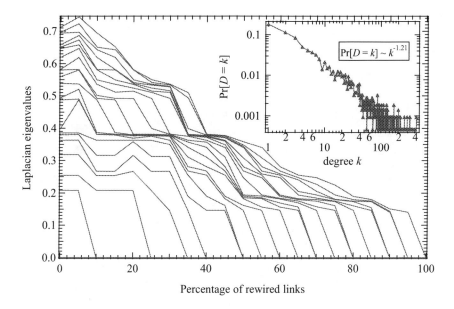

Fig. 8.8. The twenty smallest eigenvalues of the Laplacian matrix of the US air transportation network versus the percentage of rewired links. The insert shows the degree distribution that is maintained in each degree-preserving rewiring step.

8.6 Reconstructability of complex networks

In this section, we investigate, given the set of eigenvectors $x_1, x_2, \ldots, x_N$, how many eigenvalues of the adjacency matrix A are needed to reconstruct A exactly. Specifically, we perturb the spectrum by omitting the j smallest eigenvalues in absolute value of A and we determine the maximal value of j such that the matrix A can be exactly reconstructed. **Art.** 97 shows that if the orthogonal matrix X of the adjacency matrix A is known and if rank$(\Xi) = N - 1$, where the Hadamard product is $\Xi = X \circ X$, then the adjacency matrix A can be reconstructed exactly, without needing the eigenvalue vector λ!

Since $\sum_{j=0}^{N} \lambda_j = 0$ (**art.** 46), on average half of the eigenvalues of the adjacency matrix A are negative. Therefore, we reorder the eigenvalues as $\left|\lambda_{(1)}\right| \leq \left|\lambda_{(2)}\right| \leq \cdots \leq \left|\lambda_{(N)}\right|$ such that $\lambda_{(j)}$ is the j-th smallest (in absolute value) eigenvalue corresponding to the eigenvector $x_{(j)}$. Let us define the $N \times N$ matrices

$$\Lambda_{(j)} = \text{diag}\left(0, \ldots, 0, \lambda_{(j+1)}, \lambda_{(j+2)}, \cdots, \lambda_{(N)}\right)$$

and

$$A_{(j)} = \tilde{X}\Lambda_{(j)}\tilde{X}^T$$

where $\tilde{X} = \begin{bmatrix} x_{(1)} & x_{(2)} & \cdots & x_{(N)} \end{bmatrix}$ is the reordered version of the orthogonal matrix X in (1.2) corresponding to the eigenvalues ranked in absolute value. Thus, $\Lambda_{(j)}$ is the diagonal matrix where the j smallest (in absolute value) eigenvalues are

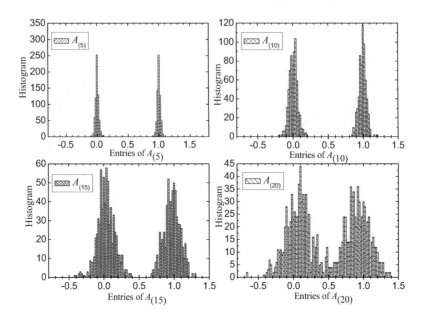

Fig. 8.9. The histograms of the entries of $A_{(5)}$, $A_{(10)}$, $A_{(15)}$ and $A_{(20)}$. The matrix A ($A = A_{(0)}$) is the adjacency matrix of an Erdős-Rényi random graph with $N = 36$ nodes and link density $p = 0.5$.

put equal to zero, or, equivalently, are removed from the spectrum of A. The spectral perturbation here considered consists of consecutively removing more eigenvalues from the spectrum until we can no longer reconstruct the adjacency matrix A. Clearly, when $j = 0$, we have that $A_{(0)} = A$ and that, for any other $j > 0$, $A_{(j)} \neq A$. Moreover, when $j > 0$, $A_{(j)}$ is not a zero-one matrix anymore. Fig. 8.9 plots the histograms of the entries of $A_{(5)}$, $A_{(10)}$, $A_{(15)}$ and $A_{(20)}$ for an Erdős-Rényi random graph with $N = 36$ nodes and link density of $p = 0.5$. The removal of a part of the eigenvalues causes roughly the same impact on the 1 and 0 elements of the adjacency matrix A, as shown in Fig. 8.9. This means that the deviations on 1s and 0s are almost the same, and that the distribution of values around 1 and 0 will reach $1/2$ roughly simultaneously, when the number of removed eigenvalues increases gradually. Using Heaviside's step function $h(x)$,

$$h(x) = \begin{cases} 0 & \text{if } x < 0 \\ \frac{1}{2} & \text{if } x = 0 \\ 1 & \text{if } x > 0 \end{cases}$$

we truncate the elements of $A_{(j)}$ as $h\left(\left(A_{(j)}\right)_{ij} - \frac{1}{2}\right)$. If we now define the operator

$\mathcal{H}$ applied to a matrix $A_{(j)}$ that replaces each element of $A_{(j)}$ by $h\left(\left(A_{(j)}\right)_{ij} - \frac{1}{2}\right)$, then $\widetilde{A_j} = \mathcal{H}\left(A_{(j)}\right)$ is a zero-one matrix, with the possible exception of elements $\frac{1}{2}$. The interesting observation from extensive simulation is that there seems to exist a maximal number θ, for which $\widetilde{A_j} = A$, if $j \leq \theta$ and $\widetilde{A_j} \neq A$, if $j > \theta$. In other words, θ is the maximum number of eigenvalues that can be removed from the spectrum of the graph such that the graph can still be reconstructed precisely, given the matrix X. We therefore call θ the *reconstructability coefficient*.

8.6.1 Theory

Art. 254 shows that any real, symmetric matrix A can be rewritten as (A.138),

$$A = \sum_{k=1}^{N} \lambda_k x_k x_k^T = \sum_{k=1}^{N} \lambda_k E_k$$

where the matrix $E_k = x_k x_k^T$ is the outer product of x_k by itself. Any element of A can be written, with the above relabeling of the eigenvectors according to a ranking in absolute values of the eigenvalues $\left|\lambda_{(1)}\right| \leq \left|\lambda_{(2)}\right| \leq \cdots \leq \left|\lambda_{(N)}\right|$ as

$$a_{ij} = \sum_{k=1}^{m} \lambda_{(k)} \left(E_{(k)}\right)_{ij} + \sum_{k=m+1}^{N} \lambda_{(k)} \left(E_{(k)}\right)_{ij} \tag{8.13}$$

where $m \in [1, N]$ is, for the time being, an integer. As shown in **art.** 255, the 2-norm of E_k is not larger than 1, so that $\left|\left(E_{(k)}\right)_{ij}\right| \leq 1$ for any $1 \leq k \leq N$, which implies that $-1 \leq \left(E_{(k)}\right)_{ij} \leq 1$. Relation (A.138) also explains why an ordering in absolute value is most appropriate for our spectral perturbation: the usual ordering $\lambda_1 \geq \lambda_2 \geq \cdots \geq \lambda_{N-1} \geq \lambda_N$ in algebraic graph theory would first remove $\lambda_N < 0$, then λ_{N-1} and so on. However, $|\lambda_N|$ can be large and its omission from the spectrum is likely to cause too big an impact.

The reconstructability of a graph is now reformulated as follows. Since a_{ij} is either zero or one, it follows from (8.13) that, if

$$\left| a_{ij} - \sum_{k=m+1}^{N} \lambda_{(k)} \left(E_{(k)}\right)_{ij} \right| < \frac{1}{2} \tag{8.14}$$

we can reconstruct the element a_{ij} as $a_{ij} = 1_{\left\{\sum_{k=m+1}^{N} \lambda_{(k)} \left(E_{(k)}\right)_{ij} > \frac{1}{2}\right\}}$. The reconstructability requirement (8.14) determines the values of m that satisfy the inequality. The largest value of m obeying (8.14) is the reconstructability coefficient θ of a graph. Using (8.13), the reconstructability requirement (8.14) is equivalent to

$$\left| \sum_{k=1}^{\theta} \lambda_{(k)} \left(E_{(k)}\right)_{ij} \right| < \frac{1}{2}$$

8.6.2 The average reconstructability coefficient $E[\theta]$

Via extensive simulations, Liu *et al.* (2010) investigated the properties of the reconstructability coefficient θ for several important types of complex networks introduced in Section 1.5, such as Erdős-Rényi random graphs, Barabási-Albert scale-free networks and Watts-Strogatz small-world networks, and also other special deterministic types of graphs. A general linear scaling law was found:

$$E[\theta] = aN \tag{8.15}$$

where the real number $a \in [0,1]$ depends on the graph G. Moreover, the variance $\text{Var}[\theta]$ was sufficiently smaller than the mean $E[\theta]$ such that $E[\theta]$ serves as an excellent estimate for θ. For sufficiently large N, a portion a of the smallest eigenvalues (in absolute value) can be removed from the spectrum and the adjacency matrix is still reconstructible with its original eigenvectors. The magnitude of a for different types of complex networks with different parameters was found to vary from 39% to 76%, which is surprisingly high.

The reconstructability coefficient θ or the scaled coefficient $a = \frac{E[\theta]}{N}$ in (8.15) can be regarded as a spectral metric of the graph that expresses how many dimensions or orthogonal eigenvectors of the N-dimensional space are needed to represent or reconstruct the graph. Roughly, a high reconstructability coefficient θ reflects a "geometrically simple" graph that only needs a few orthogonal dimensions to be described.

8.7 Spectral graph metrics

Most of the graph metrics, such as the hopcount, diameter, clustering coefficient, and many more listed in Section 1.6, are defined in the topology domain. In this section, we briefly mention graph metrics that are defined in the spectral domain, but we study the effective graph resistance in depth in Section 5.2. We have already encountered some spectral graph metrics such as the algebraic connectivity μ_{N-1} in Section 4.3 and the reconstructability coefficient θ in Section 8.6.

8.7.1 Eigenvector centrality

The per component eigenvalue equation (1.4) of the k-th eigenvector,

$$(x_k)_j = \frac{(Ax_k)_j}{\lambda_k} = \frac{1}{\lambda_k} \sum_{l \in \text{ neighbors}(j)} (x_k)_l \tag{8.16}$$

is called the eigenvector centrality of node j according to the eigenvector x_k of the adjacency matrix A of a graph G. This "centrality" measure reflects the importance with respect to the eigenvector x_k of a node in a network and provides a ranking of the nodes in the network according to the eigenvector x_k. Since the eigenvector x_1 has non-zero components (**art.** 41), this largest eigenvector is considered most often. Perhaps the best known example of this spectral graph metric is Google's Page Rank, explained in Van Mieghem (2014, Section 11.5), where the importance of webpages is ranked according to the components of the largest eigenvector of a weighted adjacency matrix, actually the stochastic matrix $P = \Delta^{-1} A$ of the web.

Van Mieghem (2015a) advocates the squared eigenvector components $(x_k)_i^2$ of e.g. the adjacency matrix as nodal centrality metrics.

8.7.2 Graph energy

The graph energy E_G is defined as

$$E_G = \sum_{j=1}^{N} |\lambda_j (A)| \tag{8.17}$$

The graph energy (8.17) is inspired by the energy eigenstates of the Hamiltonian applied to molecules (such as hydrocarbons) and was first proposed by Gutman (Dehmer and Emmert-Streib, 2009, Chapter 7). The chemical origin does not directly help to interpret the notion of graph energy, so that the graph energy is best considered as one of the spectral metrics of a graph.

The absolute sign in the definition (8.17) complicates exact computations, but a large number of bounds exist. A direct application of the inequality (B.75) and **art.** 243 gives

$$(E_G - \lambda_1 (A))^{2m} \leq (N-1)^{2m-1} \left\{ \text{trace} \left(A^{2m} \right) - \lambda_1^{2m} (A) \right\}$$

Rewritten with definition of $W_k = \text{trace}(A^k)$ in **art.** 65, we obtain, for any integer $m > 0$, the upper bound

$$E_G \leq \lambda_1 (A) + (N-1)^{1-1/(2m)} \sqrt[2m]{W_{2m} - \lambda_1^{2m} (A)}$$

Other upper bounds are found in Dehmer and Emmert-Streib (2009, Chapter 7). A lower bound is deduced from

$$E_G^2 = \sum_{j=1}^{N} \lambda_j^2 (A) + \sum_{j=1}^{N} \sum_{k=1; k \neq j}^{N} |\lambda_j (A)| \, |\lambda_k (A)|$$

We apply the product in the harmonic, geometric and arithmetic mean inequality (6.38) to the last sum and find

$$E_G \geq \sqrt{2L + N (N-1) \left(|\det (A)| \right)^{2/N}}$$

which shows that $E_G \geq \sqrt{2L}$.

The determination of graphs that maximize the graph energy E_G is an active domain of research. We content ourselves here to list a few results and refer to Dehmer and Emmert-Streib (2009, Chapter 7) for a detailed review. The graph with minimum energy is the empty graph consisting of isolated nodes, i.e., the complement of the complete graph $(K_N)^c$. Among the trees, the star $K_{1,n}$ has minimal graph energy, whereas the path possesses maximum energy. Simulations, as mentioned by Gutman *et al.* in Dehmer and Emmert-Streib (2009, Chapter 7), show that E_G seems to decrease almost linearly in the multiplicity of the zero eigenvalue for a certain class of graphs.

8.7.3 Delft graph metrics

Delft graph metrics, created in my NAS group in Delft, are defined as the quotient of quadratic forms for positive integers k,

$$_kD = \frac{u^T A^k \Omega A^k u}{u^T A^{2k} u} \tag{8.18}$$

where the denominator $N_{2k} = u^T A^{2k} u$ is the total number of walks (**art. 59**) in the graph G with $2k$ hops or of length $2k$. If $k = 0$, then with $u^T u = N$, the definition (8.18) leads to the effective graph resistance $R_G = \frac{1}{2} u^T \Omega u$ in (5.8)

$$_0D = \frac{u^T \Omega u}{u^T u} = \frac{2R_G}{N}$$

and if $k = 1$, the definition (8.18) relates to the Kemeny constant $K_G = \frac{d^T \Omega d}{4L}$ in Wang *et al.* (2017)

$$_1D = \frac{d^T \Omega d}{d^T d} = \frac{4L}{d^T d} K_G$$

The definition (8.18) itself is an instance of the Rayleigh quotient $\frac{y^T \Omega y}{y^T y}$, where the vector y is chosen as $y = A^k u$. Rayleigh's principle (**art. 251**) states that, for any k,

$$_kD \le \rho_1$$

where ρ_1 in Section 5.5 is the largest eigenvalue of the effective resistance matrix Ω. The normalization $N_{2k} = u^T A^{2k} u$ in (8.18) is algebraically more convenient than the current definitions of the effective graph resistance $R_G = \frac{1}{2} u^T \Omega u$ and the Kemeny constant $K_G = \frac{d^T \Omega d}{4L}$, where the quotient $\frac{u^T A^k \Omega A^k u}{u^T A^k u}$ is implicitly chosen. The power method (**art. 244**) shows for sufficiently large k that the vector $y = A^k u$ tends to cx_1, where x_k is the normalized eigenvector of the adjacency matrix A belonging to the eigenvalue λ_k and where c is a constant. Hence, it can be proved that

$$\lim_{k \to \infty} {}_kD = x_1^T \Omega x_1$$

After writing $y = A^k u$ as a linear combination (**art. 251**) of the eigenvectors $\{v_k\}_{1 \le k \le N}$ of Ω, Theorem 34 shows that

$$_kD = \frac{\sum_{k=1}^N \left(v_k^T A^k u\right)^2 \rho_k}{u^T A^{2k} u} = \frac{\left(v_1^T A^k u\right)^2}{u^T A^{2k} u} \rho_1 - \sum_{k=2}^N \frac{\left(v_k^T A^k u\right)^2}{u^T A^{2k} u} |\rho_k| \le \frac{\left(v_1^T A^k u\right)^2}{\sum_{k=1}^N \left(v_k^T A^k u\right)^2} \rho_1$$

The class of regular graphs with degree r indeed demonstrates, since u is the eigenvector (**art.** 55) belonging to the largest eigenvalue $\lambda_1 = r$, that

$$_k D_{\text{regular graph}} = \frac{r^{2k} u^T \Omega u}{r^{2k} u^T u} = \frac{2R_G}{N} \leq \rho_1$$

which is independent from k. Equality in the last inequality occurs when $\zeta = \frac{R_G}{N^2} u$ (see Theorem 35, which appears in so-called vertex-transitive graphs, that are a subclass of regular graphs.

Since all involved matrices and vectors in (8.18) are non-negative, the graph metrics $_k D$ are positive and reflect a global graph property. The electric power $\mathcal{P} = -\frac{1}{2} x^T \Omega x$ in (5.12), subject to $u^T x = 0$ and at least one vector component of the injected x is negative, indicates by decomposing $y = A^k u = au + bx$ that the Delft graph metrics $_k D$ only contains a small contribution of the electric power $\mathcal{P}$.

8.8 Laplacian spectrum of interdependent networks

An interdependent network, also called interconnected or multi-layer network, is a network consisting of different types of networks, that depend upon each other for their functioning (Buldyrev *et al.*, 2010). For example, a power grid is steered by a computer network, that in turn needs electricity to function. Van Mieghem (2016) shows that *regularity* in the interconnection pattern features attractive properties, that provide engineers with handles to control or uncouple the network's dynamics by changing the strength of the interconnectivity as well as by balancing or distributing that total strength over several inter-links, that connect nodes in different networks or layers.

A two-fold interconnected network G has an adjacency matrix

$$A = \begin{bmatrix} (A_1)_{n \times n} & B_{n \times m} \\ (B^T)_{m \times n} & (A_2)_{m \times m} \end{bmatrix} \tag{8.19}$$

where A_1 is the $n \times n$ adjacency matrix of the graph G_1 with n nodes, A_2 is the $m \times m$ adjacency matrix of the graph G_2 with m nodes and B is the $n \times m$ weighted matrix interconnecting G_1 and G_2, whose elements are real, non-negative numbers. The total number of nodes in G is $N = n+m$. The Laplacian Q of G, corresponding to (8.19) generalizes that of the cone in **art.** 166 and equals

$$Q = \begin{bmatrix} (Q_1)_{n \times n} + \text{diag}\left((Bu_m)\right) & -B_{n \times m} \\ -(B^T)_{m \times n} & (Q_2)_{m \times m} + \text{diag}\left((B^T u_n)\right) \end{bmatrix} \tag{8.20}$$

where $Q_1 = \Delta_1 - A_1$, $Q_2 = \Delta_2 - A_2$ and $\Delta_k = \text{diag}\left((A_k u_k)\right) = \text{diag}(d(G_k))$ for $k = 1, 2$ and where $d_i(G_k)$ denotes the degree of node i in the graph G_k. Only if B is a zero-one matrix, the total number of links in G equals $L = L_{G_2} + L_{G_1} + u_n^T B_{n \times m} u_m$, where $L_{G_k} = \frac{1}{2} u^T A_k u$ is the number of links in G_k. The block matrix (8.20) illustrates that a submatrix such as $Q_S = (Q_1)_{n \times n} + \text{diag}((Bu_m))$ of a Laplacian

Q of a connected graph is *positive definite*, i.e. $z^T Q_S z > 0$. Indeed, if $z \neq u_n$, then $z^T Q_1 z > 0$, else $u^T \mathrm{diag}((Bu_m)) u > 0$; hence the inverse Q_S^{-1} exists.

Any eigenvector $x_k = x = \left(x_1^T, x_2^T\right)^T$ of the Laplacian Q of the interconnected network G with $1 \leq k < N$, thus excluding $x_N = u$, obeys

$$u_n^T x_1 + u_m^T x_2 = 0 \tag{8.21}$$

while the normalization $x^T x = 1$ of the eigenvector x translates to

$$x_1^T x_1 + x_2^T x_2 = 1 \tag{8.22}$$

The Laplacian eigenvalue equation for the eigenvector $x = \left(x_1^T, x_2^T\right)^T$ belonging to the eigenvalue μ,

$$\begin{bmatrix} (Q_1)_{n \times n} + \mathrm{diag}\left((Bu_m)\right) & -B_{n \times m} \\ -\left(B^T\right)_{m \times n} & (Q_2)_{m \times m} + \mathrm{diag}\left(\left(B^T u_n\right)\right) \end{bmatrix} \begin{bmatrix} x_1 \\ x_2 \end{bmatrix} = \mu \begin{bmatrix} x_1 \\ x_2 \end{bmatrix}$$

is equivalent to the set

$$\begin{cases} Q_1 x_1 + \mathrm{diag}\left((Bu_m)_i\right) x_1 - B x_2 = \mu x_1 \\ Q_2 x_2 + \mathrm{diag}\left(\left(B^T u_n\right)_i\right) x_2 - B^T x_1 = \mu x_2 \end{cases} \tag{8.23}$$

The quadratic form of Q has the following property, proven[5] in Van Mieghem (2016):

Theorem 56 *Let $y = \left(y_1^T, y_2^T\right)^T$ be any real vector, then the quadratic form $y^T Q y$, where the Laplacian matrix Q is defined in (8.20), equals*

$$y^T Q y = y_1^T Q_1 y_1 + y_2^T Q_2 y_2 + R_{(y_1, y_2)} \tag{8.24}$$

where

$$R_{(y_1, y_2)} = \sum_{i=1}^{n} \sum_{j=1}^{m} B_{ij} \left((y_1)_i - (y_2)_j\right)^2 \tag{8.25}$$

which is always non-negative because $B_{ij} \geq 0$.

Since any Laplacian is positive semidefinite, each term in the Laplacian quadratic form (8.24) is non-negative. Excluding uncoupled networks, $B \neq O$, then (8.25) shows that $R_{(y_1, y_2)} = 0$ only if $(y_1)_i = (y_2)_j$ for all possible pairs (i, j) of nodal interconnections with positive coupling strength $B_{ij} > 0$. In particular, when $y = u = \left(u_n^T, u_m^T\right)^T$, then $R_{(u_n, u_m)} = 0$ independently of the structure of B (as also follows from (8.24) because $x_N = u$ is the eigenvector belonging to the zero Laplacian eigenvalue $\mu_N = 0$). As a consequence of $R_{(y_1, y_2)} \geq 0$, we find with $y = x$ in (8.24) that any eigenvalue μ of Q belonging to eigenvector $x = \left(x_1^T, x_2^T\right)^T$ is lower bounded (**art.** 163) by

$$\mu \geq x_1^T Q_1 x_1 + x_2^T Q_2 x_2$$

---

[5] Theorem 56 is generalized to an m-fold interconnected network with $m \geq 2$ in Van Mieghem (2016).

We may interpret $R_{(y_1, y_2)}$ in (8.25) as the total interconnection energy between G_1 and G_2 due to the vector $y = \left(y_1^T, y_2^T\right)^T$. In such an interpretation, $y^T Q y$ represents the total network energy for a state vector y.

If the interconnected network G with $N = m + n$ nodes has a special regular structure, we can determine at least two eigenmodes of Q:

Theorem 57 *Only if the $n \times m$ interconnection matrix B has a constant row sum equal to $\frac{\mu^*}{N} m$ and a constant column sum equal to $\frac{\mu^*}{N} n$, which we call the regularity condition for $B_{n \times m}$,*

$$\begin{cases} B u_m = \mu^* \frac{m}{m+n} u_n \\ B^T u_n = \mu^* \frac{n}{m+n} u_m \end{cases} \tag{8.26}$$

then is

$$x = \frac{1}{\sqrt{N}} \left[\sqrt{\tfrac{m}{n}} u_n^T \quad -\sqrt{\tfrac{n}{m}} u_m^T \right]^T \tag{8.27}$$

an eigenvector of Q, defined in (8.20), belonging to the eigenvalue

$$\mu^* = \left(\frac{1}{n} + \frac{1}{m}\right) u_n^T B_{n \times m} u_m \tag{8.28}$$

and $u_n^T B_{n \times m} u_m = \sum_{i=1}^n \sum_{j=1}^m B_{ij}$ equals the sum of the elements in B, specifying to the total strength of the interconnection between G_1 and G_2.

The proof is a consequence of equitable partitions in Section 2.5. Since each element $B_{ij} \geq 0$, the eigenvalue $\frac{\mu^*}{N} = \frac{u_n^T B_{n \times m} u_m}{nm}$ in (8.28) represents the average "coupling strength" per element in B and can only be zero if $B = O$, in which case the two networks G_1 and G_2 are disconnected. The eigenvector and eigenvalue in Theorem 57 are only determined by the interconnection matrix B and are independent of the structure of G_1 and of G_2, because each eigenvector component of x in (8.27) satisfies $Q_1 x_1 = Q_1 u_n = 0$ and, similarly, $Q_2 x_2 = Q_2 u_m = 0$.

By choosing y_1 equal to the k-th normalized eigenvector (i.e. $y_1^T y_1 = 1$, while $y^T y = y_1^T y_1 + y_2^T y_2$) of Q_1 and y_2 equal to l-th normalized eigenvector of Q_2, the quadratic form (8.24) reads

$$y^T Q y = \mu_k (Q_1) + \mu_l (Q_2) + R_{(y_1, y_2)}$$

In particular, confining to the algebraic connectivity where $y_1 = x_{n-1}$ and $y_2 = x_{m-1}$ are the eigenvectors belonging to the respective algebraic connectivity μ_{n-1} in G_1 and μ_{m-1} in G_2, leads to

$$y^T Q y = \mu_{n-1} (Q_1) + \mu_{m-1} (Q_2) + R_{(x_{n-1}, x_{m-1})}$$

where $u_n^T x_{n-1} = 0$ and $u_m^T x_{m-1} = 0$, so that $y^T u_N = 0$. In that case, the Rayleigh inequality $y^T Q y \geq \mu_{m+n-1} (Q) y^T y = 2\mu_{m+n-1} (Q)$ leads to the upper bound for the algebraic connectivity of Q,

$$\mu_{N-1} (Q) \leq \frac{1}{2} \left(\mu_{n-1} (Q_1) + \mu_{m-1} (Q_2) + R_{(x_{n-1}, x_{m-1})} \right) \tag{8.29}$$

Let us investigate the smallest, non-zero eigenvalue μ, corresponding to any eigenvector $x = \left(x_1^T, x_2^T\right)^T$ of Q obeying $u_n^T x_1 = u_m^T x_2 = 0$, a necessary condition for regularity of B. The Rayleigh inequality demonstrates for $u_n^T x_1 = 0$ that $x_1^T Q_1 x_1 \geq \mu_{n-1}(Q_1) x_1^T x_1$ and, similarly for $u_m^T x_2 = 0$, $x_2^T Q_2 x_2 \geq \mu_{m-1}(Q_2) x_2^T x_2$ with equality only if x_1 and x_2 are the eigenvectors of Q_1 and Q_2 belonging to the algebraic connectivity, eigenvalues $\mu_{n-1}(Q_1)$ and $\mu_{m-1}(Q_2)$, respectively. Complementary to the *upper* bound (8.29), the quadratic form (8.24) leads to the *lower* bound (8.30) for the algebraic connectivity of G with regular interconnection matrix B,

$$\mu_{N-1}(Q) \geq \mu_{n-1}(Q_1) x_1^T x_1 + \mu_{m-1}(Q_2) x_2^T x_2 + R_{(x_1,x_2)} \geq 0 \qquad (8.30)$$

with $0 < x_2^T x_2 = 1 - x_1^T x_1 < 1$. Combining the *upper* bound (8.29) and the *lower* bound (8.30) yields, for a regular interconnection matrix B,

$$\frac{\mu_{n-1}(Q_1)}{\frac{1}{x_1^T x_1}} + \frac{\mu_{m-1}(Q_2)}{\frac{1}{1-x_1^T x_1}} + R_{(x_1,x_2)} \leq \mu_{N-1}(Q) \leq \frac{\mu_{n-1}(Q_1)}{2} + \frac{\mu_{m-1}(Q_2)}{2} + \frac{R_{(x_{n-1},x_{m-1})}}{2}$$

$$(8.31)$$

Even when both G_1 and G_2 are disconnected ($\mu_{n-1}(Q_1) = \mu_{m-1}(Q_2) = 0$), a positive interconnection energy $R_{(x_1,x_2)} > 0$ results in a connected interdependent network G (i.e. $\mu_{N-1}(Q) > 0$). We observe from (8.31) that, if $\mu_{n-1}(Q_1) = \mu_{m-1}(Q_2)$, then we obtain the curious inequality

$$\mu_{n-1}(Q_1) + R_{(x_1,x_2)} \leq \mu_{N-1}(Q) \leq \mu_{n-1}(Q_1) + \frac{R_{(x_{n-1},x_{m-1})}}{2}$$

illustrating that $0 \leq R_{(x_1,x_2)} \leq \mu_{N-1}(Q) - \mu_{n-1}(Q) \leq \frac{R_{(x_{n-1},x_{m-1})}}{2}$. The interconnection energy $R_{(x_1,x_2)}$ for the Fiedler eigenvector x of Q is smaller than half the interconnection energy $R_{(x_{n-1},x_{m-1})}$ of the individual Fiedler vectors x_{n-1} of Q_1 and x_{m-1} of Q_2, both belonging to a same algebraic connectivity $\mu_{n-1}(Q_1) = \mu_{m-1}(Q_2)$.

The scaling of elements in B also causes that the eigenvalue μ^* in (8.28) is not necessarily equal to the second smallest eigenvalue μ_{N-1}. Indeed, by lowering the total coupling strength $u_n^T B_{n \times m} u_m$, we can always force μ^* to be lower than $\mu_{N-1}(Q)$. The possibility of modifying the total coupling strength $u_n^T B_{n \times m} u_m$ leads to consequences elaborated in Sahneh *et al.* (2015), where the coupling strength w in $B = wI$ was computed so that $\mu^* = \mu_{N-1}(Q)$.

8.9 Graph sparsification

The goal of sparsification is to approximate a given weighted graph G by a sparse weighted graph H with less links and potentially different link weights, but on the same set of nodes, by trying to preserve in H some property of G. For example, cut-sparsifiers approximately preserve the sizes of all cuts, while a spectral sparsifier approximately preserves eigenvalues (see e.g. Chu *et al.* (2020)). Spielman and

Srivastava (2011) have proposed an elegant spectral sparsification algorithm for any graph G that samples a link $l = (i, j)$ in G with probability p_l proportional to the relative resistance $w_l \omega_l$, where w_l is the weight of link l and $\omega_l = \omega_{ij}$ is its effective resistance.

Before specifying the Spielman-Srivastava sparsification algorithm in Batson *et al.* (2013), we define *spectral similarity*. Two symmetric $N \times N$ matrices A and B are ε-spectrally similar if

$$(1 - \varepsilon) \, x^T B x \leq x^T A x \leq (1 + \varepsilon) \, x^T B x \qquad \text{for all } N \times 1 \text{ vectors } x$$

If the components of the vector x are restricted to $x_i \in \{-1, 1\}$ for all $i \in \mathcal{N}$, one obtains cut-sparsifiers (**art. 143**). The Courant-Fischer Theorem, $\lambda_i(A) = \max_{\dim \mathcal{V} = i} \min_{x \in \mathcal{V}} \frac{x^T A x}{x^T x}$ in (A.132), implies that

$$(1 - \varepsilon) \, \lambda_i(B) \leq \lambda_i(A) \leq (1 + \varepsilon) \, \lambda_i(B)$$

The notion of ε-spectral similarity is written with the inequality sign "$\preccurlyeq$" as

$$(1 - \varepsilon) \, B \preccurlyeq A \preccurlyeq (1 + \varepsilon) \, B$$

because it allows matrix operations as with the usual inequality sign $\leq$, in particular for positive semidefinite matrices. Indeed, let $\widetilde{Q}_G$ and $\widetilde{Q}_H$ denote the weighted Laplacian of the graph G and its sparsifier H, respectively. Since all Laplacian eigenvalues are non-negative, we have that $\widetilde{Q} = \left(\widetilde{Q}\right)^{\frac{1}{2}} \left(\widetilde{Q}\right)^{\frac{1}{2}}$ and left- and right-multiplying $(1 - \varepsilon) \widetilde{Q}_G \preccurlyeq \widetilde{Q}_H \preccurlyeq (1 + \varepsilon) \widetilde{Q}_G$ with $\left(Q_G^\dagger\right)^{\frac{1}{2}} = \left(Q_G^{\frac{1}{2}}\right)^\dagger$ (see **art. 128**), using $\left(Q_G^\dagger\right)^{\frac{1}{2}} \widetilde{Q}_G \left(Q_G^\dagger\right)^{\frac{1}{2}} = I - \frac{1}{N} J$, yields

$$(1 - \varepsilon) \left(I - \frac{1}{N} J\right) \preccurlyeq \left(Q_G^\dagger\right)^{\frac{1}{2}} \widetilde{Q}_H \left(Q_G^\dagger\right)^{\frac{1}{2}} \preccurlyeq (1 + \varepsilon) \left(I - \frac{1}{N} J\right) \qquad (8.32)$$

Invoking $\widetilde{Q} = \sum_{l \in \mathcal{L}} w_l \, (e_{l+} - e_{l-}) \, (e_{l+} - e_{l-})^T$ in (4.5) indicates that

$$\left(Q_G^\dagger\right)^{\frac{1}{2}} \widetilde{Q}_H \left(Q_G^\dagger\right)^{\frac{1}{2}} = \sum_{l \in \mathcal{L}_G} w_{l;H} \left(Q_G^\dagger\right)^{\frac{1}{2}} (e_{l+} - e_{l-}) (e_{l+} - e_{l-})^T \left(Q_G^\dagger\right)^{\frac{1}{2}} \qquad (8.33)$$

where the link weight $w_{l;H}$ is crucial and determined below.

Batson *et al.* (2013, Theorem 5) state the algorithm: "Set $q = 8N \log \frac{N}{\varepsilon^2}$. Choose a random link l of G with probability p_l proportional to $w_l \omega_l$. Add l to H with link weight $\frac{w_l}{q p_l}$. Take q samples independently with replacement, summing weights if a link is chosen more than once." With probability at least $1/2$, H is a $(1 + \varepsilon)$-spectral approximation of G. We first formulate a variant without link replacements, which is more natural for sparsification and akin to random matrix theory in Section 7.5. Limitations of sampling without replacement leads to the Spielman-Srivastava sparsification algorithm with link replacements.

8.9.1 Sampling links without replacement

The link weight in the sparsifier H is written as $w_{l;H} = s_l u_l w_l$, where the positive real number u_l is a link weight scaler and the indicator $s_l = 1_{\{l \in \mathcal{L}_H\}}$ is $s_l = 1$ with probability $p_l = \Pr[l \in \mathcal{L}_H]$ if link $l \in \mathcal{L}_G$ is maintained as a link in the sparsifier H, else $s_l = 0$. The mean of an indicator or Bernoulli random variable is $E[s_l] = p_l$, but we require that the number of links $L_H = \sum_{l \in \mathcal{L}_G} s_l$ is fixed[6] and not a random variable. Thus, the expected number of links in H is $E[\mathcal{L}_H] = \sum_{l \in \mathcal{L}_G} p_l = L_H$. The expectation

$$E\left[\widetilde{Q}_H\right] = E\left[\sum_{l \in \mathcal{L}_G} s_l u_l w_l \left(e_{l+} - e_{l-}\right)\left(e_{l+} - e_{l-}\right)^T\right]$$

$$= \sum_{l \in \mathcal{L}_G} p_l u_l w_l \left(e_{l+} - e_{l-}\right)\left(e_{l+} - e_{l-}\right)^T$$

suggests to choose $u_l = \frac{1}{p_l}$ for any link $l \in \mathcal{L}_G$ so that $E\left[\widetilde{Q}_H\right] = \widetilde{Q}_G$, meaning that the average link weight in the sparsifier H and the original graph G is the same. Since $p_l < 1$ as $L_H < L_G$, the link weight scaler $u_l > 1$ and any sampled link of G possesses a higher weight in the sparsifier H.

The expectation of the random matrix in (8.33) becomes

$$E\left[\left(Q_G^\dagger\right)^{\frac{1}{2}} \widetilde{Q}_H \left(Q_G^\dagger\right)^{\frac{1}{2}}\right] = E\left[\sum_{l \in \mathcal{L}_G} w_{l;H} \left(Q_G^\dagger\right)^{\frac{1}{2}} \left(e_{l+} - e_{l-}\right)\left(e_{l+} - e_{l-}\right)^T \left(Q_G^\dagger\right)^{\frac{1}{2}}\right]$$

$$= \sum_{l \in \mathcal{L}_G} E\left[s_l u_l w_l\right] \left(Q_G^\dagger\right)^{\frac{1}{2}} \left(e_{l+} - e_{l-}\right)\left(e_{l+} - e_{l-}\right)^T \left(Q_G^\dagger\right)^{\frac{1}{2}}$$

Since the choice $u_l = \frac{1}{p_l}$ implies that $E[s_l u_l w_l] = u_l w_l E[s_l] = u_l w_l p_l = w_l$, we arrive at

$$E\left[\left(Q_G^\dagger\right)^{\frac{1}{2}} \widetilde{Q}_H \left(Q_G^\dagger\right)^{\frac{1}{2}}\right] = \sum_{l \in \mathcal{L}_G} \left(Q_G^\dagger\right)^{\frac{1}{2}} w_l \left(e_{l+} - e_{l-}\right)\left(e_{l+} - e_{l-}\right)^T \left(Q_G^\dagger\right)^{\frac{1}{2}}$$

$$= \left(Q_G^\dagger\right)^{\frac{1}{2}} \widetilde{Q}_G \left(Q_G^\dagger\right)^{\frac{1}{2}} = I - \frac{1}{N}J$$

and at the random matrix

$$\left(Q_G^\dagger\right)^{\frac{1}{2}} \widetilde{Q}_H \left(Q_G^\dagger\right)^{\frac{1}{2}} = \sum_{l \in \mathcal{L}_G} \left\{s_l \sqrt{\frac{w_l}{p_l}} \left(Q_G^\dagger\right)^{\frac{1}{2}}(e_{l+} - e_{l-})\right\}\left\{s_l \sqrt{\frac{w_l}{p_l}} \left(Q_G^\dagger\right)^{\frac{1}{2}}(e_{l+} - e_{l-})\right\}^T$$

Denoting the random vector $g_l = s_l \sqrt{\frac{w_l}{p_l}} \left(Q_G^\dagger\right)^{\frac{1}{2}} (e_{l+} - e_{l-})$, the norm of the

---

[6] Similar to the variant of Erdős-Rényi graphs (see e.g. Van Mieghem (2014, p. 376) and Bollobás (2001, Section 2.1)), where a link l occurs with probability p_l and the total number of links is fixed. Links in H are weakly dependent and $E[s_l s_m] \neq E[s_l] E[s_m]$.

random matrix using (A.18) is

$$\left\|\left(Q_G^\dagger\right)^{\frac{1}{2}}\widetilde{Q}_H\left(Q_G^\dagger\right)^{\frac{1}{2}}\right\|_2 = \left\|\sum_{l\in\mathcal{L}_G} g_l g_l^T\right\|_2 \leq \sum_{l\in\mathcal{L}_G} \left\|g_l g_l^T\right\|_2 = \sum_{l\in\mathcal{L}_G} \|g_l\|_2^2$$

and

$$\|g_l\|_2^2 = \left\|\sqrt{s_l\frac{w_l}{p_l}}\left(Q_G^\dagger\right)^{\frac{1}{2}}(e_{l+}-e_{l-})\right\|_2^2 = s_l\frac{w_l}{p_l}\,(e_{l+}-e_{l-})^T Q_G^\dagger(e_{l+}-e_{l-}) = s_l\frac{w_l}{p_l}\omega_l \tag{8.34}$$

where the last equality follows from the definition (5.2). Thus,

$$\left\|\left(Q_G^\dagger\right)^{\frac{1}{2}}\widetilde{Q}_H\left(Q_G^\dagger\right)^{\frac{1}{2}}\right\|_2 \leq \sum_{l\in\mathcal{L}_G} s_l\frac{w_l}{p_l}\omega_l = \sum_{l\in\mathcal{L}_H} \frac{w_l}{p_l}\omega_l$$

and Foster's rule $\sum_{l\in\mathcal{L}} w_l\omega_l = N-1$ in (5.20) with $w_l = \frac{1}{r_l}$ illustrates that

$$E\left[\left\|\left(Q_G^\dagger\right)^{\frac{1}{2}}\widetilde{Q}_H\left(Q_G^\dagger\right)^{\frac{1}{2}}\right\|_2\right] \leq \sum_{l\in\mathcal{L}_G} w_l\omega_l = N-1$$

If $\|g_l\|_2^2 = \alpha$ in (8.34), then $\frac{w_l}{p_l}\omega_l = \alpha$ for a link l in the sparsifier H, where the relative resistance $w_l\omega_l \leq 1$ for any link l as follows from the parallel resistor formula (5.6). Sampling without replacements leads to the number of links $L_H = \sum_{l\in\mathcal{L}_G} p_l = \frac{1}{\alpha}\sum_{l\in\mathcal{L}_G} w_l\omega_l$ in the sparsifier H, from which follows that $\alpha = \frac{N-1}{L_H}$ and $p_l = \frac{L_H}{N-1}w_l\omega_l$. Then, the norm of the random matrix is bounded by

$$\left\|\left(Q_G^\dagger\right)^{\frac{1}{2}}\widetilde{Q}_H\left(Q_G^\dagger\right)^{\frac{1}{2}}\right\|_2 \leq \sum_{l\in\mathcal{L}_G} \|g_l\|_2^2 = \frac{L_G}{L_H}(N-1)$$

while the norm of its average $\left\|E\left[\left(Q_G^\dagger\right)^{\frac{1}{2}}\widetilde{Q}_H\left(Q_G^\dagger\right)^{\frac{1}{2}}\right]\right\|_2 = \left\|I - \frac{1}{N}J\right\|_2 = 1$ is substantially lower! Since $p_l = \frac{L_H}{N-1}w_l\omega_l \leq 1$ for any link $l \in \mathcal{L}_G$, it holds that

$$L_H \leq \min_{l\in\mathcal{L}_G} \frac{N-1}{w_l\omega_l}$$

which confines the number of links L_H in the sparsifier H. This constraint is absent in Spielman and Srivastava (2011). More importantly, the stringency ε in (8.32) cannot be tuned. In order to create a sparsifier H whose Laplacian eigenvalues are within stringency ε from those of the original graph G, Spielman's sampling *with* replacement is needed.

8.9.2 Sampling links with replacement

Instead of selecting a link in the original graph G only once, sampling with replacement allows to choose a link several times. The number of link samplings is q and the random variable $\widetilde{s}_l = qs_l = q.1_{\{l\in\mathcal{L}_H\}}$ denotes the number of times link l is sampled, leading to the probability that link $l \in \mathcal{L}_G$ appears in the sparsifier H is now qp_l. The link weight is $w_{l;H} = \widetilde{s}_l \widetilde{u}_l w_l$, where the link weight scaler is $\widetilde{u}_l = \frac{1}{qp_l}$ to obey $E\left[\widetilde{Q}_H\right] = \widetilde{Q}_G$. Thus, each time the same link is chosen, its weight is increased with $\frac{w_l}{qp_l}$, but $w_{l;H} = \widetilde{s}_l \widetilde{u}_l w_l = s_l u_l w_l$ remains the same as in sampling without replacement. Instead of imposing that $E[\mathcal{L}_H] = L_H$, which creates dependence, only the average number of links $E[\mathcal{L}_H] = q\sum_{l \in \mathcal{L}_G} p_l$ can be determined and each sampled link is independent from all q others.

In order to bound the deviations in the random matrix $\left(Q_G^\dagger\right)^{\frac{1}{2}} \widetilde{Q}_H \left(Q_G^\dagger\right)^{\frac{1}{2}}$ from its mean $I - \frac{1}{N}J$, Spielman and Srivastava (2011) rewrite (8.33) as

$$\left(Q_G^\dagger\right)^{\frac{1}{2}} \widetilde{Q}_H \left(Q_G^\dagger\right)^{\frac{1}{2}} = \sum_{l \in \mathcal{L}_G} \widetilde{s}_l \widetilde{u}_l w_l \left(Q_G^\dagger\right)^{\frac{1}{2}} (e_{l+} - e_{l-})(e_{l+} - e_{l-})^T \left(Q_G^\dagger\right)^{\frac{1}{2}} = \frac{1}{q}\sum_{i=1}^q y_i y_i^T$$

where y_i are i.i.d. random vectors drawn with probability p_l from the distribution $y_l = \sqrt{\frac{w_l}{p_l}} \left(Q_G^\dagger\right)^{\frac{1}{2}} (e_{l+} - e_{l-})$ with link $l \in \mathcal{L}_G$ and they apply[7]

Theorem 58 (Rudelson and Vershynin (2007)) *Let* $y_1, y_2, \ldots, y_q$ *be independent random vectors with* $\max \|y_i\|_2 \leq m$ *and* $\|E[yy^T]\|_2 \leq 1$, *then*

$$E\left[\left\|\frac{1}{q}\sum_{i=1}^q y_i y_i^T - E[yy^T]\right\|_2\right] \leq \min\left(8m\sqrt{\frac{\log q}{q}}, 1\right)$$

and, for $0 < t < 1$ *and for a constant* c,

$$\Pr\left[\left\|\frac{1}{q}\sum_{i=1}^q y_i y_i^T - E[yy^T]\right\|_2 > t\right] \leq \exp\left(-c\frac{q}{m\log q}t^2\right) \qquad (8.35)$$

Since the definition of the matrix norm in (A.23) shows, with **art. 193**, that $\|E[yy^T]\|_2 = \left\|E\left[\left(Q_G^\dagger\right)^{\frac{1}{2}}\widetilde{Q}_H\left(Q_G^\dagger\right)^{\frac{1}{2}}\right]\right\|_2 = \|I - \frac{1}{N}J\|_2 = 1$, only the condition $\max \|y_i\|_2 \leq m$ in Theorem 58 needs to be obeyed. Spielman and Srivastava (2011) choose the norms of all random vectors y_i equal and deduce $m^2 = \max_{1 \leq i \leq q} \|y_i\|_2^2 = N - 1$, which is optimal, because the lowest right-hand side bound in Rudelson's theorem 58 is attained. Finally, they find that $p_l = \frac{1}{N-1}w_l\omega_l$, which demonstrates the key idea in Spielman and Srivastava (2011) of sampling links in G with a probability p_l proportional to the relative resistance $w_l\omega_l = \frac{\omega_l}{r_l}$.

It remains to estimate the number q of link samplings. Together with ε-spectral

---

[7] Rudelson and Vershynin (2007, remark 3.4) state that the estimates in Theorem 58 are generally best possible.

similarity, $\left(Q_G^\dagger\right)^{\frac{1}{2}} \tilde{Q}_H \left(Q_G^\dagger\right)^{\frac{1}{2}} \preccurlyeq (1+\varepsilon)\left(I - \frac{1}{N}J\right)$, Rudelson's theorem

$$E\left[\left\|\frac{1}{q}\sum_{i=1}^{q} y_i y_i^T - E\left[yy^T\right]\right\|_2\right] = E\left[\left\|\left(Q_G^\dagger\right)^{\frac{1}{2}} \tilde{Q}_H \left(Q_G^\dagger\right)^{\frac{1}{2}} - \left(I - \frac{1}{N}J\right)\right\|_2\right]$$

$$\leq \varepsilon \left\|\left(I - \frac{1}{N}J\right)\right\|_2 = \varepsilon$$

then states that $8\sqrt{\frac{(N-1)\log q}{q}} \leq \varepsilon$, which leads to $q = O\left(\frac{N}{\varepsilon^2}\log\frac{N}{\varepsilon^2}\right)$ for large N. Indeed[8], a few iterations in the recusion inequality $q_n \geq \frac{64(N-1)}{\varepsilon^2}\log q_{n-1}$, with $q_0 = 64\frac{(N-1)}{\varepsilon^2}$, shows that $q \approx \frac{64}{\varepsilon^2}N\left(\log\frac{N}{\varepsilon^2} + \log\log\frac{N}{\varepsilon^2} + \cdots\right)$. The probability (8.35) then demonstrates that the deviations in the random matrix $\left(Q_G^\dagger\right)^{\frac{1}{2}} \tilde{Q}_H \left(Q_G^\dagger\right)^{\frac{1}{2}}$ from its mean $I - \frac{1}{N}J$ are small.

Sampling with replacements allows to incorporate a stringent $\varepsilon > \varepsilon_{\min}$ accuracy, at the expense of a large number q of samplings. The accuracy ε that specifies the deviations of the Laplacian eigenvalues of the sparsifier H from those in the orginal graph G and the sparseness of H remains a trade-off: if $\varepsilon < \varepsilon_{\min}$, then $H = G$. Since the Spielman-Srivastava sparsification algorithm is stochastic, more than one sparsifier H can be found that meets the accuracy range in (8.32).

8.10 Machine learning: Assigning labels to nodes

A classical task in machine learning consists of classifying objects such as images. Suppose that there are N objects represented by points in an m-dimensional space. For example, an image with 256 pixels is represented by an $m = 256$ dimensional vector, also called a feature vector. Precisely l objects are labeled or classified by labels $y_1, y_2, \ldots, y_l$. For example, an image i of a written letter $\mathcal{A}$ is recognized precisely and labeled as $y_i = 1$. The remaining $u = N - l$ objects, typically $l \ll u$, are unlabeled and the aim is to label them. This problem is called semi-supervised learning.

All N points or objects are considered as nodes in a graph G. The labeled nodes belong to the set $\mathcal{S} \subset \mathcal{N}$ and the unlabeled nodes to its complement $\mathcal{S}^c = \mathcal{N}\backslash\mathcal{S}$. The links are specified by an $N \times N$ symmetric, weighted matrix W, which we assume is given[9]. For example, the link weight between node i at position r_i and node j at position r_j is $w_{ij} = g\left(\|r_i - r_j\|\right)$ and $g(x)$ is a decreasing function of the distance $x = \|r_i - r_j\|$ in the m-dimensional Euclidean space. The corresponding weighted adjacency matrix (**art.** 5) is $\tilde{A} = W$.

The labeling requires to construct a real-valued function that maps a node $i \in \mathcal{N}$

---

[8] The exact solution of $\frac{\log q}{q} = \frac{\varepsilon^2}{64(N-1)}$ is $q = \exp\left(-W\left(-\frac{\varepsilon^2}{64(N-1)}\right)\right)$, where $W(x)$ is the Lambert function with inverse $W^{-1}(x) = xe^x$.

[9] We refer to Zhu *et al.* (2003, Section 6) for a method to estimate W if the weighted matrix W is neither given nor fixed.

to a label, which is a real number. Inspired by the elegant properties of an electric resistor network (**art. 14**), Zhu *et al.* (2003) propose as a real-valued function to consider the potential v, which is a harmonic function (**art. 15**). Consequently, the labels reflect injected currents, provided that the current conservation law $u^T x = 0$ is obeyed, which can be met by a current vector $x = y - \frac{u^T y}{N} u$. The fundamental current-voltage relation $x = \widetilde{Q} v$ in (2.15) shows that $\left(\widetilde{Q} v \right)_{j \in \mathcal{S}^c} = 0$ for an unlabeled node j, whereas $x_k = \left(\widetilde{Q} v \right)_k$ for a labeled node $k \in \mathcal{S}$. The inverse relation $v = Q^\dagger x$ in (4.32) specifies the labeling function v at each node completely, with average label $u^T v = 0$. Since v is harmonic, its maximum and minimum value is attained at the boundary $\mathcal{S}$ and all unlabeled nodes in $\mathcal{S}^c$ will receive a label between maximum and minimum. Examples of this "harmonic" label assignment and a comparison with other methods are given in Zhu *et al.* (2003). Harmonic label assignment assumes that direct neighbors in G likely share the same label.

8.11 Graph neural networks

We only review the theory and concepts and refer for examples and applications to Ortega (2022) and Gama *et al.* (2020).

8.11.1 Graph signal processing

A process on a graph G with N nodes, such as an epidemic in a contact network, power transport in an electric network, etc., can generate an outcome of that process, called "data", at each node of the graph. In graph signal processing (see e.g. Gama *et al.* (2020)), the datum x_i at node i is the i-th component of a graph signal x, which is an $N \times 1$ vector. The graph signal x can be shifted over the nodes by the graph shift operator S, which is an $N \times N$ real symmetric graph-related matrix, such as the adjacency or Laplacian matrix. After a shift Sx, the datum at node i is $(Sx)_i = \sum_{j=1}^N s_{ij} x_j$, which is a linear combination of the data x_j at neighbors $j \in \mathcal{N}_i$ of node i. Given the $K \times 1$ parameter vector $h = \begin{bmatrix} h_1 & h_2 & \cdots & h_K \end{bmatrix}^T$, a graph convolution is defined as

$$H(S) x = \sum_{k=0}^K h_k S^k x \qquad (8.36)$$

The k-shifted signal $S^k x$ contains a summary of the information located in the k-hop neighborhood, akin to $A^k x$ in Section 1.3, and h_k weighs this summary. The graph convolution (8.36) is said to filter a graph signal x with a finite impulse response (FIR) graph filter $H(S)$ and the weights h_k are called filter taps or filter weights (Gama *et al.* (2020)). The spectral domain, called the graph frequency domain in signal processing, is specified by the eigendecomposition of the shift operator $S = V \Lambda V^T$, similar to $A = X \Lambda X^T$. The set of eigenvectors $\{v_j\}_{1 \leq j \leq N}$ form the graph frequency basis of graph G and can be interpreted as signals representing the graph eigenmodes or graph oscillating modes, while the set of eigenvalues $\{\lambda_j\}_{1 \leq j \leq N}$ are called graph frequencies. The graph Fourier transform (GFT) of the graph signal x is $\widetilde{x} = V^T x$ and the component $\widetilde{x}_i$ is the Fourier coefficient associated to graph frequency λ_i, which quantifies the contribution of mode v_i to the signal x. Since $V^T V = V V^T = I$, the inverse GFT is $x = V\widetilde{x}$. The GFT of the graph convolution $y = H(S) x$ in (8.36) is

$$\widetilde{y} = V^T y = \sum_{k=0}^{K} h_k V^T V \Lambda^k V^T x = \sum_{k=0}^{K} h_k \Lambda^k \widetilde{x} = H(\Lambda)\widetilde{x}$$

where $H(\Lambda)$ is an $N \times N$ diagonal matrix with i-th diagonal element $H_{ii}(\Lambda) = \sum_{k=0}^{K} h_k \lambda_i^k = h(\lambda_i)$, where $h(z) = \sum_{k=0}^{K} h_k z^k$ is the frequency response of the graph filter $H(\Lambda)$ and solely determined by the filter tap h. The i-th frequency content of the GFT $\widetilde{y}$ is $\widetilde{y}_i = h(\lambda_i)\widetilde{x}_i$ and the graph convolution (8.36) modifies the i-th frequency content $\widetilde{x}_i$ of the input signal x according to the filter value $h(\lambda_i)$ at frequency λ_i.

8.11.2 Graph convolutional neural networks

Learning from graph data x requires a map Φ between the data x and the target representation y that incorporates the graph structure: $y = \Phi(x; S)$. The image of the map Φ is known as the representation space and determines the space of all possible representations y for a given graph shift-operator S and any graph signal x. One example of a representation map is the graph convolution $\Phi(x; S, \mathcal{H}) = H(S) x$ in (8.36), where the set $\mathcal{H} = \{h\}$ contains the filter taps that characterize the representation space. In order to learn this map, a cost function J is optimized, given a training set $\mathcal{T} = \{x_1, x_2, \ldots, x_T\}$ with $|\mathcal{T}|$ samples. The learned map is then $\Phi(x; S, \mathcal{H}^*)$ with

$$\mathcal{H}^* = \arg \min_{\mathcal{H}} \frac{1}{|\mathcal{T}|} \sum_{x \in \mathcal{T}} J(\Phi(x; S, \mathcal{H})) \tag{8.37}$$

Typical cost functions are the mean squared error, the L_1 norm for regression, cross-entropy for classification and the maximum likelihood for stochastic processes. If $\Phi(x; S, \mathcal{H}) = H(S) x$, then the optimization in (8.37) returns the $K + 1$ best filter taps $\mathcal{H}^* = \{h^*\}$ that best fit the training data with respect to the cost function J and where K can be considered as a known design parameter or as a variable to be optimized. However, graph convolutions limit the representation problem to a linear map. Non-linear maps lead to the concept of a graph perceptron $\sigma(y)$, where each component y_i in the vector y is transformed to $\sigma(y_i)$, thus the vector $\sigma(y) = (\sigma(y_1), \sigma(y_2), \ldots, \sigma(y_N))$. In other words, if x is the input signal to a graph convolution $y = H(S) x$, then the graph perceptron $z = \sigma(y)$ creates a non-linear output signal, where $\sigma(t)$ is usually a sigmoid function of the parameter t such as $\sigma(t) = \tanh(\beta t)$ or $\sigma(t) = \max(t, 0)$.

Instead of one non-linear perceptron that maps each input component x_i to an output component $z_i = \sigma\left(\left(H\left(S\right)x\right)_i\right)$, a cascade of L graph perceptrons is built that forms a multi-layer graph perception, also called a graph convolutional neural network, where layer $l-1$ feeds the next layer l by

$$x_l = \sigma_l\left(H_l\left(S\right)x_{l-1}\right) \qquad \text{where } l = 1,\dots,L$$

Solving this recursion indicates that an input signal x_0 of an L layer graph perceptron is eventually transformed to an output signal

$$x_L = \sigma_L\left(H_L\sigma_{L-1}\left(H_{L-1}\sigma_{L-2}\left(H_{L-2}\cdots\sigma_1\left(H_1 x_0\right)\right)\right)\right)$$

which is a nested function evaluation of sigmoid functions σ_l and where we have omitted the dependence of H_l on the underlying graph via the shift-operator S. The L layer graph perceptron is actually a generalized, directed bipartite graph, where odd layers are placed on the left side and even layers are the right-side nodes of the bipartite graph and links are directed from layer l to layer $l+1$. Graph theoretically, the implicit underlying bipartite structure of a multi-layer perceptron neural network is perhaps one of its major limitations, because it prevents that layer l can interact directly with layer $l' \neq l+1$. The corresponding representation problem $\Phi\left(x;S,\mathcal{H}\right) = \sigma\left(H_l\sigma\left(H_{l-1}\sigma\left(H_{l-2}\cdots\sigma\left(H_1 x_0\right)\right)\right)\right)$ in (8.37) then finds the best filter coefficients $\mathcal{H}^* = \{h_l^*\}_{1\leq l\leq L}$, given the sigmoid functions $\{\sigma_l\}_{\leq l\leq L}$ and the order K_l of each graph convolution in (8.36). The number L of layers is either fixed and given or it can be iteratively increased to enhance accuracy.

We refer to the specific and rapidly evolving literature (see e.g. Gama *et al.* (2020), Ruiz *et al.* (2021) and references therein) for implementation details, algorithms to find the optimal filter coefficients $\mathcal{H}^*$, such as stochastic gradient descent, and applications.

Part II

Eigensystem

9

Topics in linear algebra

This chapter reviews a few, general results from linear algebra. In-depth analyses are found, among many others, in the books by Gantmacher (1959a,b), Wilkinson (1965), Shilov (1977), Mirsky (1982) and Meyer (2000). We refer to Golub and Van Loan (1996) for matrix computational methods and algorithms.

In this chapter, A is a general matrix, not the adjacency matrix.

9.1 Matrix transformations

Any linear transformation, that maps a vector x to a vector y in an n-dimensional vector space, can be represented by $y = Ax$, where A is an $n \times n$ matrix. The relation $y = Ax$ links matrix theory to geometric mappings and vector spaces. After reviewing the basic concept of a coordinate of a point in an n-dimensional space, we discuss here matrix transformations, that can make certain components of a vector x zero and that play a role in the transformation of a matrix A to a triangular form.

191. *Coordinates.* Let us consider an n-dimensional space and n basic vectors $e_1, e_2, \ldots, e_n$ with vector components $(e_k)_j = \delta_{kj}$, where δ_{kj} is the Kronecker delta, i.e. $(e_k)_j = 0$ if $k \neq j$ and $(e_k)_j = 1$ if $k = j$. This set of basic vectors $e_1, e_2, \ldots, e_n$ is orthonormal, because $e_m^T e_k = \delta_{km}$, and is said to span the n-dimensional space, because any $n \times 1$ vector $y = (y_1, y_2, \ldots, y_n)$ can be written as a linear combination of the basic vectors $y = \sum_{j=1}^{n} \alpha_j e_j$, where α_j is a real number. If we explicitly use the definition of the basic vectors $e_1, e_2, \ldots, e_n$, then we can write this expression in matrix form as

$$y = \alpha_1 \begin{bmatrix} 1 \\ 0 \\ \vdots \\ 0 \end{bmatrix} + \alpha_2 \begin{bmatrix} 0 \\ 1 \\ \vdots \\ 0 \end{bmatrix} + \alpha_n \begin{bmatrix} 0 \\ 0 \\ \vdots \\ 1 \end{bmatrix} = \begin{bmatrix} 1 & 0 & \cdots & 0 \\ 0 & 1 & \cdots & 0 \\ \vdots & \vdots & \ddots & \vdots \\ 0 & 0 & \cdots & 1 \end{bmatrix} \begin{bmatrix} \alpha_1 \\ \alpha_2 \\ \vdots \\ \alpha_n \end{bmatrix}$$

In matrix notation, we have that $y = I\alpha = \alpha$, where I is the $n \times n$ identity matrix and where the $n \times 1$ vector $\alpha = (\alpha_1, \alpha_2, \ldots, \alpha_n)$, and we observe from $y = \alpha$ that the vector components $y_k = \alpha_k$. The projection of the vector y on a basic vector e_k is equal to $y^T e_k = \sum_{j=1}^n y_j \left(e_j^T e_k\right) = y_k$, where we have invoked orthogonality. A point p in the n-dimensional space can be represented by a vector y and the k-th vector component y_k is called the coordinate of the point p in the basis determined by the set of vectors $e_1, e_2, \ldots, e_n$. Finally, the Euclidean norm of the vector y equals $y^T y$, which is the distance of the point p to the origin, the point o with zero coordinates and norm.

192. *Orthogonal transformation.* Suppose that X is an $n \times n$ orthogonal matrix (**art.** 247, 248 and 249) satisfying $X^T X = X X^T = I$, from which $X^{-1} = X^T$. This means that the k-th column vector of X, which we denote by x_k, is orthogonal to x_m if $k \neq m$. Thus, the set of orthogonal vectors $x_1, x_2, \ldots, x_n$ can span the n-dimensional space and, similarly as in **art.** 191, we can write the vector $y = \sum_{j=1}^n \beta_j x_j$ as a linear combination of those orthogonal vectors and orthogonality shows that $\beta_k = y^T x_k$. In matrix notation, $y = X\beta$, where the $n \times 1$ vector $\beta = (\beta_1, \beta_2, \ldots, \beta_n)$ are the coordinates of the point p, represented by the vector y, in the basis specified by the orthogonal vectors $x_1, x_2, \ldots, x_n$. The Euclidean norm of the vector y in the coordinate system determined by the orthogonal vectors $x_1, x_2, \ldots, x_n$ is

$$y^T y = \sum_{j=1}^n \beta_j x_j^T \sum_{m=1}^n \beta_m x_m = \sum_{j=1}^n \sum_{m=1}^n \beta_j \beta_m \left(x_j^T x_m\right) = \sum_{m=1}^n \beta_j^2$$

or, in matrix notation, $y^T y = (X\beta)^T X\beta = \beta^T X^T X\beta = \beta^T \beta$. The equality $y^T y = \beta^T \beta$ implies that the Euclidean norm is preserved after an orthogonal transformation: the distance of the point p to the origin is unaltered in the coordinate system $e_1, e_2, \ldots, e_n$ and in the coordinate system $x_1, x_2, \ldots, x_n$, although the corresponding coordinates $y_k = y^T e_k$ and $\beta_k = y^T x_k$ are generally different.

The relation between the coordinates $(y_1, y_2, \ldots, y_n)$ of the point p in the basis $e_1, e_2, \ldots, e_n$ and its coordinates $(\beta_1, \beta_2, \ldots, \beta_n)$ in the basis $x_1, x_2, \ldots, x_n$ is the linear transformation $y = X\beta$ and its inverse transformation $\beta = X^{-1} y = X^T y$. Hence, an orthogonal transformation, characterized by an orthogonal matrix X, preserves the Euclidean distance and provides an easy way to interrelate the coordinates of a same point p, without requiring the computation of the inverse matrix. Geometrically, any orthogonal transformation is a rotation of the vector y around the origin to a vector β, and both y and β have equal Euclidean distance or norm (see **art.** 201).

193. *Elementary orthogonal projector.* A matrix of the form

$$S = I - \frac{1}{v^T v} v.v^T \tag{A.1}$$

is called an elementary orthogonal projector onto the hyperplane through the origin that is orthogonal to the vector v. That hyperplane contains all the vectors y that are orthogonal to the vector v, i.e. that satisfy $y^T v = 0$. Indeed, the vector $Sx =$

$x - \left(\frac{v^T x}{v^T v}\right) v$ obeys $v^T (Sx) = v^T x - \left(\frac{v^T x}{v^T v}\right) v^T v = 0$ and thus lies in the hyperplane orthogonal to v. Moreover, $(I - S) x = \left(\frac{v^T x}{v^T v}\right) v$ is the vector proportional to v or lying on the line determined by the vector v.

The elementary orthogonal projection S is symmetric, because the matrix $v.v^T$ is symmetric. However, S is not orthogonal because $S^T S = I - \frac{v.v^T}{v^T v} \neq I$. Hence, since $S^T S = S^2$, we observe that $S^2 = S$, and $S^k = S$ for $k \in \mathbb{N}$. The squared length $\|w\|_2^2 = w^T w$ of the projected vector $w = Sx$ equals $x^T S^T Sx = x^T S^2 x = x^T Sx = x^T x - \frac{(v^T x)^2}{v^T v}$, which can be viewed as the higher dimensional analogue of Pythagoras' theorem.

194. *Spectrum of the elementary orthogonal projector.* The functional equation $S^2 = S$ in **art.** 193 means that the vector x is transformed to the vector $w = Sx$, which is orthogonal to v, and that any subsequent set of projections, $S^k w$, keeps the vector w unchanged in the hyperplane orthogonal to the vector v. Indeed, if w is orthogonal to v, then $v^T w = 0$ and $Sw = w$, which shows that $\lambda = 1$ is an eigenvalue of S for all vectors orthogonal to v. Each vector x that is not orthogonal to v is of the form $x = \alpha v + \beta y$, where $y^T v = 0$, and is transformed into a vector $w = Sx = \beta y$ that is not proportional anymore to itself. The eigenvalue equation $Sx = \lambda x$ only has a solution if $\lambda = 0$ and $y = 0$, in which case $Sv = 0$. Hence, S is singular, i.e. $\det S = 0$.

Similar to the spectrum of the complete graph in Section 6.1, the eigenvalues of S also follow from (A.66): the eigenvalues of S are $\lambda = 1$ with multiplicity $N - 1$ and $\lambda = 0$ with multiplicity 1.

195. *General Projector.* The elementary orthogonal projector S can be generalized. Let $\mathcal{X}$ and $\mathcal{Y}$ be complementary subspaces of a vector space $\mathcal{V}$ so that every vector $v \in \mathcal{V}$ can be uniquely resolved as $v = x + y$, where $x \in \mathcal{X}$ and $y \in \mathcal{Y}$. The unique linear operator P defined by $Pv = x$ is called the projector onto $\mathcal{X}$ along $\mathcal{Y}$ and P has the following properties: (a) $P^2 = P$ (i.e., P is idempotent), (b) $I - P$ is the complementary projector onto $\mathcal{Y}$ along $\mathcal{X}$, (c) If $\mathcal{V}$ is $\mathbb{R}^n$, then

$$P = [\; X \;\; Y \;] \begin{bmatrix} I & O \\ O & O \end{bmatrix} [\; X \;\; Y \;]^{-1}$$

where the columns of X and Y are respective bases for $\mathcal{X}$ and $\mathcal{Y}$. These results are proved in Meyer (2000, p. 386). If $\mathcal{Y} = \mathcal{X}^{\perp}$, then P is the orthogonal projector onto $\mathcal{X}$. In that case, Meyer (2000, p. 430) shows that $P = X (X^T X)^{-1} X^T$. Moreover, if the basis vectors of $\mathcal{X}$ are orthogonal, i.e., $X^T X = I = X X^T$, we have $P = X X^T = I$.

196. *Gauss transformation.* Gaussian elimination is the key technique to solve a set of n linear equations of the form $Ax = b$. A characteristic step in the Gaussian elimination is illustrated for $n = 2$ variables and transforms the set

$$\begin{cases} a_{11} x_1 + a_{12} x_2 = b_1 \\ a_{21} x_1 + a_{22} x_2 = b_2 \end{cases}$$

after multiplying the first equation by $\frac{a_{21}}{a_{11}}$, assuming $a_{11} \neq 0$, and subtracting from the second equation to the equivalent set

$$\begin{cases} a_{11}x_1 + a_{12}x_2 = b_1 \\ \left(a_{22} - a_{21}\frac{a_{12}}{a_{11}}\right)x_2 = b_2 - \frac{a_{21}}{a_{11}}b_1 \end{cases}$$

In matrix form, this transformation is written as

$$\begin{bmatrix} 1 & 0 \\ -\frac{a_{21}}{a_{11}} & 1 \end{bmatrix} \begin{bmatrix} a_{11} & a_{12} \\ a_{21} & a_{22} \end{bmatrix} = \begin{bmatrix} a_{11} & a_{12} \\ 0 & a_{22} - a_{21}\frac{a_{12}}{a_{11}} \end{bmatrix}$$

Multiplying both sides with the inverse transformation yields

$$\begin{bmatrix} a_{11} & a_{12} \\ a_{21} & a_{22} \end{bmatrix} = \begin{bmatrix} 1 & 0 \\ \frac{a_{21}}{a_{11}} & 1 \end{bmatrix} \begin{bmatrix} a_{11} & a_{12} \\ 0 & a_{22} - a_{21}\frac{a_{12}}{a_{11}} \end{bmatrix}$$

which is called the LU decomposition of $A = LU$, where L is a unit lower triangular and U is an upper triangular matrix. The solution $Ax = b$ is found in two steps by solving triangular matrices that have an easy solution (see e.g. Golub and Van Loan (1996)): (a) solve y in $Ly = b$, (b) solve x in $Ux = y$. Indeed, we verify that x is the solution, because $Ax = LUx = Ly = b$.

A crucial step in Gaussian elimination is the transformation G_k of a vector $x^T = \begin{bmatrix} x_1 & \cdots & x_k & x_{k+1} & \cdots & x_n \end{bmatrix}$ to a vector $(G_k x)^T = \begin{bmatrix} x_1 & \cdots & x_k & 0 & \cdots & 0 \end{bmatrix}$. The case $n = 2$ above has shown, for $\tau = \frac{x_2}{x_1}$ (and $x_1 \neq 0$), that

$$\begin{bmatrix} 1 & 0 \\ -\tau & 1 \end{bmatrix} \begin{bmatrix} x_1 \\ x_2 \end{bmatrix} = \begin{bmatrix} x_1 \\ 0 \end{bmatrix}$$

In general, the rank-one update

$$G_k = I - t_k e_k^T \tag{A.2}$$

where the vector $t_k^T = \begin{bmatrix} 0 & \cdots & 0 & \tau_{k+1} & \cdots & \tau_n \end{bmatrix}$ and $\tau_j = \frac{x_j}{x_k}$ for $k+1 \leq j \leq n$ transforms x into $G_k x = \left(I - t_k e_k^T\right) x = x - t_k x_k$ with the desired property that only the first k components of $G_k x$ are non-zero and equal to those of the first k components of x. The transform G_k in (A.2) is called the Gauss transformation with Gauss vector t_k. The Gauss transformation is a unit lower triangular matrix,

$$G_k = \begin{bmatrix} 1 & \cdots & 0 & 0 & 0 & \cdots & 0 \\ \vdots & \ddots & \vdots & \vdots & \vdots & \cdots & \vdots \\ 0 & \cdots & 1 & 0 & 0 & \cdots & 0 \\ 0 & \cdots & 0 & 1 & 0 & \cdots & 0 \\ 0 & \cdots & 0 & -\tau_{k+1} & 1 & \cdots & 0 \\ \vdots & \vdots & \vdots & \vdots & \vdots & \ddots & \vdots \\ 0 & \cdots & 0 & -\tau_n & 0 & \cdots & 1 \end{bmatrix}$$

The inverse Gauss transform follows from the inverse (A.67) of the rank-one update with $e_k^T t_k = 0$ as

$$G_k^{-1} = \left(I - t_k e_k^T\right)^{-1} = I + t_k e_k^T \tag{A.3}$$

which is a unit lower triangular matrix. In general, after $n - 1$ Gauss transforms $G_1, G_2, \ldots, G_{n-1}$, the matrix A is transformed into an upper triangular matrix $U = G_{n-1} \ldots G_1 A$ so that

$$A = (G_{n-1} \ldots G_1)^{-1} U = LU$$

where, using (A.3),

$$L = (G_{n-1} \ldots G_1)^{-1} = \prod_{j=1}^{n-1} G_j^{-1} = G_1^{-1} \ldots G_{n-1}^{-1} = I + \sum_{j=1}^{n-1} t_k e_k^T$$

is a unit lower triangular matrix. Numerical aspects (such as pivoting, error analyses, algorithms and numerical complexity) of the LU decomposition $A = LU$ are discussed in Golub and Van Loan (1996).

197. *Householder reflections.* The $n \times n$ real[1] matrix R of the form

$$R = I - \frac{2}{v^T v} v.v^T \tag{A.4}$$

is called a Householder reflection or transformation with Householder vector v. The Householder transformation is symmetric, because $\left(v.v^T\right)^T = v.v^T$, and orthogonal, because $R^T R = I$. Moreover, the Householder transformation (A.4) is the only orthogonal rank-one update transformation. Indeed, a rank-one update transformation $V = I - cd^T$, that is orthogonal, must satisfy

$$I = V^T V = I - cd^T - dc^T + \left(c^T c\right) dd^T$$

so that $cd^T + dc^T = \left(c^T c\right) dd^T$. Let $c = \alpha d + \beta y$ and $y^T d = 0$, then the requirement is $2\alpha dd^T + \beta \left(yd^T + dy^T\right) = \left(\alpha^2 d^T d + \beta^2 y^T y\right) dd^T$, which shows that $y = 0$ and $2\alpha = \alpha^2 d^T d$ or $\alpha = \frac{2}{d^T d}$. Hence, an orthogonal rank-one update transformation is of the form $V = I - \frac{2}{d^T d} dd^T$.

Let x be a non-zero vector, then

$$Rx = \left(I - \frac{2}{v^T v} v.v^T\right) x = x - \frac{2v^T x}{v^T v} v \tag{A.5}$$

Since R is an orthogonal transformation, $(Rx)^T Rx = x^T x$, the length of the vector x is preserved after transformation, implying that both x and Rx lie on the same hypersphere with center at the origin and radius $\sqrt{x^T x}$. Interestingly, after applying the elementary orthogonal projector S, defined in (A.1) and **art. 193**, to $y = Rx$, we find that

$$Sy = \left(I - \frac{vv^T}{v^T v}\right) \left(x - \frac{2v^T x}{v^T v} v\right) = x - \frac{v^T x}{v^T v} v = Sx$$

---

[1] The Householder reflection with a complex Householder vector v is defined as $R = I - \frac{2}{v^H v} v.v^H$.

Hence, both the orthogonal projection onto the hyperplane orthogonal to v of the vector x and the vector $y = Rx$ are the same. This means that the Householder transformation (A.4) creates a reflection of the vector x with respect to the hyperplane orthogonal to v. Since $R^{-1} = R^T = R$, we have that $R^2 = I$.

Since the Householder transformation R in (A.4) is orthogonal, in contrast to the elementary orthogonal projector S, a useful application is to find the vector v so that $Rx = e_1$, where x is a given vector. Equation (A.5) indicates that v is a linear combination of x and e_1. Thus, setting $v = x + \alpha e_1$ gives, with $x_1 = e_1^T x$,

$$e_1 = Rx = \left(1 - \frac{2\left(x^T x + \alpha x_1\right)}{x^T x + 2\alpha x_1 + \alpha^2}\right) x - \alpha \frac{2\left(x^T x + \alpha x_1\right)}{x^T x + 2\alpha x_1 + \alpha^2} e_1$$

which requires that the coefficient of x must be zero, $1 - \frac{2\left(x^T x + \alpha x_1\right)}{x^T x + 2\alpha x_1 + \alpha^2} = 0$ or $\alpha^2 = x^T x$. Hence, for a Householder vector $v = x \pm \|x\|_2 e_1$, it follows that[2]

$$Rx = \mp \|x\|_2 e_1 \tag{A.6}$$

The orthogonality and symmetry of the Householder transformation implies that $R^{-1} = R^T = R$, so that $x = \mp \|x\|_2 Re_1$, which illustrates that the orthogonal matrix R contains the vector x in the first column (which is Re_1).

In summary, the columns of the Householder matrix R with Householder vector $v = x \pm \|x\|_2 e_1$ represent a set of orthogonal vectors of which the first equals x.

198. *Householder reduction.* Gaussian elimination (**art.** 196) is a technique to transform a matrix A to an upper triangular matrix U. Consecutive application of the Householder transformation (**art.** 197) can also achieve a similar result as we will show here.

Consider a real $m \times n$ matrix $A = \begin{bmatrix} a_1 & a_2 & \cdots & a_n \end{bmatrix}$ where a_k is the $m \times 1$ vector of the k-th column. We invoke the $m \times m$ Householder reflection R_1 with Householder vector $v_1 = a_1 \pm \|a_1\|_2 e_1$ to transform the first column a_1 to $u_{11} (e_1)_{m \times 1}$, with $u_{11} = \pm \|a_1\|_2$ according to (A.6). Hence,

$$R_1 A = \begin{bmatrix} u_{11} (e_1)_{m \times 1} & R_1 a_2 & \cdots & R_1 a_n \end{bmatrix} = \begin{bmatrix} u_{11} & u_1^T \\ 0 & B \end{bmatrix}$$

where the real $(m-1) \times (n-1)$ matrix $B = \begin{bmatrix} b_1 & b_2 & \cdots & b_{n-1} \end{bmatrix}$ has $(m-1) \times 1$ column vectors.

After $(m-1) \times (m-1)$ Householder reflection $\widetilde{R}_2$ with $(m-1) \times 1$ Householder vector $\widetilde{v}_2 = b_1 \pm \|b_1\|_2 e_1$ of the matrix B, (A.6) shows that

$$\widetilde{R}_2 B = \begin{bmatrix} u_{22} (e_1)_{(m-1) \times 1} & R_2 b_2 & \cdots & R_2 b_n \end{bmatrix}$$

where $u_{22} = \pm \|b_1\|_2$. Now, if U and V are orthogonal matrices, so is $\begin{bmatrix} U & O \\ O & V \end{bmatrix}$.

---

[2] Besides the Householder reflection, also Givens rotations, explained in Meyer (2000); Golub and Van Loan (1996), can easily map an arbitrary vector x to the basic vector e_1.

Hence, the matrix $R_2 = \begin{bmatrix} 1 & 0 \\ 0 & \tilde{R}_2 \end{bmatrix}$ is orthogonal and, moreover, R_2 is also a Householder reflection with Householder vector $v^T = \left(0, \tilde{v}_2^T\right)$. Thus,

$$R_2 R_1 A = \begin{bmatrix} u_{11} & u_1^T \\ 0 & \tilde{R}_2 B \end{bmatrix} = \begin{bmatrix} u_{11} & * & * \\ 0 & u_{22} & u_2^T \\ 0 & 0 & C \end{bmatrix}$$

After the k-th iteration, the result is

$$R_k \ldots R_2 R_1 A = \begin{bmatrix} (U_k)_{k \times k} & (W_k)_{k \times (n-k)} \\ O_{(m-k) \times k} & K_{(m-k) \times (n-k)} \end{bmatrix}$$

where U_k is upper triangular and $R_j = \begin{bmatrix} I_{j-1} & 0 \\ 0 & \tilde{R}_j \end{bmatrix}$ is a Householder reflection. The final result after $k = m-1$ steps is $RA = U$, where $R = R_{m-1} \ldots R_1$ is an $m \times m$ orthogonal matrix[3] and U is an $m \times n$ upper triangular matrix, with diagonal element u_{jj} equal to plus or minus the Euclidean norm of an $(m-j) \times 1$ vector. In fact, the above method shows that we can always choose the sign of u_{jj}. In summary, Householder reduction[4] results in the factorization $A = R^T U$ of any $m \times n$ matrix, where R^T is an orthogonal (unitary) matrix and U is an upper triangular matrix.

199. *Quadratic form.* To a real symmetric matrix A, a bilinear form $x^T A y$ is associated, which is a scalar defined as

$$x^T A y = x A y^T = \sum_{i=1}^{n} \sum_{j=1}^{n} a_{ij} x_i y_j$$

We call a bilinear form a quadratic form if $y = x$. A necessary and sufficient condition for a quadratic form to be positive definite, i.e., $x^T A x > 0$ for all $x \neq 0$, is that all eigenvalues of A should be positive. Indeed, for a real symmetric matrix A, **art.** 247 shows the existence of an orthogonal matrix U that transforms A to a diagonal form. Let $x = Uz$, then

$$x^T A x = z^T U^T A U z = \sum_{k=1}^{n} \lambda_k z_k^2 \tag{A.7}$$

which is only positive for any vector component z_k provided $\lambda_k > 0$ for all k. From $\det A = \prod_{k=1}^{n} \lambda_k$ in (A.98), a positive definite quadratic form $x^T A x$ possesses a positive determinant $\det A > 0$. The problem of determining an orthogonal matrix U or the eigenvectors of A is equivalent to the geometrical problem of determining

---

[3] Although each Householder reflection $R_j = R_j^T$ is symmetric, the product is, in general, not symmetric, because $(R_{m-1} \ldots R_1)^T = R_1 \ldots R_{m-1}$.

[4] Householder reduction is only one of the techniques to obtain this type of matrix factorization, also known as QR factorization or decomposition, which is treated in depth by Golub and Van Loan (1996).

the principal axes of the hyper-ellipsoid $\sum_{i=1}^{n} \sum_{j=1}^{n} a_{ij} x_i y_j = 1$. Relation (A.7) illustrates that the inverse eigenvalue λ_k^{-1} is the square of the principal axis along the z_k vector. A multiple eigenvalue refers to an indeterminacy of the principal axes. For example if $n = 3$, an ellipsoid with two equal principal axis means that any section along the third axis is a circle. Any two perpendicular diameters of the largest circle orthogonal to the third axis are principal axes of that ellipsoid.

For additional properties of quadratic forms, such as Sylvester's law[5] of inertia in **art.** 266, we refer to Courant and Hilbert (1953) and Gantmacher (1959a).

200. *Taylor series of a multivariable function and a quadratic form.* The Taylor expansion of a differentiable function $F(x)$ of the vector x around the vector h is

$$F(x) = F(h) + (\nabla F(h))^T (x - h) + \frac{1}{2} (x - h)^T H(h) (x - h) + R$$

where R is the remainder of the order of $O\left(\|x - h\|^3\right)$, the gradient vector $\nabla F(h) = \left(\left.\frac{\partial F(x)}{\partial x_1}\right|_{x=h}, \left.\frac{\partial F(x)}{\partial x_2}\right|_{x=h}, \ldots, \left.\frac{\partial F(x)}{\partial x_n}\right|_{x=h}\right)$ and the $n \times n$ Hessian matrix is

$$H(h) = \begin{bmatrix} \left.\frac{\partial^2 F(x)}{\partial x_1^2}\right|_{x=h} & \left.\frac{\partial^2 F(h)}{\partial x_1 \partial x_2}\right|_{x=h} & \cdots & \left.\frac{\partial^2 F(h)}{\partial x_1 \partial x_n}\right|_{x=h} \\ \left.\frac{\partial^2 F(h)}{\partial x_2 \partial x_1}\right|_{x=h} & \left.\frac{\partial^2 F(x)}{\partial x_2^2}\right|_{x=h} & \cdots & \left.\frac{\partial^2 F(h)}{\partial x_2 \partial x_n}\right|_{x=h} \\ \vdots & \vdots & \ddots & \vdots \\ \left.\frac{\partial^2 F(h)}{\partial x_n \partial x_1}\right|_{x=h} & \left.\frac{\partial^2 F(h)}{\partial x_n \partial x_2}\right|_{x=h} & \cdots & \left.\frac{\partial^2 F(x)}{\partial x_n^2}\right|_{x=h} \end{bmatrix}$$

If $F(x) = x^T A x$ is a quadratic form and $y = x - h$, then

$$\begin{aligned} F(x) = F(h + y) &= \left(h^T + y^T\right) A (h + y) \\ &= h^T A h + h^T A y + y^T A h + y^T A y \end{aligned}$$

Since all terms are scalars, $y^T A h = \left(y^T A h\right)^T = h^T A^T y$, we have

$$F(x) = F(h) + h^T \left(A + A^T\right) (x - h) + F(x - h)$$

Comparison with the above general Taylor series indicates that the remainder $R = 0$ and that $\nabla F(h) = h^T \left(A + A^T\right)$ and $H(h) = 2A$. After putting $x \to x + h$ and $h \to x$ in the Taylor series of a quadratic function $F(x) \equiv F(x_1, x_2, \ldots, x_n) = x^T A x$ yields

$$F(x_1 + h_1, x_2 + h_2, \ldots, x_n + h_n) = F(x_1, x_2, \ldots, x_n) + \sum_{j=1}^{n} \frac{\partial F(x_1, x_2, \ldots, x_n)}{\partial x_j} h_j$$

$$+ \frac{1}{2} \sum_{i=1}^{n} \sum_{j=1}^{n} \frac{\partial^2 F(x_1, x_2, \ldots, x_n)}{\partial x_i \partial x_j} h_i h_j \qquad \text{(A.8)}$$

where $\frac{\partial^2 F(x_1, x_2, \ldots, x_n)}{\partial x_i \partial x_j} = 2a_{ij}$ and $\frac{\partial F(x_1, x_2, \ldots, x_n)}{\partial x_j} = \sum_{k=1}^{n} (a_{jk} + a_{kj}) x_k$.

---

[5] The number of positive and negative coefficients in a quadratic form reduced to the form (A.7) by a non-singular real linear transformation does not depend on the particular transformation.

9.2 Vector and matrix norms

201. Vector and matrix norms, denoted by $\|x\|$ and $\|A\|$ respectively, provide a single number reflecting a "size" of the vector or matrix and may be regarded as an extension of the concept of the modulus of a complex number. A norm is a certain function of the vector components or matrix elements. All norms, vector as well as matrix norms, satisfy the three "distance" relations:

  (i) $\|x\| > 0$ unless $x = 0$;
  (ii) $\|\alpha x\| = |\alpha|\,\|x\|$ for any complex number α;
  (iii) $\|x + y\| \leq \|x\| + \|y\|$ (triangle inequality)

An example of a non-homogeneous vector norm is the quadratic form

$$\sqrt{\|x\|_A} = \sqrt{x^T A x}$$

provided A is positive definite. Relation (A.7) shows that, if not all eigenvalues λ_j of A are the same, then not all components of the vector x are weighted similarly and, thus, in general, $\sqrt{\|x\|_A}$ is a non-homogeneous norm. The quadratic form $\|x\|_I$ equals the homogeneous Euclidean norm $\|x\|_2^2$.

202. *Hölder q-norm.* The Hölder q-norm of a vector x is defined as

$$\|x\|_q = \left(\sum_{j=1}^{n} |x_j|^q \right)^{1/q} \tag{A.9}$$

The well-known Euclidean norm or length of the vector x is found for $q = 2$ and $\|x\|_2^2 = x^H x$. In probability theory where x denotes a discrete probability density function, the law of total probability states that $\|x\|_1 = \sum_{j=1}^{n} x_j = 1$ and we will write $\|x\|_1 = \|x\|$. The extreme case $q \to \infty$ follows from (A.9) as $\max |x_j| = \lim_{q \to \infty} \|x\|_q = \|x\|_\infty = \max_{1 \leq j \leq n} |x_j|$. The unit-spheres $S_q = \{x : \|x\|_q = 1\}$ are, in three dimensions $n = 3$, for $q = 1$ an octahedron; for $q = 2$ a ball; and for $q = \infty$ a cube. Furthermore, S_1 fits into S_2, which in turn fits into S_∞, and this implies (**art.** 203) that $\|x\|_1 \geq \|x\|_2 \geq \|x\|_\infty$ for any x.

The Hölder inequality, proved in e.g. Van Mieghem (2014, p. 106), states that, for $\frac{1}{p} + \frac{1}{q} = 1$ and real $p > 1$,

$$\sum_{j=1}^{n} |x_j y_j| \leq \left(\sum_{j=1}^{n} |x_j|^p \right)^{\frac{1}{p}} \left(\sum_{j=1}^{n} |y_j|^q \right)^{\frac{1}{q}} \tag{A.10}$$

and in vector form

$$\left| x^H y \right| \leq \|x\|_p \|y\|_q \tag{A.11}$$

A special case of the Hölder inequality where $p = q = 2$ is the Cauchy-Schwarz inequality

$$\left| x^H y \right| \leq \|x\|_2 \|y\|_2 \tag{A.12}$$

The Cauchy-Schwarz inequality (A.12) follows immediately from the Cauchy identity (A.71) as shown in **art. 222**. The $q = 2$ norm is invariant under a unitary, hence also orthogonal, transformation U, where $U^H U = I$, because $\|Ux\|_2^2 = x^H U^H U x = x^H x = \|x\|_2^2$ (see **art. 192**).

203. *Norm inequalities.* All norms are equivalent in the *finite* dimensional case[6]: there exist positive real numbers c_1 and c_2 such that, for all vectors x,

$$c_1 \|x\|_p \leq \|x\|_q \leq c_2 \|x\|_p \tag{A.13}$$

For example,

$$\|x\|_2 \leq \|x\|_1 \leq \sqrt{n} \, \|x\|_2$$
$$\|x\|_\infty \leq \|x\|_1 \leq n \, \|x\|_\infty$$
$$\|x\|_\infty \leq \|x\|_2 \leq \sqrt{n} \, \|x\|_\infty$$

By choosing $x_j \to \alpha_j x_j^{\frac{s}{p}}$ for real $s > 0$, $y_j \to \alpha_j^{\frac{1}{q}} > 0$ and $p = \frac{1}{\theta}$ in the Hölder inequality (A.10), we obtain with $0 < \theta < 1$ the inequality

$$\left(\frac{\sum_{j=1}^n \alpha_j^{2-\theta} x_j^{s\theta}}{\sum_{j=1}^n \alpha_j} \right)^{\frac{1}{s\theta}} \leq \left(\frac{\sum_{j=1}^n \alpha_j^{\frac{1}{\theta}} |x_j|^s}{\sum_{j=1}^n \alpha_j} \right)^{\frac{1}{s}}$$

For $\alpha_j = 1$, the weights α_j disappear such that the inequality for the Hölder q-norm becomes $\|x\|_{s\theta} \leq \|x\|_s \, n^{\frac{1}{s}(\frac{1}{\theta}-1)}$, where $n^{\frac{1}{s}(\frac{1}{\theta}-1)} \geq 1$. On the other hand, with $0 < \theta < 1$ and for real $s > 0$,

$$\frac{\|x\|_s}{\|x\|_{s\theta}} = \frac{\left(\sum_{j=1}^n |x_j|^s \right)^{\frac{1}{s}}}{\left(\sum_{k=1}^n |x_k|^{s\theta} \right)^{\frac{1}{s\theta}}} = \left(\sum_{j=1}^n \frac{|x_j|^s}{\left(\sum_{k=1}^n |x_k|^{s\theta} \right)^{\frac{1}{\theta}}} \right)^{\frac{1}{s}} = \left(\sum_{j=1}^n \left(\frac{|x_j|^{s\theta}}{\sum_{k=1}^n |x_k|^{s\theta}} \right)^{\frac{1}{\theta}} \right)^{\frac{1}{s}}$$

Since $y = \frac{|x_j|^{s\theta}}{\sum_{k=1}^n |x_k|^{s\theta}} \leq 1$ and $\frac{1}{\theta} > 1$, it holds that $y^{\frac{1}{\theta}} \leq y$ and

$$\left(\sum_{j=1}^n \left(\frac{|x_j|^{s\theta}}{\sum_{k=1}^n |x_k|^{s\theta}} \right)^{\frac{1}{\theta}} \right)^{\frac{1}{s}} \leq \left(\sum_{j=1}^n \frac{|x_j|^{s\theta}}{\sum_{k=1}^n |x_k|^{s\theta}} \right)^{\frac{1}{s}} = \left(\frac{\sum_{j=1}^n |x_j|^{s\theta}}{\sum_{k=1}^n |x_k|^{s\theta}} \right)^{\frac{1}{s}} = 1$$

which leads to an opposite inequality $\|x\|_s \leq \|x\|_{s\theta}$.

In summary, if $p > q > 0$, then the general inequality for Hölder q-norm is

$$\|x\|_p \leq \|x\|_q \leq \|x\|_p \, n^{\frac{1}{q}-\frac{1}{p}} \tag{A.14}$$

The Minkowski inequality for the elements $a_{ij} \geq 0$ of an $m \times n$ non-negative

---

[6] For a finite dimensional vector space, the inequality (A.13) shows that the concept of "convergence of a sequence $\{x_n\}_{n \geq 1}$ to a point x^*" does not depend on the particular norm. For infinite dimensional vector spaces, however, the choice of the norm matters: for example, a Fourier series converges to the function with respect to the L_2-norm (**art. 350**), but the Gibbs phenomenon illustrates that the Fourier series does not converge uniformly.

matrix A, proved in e.g. Van Mieghem (2014, p. 108), is

$$\left(\sum_{j=1}^{n} \left(\sum_{k=1}^{m} a_{kj} \right)^{p} \right)^{\frac{1}{p}} \leq \sum_{i=1}^{m} \left(\sum_{j=1}^{n} a_{ij}^{p} \right)^{\frac{1}{p}} \tag{A.15}$$

and reduces for $m = 2$ to

$$\left(\sum_{j=1}^{n} |x_j + y_j|^q \right)^{\frac{1}{q}} \leq \left(\sum_{j=1}^{n} |x_j|^q \right)^{\frac{1}{q}} + \left(\sum_{j=1}^{n} |y_j|^q \right)^{\frac{1}{q}} \tag{A.16}$$

which is also known as the "triangle inequality", $\|x + y\|_q \leq \|x\|_q + \|y\|_q$, for the vector q-norm (A.9).

204. *Matrix norms.* For $m \times n$ matrices A, the most frequently used norms are the Euclidean or Frobenius norm

$$\|A\|_F = \left(\sum_{i=1}^{m} \sum_{j=1}^{n} |a_{ij}|^2 \right)^{1/2} \tag{A.17}$$

and the q-norm

$$\|A\|_q = \sup_{x \neq 0} \frac{\|Ax\|_q}{\|x\|_q} \tag{A.18}$$

The second distance relation in **art. 201**, $\frac{\|Ax\|_q}{\|x\|_q} = \left\| A \frac{x}{\|x\|_q} \right\|_q$, shows that

$$\|A\|_q = \sup_{\|x\|_q = 1} \|Ax\|_q \tag{A.19}$$

Furthermore, the matrix q-norm (A.18) implies that

$$\|Ax\|_q \leq \|A\|_q \|x\|_q \tag{A.20}$$

Since the vector norm is a continuous function of the vector components and since the domain $\|x\|_q = 1$ is closed, there must exist a vector x for which equality $\|Ax\|_q = \|A\|_q \|x\|_q$ holds. The i-th vector component of Ax is $(Ax)_i = \sum_{j=1}^{n} a_{ij} x_j$ and the Hölder q-norm (A.9) indicates that

$$\|Ax\|_q = \left(\sum_{i=1}^{m} \left| \sum_{j=1}^{n} a_{ij} x_j \right|^q \right)^{1/q}$$

For example, for all x with $\|x\|_1 = 1$, we have that

$$\|Ax\|_1 = \sum_{i=1}^{m} \left| \sum_{j=1}^{n} a_{ij} x_j \right| \leq \sum_{i=1}^{m} \sum_{j=1}^{n} |a_{ij}| |x_j| = \sum_{j=1}^{n} |x_j| \sum_{i=1}^{m} |a_{ij}|$$

$$\leq \sum_{j=1}^{n} |x_j| \left(\max_k \sum_{i=1}^{m} |a_{ik}| \right) = \|x\|_1 \max_k \sum_{i=1}^{m} |a_{ik}| = \max_k \sum_{i=1}^{m} |a_{ik}|$$

There exists a vector x for which equality holds, namely, if k is the column in A with maximum absolute sum, then $x = e_k$, the k-th basis vector with $(e_k)_m = \delta_{km}$. Similarly, for all x with $\|x\|_\infty = 1$,

$$\|Ax\|_\infty = \max_i \left| \sum_{j=1}^n a_{ij} x_j \right| \leq \max_i \sum_{j=1}^n |a_{ij}| \, |x_j| \leq \max_i \sum_{j=1}^n |a_{ij}|$$

If r is the row with maximum absolute sum and $x_j = 1.\text{sign}(a_{rj})$ such that $\|x\|_\infty = 1$, then $(Ax)_r = \sum_{j=1}^n |a_{rj}| = \max_i \sum_{j=1}^n |a_{ij}| = \|Ax\|_\infty$. Hence, we have proved that

$$\|A\|_\infty = \max_i \sum_{j=1}^n |a_{ij}| \tag{A.21}$$

$$\|A\|_1 = \max_j \sum_{i=1}^m |a_{ij}| \tag{A.22}$$

from which $\left\|A^H\right\|_\infty = \|A\|_1$.

205. The $q = 2$ matrix norm $\|Ax\|_2$ is obtained differently. Consider

$$\|Ax\|_2^2 = (Ax)^H Ax = x^H A^H Ax$$

Since $A^H A$ is a Hermitian matrix, **art. 247** shows that all eigenvalues are real and non-negative because a norm $\|Ax\|_2^2 \geq 0$. These ordered eigenvalues of $A^H A$ are denoted as $\sigma_1^2 \geq \sigma_2^2 \geq \cdots \geq \sigma_n^2 \geq 0$. Theorem 68 in **art. 247** states that there exists a unitary matrix U such that $x = Uz$ yields

$$x^H A^H Ax = z^H U^H A^H A U z = z^H \text{diag}\left(\sigma_j^2\right) z \leq \sigma_1^2 z^H z = \sigma_1^2 \|z\|_2^2$$

Since the $q = 2$ norm is invariant under a unitary and orthogonal transform $\|x\|_2 = \|z\|_2$, the definition (A.18) shows that

$$\|A\|_2 = \sup_{x \neq 0} \frac{\|Ax\|_2}{\|x\|_2} = \sigma_1 \tag{A.23}$$

where the supremum is achieved if x is the eigenvector of $A^H A$ belonging to σ_1^2. Meyer (2000, p. 279) proves the corresponding result for the minimum eigenvalue provided that A is non-singular,

$$\left\|A^{-1}\right\|_2 = \frac{1}{\min_{\|x\|_2 = 1} \|Ax\|_2} = \sigma_n^{-1}$$

The non-negative quantity σ_j is called the j-th singular value of the $n \times m$ matrix A and σ_1 is the largest singular value of A. An extension of the eigenvalue problem (1.3) to non-square matrices is called the singular value decomposition. A detailed discussion is found in Golub and Van Loan (1996) and Horn and Johnson (2013, Chapter 2).

206. The Frobenius norm $\|A\|_F^2 = \text{trace}(A^H A)$. With the trace-formula (A.99) and the analysis of $A^H A$ above,

$$\|A\|_F^2 = \sum_{k=1}^{n} \sigma_k^2 \tag{A.24}$$

In view of (A.23), the bounds $\|A\|_2 \le \|A\|_F \le \sqrt{n}\,\|A\|_2$ may be attained.

207. *Additional norm inequalities.* Since $\left\|A^k\right\| = \left\|AA^{k-1}\right\| \le \|A\|\left\|A^{k-1}\right\|$, by induction, we have for any integer k, that

$$\left\|A^k\right\| \le \|A\|^k$$

and

$$\lim_{k \to \infty} A^k = 0 \text{ if } \|A\| < 1$$

We apply the norm inequality (A.20) twice to the product AB

$$\|ABx\|_q \le \|A\|_q \|Bx\|_q \le \|A\|_q \|B\|_q \|x\|_q$$

The q-norm definition $\|A\|_q = \sup_{x \ne 0} \frac{\|Ax\|_q}{\|x\|_q}$ in (A.18) then leads to

$$\|AB\|_q \le \|A\|_q \|B\|_q \tag{A.25}$$

The norm $\|Ax\| = |\lambda|\,\|x\|$ of the eigenvalue equation (1.3) leads with $\|Ax\|_q \le \|A\|_q \|x\|_q$ in (A.20) to

$$|\lambda| \le \|A\|_q \tag{A.26}$$

Hence, the largest in absolute value eigenvalue of a matrix A does not exceed any matrix q-norm in (A.18). Applied to $A^H A$, for any q-norm,

$$\sigma_1^2 \le \left\|A^H A\right\|_q \le \left\|A^H\right\|_q \|A\|_q$$

Choose $q = 1$ and with (A.23),

$$\|A\|_2^2 \le \left\|A^H\right\|_1 \|A\|_1 = \|A\|_\infty \|A\|_1$$

Any matrix A can be transformed (**art. 239**) by a similarity transform H to a Jordan canonical form C as $A = HCH^{-1}$, from which $A^k = HC^k H^{-1}$. A typical Jordan submatrix $(C_m(\lambda))^k = \lambda^{k-2} B$, where B is independent of k. Hence, for large k, $A^k \to 0$ if and only if $|\lambda| < 1$ for all eigenvalues.

9.3 Formulae of determinants

The theory of determinants is discussed in historical order up to 1920 by Muir (1930) in five impressive volumes. Muir claims to be comprehensive. His treatise summarizes each paper and relates that paper to others. A remarkably large amount of papers are by his hand. Many papers deal with specially structured determinants that sometimes possess a nice, closed form[7].

208. *Definition.* A determinant of an $n \times n$ matrix A is defined by

$$\det A = \sum_p (-1)^{\sigma(p)} \prod_{j=1}^{n} a_{jk_j} \tag{A.27}$$

where the sum is over all the $n!$ permutations $p = (k_1, k_2, \ldots, k_n)$ of $(1, 2, \ldots, n)$ and $\sigma(p)$ is the number of interchanges between p and the natural order $(1, 2, \ldots, n)$. For example, $p = (1, 3, 2, 4)$ has 1 interchange, $\sigma(p) = 1$, while $p = (4, 3, 2, 1)$ has $\sigma(p) = 2$. Thus, $\sigma(p)$ is the number of interchanges to bring p back to the natural order. The determinant of a non-square matrix is not defined.

An important observation from the definition (A.27) of a determinant is that the product $\prod_{j=1}^{n} a_{jk_j}$ contains precisely one element from each row and one element of each column. Hence, if the matrix A contains a zero row or zero column, then its determinant $\det A = 0$. Any $n \times n$ diagonal matrix $A = \mathrm{diag}(a)$ possesses only one non-zero product $\prod_{j=1}^{n} a_{jk_j} = \prod_{j=1}^{n} a_{jj}$ and (A.27) with $\sigma(p) = 0$ reduces to

$$\det(\mathrm{diag}(a)) = \prod_{j=1}^{n} a_{jj} \tag{A.28}$$

The same argument shows that any $n \times n$ triangular matrix T, with all zero elements $t_{ij} = 0$ below (or above) the main diagonal, has a determinant equal to $\det T = \prod_{j=1}^{n} t_{jj}$, which generalizes the result (A.28) for the diagonal matrix.

209. *Elementary properties.* From the definition (A.27) of a determinant in **art. 208**, the following elementary properties can be derived (see e.g. Meyer (2000) or Mirsky (1982)).

(a) The transpose of a square matrix does not alter the determinant:

$$\det(A^T) = \det A \tag{A.29}$$

Hence, a sequence of row manipulations performed on a matrix results in the same determinant after performing the same sequence of corresponding column manipulations.

(b) If two rows (or columns) of a matrix A are interchanged, then the determinant

---

[7] We mention as an example the following $n \times n$ determinant of Scott (1880) in Muir (1930, vol. IV, p. 124), which involves all players of the harmonic, geometric and arithmetic mean inequality (6.38),

$$\begin{vmatrix} 0 & a_1 + a_2 & a_1 + a_3 & \cdots \\ a_1 + a_2 & 0 & a_2 + a_3 & \cdots \\ a_1 + a_3 & a_2 + a_3 & \ddots & \\ \vdots & \vdots & & \ddots \end{vmatrix} = \frac{(-2)^{n-1}}{2} \prod_{j=1}^{n} a_j \left\{ \left(\sum_{j=1}^{n} a_j \right) \left(\sum_{j=1}^{n} \frac{1}{a_j} \right) - (n-2)^2 \right\}$$

of the resulting matrix B equals

$$\det B = -\det A \qquad (A.30)$$

An immediate consequence of property (A.30) is that, when a matrix A contains two identical rows (or columns), its determinant is zero. Indeed, after interchanging these identical rows in A, property (A.30) indicates that the sign of the determinant must change, but the matrix A is unchanged! Hence, $\det A = -\det A$ implies that $\det A = 0$.

(c) If a row (or column) in a matrix A is multiplied by a complex number z, then the determinant of the resulting matrix B equals

$$\det B = z \det A \qquad (A.31)$$

Clearly, if $z = 0$, then the matrix B has one zero row and we obtain again the property, deduced in **art.** 208, that directly follows from the definition (A.27).

(d) The column (or row) addition property for determinants states that

$$
\det A = \begin{vmatrix} a_{11} & \cdots & b_{1k}+c_{1k} & \cdots & a_{1n} \\ a_{12} & \cdots & b_{2k}+c_{2k} & \cdots & a_{2n} \\ \vdots & & \vdots & & \vdots \\ a_{n1} & \cdots & b_{nk}+c_{nk} & \cdots & a_{nn} \end{vmatrix}
$$

$$
= \begin{vmatrix} a_{11} & \cdots & b_{1k} & \cdots & a_{1n} \\ a_{12} & \cdots & b_{2k} & \cdots & a_{2n} \\ \vdots & & \vdots & & \vdots \\ a_{n1} & \cdots & b_{nk} & \cdots & a_{nn} \end{vmatrix} + \begin{vmatrix} a_{11} & \cdots & c_{1k} & \cdots & a_{1n} \\ a_{12} & \cdots & c_{1k} & \cdots & a_{2n} \\ \vdots & & \vdots & & \vdots \\ a_{n1} & \cdots & c_{1k} & \cdots & a_{nn} \end{vmatrix}
$$

$$= \det A_1 + \det A_2 \qquad (A.32)$$

In other words, when the matrix A is written in terms of its column vectors $a_k = (a_{1k}, a_{2k}, \ldots, a_{nk})$ as

$$A = \begin{bmatrix} a_1 & \cdots & a_{k-1} & a_k & a_{k+1} & \cdots & a_n \end{bmatrix}$$

and $a_k = b_k + c_k$ so that

$$A_1 = \begin{bmatrix} a_1 & \cdots & a_{k-1} & b_k & a_{k+1} & \cdots & a_n \end{bmatrix}$$

and

$$A_2 = \begin{bmatrix} a_1 & \cdots & a_{k-1} & c_k & a_{k+1} & \cdots & a_n \end{bmatrix}$$

then (A.32) holds, but clearly $A \neq A_1 + A_2$. A consequence of property (A.32) and (A.30) is that the determinant is unaltered if a multiple of a column (row) is added to another column (row). Thus, if we add to the column vector b_k in A_1 the column vector $c_k = za_j$ for any $1 \leq j \neq k \leq n$ (thus $a_j \neq b_k$) and any complex number z, then the matrix A_2 consists of two identical columns after applying the column multiplication property (A.31) and its determinant vanishes, so that $\det A = \det A_1$.

(e) Product rule: If A, B, C are $n \times n$ matrices and $C = AB$, then

$$\det C = \det A. \det B \qquad (A.33)$$

210. *Explicit form of* $\det(A + B)$. We illustrate the definition (A.27) of a determinant by computing $\det(A + B)$, the determinant of a sum of two matrices, which can be recursively obtained from the column addition property (A.32). Here, we present a direct computation of

$$\det(A + B) = \sum_p (-1)^{\sigma(p)} \prod_{j=1}^n \left(a_{jk_j} + b_{jk_j}\right)$$

We first compute the product, rewritten as $\prod_{j=1}^n (x_j + y_j) = \prod_{j=1}^n x_j \prod_{j=1}^n \left(1 + \frac{y_j}{x_j}\right)$ and the latter product is a special case of the polynomial $\prod_{j=1}^n (z - z_j) = \sum_{k=0}^n a_k z^k$ in (B.1) with $a_n = 1$, where $z_j = -\frac{y_j}{x_j}$ and $z = 1$. We invoke Vieta's formula (B.11) and find that

$$\prod_{j=1}^n (z - z_j) = \sum_{k=0}^n \left((-1)^{n-k} \sum_{j_1=1}^n \sum_{j_2=j_1+1}^n \cdots \sum_{j_{n-k}=j_{n-k-1}+1}^n \prod_{i=1}^{n-k} z_{j_i}\right) z^k$$

We return to the original product

$$\prod_{j=1}^n (x_j + y_j) = \sum_{k=0}^n \left(\sum_{j_1=1}^n \sum_{j_2=j_1+1}^n \cdots \sum_{j_{n-k}=j_{n-k-1}+1}^n \prod_{i=1}^{n-k} y_{j_i} \prod_{i=n-k+1}^n x_{j_i}\right)$$

which, introduced into $\det(A + B) = \sum_p (-1)^{\sigma(p)} \prod_{j=1}^n \left(a_{jk_j} + b_{jk_j}\right)$, yields

$$\det(A + B) = \sum_{k=0}^n \left(\sum_{j_1=1}^n \sum_{j_2=j_1+1}^n \cdots \sum_{j_{n-k}=j_{n-k-1}+1}^n \sum_{p_n} (-1)^{\sigma(p_n)} \prod_{i=1}^{n-k} a_{j_i l_{j_i}} \prod_{i=n-k+1}^n b_{j_i l_{j_i}}\right)$$
$$(A.34)$$

where the p_n sum is over all the $n!$ permutations $p_n = (l_{j_1}, l_{j_2}, \ldots, l_{j_n})$ of the n-tuple $(j_1, j_2, \ldots, j_n)$. We observe that

$$\sum_{p_n} (-1)^{\sigma(p_n)} \prod_{i=1}^{n-k} a_{j_i l_{j_i}} \prod_{i=n-k+1}^n b_{j_i l_{j_i}}$$

is the determinant of the matrix with rows $j_1, j_2, \ldots, j_{n-k}$ consisting of elements of the matrix A and the remaining rows $j_{n-k+1}, \ldots, j_n$ with elements of the matrix B.

211. *Explicit form of* $\det(A - \lambda I)$. The special case of (A.34), where $B = -\lambda I$, is of particular interest (see **art. 235**). If $B = -\lambda I$, then the factor $\prod_{i=n-k+1}^n b_{j_i l_{j_i}}$

is non-zero and equal to $(-\lambda)^k$, only if $l_{j_m} = j_m$ for $n - k + 1 \le m \le n$, so that, using the notation of (A.95) in **art. 235**,

$$\det(A - \lambda I) = \sum_{k=0}^{n} c_k \lambda^k$$

with

$$c_k = (-1)^k \sum_{j_1=1}^{n} \sum_{j_2=j_1+1}^{n} \cdots \sum_{j_{n-k}=j_{n-k-1}+1}^{n} \sum_{p_{n-k}} (-1)^{\sigma(p_{n-k})} \prod_{i=1}^{n-k} a_{j_i l_{j_i}} \qquad \text{(A.35)}$$

where the last sum is over all $(n-k)!$ permutations of $p_{n-k} = \left(l_{j_1}, l_{j_2}, \ldots, l_{j_{n-k}}\right)$ of $(j_1, j_2, \ldots, j_{n-k})$. The latter determinant, called a principal minor, is thus obtained from the matrix A by selecting only $(j_1, j_2, \ldots, j_{n-k})$ rows and the same columns. For example, the case $k = n - 1$ in (A.35) equals $c_{n-1} = (-1)^{n-1} \sum_{j=1}^{n} a_{jj}$, presented in (A.99). For $k = n - 2$, (A.35) becomes

$$c_{n-2} = (-1)^n \sum_{i=1}^{n} \sum_{j=i+1}^{n} (a_{ii}a_{jj} - a_{ij}a_{ji}) = (-1)^n \sum_{i=1}^{n} \sum_{j=i+1}^{n} \begin{vmatrix} a_{ii} & a_{ij} \\ a_{ji} & a_{jj} \end{vmatrix}$$

212. *Expansion of the determinant in cofactors.* A cofactor of the element (i, j) in the $n \times n$ matrix A is defined as

$$\mathring{A}_{ij} = (-1)^{i+j} \det A_{\backslash \, \text{row } i \backslash \text{col } j} \qquad \text{(A.36)}$$

where $A_{\backslash \, \text{row } i \backslash \text{col } j}$ is the $(n-1) \times (n-1)$ matrix obtained from A by deleting the i-th row and the j-th column. The determinant $M_{ij} = \det A_{\backslash \, \text{row } i \backslash \text{col } j}$ is also called the minor of element a_{ij} in the matrix A. The adjugate of the matrix A is the transpose of the matrix of cofactors,

$$\text{adj}A = \mathring{A}^T \qquad \text{(A.37)}$$

and

$$(\text{adj}A)_{ij} = \mathring{A}_{ji} = (-1)^{i+j} \det A_{\backslash \, \text{row } j \backslash \text{col } i} \qquad \text{(A.38)}$$

Theorem 59 (Cofactor Expansion) *If $\mathring{A}_{ij}$ is the cofactor of a_{ij} in the $n \times n$ matrix A and δ_{ij} is the Kronecker delta, then for $1 \le i \le n$ and $1 \le j \le n$, the i-th row expansion of the determinant of A equals*

$$\sum_{k=1}^{n} a_{ik} \mathring{A}_{jk} = \delta_{ij} \det A \qquad \text{(A.39)}$$

while the i-th column expansion is

$$\sum_{k=1}^{n} a_{ki} \mathring{A}_{kj} = \delta_{ij} \det A \qquad \text{(A.40)}$$

Proof: See, e.g., Mirsky (1982, pp. 15-20).   □

When $a_k = (a_{1k}, a_{2k}, \ldots, a_{nk})$ is the k-th column vector in the matrix A and $\mathring{A}_k = \left(\mathring{A}_{1k}, \mathring{A}_{2k}, \ldots, \mathring{A}_{nk}\right)$ is the vector of cofactors of the components of the vector a_k, then (A.40) in the cofactor expansion Theorem 59 is rewritten as the scalar product $(a_i)^T \mathring{A}_j = \delta_{ij} \det A$. Only when each element in column i is multiplied by its corresponding cofactor, we obtain the value of the determinant of A, else $(a_i)^T \mathring{A}_j = 0$. In other words, the vectors a_i and $\mathring{A}_j$ are orthogonal if $i \neq j$.

An interesting application of the cofactor expansion Theorem 59 is Cauchy's formula

$$\det(\text{adj}A) = (\det A)^{n-1} \tag{A.41}$$

Consider the matrix $C = A.\text{adj}A$, where $c_{ij} = \sum_{k=1}^{n} a_{ik}\mathring{A}_{jk}$. Invoking (A.39) shows that the matrix C is a diagonal matrix with the same diagonal elements, $c_{ii} = \det A$. Similarly, using (A.40) shows that $\text{adj}A.A = I \det A = C$. Hence, by (A.28), we have that $\det C = (\det A)^n$ and by the product rule (A.33), we arrive at (A.41).

Moreover, from the basic property of the adjugate

$$A.\text{adj}A = I \det A = \text{adj}A.A \tag{A.42}$$

we find that the inverse of a matrix A equals

$$A^{-1} = \frac{\text{adj}A}{\det A} \tag{A.43}$$

and with (A.38)

$$\left(A^{-1}\right)_{ij} = (-1)^{i+j} \frac{\det A_{\setminus \text{row } j \setminus \text{col } i}}{\det A} \tag{A.44}$$

213. *Derivative of a determinant.* Suppose that the elements a_{ij} of a matrix A are differentiable functions of t. Then, the derivative of $\det A$ is computed from the definition (A.27) of the determinant of $A(t)$ as

$$\frac{d \det A(t)}{dt} = \sum_p (-1)^{\sigma(p)} \frac{d}{dt} \prod_{j=1}^{n} a_{jk_j}(t)$$

Since

$$\frac{d}{dt} \prod_{j=1}^{n} a_{jk_j}(t) = \sum_{l=1}^{n} \frac{da_{lk_l}(t)}{dt} \prod_{j=1; j \neq l}^{n} a_{jk_j}(t)$$

we have

$$\frac{d \det A(t)}{dt} = \sum_{l=1}^{n} \sum_p (-1)^{\sigma(p)} \frac{da_{lk_l}(t)}{dt} \prod_{j=1; j \neq l}^{n} a_{jk_j}(t)$$

The definition (A.27) shows that $\sum_p (-1)^{\sigma(p)} \frac{da_{lk_l}(t)}{dt} \prod_{j=1; j \neq l}^{n} a_{jk_j}(t) = \det A_l$, where the matrix A_l is equal to the matrix A, except that the l-th row in A_l is replaced by the derivatives $\frac{da_{lj}(t)}{dt}$, for $1 \leq j \leq n$. Hence,

$$\frac{d \det A(t)}{dt} = \sum_{l=1}^{n} \det A_l \qquad (A.45)$$

We compute the derivative of the characteristic polynomial $c_A(\lambda) = \det(A - \lambda I)$ with respect to λ using (A.45). Since $\frac{da_{lj}(\lambda)}{d\lambda} = -\delta_{lj}$, the cofactor Theorem 59 and the definition of a cofactor (A.36) in **art. 212** indicate that $\det A_l = \det A_{\backslash\{l\}}$, where $A_{\backslash\{l\}}$ is the $(n-1) \times (n-1)$ matrix deduced from A by deleting the l-th row and l-th column. Thus, we find that

$$\frac{dc_A(\lambda)}{d\lambda} = \frac{d \det(A - \lambda I)}{d\lambda} = -\sum_{l=1}^{n} \det\left(A_{\backslash\{l\}} - \lambda I\right) = -\sum_{l=1}^{n} c_{A_{\backslash\{l\}}}(\lambda) \qquad (A.46)$$

Invoking $c_1 = \frac{dc_A(\lambda)}{d\lambda}\Big|_{\lambda=0} = \det A \sum_{k=1}^{n} \frac{1}{\lambda_k(A)}$ in (A.100), the case for $\lambda = 0$ in (A.46) leads to

$$\sum_{j=1}^{n} \det\left(A_{\backslash\{j\}}\right) = \det A \sum_{k=1}^{n} \frac{1}{\lambda_k(A)}$$

214. *Generalized expansion of the determinant.* In 1772, Laplace has presented a generalization of the cofactor expansion Theorem 59 in **art. 212**. Before stating Laplace's theorem, the definition of the cofactor needs to be generalized. We denote by $A(i_1 \cdots i_k | j_1 \cdots j_k)$ the $k \times k$ submatrix of the $n \times n$ matrix A formed by the rows $i_1, i_2, \ldots, i_k$ intersected by the column $j_1, j_2, \ldots, j_k$. The corresponding minor $M(i_1 \cdots i_k | j_1 \cdots j_k)$ is the $(n-k) \times (n-k)$ determinant of the submatrix of A obtained by deleting the rows $i_1, i_2, \ldots, i_k$ and the column $j_1, j_2, \ldots, j_k$ from A. The cofactor of $A(i_1 \cdots i_k | j_1 \cdots j_k)$ is defined as

$$\mathring{A}(i_1 \cdots i_k | j_1 \cdots j_k) = (-1)^{\sum_{m=1}^{k} i_m + j_m} M(i_1 \cdots i_k | j_1 \cdots j_k) \qquad (A.47)$$

When $k = 1$, then $A(i_1 \cdots i_k | j_1 \cdots j_k) = A(i|j) = a_{ij}$ and $\mathring{A}(i_1 \cdots i_k | j_1 \cdots j_k) = \mathring{A}(i|j) = (-1)^{i+j} M_{ij} = \mathring{A}_{ij}$, consistent with the definition (A.36) of the cofactor.

Theorem 60 (Laplace) *For each fixed set of row indices* $1 \leq i_1 < i_2 < \cdots < i_k \leq n$, *it holds that*

$$\det A = \sum_{1 \leq j_1 < \cdots < j_k \leq n} \det A(i_1 \cdots i_k | j_1 \cdots j_k) \, \mathring{A}(i_1 \cdots i_k | j_1 \cdots j_k) \qquad (A.48)$$

where the sum is over all $\binom{n}{k}$ *ways in which a set* $j_1, j_2, \ldots, j_k$ *of* k *columns can be chosen.*

Proof: See, e.g., Mirsky (1982, pp. 22-23).   □

Due to property (A.29), we have similarly, for each fixed set of column indices $1 \le j_1 < j_2 < \cdots < j_k \le n$, that

$$\det A = \sum_{1 \le i_1 < \cdots < i_k \le n} \det A \left(i_1 \cdots i_k | j_1 \cdots j_k \right) \mathring{A} \left(i_1 \cdots i_k | j_1 \cdots j_k \right) \qquad \text{(A.49)}$$

Mirsky (1982) remarks that Laplace's expansion Theorem 60 can be obtained from the column or row addition property (A.32): select the $i_1, i_2, \ldots i_k$ rows in the matrix A and write each element in those rows as $a_{ij} + 0$, while every other element in each remaining row as $0 + a_{pq}$. After repeatedly invoking the row addition property (A.32), we obtain a sum of 2^n determinants. The non-zero of those 2^n determinants can be written as a product of two determinants, corresponding to a $k \times k$ submatrix and its corresponding cofactor.

The next generalization is a famous theorem of Jacobi from 1833.

Theorem 61 (Jacobi) *For* $1 \le k \le n$, *it holds that*

$$\det \left(\text{adj}A \left(i_1 \cdots i_k | j_1 \cdots j_k \right) \right) = \left(\det A \right)^{k-1} \mathring{A} \left(i_1 \cdots i_k | j_1 \cdots j_k \right) \qquad \text{(A.50)}$$

Proof: See, e.g., Mirsky (1982, pp. 25-27). □

For $k = 1$, (A.50) reduces to an identity. If $k = n$, then $\text{adj}A \left(i_1 \cdots i_n | j_1 \cdots j_n \right) = \text{adj}A$ and (A.50) reduces to Cauchy's formula (A.41) when we define

$$\mathring{A} \left(i_1 \cdots i_n | j_1 \cdots j_n \right) = 1$$

consistent with (A.48). If $k = n - 1$, then $\mathring{A} \left(i_1 \cdots i_k | j_1 \cdots j_k \right)$ equals an element in A, say $(-1)^{r+s} a_{rs}$, and $\det \left(\text{adj}A \left(i_1 \cdots i_k | j_1 \cdots j_k \right) \right)$ is the cofactor of element (r, s) in the adjugate matrix $\text{adj}A$, which is, by (A.50), equal to $(-1)^{r+s} a_{rs} \left(\det A \right)^{n-2}$. If $k = 2$ and let $i_1 = i$, $i_2 = m$, $j_1 = j$ and $j_2 = l$, then

$$\det \left(\text{adj}A \left(i_1 i_2 | j_1 j_2 \right) \right) = \det \begin{bmatrix} \mathring{A}_{ij} & \mathring{A}_{il} \\ \mathring{A}_{mj} & \mathring{A}_{ml} \end{bmatrix} = \mathring{A}_{ij} \mathring{A}_{ml} - \mathring{A}_{il} \mathring{A}_{mj}$$

$$= (-1)^{i+j+m+l} \det A_{\backslash \text{row } i \backslash \text{col } j} \det A_{\backslash \text{row } m \backslash \text{col } l}$$
$$- (-1)^{i+j+m+l} \det A_{\backslash \text{row } i \backslash \text{col } l} \det A_{\backslash \text{row } m \backslash \text{col } j}$$

and

$$\mathring{A} \left(i_1 i_2 | j_1 j_2 \right) = (-1)^{i+m+j+l} \det \left(A_{\backslash \text{row } i \backslash \text{row } m \backslash \text{col } j \backslash \text{col } l} \right)$$

If the latter is non-zero, Jacobi's formula (A.50) becomes

$$\det A = \frac{\det A_{\backslash \text{row } i \backslash \text{col } j} \det A_{\backslash \text{row } m \backslash \text{col } l} - \det A_{\backslash \text{row } i \backslash \text{col } l} \det A_{\backslash \text{row } m \backslash \text{col } j}}{\det \left(A_{\backslash \text{row } i \backslash \text{row } m \backslash \text{col } j \backslash \text{col } l} \right)}$$

$$\text{(A.51)}$$

215. *Resolvent and Jacobi's trace formula.* The diagonal element of the matrix

$(xI - A)^{-1}$, called the resolvent (**art. 262**) of matrix A, follows from (A.44) as

$$(xI - A)^{-1}_{jj} = \frac{\det\left(xI - A_{\setminus\{j\}}\right)}{\det(xI - A)} \tag{A.52}$$

where $A_{\setminus\{j\}}$ is the $(n-1)\times(n-1)$ matrix obtained from A by deleting the j-th row and column. The expression $\frac{d}{dx}\det(xI - A) = \sum_{j=1}^{n}\det\left(xI - A_{\setminus\{j\}}\right)$ in (A.46) in **art. 213**) shows that

$$\sum_{j=1}^{n}(xI - A)^{-1}_{jj} = \frac{\frac{d}{dx}\det(xI - A)}{\det(xI - A)} = \frac{d}{dx}\log\det(xI - A)$$

which is rewritten as

$$\operatorname{trace}\left((xI - A)^{-1}\right) = \frac{d}{dx}\log\det(xI - A) \tag{A.53}$$

Integrating both sides with respect to x yields (**art. 231**)

$$\operatorname{trace}\left(\log(xI - A)\right) = \log\det(xI - A)$$

By substitution of $B = \log(xI - A)$, we find Jacobi's expression, valid for any matrix B,

$$e^{\operatorname{trace}(B)} = \det e^{B} \tag{A.54}$$

After taking the logarithm in (A.54), the trace is expressed in terms of the determinant,

$$\operatorname{trace}(A) = \log\det e^{A}$$

while by substituting $A = e^{B}$ in (A.54), Jacobi's identity expresses a determinant as a function of the trace

$$\det A = e^{\operatorname{trace}(\log A)}$$

Expanding the last expression in a Taylor series shows the relation with the Newton identities (B.4) as demonstrated in **art. 65**.

216. *Christoffel-Darboux formula for resolvents.* The resolvent $(xI - A)^{-1}$ of matrix A obeys

$$(xI - A)^{-1} - (yI - A)^{-1} = (y - x)(xI - A)^{-1}(yI - A)^{-1}$$

which is verified by left-multiplication by $(yI - A)$ and right-multiplication by $(xI - A)$. With the inverse matrix $B^{-1} = \frac{\operatorname{adj} B}{\det B}$ in (A.43),

$$\frac{\operatorname{adj}(xI - A)}{\det(xI - A)} - \frac{\operatorname{adj}(yI - A)}{\det(yI - A)} = (y - x)\frac{\operatorname{adj}(xI - A)}{\det(xI - A)}\frac{\operatorname{adj}(yI - A)}{\det(yI - A)}$$

After multiplying both sides by $(y - x)^{-1}\det(xI - A)\det(yI - A)$, the element ij

of the resulting matrix $T = \mathrm{adj}(xI - A)\mathrm{adj}(yI - A)$ is

$$t_{ij} = \frac{\mathrm{adj}_{ij}(xI - A)\det(yI - A) - \mathrm{adj}_{ij}(yI - A)\det(xI - A)}{y - x}$$

$$= \sum_{k=1}^{n} \mathrm{adj}_{ik}(xI - A)\,\mathrm{adj}_{kj}(yI - A)$$

Since both the adjugate and the determinant of $xI - A$ are polynomials in x of degree $n - 1$ and n, respectively, the Christoffel-Darboux identity reflects a polynomial identity, whose strength is applied in the study of orthogonal polynomials (see **art. 358**). In particular, the limit $y \to x$ results in

$$\lim_{y \to x} t_{ij} = \mathrm{adj}_{ij}(xI - A)\frac{d}{dx}\det(xI - A) - \frac{d}{dx}\mathrm{adj}_{ij}(xI - A)\det(xI - A)$$

$$= \sum_{k=1}^{n} \mathrm{adj}_{ik}(xI - A)\,\mathrm{adj}_{kj}(xI - A)$$

If $i = j$ and A is symmetric, then $\mathrm{adj}A$ is symmetric and

$$\mathrm{adj}_{ii}(xI - A)\frac{d\det(xI - A)}{dx} - \frac{d\,\mathrm{adj}_{ii}(xI - A)}{dx}\det(xI - A) = \sum_{k=1}^{n}(\mathrm{adj}_{ik}(xI - A))^2 \tag{A.55}$$

By using the same arguments as in **art. 364**, the above expression implies that the zeros of the polynomial $p(x) = \mathrm{adj}_{ii}(xI - A)$ and the polynomial $q(x) = \det(xI - A)$ interlace.

217. *Schur complements.* From the Schur identity

$$\begin{bmatrix} A & B \\ C & D \end{bmatrix} = \begin{bmatrix} I & O \\ CA^{-1} & I \end{bmatrix}\begin{bmatrix} A & B \\ O & D - CA^{-1}B \end{bmatrix} \tag{A.56}$$

which is a block Gaussian elimination in **art. 196** to construct an upper block triangular matrix, we find that

$$\det\begin{bmatrix} A & B \\ C & D \end{bmatrix} = \det A \det(D - CA^{-1}B) \tag{A.57}$$

and $D - CA^{-1}B$ is called the Schur complement of A. A similar identity

$$\begin{bmatrix} A & B \\ C & D \end{bmatrix} = \begin{bmatrix} A - BD^{-1}C & B \\ O & D \end{bmatrix}\begin{bmatrix} I & O \\ D^{-1}C & I \end{bmatrix} \tag{A.58}$$

leads to

$$\det\begin{bmatrix} A & B \\ C & D \end{bmatrix} = \det D \det(A - BD^{-1}C) \tag{A.59}$$

with Schur complement $A - BD^{-1}C$.

Applying $\det(AB) = \det(A)\det(B)$ in (A.33) to the right-hand side of (A.57), provided $A_{n\times n}$ and $D_{m\times m}$ have the same dimensions $m = n$, results in

$$\det A \det\left(D - CA^{-1}B\right) = \det\left(AD - ACA^{-1}B\right) = \det\left(DA - CA^{-1}BA\right)$$

which illustrates that, if A and C or A and B commute (i.e. $AC = CA$ or $AB = BA$), then the Schur determinant simplifies to

$$\det\begin{bmatrix} A & B \\ C & D \end{bmatrix} = \det(AD - CB) = \det(DA - CB)$$

which is formally equal to the determinant of a 2×2 matrix.

We can further reduce the block triangular matrices to block diagonal matrices as

$$\begin{bmatrix} A & B \\ O & D - CA^{-1}B \end{bmatrix} = \begin{bmatrix} A & O \\ O & D - CA^{-1}B \end{bmatrix}\begin{bmatrix} I & A^{-1}B \\ O & I \end{bmatrix}$$

and

$$\begin{bmatrix} A - BD^{-1}C & B \\ O & D \end{bmatrix} = \begin{bmatrix} I & BD^{-1} \\ O & I \end{bmatrix}\begin{bmatrix} A - BD^{-1}C & O \\ O & D \end{bmatrix}$$

so that the first (A.56) and second (A.58) Schur identities become

$$\begin{bmatrix} A & B \\ C & D \end{bmatrix} = \begin{bmatrix} I & O \\ CA^{-1} & I \end{bmatrix}\begin{bmatrix} A & O \\ O & D - CA^{-1}B \end{bmatrix}\begin{bmatrix} I & A^{-1}B \\ O & I \end{bmatrix}$$

and

$$\begin{bmatrix} A & B \\ C & D \end{bmatrix} = \begin{bmatrix} I & BD^{-1} \\ O & I \end{bmatrix}\begin{bmatrix} A - BD^{-1}C & O \\ O & D \end{bmatrix}\begin{bmatrix} I & O \\ D^{-1}C & I \end{bmatrix}$$

From the identity

$$\begin{bmatrix} I & X \\ O & I \end{bmatrix}\begin{bmatrix} I & -X \\ O & I \end{bmatrix} = \begin{bmatrix} I & O \\ O & I \end{bmatrix}$$

it follows that

$$\begin{bmatrix} I & X \\ O & I \end{bmatrix}^{-1} = \begin{bmatrix} I & -X \\ O & I \end{bmatrix}$$

while the inverse of a diagonal block matrix equals

$$\begin{bmatrix} A & O \\ O & D \end{bmatrix}^{-1} = \begin{bmatrix} A^{-1} & O \\ O & D^{-1} \end{bmatrix}$$

With $(ABC)^{-1} = C^{-1}B^{-1}C^{-1}$, we find two expressions for the inverse of the block matrix:

$$
\begin{bmatrix} A & B \\ C & D \end{bmatrix}^{-1} = \begin{bmatrix} I & -A^{-1}B \\ 0 & I \end{bmatrix} \begin{bmatrix} A^{-1} & O \\ O & (D-CA^{-1}B)^{-1} \end{bmatrix} \begin{bmatrix} I & O \\ -CA^{-1} & I \end{bmatrix}
$$
$$
= \begin{bmatrix} A^{-1}+A^{-1}B(D-CA^{-1}B)^{-1}CA^{-1} & -A^{-1}B(D-CA^{-1}B)^{-1} \\ -(D-CA^{-1}B)^{-1}CA^{-1} & (D-CA^{-1}B)^{-1} \end{bmatrix}
$$
(A.60)

and

$$
\begin{bmatrix} A & B \\ C & D \end{bmatrix}^{-1} = \begin{bmatrix} I & O \\ -D^{-1}C & I \end{bmatrix} \begin{bmatrix} (A-BD^{-1}C)^{-1} & O \\ O & D^{-1} \end{bmatrix} \begin{bmatrix} I & -BD^{-1} \\ O & I \end{bmatrix}
$$
$$
= \begin{bmatrix} (A-BD^{-1}C)^{-1} & -(A-BD^{-1}C)^{-1}BD^{-1} \\ -D^{-1}C(A-BD^{-1}C)^{-1} & D^{-1}+D^{-1}C(A-BD^{-1}C)^{-1}BD^{-1} \end{bmatrix}
$$
(A.61)

Equating corresponding blocks at the right-hand side of (A.60) and (A.61) returns the formulae

$$
A^{-1}B\left(D-CA^{-1}B\right)^{-1} = \left(A-BD^{-1}C\right)^{-1}BD^{-1}
$$

and

$$
\left(A-BD^{-1}C\right)^{-1} = A^{-1} + A^{-1}B\left(D-CA^{-1}B\right)^{-1}CA^{-1}
$$
(A.62)

where the latter (A.62) is known as the Sherman-Morrison-Woodbury formula.

218. *Schur's complement extended to a general block matrix.* Powell (2011) applied block Gaussian elimination to the $nN \times nN$ block matrix A, which is partitioned into N^2 blocks, each of size $n \times n$,

$$
A = \begin{bmatrix} A_{11} & A_{12} & \cdots & A_{1N} \\ A_{21} & A_{22} & \cdots & A_{2N} \\ \vdots & \vdots & \ddots & \vdots \\ A_{N1} & A_{N2} & \cdots & A_{NN} \end{bmatrix}
$$

The determinant of A is

$$
\det A = \prod_{k=1}^{N} \det\left(S_{kk}^{(N-k)}\right)
$$
(A.63)

where the $n \times n$ matrices $S_{ij}^{(k)}$ obey the recursion

$$
\begin{cases} S_{ij}^{(0)} = A_{ij} \\ S_{ij}^{(k+1)} = S_{ij}^{(k)} - S_{i,N-k}^{(k)}\left(S_{N-k,N-k}^{(k)}\right)^{-1}S_{N-k,j}^{(k)} & \text{for } 0 \le k < N \end{cases}
$$
(A.64)

In the case $N = 2$, (A.63) reduces to $\det A = \det S_{11}^{(1)} \det S_{22}^{(0)} = \det S_{11}^{(1)} \det A_{22}$ and the recursion (A.64) for $S_{11}^{(1)}$ becomes

$$S_{11}^{(1)} = S_{11}^{(0)} - S_{12}^{(0)} \left(S_{2,2}^{(0)}\right)^{-1} S_{21}^{(0)} = A_{11} - A_{12}A_{22}^{-1}A_{21}$$

which is precisely equal to Schur's block determinant (A.59).

219. *Rank one update formulae.* An interesting application of **art. 217** is

$$\det \begin{bmatrix} A_{n\times n} & -C_{n\times k} \\ D_{k\times n}^T & I_k \end{bmatrix} = \det \left(A_{n\times n} + C_{n\times k}D_{k\times n}^T\right) = \det A \det \left(I_k + D^T A^{-1} C\right)$$
(A.65)

which follows by applying both (A.59) and (A.57). For $k = 1$ and $A = I$ in (A.65), we obtain the "rank one update" formula

$$\det \left(I + cd^T\right) = 1 + d^T c \qquad (A.66)$$

This example shows that interesting relations can be obtained when the inverse of either A or D or both in (A.57) and (A.59) are explicitly known.

The inverse of $\left(A_{n\times n} + C_{n\times k}D_{k\times n}^T\right)$ follows from formula (A.62) as

$$\left(A_{n\times n} + C_{n\times k}D_{k\times n}^T\right)^{-1} = A^{-1} - A^{-1}C\left(I + D^T A^{-1} C\right)^{-1} D^T A^{-1}$$

from which the special case $k = 1$ of the "rank one update" follows as

$$\left(A + cd^T\right)^{-1} = A^{-1} - \frac{A^{-1}cd^T A^{-1}}{1 + d^T A^{-1}c} \qquad (A.67)$$

and, in particular for $A = I$,

$$\left(I + cd^T\right)^{-1} = I - \frac{cd^T}{1 + d^T c}$$

The classical example of (A.67) is the case where one element a_{ij} in an $n \times n$ matrix A is increased by a number x, which is established if $c = e_i$ and $d = xe_j$ so that $cd^T = xe_ie_j^T$ and

$$\left(A + xe_ie_j^T\right)^{-1} = A^{-1} - \frac{xA^{-1}e_ie_j^T A^{-1}}{1 + xe_j^T A^{-1}e_i} = A^{-1} - \frac{x\sum_{k=1}^n \left(A^{-1}\right)_{jk}\left(A^{-1}\right)_{ki}}{1 + x\left(A^{-1}\right)_{ji}}$$

$$= A^{-1} - \frac{x\left(A^{-2}\right)_{ji}}{1 + x\left(A^{-1}\right)_{ji}}$$

Hence, if the inverse A^{-1} is known, the inverse $\left(A + xe_ie_j^T\right)^{-1}$ is efficiently computed in terms of the elements of A^{-1}, which is useful for perturbation or sensitivity analyses.

220. *Cramer's rule.* The linear set of equations, $Ax = b$, has a unique solution

$x = A^{-1}b$ provided $\det A \neq 0$. If we write the matrix A in terms of its column vectors $a_k = (a_{1k}, a_{2k}, \ldots, a_{nk})$, then

$$A = \begin{bmatrix} a_1 & \cdots & a_{k-1} & a_k & a_{k+1} & \cdots & a_n \end{bmatrix}$$

Cramer's rule expresses the solution of $x = (x_1, x_2, \ldots, x_n)$ per component as

$$x_k = \frac{\det \begin{bmatrix} a_1 & \cdots & a_{k-1} & b & a_{k+1} & \cdots & a_n \end{bmatrix}}{\det A} \tag{A.68}$$

Indeed, the matrix A with the k-th column replaced by the vector b is

$$A_k = A + (b - a_k)\, e_k^T$$

where e_k is the k-th basis vector. Hence, ye_k^T equals the zero matrix with the k-th column replaced by the vector y and it has rank 1. Then,

$$\det A_k = \det \left(A + (b - a_k)\, e_k^T \right) = \det A \det \left(I + A^{-1} (b - a_k)\, e_k^T \right)$$

The "rank one update" formula (A.66), with $Ae_k = a_k$ and $e_k = A^{-1}a_k$, produces

$$\det \left(I + A^{-1} (b - a_k)\, e_k^T \right) = 1 + e_k^T A^{-1} (b - a_k) = 1 + e_k^T \left(A^{-1}b - A^{-1}a_k \right)$$
$$= 1 + e_k^T (x - e_k) = x_k$$

which demonstrates Cramer's formula (A.68).

221. *Expansion of the determinant of a product.*

Theorem 62 (Binet-Cauchy) *Let $C = AB$ where $A_{m \times n}$ and $B_{n \times m}$. Then,*

$$\det C = \sum_{1 \leq k_1 < k_2 < \cdots < k_m \leq n} \begin{vmatrix} a_{1k_1} & \cdots & a_{1k_m} \\ \vdots & \cdots & \vdots \\ a_{mk_1} & \cdots & a_{mk_m} \end{vmatrix} \begin{vmatrix} b_{k_1 1} & \cdots & b_{k_m 1} \\ \vdots & \cdots & \vdots \\ b_{k_1 m} & \cdots & b_{k_m m} \end{vmatrix} \tag{A.69}$$

Proof: See, e.g., Gantmacher (1959a, pp. 9-10). $\qquad\square$

If $B_{n \times m} = (A_{m \times n})^T$ (thus $b_{ij} = a_{ji}$), then the Binet-Cauchy formula (A.69) reduces to

$$\det AA^T = \sum_{k_1=1}^{n} \sum_{k_2=k_1+1}^{n} \cdots \sum_{k_m=k_{m-1}+1}^{n} \begin{vmatrix} a_{1k_1} & \cdots & a_{1k_m} \\ \vdots & \cdots & \vdots \\ a_{mk_1} & \cdots & a_{mk_m} \end{vmatrix}^2 \tag{A.70}$$

222. *The Cauchy identity.* The Cauchy identity

$$\sum_{j=1}^{n} x_j^2 \sum_{j=1}^{n} y_j^2 - \left(\sum_{j=1}^{n} x_j y_j \right)^2 = \frac{1}{2} \sum_{j=1}^{n} \sum_{k=1}^{n} (x_j y_k - x_k y_j)^2 \tag{A.71}$$

is the special case for the dimension $m = 2$ in the Binet-Cauchy Theorem 62.

Specifically[8], (A.70) reduces to (A.71) for the matrix $A_{2 \times n} = \begin{bmatrix} x^T \\ y^T \end{bmatrix}$, where x and y are $n \times 1$ vectors. Since the right-hand side in the Cauchy *identity* (A.71) is non-negative for real vectors x and y, the Cauchy-Schwarz *inequality* (A.12) is

$$\sum_{j=1}^{n} x_j^2 \sum_{j=1}^{n} y_j^2 \geq \left(\sum_{j=1}^{n} x_j y_j \right)^2 \tag{A.72}$$

The equality sign is only possible if and only if all $x_j = x$ and all $y_j = y$. With the scalar product $x^T y = \|x\|_2 \|y\|_2 \cos \theta_{x,y}$, where $\theta_{x,y}$ is the angle between the vector x and y, the Cauchy identity (A.71) is represented as

$$\|x\|_2 \|y\|_2 |\sin \theta_{x,y}| = \sqrt{\frac{1}{2} \sum_{j=1}^{n} \sum_{k=1}^{n} (x_j y_k - x_k y_j)^2}$$

Since $\text{Var}[X] = E\left[X^2\right] - (E\left[X\right])^2$, Cauchy's equality (A.71) shows for any random variable X in a graph, such as the degree D, that the variance equals

$$\text{Var}\left[X\right] = \frac{1}{n} \sum_{j=1}^{n} x_j^2 - \left(\frac{1}{n} \sum_{j=1}^{n} x_j \right)^2 = \sum_{j=2}^{n} \sum_{k=1}^{j-1} \left(\frac{x_j - x_k}{n} \right)^2 \tag{A.73}$$

where the last term sums the square of the difference in realizations of X over all pairs of nodes in the graph.

223. *The de Bruijn inequality.* If $a_1, a_2, \ldots, a_n$ are real numbers and $z_1, z_2, \ldots, z_n$ are complex numbers, then de Bruijn (1960) found the interesting inequality

$$\left| \sum_{j=1}^{n} a_j z_j \right|^2 \leq \frac{1}{2} \sum_{j=1}^{n} a_j^2 \left(\sum_{j=1}^{n} |z_j|^2 + \left| \sum_{j=1}^{n} z_j^2 \right| \right) \tag{A.74}$$

Since $\left| \sum_{j=1}^{n} z_j^2 \right| \leq \sum_{j=1}^{n} |z_j|^2$, the de Bruijn inequality (A.74) is sharper than

$$\left| \sum_{j=1}^{n} a_j z_j \right|^2 \leq \sum_{j=1}^{n} a_j^2 \sum_{j=1}^{n} |z_j|^2$$

which follows from the Cauchy-Schwarz inequality (A.12), because $\left| \sum_{j=1}^{n} a_j z_j \right| \leq \sum_{j=1}^{n} a_j |z_j|$.

Proof of (A.74): Let $S = \sum_{j=1}^{n} a_j z_j$ and denote $z_j = x_j + i y_j$. de Bruijn (1960) observes that $S e^{i\theta}$ and S have the same modulus $|S|$, so that we may assume that $\sum_{j=1}^{n} a_j z_j = \sum_{j=1}^{n} a_j x_j \geq 0$ and $\sum_{j=1}^{n} a_j y_j = 0$, which corresponds to

---

[8] The case $n = 2$ in the Cauchy identity (A.71), $\left(x_1^2 + x_2^2\right)\left(y_1^2 + y_2^2\right) = \left(x_1 y_1 + x_2 y_2\right)^2 + \left(x_1 y_2 - x_2 y_1\right)^2$, has played (Weil, 1984, p. 67-69) a role in Fermat's "Christmas 1640" Theorem, that every prime of the form $p = 4m + 1$, where m is a positive integer, can be written in one and only one way as sum of two squares.

a simultaneous rotation of $z_1, z_2, \ldots, z_n$ around the origin of the complex plane. Invoking the Cauchy-Schwarz inequality (A.12) then yields

$$\left| \sum_{j=1}^{n} a_j z_j \right|^2 = \left(\sum_{j=1}^{n} a_j x_j \right)^2 \le \sum_{j=1}^{n} a_j^2 \sum_{j=1}^{n} x_j^2$$

Since $|z_k|^2 = x_k^2 + y_k^2$ and $\operatorname{Re}\left(z_k^2\right) = x_k^2 - y_k^2$, we have that $x_k^2 = \frac{1}{2}\left(|z_k|^2 + \operatorname{Re}\left(z_k^2\right)\right)$. Together with $\sum_{k=1}^{n} \operatorname{Re}\left(z_k^2\right) = \operatorname{Re}\left(\sum_{k=1}^{n} z_k^2\right) \le \left|\sum_{k=1}^{n} z_k^2\right|$, the de Bruijn inequality (A.74) is proved. $\qquad\square$

224. *Vandermonde matrix.* The $n \times n$ Vandermonde matrix of the vector x is defined as[9]

$$V_n(x) = \begin{bmatrix} 1 & x_1 & x_1^2 & x_1^3 & \cdots & x_1^{n-1} \\ 1 & x_2 & x_2^2 & x_2^3 & \cdots & x_2^{n-1} \\ 1 & x_3 & x_3^2 & x_3^3 & \cdots & x_3^{n-1} \\ \vdots & \vdots & \vdots & \vdots & \vdots & \vdots \\ \vdots & \vdots & \vdots & \vdots & \vdots & \vdots \\ 1 & x_n & x_n^2 & x_n^3 & \cdots & x_n^{n-1} \end{bmatrix} \tag{A.75}$$

The Vandermonde determinant obeys the recursion

$$\det V_n(x) = \det V_{(n-1)}(x) \prod_{j=1}^{n-1} (x_n - x_j) \tag{A.76}$$

with $\det V_2(x) = x_2 - x_1$. Indeed, subtracting the last row from all previous rows and using the algebraic formula $x^k - y^k = (x - y)\sum_{j=0}^{k-1} x^{k-1-j} y^j$ yields

$$\det V_n(x) = \begin{vmatrix} 0 & x_1 - x_n & (x_1 - x_n)(x_1 + x_n) & \cdots & x_1^{n-1} - x_n^{n-1} \\ 0 & x_2 - x_n & (x_2 - x_n)(x_2 + x_n) & \cdots & x_2^{n-1} - x_n^{n-1} \\ 0 & x_3 - x_n & (x_3 - x_n)(x_3 + x_n) & \cdots & x_3^{n-1} - x_n^{n-1} \\ \vdots & \vdots & \vdots & \vdots & \vdots \\ 0 & x_{n-1} - x_n & (x_{n-1} - x_n)(x_{n-1} + x_n) & \vdots & \vdots \\ 1 & x_n & x_n^2 & \cdots & x_n^{n-1} \end{vmatrix}$$

After expanding the determinant as $(-1)^n$ times the cofactor of the last element of the first column, the resulting determinant is, after dividing each row r by the

---

[9] There are different ways to define the Vandermonde matrix, for instance, by organizing the powers of the vector x in rows (as in **art.** 242) instead of in columns, and by choosing the sequence of powers in either decreasing or increasing order.

factor $x_r - x_n$,

$$\det W_{n-1} = \begin{vmatrix} 1 & x_1 + x_n & x_1^2 + x_1 x_n + x_n^2 & \cdots & \sum_{j=0}^{n-2} x_n^{n-2-j} x_1^j \\ 1 & x_2 + x_n & x_2^2 + x_2 x_n + x_n^2 & \cdots & \sum_{j=0}^{n-2} x_n^{n-2-j} x_2^j \\ 1 & x_3 + x_n & x_3^2 + x_2 x_n + x_n^2 & \cdots & \sum_{j=0}^{n-2} x_n^{n-2-j} x_3^j \\ \vdots & \vdots & \vdots & \vdots & \vdots \\ 1 & x_{n-2} + x_n & x_{n-2}^2 + x_{n-2} x_n + x_n^2 & \vdots & \vdots \\ 1 & x_{n-1} + x_n & x_{n-1}^2 + x_{n-1} x_n + x_n^2 & \cdots & \sum_{j=0}^{n-2} x_n^{n-2-j} x_{n-1}^j \end{vmatrix}$$

A determinant remains unchanged by adding a column multiplied by some number α to another column. Since $\sum_{j=0}^{k-1} x^{k-1-j} y^j = x^{k-1} + y \sum_{j=0}^{k-2} x^{k-2-j} y^j$, we can subsequently multiply each but the last column k by x_n and subtract the result from the column $k+1$ to arrive at $W_{n-1} = V_{n-1}(x)$. This establishes the recursion (A.76). Iterating the recursion (A.76) results in

$$\det V_n(x) = \prod_{1 \leq i < j \leq n} (x_j - x_i) = \prod_{i=1}^{n} \prod_{j=i+1}^{n} (x_j - x_i) \tag{A.77}$$

The cofactor of the Vandermonde matrix $V_n(x)$ is elegantly derived as (B.26) in **art.** 305 using the Lagrange interpolation polynomial.

225. *Hadamard's inequality.* Consider the matrix $A = \begin{bmatrix} a_1 & a_2 & \cdots & a_n \end{bmatrix}$, with the vectors $\{a_k\}_{1 \leq k \leq n}$ as columns. The Hadamard inequality for the determinant, proved in Meyer (2000, p. 469), is

$$|\det A| \leq \prod_{k=1}^{n} \|a_k\|_2 = \prod_{k=1}^{n} \sqrt{\sum_{j=1}^{n} |a_{kj}|^2} \tag{A.78}$$

with equality only if all the vectors $a_1, a_2, \ldots, a_n$ are mutually orthonormal, i.e., if $(a_k)^T a_j = \delta_{kj}$ or, when complex $(a_k)^H a_j = \delta_{kj}$, for all pairs (k, j). As proved by Meyer (2000, p. 469), the volume v_n of an n-dimensional parallelepiped, a possibly *skewed* rectangular box generated by n independent vectors $a_1, a_2, \ldots, a_n$, equals $v_n = |\det A|$. This relation provides a geometrical interpretation of the determinant. Hadamard's inequality (A.78) asserts that the volume of an n-dimensional parallelepiped generated by the columns of A cannot exceed the volume of a rectangular box whose sides have length $\|a_k\|_2$. In general, an n-dimensional parallelepiped is skewed, i.e., its n independent, generating vectors $a_1, a_2, \ldots, a_n$ are not orthogonal, which geometrically explains Hadamard's inequality (A.78).

We apply the Hadamard inequality (A.78) to the Vandermonde determinant in **art.** 224, where the components of the vector x are ordered as $|x_1| > |x_2| > \ldots > |x_m| > 1 \geq |x_{m+1}| > \ldots > |x_n|$. After dividing the first m rows, corresponding to the components with absolute value larger than 1, by x_j^{n-1} for $1 \leq j \leq m$, we obtain

$$\det V_n\left(x\right) = \prod_{j=1}^{m} x_j^{n-1} \begin{vmatrix} x_1^{-(n-1)} & x_1^{-(n-2)} & x_1^{-(n-3)} & x_1^{-(n-4)} & \cdots & 1 \\ \vdots & \vdots & \vdots & \vdots & \vdots & \vdots \\ x_m^{-(n-1)} & x_m^{-(n-2)} & x_m^{-(n-3)} & x_m^{-(n-4)} & \cdots & 1 \\ 1 & x_{m+1} & x_{m+1}^2 & x_{m+1}^3 & \cdots & x_{m+1}^{n-1} \\ \vdots & \vdots & \vdots & \vdots & \vdots & \vdots \\ 1 & x_n & x_n^2 & x_n^3 & \cdots & x_n^{n-1} \end{vmatrix}$$

Since none of the elements in this determinant exceeds in absolute value unity, Hadamard's inequality (A.78) shows that $|\det V_n\left(x\right)| \le n^{\frac{n}{2}} \prod_{j=1}^{m} |x_j|^{n-1}$ with equality if and only if the row vectors are orthogonal. **Art.** 242 shows that orthogonality is only possible if all $x_j = e^{2\pi i \frac{j}{n}}$ corresponding to the zeros of $p_n\left(z\right) = a_n\left(z^n \pm 1\right)$. Using (A.77) and $|x_j| = 1$ yields the identity

$$\prod_{k=1}^{n} \prod_{j=k+1}^{n} \left| e^{\frac{2\pi i j}{n}} - e^{\frac{2\pi i k}{n}} \right| = n^{\frac{n}{2}} \tag{A.79}$$

226. *A Hadamard matrix.* An $n \times n$ Hadamard matrix H_n contains as elements either -1 and 1 and obeys $H_n H_n^T = n I_n$. The normalized matrix $X_n = \frac{1}{\sqrt{n}} H_n$ is an orthogonal matrix (**art.** 248), from which it follows that $\det H_n = n^{\frac{n}{2}}$. **Art.** 225 demonstrates that $\det H_n$ is maximal among all $n \times n$ matrices with elements in absolute value less than or equal to 1, which includes all orthogonal matrices. Any relabeling (permutation of rows and columns, **art.** 31) of a Hadamard matrix is again a Hadamard matrix; multiplying any row or column by -1 preserves the Hadamard properties.

Sylvester found a construction for *symmetric* Hadamard matrices $H_{2^k} = H_{2^{k-1}} \otimes H_2$, where $\otimes$ is the Kronecker product (**art.** 286) and $H_2 = \begin{bmatrix} 1 & 1 \\ 1 & -1 \end{bmatrix}$, that contain the u vector in the first column.

9.4 Function of a matrix

227. *Bézout's Theorem.* Consider an arbitrary matrix polynomial in λ,

$$F(\lambda) = \sum_{k=0}^{m} F_k \lambda^k$$

where all F_k are $n \times n$ matrices and $F_m \ne O$. Hence, any element of the $n \times n$ matrix $F\left(\lambda\right)$ is a polynomial $F_{ij}\left(\lambda\right) = \sum_{k=0}^{m} \left(F_k\right)_{ij} \lambda^k$ of at most order m in λ.

Any matrix polynomial $F(\lambda)$ can be right and left divided by another (non-zero) matrix polynomial $B(\lambda)$ in a unique way as proved in Gantmacher (1959a, Chapter IV). Hence, the left-quotient and left-remainder $F(\lambda) = B(\lambda)Q_L(\lambda) + R_L(\lambda)$ and the right-quotient and right-remainder $F(\lambda) = Q_R(\lambda)B(\lambda) + R_R(\lambda)$ are unique. Let us concentrate on the right-remainder in the case where $B(\lambda) = \lambda I - A$ is a linear polynomial in λ. Using Euclid's division scheme for polynomials (**art.** 309),

$$F(\lambda) = F_m \lambda^{m-1} (\lambda I - A) + (F_m A + F_{m-1}) \lambda^{m-1} + \sum_{k=0}^{m-2} F_k \lambda^k$$

$$= \left[F_m \lambda^{m-1} + (F_m A + F_{m-1}) \lambda^{m-2} \right] (\lambda I - A)$$

$$+ \left(F_m A^2 + F_{m-1} A + F_{m-2} \right) \lambda^{m-2} + \sum_{k=0}^{m-3} F_k \lambda^k$$

and continuing, we arrive at

$$F(\lambda) = \left[F_m \lambda^{m-1} + \cdots + \lambda^{k-1} \sum_{j=k}^{m} F_j A^{j-k} + \cdots + \sum_{j=1}^{m} F_j A^{j-1} \right] (\lambda I - A) + \sum_{j=0}^{m} F_j A^j$$

In summary, $F(\lambda) = Q_R(\lambda) (\lambda I - A) + R(\lambda)$ and similarly for the left-quotient and left-remainder with

$$Q_R(\lambda) = \sum_{k=1}^{m} \lambda^{k-1} \left(\sum_{j=k}^{m} F_j A^{j-k} \right) \quad Q_L(\lambda) = \sum_{k=1}^{m} \lambda^{k-1} \left(\sum_{j=k}^{m} A^{j-k} F_j \right)$$
$$R_R(\lambda) = \sum_{j=0}^{m} F_j A^j = F(A) \qquad R_L(\lambda) = \sum_{j=0}^{m} A^j F_j$$

$$\text{(A.80)}$$

and where the left- and right-remainder is independent of λ. The Generalized Bézout Theorem states that the polynomial $F(\lambda)$ is divisible by $(\lambda I - A)$ on the right (left) if and only if $F(A) = R_R(\lambda) = O$ (or $R_L(\lambda) = O$).

228. *The Cayley-Hamilton Theorem.* Operations with matrices are different from operations with scalars. Well-known examples of the difference are the non-commutativity of the matrix product and the fact that AB can be the null matrix O, although both $A \neq O$ and $B \neq O$. The Cayley-Hamilton Theorem is another example that leads to a remarkable consequence discussed in **art.** 233.

Theorem 63 (Cayley-Hamilton) *An $n \times n$ matrix A satisfies its own characteristic polynomial*

$$c_A(A) = O \tag{A.81}$$

where the characteristic polynomial is $c_A(\lambda) = \det(A - \lambda I) = \sum_{k=0}^{n} c_k \lambda^k$.

There exist several proofs of the Cayley-Hamilton Theorem. Due to the importance of the Cayley-Hamilton Theorem, valid for any $n \times n$ matrix A, we provide a general proof.

Proof: Applying the basic property of the adjugate (A.42) in **art.** 212 to the matrix $A - \lambda I$, yields, with $c_A(\lambda) = \det(A - \lambda I)$,

$$(A - \lambda I) \operatorname{adj}(A - \lambda I) = c_A(\lambda) I_n = \operatorname{adj}(A - \lambda I)(A - \lambda I) \tag{A.82}$$

Since the characteristic polynomial $c_A(\lambda) = \sum_{k=0}^{n} c_k \lambda^k$ is a polynomial of degree

n in λ, **art.** 227 demonstrates that[10] $Q(\lambda) = \mathrm{adj}(A - \lambda I) = \sum_{k=0}^{n-1} C_k \lambda^k$ must be a polynomial of at most degree $n-1$ in λ with $n \times n$ matrix coefficients. Hence,

$$(A - \lambda I)\,\mathrm{adj}\,(A - \lambda I) = (A - \lambda I)\sum_{k=0}^{n-1} C_k \lambda^k = \sum_{k=0}^{n-1} AC_k \lambda^k - \sum_{k=1}^{n} C_{k-1}\lambda^k$$

$$= AC_0 + \sum_{k=1}^{n-1}\left(AC_k - C_{k-1}\right)\lambda^k - C_{n-1}\lambda^n$$

Equating corresponding powers of λ in (A.82) yields

$$c_0 I_n = AC_0$$
$$c_k I_n = (AC_k - C_{k-1}) \qquad \text{for } 1 \le k \le n-1$$
$$c_n I_n = C_{n-1}$$

After multiplying the above equation of the coefficients of λ^k from the left by A^k, we obtain

$$\sum_{k=0}^{n} c_k A^k = AC_0 + \sum_{k=1}^{n-1}\left(A^{k+1}C_k - A^k C_{k-1}\right) - A^n C_{n-1} = 0$$

Hence, $c_A(A) = \sum_{k=0}^{n} c_k A^k = O$, which completes the proof. $\qquad\square$

229. *The minimal polynomial of a square matrix.* Let $m_{c_A}(z) = \sum_{k=0}^{l} b_k z^k$ denote the minimal polynomial, defined in **art.** 310, of the characteristic polynomial $c_A(z)$ of a matrix A and the degree of the minimal polynomial obeys $l \le n$, where l is the number of different eigenvalues of A.

A polynomial $f_a(z) = \sum_{k=0}^{a} f_k z^k$ is called an *annihilating polynomial* of the square matrix A if $f_a(A) = O$. The minimal polynomial $m_{c_A}(z) = \sum_{k=0}^{l} b_k z^k$ of degree l is the annihilating polynomial of A of least degree with highest coefficient $b_l = 1$. Consider the division

$$f_a(z) = m_{c_A}(z)\,q(z) + r(z)$$

where $r(z)$ is a polynomial in z of degree less than l and consider the corresponding matrix division

$$f_a(A) = m_{c_A}(A)\,q(A) + r(A)$$

Since $f_a(A) = O$ and $m_{c_A}(A) = O$, we conclude that $r(A) = O$ and that $r(z)$

---

[10] Given the matrix A, Gantmacher (1959a) describes a method due to Faddeev that simultaneously computes the coefficients p_k of the characteristic polynomial $\det(\lambda I - A) = \lambda^n - \sum_{k=0}^{n-1} p_{n-k}\lambda^k$ as well as the matrix coefficients B_k of the adjoint matrix $\widetilde{Q}(\lambda) = \sum_{k=0}^{n-1} B_{n-1-k}\lambda^k$, differently defined than ours in (A.84). Faddeev defines, for $1 \le k \le n$, the system $r_k = \frac{1}{k}\mathrm{trace}(A_k)$ and $B_k = A_k - r_k I_n$, where $A_k = AB_k$ and $A_1 = A$. A check is $B_n = A_n - r_n I_n = O$. The solution is $A_k = A^k - \sum_{j=1}^{k-1} r_{k-j} A^j$. After taking the trace of both sides of this solution and comparing the result with the Newton identities (B.9), we find that $r_k = p_k$ are the coefficients of the characteristic polynomial $\det(\lambda I - A) = \lambda^n - \sum_{k=0}^{n-1} p_{n-k}\lambda^k$.

is also an annihilating polynomial of A. But, the degree of $r(z)$ is lower than the degree of the minimal polynomial $m_{c_A}(z)$, which is impossible, else $r(z)$ should be the minimal polynomial. Hence, $r(z)$ must be zero. In conclusion, there holds: $f_a(A) = m_{c_A}(A) q(A)$ or *every annihilating polynomial $f_a(z)$ of a matrix A, obeying $f_a(A) = O$, is divisible, i.e. without remainder, by its minimal polynomial,* $m_{c_A}(z) | f_a(z)$. Consequently, since the characteristic polynomial is an annihilating polynomial by the Cayley-Hamilton Theorem 63, it holds that $m_{c_A}(z) | c_A(z)$.

Moreover, given the matrix A, its corresponding minimal polynomial $m_{c_A}(z)$ is unique. The uniqueness of the minimal polynomial also follows from the above argument. Indeed, if $n_{c_A}(z)$ were another minimal polynomial of A, then $n_{c_A}(z) | m_{c_A}(z)$ as well as $m_{c_A}(z) | n_{c_A}(z)$. Hence, $m_{c_A}(z) = \beta n_{c_A}(z)$, but the constant β must be one since the highest coefficient of a minimal polynomial is 1.

We remark that the coefficients of an annihilating polynomial are scalars. A general matrix polynomial $F(z)$, satisfying $F(A) = \sum_{k=0}^{m} F_k A^k = O$ where the coefficients F_k are $n \times n$ matrices, can be of lower degree than the minimal polynomial $m_{c_A}(z) = \sum_{k=0}^{l} b_k z^k$, i.e. $m < l$, as exemplified in **art.** 55.

230. *The adjoint matrix.* By the Generalized Bézout Theorem, the polynomial $F(\lambda) = g(\lambda)I - g(A)$ is divisible by $(\lambda I - A)$ because $F(A) = g(A)I - g(A) = O$. If $F(\lambda)$ is an ordinary polynomial (i.e. all coefficients F_k are scalars), then the right- and left-quotient and the remainders are equal, $Q_R(\lambda) = Q_L(\lambda) = Q(\lambda)$ and $R_R(\lambda) = R_L(\lambda) = R(\lambda)$,

$$F(\lambda) = Q(\lambda)(\lambda I - A) + R(\lambda) = (\lambda I - A)Q(\lambda) + R(\lambda)$$

Let $g(\lambda) = c_A(\lambda)$, then

$$c_A(\lambda)I - c_A(A) = Q(\lambda)(\lambda I - A) = (\lambda I - A)Q(\lambda) \tag{A.83}$$

The Cayley-Hamilton Theorem 63 states that $c_A(A) = O$, which indicates that $c_A(\lambda)I = Q(\lambda)(\lambda I - A)$ and also $c_A(\lambda)I = (\lambda I - A)Q(\lambda)$. Incidentally, the relation (A.83) also proves the Cayley-Hamilton Theorem 63, based on the property (A.82) of the adjugate matrix (**art.** 212). The two proofs illustrate the intimate relation between the adjugate and the Cayley-Hamilton Theorem 63.

The matrix

$$Q(\lambda) = \operatorname{adj}(A - \lambda I) = (\lambda I - A)^{-1} c_A(\lambda) \tag{A.84}$$

is called the adjoint matrix of A. Explicitly, from (A.80),

$$Q(\lambda) = \sum_{k=1}^{n} \lambda^{k-1} \left(\sum_{j=k}^{n} c_j A^{j-k} \right)$$

With $c_0 = \det A$ in (A.98), it holds that $Q(0) = -A^{-1} \det A = \sum_{j=1}^{n} c_j A^{j-1}$. The Cayley-Hamilton Theorem (A.81) and (A.98), $O = \sum_{j=1}^{n} c_j A^j + I \det A$, directly lead to the above polynomial form for the inverse matrix A^{-1} of a non-singular matrix A.

The main theoretical interest of the adjoint matrix $Q(\lambda)$ stems from its definition,

$$c_A(\lambda)I = Q(\lambda)(\lambda I - A) = (\lambda I - A)Q(\lambda)$$

In case $\lambda = \lambda_k$ is an eigenvalue of A, then $(\lambda_k I - A)Q(\lambda_k) = O$, which indicates by (1.3) and the commutative property $(\lambda I - A)Q(\lambda) = Q(\lambda)(\lambda I - A)$ that every non-zero column(row) of the adjoint matrix $Q(\lambda_k)$ is a right(left)-eigenvector belonging to the eigenvalue λ_k. In addition, by differentiation with respect to λ, we obtain

$$c'_A(\lambda)I = (\lambda I - A)Q'(\lambda) + Q(\lambda)$$

This demonstrates that, if $Q(\lambda_k) \neq O$, the eigenvalue λ_k is a simple root of $c_A(\lambda)$ and, conversely, if $Q(\lambda_k) = O$, the eigenvalue λ_k has higher multiplicity.

The adjoint matrix $Q(\lambda) = (\lambda I - A)^{-1} c_A(\lambda)$ is computed by observing that, on the Generalized Bézout Theorem, $r(\lambda, \mu) = \frac{c_A(\lambda) - c_A(\mu)}{\lambda - \mu}$ is divisible without remainder. By replacing λ and μ in this polynomial $r(\lambda, \mu)$ by λI and A respectively, $Q(\lambda) = r(\lambda I, A)$ readily follows.

231. Consider the arbitrary polynomial of degree l,

$$g(x) = g_0 \prod_{j=1}^{l}(x - \mu_j)$$

Substitute x by A, then $g(A) = g_0 \prod_{j=1}^{l}(A - \mu_j I)$. Since $\det(AB) = \det A \det B$ and $\det(kA) = k^n \det A$, we have $\det(g(A)) = g_0^n \prod_{j=1}^{l} \det(A - \mu_j I) = g_0^n \prod_{j=1}^{l} c_A(\mu_j)$. With $c_A(\lambda) = \prod_{k=1}^{n}(\lambda_k - \lambda)$ in (A.97),

$$\det(g(A)) = g_0^n \prod_{j=1}^{l}\prod_{k=1}^{n}(\lambda_k - \mu_j) = \prod_{k=1}^{n} g_0 \prod_{j=1}^{l}(\lambda_k - \mu_j) = \prod_{k=1}^{n} g(\lambda_k)$$

Let $g(x) = h(x) - \lambda$, then we arrive at the general result: for any polynomial $h(x)$, the eigenvalues of $h(A)$ are $h(\lambda_1), \ldots, h(\lambda_n)$ and the characteristic polynomial is

$$\det(h(A) - \lambda I) = \prod_{k=1}^{n}(h(\lambda_k) - \lambda) \tag{A.85}$$

which is a polynomial in λ of degree at most n. Since the result holds for an arbitrary polynomial, it should not surprise that, under appropriate conditions of convergence, it can be extended to infinite polynomials, in particular to the Taylor series of a complex function.

232. *A function of a matrix.* As proved in Gantmacher (1959a, Chapter V), if the power series of a function $f(z)$ around $z = z_0$,

$$f(z) = \sum_{j=0}^{\infty} f_j(z_0)(z - z_0)^j \quad \text{where} \quad f_j(z_0) = \frac{1}{j!}\left.\frac{d^j f(z)}{dz^j}\right|_{z=z_0} \tag{A.86}$$

converges for all z in the disc $|z - z_0| < R$, then

$$f(A) = \sum_{j=0}^{\infty} f_j(z_0)(A - z_0 I)^j \qquad (A.87)$$

provided that all eigenvalues of A lie within the region of convergence of (A.86), i.e., $|\lambda - z_0| < R$. For example,

$$
\begin{array}{ll}
(I - zA)^{-1} = \sum_{k=0}^{\infty} z^k A^k & \text{for } |z\lambda_k| < 1, \text{ all } 1 \le k \le n \\
e^{Az} = \sum_{k=0}^{\infty} \frac{z^k A^k}{k!} & \text{for all } A \\
\log A = \sum_{k=1}^{\infty} \frac{(-1)^{k-1}}{k}(A - I)^k & \text{for } |\lambda_k - 1| < 1, \text{ all } 1 \le k \le n
\end{array}
$$

The Taylor series of an analytic function can be differentiated and integrated within the region of convergence, which leads us to define other matrix functions. For example, when $|z\lambda_k| < 1$ for all $1 \le k \le n$,

$$\int (I - zA)^{-1}\, dz = \sum_{k=0}^{\infty} \frac{z^{k+1}}{k+1} A^k = A^{-1} \sum_{k=1}^{\infty} \frac{z^k}{k} A^k = A^{-1} \log(I - zA)$$

from which $\log(I - zA) = \int A(I - zA)^{-1}\, dz$, while, for all $\left|\frac{\lambda_k}{z}\right| < 1$, we find $\frac{d\log(zI - A)}{dz} = (zI - A)^{-1}$.

Expression (A.85) shows that the eigenvalues of e^{Az} are $e^{z\lambda_1}, \ldots, e^{z\lambda_1}$. Hence, the knowledge of the eigenstructure of a matrix A allows us to compute any function of A under the same convergence restrictions as complex numbers z.

233. *A function of a matrix is a polynomial.* Any function $f(z)$, that has a Taylor series (A.86) around some point z_0, can define (**art. 232**) the function $f(A)$ as a Taylor series (A.87) for any $n \times n$ matrix A, provided that the Taylor series (A.87) converges. In that case, the Taylor series of $f(z)$ consists of an infinite number of terms, except when $f(z) = p_m(z)$ is a polynomial of degree m.

The situation for $f(A)$ is surprisingly different: *if the Taylor series (A.87) of $f(A)$ converges and, hence, defines $f(A)$, then there is a polynomial $p_k(z)$ of degree $k \le n - 1$ such that $f(A) = p_k(A)$.* This remarkable property is a direct consequence of the Cayley-Hamilton Theorem (A.81): $A^n = -(-1)^n \sum_{k=0}^{n-1} c_k A^k$ so that each matrix A^{n+k} for any integer $k \ge 0$ and similarly each term $(A - z_0 I)^j$, for $j \ge n$, in the Taylor series (A.87) can be expressed as a polynomial in A of degree not exceeding $n - 1$, as illustrated in **art. 234** below.

234. *The function of a symmetric real matrix.* Using the vector notation (A.138) of the eigenvalue decomposition of a symmetric matrix A and (A.117), we have that

$$(A - z_0 I)^j = \sum_{k=1}^{n} (\lambda_k - z_0)^j x_k x_k^T$$

For any analytic function f that possesses a converging Taylor series around some point z_0, the function $f(A)$ is, with (A.87),

$$f(A) = \sum_{j=0}^{\infty} f_j(z_0)(A - z_0 I)^j = \sum_{j=0}^{\infty} f_j(z_0) \sum_{k=1}^{n} (\lambda_k - z_0)^j \, x_k x_k^T$$

$$= \sum_{k=1}^{n} \left(\sum_{j=0}^{\infty} f_j(z_0) \, (\lambda_k - z_0)^j \right) x_k x_k^T$$

Hence, provided that all eigenvalues of A lie within the radius of convergence of the Taylor series (A.86) around z_0, we find that

$$f(A) = \sum_{k=1}^{n} f(\lambda_k) x_k x_k^T \tag{A.88}$$

which indicates that the function f cannot map a symmetric matrix A into an asymmetric matrix, where $f(A) \neq (f(A))^T$.

Art. 233 demonstrates that $f(A) = p_{n-1}(A)$ for any $n \times n$ matrix A, if the minimal polynomial $m_{c_A}(z)$ is of degree $n - 1$, in which case all eigenvalues are distinct. Hence, if there exists a polynomial for which $f(\lambda_k) = p_{n-1}(\lambda_k)$, for all $1 \leq k \leq n$ eigenvalues of A, then

$$f(A) = \sum_{k=1}^{n} f(\lambda_k) x_k x_k^T = \sum_{k=1}^{n} p_{n-1}(\lambda_k) x_k x_k^T = p_{n-1}(A) \tag{A.89}$$

A polynomial $p_{n-1}(z)$ of degree $n-1$ that passes through a set of n different points $\{(\lambda_k, f(\lambda_k))\}_{1 \leq k \leq n}$ is precisely the Lagrange interpolation polynomial (B.20),

$$p_{n-1}(x) = \sum_{k=1}^{n} f(\lambda_k) \prod_{j=1; j \neq k}^{n} \frac{x - \lambda_j}{\lambda_k - \lambda_j}$$

studied in **art.** 303, and thus,

$$p_{n-1}(A) = \sum_{k=1}^{n} f(\lambda_k) \prod_{j=1; j \neq k}^{n} \frac{A - \lambda_j I}{\lambda_k - \lambda_j} \tag{A.90}$$

Substituting the relations in **art.** 303 into (A.90), the function $f(A)$ in (A.89) can be written explicitly as a polynomial of degree $n - 1$ in A,

$$f(A) = \sum_{k=0}^{n-1} c_k[f] A^k \tag{A.91}$$

where the coefficient $c_k[f]$, which depends on the function f and on the eigenvalues of A, is

$$c_k[f] = \frac{1}{k!} \sum_{m=1}^{n} \frac{f(\lambda_m)}{\prod_{j=1; j \neq m}^{n} (\lambda_m - \lambda_j)} \frac{d^k}{dx^k} \prod_{j=1; j \neq m}^{n} (x - \lambda_j) \Bigg|_{x=0}$$

Since both expressions (A.89) and (A.90) hold for any function f, we conclude that

$$x_k x_k^T = \prod_{j=1; j \neq k}^{n} \frac{A - \lambda_j I}{\lambda_k - \lambda_j} \tag{A.92}$$

Another proof of (A.92) follows from (A.88) with $f(x) = \frac{c_A(x)}{x - \lambda_k} = \prod_{j=1; j \neq k}^{n} (x - \lambda_j)$, where $c_A(x)$ is the characteristic polynomial of A,

$$\prod_{j=1; j \neq k}^{n} (A - \lambda_j) = \sum_{j=1}^{n} \frac{c_A(\lambda_j)}{\lambda_j - \lambda_k} x_j x_j^T = \lim_{x \to \lambda_k} \frac{c_A(x) - c_A(\lambda_k)}{x - \lambda_k} x_k x_k^T = c_A'(\lambda_k) x_k x_k^T$$

because $c_A'(\lambda_k) = \prod_{j=1; j \neq k}^{n} (\lambda_k - \lambda_j)$.

The above discussion has assumed that the eigenvalues of A are distinct, in order to straightforwardly apply the Lagrange interpolation (**art. 303**). However, when A has eigenvalues with multiplicity larger than one or when A is not symmetric nor diagonalizable, but has the Jordan form (**art. 239**), matrix polynomials $p_{n-1}(A)$ based on the spectrum of A can still be deduced. The analysis (see e.g. Gantmacher (1959a, Chapter V), Meyer (2000, Section 7.9)) becomes more complicated and is here omitted.

10

Eigensystem of a matrix

This chapter reviews general results about the eigensystem or spectrum of a square matrix A, the set of eigenvalues with their corresponding eigenvectors. The emphasis lies on symmetric matrices, $A = A^T$, for whom the spectral theory belongs to the pearls of linear algebra.

10.1 Eigenvalues and eigenvectors

235. The algebraic eigenproblem $Ax = \lambda x$ in (1.3) asks for the determination of the eigenvalue λ, a complex number, and the corresponding $n \times 1$ eigenvector x of an $n \times n$ matrix A for which the set of n homogeneous linear equations

$$
\begin{bmatrix}
a_{11} - \lambda & a_{12} & a_{13} & \cdots & a_{1n} \\
a_{21} & a_{22} - \lambda & a_{23} & \cdots & a_{2n} \\
a_{31} & a_{32} & a_{33} - \lambda & \cdots & a_{3n} \\
\vdots & \vdots & \vdots & \ddots & \vdots \\
a_{n1} & a_{n2} & a_{n3} & \cdots & a_{nn} - \lambda
\end{bmatrix}
\begin{bmatrix}
x_1 \\
x_2 \\
x_3 \\
\vdots \\
x_n
\end{bmatrix} = 0
\tag{A.93}
$$

in n unknowns $x_1, x_2, \ldots, x_n$ has a non-zero solution. Clearly, the zero vector $x = 0$ is always a solution of (1.3). A non-zero solution of eigenvalue equation $Ax = \lambda x$ is only possible if and only if the matrix $A - \lambda I$ is singular, that is,

$$
\det (A - \lambda I) = 0
\tag{A.94}
$$

As shown[1] in **art.** 211, this determinant $c_A(\lambda) = \det (A - \lambda I)$ can be expanded as a polynomial in λ of degree n,

$$
c_A(\lambda) = \sum_{k=0}^{n} c_k \lambda^k = c_n \lambda^n + c_{n-1} \lambda^{n-1} + \cdots + c_1 \lambda + c_0 = 0
\tag{A.95}
$$

---

[1] Another proof is given in Meyer (2000, p. 495).

which is called the characteristic or eigenvalue polynomial of the matrix A. Apart from $c_n = (-1)^n$, the coefficients for $0 \leq k < n$ are

$$c_k = (-1)^k \sum_{all} M_{n-k} \tag{A.96}$$

and M_k is a principal minor[2], given explicitly in (A.35) in **art. 211**. Meyer (2000, p. 504) mentions the Leverrier–Souriau–Frame algorithm that computes the coefficients c_k of characteristic polynomial $c_A(\lambda)$ in (A.95) as

$$c_k = -\frac{\text{trace}(AB_{k-1})}{k}$$

where the matrix B_k obeys $B_0 = I$ and the recursion $B_k = -\frac{\text{trace}(AB_{k-1})}{k}I + AB_{k-1}$ for $k = 1, 2, \ldots, n$.

Since a polynomial of degree n has n complex zeros (**art. 291**), the $n \times n$ square matrix A possesses n eigenvalues $\lambda_1, \lambda_2, \ldots, \lambda_n$, not all necessarily distinct. In general, the characteristic polynomial can be written in product form (B.1),

$$c_A(\lambda) = \prod_{k=1}^{n} (\lambda_k - \lambda) \tag{A.97}$$

Since $c_A(\lambda) = \det(A - \lambda I)$, it follows from (A.95) and (A.97) that, for $\lambda = 0$,

$$\det A = c_0 = \prod_{k=1}^{n} \lambda_k \tag{A.98}$$

Hence, if $\det A = 0$, there is at least one zero eigenvalue. Also (see **art. 211**),

$$(-1)^{n-1} c_{n-1} = \sum_{k=1}^{n} \lambda_k = \text{trace}(A) \tag{A.99}$$

and

$$c_1 = -\sum_{k=1}^{n} \prod_{j=1; j \neq k}^{n} \lambda_j = -\det A \sum_{k=1}^{n} \frac{1}{\lambda_k} \tag{A.100}$$

For any eigenvalue λ, the linear set (A.93) has at least one non-zero eigenvector x. Furthermore, if x is a non-zero eigenvector, also kx is a non-zero eigenvalue. Therefore, eigenvectors are often normalized, for instance, a probabilistic eigenvector has the sum of its components equal to 1 or a norm $\|x\|_1 = 1$ as defined in (A.9). The most common normalization is the Euclidean norm $\|x\|_2^2 = x^T x = 1$.

236. *Multiplicity of eigenvalues.* If the same eigenvalue λ_k reappears m_k times as

---

[2] A principal minor M_k is the determinant of a principal $k \times k$ submatrix $M_{k \times k}$ obtained by deleting the same $n - k$ rows and columns in A. Hence, the main diagonal elements $(M_{k \times k})_{ii}$ are k elements of main diagonal elements $\{a_{ii}\}_{1 \leq i \leq n}$ of A.

a zero of the characteristic polynomial $c_A(\lambda)$ in (A.95), then (A.97) can be written as

$$c_A(\lambda) = \prod_{j=1}^{l} (\lambda_j - \lambda)^{m_j} \qquad \text{with} \quad \sum_{j=1}^{l} m_j = n$$

and m_k is called the *algebraic multiplicity* of the eigenvalue λ_k of the $n \times n$ matrix A.

If the rank of $A - \lambda I$ is less than $n - 1$, there will be more than one independent eigenvector belonging to the eigenvalue λ. The *geometric multiplicity* of the eigenvalue λ_k of the $n \times n$ matrix A is defined as $n-$ rank$(A - \lambda_k I)$, which equals the number of linearly independent eigenvectors associated with the eigenvalue λ_k. For any $n \times n$ complex matrix, the algebraic multiplicity of an eigenvalue is larger than or equal to its geometric multiplicity (Meyer, 2000, p. 511). If a matrix is diagonalizable as any symmetric matrix; **art.** 247, then the algebraic and geometric multiplicity of any eigenvalue are equal (Meyer, 2000, p. 512).

Multiplicity of eigenvalues seriously complicates the eigenvalue problem. In the sequel, we omit a detailed discussion on multiple eigenvalues and refer to Wilkinson (1965).

237. *Eigenproblem of the transpose A^T.* The eigenvalue equation (1.3) of the transposed matrix A^T,

$$A^T y = \lambda y \tag{A.101}$$

is of singular importance. The determinant of a matrix is equal to the determinant of its transpose (**art.** 209). This property $\det(A^T - \lambda I) = \det(A - \lambda I)$ shows that the eigenvalues of A and A^T are the same.

However, the eigenvectors are, in general, different. Transposing (A.101) yields

$$y^T A = \lambda y^T \tag{A.102}$$

The vector y_j^T is therefore called the left-eigenvector of A belonging to the eigenvalue λ_j, whereas x_j in $Ax_j = \lambda_j x_j$ is called the right-eigenvector belonging to the same eigenvalue λ_j. An important relation between the left- and right-eigenvectors of a matrix A is, for $\lambda_j \neq \lambda_k$,

$$y_j^T x_k = 0 \tag{A.103}$$

Indeed, left-multiplying $Ax_k = \lambda_k x_k$ in (1.3) by y_j^T, $y_j^T A x_k = \lambda_k y_j^T x_k$, and similarly right-multiplying $y_j^T A = \lambda_j y_j^T$ in (A.102) by x_k, $y_j^T A x_k = \lambda_j y_j^T x_k$, leads, after subtraction, to $0 = (\lambda_k - \lambda_j) y_j^T x_k$ and (A.103) follows.

Since eigenvectors can be complex and since $y_j^T x_k = x_k^T y_j$, the expression $y_j^T x_k$ is not an inner-product that is always real and for which $y_j^T x_k = (x_k^T y_j)^*$ holds. However, (A.103) expresses that the sets of left- and right-eigenvectors are orthogonal if $\lambda_j \neq \lambda_k$.

238. If A has n distinct eigenvalues, then the n eigenvectors are linearly independent and span the whole n-dimensional space. The proof is by *reductio ad absurdum*. Assume that s is the smallest number of linearly dependent eigenvectors labeled by the first s smallest indices. Linear dependence then means that

$$\sum_{k=1}^{s} \alpha_k x_k = 0 \tag{A.104}$$

where $\alpha_k \neq 0$ for $1 \leq k \leq s$. Left-multiplying by A and invoking the eigenvalue equation (1.3) yields

$$\sum_{k=1}^{s} \alpha_k \lambda_k x_k = 0 \tag{A.105}$$

On the other hand, multiplying (A.104) by λ_s and subtracting from (A.105) leads to

$$\sum_{k=1}^{s-1} \alpha_k \left(\lambda_k - \lambda_s \right) x_k = 0,$$

which, because all eigenvalues are distinct, implies that there is a smaller set of $s-1$ linearly depending eigenvectors. This contradicts the initial hypothesis.

This important property has a number of consequences. First, it applies to left- as well as to right-eigenvectors. Relation (A.103) then shows that the sets of left- and right-eigenvectors form a bi-orthogonal system with $y_k^T x_k \neq 0$. For, if x_k were orthogonal to y_k, thus $y_k^T x_k = 0$, then (A.103) demonstrates that x_k would be orthogonal to all left-eigenvectors y_j. Since the set of left-eigenvectors span the n dimensional vector space, it would mean that the n-dimensional vector x_k would be orthogonal to the whole n-space, which is impossible because x_k is not the null vector. Second, any n-dimensional vector can be written in terms of either the left- or right-eigenvectors.

239. Let us denote by X the matrix with the right-eigenvector x_j in column j and by Y^T the matrix with the left-eigenvector y_k^T in row k. If the right- and left-eigenvectors are scaled such that $y_k^T x_k = 1$, for all $1 \leq k \leq n$, then (A.103) leads to

$$Y^T X = I \tag{A.106}$$

Thus, the matrix Y^T is the inverse of the matrix X. Furthermore, for any right-eigenvector, the eigenvalue equation $Ax = \lambda x$ in (1.3) holds, rewritten in matrix form,

$$AX = X \, \text{diag}(\lambda) \tag{A.107}$$

where the $n \times 1$ eigenvalue vector is $\lambda = (\lambda_1, \lambda_1, \ldots, \lambda_n)$. Left-multiplying by $X^{-1} = Y^T$ yields the similarity transform of matrix A,

$$X^{-1}AX = Y^T AX = \text{diag}(\lambda) \tag{A.108}$$

Thus, when the eigenvalues of A are distinct, there exists a similarity transform $H^{-1}AH$ that reduces A to diagonal form.

In many applications, similarity transforms by a matrix H are applied to simplify matrix problems. The only condition is that the inverse H^{-1} of the matrix H must exist. Indeed, if $Ax = \lambda x$, then $\lambda H^{-1}x = H^{-1}Ax = (H^{-1}AH)H^{-1}x$. Thus, a similarity transform preserves the eigenvalues; the matrix $H^{-1}AH$ possesses the same eigenvalues as A, while the eigenvectors x of A are transformed to $H^{-1}x$.

When A has multiple eigenvalues, it may be impossible to reduce A to a diagonal form by similarity transforms. Instead of a diagonal form, the most compact form when A has r distinct eigenvalues each with multiplicity m_j such that $\sum_{j=1}^{r} m_j = n$ is the Jordan canonical form C,

$$
C = \begin{bmatrix}
C_{m_1}(\lambda_1) \\
& C_{m_2}(\lambda_1) \\
& & \vdots \\
& & & C_{m_{r-1}}(\lambda_{r-1}) \\
& & & & C_{m_r}(\lambda_r)
\end{bmatrix}
$$

where $C_m(\lambda)$ is an $m \times m$ submatrix of the form

$$
C_m(\lambda) = \begin{bmatrix}
\lambda & 1 & 0 & \cdots & 0 \\
0 & \lambda & 1 & 0 & \cdots \\
\vdots & \vdots & \vdots & \vdots & \vdots \\
0 & \cdots & 0 & \lambda & 1 \\
0 & \cdots & 0 & 0 & \lambda
\end{bmatrix}
$$

The number of independent eigenvectors is equal to the number of submatrices in C. If an eigenvalue λ has multiplicity m, there can be one large submatrix $C_m(\lambda)$, but also a number k of smaller submatrices $C_{b_j}(\lambda)$ such that $\sum_{j=1}^{k} b_j = m$. This illustrates, as mentioned in **art. 235**, the much higher complexity of the eigenproblem in case of multiple eigenvalues.

240. *Frequency interpretation of the eigenvalue equation.* The dependence on the parameter λ in the eigenvalue equation (1.3) in **art. 235** is made explicit in

$$
Ax(\lambda) = \lambda x(\lambda) \tag{A.109}
$$

where a non-zero vector $x(\lambda)$ only satisfies this linear equation if λ is an eigenvalue of A such that the eigenvector $x_j = x(\lambda_j)$. We can interpret λ as a frequency that ranges continuously over all real numbers. This "frequency" interpretation of the eigenvalue equation will be exploited in **art. 241**, where the application of calculus to (A.109) is illustrated, and in **art. 249**.

241. *Principal vector of grade m.* Invoking Leibniz' rule to the parameterized eigenvalue equation (A.109) of the $n \times n$ matrix A in **art. 240**, the m-th derivative of both sides of $Ax(\lambda) = \lambda\, x(\lambda)$ with respect to λ is

$$A\frac{d^l x(\lambda)}{d\lambda^l} = \sum_{k=0}^{l} \binom{l}{k}\frac{d^k}{d\lambda^k}(\lambda)\frac{d^{l-k}}{d\lambda^{l-k}}x(\lambda) = \lambda\frac{d^l x(\lambda)}{d\lambda^l} + l\frac{d^{l-1}x(\lambda)}{d\lambda^{l-1}}$$

so that, for any integer $l \geq 1$,

$$(A - \lambda I)\frac{d^l x(\lambda)}{d\lambda^l} = l\frac{d^{l-1}x(\lambda)}{d\lambda^{l-1}} \tag{A.110}$$

Explicitly, denoting $x^{(l)}(\lambda) = \frac{d^l x(\lambda)}{d\lambda^l}$ and $x^{(0)}(\lambda) = x(\lambda)$, we obtain the sequence for $l = 0, 1, \ldots, m$,

$$(A - \lambda I)\,x(\lambda) = 0, \quad (A - \lambda I)\,x^{(1)}(\lambda) = x(\lambda), \quad (A - \lambda I)\,x^{(2)}(\lambda) = 2x^{(1)}(\lambda), \quad \cdots$$

up to

$$(A - \lambda I)\,x^{(m)}(\lambda) = mx^{(m-1)}(\lambda) \tag{A.111}$$

Multiplying both sides in (A.111) with $(A - \lambda I)^{k-1}$,

$$(A - \lambda I)^k\,x^{(m)}(\lambda) = m\,(A - \lambda I)^{k-1}\,x^{(m-1)}(\lambda)$$

using (A.111) to the right-hand side iteratively p-times

$$(A - \lambda I)^k\,x^{(m)}(\lambda) = m\,(A - \lambda I)^{k-1}\,x^{(m-1)}(\lambda) = m\,(m-1)\,(A - \lambda I)^{k-2}\,x^{(m-2)}(\lambda)$$
$$= m\,(m-1)\,(m-2)\,(A - \lambda I)^{k-3}\,x^{(m-3)}(\lambda) = \cdots$$

yields

$$(A - \lambda I)^k\,x^{(m)}(\lambda) = \frac{m!}{(m-p)!}\,(A - \lambda I)^{k-p}\,x^{(m-p)}(\lambda) \tag{A.112}$$

Choose $p = m$ and $x^{(0)}(\lambda) = x(\lambda)$, then $(A - \lambda I)^k\,x^{(m)}(\lambda) = m!\,(A - \lambda I)^{k-m}\,x(\lambda)$. Subsequently with $k = m$, we find that

$$(A - \lambda I)^m\,x^{(m)}(\lambda) = m!x(\lambda) \tag{A.113}$$

and, from $(A - \lambda I)\,x(\lambda) = 0$, that

$$(A - \lambda I)^{m+1}\,x^{(m)}(\lambda) = 0 \tag{A.114}$$

If λ is not an eigenvalue so that $A - \lambda I$ is of rank n and invertible, then (A.114) and (A.113) show that $x^{(m)}(\lambda) = 0$ and $x(\lambda) = 0$, while the recursion (A.111) further tells that all higher order derivatives vanish, $x^{(j)}(\lambda) = 0$ for $0 \leq j \leq m$. If λ is an eigenvalue, the vector $x^{(m)}(\lambda)$ can be different from the zero vector and orthogonal to all the row vectors of $(A - \lambda I)^{m+1}$.

Theorem 64 *The set of vectors $\{x(\lambda), x^{(1)}(\lambda), x^{(2)}(\lambda), \ldots, x^{(m)}(\lambda)\}$ is linearly independent.*

Proof: Assume, on the contrary, that these vectors are dependent, then

$$b_0 x(\lambda) + b_1 x^{(1)}(\lambda) + b_2 x^{(2)}(\lambda) + \ldots + b_m x^{(m)}(\lambda) = \sum_{j=0}^{m} b_j x^{(j)}(\lambda) = 0$$

and not all b_j are zero. Left-multiplying both sides with $(A - \lambda I)^m$ and taking (A.114) into account that $(A - \lambda I)^{m+j} x^{(j)}(\lambda) = 0$ for any $j \geq 1$ and $m \geq 1$ leads to

$$b_m (A - \lambda I)^m x^{(m)}(\lambda) = 0$$

and (A.113) indicates that b_m must be zero. Next, we repeat the argument and left-multiply both sides with $(A - \lambda I)^{m-1}$, which leads us to conclude that $b_{m-1} = 0$. Continuing in this way shows that each coefficient $b_j = 0$ for $0 \leq j \leq n$, which contradicts the assumption and proves the Theorem 64. □

Let us now consider the integer $m = n$, equal to the dimensions of the matrix A. From (A.112), we define for $1 \leq k \leq n$ the vectors

$$y_k = (A - \lambda I)^k x^{(n)}(\lambda)$$

that satisfy

$$y_k = \frac{n!}{(n-p)!} (A - \lambda I)^{k-p} x^{(n-p)}(\lambda)$$

while relation (A.113) shows that $y_n = n! x(\lambda)$. Hence, any vector y_k is generated by the vector $x^{(n)}(\lambda)$ and Theorem 64 states that the set $\{y_1, y_2, \ldots, y_n\}$ is linearly independent and thus spans the n-dimensional space. In the classical eigenvalue theory (Wilkinson, 1965, p. 43), the vector z satisfying $(A - \lambda_k I)^{n+1} z = 0$ is called a *principal vector of grade* $n+1$ corresponding to eigenvalue λ_k. Theorem 64 and (A.114) show that $z = \beta x^{(n)}(\lambda)$, for any non-zero number β.

We now concentrate on eigenvalues. Left-multiplying (A.113) by $x^T(\xi)$ yields

$$n! x^T(\xi) x(\lambda) = x^T(\xi) (A - \lambda I)^n x^{(n)}(\lambda)$$

If A is a symmetric matrix and ξ is an eigenvalue of A, then $x^T(\xi) (A - \lambda I)^n = x^T(\xi) (\xi - \lambda)^n$, so that

$$\delta_{\xi\lambda} = x^T(\xi) x(\lambda) = \frac{(\xi - \lambda)^n}{n!} x^T(\xi) x^{(n)}(\lambda)$$

Hence, $x^T(\xi) x^{(n)}(\lambda) = 0$ for all $n \geq 0$, if the eigenvalue ξ is different from the eigenvalue λ. However, if $\xi = \lambda$, an inconsistency appears when $n > 0$, which implies that a *principal vector* $x^{(n)}(\lambda)$ *of grade* $n+1$ with $n > 0$ does not exist for symmetric matrices. Indeed, for symmetric matrices, the set of eigenvectors $\{x_m\}_{1 \leq m \leq N}$ spans the entire space, as demonstrated in art. 247 below, so that $x^{(n)}(\lambda) = 0$ for $n \geq 1$, because a non-zero vector cannot be orthogonal to all eigenvectors. Hence, a *principal vector* $x^{(m)}(\lambda)$ *of grade* $m+1$ with $m > 0$ only exists for asymmetric matrices and may be helpful to construct an orthogonal set of vectors when degeneracy occurs as in Jordan forms in **art**. 239.

242. *Companion matrix.* The companion matrix of a polynomial $p_n(z) = \sum_{k=0}^{n} a_k z^k$ is defined[3] as

$$
C = \begin{bmatrix}
-\frac{a_{n-1}}{a_n} & -\frac{a_{n-2}}{a_n} & \cdots & -\frac{a_1}{a_n} & -\frac{a_0}{a_n} \\
1 & 0 & \cdots & 0 & 0 \\
0 & 1 & \cdots & 0 & 0 \\
\vdots & \vdots & \vdots & \vdots & \vdots \\
0 & 0 & \cdots & 1 & 0
\end{bmatrix}
$$

The basic property of the companion matrix C is

$$
\det(C - \lambda I) = (-1)^n \frac{p_n(\lambda)}{a_n} \tag{A.115}
$$

Indeed, in

$$
\det(C - \lambda I) = \begin{vmatrix}
-\frac{a_{n-1}}{a_n} - \lambda & -\frac{a_{n-2}}{a_n} & \cdots & -\frac{a_1}{a_n} & -\frac{a_0}{a_n} \\
1 & -\lambda & \cdots & 0 & 0 \\
0 & 1 & \cdots & 0 & 0 \\
\vdots & \vdots & \vdots & \vdots & \vdots \\
0 & 0 & \cdots & 1 & -\lambda
\end{vmatrix}
$$

multiply the first column by λ^{n-1}, the second column by λ^{n-2}, and so on, and add them to the last column. The resulting last column elements are zero, except for that in the first row, which is $-\frac{p_n(\lambda)}{a_n}$. The corresponding cofactor is one, which proves (A.115). The inverse C^{-1} of the companion matrix C is

$$
C^{-1} = \begin{bmatrix}
0 & 1 & 0 & \cdots & 0 \\
0 & 0 & 1 & \cdots & 0 \\
\vdots & \vdots & \vdots & \vdots & \vdots \\
0 & 0 & 0 & \cdots & 1 \\
-\frac{a_n}{a_0} & -\frac{a_{n-1}}{a_0} & -\frac{a_{n-2}}{a_0} & \cdots & -\frac{a_1}{a_0}
\end{bmatrix}
$$

The companion matrix of the characteristic polynomial (A.95) of A is defined as

$$
C = \begin{bmatrix}
(-1)^{n-1}c_{n-1} & (-1)^{n-1}c_{n-2} & \cdots & (-1)^{n-1}c_1 & (-1)^{n-1}c_0 \\
1 & 0 & \cdots & 0 & 0 \\
0 & 1 & \cdots & 0 & 0 \\
\vdots & \vdots & \vdots & \vdots & \vdots \\
0 & 0 & \cdots & 1 & 0
\end{bmatrix}
$$

such that $\det(C - \lambda I) = c_A(\lambda)$. If A has distinct eigenvalues, A as well as C are similar to $\mathrm{diag}(\lambda_i)$. It has been shown in **art**. 239 that the similarity transform H

---

[3] Other variants with the first row replaced to the last column also appear in the literature.

for A equals $H = X$. The similarity transform for C is the Vandermonde matrix $V_n(\lambda)$ in **art.** 224, where

$$
V_n(x) = \begin{bmatrix}
x_1^{n-1} & x_2^{n-1} & \cdots & x_{n-1}^{n-1} & x_n^{n-1} \\
x_1^{n-2} & x_2^{n-2} & \cdots & x_{n-1}^{n-2} & x_n^{n-2} \\
\vdots & \vdots & \cdots & \vdots & \vdots \\
x_1 & x_2 & \vdots & x_{n-1} & x_n \\
1 & 1 & \cdots & 1 & 1
\end{bmatrix}
$$

Indeed,

$$
V_n(\lambda)\text{diag}(\lambda_i) = \begin{bmatrix}
\lambda_1^n & \lambda_2^n & \cdots & \lambda_{n-1}^n & \lambda_n^n \\
\lambda_1^{n-1} & \lambda_2^{n-1} & \cdots & \lambda_{n-1}^{n-1} & \lambda_n^{n-1} \\
\vdots & \vdots & \cdots & \vdots & \vdots \\
\lambda_1^2 & \lambda_2^2 & \vdots & \lambda_{n-1}^2 & \lambda_n^2 \\
\lambda_1 & \lambda_2 & \cdots & \lambda_{n-1} & \lambda_n
\end{bmatrix}
$$

while

$$
CV_n(\lambda) = \begin{bmatrix}
(-1)^{n-1}c_A(\lambda_1) + \lambda_1^n & (-1)^{n-1}c_A(\lambda_2) + \lambda_2^n & \cdots & (-1)^{n-1}c_A(\lambda_n) + \lambda_n^n \\
\lambda_1^{n-1} & \lambda_2^{n-1} & \cdots & \lambda_n^{n-1} \\
\vdots & \vdots & \cdots & \vdots \\
\lambda_1^2 & \lambda_2^2 & \vdots & \lambda_n^2 \\
\lambda_1 & \lambda_2 & \cdots & \lambda_n
\end{bmatrix}
$$

Since $c_A(\lambda_j) = 0$, it follows that $CV_n(\lambda) = V_n(\lambda)\text{diag}(\lambda_i)$, which demonstrates the claim. Hence, the eigenvector x_k of C belonging to eigenvalue λ_k is

$$
x_k^T = \begin{bmatrix} \lambda_k^{n-1} & \lambda_k^{n-2} & \cdots & \lambda_k & 1 \end{bmatrix}
$$

The Vandermonde matrix $V_n(\lambda)$ is non-singular if all eigenvalues are distinct (see also **art.** 224). In the case that $\det V_n(\lambda) \neq 0$, the matrix $V_n(\lambda)$ is of rank n, implying that all eigenvectors are linearly independent. The eigenvectors are only orthogonal if $x_k^T x_m = 0$ for each pair (k, m) with $k \neq m$. In other words, if

$$
0 = \sum_{j=1}^{n} (\lambda_k)^j (\lambda_m)^j = \lambda_k \lambda_m \frac{(\lambda_k \lambda_m)^n - 1}{\lambda_k \lambda_m - 1}
$$

The solution is $\lambda_k \lambda_m = e^{\frac{2\pi i l}{n}}$ for $l = 1, 2, \ldots, n-1$, which implies that each of the n eigenvalues $\{\lambda_k\}_{1 \leq k \leq n}$ must be an n-th distinct root of unity and that the associated polynomial to the companion matrix is $p_n(z) = a_n(z^n \pm 1)$.

The first component or row in the eigenvalue equation $Cx = \lambda x$ expresses explicitly the root equation $c_A(\lambda) = 0$ of the polynomial

$$
(Cx_k)_1 - (\lambda_k x_k)_1 = -c_A(\lambda_k) = 0 \tag{A.116}
$$

and any other row is an identity. If C has an eigenvalue λ_k of multiplicity m, then λ_k satisfies $c_A(\lambda_k) = c'_A(\lambda_k) = \ldots = c_A^{(m-1)}(\lambda_k) = 0$. The first equality is equivalent to (A.116). The others are similarly derived by differentiating (A.116) with respect to λ_k such that, for $1 \le j \le m-1$,

$$\left(Cx_k^{(j)}\right)_1 - \left(jx_k^{(j-1)}\right)_1 - \left(\lambda_k x_k^{(j)}\right)_1 = -c_A^{(j)}(\lambda_k) = 0$$

Hence, if λ_k is a zero with multiplicity m, then $Cx_k = \lambda_k x_k$, where x_k is the eigenvector and the other $2 \le j \le m$ equations are $Cy_j = \lambda_k y_j + y_{j-1}$, where $y_j = \frac{x_k^{(j-1)}}{(j-1)!}$ is a generalized eigenvector (**art. 241**),

$$y_j^T = \begin{bmatrix} 0 & 0 & \ldots & 1 & \binom{j}{j-1}\lambda_k & \ldots & \binom{n-1}{j-1}\lambda_k^{n-j} \end{bmatrix}$$

which has a 1 in the j-th position. Clearly, with this notation, $x_k = y_1$. Moreover, the set of the eigenvectors and $m-1$ generalized eigenvectors are independent because the $m \times n$ matrix formed by their components has rank m.

243. *Powers of a matrix and eigenvalues.* When left-multiplying (1.3), we obtain

$$A^2 x = \lambda A x = \lambda^2 x$$

and, in general, for any integer $k \ge 0$,

$$A^k x = \lambda^k x \tag{A.117}$$

Since an eigenvalue λ satisfies its characteristic polynomial $c_A(\lambda) = \sum_{k=0}^{n} c_k \lambda^k = 0$, we directly find from (A.117) that $c_A(A)x = 0$. Only if the set of all eigenvectors $x_1, x_2, \ldots, x_n$ spans the n-dimensional space and forms a basis, thus only for $n \times n$ diagonalizable matrices, then $c_A(A) \begin{bmatrix} x_1 & x_2 & \cdots & x_n \end{bmatrix} = 0$ implies $c_A(A) = O$, which demonstrates the Cayley-Hamilton Theorem 63 (**art. 228**).

If A has no zero eigenvalue, i.e., A^{-1} exists, then left-multiplying (1.3) with A^{-1} yields $A^{-1}x = \lambda^{-1}x$. We apply (A.117) to the matrix A^{-1} and conclude that

$$A^{-k} x = \lambda^{-k} x$$

In other words, if the inverse matrix A^{-1} exists, then equation (A.117) is valid for any integer, positive as well as negative.

Combining (A.117) and (A.99) implies that

$$\text{trace}\left(A^k\right) = \sum_{j=1}^{n} \lambda_j^k \tag{A.118}$$

244. *Power method.* Let $x_1, x_2, \ldots, x_n$ denote the complete set of eigenvectors of A. For example, if A has *distinct* eigenvalues $\lambda_1, \lambda_2, \ldots, \lambda_n$, then **art. 238** demonstrates that the set $x_1, x_2, \ldots, x_n$ of linearly independent vectors spans the n-dimensional space. Any symmetric matrix possesses an orthogonal set of eigenvectors by Theorem 68 in **art. 247**. For such matrices, any other vector w can be written as a linear combination,

$$w = \sum_{j=1}^{n} \alpha_j x_j$$

Then, for any integer k and using (A.117) yields

$$A^k w = \sum_{j=1}^{n} \alpha_j A^k x_j = \sum_{j=1}^{n} \alpha_j \lambda_j^k x_j$$

If the largest eigenvalue obeys that $|\lambda_1| > |\lambda_2|$ and $|\lambda_2| \geq |\lambda_j|$ for any $3 \leq j \leq n$, then, for large k and assuming that $\alpha_1 \neq 0$ nor too small, we observe that

$$A^k w = \alpha_1 \lambda_1^k x_1 \left(1 + O\left(\left(\frac{|\lambda_2|}{|\lambda_1|} \right)^k \right) \right)$$

This shows that, after subsequent multiplications with A, an arbitrary vector w (not orthogonal to x_1, i.e. $\alpha_1 = w^T x_1 \neq 0$) aligns increasingly more towards the largest eigenvector x_1. This so-called power method lies at the basis of the computations of the largest eigenvector, especially in large and sparse matrices. In particular, the sequence $Aw, A^2 w, A^4 w, \ldots, A^{2^m} w$, tends exponentially fast to a vector, proportional to the largest eigenvector x_1 of A under the very mild condition that $|\lambda_1| > |\lambda_2|$.

10.2 Locations of eigenvalues

245. *General bounds on the position of eigenvalues.* Marcus and Ming (1964) overview the historic achievements on the localization of eigenvalues of a complex $n \times n$ matrix A from the early beginning up to 1964. Gerschgorin's Theorem 65 has been central, although several other scholars have earlier rephrased variants of Theorem 65.

Theorem 65 (Gerschgorin) *Every eigenvalue of a matrix A lies in at least one of the circular disks with center a_{jj} and radii $R_j = \sum_{k=1; k\neq j} |a_{jk}|$ or $r_j = \sum_{k=1; k\neq j} |a_{kj}|$*

Proof: Suppose that the j-th component of the eigenvector x of A belonging to eigenvalue λ has the largest modulus. An eigenvector can always be scaled and we normalize such that

$$x^T = (x_1, x_2, \ldots, x_{j-1}, 1, x_{j+1}, \ldots, x_n)$$

where $|x_k| \leq 1$, for all $1 \leq k \leq n$. Equating the j-th component on both sides of the eigenvalue equation $Ax = \lambda x$ gives $\sum_{k=1}^{n} a_{jk} x_k = \lambda x_j = \lambda$. Hence,

$$|a_{jj} - \lambda| \leq \sum_{k=1; k\neq j}^{n} |a_{jk} x_k| \leq \sum_{k=1; k\neq j}^{n} |a_{jk}| |x_k| \leq \sum_{k=1; k\neq j}^{n} |a_{jk}|$$

which shows that λ lies in a circular disk in the complex plane centered at a_{jj} with a radius not larger than $\sum_{k=1;k\neq j}^{n} |a_{jk}|$. The other radius mentioned follows from the fact that A and A^T have the same eigenvalues as shown in **art. 237.** □

A real $n \times n$ matrix A has a characteristic polynomial with real coefficients as follows from (A.96), so that its n zeros are either complex conjugate or real (**art.** 314). A consequence of Gerschgorin's Theorem 65 is

Corollary 3 (Gerschgorin) *If all n Gerschgorin disks, each centered around a diagonal element a_{jj} of a real matrix A, are disconnected (i.e. not overlapping), then all eigenvalues of A are real.*

Marcus and Ming (1964, p. 150) mention a generalization of Gerschgorin's Theorem 65 due to Ostrowski:

Theorem 66 (Ostrowski) *Let* $0 \leq \theta \leq 1$, $R_j = \sum_{k=1;k\neq j} |a_{jk}|$ *and* $r_j = \sum_{k=1;k\neq j} |a_{kj}|$, *then* $\det A \neq 0$, *provided* $|a_{jj}| > R_j^\theta r_j^{1-\theta}$ *for each* $1 \leq j \leq n$.

Proof: The j-th row of the eigenvalue equation $Ax = \lambda x$ is $(a_{jj} - \lambda)\, x_j = -\sum_{k=1;k\neq j}^{n} a_{jk} x_k$, from which

$$|a_{jj} - \lambda|\,|x_j| \leq \sum_{k=1;k\neq j}^{n} |a_{jk}|\,|x_k| = \sum_{k=1;k\neq j}^{n} |a_{jk}|^\theta\, |a_{jk}|^{1-\theta}\, |x_k|$$

After applying the Hölder inequality (A.10) with $p = \frac{1}{\theta} \geq 1$,

$$\sum_{k=1;k\neq j}^{n} |a_{jk}|^\theta\, |a_{jk}|^{1-\theta}\, |x_k| \leq \left(\sum_{k=1;k\neq j}^{n} |a_{jk}| \right)^\theta \left(\sum_{k=1;k\neq j}^{n} |a_{jk}|\,|x_k|^{\frac{1}{1-\theta}} \right)^{1-\theta}$$

and the definition $R_j = \sum_{k=1;k\neq j} |a_{jk}|$, we obtain for each $1 \leq j \leq n$,

$$|a_{jj} - \lambda|\,|x_j| \leq R_j^\theta \left(\sum_{k=1;k\neq j}^{n} |a_{jk}|\,|x_k|^{\frac{1}{1-\theta}} \right)^{1-\theta} \tag{A.119}$$

We deduce a contradiction by supposing that $\det A = 0$, which implies that $Ax = 0$ has a non-zero vector x belonging to $\lambda = 0$ as solution. Ostrowski's theorem states that $R_j^\theta r_j^{1-\theta} < |a_{jj}|$ for all $1 \leq j \leq n$, so that the general inequality (A.119) becomes

$$r_j\,|x_j|^{\frac{1}{1-\theta}} \leq \sum_{k=1;k\neq j}^{n} |a_{jk}|\,|x_k|^{\frac{1}{1-\theta}}$$

Summing over all j, recalling that the inequality is strict for at least one j,

$$\sum_{j=1}^{n} r_j\,|x_j|^{\frac{1}{1-\theta}} < \sum_{j=1}^{n} \sum_{k=1;k\neq j}^{n} |a_{jk}|\,|x_k|^{\frac{1}{1-\theta}} = \sum_{j=1}^{n} \left(\sum_{k=1}^{n} |a_{jk}|\,|x_k|^{\frac{1}{1-\theta}} - |a_{jj}|\,|x_j|^{\frac{1}{1-\theta}} \right)$$

and reversing the j- and k-sum, leads to

$$\sum_{j=1}^{n} r_j |x_j|^{\frac{1}{1-\theta}} < \sum_{k=1}^{n} \sum_{j=1;j\neq k}^{n} |a_{jk}| |x_k|^{\frac{1}{1-\theta}} = \sum_{k=1}^{n} |x_k|^{\frac{1}{1-\theta}} \sum_{j=1;j\neq k}^{n} |a_{jk}| = \sum_{k=1}^{n} r_k |x_k|^{\frac{1}{1-\theta}}$$

which is a contradiction. $\qquad\square$

Since each eigenvalue λ obeys $\det(A - \lambda I) = 0$, Ostrowski's Theorem 66 implies that each eigenvalue of A lies on or inside at least one of the disks

$$|a_{jj} - \lambda| \leq R_j^{\theta} r_j^{1-\theta} \qquad (0 \leq \theta \leq 1) \qquad\qquad (A.120)$$

that reduces to that of Gerschgorin for $\theta = 1$.

The idea in the proof of Ostrowski's Theorem 66 can be pushed a little further, by showing that at least two components in the vector x, satisfying $Ax = 0$, should not be zero. Indeed, let x_r and x_s be the components of the vector x such that $|x_r| \geq |x_s| \geq |x_k|$ for any $k \neq r$. Now, suppose that $x_s = 0$, then all components of the vector x, except for x_r, are zero. But, row r in $Ax = 0$, $\sum_{k=1}^{n} a_{rk} x_k = 0$, shows that $a_{rr} = 0$, which contradicts the condition $|a_{jj}| > R_j^{\theta} r_j^{1-\theta}$ for each row j. Hence, there are at least two equations for which $\det A = 0$ implies that

$$|a_{rr}| |x_r| \leq \sum_{k=1;k\neq r}^{n} |a_{rk}| |x_k| \leq |x_s| \sum_{k=1;k\neq r}^{n} |a_{rk}| = R_r |x_s|$$

and

$$|a_{ss}| |x_s| \leq \sum_{k=1;k\neq s}^{n} |a_{sk}| |x_k| \leq |x_r| \sum_{k=1;k\neq s}^{n} |a_{sk}| = R_s |x_r|$$

leading to $|a_{rr}| |a_{ss}| \leq R_r R_s$ and to

Theorem 67 (Ovals of Cassini) *If, for each $i, j \in \{1, 2, \ldots, n\}$ and $i \neq j$, it holds that*

$$|a_{ii}| |a_{jj}| > R_i R_j$$

then $\det A \neq 0$.

A direct consequence of Theorem 67 is that each eigenvalue of a complex $n \times n$ matrix A lies in at least one of the $\binom{n}{2}$ ovals $|\lambda - a_{ii}| |\lambda - a_{jj}| \leq R_i R_j$ of Cassini. A combination of Ostrowski's Theorem 66 and Theorem 67 leads to the generalization that each eigenvalue of A lies in at least one of the $\frac{n(n-1)}{2}$ ovals of Cassini, specified by

$$|\lambda - a_{ii}| |\lambda - a_{jj}| \leq R_i^{\theta} r_i^{1-\theta} R_j^{\theta} r_j^{1-\theta} \qquad (0 \leq \theta \leq 1)$$

10.3 Hermitian and real symmetric matrices

246. *A Hermitian matrix.* A Hermitian matrix A is a complex matrix that obeys $A^H = \left(A^T\right)^* = A$, where $a^H = (a_{ij})^*$ is the complex conjugate of a_{ij}. The superscript H, in honor of Charles Hermite, means to take the complex conjugate and then a transpose. Hermitian matrices possess a number of attractive properties. A particularly interesting subclass of Hermitian matrices are real, symmetric matrices that obey $A^T = A$. The inner-product of vector y and x is defined as $y^H x$ and obeys $\left(y^H x\right)^* = \left(y^H x\right)^H = x^H y$. The inner-product $x^H x = \sum_{j=1}^{n} |x_j|^2$ is real and positive for all vectors except for the null vector.

247. The eigenvalues of a Hermitian matrix are all real. Indeed, left-multiplying the eigenvalue equation $Ax = \lambda x$ in (1.3) by x^H yields

$$x^H A x = \lambda x^H x$$

Since $\left(x^H A x\right)^H = x^H A^H x = x^H A x$, it follows that $\lambda x^H x = \lambda^H x^H x$ or $\lambda = \lambda^H$, because $x^H x$ is a positive real number. Furthermore, with $A = A^H$, we have $A^H x = \lambda x$. Taking the complex conjugate, yields

$$A^T x^* = \lambda x^*$$

In general, the eigenvectors of a Hermitian matrix are complex, but real for a real symmetric matrix, because $A^H = A^T$. Moreover, the left-eigenvector y^T is the complex conjugate of the right-eigenvector x. Hence, the orthogonality relation (A.103) reduces, after normalization, to the inner-product

$$x_k^H x_j = \delta_{kj} \tag{A.121}$$

where δ_{kj} is the Kronecker delta, which is zero if $k \neq j$ and else $\delta_{kk} = 1$. Consequently, (A.106) reduces to

$$X^H X = I \tag{A.122}$$

which implies that the matrix X formed by the eigenvectors is a unitary matrix obeying $X^{-1} = X^H$.

For a real symmetric matrix A, the corresponding relation $X^T X = I$ implies that X is an orthogonal matrix obeying $X^{-1} = X^T$ and $X^T X = XX^T = I$, where the first equality follows from the commutativity of the inverse of a matrix, $X^{-1} X = XX^{-1}$. Hence, all eigenvectors of a symmetric matrix are orthogonal.

Although the arguments in Section 10.1 for a complete set of eigenvectors that spans the n-dimensional space have assumed that the eigenvalues of A are distinct, Theorem 68 for Hermitian and real matrices, proved in Wilkinson (1965, Section 47), applies to eigenvalues of any multiciplcty:

Theorem 68 (Hermitian and real symmetric diagonalizability) *For any*

Hermitian matrix A, there exists a unitary matrix U such that, for real λ_j,

$$U^H AU = diag\,(\lambda_j)$$

and for any real symmetric matrix A, there exists an orthogonal matrix X such that, for real λ_j,

$$X^T AX = diag\,(\lambda_j)$$

Due to the fundamental character of these diagonalizations, we will prove the last case in which A is real and symmetric. Transformations of a matrix A to another matrix B that contains more zero elements are key operations in linear algebra, that, as mentioned by Meyer (2000), have started with Gauss, who frequently used the technique of Gaussian elimination. In 1909, Schur has proved that every square matrix is unitarily similar to a triangular matrix:

Theorem 69 (Shur's Triangularization) *For any square matrix A, there exists a unitary matrix U, which is not necessarily unique, and a possibly non-unique triangular matrix T such that $U^H AU = T$. The diagonal entries of T are the eigenvalues of A.*

The proof of Schur's Triangularization Theorem 69 (see e.g. Meyer (2000), Mirsky (1982, p. 307-308)) is similar in nature to that of Theorem 68. The proof relies on interesting properties of the Householder transformation (**art.** 197) and reduction (**art.** 198). The proof is here for an upper-triangular matrix, but Theorem 68 also holds for a lower-triangular matrix.

Proof of Theorem 68: Let x_1 be the real eigenvector of an $n \times n$ symmetric matrix $A = A^T$ belonging to the real eigenvalue λ_1 and normalized such that $x_1^T x_1 = 1$. We invoke the Householder reflection $R = R^T$ in **art.** 197 with Householder vector $v = x_1 + e_1$, such that $Rx_1 = e_1$ and also $x_1 = R^T e_1 = R\,e_1$. From the eigenvalue equation $Ax_1 = \lambda_1 x_1$, we obtain, after Householder reflection, that $\lambda_1 e_1 = RAx_1 = R^T AR\,e_1$, which indicates that the first column of the matrix $R^T AR$ is a multiple of e_1. But, since $R^T AR$ is symmetric, it must be of the form

$$R^T AR = \begin{bmatrix} \lambda_1 & 0 \\ 0 & B \end{bmatrix}$$

where B is an $(n-1) \times (n-1)$ symmetric matrix. We can proceed to iteratively apply the above recipe to the matrix B as in **art.** 198. Here, we use induction and assume that there exists an orthogonal $(n-1) \times (n-1)$ matrix V such that $V^T BV = diag(\nu)$. The existence is clearly true for $n = 2$, since V is then a scalar. Next,

$$\begin{bmatrix} 1 & 0 \\ 0 & V^T \end{bmatrix} R^T AR \begin{bmatrix} 1 & 0 \\ 0 & V \end{bmatrix} = \begin{bmatrix} 1 & 0 \\ 0 & V^T \end{bmatrix}\begin{bmatrix} \lambda_1 & 0 \\ 0 & B \end{bmatrix}\begin{bmatrix} 1 & 0 \\ 0 & V \end{bmatrix} = \begin{bmatrix} \lambda_1 & 0 \\ 0 & diag\,(\nu) \end{bmatrix}$$

Since $X = R\begin{bmatrix} 1 & 0 \\ 0 & V \end{bmatrix}$ is an orthogonal matrix, the induction assumption also holds for the $n \times n$ matrix A, which establishes the diagonalization for any real symmetric matrix by a real orthogonal matrix. After comparison with the eigenvalue equation $AX = X\Lambda$, we conclude that $\nu_j = \lambda_j$. Hence, we have demonstrated that $X^T A X = \text{diag}(\lambda)$ for any real symmetric matrix A. □

248. *The orthogonal matrix X of a real symmetric matrix.* We elaborate on the results of **art.** 247 for a real symmetric matrix and point to the notion of double orthogonality. We denote by x_k the eigenvector of the $n \times n$ symmetric matrix A belonging to the eigenvalue λ_k, normalized so that $x_k^T x_k = 1$. The eigenvalues of a symmetric matrix $A = A^T$ are real and can be ordered as $\lambda_1 \geq \lambda_2 \geq \ldots \geq \lambda_n$. The orthogonal matrix X with the eigenvectors of A in the columns,

$$X = \begin{bmatrix} x_1 & x_2 & x_3 & \cdots & x_n \end{bmatrix}$$

is explicitly written in terms of the m-th component $(x_j)_m$ of eigenvector x_j,

$$X = \begin{bmatrix} (x_1)_1 & (x_2)_1 & (x_3)_1 & \cdots & (x_n)_1 \\ (x_1)_2 & (x_2)_2 & (x_3)_2 & \cdots & (x_n)_2 \\ (x_1)_3 & (x_2)_3 & (x_3)_3 & \cdots & (x_n)_3 \\ \vdots & \vdots & \vdots & \ddots & \vdots \\ (x_1)_n & (x_2)_n & (x_3)_n & \cdots & (x_n)_n \end{bmatrix} \tag{A.123}$$

where the element $X_{ij} = (x_j)_i$. For a graph related matrix, the row i of X details the eigenvalue components of the node i over all eigenfrequencies/eigenvalues (**art.** 240), while the column k of X equals the eigenvector x_j belonging to the j-th largest eigenfrequency/eigenvalue λ_j.

The relation $X^T X = I = X X^T$ (**art.** 247) expresses, in fact, *double orthogonality*. The first equality $X^T X = I$ translates to the orthogonality relation – the real companion of (A.121) –

$$x_k^T x_m = \sum_{j=1}^{n} (x_k)_j (x_m)_j = \delta_{km} \tag{A.124}$$

stating that the eigenvector x_k belonging to eigenvalue λ_k is orthogonal to any other eigenvector belonging to a different eigenvalue. The second equality $X X^T = I$, which arises from the commutativity of the inverse matrix $X^{-1} = X^T$ with the matrix X itself, can be written as $\sum_{j=1}^{n} (x_j)_m (x_j)_k = \delta_{mk}$ and suggests us to define the row vector in X as

$$v_m = ((x_1)_m, (x_2)_m, \ldots, (x_N)_m) \tag{A.125}$$

Then, the second orthogonality condition $X X^T = I$ implies orthogonality of the vectors

$$v_l^T y_j = \sum_{k=1}^{n} (x_k)_l (x_k)_j = \delta_{lj} \tag{A.126}$$

Beside the first (A.124) and second (A.126) orthogonality relations, the third combination equals

$$v_j^T x_k = \sum_{l=1}^{n} (x_l)_j (x_k)_l = \sum_{l=1}^{n} X_{jl} X_{lk} = (X^2)_{jk} \tag{A.127}$$

Each $n \times n$ symmetric matrix A possesses 2^n different orthogonal matrices, because each column in (A.123), thus each eigenvector, can be multiplied by -1 without violating the orthogonality conditions (A.124) and (A.126).

249. *Continuous form of the orthogonality relations.* Invoking the frequency interpretation of the eigenvalue equation (A.109) in **art.** 240 and the Dirac delta-function $\delta(t)$ in **art.** 172, the first orthogonality relation (A.124) becomes

$$\sum_{j=1}^{n} (x_j)_m (x_j)_k = \sum_{\lambda \in \{\lambda_1, \lambda_2, \ldots, \lambda_n\}} (x(\lambda))_m (x(\lambda))_k$$

$$= \sum_{j=1}^{n} \int_{-\infty}^{\infty} \delta(\lambda - \lambda_j) (x(\lambda))_m (x(\lambda))_k \, d\lambda$$

Using the non-negative weight function in **art.** 350

$$w(\lambda) = \sum_{j=1}^{n} \delta(\lambda - \lambda_j) = \delta(\det(A - \lambda I)) \left| \frac{d \det(A - xI)}{dx} \right|_{x=\lambda}$$

shows that

$$\sum_{j=1}^{n} (x_j)_m (x_j)_k = \int_{-\infty}^{\infty} w(\lambda) (x(\lambda))_m (x(\lambda))_k \, d\lambda = \delta_{mk} \tag{A.128}$$

The right-hand side in (A.128) is the continuous variant of (A.126) that expresses orthogonality between functions with respect to the weight function w in **art.** 351. Specifically, the orthogonality property (A.128) applied to a general tri-diagonal matrix (see e.g. Van Mieghem (2013, Appendix) and **art.** 370) shows that the set $\{(x(\lambda))_m\}_{1 \leq m \leq N}$ is a set of n orthogonal polynomials in λ, that are further studied in Chapter 12.

250. *The spectrum of a unitary matrix.* We denote the eigenvalues of the $n \times n$ unitary matrix U by $\mu_1, \mu_2, \ldots, \mu_n$.

Theorem 70 *All eigenvalues of a unitary matrix have absolute value 1, i.e. $|\mu_k| = 1$ for all $1 \leq k \leq n$.*

Proof: The orthogonality relation (A.121) for $k = j$ or the matrix product of the j-th diagonal element in I in the orthogonality relation (A.122) equals $\sum_{i=1}^{n} |U_{ij}|^2 = 1$, which implies that the elements U_{ij} of a unitary matrix cannot exceed unity in absolute value. Therefore, the absolute value of the coefficients c_k in (A.96) of the characteristic polynomial is bounded for any $n \times n$ unitary matrix U.

Taking the determinant of the orthogonality relation (A.122) gives

$$1 = \det\left(U^H U\right) = \det\left(U^H\right)\det U = |\det U|^2$$

while (A.98) then leads to $\prod_{k=1}^{n} |\mu_k|^2 = 1$. Hence, a unitary matrix cannot have a zero eigenvalue. In addition, it shows, together with the bounds on $|c_k|$ that are only function of n, that all eigenvalues must lie between some lower and upper bound for any $n \times n$ unitary matrix U and these bounds are not dependent on the unitary matrix elements considered. **Art. 243** shows that any integer power m, positive as well as negative, of U^m has the same eigenvalues of U raised to the power m. In addition, U^m is also an $n \times n$ unitary matrix obeying $(U^m)^H U^m = I$ as follows by induction on

$$(U^m)^H U^m = \left(U^{m-1}\right)^H U^H U U^{m-1} = \left(U^{m-1}\right)^H U^{m-1}$$

Hence, $|\det U^m|^2 = 1$ and, by (A.98), we have, for any $m \in \mathbb{Z}$, that $\prod_{k=1}^{n} |\mu_k|^{2m} = 1$. But, the absolute value of these powers, positive as well as negative, can only remain below a bound independent of m provided $|\mu_k| = 1$ for all k. $\qquad\square$

A unitary matrix $U = U_R + iU_I$ obeys $U^H U = I$,

$$U^H U = (U_R - iU_I)^T (U_R + iU_I) = U_R^T U_R + U_I^T U_I + i\left(U_R^T U_I - U_I^T U_R\right)$$

which requires that $U_R^T U_R + U_I^T U_I = I$ and $U_R^T U_I = U_I^T U_R$. The latter, written as $\left(U_I^T U_R\right)^T = U_I^T U_R$, shows that $U_I^T U_R$ is a symmetric matrix. An orthogonal matrix U_R that obeys $U_R^T U_R = I$ can be regarded as a unitary matrix $U = U_R + iU_I$ with imaginary part $U_I = 0$. Theorem 70 states that the j-th eigenvector $z_j = x_j + iy_j$ obeys the eigenvalue equation $U z_j = e^{i\theta_j} z_j$ for real θ_j, explicitly,

$$(U_R + iU_I)(x_j + iy_j) = (\cos\theta_j + i\sin\theta_j)(x_j + iy_j)$$

Thus, in general, the eigenvalues $e^{i\theta_j}$ and eigenvector z_j of an orthogonal matrix U_R with $U_I = 0$ are complex with unit modulus. An alternative proof starts from the eigenvalue equation of a real, orthogonal matrix $U_R z_j = \lambda_j z_j$, whose complex conjugate is $U_R z_j^* = \lambda_j^* z_j^*$ from which transposing follows as $\left(z_j^*\right)^T U_R^T = \lambda_j^* \left(z_j^*\right)^T$. After multiplication, we obtain

$$\lambda_j \lambda_j^* \left(z_j^*\right)^T z_j = \left(z_j^*\right)^T U_R^T U_R z_j = \left(z_j^*\right)^T z_j$$

Since an eigenvector is different from the zero vector, we find that $\lambda_j \lambda_j^* = |\lambda_j|^2 = 1$. Only if the orthogonal matrix U_R is symmetric, i.e. when $U_R^{-1} = U_R$, the eigenvectors and eigenvalues are real (**art. 247**). Hence, an eigenvalue of a symmetric orthogonal matrix $U_R = U_R^T$ is either 1 or -1.

251. The Rayleigh inequalities. Art. 247 tells that the normalized eigenvectors x_k and x_m of real symmetric[4] A obey $x_k^T A x_m = 0$ if $k \neq m$ and $x_k^T A x_k = \lambda_k$. These

---

[4] The extension to a Hermitian matrix is straightforward and omitted.

n eigenvectors span the n-dimensional space. Let w be an $n \times 1$ vector that can be written as a linear combination of the first j eigenvectors of A,

$$w = c_1 x_1 + c_2 x_2 + \cdots + c_j x_j$$

where all $c_k = w^T x_k \in \mathbb{R}$. Equivalently, $w \in \mathcal{X}_j$, where $\mathcal{X}_j$ is the space spanned by the vectors $\{x_1, x_2, \ldots, x_j\}$. Then $w^T w = \sum_{k=1}^{j} \sum_{m=1}^{j} c_k c_m x_k^T x_m = \sum_{k=1}^{j} c_k^2$ and

$$w^T A w = \sum_{k=1}^{j} \sum_{m=1}^{j} c_k c_m x_k^T A x_m = \sum_{k=1}^{j} c_k^2 \lambda_k$$

Since A has real eigenvalues $\lambda_1 \geq \lambda_2 \geq \cdots \geq \lambda_n$, this ordering of the eigenvalues leads to the bound $\lambda_j \sum_{k=1}^{j} c_k^2 \leq \sum_{k=1}^{j} c_k^2 \lambda_k \leq \lambda_1 \sum_{k=1}^{j} c_k^2$ from which the Rayleigh inequalities for $w \in \mathcal{X}_j$ follow as

$$\lambda_j \leq \frac{w^T A w}{w^T w} \leq \lambda_1 \tag{A.129}$$

Equality in $\frac{w^T A w}{w^T w} = \lambda_k$ is only attained provided $w \in \mathcal{X}_j$ is an eigenvector of A belonging to eigenvalue λ_k with $\lambda_j \leq \lambda_k \leq \lambda_1$. If w is a vector that is orthogonal to the first j eigenvectors of A, which means that $w = c_{j+1} x_{j+1} + c_{j+2} x_{j+2} + \cdots + c_n x_n$ can be written as a linear combination of the last $n-j$ eigenvectors or that $w \in \mathcal{X}_j^{\perp}$, then $\lambda_n \leq \frac{w^T A w}{w^T w} \leq \lambda_{j+1}$.

The two extreme eigenvalues can thus be written as

$$\lambda_1 = \sup_{y \neq 0} \frac{y^T A y}{y^T y} \tag{A.130}$$

$$\lambda_n = \inf_{y \neq 0} \frac{y^T A y}{y^T y} \tag{A.131}$$

The Courant-Fischer Theorem, proved in (Meyer, 2000, p. 550), is the generalization to Hermitian matrices and infinite-dimensional operators,

$$\lambda_i = \max_{\dim \mathcal{V} = i} \min_{y \in \mathcal{V}} \frac{y^H A y}{y^H y} \tag{A.132}$$

$$\lambda_i = \min_{\dim \mathcal{V} = n+1-i} \max_{y \in \mathcal{V}} \frac{y^H A y}{y^H y} \tag{A.133}$$

252. *Field of values.* The field of values $\Phi(.)$ is a set of complex numbers associated to an $n \times n$ matrix A,

$$\Phi(A) = \{x^H A x : x \in \mathbb{C}^n, x^H x = 1\} \tag{A.134}$$

While the spectrum of a matrix is a discrete set, the field of values $\Phi(.)$, of which an instance appeared in **art.** 251, can be a continuum. However, $\Phi(A)$ is always a convex subset of $\mathbb{C}$ for any matrix A, a fundamental fact known as the Toeplitz-Hausdorff Theorem and proved in Horn and Johnson (1991, Section 1.3). Another property is the subadditivity, $\Phi(A+B) \subset \Phi(A) + \Phi(B)$, which follows from the definition (A.134). Indeed,

$$\Phi(A+B) = \left\{ x^H Ax + x^H Bx : x \in \mathbb{C}^n, x^H x = 1 \right\}$$
$$\subset \left\{ x^H Ax : x \in \mathbb{C}^n, x^H x = 1 \right\} + \left\{ y^H By : y \in \mathbb{C}^n, y^H y = 1 \right\}$$
$$= \Phi(A) + \Phi(B)$$

Since the set $\lambda(A)$ of eigenvalues of A belongs to $\Phi(A)$, it holds that

$$\lambda(A+B) \subset \Phi(A+B) \subset \Phi(A) + \Phi(B)$$

which can provide possible information about the eigenvalues of $A+B$, given $\Phi(A)$ and $\Phi(B)$.

In general, given the spectrum $\lambda(A)$ and $\lambda(B)$, surprisingly little can be said about $\lambda(A+B)$ (see also **art. 267** and **284**). For example, even if the eigenvalues of A and B are known and bounded, the largest eigenvalue of $A+B$ can be unbounded, as deduced from the example inspired by Horn and Johnson (1991, p. 5), where

$$A = \begin{bmatrix} 1-x & 1 \\ f(x) & x \end{bmatrix} \text{ and } B = \begin{bmatrix} 1+x & 1 \\ g(x) & -x \end{bmatrix}$$

Clearly, the eigenvalues of A, B and $A+B$ are

$$\lambda_{1,2}(A) = \frac{1}{2}\left(1 \pm \sqrt{1 + 4\left(x^2 - x + f(x)\right)}\right)$$
$$\lambda_{1,2}(B) = \frac{1}{2}\left(1 \pm \sqrt{1 + 4\left(x^2 + x + g(x)\right)}\right)$$
$$\lambda_{1,2}(A+B) = \frac{1}{2}\left(1 \pm \sqrt{1 + 4\left(f(x) + g(x)\right)}\right)$$

It suffices to choose $f(x) = -x^2 + x + c_1$ and $g(x) = -x^2 - x + c_2$ for arbitrary constants c_1 and c_2 to have bounded eigenvalues, independent of x, while $\lim_{x \to \infty} |\lambda_{1,2}(A+B)| = \infty$.

253. *Weyl's problem.* Knutson and Tao (2001) discuss the problem of Weyl (1912): Given the eigenvalues of two $n \times n$ Hermitian matrices A and B, determine all the possible sets of eigenvalues of $A+B$. Apart from the trace equality (A.99),

$$\sum_{j=1}^{n} \lambda_j(A+B) = \sum_{j=1}^{n} \lambda_j(A) + \sum_{j=1}^{n} \lambda_j(B)$$

Horn (1962) has shown that a finite number of inequalities of the form

$$\sum_{q=1}^{r} \lambda_{k_q + (r-q+1)}(A+B) \leq \sum_{q=1}^{r} \lambda_{i_q + (r-q+1)}(A) + \sum_{q=1}^{n} \lambda_{j_q + (r-q+1)}(B) \qquad \text{(A.135)}$$

needs to be obeyed, where $1 \leq r < n$ and all triplets of indices $1 \leq i_1 < \cdots < i_r$, $1 \leq j_1 < \cdots < j_r$ and $1 \leq k_1 < \cdots < k_r$ belong to a certain finite set $T_{r,n}$.

Horn conjectured in 1962, but Knutson and Tao proved in 1999, that the set $T_{r,n}$ is generated by the triplets of indices that obey

$$\sum_{q=1}^{r} i_q + \sum_{q=1}^{r} j_q = \sum_{q=1}^{r} k_q + \frac{r(r+1)}{2} \tag{A.136}$$

and the inequalities

$$\sum_{l=1}^{s} i_{a_l} + \sum_{l=1}^{s} j_{b_l} \geq \sum_{l=1}^{s} k_{c_l} + \frac{s(s+1)}{2} \tag{A.137}$$

for $1 \leq s < r$ and all triplets of indices $1 \leq a_1 < \cdots < a_s$, $1 \leq b_1 < \cdots < b_s$ and $1 \leq c_1 < \cdots < c_s$ in $T_{s,r}$. The above equations are a highly recursive algorithm to generate the sets $T_{r,n}$ in terms of earlier generations $T_{s,r}$ and give the complete solution to Weyl's problem for any dimension n. The complicated proof of Knutsen and Tao relies on their discovery of an equivalence between Weyl's problem and a planar graph, called the honeycomb, that is further explained in Knutson and Tao (2001).

For the case that $r = 1$, (A.135) reduces to

$$\lambda_{k_1+1}(A+B) \leq \lambda_{i_1+1}(A) + \lambda_{j_1+1}(B)$$

while the indices (A.136) satisfy $i_1 + j_1 = k_1 + 1$. Hence, for the $n \times n$ Hermitian matrices A and B, the Weyl inequality in Weyl (1912, Sec. 1), for $2 \leq i+j \leq n+1$, is

$$\lambda_{i+j-1}(A+B) \leq \lambda_i(A) + \lambda_j(B)$$

while, for $2n \geq i + j \geq n + 1$, the dual Weyl inequality is

$$\lambda_{i+j-n}(A+B) \geq \lambda_i(A) + \lambda_j(B)$$

254. *The eigenvalue decomposition of a symmetric matrix.* **Art.** 247 shows that any real, symmetric matrix $A_{n \times n}$ can be written as $A = X \Lambda X^T$, where $\Lambda = \text{diag}(\lambda)$ with eigenvalue vector $\lambda = (\lambda_1, \lambda_2, \ldots, \lambda_n)$ and where $X = \begin{bmatrix} x_1 & x_2 & \ldots & x_n \end{bmatrix}$ is an orthogonal matrix, obeying $X^T X = X X^T = I$, formed by the real and normalized eigenvectors $x_1, x_2, \ldots, x_n$ of A corresponding to the eigenvalues $\lambda_1 \geq \lambda_2 \geq \cdots \geq \lambda_n$. In vector notation,

$$A = \sum_{k=1}^{n} \lambda_k x_k x_k^T \tag{A.138}$$

where the matrix $E_k = x_k x_k^T$ is the outer product of x_k by itself.

255. *Properties of the matrix $E_k = x_k x_k^T$.* The definition $E_k = x_k x_k^T$ shows that

$E_k = E_k^T$ is symmetric. The explicit form of the $n \times n$ matrix E_k is

$$E_k = x_k x_k^T = \begin{bmatrix} (x_k)_1^2 & (x_k)_1 (x_k)_2 & (x_k)_1 (x_k)_3 & \cdots & (x_k)_1 (x_k)_n \\ (x_k)_2 (x_k)_1 & (x_k)_2^2 & (x_k)_2 (x_k)_3 & \cdots & (x_k)_2 (x_k)_n \\ (x_k)_3 (x_k)_1 & (x_k)_3 (x_k)_2 & (x_k)_3^2 & \cdots & (x_k)_3 (x_k)_n \\ \vdots & \vdots & \vdots & \vdots & \vdots \\ (x_k)_n (x_k)_1 & (x_k)_n (x_k)_2 & (x_k)_n (x_k)_3 & \cdots & (x_k)_n^2 \end{bmatrix}$$

which shows that the diagonal element $(E_k)_{ii} = (x_{ki})^2$ equals the square of the i-th vector component of the eigenvector x_k. Hence,

$$\text{trace}\,(E_k) = \sum_{i=1}^{n} (x_k)_i^2 = x_k^T x_k = 1 \tag{A.139}$$

We write E_k in terms of the elementary orthogonal projector $S = I - \frac{1}{v^T v} v.v^T$ onto the hyperplane through the origin that is orthogonal to the vector v in **art. 193** as

$$E_k = I - S_k = x_k x_k^T$$

which represents the orthogonal projection onto the eigenspace of λ_k. Any vector w is projected by $E_k w = (x_k^T w)\, x_k$ onto the vector x_k.

The orthogonality property (A.121) of eigenvectors x_k of a symmetric matrix indicate that $E_k^2 = E_k$ and $E_k E_m = 0$ for $k \neq m$. The eigenvalue equation $E_k y_j = \xi_j y_j$ of the symmetric matrix E_k has (**art. 194**) a zero eigenvalue $\lambda_k = 0$ with multiplicity $n - 1$ and one eigenvalue $\lambda_k = 1$, such that $\|E_k\|_2 = 1$, which follows from (A.23). The zero eigenvalues imply that $\det(E_k) = 0$ and that the inverse of E_k does not exist. Geometrically, this is understood because, by projecting, information is lost and the inverse cannot create information.

The notation of E_k so far has implicitly assumed that all eigenvalues of the symmetric matrix A are different. If the eigenvalue λ_k has multiplicity m_k, then there are m_k eigenvectors belonging to λ_k that form an orthonormal basis for the eigenspace belonging to λ_k. Let U_k denote the $n \times m_k$ matrix with its columns equal to the m_k eigenvectors belonging to λ_k, then the matrix $E_k = U_k U_k^T$ generalizes $E_k = x_k x_k^T$. Thus, with (A.99), trace(E_k) = m_k is equal to the rank of E_k, which is the dimension of the eigenspace associated with λ_k.

Consider now the $n \times n$ matrix $Y = \sum_{k=1}^{l} E_k$, where the k index ranges over all distinct $l \leq n$ eigenvalues $\{\lambda_k\}_{1 \leq k \leq l}$ of A. Since $E_k^2 = E_k$ and $E_k E_m = 0$ for $k \neq m$, we find that $Y^2 = \sum_{k=1}^{l} E_k \sum_{j=1}^{l} E_j = \sum_{k=1}^{l} E_k^2 + 2 \sum_{k=1}^{l} \sum_{j=1}^{k-1} E_k E_j = Y$ such that all eigenvalues of the symmetric (Hermitian) matrix Y are either 1 or 0. But, trace(Y) = $\sum_{k=1}^{l} \text{trace}(E_k) = \sum_{k=1}^{l} m_k = n$ implies that all eigenvalues of Y must be equal to 1, and thus that $Y = I$. The fact that $\sum_{k=1}^{l} E_k = I_{n \times n}$ follows directly from (A.89) for $f(x) = e^{zx}$, after letting $z = 0$. Moreover, this relation is rewritten as $X X^T = I$, which, combined with the normalization (A.122), implies that X is an orthogonal matrix, which we already knew from **art. 247**. It means that the sum of the orthogonal projections onto all eigenspaces of A spans again the total $n \times n$ space.

256. *Hadamard product form of the decomposition of two symmetric matrices.* Let $A = A^T = X\Lambda X^T$ and $B = B^T = VMV^T$, then their Hadamard product is written with (A.138) as

$$(A \circ B)_{ij} = a_{ij}b_{ij} = \sum_{k=1}^{n}\sum_{m=1}^{n} \lambda_k \mu_m \left(x_k x_k^T\right)_{ij} \left(v_m v_m^T\right)_{ij}$$

Since

$$\left(x_k w_l^T\right)_{ij} \left(v_m y_q^T\right)_{ij} = (x_k)_i\,(w_l)_j\,(v_m)_i\,(y_q)_j = (x_k \circ v_m)_i\,(w_l \circ y_q)_j$$
$$= \left((x_k \circ v_m)(w_l \circ y_q)^T\right)_{ij}$$

we obtain $(A \circ B)_{ij} = \left(\sum_{k=1}^{n}\sum_{m=1}^{n} \lambda_k \mu_m (x_k \circ v_m)(x_k \circ v_m)^T\right)_{ij}$. Hence, we find that the Hadamard product of two symmetric matrices is

$$A \circ B = \sum_{k=1}^{n}\sum_{m=1}^{n} \lambda_k \mu_m (x_k \circ v_m)(x_k \circ v_m)^T \qquad \text{(A.140)}$$

257. *The eigenvalue decomposition of a function of a symmetric matrix.* **Art.** 234 generalizes the spectral decomposition (A.138) of a real symmetric matrix A to a function $f(A)$ in (A.88), whose elements are

$$(f(A))_{ij} = \sum_{k=1}^{n} f(\lambda_k)(x_k)_i\,(x_k)_j \qquad \text{(A.141)}$$

Using the identity $ab = \frac{(a+b)^2 - (a^2+b^2)}{2}$ in (A.141) yields

$$(f(A))_{ij} = \frac{1}{2}\sum_{k=1}^{n} f(\lambda_k)\left((x_k)_i + (x_k)_j\right)^2 - \frac{1}{2}\sum_{k=1}^{n} f(\lambda_k)(x_k)_i^2 - \frac{1}{2}\sum_{k=1}^{n} f(\lambda_k)(x_k)_j^2$$

and

$$(f(A))_{ij} = \frac{1}{2}\sum_{k=1}^{n} f(\lambda_k)\left((x_k)_i + (x_k)_j\right)^2 - \frac{(f(A))_{ii} + (f(A))_{jj}}{2} \qquad \text{(A.142)}$$

Similarly, using $ab = \frac{(a^2+b^2) - (a-b)^2}{2}$ leads to

$$(f(A))_{ij} = \frac{(f(A))_{ii} + (f(A))_{jj}}{2} - \frac{1}{2}\sum_{k=1}^{n} f(\lambda_k)\left((x_k)_i - (x_k)_j\right)^2 \qquad \text{(A.143)}$$

After addition, we obtain

$$(f(A))_{ij} = \frac{1}{4}\sum_{k=1}^{n} f(\lambda_k)\left((x_k)_i + (x_k)_j\right)^2 - \frac{1}{4}\sum_{k=1}^{n} f(\lambda_k)\left((x_k)_i - (x_k)_j\right)^2 \qquad \text{(A.144)}$$

Since $\sum_{k=1}^{n} \left((x_k)_i \pm (x_k)_j \right)^2 = 2 \left(1 \pm \delta_{ij} \right)$, upper and lower bounding results in

$$(1 + \delta_{ij}) \min_{1 \leq k \leq n} f(\lambda_k) \leq \frac{(f(A))_{ii} + (f(A))_{jj}}{2} + (f(A))_{ij} \leq (1 + \delta_{ij}) \max_{1 \leq k \leq n} f(\lambda_k) \tag{A.145}$$

and

$$(1 - \delta_{ij}) \min_{1 \leq k \leq n} f(\lambda_k) \leq \frac{(f(A))_{ii} + (f(A))_{jj}}{2} - (f(A))_{ij} \leq (1 - \delta_{ij}) \max_{1 \leq k \leq n} f(\lambda_k) \tag{A.146}$$

while the addition formula (A.144) leads to the bound

$$\left| (f(A))_{ij} - \frac{\delta_{ij}}{2} \left(\min_{1 \leq k \leq n} f(\lambda_k) + \max_{1 \leq k \leq n} f(\lambda_k) \right) \right| \leq \frac{1}{2} \left(\max_{1 \leq k \leq n} f(\lambda_k) - \min_{1 \leq k \leq n} f(\lambda_k) \right) \tag{A.147}$$

Each non-diagonal element in (A.147) is bounded by

$$\left| (f(A))_{ij} \right| \leq \frac{1}{2} \left(\max_{1 \leq k \leq n} f(\lambda_k) - \min_{1 \leq k \leq n} f(\lambda_k) \right)$$

while each diagonal element is bounded by (A.145) as $\min_{1 \leq k \leq n} f(\lambda_k) \leq (f(A))_{jj} \leq \max_{1 \leq k \leq n} f(\lambda_k)$. If $\min_{1 \leq k \leq n} f(\lambda_k) \geq 0$, then $f(A)$ is positive semidefinite and (A.146) shows, for $i \neq j$, that $(f(A))_{ij} \leq \frac{(f(A))_{ii} + (f(A))_{jj}}{2}$, which is a well-known bound for the off-diagonal elements (**art. 279**).

10.4 Recursive eigenvalue equation of a symmetric matrix

258. *The eigenvalue equation of a symmetric block matrix.* We write the $n \times n$ symmetric matrix A as a block matrix

$$A = \begin{bmatrix} A_1 & B \\ B^T & A_2 \end{bmatrix} \tag{A.148}$$

where A_1 is an $(n - m) \times (n - m)$ symmetric matrix and A_2 is an $m \times m$ symmetric matrix. The eigenvalue equation $Ax = \lambda(A)x$ with the block eigenvector $x^T = \begin{bmatrix} y & z \end{bmatrix}^T$,

$$\begin{bmatrix} A_1 & B \\ B^T & A_2 \end{bmatrix} \begin{bmatrix} y \\ z \end{bmatrix} = \lambda(A) \begin{bmatrix} y \\ z \end{bmatrix}$$

is written as the linear block set,

$$\begin{cases} A_1 y + B z = \lambda(A) y \\ B^T y + A_2 z = \lambda(A) z \end{cases} \tag{A.149}$$

where the normalization $x^T x = 1$ is equivalent to $y^T y + z^T z = 1$.

After left-multiplying the first block equation in (A.149) with y^T and the second block equation with z^T, we obtain

$$\begin{cases} y^T A_1 y + y^T B z = \lambda(A) y^T y \\ z^T B^T y + z^T A_2 z = \lambda(A) z^T z \end{cases}$$

Adding and subtracting these two scalar equations and using $y^T B z = z^T B^T y$ and $y^T y + z^T z = 1$ yields

$$\lambda(A) = y^T A_1 y + 2 y^T B z + z^T A_2 z \tag{A.150}$$

and

$$y^T A_1 y = \lambda(A)\left(1 - 2z^T z\right) + z^T A_2 z \tag{A.151}$$

which only contains two quadratic forms after the elimination of the $(n-m) \times m$ matrix B.

Applying the Rayleigh inequality (A.129) to $v^T A_1 v$, $\lambda_1(A_1) \geq \frac{y^T A_1 y}{y^T y} \geq \lambda_n(A_1)$ leads, after substituting (A.151) and using the corresponding eigenvector, to a lower bound for the spectral radius of A_1,

$$\lambda_1(A_1) \geq \frac{\lambda_1(A)\left(1 - 2z_1^T z_1\right) + z_1^T A_2 z_1}{1 - z_1^T z_1} \tag{A.152}$$

and an upper bound for the smallest eigenvalue of A_1,

$$\lambda_n(A_1) \leq \frac{\lambda_n(A)\left(1 - 2z_n^T z_n\right) + z_n^T A_2 z_n}{1 - z_n^T z_n} \tag{A.153}$$

259. *The eigenvalue equation approached recursively.* We revisit the eigenvalue equation (1.3) for a real symmetric $n \times n$ matrix A_n, which we write in terms of the $(n-1) \times (n-1)$ symmetric matrix A_{n-1} by adding the last column and row as

$$A_n = \begin{bmatrix} A_{n-1} & v_{(n-1)\times 1} \\ \left(v^T\right)_{1\times(n-1)} & a_{nn} \end{bmatrix} \tag{A.154}$$

where the $(n-1) \times 1$ vector $v = (a_{1n}, a_{2n}, \ldots, a_{n-1,n})$. Let $x_k(A_n) = \begin{bmatrix} y & z \end{bmatrix}^T$ be the eigenvector of A_n belonging to the eigenvalue $\lambda_k(A_n)$, normalized in the usual way as $x_k^T(A_n) x_k(A_n) = 1$ and y is an $(n-1) \times 1$ vector, while z is a real number. With $\lambda = \lambda_k(A_n)$ to simplify the notation, the block eigenvalue equations in (A.149) in **art.** 258 reduce to

$$\begin{cases} (A_{n-1} - \lambda I) y + zv = 0 \\ v^T y + z(a_{nn} - \lambda) = 0 \end{cases}$$

Assuming that $(A_{n-1} - \lambda I)^{-1}$ exists, which implies that λ is *not* an eigenvalue of A_{n-1}, the first equation is solved for the $(n-1) \times 1$ vector y as

$$y = -z(A_{n-1} - \lambda I)^{-1} v$$

Introduced into the second equation, $z = \frac{1}{a_{nn}-\lambda}\left(-v^T y\right) = \frac{zv^T\left(A_{n-1}-\lambda I\right)^{-1}v}{a_{nn}-\lambda}$ leads to an equation for λ,

$$v^T\left(A_{n-1}-\lambda I\right)^{-1}v = a_{nn}-\lambda$$

which is consistent with (A.157) in **art. 264**, because λ is not an eigenvalue of A_{n-1}. The normalization of $x_k\left(A_n\right)$ shows that $y^T y + z^2 = 1$ and, explicitly,

$$1 = z^2 + \left(v^T\left(A_{n-1}-\lambda I\right)^{-1}z\right)z\left(A_{n-1}-\lambda I\right)^{-1}v = z^2\left(1 + v^T\left(A_{n-1}-\lambda I\right)^{-2}v\right)$$

With $z = \left(x_k\left(A_n\right)\right)_n$ and provided that λ is *not* an eigenvalue of A_{n-1}, the n-th component of the eigenvector $x_k\left(A_n\right)$ obeys

$$z^2 = \frac{1}{1 + v^T\left(A_{n-1}-\lambda I\right)^{-2}v} = \frac{1}{1 + \left\|\left(A_{n-1}-\lambda I\right)^{-1}v\right\|_2^2} \tag{A.155}$$

Since rows can be interchanged, a similar type of expression can be deduced for each eigenvector component of A_n. Invoking $f\left(A\right) = \sum_{k=1}^n f\left(\lambda_k\right)x_k x_k^T$ in (A.88) to the symmetric matrix A_{n-1} yields

$$\left(A_{n-1}-\lambda I\right)^{-2} = \sum_{l=1}^{n-1}\frac{x_l\left(A_{n-1}\right)x_l^T\left(A_{n-1}\right)}{\left(\lambda - \lambda_l\left(A_{n-1}\right)\right)^2}$$

Finally, the square in (A.155) of any component of an eigenvector $x_k\left(A_n\right)$ of A_n, belonging to the eigenvalue $\lambda = \lambda_k\left(A_n\right)$, can be written in terms of the eigenstructure of a submatrix A_{n-1} with corresponding vector v,

$$z^2 = \left(1 + \sum_{l=1}^{n-1}\frac{\left(v^T x_l\left(A_{n-1}\right)\right)^2}{\left(\lambda_k\left(A_n\right) - \lambda_l\left(A_{n-1}\right)\right)^2}\right)^{-1}$$

Using the norm inequality (A.12): $\left\|\left(A_{n-1}-\lambda\right)^{-1}v\right\|_2^2 \le \left\|\left(\lambda I - A_{n-1}\right)^{-1}\right\|_2^2\|v\|_2^2$. Since $\left(A_{n-1}-\lambda I\right)^{-1}$ is symmetric, **art. 205** illustrates that $\left\|\left(\lambda I - A_{n-1}\right)^{-1}\right\|_2 = \frac{1}{\lambda_{\min}\left(\lambda I - A_{n-1}\right)}$. If $\lambda = \lambda_1\left(A_n\right)$ is the largest eigenvalue, the interlacing Theorem 71 shows that $\lambda_1\left(A_n\right) \ge \lambda_1\left(A_{n-1}\right)$. By assumption that λ is not an eigenvalue of A_{n-1}, we conclude that $\lambda_1\left(A_n\right) > \lambda_1\left(A_{n-1}\right)$ in which case the minimum eigenvalue of the matrix $\lambda I - A_{n-1}$ equals $\lambda - \lambda_1\left(A_{n-1}\right)$. Hence, as follows from (A.155) and if $\lambda_1\left(A_n\right) > \lambda_1\left(A_{n-1}\right)$, then any component of a principal vector of a symmetric matrix can be lower bounded by

$$z^2 \ge \frac{1}{1 + \frac{\|v\|_2^2}{\left(\lambda_1(A_n)-\lambda_1(A_{n-1})\right)^2}} \tag{A.156}$$

260. *Determinant for the eigenvalue equation of blockmatrix (A.154).* The characteristic polynomial of the recursive, symmetric block matrix A_n in (A.154) is, invoking the Schur complement (A.57),

$$\det\left(A_n - \lambda I\right) = \det \begin{bmatrix} A_{n-1} - \lambda I & v \\ v^T & a_{nn} - \lambda \end{bmatrix}$$

$$= \det\left(A_{n-1} - \lambda I\right)\det\left(a_{nn} - \lambda - \left(v^T\left(A_{n-1} - \lambda I\right)^{-1} v\right)_{1\times 1}\right)$$

and

$$\det\left(A_n - \lambda I\right) = \left(a_{nn} - \lambda - v^T\left(A_{n-1} - \lambda I\right)^{-1} v\right)\det\left(A_{n-1} - \lambda I\right) \qquad \text{(A.157)}$$

For any symmetric matrix A_{n-1}, the resolvent in (A.162) in **art. 262** is

$$\left(A_{n-1} - \lambda I\right)^{-1} = \sum_{m=1}^{n-1} \frac{x_m x_m^T}{\xi_m - \lambda}$$

where $x_1, x_2, \ldots, x_{n-1}$ are the orthogonal eigenvectors of A_{n-1}, belonging to the eigenvalues $\xi_1 \geq \xi_2 \geq \ldots \geq \xi_{n-1}$, respectively. Hence,

$$v^T\left(A_{n-1} - \lambda I\right)^{-1} v = \sum_{m=1}^{n-1} \frac{\left(v^T x_m\right)^2}{\xi_m - \lambda}$$

and (A.157) is written with the projection[5] $c_m = v^T x_m$ of the vector v on the eigenvector x_m as

$$\frac{\det\left(A_n - \lambda I\right)}{\det\left(A_{n-1} - \lambda I\right)} = a_{nn} - \lambda - \sum_{m=1}^{n-1} \frac{c_m^2}{\xi_m - \lambda}$$

Since $\det\left(A_{n-1} - \lambda I\right) = \prod_{k=1}^{n-1}\left(\xi_k - \lambda\right)$ as shown in **art. 235**, we find the characteristic polynomial of A_n, written in terms of the eigenvalues $\{\xi_k\}_{1\leq k\leq n-1}$ of A_{n-1},

$$c_{A_n}(\lambda) = \det\left(A_n - \lambda I\right) = \left(a_{nn} - \lambda\right)\prod_{k=1}^{n-1}\left(\xi_k - \lambda\right) - \sum_{m=1}^{n-1} c_m^2 \prod_{k=1;k\neq m}^{n-1}\left(\xi_k - \lambda\right)$$

$$\text{(A.158)}$$

Equation (A.158) shows that

$$c_{A_n}(\xi_q) = (-1)^{n-q} c_q^2 \prod_{k=1}^{q-1}|\xi_k - \xi_q| \prod_{k=q+1}^{n-1}|\xi_k - \xi_q| \qquad \text{(A.159)}$$

The consequences of (A.158) and (A.159) are analyzed in **art. 264**.

261. *Coefficients of the characteristic polynomial of A_n in terms of those of A_{n-1}.*

---

[5] **Art**. 192 regards the values $(c_1, c_2, \ldots, c_{n-1})$ as the coordinates of the point $v = \sum_{k=1}^{n-1} c_m x_m$ in the $n-1$ dimensional space with respect to the coordinate axes generated by the orthogonal eigenvectors $x_1, x_2, \ldots, x_{n-1}$ of A_{n-1}.

Invoking the expansion of the resolvent of a matrix A, valid for $\lambda > \lambda_1(A)$,

$$(A - \lambda I)^{-1} = -\frac{1}{\lambda}\left(I - \lambda^{-1}A\right)^{-1} = -\frac{1}{\lambda}\sum_{k=0}^{\infty}\frac{A^k}{\lambda^k} = -\frac{1}{\lambda}\left(I + \frac{A}{\lambda} + \frac{A^2}{\lambda^2} + \cdots\right)$$

(A.160)

in (A.157) yields

$$v^T\left(A_{n-1} - \lambda I\right)^{-1}v = -\frac{1}{\lambda}\sum_{k=0}^{\infty}\frac{v^T A_{n-1}^k v}{\lambda^k}$$

We translate (A.157) in the polynomial[6] form $c_{A_n}(\lambda) = \sum_{k=0}^{n}c_k(n)\lambda^k$ in (A.95) of the characteristic polynomial $c_{A_n}(\lambda) = \det(A_n - \lambda I)$,

$$\sum_{k=0}^{n}c_k(n)\lambda^k = \left(a_{nn} - \lambda + \sum_{k=0}^{\infty}\frac{v^T A_{n-1}^k v}{\lambda^{k+1}}\right)\sum_{k=0}^{n-1}c_k(n-1)\lambda^k$$

$$= (a_{nn} - \lambda)\sum_{k=0}^{n-1}c_k(n-1)\lambda^k + \sum_{k=0}^{\infty}\sum_{j=0}^{n-1}\left(v^T A_{n-1}^k v\right)c_j(n-1)\lambda^{j-k-1}$$

The substitution of $m = j - k - 1$ changes the summation in

$$\sum_{k=0}^{\infty}\sum_{j=0}^{n-1}\left(v^T A_{n-1}^k v\right)c_j(n-1)\lambda^{j-k-1} = \sum_{m=-\infty}^{n-2}\sum_{j=\max(0,m+1)}^{n-1}\left(v^T A_{n-1}^{j-m-1}v\right)c_j(n-1)\lambda^m$$

For $m < 0$, the last sum is $v^T\left(\sum_{j=0}^{n-1}c_j(n-1)A_{n-1}^j\right)A_{n-1}^{-m-1}v = 0$ by the Cayley-Hamilton Theorem (**art. 228**), stating that $c_{A_n}(A_n) = \sum_{j=0}^{n}c_j(n)A_n^j = 0$. Hence, we obtain

$$\sum_{k=0}^{n}c_k(n)\lambda^k = (a_{nn} - \lambda)\sum_{k=0}^{n-1}c_k(n-1)\lambda^k + \sum_{k=0}^{n-2}\sum_{j=k+1}^{n-1}c_j(n-1)\left(v^T A_{n-1}^{j-k-1}v\right)\lambda^k$$

Equating corresponding powers in λ yields, for[7] $1 \leq k \leq n-1$, a recursion that expresses the coefficients of the characteristic polynomial of A_n in terms of those

---

[6] The first equation only holds for $\lambda > \lambda_1(A)$, but since the final result is polynomial, by analytic continuation (see e.g. Titchmarsh (1964)), it holds for all complex λ.

[7] For $k = 0$, we find with Cayley-Hamilton's Theorem that

$$c_0(n) = a_{nn}c_0(n-1) + v^T\left(\sum_{j=1}^{n-1}c_j(n-1)A_{n-1}^{j-1}\right)v = c_0(n-1)\left(a_{nn} - v^T A_{n-1}^{-1}v\right)$$

which, indeed, equals (A.157) when $\lambda = 0$, while for $k = n$, $c_n(n) = -c_{n-1}(n-1)$ leads to $c_n(n) = (-1)^n$, a requirement for any characteristic polynomial $c_A(\lambda) = \det(A - \lambda I)$. For $k = n - 1$, it holds that

$$c_{n-1}(n) = a_{nn}c_{n-1}(n-1) - c_{n-2}(n-1) = (-1)^{n-1}a_{nn} - c_{n-2}(n-1)$$

which we write as $(-1)^{n-1}c_{n-1}(n) = a_{nn} + (-1)^{n-2}c_{n-2}(n-1)$. Let $t_n = (-1)^{n-1}c_{n-1}(n)$, then the recursion $t_n = a_{nn} + t_{n-1}$ yields $t_n = \sum_{j=1}^{n}a_{jj}$, which is (A.99).

of A_{n-1},

$$c_k(n) = a_{nn}c_k(n-1) - c_{k-1}(n-1) + \sum_{j=k+1}^{n-1} c_j(n-1)\left(v^T A_{n-1}^{j-(k+1)} v\right) \quad \text{(A.161)}$$

Given the coefficients $\{c_k(n-1)\}_{0 \le k < n}$, the quadratic forms $r_k(v) = v^T A_{n-1}^k v$ for $1 \le k < n$ constitute the major computational effort in (A.161) to determine the coefficients $\{c_k(n)\}_{0 \le k < n}$. Starting with $n = 2$, the set can be iterated up to any size n and any structure in A_n, each n producing the set of coefficients $\{c_k(n)\}_{0 \le k < n}$ of a polynomial with integer coefficients and real zeros. Moreover, **art. 263** shows that all eigenvalues of A_{n-1} interlace with those of A_n. These properties are also shared by orthogonal polynomials (see Chapter 12).

10.5 Interlacing

262. *The resolvent.* The resolvent $(xI - A)^{-1}$ of matrix A is defined in **art. 215** and is related in **art. 230** to the adjoint matrix $Q(x) = (xI - A)^{-1} c_A(x)$, where $c_A(x)$ is the characteristic polynomial of A. Applying $f(A) = \sum_{k=1}^{n} f(\lambda_k) x_k x_k^T$ in (A.88) for a symmetric matrix A to the function $f(y) = \frac{1}{z-y}$, which is everywhere analytic except for $z = y$, yields the spectral decomposition

$$(zI - A)^{-1} = \sum_{k=1}^{n} \frac{x_k x_k^T}{z - \lambda_k} \quad \text{(A.162)}$$

Since $\text{trace}\left(x_k x_k^T\right) = \sum_{j=1}^{n} (x_k)_j^2 = 1$ by orthogonality (A.124), we have

$$\text{trace}\,(zI - A)^{-1} = \sum_{k=1}^{n} \frac{1}{z - \lambda_k} \quad \text{(A.163)}$$

263. *Interlacing.* For any $n \times 1$ real vector y, we obtain from (A.162) for a symmetric matrix A that

$$\phi_y(z) = y^T(zI - A)^{-1} y = \sum_{k=1}^{n} \frac{\left(y^T x_k\right)^2}{z - \lambda_k}$$

which implies that the rational function $\phi_y(z)$ has simple poles at the real eigenvalues of A. Differentiating with respect to z yields for a real vector y

$$\frac{d\phi_y(z)}{dz} = -y^T(zI - A)^{-2} y = -\sum_{k=1}^{n} \frac{\left(y^T x_k\right)^2}{(z - \lambda_k)^2}$$

Since $A = A^T$ is symmetric

$$y^T(zI - A)^{-2} y = y^T\left((zI - A)^{-1}\right)^T (zI - A)^{-1} y = \left\|(zI - A)^{-1} y\right\|_2^2 \ge 0$$

we observe that $\frac{d\phi_y(z)}{dz}$ is always strictly negative whenever z is real and not a pole of $\phi_y(z)$. This implies that each zero of $\phi_y(z)$ must be simple and lying between two consecutive poles of $\phi_y(z)$. By choosing $y = e_j$ equal to the base vector e_j, we find that

$$\phi_{e_j}(z) = (zI - A)_{jj}^{-1} = \frac{\det\left(zI - A_{\backslash\{j\}}\right)}{\det(zI - A)}$$

where the last equality is (A.52) in **art. 215**. Thus, the polynomial $\det\left(zI - A_{\backslash\{j\}}\right)$ has simple zeros that lie between the zeros of $\det(zI - A)$.

In summary, all eigenvalues of the symmetric matrix $A_{\backslash\{j\}}$ lie in between eigenvalues of $A = A^T$,

$$\lambda_{i+1}(A) \le \lambda_i\left(A_{\backslash\{j\}}\right) \le \lambda_i(A)$$

for any $1 \le i \le n-1$. This property is called[8] *interlacing*. Repeating the argument to a principal submatrix $A_{\backslash\{j,k\}}$ of $A_{\backslash\{j\}}$, obtained by deleting a same row k and column k, we arrive at the general interlacing theorem:

Theorem 71 (Interlacing) *For a real symmetric matrix $A_{n \times n}$ and any principal submatrix $B_{m \times m}$ of A obtained by deleting $n - m$ same rows and columns in A, the eigenvalues of B interlace with those of A as*

$$\lambda_{n-m+i}(A) \le \lambda_i(B) \le \lambda_i(A) \tag{A.164}$$

for any $1 \le i \le m$.

Also the zeros of orthogonal polynomials (**art. 364**) are interlaced. There is an interesting corollary of Theorem 71:

Corollary 4 *Let A be a real symmetric $n \times n$ matrix with eigenvalues $\lambda_n(A) \le \lambda_{n-1}(A) \le \cdots \le \lambda_1(A)$ and ordered diagonal elements $d_n \le d_{n-1} \le \cdots \le d_1$ then, for any $1 \le k \le n$, it holds that*

$$\sum_{j=1}^{k} d_j \le \sum_{j=1}^{k} \lambda_j(A)$$

Proof: Let B denote the principal submatrix of A obtained by deleting the rows and columns containing the $n - k$ smallest diagonal elements $d_{k+1}, d_{k+2}, \ldots, d_n$. The trace formula (A.99) indicates that $\text{trace}(B) = \sum_{j=1}^{k} \lambda_j(B)$ and, by construction of B, $\text{trace}(B) = \sum_{j=1}^{k} d_j$. The Interlacing Theorem provides the inequality (A.164) from which $\sum_{j=1}^{k} \lambda_j(B) \le \sum_{j=1}^{k} \lambda_j(A)$. Combining the relations proves the corollary. $\qquad\square$

Corollary 4 is differently proved in (A.181) in **art. 275** based on properties of

---

[8] A sequence of real numbers $b_m \le \cdots \le b_2 \le b_1$ is said to interlace another sequence of real numbers $a_n \le \cdots \le a_2 \le a_1$ with $n > m$, if $a_{n-m+i} \le b_i \le a_i$ for $i = 1, \ldots, m$. The *interlacing is called tight* if there is an integer k, with $0 \le k \le m$, such that $a_i = b_i$ for $i = 1, \ldots, k$, and $a_{n-m+i} = b_i$ for $i = k+1, \ldots, m$.

the $n \times n$ doubly stochastic matrix Ξ, defined in (A.178) and associated to the real symmetric matrix A.

264. *Strict interlacing.* A number of interesting conclusions can be derived from (A.158) and (A.159).

(a) If $\xi_q = \xi_{q+1}$ is an eigenvalue of A_{n-1} with multiplicity larger than 1, then (A.159) indicates that $c_{A_n}(\xi_q) = 0$, implying that ξ_q is also an eigenvalue of A_n. Eigenvalues of A_{n-1} with multiplicity exceeding 1 are found as a degenerate case of the simple eigenvalue situation when $\xi_q \to \xi_{q+l}$ with $l > 1$. Thus, we assume next that all eigenvalues of A_{n-1} are distinct and simple such that the product of the absolute values of the differences of eigenvalues in (A.159) is strict positive.

Then, (A.159) shows that the eigenvalue ξ_q of A_{n-1} cannot be an eigenvalue of A_n, unless $c_q = 0$, which means that v is orthogonal to the eigenvector x_q. If v is not orthogonal to any eigenvector of A_{n-1}, then $c_m \neq 0$ for $1 \leq m \leq n-1$ and $c_{A_n}(\xi_q) \neq 0$ for $1 \leq q \leq n-1$. Moreover, $c_{A_n}(\xi_q)$ is alternatingly negative, starting from $q = n-1$, then positive for $q = n-2$, again negative for $q = n-3$, etc. Since the polynomial $c_{A_n}(x) = (-x)^n + O(x^{n-1})$ for large x as follows from (A.158), there is a zero smaller than ξ_{n-1} (because $c_{A_n}(\xi_{n-1}) < 0$ and $\lim_{x \to -\infty} c_{A_n}(x) > 0$) and a zero larger than ξ_1 (because $(-1)^{n-1} c_{A_n}(\xi_1) > 0$ and $\lim_{x \to \infty}(-1)^{n-1} c_{A_n}(x) < 0$). Since all zeros of $c_{A_n}(x)$ are real (**art. 247**) and the total number of zeros is n (**art. 291**), all zeros of $c_{A_n}(x)$ are simple and there is precisely one zero of $c_{A_n}(x)$ in between two consecutive zeros of $c_{A_{n-1}}(x)$.

This argument presents another derivation of the interlacing principle in **art. 263**. But, the conclusion is more precise and akin to interlacing for orthogonal polynomials (**art. 364**): if the vector v is not orthogonal to any eigenvector of A_{n-1}, which is equivalent to the requirement that $c_m \neq 0$ for $1 \leq m \leq n-1$, then the interlacing is strict in the sense that

$$\lambda_n(A_n) < \lambda_{n-1}(A_{n-1}) < \lambda_{n-1}(A_n) < \ldots < \lambda_1(A_{n-1}) < \lambda_1(A_n)$$

Only if v is orthogonal to some eigenvectors, the corresponding eigenvalues are the same for A_n and A_{n-1}.

(b) If v is proportional to an eigenvector, say $v = c_q x_q$, then $c_m = 0$ for all $1 \leq m \leq n-1$, except when $m = q$, such that (A.158) reduces to

$$\det(A_n - \lambda I) = \{(a_{nn} - \lambda)(\xi_q - \lambda) - c_q^2\} \prod_{k=1; k \neq q}^{n-1} (\xi_k - \lambda)$$

which shows that $\det(A_n - \lambda I)$ and $\det(A_{n-1} - \lambda I)$ have $n-2$ eigenvalues in common and only ξ_q and the zeros of the quadratic equation are different. Indeed, ξ_q is not a zero of $p_2(\lambda) = (a_{nn} - \lambda)(\xi_q - \lambda) - c_q^2$ because $p_2(\xi_q) = -c_q^2 \neq 0$, by construction. An example is given in **art. 85**. The observation is readily extended:

if v is a linear combination of l eigenvectors, then there are $n-1-l$ eigenvalues in common. From (A.158) with $c_{l+1} = c_{l+2} = \ldots = c_{n-1} = 0$, we have

$$\det(A_n - \lambda I) = (a_{nn} - \lambda) \prod_{k=1}^{n-1} (\xi_k - \lambda) - \sum_{m=1}^{l} c_m^2 \prod_{k=1; k \neq m}^{l} (\xi_k - \lambda) \prod_{k=l+1}^{n-1} (\xi_k - \lambda)$$

$$= p_{l+1}(\lambda) \prod_{k=l+1}^{n-1} (\xi_k - \lambda)$$

where $p_{l+1}(\lambda) = (a_{nn} - \lambda) \prod_{k=1}^{l} (\xi_k - \lambda) - \sum_{m=1}^{l} c_m^2 \prod_{k=1; k \neq m}^{l} (\xi_k - \lambda)$. We see that $p_{l+1}(\xi_q) = -c_q^2 \neq 0$, for $1 \leq q \leq l$, by construction and that the real $l + 1$ zeros of the polynomial $p_{l+1}(\lambda)$ determine the zeros of $c_{A_n}(x)$, that are different from those of $c_{A_{n-1}}(x)$.

In summary, if we build up the matrix A_n in (A.154) by iterating from $n = 2$ and requiring in each iteration i that the corresponding $(i - 1) \times 1$ vector v is not orthogonal to any eigenvector of A_{i-1}, then each matrix in the sequence $A_2, A_3, \ldots,$ A_n has only simple eigenvalues, that all interlace over $2 \leq i \leq n$. Their associated characteristic polynomials are very likely a set of orthogonal polynomials.

265. *Symmetric matrix with simple, distinct eigenvalues.* In order to have simple, distinct eigenvalues, it is sufficient for A_2 in (A.154) that all elements in the upper triangular part including the diagonal are different. However, the statement that "the symmetric matrix A_n has only real, simple eigenvalues provided all its upper triangular (including the diagonal) elements are different" is not correct for $n > 2$ as follows from the counter example[9]

$$A_3 = \begin{bmatrix} 9 & 3 & 6 \\ 3 & 1 & 2 \\ 6 & 2 & 4 \end{bmatrix} = A_3^T$$

because the eigenvalues of A_3 are $14, 0, 0$. In fact, A_3 is the $n = 3$ case of a Fibonacci product matrix A_n, with elements $a_{ij} = F_{i+1} F_{j+1}$, where F_i denotes the i-th Fibonacci number. Since $F_{i+1} F_{j+1}$ and $F_{n+1} F_{m+1}$ are only equal if $i = m$ and $j = n$, all elements in the upper triangular part are different. Since all rows are dependent, we have $n - 1$ eigenvalues equal to 0 and one eigenvalue equal to the sum of the diagonal elements, $\sum_{j=1}^{n} F_{j+1}^2$.

Although not correct for $n > 2$, we provide a probabilistic argument that the statement is *in most, but not all cases* correct. A random vector v has almost surely all real elements (components) different. In additional, such a random vector v is almost never orthogonal to any of the $n - 1$ given orthogonal eigenvectors of A_{n-1}, that span the $n - 1$ dimensional space. Intuitively, one may think of a unit sphere in $n = 3$ dimensions in which the eigenvectors form an orthogonal coordinate axis. The (normalized) vector v is a point on the surface of that unit sphere. The orthogonality requirement translates to three circles on the sphere's surface, each of them passing through two orthogonal eigenvector points. The vector v is not

---

[9] This example is due to F.A. Kuipers.

allowed to lie on such a circle. These circles occupy a region with negligible area, thus they have Lesbegue measure zero on that surface. Hence, the probability that v coincides with such a forbidden region is almost zero. Geometric generalizations to higher dimensions are difficult to imagine, but the argument, that the forbidden "orthogonality" regions have a vanishingly small probability to be occupied by a random vector, also holds for $n > 3$. In practice, most matrices A_n that obey the statement have distinct eigenvalues.

266. *General interlacing.* Theorem 71 in **art.** 263 is extended by Haemers (1995):

Theorem 72 (Generalized Interlacing) *Let A be a real symmetric $n \times n$ matrix and S be a real $n \times m$ orthogonal matrix satisfying $S^T S = I$. Denote the eigenvector v_k belonging to the eigenvalue $\lambda_k(B)$ of the $m \times m$ matrix $B = S^T A S$. Then,*

  (i) *the eigenvalues of B interlace with those of A;*
  (ii) *if $\lambda_k(B) = \lambda_k(A)$ or $\lambda_k(B) = \lambda_{n-m+k}(A)$ for some $k \in [1, m]$, then Sv_k is an eigenvector of A belonging to $\lambda_k(A)$;*
  (iii) *if there exists an integer $k \in [0, m]$ such that $\lambda_j(B) = \lambda_j(A)$ for $1 \le j \le k$ and $\lambda_j(B) = \lambda_{n-m+j}(A)$ for $k+1 \le j \le m$, then $SB = AS$.*

Proof: The Rayleigh inequalities (A.129) in **art.** 251, applied to an $m \times 1$ vector s_j being a vector belonging to the space spanned by the eigenvectors $\{v_1, v_2, \ldots, v_j\}$, are $\frac{s_j^T B s_j}{s_j^T s_j} \ge \lambda_j(B)$. Since $\frac{s_j^T B s_j}{s_j^T s_j} = \frac{(Ss_j)^T A S s_j}{(Ss_j)^T S s_j}$, Rayleigh's principle, now applied to the vector Ss_j, states that the right-hand side is smaller than $\lambda_j(A)$ provided Ss_j belongs to the space spanned by $\{x_j, x_j, \ldots, x_n\}$, the last $n+1-j$ eigenvectors of A. In that case, Ss_j can be written as a linear combination, $Ss_j = \sum_{k=j}^{n} c_k x_k$. Using the orthogonality $S^{-1} = S^T$, we have $s_j = \sum_{k=j}^{n} c_k S^T x_k$. Hence, if we choose s_j belonging to the space spanned by $\{v_1, v_2, \ldots, v_j\}$ and orthogonal to the space spanned by $\{S^T x_1, S^T x_2, \ldots, S^T x_{j-1}\}$, then $\lambda_{j+1}(A) \le \lambda_j(B) \le \lambda_j(A)$ for any $1 \le j \le m$. If the same reasoning is applied to $-A$ and $-B$, we obtain $\lambda_j(B) \ge \lambda_{n-m+j}(A)$, thereby proving (i). Equality, occurring in the Rayleigh inequalities, $\lambda_j(B) = \lambda_j(A)$, means that the $s_j = v_j$ is an eigenvector of B belonging to the eigenvalue $\lambda_j(B)$ *and* that $Ss_j = Sv_j = x_j$ is an eigenvector of A belonging to the eigenvalue $\lambda_j(A)$. This proves (ii). The last point (iii) implies, using (ii), that $Sv_1, Sv_2, \ldots, Sv_m$ is an orthonormal set of eigenvectors of A belonging to the eigenvalues $\lambda_1(B), \lambda_2(B), \ldots, \lambda_m(B)$. Left-multiplying the eigenvalue equation $Bv_j = \lambda_j(B)v_j$ by S yields $SBv_j = \lambda_j(B)Sv_j = \lambda_j(A)x_j = Ax_j = ASv_j$ from which $SB = AS$ follows because all $1 \le j \le m$ eigenvectors span the m-dimensional space. $\square$

By choosing $S = \begin{bmatrix} I_{m \times m} & O_{m \times (n-m)} \end{bmatrix}^T$, we find that B is just a principal submatrix of A. This observation shows that Theorem 71 is a special case of the general interlacing Theorem 72, which was already known to Cauchy. Another useful choice is the community matrix S in **art.** 36 that defines the quotient matrix (2.43) of a matrix in **art.** 37. Haemer's Theorem 72 may be compared to

Theorem 73 (Sylvester's law of inertia) *Let A be a real symmetric $n \times n$ matrix and B be a real non-singular $n \times n$ matrix. Then, the matrix $B^T A B$ has the same number of positive, negative and zero eigenvalues as A.*

Sylvester's theorem also holds for Hermitian matrices A and B and follows from the Courant-Fischer expression (A.132) or (A.133) with $y = Bx$. If B is an orthogonal matrix, then Theorem 68 shows that $B^T A B$ is a diagonal matrix D. Sylvester's law of inertia states that the number of positive, negative and zero diagonal elements of D is an invariant of A, that does not depend on the matrix B. Bapat (2013) computes the inertia of threshold graphs (**art.** 114).

267. *Interlacing and the sum $A + B$.*

Lemma 7 *For symmetric $n \times n$ matrices A, B, it holds that*

$$\lambda_n (B) + \lambda_k (A) \le \lambda_k (A + B) \le \lambda_k (A) + \lambda_1 (B) \qquad \text{(A.165)}$$

Proof: The proof is based on the Rayleigh inequalities (**art.** 251) of eigenvalues (see, e.g., Wilkinson (1965, pp. 101-102)). $\qquad \square$

An extension of Lemma 7 is, for $i + j - 1 \le n$,

$$\lambda_{i+j-1} (A + B) \le \lambda_i (A) + \lambda_j (B)$$

which is also called an *interlacing* property. These inequalities are also known as the Courant-Weyl inequalities and also hold for Hermitian matrices (see **art.** 253).

Lemma 8 *If $X = \begin{bmatrix} A & C \\ C^T & B \end{bmatrix}$ is a real symmetric matrix, where A and B are square, and consequently symmetric, matrices, then*

$$\lambda_{\max} (X) + \lambda_{\min} (X) \le \lambda_{\max} (A) + \lambda_{\max} (B) \qquad \text{(A.166)}$$

Proof: See, e.g., Biggs (1996, p. 56). $\qquad \square$

Theorem 74 (Wielandt-Hoffman) *For symmetric matrices A and B, it holds that*

$$\sum_{k=1}^{n} (\lambda_k (A + B) - \lambda_k (A))^2 \le \sum_{k=1}^{n} \lambda_k^2 (B) \qquad \text{(A.167)}$$

Proof: See, e.g., Wilkinson (1965, pp. 104-108). $\qquad \square$

We can rewrite (A.167) with $C = A + B$ and $B = C - A$ as

$$\sum_{k=1}^{n} \lambda_k^2 (A) + \sum_{k=1}^{n} \lambda_k^2 (C) - \sum_{k=1}^{n} \lambda_k^2 (C - A) \le 2 \sum_{k=1}^{n} \lambda_k (A) \lambda_k (C)$$

Using (A.118), we have

$$2 \sum_{k=1}^{n} \lambda_k(A) \lambda_k(C) \geq \text{trace}\left(A^2\right) + \text{trace}\left(C^2\right) - \text{trace}\left(\left(C - A\right)^2\right)$$

$$= \text{trace}\left(CA + AC\right) = 2 \sum_{i=1}^{n} \sum_{j=1}^{n} a_{ij} c_{ij}$$

Hence, an equivalent form of the Wielandt-Hoffman Theorem 74 for symmetric matrices A and B is

$$\text{trace}\left(AB\right) \leq \sum_{k=1}^{n} \lambda_k(A) \lambda_k(B) \tag{A.168}$$

10.6 Non-negative matrices

268. *Reducibility.* A matrix A is reducible if there is a relabeling that leads to

$$\widetilde{A} = \begin{bmatrix} A_1 & B \\ O & A_2 \end{bmatrix}$$

where A_1 and A_2 are square matrices. Otherwise A is irreducible. Relabeling amounts to permuting rows and columns in the same fashion. Thus, there exists a similarity transform H such that $A = H\widetilde{A}H^{-1}$.

For doubly stochastic matrices, where $\sum_{k=1}^{n} a_{kj} = \sum_{k=1}^{n} a_{jk} = 1$, Fiedler (1972) has proposed the "measure $r(A)$ of irreducibility" of A defined as

$$r(A) = \min_{\mathcal{M} \subset \mathcal{N}} \sum_{i \in \mathcal{M}, k \notin \mathcal{M}} a_{ik} \tag{A.169}$$

because A is reducible if there exists a non-empty subset $\mathcal{M}$ of the set of all indices in $\mathcal{N}$ such that $a_{ik} = 0$ for all $i \in \mathcal{M}$ and $k \notin \mathcal{M}$. Hence, if A is reducible, then $r(A) = 0$. Since $\sum_{i \in \mathcal{M}, k \notin \mathcal{M}} a_{ik} \leq \sum_{k=2}^{n} a_{1k} \leq 1$ for a doubly stochastic matrix, the measure of irreducibility lies between $0 \leq r(A) \leq 1$.

269. *The famous Perron-Frobenius theorem for non-negative matrices.*

Theorem 75 (Perron-Frobenius) *An irreducible, non-negative $n \times n$ matrix A always has a real, positive eigenvalue $\lambda_1 = \lambda_{\max}(A)$ and the modulus of any other eigenvalue does not exceed $\lambda_{\max}(A)$, i.e., $|\lambda_k(A)| \leq \lambda_{\max}(A)$ for $k = 2, \ldots, n$. Moreover, λ_1 is a simple zero of the characteristic polynomial $\det(A - \lambda I)$. The eigenvector belonging to λ_1 has positive components.*

If A has h eigenvalues $\lambda_1, \lambda_2, \ldots, \lambda_h$ with $|\lambda_h| = \lambda_1$, then all these equal-moduli eigenvalues satisfy the polynomial $\lambda^h - \lambda_1^h = 0$, i.e., $\lambda_k = \lambda_1 e^{\frac{2\pi i(k-1)}{h}}$ for $k = 1, \ldots, h$.

Proof: See, e.g., Gantmacher (1959b, Chapter XIII). $\square$

If a non-negative $n \times n$ matrix A is reducible, then A has always a *non-negative* eigenvalue $\lambda_{\max}(A)$ and no other eigenvalue has a larger modulus than $\lambda_{\max}(A)$. The corresponding eigenvector belonging to $\lambda_{\max}(A)$ has *non-negative* components. Hence, reducibility removes the positivity of the largest eigenvalue and that of the components of its corresponding eigenvector. An essential Lemma in Frobenius' proof, beside the variational property of the largest eigenvalue

$$\lambda_{\max}(A) = \max_{x \neq 0} \min_{1 \leq j \leq n} \frac{(Ax)_j}{x_j} \tag{A.170}$$

akin to Rayleigh's inequality (A.130) and a consequence of (A.171) in **art.** 270 for a symmetric matrix, is:

Lemma 9 *If A is an $n \times n$ non-negative, irreducible matrix and C is an $n \times n$ complex matrix, in which each element obeys $|c_{ij}| \leq a_{ij}$, then every eigenvalue $\lambda(C)$ of C satisfies the inequality $|\lambda(C)| \leq \lambda_{\max}(A)$.*

Proof: See, e.g., Gantmacher (1959b, Chapter XIII). $\square$

An application of Lemma 9 is Lemma 10 for non-negative matrices, which is useful in assessing the largest eigenvalue of the adjacency matrix of a graph:

Lemma 10 *If one element in a non-negative matrix A is increased, then the largest eigenvalue is also increased. The increase is strict for irreducible matrices.*

Proof: Consider the non-negative matrix C and $A = C + \varepsilon e_i e_j^T$, where $\varepsilon > 0$, e_i and e_j are the basic vectors and $\tilde{O} = e_i e_j^T$ is the zero matrix, except for the element $\tilde{O}_{ij} = 1$. Lemma 9 shows that $\lambda_{\max}(A) \geq \lambda(C)$. We now demonstrate the strict inequality $\lambda_{\max}(A) > \lambda(C)$ for irreducible matrices. If x denotes the eigenvector belonging to the largest eigenvalue of C, then the variational property (A.170) implies

$$\lambda_{\max}(A) \geq \frac{x^T A x}{x^T x} = \frac{x^T C x}{x^T x} + \varepsilon \frac{x^T e_i e_j^T x}{x^T x} = \lambda_{\max}(C) + \varepsilon \frac{x_i x_j}{x^T x}$$

Since all components of the largest eigenvector x are non-negative and even positive if C is irreducible, the lemma is proved. $\square$

270. *Bounds for the largest eigenvalue of symmetric, irreducible, non-negative matrices.* If the irreducible, non-negative matrix is symmetric, we can exploit symmetry to deduce bounds for the largest eigenvalue by considering the quadratic form $y^T A x_1 = x_1^T A y$, where x_1 is the eigenvector with positive components (Perron-Frobenius Theorem 75) belonging to the largest eigenvalue λ_1 and y is a vector with positive components. Using the eigenvalue equation $A x_1 = \lambda_1 x_1$, we obtain

$y^T A x_1 = \lambda_1 y^T x_1$. On the other hand, we have

$$x_1^T A y = \sum_{i=1}^{n} \sum_{j=1}^{n} a_{ij} (x_1)_i y_j = \sum_{i=1}^{n} (x_1)_i y_i \left(\sum_{j=1}^{n} \frac{a_{ij} y_j}{y_i} \right)$$

and, since the components of y and x_1 are positive,

$$\left(\min_{1 \le i \le n} \sum_{j=1}^{n} \frac{a_{ij} y_j}{y_i} \right) \sum_{i=1}^{n} (x_1)_i y_i \le x_1^T A y \le \left(\max_{1 \le i \le n} \sum_{j=1}^{n} \frac{a_{ij} y_j}{y_i} \right) \sum_{i=1}^{n} (x_1)_i y_i$$

By combining both expressions, taking into account that $y^T x_1 = \sum_{i=1}^{n} (x_1)_i y_i > 0$, we obtain, for a positive vector y and for any symmetric, irreducible, non-negative matrix A, the bounds

$$\min_{1 \le i \le n} \frac{(Ay)_i}{y_i} \le \lambda_1 \le \max_{1 \le i \le n} \frac{(Ay)_i}{y_i} \tag{A.171}$$

271. *Maximum ratio of principal eigenvector components of a positive matrix.* Ostrowski (1960) considered the maximum ratio

$$\gamma = \max_{1 \le i,j \le n} \frac{(x_1)_i}{(x_1)_j} \tag{A.172}$$

of components of the principal eigenvector x_1 of an $n \times n$ positive matrix M, that is irreducible, hence, $(x_1)_j > 0$ for all $1 \le j \le n$. Minc (1970) briefly overviews the results and proves the following theorem:

Theorem 76 (Minc) *Let M be a positive $n \times n$ matrix with principal eigenvector x_1, then*

$$\gamma \le \max_{1 \le s,r,j \le n} \frac{m_{sj}}{m_{rj}} \tag{A.173}$$

Equality in (A.173) holds if and only if the p-th row of M is a multiple of the q-th row, for some pairs of indices p and q satisfying $\frac{m_{ph}}{m_{qh}} = \max_{1 \le s,r,j \le n} \frac{m_{sj}}{m_{rj}}$.

The main idea of Minc's proof is as follows. We consider the eigenvalue equation for both the minimum $(x_1)_\nu = \min_{1 \le j \le n} (x_1)_j$ and the maximum $(x_1)_\mu = \max_{1 \le j \le n} (x_1)_j$ principal eigenvector component:

$$\begin{cases} \lambda_1 (M) (x_1)_\mu = \sum_{j=1}^{n} m_{\mu j} (x_1)_j \\ \lambda_1 (M) (x_1)_\nu = \sum_{j=1}^{n} m_{\nu j} (x_1)_j \end{cases}$$

Their ratio equals

$$\gamma = \frac{(x_1)_\mu}{(x_1)_\nu} = \frac{\sum_{j=1}^{n} m_{\mu j} (x_1)_j}{\sum_{j=1}^{n} m_{\nu j} (x_1)_j} \tag{A.174}$$

Since $(x_1)_j > 0$ and $m_{ij} > 0$, invoking the inequality (3.88) yields

$$\gamma \leq \max_{1 \leq j \leq n} \frac{m_{\mu j}}{m_{\nu j}} \leq \max_{1 \leq s,r,j \leq n} \frac{m_{sj}}{m_{rj}}$$

If equality holds, then $\gamma = \frac{m_{\mu j}}{m_{\nu j}}$ for $1 \leq j \leq n$, illustrating that row μ and ν are linearly dependent and that $\det M = 0$ in **art. 209**. We refer to Minc (1970) for the converse, which then proves Theorem 76. Theorem 76 also applies to the matrix $M + yI$, because eigenvectors are the same for M and $M + yI$. The ratio (A.173) can be sharpened by choosing the optimal value for y.

272. *Eigenvector components of a non-negative matrix.* Fiedler (1975) found a nice property regarding the signs of eigenvector components of a non-negative symmetric matrix, that have a profound impact on graph partitioning (**art. 150**).

Theorem 77 (Fiedler) *Let A be an irreducible, non-negative symmetric $n \times n$ matrix with eigenvalues $\lambda_1(A) \geq \lambda_2(A) \geq \ldots \lambda_n(A)$ and z be a vector such that $Az \geq \lambda_k(A)z$ with $k \geq 2$. Then, the set of indices (nodes) $\mathcal{M} = \{j \in \mathcal{N} : z_j \geq 0\}$ is not empty and the number of connected components of the principal submatrix $A(\mathcal{M})$, with indices of rows and columns belonging to $\mathcal{M}$, is not larger than $k - 1$.*

Before proving the theorem, we rephrase the theorem when A is the adjacency matrix of a graph G. The non-negative vector components of z correspond to nodes, that induce a subgraph, specified by the adjacency matrix $A(\mathcal{M})$, with at most $k - 1$ distinct connected components.

Proof[10]: The set $\mathcal{M}$ cannot be empty. For, if $\mathcal{M}$ were empty, then all components of z would be negative such that $v = -z$ satisfies $Av \leq \lambda_k(A)v$. Since A is irreducible, Perron-Frobenius Theorem 75 demonstrates that $Ax_1 = \lambda_1(A)x_1$ and $\lambda_1(A) > \max(\lambda_2(A), \lambda_n(A))$. Thus, $x_1^T Av = \lambda_1(A)x_1^T v > \lambda_k(A)x_1^T v$, while the hypothesis implies that $x_1^T Av \leq \lambda_k(A)x_1^T v$, which leads to a contradiction. If $\mathcal{M} = \mathcal{N}$, the theorem is true by the Perron-Frobenius Theorem 75. Suppose now that $\mathcal{M} \neq \mathcal{N}$. Then, we can always write the matrix A as $A = \begin{bmatrix} \widetilde{A} & C \\ C^T & D \end{bmatrix}$,

where $\widetilde{A}$ consists of r distinct connected or irreducible matrices A_j, subject to $\sum_{j=1}^{r} \dim(A_j) = \dim \mathcal{M} < n$ with structure

$$\widetilde{A} = \begin{bmatrix} A_1 & O & \cdots & O \\ O & A_2 & \ddots & \vdots \\ \vdots & \ddots & \ddots & O \\ O & \cdots & O & A_r \end{bmatrix} \text{ and } C = \begin{bmatrix} C_1 \\ C_2 \\ \vdots \\ C_r \end{bmatrix}$$

We partition the vector z conformally,

$$z = \begin{bmatrix} x \\ -y \end{bmatrix}$$

---

[10] We have combined Fiedler's proof with that of Powers (1988).

where the vector $x^T = \begin{bmatrix} x_1^T & x_2^T & \cdots & x_r^T \end{bmatrix}$ has subvectors x_j all with non-negative components whose indices belong to $\mathcal{M}$. This implies that y contains only positive components, otherwise a component of y would belong to $\mathcal{M}$. The condition $Az \geq \lambda_k(A)z$ implies that $A_j x_j - C_j y \geq \lambda_k(A)x_j$. Since A is irreducible, none of the block matrices C_j can be zero such that $C_j y \geq 0$, with inequality in some component because all components of y are strictly positive. Hence, $A_j x_j \geq \lambda_k(A)x_j$ holds with strict inequality in some component which implies that $x_j^T A_j x_j > \lambda_k(A) x_j^T x_j$ for $1 \leq j \leq r$. By construction, each A_j is irreducible. The Perron-Frobenius Theorem 75 and the Rayleigh inequality (**art.** 251) for the largest eigenvalue state that $\lambda_1(A_j) x_j^T x_j \geq x_j^T A_j x_j$ such that $\lambda_1(A_j) > \lambda_k(A)$. Finally, the interlacing Theorem 71 shows that, if $\lambda_1(A_j) > \lambda_k(A)$ for all $1 \leq j \leq r$, then $\lambda_r(A) > \lambda_k(A)$ and $r \leq k - 1$. This proves the theorem. $\qquad\square$

An immediate consequence is that the vector $z = \alpha v_1 + v_2$, where v_1 is the largest eigenvector with all positive components and v_2 is the second largest eigenvector of A, satisfies, for $\alpha \geq 0$, $A(\alpha v_1 + v_2) = \alpha \lambda_1(A) v_1 + \lambda_2(A) v_2 \geq \lambda_2(A)(\alpha v_1 + v_2)$ and thus the inequality in Theorem 77 for $k = 2$. Hence, the index set $\mathcal{M} = \left\{ j \in \mathcal{N} : \alpha(v_1)_j + (v_2)_j \geq 0 \right\}$ corresponds to an irreducible submatrix of $A(\mathcal{M})$. Since A and $A(\mathcal{M})$ are irreducible, it means that $A(\mathcal{M}^c)$, where $\mathcal{M} \cup \mathcal{M}^c = \mathcal{N}$ and $\mathcal{M}^c = \left\{ j \in \mathcal{N} : \alpha(v_1)_j + (v_2)_j < 0 \right\}$, is also irreducible. This index set $\mathcal{M}$ decomposes the set of indices (nodes) into two irreducible submatrices (connected subgraphs).

273. *Bounds on eigenvalues of the adjacency matrix.* We present a consequence of Fiedler's eigenvector component Theorem 77 in **art.** 272. Consider the eigenvalue equation $Ax_k = \lambda_k(A)x_k$, where the eigenvalue $\lambda_k(A)$ is smaller than the largest eigenvalue $\lambda_1(A)$. The corresponding real eigenvector x_k is orthogonal to x_1, whose vector components are positive by virtue of the Perron-Frobenius Theorem 75. Let us denote the nodal sets

$$\mathcal{M}_+ = \left\{ j \in \mathcal{N} : (x_k)_j > 0 \right\}, \mathcal{M}_- = \left\{ j \in \mathcal{N} : (x_k)_j < 0 \right\}, \mathcal{M}_0 = \left\{ j \in \mathcal{N} : (x_k)_j = 0 \right\}$$

such that $\mathcal{M}_+ \cup \mathcal{M}_- \cup \mathcal{M}_0 = \mathcal{N}$. Since $x_k^T x_1 = 0$ by orthogonality (**art.** 247), it holds that $|\mathcal{M}_+| \geq 1$ and $|\mathcal{M}_-| \geq 1$, whence $|\mathcal{M}_0| \leq N - 2$ for any eigenvalue $\lambda_k(A)$ with index $k > 1$.

Suppose that $(x_k)_l = \min_{1 \leq j \leq N} (x_k)_j < 0$ and $(x_k)_m = \max_{1 \leq j \leq N} (x_k)_j > 0$. The eigenvalue equation (1.4) for component l is, assuming that $\lambda_k(A) > 0$,

$$\lambda_k(A)(x_k)_l = \sum_{j=1}^N a_{lj}(x_k)_j \geq (|\mathcal{M}_-| - 1)(x_k)_l$$

while that for component m is $\lambda_k(A)(x_k)_m \leq (|\mathcal{M}_+| - 1)(x_k)_m$. Thus, provided $\lambda_k(A) > 0$, we have that $\lambda_k(A) \leq (|\mathcal{M}_-| - 1)$ and $\lambda_k(A) \leq (|\mathcal{M}_+| - 1)$, from which $\lambda_k(A) \leq \min(|\mathcal{M}_-|, |\mathcal{M}_+|) - 1 \leq \frac{|\mathcal{M}_-| + |\mathcal{M}_+|}{2} - 1 = \frac{N - |\mathcal{M}_0|}{2} - 1$. Since

$|\mathcal{M}_0| \geq 0$, we find that

$$\lambda_2(A) \leq \left\lfloor \frac{N}{2} \right\rfloor - 1 \tag{A.175}$$

When $\lambda_k(A) < 0$, we obtain $\lambda_k(A)(x_k)_l = |\lambda_k(A)| \cdot |(x_k)_l| \leq |\mathcal{M}_+|(x_k)_m$ and $\lambda_k(A)(x_k)_m \geq |\mathcal{M}_-|(x_k)_l$. Thus, $|\lambda_k(A)|(x_k)_m \leq |\mathcal{M}_-|(x_k)_l$ from which we deduce, after multiplying both inequalities $|\lambda_k(A)| \leq \sqrt{|\mathcal{M}_+||\mathcal{M}_-|} \leq \frac{|\mathcal{M}_-|+|\mathcal{M}_+|}{2}$. Since $\sqrt{|\mathcal{M}_+||\mathcal{M}_-|} = \sqrt{|\mathcal{M}_+|(N - |\mathcal{M}_+| - |\mathcal{M}_0|)}$ and $|\mathcal{M}_0| \geq 0$, this quantity is maximal if $|\mathcal{M}_+| = \left\lfloor \frac{N}{2} \right\rfloor$ and $|\mathcal{M}_0| = 0$. Hence, the smallest eigenvalue of A obeys

$$\lambda_N(A) \geq -\sqrt{\left\lfloor \frac{N}{2} \right\rfloor \left\lceil \frac{N}{2} \right\rceil} \tag{A.176}$$

In addition to this bound (A.176), the Perron-Frobenius Theorem 75 as well as Theorem 109 indicate that $\lambda_N(A) \geq -\lambda_1(A)$.

We end this section on non-negative matrices by pointing to yet another nice article by Fiedler and Pták (1962), that studies the class of real square matrices with non-positive off-diagonal elements, to which the Laplacian matrix of a graph belongs. We also mention *totally positive matrices*. An $n \times m$ matrix is said to be totally positive if all its minors are non-negative. The current state of the art is treated by Pinkus (2010), who shows that the eigenvalues of square, totally positive matrices are both real and non-negative.

10.7 Doubly stochastic matrices

A non-negative matrix A is doubly stochastic if both its row and column sums are 1, i.e. $Au = u$ and $u^T A = u$. Sinkhorn (1964) has demonstrated that any matrix A with strictly positive elements can be made doubly stochastic by pre- and post-multiplication by diagonal matrices. Thus, for any strictly positive matrix A (without zero elements), there exist positive diagonal matrices D_1 and D_2 such that $D_1 A D_2$ is doubly stochastic.

274. *Diagonal elements of E_k and the doubly stochastic matrix* Ξ. It directly follows from (A.138) that $a_{jj} = \sum_{k=1}^{n} \lambda_k(E_k)_{jj} = \sum_{k=1}^{n} \lambda_k(x_k)_j^2$ for each $1 \leq j \leq n$ and by **art.** 243 that

$$(A^m)_{jj} = \sum_{k=1}^{n} \lambda_k^m (x_k)_j^2 \tag{A.177}$$

Geometrically, the scalar product of the eigenvalue vector $\lambda = (\lambda_1, \lambda_2, \ldots, \lambda_n)$ with the vectors $\xi_k = \left((x_k)_1^2, (x_k)_2^2, \ldots, (x_k)_n^2\right)$, where $(x_k)_j$ is the j-th component of the k-th eigenvector of A belonging to λ_k, equals the diagonal element a_{jj}. With the $n \times n$ non-negative matrix $\Xi = X \circ X$, where $\circ$ denotes the Hadamard product,

$$\Xi = \begin{bmatrix} (x_1)_1^2 & (x_2)_1^2 & (x_3)_1^2 & \cdots & (x_n)_1^2 \\ (x_1)_2^2 & (x_2)_2^2 & (x_3)_2^2 & \cdots & (x_n)_2^2 \\ (x_1)_3^2 & (x_2)_3^2 & (x_3)_3^2 & \cdots & (x_n)_3^2 \\ \vdots & \vdots & \vdots & \ddots & \vdots \\ (x_1)_n^2 & (x_2)_n^2 & (x_3)_n^2 & \cdots & (x_n)_n^2 \end{bmatrix} \tag{A.178}$$

and with the vector $b = (a_{11}, a_{22}, \ldots, a_{nn})$, the relation $a_{jj} = \sum_{k=1}^n \lambda_k (x_k)_j^2$ reads in matrix form

$$\Xi \lambda = b \tag{A.179}$$

If a function f of the vector $v = (v_1, v_2, \ldots, v_n)$ is denoted by

$$f(v) = (f(v_1), f(v_2), \ldots, f(v_n))$$

then $f(A) = \sum_{k=1}^n f(\lambda_k) x_k x_k^T$ in (A.88) leads to

$$\Xi f(\lambda) = f(b) \tag{A.180}$$

Since $\Xi u = u$ and $\Xi^T u = u$, by "double orthogonality" of (A.124) and (A.126), and since each element $0 \le (x_k)_j^2 \le 1$, the matrix Ξ with squared eigenvector components of a diagonalizable matrix A is doubly stochastic with largest eigenvalue equal to 1. Since Ξ is a non-negative matrix and $\Xi u = u$, the Perron-Frobenius Theorem in **art. 269** indicates that the eigenvalue 1 belonging to the eigenvector u with non-negative components is the largest one and that the absolute value of any other eigenvalue is smaller than 1. Hence, all eigenvalues of the asymmetric matrix Ξ lie within the unit circle.

275. *Partial sum inequalities and doubly stochastic matrices.* Consider the $n \times 1$ real vectors α and β, both with ordered components, i.e. $\alpha_1 \ge \alpha_2 \ge \ldots \ge \alpha_n$ and $\beta_1 \ge \beta_2 \ge \ldots \ge \beta_n$.

Theorem 78 *If P is a doubly stochastic matrix with elements $0 \le p_{ij} \le 1$ and $\alpha = P\beta$, then the partial sum inequalities hold,*

$$\sum_{i=1}^k \alpha_i \le \sum_{i=1}^k \beta_i \qquad \text{for } 1 \le k < n$$

$$\sum_{i=1}^n \alpha_i = \sum_{i=1}^n \beta_i$$

which is denoted as $\alpha \prec \beta$ and expresses that the vector β majorizes the vector α.

Proof (Marshall *et al.*, 2011, p. 31): Summing the first k components in $\alpha = P\beta$ yields $\sum_{i=1}^k \alpha_i = \sum_{i=1}^k (P\beta)_i = \sum_{i=1}^k \sum_{j=1}^n p_{ij}\beta_j = \sum_{j=1}^n \left(\sum_{i=1}^k p_{ij} \right) \beta_j$. We denote $t_j = \sum_{i=1}^{k} p_{ij}$ and t_j satisfies $0 \le t_j \le 1$ and $\sum_{j=1}^{n} t_j = \sum_{i=1}^{k} \sum_{j=1}^{n} p_{ij} = k$. Hence,

$$\sum_{i=1}^{k} \alpha_i - \sum_{i=1}^{k} \beta_i = \sum_{j=1}^{n} t_j \beta_j - \sum_{j=1}^{k} \beta_j = \sum_{j=1}^{n} t_j \beta_j - \sum_{j=1}^{k} \beta_j + \beta_k \left(k - \sum_{j=1}^{n} t_j \right)$$

Splitting the n-upper limit sums at the right-hand side,

$$\sum_{i=1}^{k} \alpha_i - \sum_{i=1}^{k} \beta_i = \sum_{j=1}^{k} t_j \beta_j + \sum_{j=k+1}^{n} t_j \beta_j - \sum_{j=1}^{k} \beta_j + \beta_k k - \beta_k \sum_{j=1}^{k} t_j - \beta_k \sum_{j=k+1}^{n} t_j$$

and recombining yields, recalling that $\beta_1 \ge \beta_2 \ge \ldots \ge \beta_n$,

$$\sum_{i=1}^{k} \alpha_i - \sum_{i=1}^{k} \beta_i = \sum_{j=1}^{k-1} (t_j - 1)(\beta_j - \beta_k) + \sum_{j=k+1}^{n} t_j (\beta_j - \beta_k) \le 0$$

The equality for $k = n$ follows from $u^T = u^T P$ as $u^T \alpha = u^T P \beta = u^T \beta$. $\square$

The proof does not use the ordering of the vector α, only that β is ordered. Except for the $k = n$ case, the proof also holds for a doubly substochastic matrix S, in which there are rows and columns that sum to a value less than 1. Indeed, denote $\tau_j = \sum_{i=1}^{k} s_{ij}$, it now holds that $\sum_{j=1}^{n} \tau_j \le k$, so that

$$\sum_{i=1}^{k} \alpha_i - \sum_{i=1}^{k} \beta_i = \sum_{j=1}^{n} \tau_j \beta_j - \sum_{j=1}^{k} \beta_j \le \sum_{j=1}^{n} \tau_j \beta_j - \sum_{j=1}^{k} \beta_j + \beta_k \left(k - \sum_{j=1}^{n} \tau_j \right)$$

and the remainder of the proof remains unchanged.

Application of Theorem 78 to $\Xi f(\lambda) = f(b)$ in (A.180), where the components of the vector $f(\lambda)$ decrease with index i, shows that $u^T f(\lambda) = u^T f(b)$ and, for $1 \le k < n$,

$$\sum_{i=1}^{k} (f(A))_{ii} \le \sum_{i=1}^{k} f(\lambda_l) \tag{A.181}$$

The partial sum inequalities (A.181) are also proved in Corollary 4 in **art. 263** based on interlacing.

276. *Convex functions and doubly stochastic matrices.* Schur (Marshall *et al.*, 2011, Chapter 3) demonstrated in 1923 another type of inequality that involves doubly stochastic matrices and a continuous convex function g. Consider the i-th component in (A.179), $b_i = \sum_{k=1}^{n} \xi_{ik} \lambda_k$ with $\sum_{k=1}^{n} \xi_{ik} = \sum_{i=1}^{n} \xi_{ik} = 1$ and $\xi_{ik} = (x_k)_i^2$, then the definition of convexity (Hardy *et al.*, 1999, art. 90, p. 74)

$$g\left(\frac{\sum_{k=1}^{n} p_k x_k}{\sum_{k=1}^{n} p_k} \right) \le \frac{\sum_{k=1}^{n} p_k g(x_k)}{\sum_{k=1}^{n} p_k} \tag{A.182}$$

with equality only if g is linear or all x_k are the same, shows that

$$g\left(b_i\right) = g\left(\sum_{k=1}^{n} \xi_{ik}\lambda_k\right) \leq \sum_{k=1}^{n} \xi_{ik}g\left(\lambda_k\right)$$

Summing over all i, $\sum_{i=1}^{n} g\left(b_i\right) \leq \sum_{i=1}^{n}\sum_{k=1}^{n} \xi_{ik}g\left(\lambda_k\right) = \sum_{k=1}^{n}\left(\sum_{i=1}^{n} \xi_{ik}\right) g\left(\lambda_k\right)$ results in Schur's inequality,

$$\sum_{i=1}^{n} g\left(b_i\right) \leq \sum_{k=1}^{n} g\left(\lambda_k\right) \tag{A.183}$$

which is, in vector form, $u^T g\left(b\right) \leq u^T g\left(\lambda\right)$.

277. The next remarkable theorem by Fiedler (1972) bounds the spectral gap for symmetric stochastic matrices, that are doubly stochastic.

Theorem 79 (Fiedler) *Let P be an $n \times n$ symmetric stochastic matrix with second largest eigenvalue $\lambda_2\left(P\right)$. Then*

$$\psi_n\left(r\left(P\right)\right) \leq 1 - \lambda_2\left(P\right) \leq \frac{n}{n-1}r\left(P\right) \tag{A.184}$$

where the measure of irreducibility $r\left(A\right)$ is defined in (A.169) and where the continuous, convex and increasing function $\psi_n\left(x\right) \in [0,1]$ is

$$\psi_n\left(x\right) = \begin{cases} 2x\left(1 - \cos\frac{\pi}{n}\right) & 0 \leq x \leq \frac{1}{2} \\ 1 - 2\left(1 - x\right)\cos\frac{\pi}{n} - \left(2x - 1\right)\cos\frac{2\pi}{n} & \frac{1}{2} < x \leq 1 \end{cases}$$

The inequality (A.184) is best possible: if $u, v \in \mathbb{R}$ satisfy $0 \leq u \leq 1$ and $\psi_n\left(u\right) \leq 1 - v \leq \frac{n}{n-1}u$, then there exists a symmetric stochastic matrix P with $r\left(P\right) = u$ and $\lambda_2\left(P\right) = v$.

Proof: The proof is rather involved and we refer to Fiedler (1972). $\qquad\square$

10.8 Positive (semi) definiteness

278. *Positive definiteness.* A matrix $A \in \mathbb{R}^{n \times n}$ is positive definite if the quadratic form $x^T A x > 0$ for all non-zero vectors $x \in \mathbb{R}^n$. This definition implies that A is non-singular for otherwise there would exist a non-zero vector x such that $x^T A x = 0$.

We start with a *basic property*: If $A \in \mathbb{R}^{n \times n}$ is positive definite and $Y \in \mathbb{R}^{k \times n}$ has rank k, then the $k \times k$ matrix $B = Y^T A Y$ is also positive definite. Indeed, suppose that the non-zero vector $z \in \mathbb{R}^k$ satisfies $0 \geq z^T B z = \left(Yz\right)^T A\left(Yz\right)$, then $Yz = 0$ by the positive definiteness of A. But Y has full column rank, which implies that $z = 0$, leading to a contradiction.

A consequence of the basic property is that all principal submatrices of A are positive definite. In particular, *all diagonal elements of a positive definite matrix*

A are positive. By choosing Y equal to the k column vectors of the identity matrix $I_{n \times n}$, any principal submatrix of A is found as $B = Y^T A Y$.

If A is positive semidefinite, then any principal submatrix of A is also positive semidefinite. This property is less stringent than the basic property for positive definiteness, because $z^T B z = (Yz)^T A (Yz) \geq 0$ for any vector Yz.

279. *Elements in a symmetric positive semidefinite matrix.* If $A \in \mathbb{R}^{n \times n}$ is symmetric positive semidefinite, then

$$|a_{ij}| \leq \frac{1}{2} (a_{ii} + a_{jj}) \tag{A.185}$$

$$|a_{ij}| \leq \sqrt{a_{ii} a_{jj}} \tag{A.186}$$

Proof: Since $\sqrt{xy} \leq \frac{x+y}{2}$ for positive real numbers x and y as follows from $\left(\sqrt{x} - \sqrt{y}\right)^2 \geq 0$, we only need to prove the geometric mean inequality (A.186). We give two different proofs.

(i) Positive semidefiniteness implies that $x^T A x \geq 0$ for any vector x. Choose now $x = e_i + \alpha e_j$, where α is a real number and e_i is a basis vector with $(e_i)_k = \delta_{ik}$. Using $a_{ij} = e_i^T A e_j$ yields

$$x^T A x = a_{ii} + 2\alpha a_{ij} + a_{jj}\alpha^2 = a_{jj} \left(\alpha + \frac{a_{ij}}{a_{jj}}\right)^2 + \frac{a_{ii} a_{jj} - a_{ij}^2}{a_{jj}}$$

and the positive semidefiniteness requires that $x^T A x \geq 0$, which is equivalent to the condition that $a_{ii} a_{jj} \geq a_{ij}^2$, because the diagonal elements of A are non-negative (**art. 278**).

(ii) Consider a principal submatrix of A, which is also positive semidefinite (**art. 278**). Without loss of generality, we can choose the principal submatrix $A_s = \begin{bmatrix} a_{11} & a_{12} \\ a_{21} & a_{22} \end{bmatrix}$ and the vector $z = \begin{bmatrix} \alpha \\ 1 \end{bmatrix}$. Then, $0 \leq z^T A_s z = a_{11}\alpha^2 + 2a_{12}\alpha + a_{22}$, which requires that the discriminant $4a_{12}^2 - 4a_{11}a_{22} \leq 0$. $\qquad \square$

Since the inequality (A.185) holds for all i and j, it also implies

$$\max_{i,j} |a_{ij}| \leq \max_j a_{jj}$$

280. The *Gram matrix* associated to the vectors $a_1, a_2, \ldots, a_n$ is defined as

$$G = A^T A, \qquad A = \begin{bmatrix} a_1 & a_2 & \cdots & a_n \end{bmatrix}$$

so that $G_{ij} = a_i^T a_j$ and $G_{ii} = a_i^T a_i = |a_i|^2$ for $i = 1, \ldots, n$. The Gram matrix $G = A^T A$ is symmetric and positive semidefinite because $x^T G x = (Ax)^T Ax = \|Ax\|_2^2 \geq 0$. **Art. 199** implies that all eigenvalues of G are real and non-negative. When a matrix G is positive semidefinite and symmetric, we can find the matrix A as the square root $A = \sqrt{G}$. Indeed, the eigenvalue decomposition is $G = U\text{diag}(\lambda_k(G))U^T$, where $U = \begin{bmatrix} u_1 & u_2 & \cdots & u_n \end{bmatrix}$ is an orthogonal matrix $\left(U^T U = U U^T = I\right)$ formed by the scaled, real eigenvectors u_k belonging to eigenvalue $\lambda_k(G)$. Since all eigenvalues are real and non-negative, it holds that $\sqrt{\lambda_k(G)}$ are real such that $G = U \text{diag}\left(\sqrt{\lambda_k(G)}\right)\text{diag}\left(\sqrt{\lambda_k(G)}\right)U^T$. Any orthogonal matrix W satisfies $W^T W = I$ (see **art. 247**) and a more general form is

$$G = AA^T = U \text{diag}(\sqrt{\lambda_k(G)})W^T W \left(\text{diag}(\sqrt{\lambda_k(G)})\right)^T U^T$$

from which

$$A = U \text{diag}(\sqrt{\lambda_k(G)})W^T$$

The matrix A is also called the *square root* matrix of G, but it is not unique, because we can choose any orthogonal matrix W, such as, for example, $W = I$. If $W = U$, we construct a *symmetric* square root matrix $A = A^T = U \text{diag}(\sqrt{\lambda_k(G)})U^T$, so that $G = A^2$. The matrix A can be found from the singular value decomposition of G or from Cholesky factorization (Press *et al.*, 1992). The Cholesky method gives a solution $A = \sqrt{G}$ that is, in general, not symmetric. Another example of a non-symmetric "square root" matrix A appears in **art. 374**.

Moreover, if R is an orthogonal matrix for which $R^T R = I$, then $\tilde{A} = RA$ has a same Gram matrix since

$$\tilde{G} = \tilde{A}^T \tilde{A} = (RA)^T RA = A^T R^T RA = A^T A = G$$

Hence, given a solution A of $G = A^T A$, all other solutions are found by orthogonal transformation.

In summary, any symmetric, positive semidefinite matrix can be considered as a Gram matrix G whose diagonal elements are non-negative, $G_{ii} \geq 0$. The non-negativeness of the diagonal elements was already demonstrated in **art. 278** and **art. 279**.

281. *Stieltjes matrix.* An $n \times n$ positive definite matrix A with non-positive off-diagonal elements, i.e. $a_{ij} \leq 0$ for all $i \neq j$, is called a Stieltjes matrix. Let $\lambda_1 \geq \lambda_2 \geq \ldots \geq \lambda_n > 0$ be the eigenvalues of a Stieltjes matrix A, then Micchelli and Willoughby (1979) demonstrate that the matrix polynomials

$$F_m(A) = \prod_{k=1}^{m}(\lambda_k I - A)$$

are non-negative matrices: any element $(F_m(A))_{ij} \geq 0$ for $1 \leq m \leq n$. The polynomials $F_m(z)$ in (B.17) arise in Lagrange (**art. 303**) and Newton interpolation (**art. 306**), as well as in function expansions of a matrix (**art. 234**).

282. If all eigenvalues are real and $\lambda_k > 0$ as in a symmetric, positive definite matrix (**art. 199**), we can apply the general theorem of the arithmetic and geometric mean in several real variables $x_k \geq 0$, which is nicely treated by Hardy *et al.* (1999),

$$\prod_{k=1}^{n} x_k^{q_k} \leq \sum_{k=1}^{n} q_k x_k \tag{A.187}$$

where $\sum_{k=1}^{n} q_k = 1$, to (A.98) and (A.99) with $q_k = \frac{\alpha_k}{\sum_{j=1}^{n} \alpha_j} \geq 0$,

$$\prod_{k=1}^{n} \lambda_k^{\alpha_k} \leq \left(\frac{\sum_{k=1}^{n} \alpha_k \lambda_k}{\sum_{j=1}^{n} \alpha_j} \right)^{\sum_{j=1}^{n} \alpha_j}$$

Choosing $\alpha_j = 1$, so that $\det A = \prod_{k=1}^{n} \lambda_k$ in (A.98), leads for an $n \times n$ symmetric, positive definite matrix A to the inequality

$$0 < \det A \leq \left(\frac{\mathrm{trace}(A)}{n} \right)^n$$

283. Let M be a symmetric and positive semidefinite $n \times n$ matrix with $Mu = 0$. Any square matrix whose n row sums are zero has an eigenvalue zero with corresponding eigenvector u. Let W denote the set of all column vectors x that satisfy $x^T x = 1$ and $x^T u = 0$. If M is positive semidefinite, then the second smallest eigenvalue

$$\lambda_{n-1}(M) = \min_{x \in W} x^T M x \qquad (A.188)$$

which follows from the Rayleigh inequalities in **art. 251** and the fact that the smallest eigenvalue is $\lambda_n(M) = 0$.

Theorem 80 (Fiedler) *The second smallest eigenvalue* $\lambda_{n-1}(M)$ *of a symmetric, positive definite* $n \times n$ *matrix* M *with* $Mu = 0$ *obeys*

$$\lambda_{n-1}(M) \leq \frac{n}{n-1} \min_{1 \leq j \leq n} m_{jj} \qquad (A.189)$$

In addition,

$$2 \max_{1 \leq j \leq n} \sqrt{m_{jj}} \leq \sum_{k=1}^{n} \sqrt{m_{kk}} \qquad (A.190)$$

and

$$2 \max_{1 \leq j \leq n} \sqrt{m_{jj} - \lambda_{n-1}(M)\left(1 - \frac{1}{n}\right)} \leq \sum_{k=1}^{n} \sqrt{m_{jj} - \lambda_{n-1}(M)\left(1 - \frac{1}{n}\right)} \quad (A.191)$$

Proof: Fiedler (1973) observes that the matrix $\widetilde{M} = M - \lambda_{n-1}(M)\left(I - \frac{1}{n}J\right)$ is also positive semidefinite. For, let y be any vector in $\mathbb{R}^n$. Then y can be written as $y = c_1 u + c_2 x$ where $x \in W$. Since $\widetilde{M}u = 0$ because $Ju = nu$, it follows with $Jx = u.u^T x = 0$ that $y^T \widetilde{M} y = c_2^2 x^T \widetilde{M} x = c_2^2 \left(x^T M x - \lambda_{n-1}(M) \right) \geq 0$ by (A.188). Since any symmetric, positive semidefinite matrix can be considered as a Gram matrix, whose diagonal elements are non-negative (**art. 280**), the minimum diagonal element of $\widetilde{M}$ is non-negative,

$$0 \leq \min_{1 \leq j \leq n} \widetilde{m}_{jj} = \min_{1 \leq j \leq n} \left(m_{jj} - \lambda_{n-1}(M)\left(1 - \frac{1}{n}\right) \right)$$

which proves (A.189).

Also M is a Gram matrix, i.e., $M = A^T A$ and $m_{ij} = a_i^T a_j$, where $M = M^T$ is symmetric. The fact that $Mu = 0$ translates to $Au = 0$. This implies that the row vectors $a_1, a_2, \ldots, a_n$ of A obey $\sum_{k=1}^{n} a_k = 0$. Hence, $a_j = -\sum_{k=1, k \neq j}^{n} a_k$, and taking the Euclidean norm of both sides leads to $|a_j| \leq \sum_{k=1; k \neq j}^{n} |a_k|$. Since this inequality holds for any $1 \leq j \leq n$, it also holds for $\max_{1 \leq j \leq n} |a_j|$,

$$2 \max_{1 \leq j \leq n} |a_j| \leq \sum_{k=1}^{n} |a_k|$$

With $m_{ii} = |a_i|^2$, we arrive at (A.190), which, when applied to $\widetilde{M}$, yields (A.191). $\square$

10.9 Eigenstructure of the matrix product AB

284. *Eigenvalues of the matrix product AB.*

Lemma 11 *For all matrices $A_{n \times m}$ and $B_{m \times n}$ with $n \geq m$, it holds that $\lambda(AB) = \lambda(BA)$ and $\lambda(AB)$ has $n - m$ extra zero eigenvalues.*

Proof: Consider the matrix identities

$$\begin{bmatrix} I_{n \times n} & O_{n \times m} \\ -B_{m \times n} & \mu I_{m \times m} \end{bmatrix} \begin{bmatrix} \mu I_{n \times n} & A_{n \times m} \\ B_{m \times n} & \mu I_{m \times m} \end{bmatrix} = \begin{bmatrix} \mu I_{n \times n} & A_{n \times m} \\ O_{m \times n} & (\mu^2 I - BA)_{m \times m} \end{bmatrix}$$

and

$$\begin{bmatrix} \mu I_{n \times n} & -A_{n \times m} \\ O_{m \times n} & I_{m \times m} \end{bmatrix} \begin{bmatrix} \mu I_{n \times n} & A_{n \times m} \\ B_{m \times n} & \mu I_{m \times m} \end{bmatrix} = \begin{bmatrix} (\mu^2 I - AB)_{n \times n} & O_{n \times m} \\ B_{m \times n} & \mu I_{m \times m} \end{bmatrix}$$

Taking the determinants of both sides of each identity and denoting

$$X = \begin{bmatrix} \mu I_{n \times n} & A_{n \times m} \\ B_{m \times n} & \mu I_{m \times m} \end{bmatrix}$$

gives respectively

$$\mu^m \det X = \mu^n \det (\mu^2 I - BA)$$
$$\mu^n \det X = \mu^m \det (\mu^2 I - AB)$$

from which it follows, with $\lambda = \mu^2$, that $\lambda^{n-m} \det (BA - \lambda I) = \det (AB - \lambda I)$, which is an equation of two polynomials in λ. Equating corresponding powers in λ proves Lemma 11. $\square$

If A and B are both $n \times n$ matrices and $\det(A) \neq 0$ so that A^{-1} exists, then $BA = (A^{-1}A) BA = A^{-1} (AB) A$. Thus, the matrix BA and AB are similar and **art.** 239 shows that their eigenvalues are the same.

Lemma 12 *If A and B are symmetric matrices and one (or both) is positive definite, then all eigenvalues of AB are real.*

Proof: See, e.g., Wilkinson (1965, pp. 34-36).   □

The eigenvalue equation $ABx = \lambda x$ under the conditions of Lemma 12 is equivalent to $Bx = \lambda A^{-1}x$ in case that A is positive definite, because the inverse of a positive definite matrix exists and is also positive definite (**art.** 280). If B is positive definite, then $ABx = \lambda x$ can be written as

$$A(Bx) = \lambda B^{-1}(Bx)$$

Hence, the corresponding characteristic polynomials, $\det(AB - \lambda I)$, $\det\left(B - \lambda A^{-1}\right)$ and $\det\left(A - \lambda B^{-1}\right)$ have the same zeros (roots). Finally, the eigenvalue problems $ABx = \lambda x$ and $Ay = \lambda Cy$ are equivalent if $C = B^{-1}$ is positive definite.

If A and B are symmetric matrices, but neither is positive definite, then the eigenvalues of AB can be complex, although all eigenvalues of A and B are real (**art.** 247). Wilkinson (1965) illustrates the importance of positive definiteness by the example

$$A = \begin{bmatrix} a & 0 \\ 0 & b \end{bmatrix}, \quad B = \begin{bmatrix} 0 & 1 \\ 1 & 0 \end{bmatrix} \quad \text{and} \quad AB = \begin{bmatrix} 0 & a \\ b & 0 \end{bmatrix}$$

where the eigenvalues of AB, being $\pm\sqrt{ab}$, are complex when a and b have different sign. Moreover, if neither A nor B is positive definite and ζ is a complex eigenvalue of the eigenvalue equation $Ax = \lambda Bx$ with corresponding non-zero eigenvector x, then it holds that

$$x^H A x = (\operatorname{Re}\zeta + i\operatorname{Im}\zeta)\, x^H B x$$

Since $x^H A x$ and $x^H B x$ are real (**art.** 247), we conclude that $x^H A x = x^H B x = 0$. In other words, both A and B must have a zero eigenvalue.

285. *Matrices A and B commute.*

Lemma 13 *If square matrices $A_{n \times n}$ and $B_{n \times n}$ commute such that $AB = BA$, then the set of eigenvectors of A is the same as the set of eigenvectors of B provided that all n eigenvectors are independent. The converse more generally holds: if any two matrices A and B have a common complete set of eigenvectors, then $AB = BA$.*

Proof: If x_k is an eigenvector of A corresponding to eigenvalue λ_k, then $Ax_k = \lambda_k x_k$. Left multiplying both sides by B and using the commutative property yields $A(Bx_k) = \lambda_k(Bx_k)$, which implies that, to any eigenvector x_k with eigenvalue λ_k, the matrix A also possesses an eigenvector Bx_k with same eigenvalue λ_k. Since eigenvectors are linearly independent and since the set of n eigenvectors $\{x_1, x_2, \ldots, x_n\}$ spans the n-dimensional space, the eigenvector $Bx_k = \mu_k x_k$, which means that x_k is also an eigenvector of B.

The converse follows from **art.** 239 since $A = X\mathrm{diag}(\lambda_k)\,X^{-1}$ and, similarly, $B = X\mathrm{diag}(\mu_k)\,X^{-1}$. Indeed,

$$AB = X\mathrm{diag}\,(\lambda_k)\,X^{-1}X\mathrm{diag}\,(\mu_k)\,X^{-1} = X\mathrm{diag}\,(\lambda_k\mu_k)\,X^{-1}$$
$$BA = X\mathrm{diag}\,(\mu_k)\,X^{-1}X\mathrm{diag}\,(\lambda_k)\,X^{-1} = X\mathrm{diag}\,(\lambda_k\mu_k)\,X^{-1}$$

shows that $AB = BA$. $\hfill\square$

If all eigenvalues are distinct, all eigenvectors are independent (**art.** 238). However, in case of multiple eigenvalues, the situation can be more complex such that there are fewer than n independent eigenvectors. In that case, the Lemma 13 is not applicable.

A direct consequence of Lemma 13 is that, for commuting matrices A and B, the eigenvalues of $A + B$ are $\lambda_k + \mu_k$ and both eigenvalues belong to the same eigenvector x_k. If matrices are not commuting, remarkably little can be said about the eigenvalues of $A + B$, given the spectra of A and B (see also **art.** 252).

286. *Kronecker product.* The Kronecker product of the $n \times m$ matrix A and the $p \times q$ matrix B is the $np \times mq$ matrix $A \otimes B$, where

$$A \otimes B = \begin{bmatrix} a_{11}B & a_{12}B & \cdots & a_{1m}B \\ a_{21}B & a_{22}B & \cdots & a_{2m}B \\ \vdots & \vdots & \ddots & \vdots \\ a_{n1}B & a_{n2}B & \cdots & a_{nm}B \end{bmatrix}$$

The Kronecker product $A \otimes B$ features many properties (Meyer, 2000, p. 597). The eigenvalues of $A_{n\times n} \otimes B_{m\times m}$ are the nm numbers $\{\lambda_j\,(A)\,\lambda_k\,(B)\}_{1\leq j\leq n,1\leq k\leq m}$. Likewise, the set of eigenvalues of $I_m \otimes A_{n\times n} + B_{m\times m} \otimes I_n$ equals the set of nm eigenvalues $\{\lambda_j\,(A) + \lambda_k\,(B)\}_{1\leq j\leq n,1\leq k\leq m}$.

287. *The commutator of a matrix.* Consider the matrix equation

$$A_{n\times n}X_{n\times m} + X_{n\times m}B_{m\times m} = C_{n\times m}$$

that includes the commutator equation, $AX - XA = O$, where X are all matrices that commute with A, as a special case, as well as the Lyapunov equation (Horn and Johnson, 1991, Chapter 4). The matrix equation is written in Kronecker form as

$$\left(I_m \otimes A_{n\times n} + B_{m\times m}^T \otimes I_n\right) vec\,(X) = vec\,(C) \tag{A.192}$$

where the $nm \times 1$ vector is

$$vec\,(X) = \left(x_1^T, x_2^T, \ldots, x_m^T\right) = (x_{11}, \ldots, x_{n1}, x_{12}, \ldots, x_{n2}, \ldots, x_{1m}, \ldots, x_{nm})$$

where x_j is the j-th $n \times 1$ column vector of X. The mixed-product property (Meyer, 2000, p. 597),

$$(A_1 \otimes B_1)\,(A_2 \otimes B_2) = (A_1A_2 \otimes B_1B_2)$$

shows that

$$(I_m \otimes A_{n \times n})(B_{m \times m} \otimes I_n) = (B_{m \times m} \otimes A_{n \times n}) = (B_{m \times m} \otimes I_n)(I_m \otimes A_{n \times n})$$

In other words, the square $mn \times mn$ matrices $I_m \otimes A_{n \times n}$ and $B_{m \times m} \otimes I_n$ commute. Horn and Johnson (1991) prove that, if w_k is an $n \times 1$ eigenvector of A belonging to $\lambda_k(A)$ and y_l an $m \times 1$ eigenvector of B belonging to $\lambda_l(B)$, then $y_l \otimes w_k$ is an $nm \times 1$ eigenvector of $I_m \otimes A_{n \times n} + B_{m \times m} \otimes I_n$ belonging to the eigenvalue $\lambda_k(A) + \lambda_l(B)$.

The linear equation (A.192) has a unique solution provided none of the eigenvalues $\lambda_k(A) + \lambda_l(B) = 0$ for all $1 \le k \le n$ and $1 \le l \le m$, because $\lambda_l(B^T) = \lambda_l(B)$ on **art.** 237. Likewise, if $C = O$ in (A.192), the equation $AX - XB = O$ has only a solution, provided $\{\lambda_k(A)\}_{1 \le k \le n} \cap \{\lambda_l(B)\}_{1 \le l \le m} \ne \varnothing$. Thus, when $B = A$, in which case X is the commutator of A, there are at least n zero eigenvalues of $I_n \otimes A_{n \times n} - A_{n \times n} \otimes I_n$ (and more than n if A has zero eigenvalues) illustrating that there may exist many possible commutators of a matrix A. If $C \ne O$ and $B = -A$ in (A.192), there is no solution for X. A theorem of Shoda, proved in Horn and Johnson (1991, p. 288), states that C can be written as $C = XY - YX$ for some matrices X and Y provided trace$(C) = 0$.

10.10 Perturbation theory

We confine ourselves to simple eigenvalues of a symmetric matrix A, in which case perturbation theory is relatively simple. Perturbation theory for non-symmetric matrices and for eigenvalues with higher multiplicity is more involved and omitted. We follow Wilkinson (1965, pp. 60-70), although a similar analysis, albeit a little less transparent, appears in Cvetković *et al.* (1997, Sec. 6.3).

288. *Perturbation theory around a simple eigenvalue.* Let us consider the matrix $A(\zeta) = A + \zeta B$. Perturbation theory assumes that the real number ζ is sufficiently small so that we may regard $A(\zeta)$ as the perturbation of the original $n \times n$ symmetric matrix A by an $n \times n$ matrix B, which is not necessarily symmetric. We denote by $x(\zeta)$ the $n \times 1$ eigenvector of $A(\zeta)$ belonging to the eigenvalue $\lambda(\zeta)$. As shown in Wilkinson (1965, pp. 60-70), both $x(\zeta)$ and $\lambda(\zeta)$ are analytic functions of ζ around zero and can be represented by a power series

$$x(\zeta) = x + \zeta z_1 + \zeta^2 z_2 + \cdots = \sum_{j=0}^{\infty} z_j \zeta^j \qquad (A.193)$$

$$\lambda(\zeta) = \lambda + \zeta c_1 + \zeta^2 c_2 + \cdots = \sum_{j=0}^{\infty} c_j \zeta^j \qquad (A.194)$$

where $x(0) = x = z_0$ is the eigenvector of A and $\lambda(0) = \lambda = c_0$ is its corresponding simple eigenvalue. We omit considerations about the convergence radius of the above power series. We choose $x = x_q$ as the normalized eigenvector of A corresponding to $\lambda = \lambda_q$.

The eigenvalue equation of $A(\zeta)$ is $(A + \zeta B) x(\zeta) = \lambda(\zeta) x(\zeta)$. After introducing the power series (A.193) and (A.194), we obtain

$$(A + \zeta B)\left(x_q + \sum_{j=1}^{\infty} z_j \zeta^j\right) = \sum_{j=0}^{\infty} c_j \zeta^j \sum_{j=0}^{\infty} z_j \zeta^j$$

The left-hand side equals

$$(A + \zeta B)\left(x_q + \sum_{j=1}^{\infty} z_j \zeta^j\right) = A x_q + \sum_{j=1}^{\infty} A z_j \zeta^j + \zeta B x_q + \sum_{j=1}^{\infty} B z_j \zeta^{j+1}$$

$$= \lambda_q x_q + (A z_1 + B x_q)\zeta + \sum_{j=2}^{\infty} (A z_j + B z_{j-1})\zeta^j$$

while the Cauchy product of the right-hand side gives

$$\sum_{j=0}^{\infty} c_j \zeta^j \sum_{j=0}^{\infty} z_j \zeta^j = \sum_{j=0}^{\infty}\left(\sum_{k=0}^{j} c_{j-k} z_k\right)\zeta^j = \lambda_q x_q + (c_1 x_q + \lambda_q z_1)\zeta + \sum_{j=2}^{\infty}\left(\sum_{k=0}^{j} c_{j-k} z_k\right)\zeta^j$$

Equating corresponding powers in ζ yields, for $j = 1$,

$$A z_1 + B x_q = \lambda_q z_1 + c_1 x_q \tag{A.195}$$

and, for $j > 1$,

$$A z_j + B z_{j-1} = \sum_{k=0}^{j} c_{j-k} z_k = c_j x_q + \sum_{k=1}^{j-1} c_{j-k} z_k + \lambda_q z_j \tag{A.196}$$

Relations (A.195) and (A.196) are the results of complex function theory. The solution for the $n \times 1$ vectors $\{z_j\}_{j\geq 1}$ in (A.193) and the coefficients c_k in (A.194) now requires linear algebra.

289. *Scaling of the eigenvector* $x(\zeta)$. The vector z_j can be written as a linear combination of the eigenvectors x_k of the symmetric matrix A,

$$z_j = \sum_{k=1}^{n} \beta_{jk} x_k \tag{A.197}$$

where the coefficients $\beta_{jm} = x_m^T z_j = z_j^T x_m \neq \beta_{mj}$. The particular case $j = 0$, where $z_0 = x_q$, indicates that $\beta_{0k} = \delta_{kq}$. Thus, the eigenvector in (A.193) is rewritten as

$$x(\zeta) = \sum_{j=0}^{\infty} z_j \zeta^j = \sum_{k=1}^{n}\left(\sum_{j=0}^{\infty} \beta_{jk} \zeta^j\right) x_k = \left(\sum_{j=0}^{\infty} \beta_{jq} \zeta^j\right) x_q + \sum_{k=1;k\neq q}^{n}\left(\sum_{j=0}^{\infty} \beta_{jk} \zeta^j\right) x_k$$

and

$$x(\zeta) = \left(1 + \sum_{j=1}^{\infty} \beta_{jq} \zeta^j\right) x_q + \sum_{k=1;k\neq q}^{n}\left(\sum_{j=1}^{\infty} \beta_{jk} \zeta^j\right) x_k$$

We can always scale an eigenvector by a scalar $\alpha \neq 0$, which we choose here as $\alpha = 1 + \sum_{j=1}^{\infty} \beta_{jq} \zeta^j$, assuming that the power series converges to a value different than -1. The latter condition can always be met for sufficiently small $|\zeta|$ and we arrive at

$$\alpha^{-1} x \left(\zeta \right) = x_q + \sum_{k=1; k \neq q}^{n} \left(\frac{\sum_{j=1}^{\infty} \beta_{jk} \zeta^j}{1 + \sum_{j=1}^{\infty} \beta_{jq} \zeta^j} \right) x_k$$

If we choose $\beta_{jq} = x_q^T z_j = z_0^T z_j = 0$ for $j \geq 1$ and recall that $\beta_{0q} = 1$, then $\alpha = 1$ and we simplify the computation by requiring that any "perturbation" vector z_j for $j \geq 1$ is orthogonal to the eigenvector x_q of the matrix A.

If we choose a different scaling by requiring a normalized eigenvector, such as $x^T \left(\zeta \right) x \left(\zeta \right) = 1$, then it implies that

$$1 = x^T \left(\zeta \right) x \left(\zeta \right) = \sum_{j=0}^{\infty} z_j^T \zeta^j \sum_{m=0}^{\infty} z_m \zeta^m = \sum_{j=0}^{\infty} \left(\sum_{m=0}^{j} z_m^T z_{j-m} \right) \zeta^j$$

and equating corresponding powers in ζ leads, for $j = 0$, to $z_0^T z_0 = 1$, which is satisfied for any normalized eigenvector $z_0 = x_q$ of A and, for $j > 0$, to $0 = \sum_{m=0}^{j} z_m^T z_{j-m}$. The latter condition means that $z_0^T z_1 = 0$ and furthermore that $z_0^T z_j = -\frac{1}{2} \sum_{m=1}^{j-1} z_m^T z_{j-m}$ for $j \geq 2$.

In summary, the normalization of the eigenvector $x \left(\zeta \right)$ imposes conditions on the scalar products $z_0^T z_j$ for all $j \geq 1$. Choosing a different scaling leads to a different computational scheme and the art consists of choosing the most appropriate conditions on $z_0^T z_j$.

290. *Evaluation of the power series coefficients c_k and vectors z_j.* After expressing the relations (A.195) and (A.196) with $z_j = \sum_{k=1}^{n} \beta_{jk} x_k$ in (A.197) in terms of the normalized eigenvectors $x_1, x_2, \ldots, x_n$ of the matrix A and taking the eigenvalue equation $A x_k = \lambda_k x_k$ into account, we obtain the set of linear equations

$$c_1 x_q = \sum_{k=1}^{n} \beta_{1k} \left(\lambda_k - \lambda_q \right) x_k + B x_q \tag{A.198}$$

and, for $j > 1$,

$$c_j x_q = \sum_{k=1}^{n} \beta_{jk} \left(\lambda_k - \lambda_q \right) x_k + \sum_{k=1}^{n} \beta_{j-1,k} B x_k - \sum_{l=1}^{n} \sum_{k=1}^{j-1} c_{j-k} \beta_{kl} x_l \tag{A.199}$$

in the unknown numbers $\{c_k\}_{k \geq 1}$ and $\{\beta_{jk}\}_{j \geq 1; k \geq 1}$. As eigenvector scaling, we choose $\beta_{jq} = x_q^T z_j = z_0^T z_j = 0$ for $j \geq 1$, which is computationally, the simplest choice.

Pre-multiplying (A.198) with the vector x_r^T, using $x_r^T x_q = \delta_{rq}$ yields

$$c_1 \delta_{rq} = \beta_{1r} \left(\lambda_r - \lambda_q \right) + x_r^T B x_q$$

In particular, if $r = q$, then

$$c_1 = x_q^T B x_q \tag{A.200}$$

else,

$$\beta_{1r} = \frac{x_r^T B x_q}{\lambda_q - \lambda_r} \qquad \text{for } r \neq q \tag{A.201}$$

The expression (A.201) emphasizes that the eigenvalue λ_q must be simple, which is a basic limitation of the presented perturbation method. Hence, it follows from (A.197) that $z_1 = \sum_{k=1}^N \beta_{1k} x_k = \sum_{k=1;k\neq q}^N \frac{x_k^T B x_q}{\lambda_q - \lambda_k} x_k + \beta_{1q} x_q$. With our eigenvector scaling choice $\beta_{1q} = 0$, we find the first order expansion in ζ,

$$\begin{cases} x(\zeta) = x_q + \zeta \sum_{k=1;k\neq q}^N \frac{x_k^T B x_q}{\lambda_q - \lambda_k} x_k + O\left(\zeta^2\right) \\ \lambda(\zeta) = \lambda_q + \zeta x_q^T B x_q + O\left(\zeta^2\right) \end{cases}$$

Pre-multiplying (A.199) with the vector x_r^T analogously leads, for $j > 1$, to

$$c_j \delta_{rq} = \beta_{jr} (\lambda_r - \lambda_q) + \sum_{k=1}^n \beta_{j-1,k} x_r^T B x_k - \sum_{k=1}^{j-1} c_{j-k} \beta_{kr}$$

In particular, if $r = q$, then

$$c_j = \sum_{k=1}^n \beta_{j-1,k} x_q^T B x_k - \sum_{k=1}^{j-1} c_{j-k} \beta_{kq}$$

else

$$\beta_{jr} = \frac{1}{\lambda_r - \lambda_q} \left\{ \sum_{k=1}^{j-1} c_{j-k} \beta_{kr} - \sum_{k=1}^n \beta_{j-1,k} x_r^T B x_k \right\} \qquad \text{for } r \neq q \tag{A.202}$$

With our eigenvector scaling choice $\beta_{jq} = 0$ for $j \geq 1$, the first recursive equation in the coefficients c_k simplifies considerably to

$$c_j = \sum_{k=1;k\neq q}^n \beta_{j-1,k} x_q^T B x_k \qquad \text{for } j > 1 \tag{A.203}$$

Substituting the explicit form of the coefficients c_j in (A.203) into (A.202) yields

$$\beta_{jr} = \frac{1}{\lambda_r - \lambda_q} \sum_{l=1;l\neq q}^n \left\{ \sum_{k=1}^{j-1} \beta_{kr} \beta_{j-k-1,l} x_q^T B x_l - \beta_{j-1,l} x_r^T B x_l \right\} \qquad \text{for } r \neq q$$

The scaling choice $\beta_{0,l} = \delta_{lq}$ and $\beta_{jq} = 0$ for $j \geq 1$ simplifies, for $r \neq q$, to a recursion in β_{jr}

$$\beta_{jr} = \frac{\beta_{j-1;r} x_q^T B x_q}{\lambda_r - \lambda_q} + \frac{1}{\lambda_r - \lambda_q} \sum_{l=1;l\neq q}^n \left\{ \sum_{k=1}^{j-2} \beta_{kr} \beta_{j-k-1,l} x_q^T B x_l - \beta_{j-1,l} x_r^T B x_l \right\} \tag{A.204}$$

which can be iterated up to any desired integer value of j.

For example, if $j = 2$, then (irrespective of the choice of scaling)

$$c_2 = \sum_{k=1}^{n} \beta_{1k} x_q^T B x_k - c_1 \beta_{1q} = \sum_{k=1;k\neq q}^{n} \beta_{1k} x_q^T B x_k$$

and

$$\beta_{2r} = \frac{1}{\lambda_r - \lambda_q} \left\{ \beta_{1r} x_q^T B x_q - \sum_{k=1}^{n} \beta_{1k} x_r^T B x_k \right\} \qquad \text{for } r \neq q$$

Using (A.201) results in

$$c_2 = \sum_{k=1;k\neq q}^{n} \frac{\left(x_k^T B x_q\right)^2}{\lambda_q - \lambda_k} \tag{A.205}$$

and

$$\beta_{2r} = \frac{1}{\lambda_q - \lambda_r} \sum_{k=1;k\neq q}^{n} \frac{\left(x_k^T B x_q\right)\left(x_k^T B x_r\right)}{\lambda_q - \lambda_k} - \frac{\left(x_r^T B x_q\right)\left(x_q^T B x_q\right)}{\left(\lambda_q - \lambda_r\right)^2} \qquad \text{for } r \neq q \tag{A.206}$$

Moreover, we can use β_{2r} immediately in $c_3 = \sum_{k=1;k\neq q}^{n} \beta_{2k} x_q^T B x_k$ in (A.203),

$$c_3 = \sum_{r=1;r\neq q}^{n} \frac{x_q^T B x_r}{\lambda_q - \lambda_r} \sum_{k=1;k\neq q}^{n} \frac{\left(x_k^T B x_q\right)\left(x_k^T B x_r\right)}{\lambda_q - \lambda_k} - \sum_{r=1;r\neq q}^{n} \frac{\left(x_r^T B x_q\right)^2 \left(x_q^T B x_q\right)}{\left(\lambda_q - \lambda_r\right)^2} \tag{A.207}$$

illustrating that, in general, the eigenvalue expansion (A.194) can always be computed, with the same efforts, one order higher in ζ than the eigenvector expansion (A.193). Indeed, the coefficient c_j in (A.203) only depends on $\beta_{j-1,k}$ and not on β_{jk} as z_j in (A.197).

If $\lambda_q = \lambda_1$ is the largest eigenvalue of a symmetric matrix A, then we observe that the coefficient c_2 in (A.205) is positive. Consequently, if ζ is sufficiently small so that the remainder of the series in (A.194) obeys $\left|\sum_{j=3}^{\infty} c_j \zeta^j\right| < c_2 \zeta^2$, then the first order perturbation $\lambda(\zeta) \geq \lambda_1 + \zeta x_1^T B x_1$ is a lower bound.

Part III
Polynomials

11

Polynomials with real coefficients

The characteristic polynomial of real matrices possesses real coefficients. This chapter aims to summarize general results on the location and determination of the zeros of polynomials with mainly real coefficients. The operations here are assumed to be performed over the set $\mathbb{C}$ of complex numbers. Restricting operations to other subfields of $\mathbb{C}$, such as the set $\mathbb{Z}$ of integers or finite fields (see e.g. Gilbert and Nicholson (2004)), is omitted because, in that case, we need to enter an entirely different and more complex area, which requires Galois theory, advanced group theory and number theory. A general outline for the latter is found in Govers *et al.* (2008). A nice introduction to Galois theory is written by Stewart (2004).

The study of polynomials belongs to one of the oldest researches in mathematics. The insolubility of the quintic, famously proved by Abel and extended by Galois (see **art.** 291 and Govers *et al.* (2008) for more details and for the historical context), shifted the root finding problem in polynomials from pure to numerical analysis. Numerical methods as well as matrix method based on the companion matrix (**art.** 242) are extensively treated by McNamee (2007), but omitted here. A complex function theoretic approach, covering more recent results such as self-inversive polynomials and extensions of Grace's Theorem (**art.** 331), is presented by Sheil-Small (2002) and by Milovanović *et al.* (1994) and Borwein and Erdélyi (1995), who also list many polynomial inequalities. In addition, Milovanović *et al.* (1994) treat polynomial extremal problems of the type: given that the absolute value of a polynomial is bounded in some region of the complex plane, how large can its derivative be in that region?

11.1 General properties

291. *Definition of a polynomial.* A fundamental theorem of algebra, first proved by Gauss and later by Liouville (Titchmarsh, 1964, p. 118), states that any polynomial of degree n has precisely n zeros in the complex plane. If these zeros coincide, we count zeros according to their multiplicity. Thus, if there are $l < n$ zeros and a zero z_k has multiplicity m_k, then the fundamental theorem states that $\sum_{k=1}^{l} m_k = n$.

In the sequel, we represent zeros as if they are single, however, with the convention that a coinciding zero z_k with multiplicity m_k is counted m_k times.

Let $p_n(z)$ denote a polynomial of degree n defined by

$$p_n(z) = \sum_{k=0}^{n} a_k \, z^k = a_n \prod_{k=1}^{n} (z - z_k) \tag{B.1}$$

where $\{z_k\}_{1 \leq k \leq n}$ is the set of n zeros and the coefficient a_k (for $0 \leq k \leq n$) is a finite, complex number. Moreover, we require that $a_n \neq 0$, otherwise the polynomial is not of degree n. If $a_n = 1$, which is an often used normalization, the polynomial is called "monic".

Once the set of zeros is known, the coefficients a_k can be computed by multiplying out the product in (B.1). The other direction, the determination of the set of zeros given the set of coefficients $\{a_k\}_{0 \leq k \leq n}$, proves to be much more challenging. Abel and Galois have shown that only up to degree $n = 4$ explicit relations of the zeros exist in terms of a finite number of elementary operations such as additions, subtractions, multiplications, divisions and radicals on the coefficients. The solution of the cubic ($n = 3$) and quartic ($n = 4$) can be found, for example, in Stewart (2004) and Milovanović *et al.* (1994). An important aspect of the theory of polynomials thus lies in the determination of the set of zeros.

It follows immediately from (B.1) that $p_n(0) = a_0 = a_n \prod_{k=1}^{n}(-z_k)$, which shows that the absolute value of any zero of a polynomial must be finite. If $a_0 = 0$, then at least one zero must be zero and $p_n(z) = z^m p_{n-m}(z)$ for an integer $m \geq 1$. Therefore, we often implicitly assume that $a_0 \neq 0$, otherwise the polynomial $p_n(z)$ can be trivially reduced to a lower degree polynomial. From (B.1), one readily verifies that

$$z^n \, p_n \left(\frac{1}{z} \right) = \sum_{k=0}^{n} a_{n-k} \, z^k = a_0 \prod_{k=1}^{n} \left(z - \frac{1}{z_k} \right) \tag{B.2}$$

Hence, the polynomial $\sum_{k=0}^{n} a_{n-k} \, z^k$ with the coefficients in the reverse order possesses as zeros the inverses of those of the original polynomial $\sum_{k=0}^{n} a_k \, z^k$.

292. *Polynomials with integer coefficients.* If all the coefficients a_k of $p_n(z) = \sum_{k=0}^{n} a_k z^k$ are integers and if $\xi = \frac{r}{s}$ is a rational zero (i.e. r and s are integers and coprime), then $r | a_0$ and $s | a_n$. Indeed, rewriting $p_n\left(\frac{r}{s}\right) = 0$ in (B.1) as,

$$r \sum_{k=0}^{n-1} a_{k+1} r^k s^{n-k-1} = -s^n a_0$$

shows that r divides the left-hand side and, hence, $r | s^n a_0$. Since the prime factorizations of r and s do not have a prime number in common because r and s are coprime, $r | s^n a_0$ implies that $r | a_0$. The second statement follows analogously after rewriting $p_n\left(\frac{r}{s}\right) = 0$ as $a_n r^n = -s \sum_{k=0}^{n-1} a_k r^k s^{n-1-k}$.

The zeros of a monic polynomial (i.e. $a_n = 1$) with integer coefficients are called algebraic numbers and play a fundamental role in algebraic number fields (see, e.g.,

Govers *et al.* (2008)). Since the integer s in the zero $\xi = \frac{r}{s}$ of $p_n(z)$ must divide $a_n = 1$, it must equal one so that algebraic numbers cannot be rational numbers. This fact was first observed by Gauss (1801, art. 11).

293. *Irreducibility.* While a polynomial $p_n(z) = \sum_{k=0}^{n} a_k z^k$ with complex coefficients can be factored over the complex numbers $\mathbb{C}$ as in the definition (B.1), confinement of its coefficients to rationals or integers generally also confines the factorization. If all coefficients are rational, i.e. $a_k \in \mathbb{Q}$, then by multiplying $p_n(z)$ with the least common multiple of all denominators, a polynomial with integer coefficients is obtained with the same zeros. A polynomial $p_n(z)$ is irreducible over $\mathbb{Q}$ if $p_n(z) = g_m(z) h_l(z)$ with $n = m + l$ cannot be factored into two polynomials $g_m(z)$ and $h_l(z)$ with integer coefficients. Irreducibility over $\mathbb{Q}$ means that a *monic* $p_n(z)$ with $a_n = 1$ does not have rational zeros, but the converse is not true; e.g. $(z^2 - 2)(z^2 - 3)$ does not possess rational zeros, but it is reducible. There exist criteria (e.g. due to Eisenstein and Perron) to test whether a polynomial $p_n(z)$ with integer coefficients is irreducible.

Irreducibility over $\mathbb{Z}$ occurs if $n-1$ zeros of a monic polynomial $p_n(z)$ with integer coefficients and $a_n = 1, a_0 \neq 0$ are, in absolute value, smaller than 1. In that case, the definition (B.1) shows that $p_n(z) = (z - z_1) \prod_{k=2}^{n}(z - z_k)$ where $|z_1| > 1 > |z_2| \geq |z_3| \geq |z_n|$, but $g_{n-1}(z) = \prod_{k=2}^{n}(z - z_k) = \sum_{k=0}^{n-1} b_k z^k$ cannot be a polynomial with integer coefficients $b_k \in \mathbb{Z}$, because $0 < |g_{n-1}(0)| = |b_0| = \prod_{k=2}^{n} |z_k| < 1$ and $b_0 \notin \mathbb{Z}$. This fact, combined with Perron's Theorem 85, shows that a monic polynomial $p_n(z) = \sum_{k=0}^{n} a_k z^k$ is irreducible over $\mathbb{Z}$ if its integer coefficients obey $\sum_{k=0}^{n-2} |a_k| + 1 < |a_{n-1}|$ or, by Theorem 86, if either $p_n\left(\frac{1}{2}\left(\sum_{k=0}^{n-1} |a_k| + 1\right)\right) < 0$ or $(-1)^n p_n\left(-\frac{1}{2}\left(\sum_{k=0}^{n-1} |a_k| + 1\right)\right) < 0$ holds. Perron (1907) derives several other, but more complicated irreducibility criteria.

Eisenstein's criterion, proved in Gilbert and Nicholson (2004, p. 194), is

Theorem 81 (Eisenstein) *If the coefficients a_k of $p_n(z) = \sum_{k=0}^{n} a_k z^k$ are integers and if all the following conditions hold for some prime p: (i) $p|a_k$ for all $0 \leq k < n$, (ii) $p \nmid a_n$ and (iii) $p^2 \nmid a_0$, then the polynomial $p_n(z)$ is irreducible over $\mathbb{Q}$.*

For example, $z^5 - 2$ and $2z^3 + 9z - 3$ are irreducible over the rational numbers and do not possess rational zeros.

294. *Newton identities.* The Newton identities are a recursive set of equations that relate the coefficients a_k of a polynomial $p_n(z) = \sum_{k=0}^{n} a_k z^k$ to sums of integer powers $j \in [1, n]$

$$Z_j = \sum_{k=1}^{n} z_k^j \tag{B.3}$$

of the zeros $z_1, z_2, \ldots, z_n$ of $p_n(z)$.

Theorem 82 (Newton) *For any polynomial defined by (B.1), the coefficients, for* $1 \leq l < n$, *satisfy the recursion*

$$a_l = -\frac{1}{n-l} \sum_{k=l+1}^{n} a_k Z_{k-l} \qquad (B.4)$$

Proof: The logarithmic derivative $\frac{d \log p_n(z)}{dz}$ of (B.1) is

$$p'_n(z) = p_n(z) \sum_{k=1}^{n} \frac{1}{z - z_k} \qquad (B.5)$$

For $z > \max_k z_k$, we can expand $\frac{1}{z-z_k} = \frac{1}{z\left(1-\frac{z_k}{z}\right)} = \frac{1}{z}\sum_{j=0}^{\infty} \frac{z_k^j}{z^j}$ in a geometric series $p'_n(z) = p_n(z) \sum_{k=1}^{n} \sum_{j=0}^{\infty} \frac{z_k^j}{z^{j+1}} = p_n(z) \sum_{j=0}^{\infty} \frac{Z_j}{z^{j+1}}$, where the summations can always be reversed for polynomials (finite n), but not for functions. Introducing the series representation in (B.1) of $p_n(z)$ yields

$$\sum_{k=1}^{n} k a_k z^k = \sum_{k=0}^{n} a_k z^k \sum_{j=0}^{\infty} Z_j z^{-j} = \sum_{j=0}^{\infty} \sum_{k=0}^{n} a_k Z_j z^{k-j} \qquad (B.6)$$

Let $l = k - j$, then $-\infty \leq l \leq n$. Also $j = k - l \geq 0$ such that $k \geq l$. Combined with $0 \leq k \leq n$, we have $\max(0, l) \leq k \leq n$. Thus,

$$\sum_{j=0}^{\infty} \sum_{k=0}^{n} a_k Z_j z^{k-j} = \sum_{l=-\infty}^{n} \sum_{k=\max(l,0)}^{n} a_k Z_{k-l} z^l$$

$$= \sum_{l=-\infty}^{0} \sum_{k=0}^{n} a_k Z_{k-l} z^l + \sum_{l=1}^{n} \sum_{k=l}^{n} a_k Z_{k-l} z^l$$

Equating the corresponding powers of z in (B.6) and using $Z_0 = n$ yields

$$\begin{array}{ll} \sum_{k=0}^{n} a_k Z_{k-l} = 0 & l \leq 0 \\ \sum_{k=l+1}^{n} a_k Z_{k-l} = (l-n)a_l & 1 \leq l \leq n \end{array}$$

The first set of equations, equivalent for $q \geq 0$ to $0 = \sum_{k=0}^{n} a_k Z_{k+q} = \sum_{l=1}^{n} z_l^q p_n(z_l)$ are trivial, because $p_n(z_l) = 0$. Newton's theorem thus follows from the second set of equations. □

By rewriting the Newton identities (B.4) as

$$Z_j = -\frac{1}{a_n}\left(j a_{n-j} + \sum_{k=1}^{j-1} a_{k+n-j} Z_k\right) \qquad (B.7)$$

we obtain, for $1 \leq j \leq n$, a recursion that expresses the positive powers Z_j of the zeros in terms of the coefficients a_k. Explicitly for the first few Z_j,

$$Z_1 = \sum_{k=1}^{n} z_k = -\frac{a_{n-1}}{a_n} \tag{B.8}$$

$$Z_2 = \sum_{k=1}^{n} z_k^2 = \frac{a_{n-1}^2}{a_n^2} - \frac{2a_{n-2}}{a_n}$$

$$Z_3 = \sum_{k=1}^{n} z_k^3 = -\frac{a_{n-1}^3}{a_n^3} + \frac{3a_{n-2}a_{n-1}}{a_n^2} - \frac{3a_{n-3}}{a_n}$$

$$Z_4 = \sum_{k=1}^{n} z_k^4 = \frac{a_{n-1}^4}{a_n^4} - \frac{4a_{n-2}a_{n-1}^2}{a_n^3} + \frac{2a_{n-2}^2 + 4a_{n-3}a_{n-1}}{a_n^2} - 4\frac{a_{n-4}}{a_n}$$

Applying (B.8) to the polynomial $z^n p_n\left(\frac{1}{z}\right) = \sum_{k=0}^{n} a_{n-k} z^k$ with the coefficients in reverse order (**art.** 291) gives

$$Z_{-1} = \sum_{k=1}^{n} \frac{1}{z_k} = -\frac{a_1}{a_0}$$

$$Z_{-2} = \sum_{k=1}^{n} \frac{1}{z_k^2} = \frac{a_1^2}{a_0^2} - \frac{2a_2}{a_0}$$

When changing the coefficients $a_l \to a_{n-l}$ and $Z_k \to Z_{-k}$ in the Newton identities (B.4) according to (B.2), we obtain

$$a_{n-l} = -\frac{1}{n-l} \sum_{k=l+1}^{n} a_{n-k} Z_{-k+l} = -\frac{1}{n-l} \sum_{k=1}^{n-l} a_{n-l-k} Z_{-k}$$

After letting $m = n-l$, we find, for $0 \le m \le n$, the appealing form[1] of the recursion for the sum of inverse powers of zeros,

$$m a_m = -\sum_{k=1}^{m} a_{m-k} Z_{-k} \tag{B.9}$$

$$= -a_{m-1}Z_{-1} - a_{m-2}Z_{-2} - \ldots - a_0 Z_{-m}$$

The inverse powers $Z_{-m} = \sum_{k=1}^{n} \frac{1}{z_k^m}$ and thus also the positive powers Z_m after changing $a_l \to a_{n-l}$ can be explicitly expressed for $m > 1$ as

$$Z_{-m} = m \sum_{k=1}^{m} \frac{(-1)^k}{k\, a_0^k} s[k, m]$$

where $s[k, m]$ is the characteristic coefficient (Van Mieghem, 1996),

$$s[k, m] = \sum_{\sum_{i=1}^{k} j_i = m; j_i > 0} \prod_{i=1}^{k} a_{j_i} \tag{B.10}$$

---

[1] If a polynomial is defined as $p_n(z) = \sum_{k=0}^{n} a_{n-k} z^k$ rather than our standard definition (B.1), then the Newton identities appear in this easier form.

If all zeros $\{z_k\}_{1\leq k\leq n}$ are real and positive, the harmonic, geometric and arithmetic mean inequality (6.38) shows that

$$\frac{n}{Z_{-1}} \leq \sqrt[n]{\prod_{j=1}^{n} z_j} \leq \frac{1}{n} Z_1$$

from which we find $\frac{a_{n-1}}{a_n}\frac{a_1}{a_0} \geq n^2$.

Finally, the Newton identities (B.4) are linear equations that express the coefficients a_k of a polynomial $p_n(z)$ in terms of sums of powers Z_j of zeros and illustrate that the set $\{Z_j\}_{0\leq j\leq n}$ suffices to determine the coefficients $\{a_k\}_{0\leq k\leq n}$ uniquely and, hence, the polynomial $p_n(z)$. The trace formula (A.118) relates $Z_j = \text{trace}(A^j)$, where the zero z_k equals the eigenvalue λ_k of the $n \times n$ matrix A.

295. The problem of finding the zeros z_k from the coefficients $\{a_k\}_{0\leq k\leq n}$ and the set $\{Z_j\}_{0\leq j\leq n}$ is difficult. However, from $Z_1 = \sum_{k=1}^{n} z_k = -\frac{a_{n-1}}{a_n}$ and $Z_{-1} = \sum_{k=1}^{n} \frac{1}{z_k} = -\frac{a_1}{a_0}$, two zeros z_1 and z_2 can be determined in terms of the others.

Lemma 14 *Both $A = -\frac{a_{n-1}}{a_n} - \sum_{k=3}^{n} z_k$ and $B = -\frac{a_1}{a_0} - \sum_{k=3}^{n} \frac{1}{z_k}$ determine the zeros*

$$z_1 = \frac{A}{2}\left(1 + \sqrt{1 - \frac{4}{AB}}\right) \quad and \quad z_1 = \frac{A}{2}\left(1 - \sqrt{1 - \frac{4}{AB}}\right)$$

in terms of the other $n-2$ zeros and the coefficient ratios $\frac{a_{n-1}}{a_n}$ and $\frac{a_1}{a_0}$.

Proof: Let $A = -\frac{a_{n-1}}{a_n} - \sum_{k=3}^{n} z_k$ and $B = -\frac{a_1}{a_0} - \sum_{k=3}^{n} \frac{1}{z_k}$, then it holds that $z_1 + z_2 = A$ and $\frac{1}{z_1} + \frac{1}{z_2} = B$. The last equation is rewritten as $\frac{A}{z_1 z_2} = B$. Thus, the product is $z_1 z_2 = \frac{A}{B}$, while the sum is $z_1 + z_2 = A$, leading to the quadratic equation $z^2 - Az + \frac{A}{B} = 0$, with solution $z_{1,2} = \frac{1}{2}\left(A \pm \sqrt{A^2 - 4\frac{A}{B}}\right)$. □

296. *Vieta's formulae* express the coefficients a_k of $p_n(z)$ explicitly in terms of its zeros $\{z_k\}_{1\leq k\leq n}$.

Theorem 83 (Vieta) *For any polynomial defined by (B.1), it holds, for $0 \leq k < n$, that*

$$\frac{a_k}{a_n} = (-1)^{n-k} \sum_{j_1=1}^{n} \sum_{j_2=j_1+1}^{n} \cdots \sum_{j_{n-k}=j_{n-k-1}+1}^{n} \prod_{i=1}^{n-k} z_{j_i} \qquad (j_0 = 0) \qquad (B.11)$$

$$\equiv (-1)^{n-k} \sum_{1\leq j_1<j_2<\cdots<j_{n-k}\leq n} \prod_{i=1}^{n-k} z_{j_i} \qquad (B.12)$$

Proof: The proof is based on the principle of induction. Relation (B.11) is verified for $n = 2$. Assume that it holds for $n > 2$. Consider now the polynomial of degree $n+1$ whose zeros are precisely those of $p_n(z)$, thus $z_k(n+1) = z_k(n)$ for $1 \le k \le n$ with the addition of the $n+1$-th zero, $z_{n+1}(n+1)$. Hence,

$$p_{n+1}(z) = (z - z_{n+1}(n+1)) \, p_n(z) = \sum_{k=1}^{n+1} a_{k-1}(n) \, z^k - \sum_{k=0}^{n} z_{n+1}(n+1) \, a_k(n) \, z^k$$

$$= -z_{n+1}(n+1)a_0(n) + \sum_{k=1}^{n} [a_{k-1}(n) - z_{n+1}(n+1)a_k(n)]z^k + a_n(n)z^{n+1}$$

from which the recursion

$$a_k(n+1) = a_{k-1}(n) - z_{n+1}(n+1) \, a_k(n)$$

is immediate. Since the coefficient of the highest power in (B.11) equals unity, by definition thus $a_n(n) = a_{n+1}(n+1) = 1$, and since the constant term indeed reflects $(-1)^{n+1}$ times the product of all $n+1$ zeros, we only have to verify for the coefficients $a_k(n+1)$ with $1 \le k \le n$ whether (B.11) satisfies the recursion relation. Substitution yields

$$a_k(n+1) = (-1)^{n+1-k} \sum_{j_1=1}^{n} \sum_{j_2=j_1+1}^{n} \cdots \sum_{j_{n+1-k}=j_{n-k}+1}^{n} \prod_{i=1}^{n+1-k} z_{j_i}(n)$$

$$- (-1)^{n-k} \sum_{j_1=1}^{n} \sum_{j_2=j_1+1}^{n} \cdots \sum_{j_{n-k}=j_{n-k-1}+1}^{n} \prod_{i=1}^{n-k} z_{j_i}(n) \, z_{n+1}(n+1)$$

Distributing the product of zeros over the sums and using $z_k(n+1) = z_k(n)$ for $1 \le k \le n$ leads to

$$a_k(n+1) = (-1)^{n+1-k} \sum_{j_1=1}^{n} z_{j_1}(n+1) \sum_{j_2=j_1+1}^{n} z_{j_2}(n+1) \cdots \sum_{j_{n-k}=j_{n-k-1}+1}^{n} z_{j_{n-k}}(n+1)$$

$$\left[\sum_{j_{n+1-k}=j_{n-k}+1}^{n} z_{j_{n+1-k}}(n+1) + z_{n+1}(n+1) \right]$$

$$= (-1)^{n+1-k} \sum_{j_1=1}^{n} z_{j_1}(n+1) \cdots \sum_{j_{n-k}=j_{n-k-1}+1}^{n} z_{j_{n-k}}(n+1) \sum_{j_{n+1-k}=j_{n-k}+1}^{n+1} z_{j_{n+1-k}}(n+1)$$

Since $\sum_{k=a}^{b} f(k) = 0$ if $a > b$, the last relation equals (B.11) when n is replaced by $n+1$. □

In case $z_k(n) = 1$ for all k, we have $p_n(z) = (z-1)^n = \sum_{k=0}^{n} \binom{n}{k}(-1)^{n-k} z^k$ from which the simple check

$$\sum_{j_1=1}^{n} \sum_{j_2=j_1+1}^{n} \cdots \sum_{j_{n-k}=j_{n-k-1}+1}^{n} 1 = n(n-1)\cdots(n-k)/k! \equiv \binom{n}{k}$$

follows, because only one ordering of $\{j_i\}$ out of $k!$ is allowed. Hence, the multiple sum in (B.11) consists of $\binom{n}{k}$ terms. Applying (B.11) to (B.2) yields, after substitution of $m = n - k$, the following alternative expression:

$$a_m = (-1)^m a_0 \sum_{j_1=1}^{n} \sum_{j_2=j_1+1}^{n} \cdots \sum_{j_m=j_{m-1}+1}^{n} \prod_{i=1}^{m} \frac{1}{z_{j_i}} \qquad (B.13)$$

Finally, if the multiplicity of the zeros is known, then the polynomial can be written as

$$p_n(z) = a_n \prod_{k=1}^{l} (z - z_k)^{m_k}$$

Using Newton's binomium $(z - z_k)^{m_k} = \sum_{j=0}^{m_k} \binom{m_k}{j} (-z_k)^j z^{m_k-j}$, expansion of the product yields

$$p_n(z) = a_n \sum_{j_1=0}^{m_1} \binom{m_1}{j_1} (-z_1)^{j_1} \sum_{j_2=0}^{m_2} \binom{m_2}{j_2} (-z_2)^{j_2} \cdots \sum_{j_l=0}^{m_l} \binom{m_l}{j_l} (-z_l)^{j_l} z^{n-\sum_{k=1}^{l} j_k}$$

where we have used $\sum_{k=1}^{l} m_l = n$ in **art. 291**. Let $q = \sum_{k=1}^{l} j_l$, then

$$p_n(z) = (-1)^n a_n \sum_{q=0}^{n} \left\{ \sum_{\sum_{k=1}^{l} j_k=q; j_k \geq 0} \prod_{k=1}^{l} \binom{m_k}{j_k} (-z_k)^{j_k} \right\} z^q$$

from which the coefficient a_q follows as

$$a_q = (-1)^{n-q} a_n \sum_{\sum_{k=1}^{l} j_k=q; j_k \geq 0} \prod_{k=1}^{l} \binom{m_k}{j_k} z_k^{j_k}$$

The last sum is an instance of a characteristic coefficient (B.10) of a complex function, first defined in Van Mieghem (1996) and different in form than (B.11).

297. The *elementary symmetric polynomials* of degree k in n variables $z_1, z_2, \ldots, z_n$ are defined by

$$e_k(z_1, z_2, \ldots, z_n) = (-1)^{n-k} \frac{a_{n-k}}{a_n}$$

where $\frac{a_{n-k}}{a_n}$ is given in either (B.11) or (B.12). We define $e_0(z_1, z_2, \ldots, z_n) = (-1)^n$. By Vieta's Theorem 83, any polynomial $p_n(z) = \sum_{k=0}^{n} a_k z^k = \sum_{k=0}^{n} a_{n-k} z^{n-k}$ in (B.1) can be expressed in terms of elementary symmetric polynomials as

$$p_n(z) = a_n \sum_{k=0}^{n} (-1)^k e_{n-k}(z_1, z_2, \ldots, z_n) z^k$$

which is verified by multiplying out $p_n(z) = a_n \prod_{k=1}^{n}(z - z_k)$, or easier, $p_n(-z) = (-1)^n a_n \prod_{k=1}^{n}(z + z_k)$. For example, for $n = 1$,

$$e_1(z_1) = z_1$$

for $n = 2$,

$$e_1(z_1, z_2) = z_1 + z_2$$
$$e_2(z_1, z_2) = z_1 z_2$$

for $n = 3$,

$$e_1(z_1, z_2, z_3) = z_1 + z_2 + z_3$$
$$e_2(z_1, z_2, z_3) = z_1 z_2 + z_1 z_3 + z_2 z_3$$
$$e_3(z_1, z_2, z_3) = z_1 z_2 z_3$$

for $n = 4$,

$$e_1(z_1, z_2, z_3, z_4) = z_1 + z_2 + z_3 + z_4$$
$$e_2(z_1, z_2, z_3, z_4) = z_1 z_2 + z_1 z_3 + z_1 z_4 + z_2 z_3 + z_2 z_4 + z_3 z_4$$
$$e_3(z_1, z_2, z_3, z_4) = z_1 z_2 z_3 + z_1 z_2 z_4 + z_1 z_3 z_4 + z_2 z_3 z_4$$
$$e_4(z_1, z_2, z_3, z_4) = z_1 z_2 z_3 z_4$$

For each positive integer $k \leq n$, there exists exactly one elementary symmetric polynomial $e_k(z_1, z_2, \ldots, z_n)$ of degree k in n variables, which is formed by the sum of all different products of k-tuples of the n variables. Since each such a product $\prod_{j=1}^{k} z_j$ is commutative, all linear combinations of products of the elementary symmetric polynomials constitute a commutative ring, which lies at the basis of Galois theory. For example, it can be shown that any symmetric polynomial in n variables can be expressed in a unique way in terms of the elementary symmetric polynomials $e_k(z_1, z_2, \ldots, z_n)$ for $1 \leq k \leq n$.

298. *Discriminant of a polynomial.* The discriminant of a polynomial $p_n(z)$ is defined for $n \geq 2$ as

$$\Delta(p_n) = a_n^{2n-2} \prod_{1 \leq k < j \leq n} (z_j - z_k)^2 \tag{B.14}$$

with the convention that $\Delta(p_1) = 1$. In view of (A.77), the discriminant can be written in terms of the Vandermonde determinant as

$$\Delta(p_n) = a_n^{2n-2} (\det V_n(z))^2 \tag{B.15}$$

where $z = (z_1, z_2, \ldots, z_n)$ is the vector of the zeros of $p_n(z)$. The definition (B.14) of the discriminant shows that $\Delta(p_n) = 0$ when at least one zero has a multiplicity larger than 1. In order words, $\Delta(p_n) \neq 0$ if and only if all zeros of $p_n(z)$ are simple or distinct.

Since the discriminant is a symmetric polynomial in the zeros and any symmetric polynomial can be expressed in a unique way in terms of the elementary symmetric polynomials (**art. 297**), the discriminant can also be expressed in terms of the coefficients of the polynomial. For example, for $n = 2$, we obtain the well-known discriminant of the quadratic polynomial $p_2(z) = az^2 + bz + c$ as

$$\Delta(p_2) = b^2 - 4ac$$

The discriminant of the cubic $p_3(z) = az^3 + bz^2 + cz + d$ is

$$\Delta(p_3) = b^2c^2 - 4ac^3 - 4b^3d - 27a^2d^2 + 18abcd$$

299. *Discriminant and the derivative of a polynomial.* The logarithmic derivative of the polynomial $p_n(z)$ in (B.5) shows that

$$p_n'(z) = a_n \sum_{k=1}^{n} \prod_{j=1;j\neq k}^{n} (z - z_j)$$

Evaluated at a zero z_m of $p_n(z)$ gives

$$p_n'(z_m) = a_n \prod_{j=1;j\neq m}^{n} (z_m - z_j) = a_n (-1)^{m-1} \prod_{j=1;j\neq m}^{m-1} (z_j - z_m) \prod_{j=m+1}^{n} (z_m - z_j)$$

from which we obtain $\prod_{m=1}^{n} p_n'(z_m) = a_n^n (-1)^{\frac{n(n-1)}{2}} \prod_{m=1}^{n}\prod_{j=m+1}^{n} (z_j - z_m)^2$.
By invoking the definition (B.14) of the discriminant, we arrive at

$$\Delta(p_n) = (-1)^{\frac{n(n-1)}{2}} a_n^{n-2} \prod_{m=1}^{n} p_n'(z_m) \tag{B.16}$$

which shows that, if the discriminant is non-zero, the derivative $p_n'(z)$ has all its zeros different from the simple zeros of $p_n(z)$. In cases where a differential equation for a set of polynomials is known, such as for most orthogonal polynomials, the relation (B.16) can be used to express the discriminant in closed form as shown in Milovanović *et al.* (1994, p. 67).

11.2 Transforming polynomials

300. *Linear transformation.* Any polynomial

$$q_n(z) = \sum_{j=0}^{n} b_j z^j = b_n \prod_{k=1}^{n} (z - y_k)$$

where $b_n \neq 0$ can be reduced by a linear transformation $z = x + c$ into a polynomial $p_n(x) = \sum_{k=0}^{n} a_k x^k$, where the coefficient a_{n-1} of x^{n-1} is zero. Indeed,

$$q_n(x+c) = \sum_{j=0}^{n} b_j (x+c)^j = \sum_{j=0}^{n} b_j \sum_{k=0}^{j} \binom{j}{k} x^k c^{j-k} = \sum_{k=0}^{n} \left(\sum_{j=k}^{n} b_j \binom{j}{k} c^{j-k} \right) x^k$$

$$= b_n x^n + (b_{n-1} + n b_n c) x^{n-1} + \ldots + \sum_{j=0}^{n} b_j c^j$$

shows that, if $c = -\frac{b_{n-1}}{nb_n}$, the polynomial $p_n(x) = q_n(x+c)$ possesses a zero coefficient of x^{n-1}, thus $a_{n-1} = 0$. Clearly, a linear transform shifts the zeros $\{y_k\}_{1 \le k \le n}$ of $q_n(z)$ to the zeros $\{y_k - c\}_{1 \le k \le n}$ of $p_n(z)$. The sum of zeros of $p_n(z)$ is zero by (B.8) such that c is the mean of the zeros of $q_n(z)$. Hence, without loss of generality, any polynomial $q_n(z)$ can be first transformed by $z = x + c$ with $c = -\frac{b_{n-1}}{nb_n}$ into the polynomial $p_n(x)$, where $a_k = \sum_{j=k}^n b_j \binom{j}{k} c^{j-k}$ and $a_{n-1} = 0$.

The Newton identity $Z_2 = \sum_{k=1}^n (y_k - c)^2 = -\frac{2a_{n-2}}{a_n}$ shows that the real coefficients a_{n-2} and a_n of a real polynomial $p_n(x)$, where $a_{n-1} = 0$, must have opposite signs if all zeros are real.

301. *Möbius transform or linear fractional transform.* The Möbius transform or linear fractional transform $f(z) = \frac{az+b}{cz+d}$, that maps a point in the z-plane to a point in the w-plane, is the only univalent[2] transform in the *whole* finite plane (Markushevich, 1985, Vol. II, p.116). The Möbius conformal mapping $w = \frac{1-z}{1+z}$ and $z = \frac{1-w}{1+w}$ transforms the right-half complex plane $\operatorname{Re}(z) > 0$ into the unit circle $|w| < 1$. For, let $z = re^{i\theta}$, then

$$w = \frac{1 - e^{i\theta + \ln r}}{1 + e^{i\theta + \ln r}} = \frac{e^{\frac{i\theta + \ln r}{2}}\left(e^{-\frac{i\theta + \ln r}{2}} - e^{\frac{i\theta + \ln r}{2}}\right)}{e^{\frac{i\theta + \ln r}{2}}\left(e^{-\frac{i\theta + \ln r}{2}} + e^{\frac{i\theta + \ln r}{2}}\right)} = -\tanh\left(\frac{\ln r + i\theta}{2}\right)$$

which can also be written as

$$w = \sqrt{\frac{\cosh \ln r - \cos \theta}{\cosh \ln r + \cos \theta}} e^{i\left(\arctan\left(\frac{\sin \theta}{\sinh \ln r}\right) + \pi\right)}$$

If $r = 1$, then $w = i\tan\frac{\theta}{2}$, which shows that a point z on the unit circle is mapped into a point w on the imaginary axis (and vice versa). If $\operatorname{Re}(z) < 0$ or $\theta \in \left(\frac{\pi}{2}, \frac{3\pi}{2}\right)$, then $\cos \theta < 0$ and $|w| = \sqrt{\frac{\cosh \ln r - \cos \theta}{\cosh \ln r + \cos \theta}} > 1$, while, if $\operatorname{Re}(z) > 0$ or $\theta \in \left(-\frac{\pi}{2}, \frac{\pi}{2}\right)$ and $\cos \theta > 0$, then $|w| < 1$.

If all zeros of $p_n(z)$ lie in the $\operatorname{Re}(z) > 0$ - plane and similarly, all zeros of $p_n(-z)$ lie in the left-half complex plane, then all zeros of $p_n\left(\frac{1-w}{1+w}\right)$ lie inside the unit circle $|w| < 1$. The function[3]

$$p_n\left(\frac{1-w}{1+w}\right) = (1+w)^{-n} \sum_{k=0}^n a_k (1+w)^{n-k} (1-w)^k$$

has the same (finite) zeros as the polynomial

$$q_n(w) = \sum_{k=0}^n a_k (1+w)^{n-k} (1-w)^k$$

---

[2] A single-valued function $f(z)$ is univalent (schlicht or simple) on a domain D if $f(z)$ is analytic in D, except possibly at simple poles, and if $f(z)$ takes distinct value at distinct points of D, i.e. $f(z_1) \ne f(z_2)$ for $z_1 \ne z_1$ and $z_1, z_2 \in D$.

[3] All zeros of $p_n\left(-\frac{1-w}{1+w}\right)$ lie outside the unit circle and $w^n q_n(w^{-1}) = \sum_{m=0}^n \left(\sum_{j=0}^m (-1)^j \sum_{k=0}^n a_k (-1)^k \binom{k}{j}\binom{n-k}{m-j}\right) w^m$.

Newton's binomium and the Cauchy product give

$$(1+w)^{n-k}(1-w)^k = \sum_{m=0}^n \sum_{j=0}^m (-1)^j \binom{k}{j}\binom{n-k}{m-j} w^m$$

such that the polynomial $q_n(w) = \sum_{m=0}^n b_m w^m$ has coefficients

$$b_m = \sum_{j=0}^m (-1)^j \sum_{k=0}^n a_k \binom{k}{j}\binom{n-k}{m-j} \stackrel{\text{def}}{=} \sum_{k=0}^n B_{mk} a_k$$

Defining $B_{mk} = \sum_{j=0}^m (-1)^j \binom{k}{j}\binom{n-k}{m-j}$ as matrix elements of the $(n+1)\times(n+1)$ matrix B allows us to write the coefficient vector b in terms of the coefficient vector a as $b = Ba$. Since $q_n(1) = a_0 2^n$, which is, for any set of coefficients a_k, equivalent to

$$\sum_{m=0}^n b_m = \sum_{m=0}^n \sum_{j=0}^m (-1)^j \sum_{k=0}^n a_k \binom{k}{j}\binom{n-k}{m-j} = 2^n a_0$$

we find, by equating corresponding coefficients a_k that

$$\sum_{m=0}^n \sum_{j=0}^m (-1)^j \binom{k}{j}\binom{n-k}{m-j} = 2^n 1_{\{k=0\}} = \sum_{m=0}^n B_{mk}$$

Hence, the sums over the columns of B are zero, except for the zeroth column $k=0$.

Let us consider the inverse transform $w = \frac{1-z}{1+z}$ in $p_n\left(\frac{1-w}{1+w}\right) = \frac{q_n(w)}{(1+w)^n}$,

$$p_n(z) = 2^{-n}(1+z)^n q_n\left(\frac{1-z}{1+z}\right)$$

where

$$q_n\left(\frac{1-z}{1+z}\right) = \sum_{m=0}^n b_m \left(\frac{1-z}{1+z}\right)^m = (1+z)^{-n} \sum_{m=0}^n b_m (1+z)^{n-m}(1-z)^m$$

$$= (1+z)^{-n} \sum_{m=0}^n c_m z^m$$

Since $b_k = \sum_{l=0}^k (-1)^l \sum_{q=0}^n a_q \binom{q}{l}\binom{n-q}{k-l}$, the coefficient c_m is

$$c_m = \sum_{j=0}^m (-1)^j \sum_{k=0}^n b_k \binom{k}{j}\binom{n-k}{m-j}$$

$$= \sum_{j=0}^m (-1)^j \sum_{k=0}^n \left(\sum_{l=0}^k (-1)^l \sum_{q=0}^n a_q \binom{q}{l}\binom{n-q}{k-l}\right) \binom{k}{j}\binom{n-k}{m-j}$$

$$= \sum_{q=0}^n a_q \sum_{k=0}^n \left(\sum_{l=0}^k (-1)^l \binom{q}{l}\binom{n-q}{k-l}\right) \left(\sum_{j=0}^m (-1)^j \binom{k}{j}\binom{n-k}{m-j}\right)$$

Thus, $p_n(z) = 2^{-n} \sum_{m=0}^{n} c_m z^m$ and since $p_n(z) = \sum_{m=0}^{n} a_m z^m$, we must have that $a_m = 2^{-n} c_m$ is

$$a_m = 2^{-n} \sum_{q=0}^{n} a_q \sum_{k=0}^{n} \left(\sum_{l=0}^{k} (-1)^l \binom{q}{l} \binom{n-q}{k-l} \right) \left(\sum_{j=0}^{m} (-1)^j \binom{k}{j} \binom{n-k}{m-j} \right)$$

After equating corresponding coefficients a_m, we find

$$\sum_{k=0}^{n} \left(\sum_{j=0}^{m} (-1)^j \binom{k}{j} \binom{n-k}{m-j} \right) \left(\sum_{l=0}^{k} (-1)^l \binom{q}{l} \binom{n-q}{k-l} \right) = 2^n \delta_{qm}$$

In terms of the matrix elements $B_{mk} = \sum_{j=0}^{m} (-1)^j \binom{k}{j} \binom{n-k}{m-j}$, we observe that

$$\sum_{k=0}^{n} B_{mk} B_{kq} = 2^n \delta_{qm}$$

Hence, the matrix $B^2 = 2^n I$.

If all zeros of $q_n(w)$ lie inside the unit circle, then the sequence of the power sums $Z_j = \sum_{k=1}^{n} z_k^j$ is strictly decreasing in j. **Art. 333** below gives another check. In addition, the sum of the inverses of the zeros $w_1, w_2, \ldots, w_n$ of $q_n(w)$ is (**art. 294**)

$$\sum_{k=1}^{n} \frac{1}{w_k} = -\frac{b_1}{b_0} = -\frac{\sum_{k=0}^{n} (n-2k) a_k}{\sum_{k=0}^{n} a_k}$$

$$= -\frac{\frac{d}{dw} (1+w)^n p_n \left(\frac{1-w}{1+w} \right) \Big|_{w=0}}{p_n(1)} = -n + \frac{2 p_n'(1)}{p_n(1)}$$

302. *Möbius transform of an even polynomial.* Consider an even polynomial $r_{2n}(z) = \sum_{k=0}^{n} a_{2k} z^{2k}$. Conformal mapping $w = \frac{1-z}{1+z}$ and $z = \frac{1-w}{1+w}$ leads to

$$r_{2n} \left(\frac{1-w}{1+w} \right) = (1+w)^{-2n} \sum_{k=0}^{n} a_{2k} (1+w)^{2n-2k} (1-w)^{2k} = r_{2n} \left(-\frac{1-w}{1+w} \right)$$

$$= (1+w)^{-2n} t_{2n}(w)$$

where

$$t_{2n}(w) = \sum_{m=0}^{2n} b_m^* w^m$$

with coefficients $b_m^* = \sum_{j=0}^{m} (-1)^j \sum_{k=0}^{n} a_{2k} \binom{2k}{j} \binom{2n-2k}{m-j}$. The inverse transform $w = \frac{1-z}{1+z}$ applied to $r_{2n} \left(\frac{1-w}{1+w} \right) = (1+w)^{-2n} t_{2n}(w)$ gives

$$r_{2n}(z) = 2^{-2n} (1+z)^{2n} t_{2n} \left(\frac{1-z}{1+z} \right) = 2^{-2n} (1-z)^{2n} t_{2n} \left(\frac{1+z}{1-z} \right)$$

where the latter follows from $r_{2n}(z) = r_{2n}(-z)$. Explicitly, we have that

$$(1+z)^{2n} t_{2n}\left(\frac{1-z}{1+z}\right) = \sum_{m=0}^{2n} b_m^*(1-z)^m (1+z)^{2n-m}$$

and

$$(1-z)^{2n} t_{2n}\left(\frac{1+z}{1-z}\right) = \sum_{m=0}^{2n} b_m^*(1-z)^{2n-m}(1+z)^m = \sum_{q=0}^{2n} b_{2n-m}^*(1-z)^q (1+z)^{2n-q}$$

Equating corresponding powers in $(1-z)^m (1+z)^m = (1-z^2)^m$ shows that the coefficients b_m^* are symmetric around b_n^*,

$$b_m^* = b_{2n-m}^*$$

Thus,

$$w^{2n} t_{2n}\left(\frac{1}{w}\right) = \sum_{m=0}^{2n} b_m^* w^{2n-m} = \sum_{m=0}^{2n} b_{2n-m}^* w^{2n-m} = \sum_{q=0}^{2n} b_q^* w^q = t_{2n}(w)$$

or, in symmetric form[4], $w^n t_{2n}\left(\frac{1}{w}\right) = w^{-n} t_{2n}(w)$. Hence, if the polynomial $t_{2n}(w)$ does not have a zero inside (and thus also outside) the unit circle, all $2n$ zeros of $t_{2n}(w)$ must lie on the unit circle, which is equivalent to the fact that all zeros of $r_{2n}(z)$ lie on the imaginary axis.

On the other hand, we can write

$$t_{2n}(w) = \sum_{m=0}^{n} b_m^* w^m + \sum_{m=n+1}^{2n} b_{2n-m}^* w^m = \sum_{m=0}^{n-1} b_m^* w^m + b_n^* w^n + \sum_{m=0}^{n-1} b_m^* w^{2n-m}$$

$$= w^n \left(b_n^* + \sum_{m=0}^{n-1} b_m^* \left(w^{m-n} + w^{n-m} \right) \right)$$

Let $w = re^{i\theta}$, then with $w^{m-n} + w^{n-m} = 2\cosh((n-m)(\ln r + i\theta))$, we have

$$w^{-n} t_{2n}(w) = b_n^* + 2\sum_{k=1}^{n} b_{n+k}^* \cosh(k(\ln r + i\theta))$$

If $r = 1$ and if $b_n^* < 2b_{n+1}^* < 2b_{n+2}^* < \cdots < 2b_{2n}^*$, then

$$e^{-in\theta} t_{2n}(e^{i\theta}) = b_n^* + 2\sum_{k=1}^{n} b_{n+k}^* \cos(k\theta)$$

has $2n$ distinct real roots in the interval $0 < \theta < 2\pi$, and no imaginary roots at all. The proof is given in Markushevich (1985, Vol. II, pp. 50-52).

---

[4] Polynomials $p_n(z)$ with complex coefficients that satisfy $\sum_{k=0}^{n} a_k z^k = \sum_{k=0}^{n} (a_{n-k})^* z^k$, equivalent to $p_n(z) = (z^n p_n(z^{-1}))^*$ are called self-inversive and discussed in Sheil-Small (2002, Chapter 7).

11.3 Interpolation

303. *Lagrange interpolation.* Interpolation consists of constructing a polynomial that passes through a set of n distinct points, defined by their finite, possibly complex, coordinates (x_j, y_j) for $1 \leq j \leq n$. In many cases, each ordinate $y_j = f(x_j)$ is a known value at x_j of a function $f(x)$, that we want to approximate by a polynomial. Lagrange has developed a convenient way to construct an "interpolating" polynomial.

We start by considering the polynomial of degree n,

$$F_n(x) = \prod_{j=1}^{n} (x - x_j) \tag{B.17}$$

Clearly, $\frac{F_n(x)}{x - x_k} = \prod_{j=1; j \neq k}^{n} (x - x_j)$ is a polynomial of degree $n - 1$ and, since $F_n(x_k) = 0$, we have that

$$\lim_{x \to x_k} \frac{F_n(x)}{x - x_k} = \lim_{x \to x_k} \frac{F_n(x) - F_n(x_k)}{x - x_k} = F_n'(x_k) = \prod_{j=1; j \neq k}^{n} (x_k - x_j) \tag{B.18}$$

Since all x_k are distinct, x_k is a simple zero of $F_n(x)$ such that $F_n'(x_k) \neq 0$. The polynomial of degree $n - 1$,

$$l_{n-1}(x; x_k) = \frac{F_n(x)}{(x - x_k) F_n'(x_k)} = \prod_{j=1; j \neq k}^{n} \frac{x - x_j}{x_k - x_j} \tag{B.19}$$

possesses the interesting property that, at any of the abscissae $x_1, x_2, \ldots, x_n$, it vanishes, except at $x = x_k$, where it is one. Thus, with Kronecker's delta δ_{kj}, it holds that

$$l_{n-1}(x_j; x_k) = \delta_{kj}$$

Lagrange observed that the polynomial of degree $n - 1$,

$$p_{n-1}(x) = \sum_{j=1}^{n} y_j l_{n-1}(x; x_j) \tag{B.20}$$

passes through all n points $\{(x_j, y_j)\}_{1 \leq j \leq n}$ satisfying $p_{n-1}(x_j) = y_j$ for $1 \leq j \leq n$. The polynomial (B.20) is called the Lagrange interpolation polynomial corresponding to the set of n points $\{(x_j, y_j)\}_{1 \leq j \leq n}$.

The Lagrange polynomial (B.20) is unique. Indeed, assume that there is another polynomial $q_{n-1}(x)$ that passes through the same set of n points. Then $p_{n-1}(x) - q_{n-1}(x)$ is again a polynomial of degree $n - 1$ that possesses n zeros at x_j for $1 \leq j \leq n$, which is impossible (**art. 291**). Hence, $p_{n-1}(x) = q_{n-1}(x)$, which establishes the uniqueness.

If the function $f(x)$ that generates the set of n ordinate values $\{y_j = f(x_j)\}_{1 \leq j \leq n}$ is a polynomial of degree r, then the Lagrange polynomial (B.20) returns precisely that polynomial $f(x)$ provided that $n \geq r + 1$. This property follows, similarly as the argument above, by considering the difference polynomial $f(x) - p_{n-1}(x)$, which is zero at the n different abscissa points $\{x_j\}_{1 \leq j \leq n}$. Hence, using more (different) sampling points than $r + 1$ to determine a polynomial $f(x)$ of degree r cannot lead to a Lagrange polynomial of a higher degree than r. On the other hand, using more information than necessary does not degrade the Lagrange polynomial $p_{n-1}(x)$ in the sense that $p_{n-1}(x)$ is still precisely equal to $f(x)$. This property can be useful when we possess a set of n distinct points $\{(x_j, f(x_j))\}_{1 \leq j \leq n}$ of which it is unknown whether $f(x)$ is a polynomial. If $f(x)$ is a polynomial of degree r, then after generating more than $n > r + 1$ function evaluations, the Lagrange polynomial does not change anymore and we may conclude that $f(x)$ is a polynomial of degree r. Otherwise, the degree of the Lagrange polynomial $p_{n-1}(x)$ will continue to increase with n.

304. *Approximating a function by a polynomial.* Given the set $\{x_j\}_{1 \leq j \leq n}$ of different abscissae lying in the interval $[a, b]$ and ordered $b > x_1 > x_2 > \cdots > x_n > a$, the goodness of the approximation of $f(x)$ by the Lagrange polynomial $p_{n-1}(x)$ is usually measured by the maximum deviation $\max_{a \leq x \leq b} |f(x) - p_{n-1}(x)|$ for increasing $n > 0$, while the remainder is $r_n(x) = f(x) - p_{n-1}(x)$. From (B.20), we deduce that

$$\max_{a \leq x \leq b} |f(x) - p_{n-1}(x)| = \max_{a \leq x \leq b} \left| \sum_{j=1}^{n} (f(x) - y_j) l_{n-1}(x; x_j) \right|$$

$$\leq \max_{a \leq x \leq b} |f(x) - f(x_j)| \max_{a \leq x \leq b} \sum_{j=1}^{n} |l_{n-1}(x; x_j)|$$

and the definition (B.19) shows that $\sum_{j=1}^{n} |l_{n-1}(x; x_j)|$ is independent of the function $f(x)$. The smaller $\max_{a \leq x \leq b} \sum_{j=1}^{n} |l_{n-1}(x; x_j)|$, the better the sequence of Lagrange interpolating polynomials at the set $\{x_j\}_{1 \leq j \leq n}$ approximates the function $f(x)$ uniformly over $[a, b]$. Often, the interval $[a, b]$ is transformed to $[-1, 1]$ by the linear transformation $\frac{2(x-a)}{b-a} - 1$. Finding the best set $\{x_j\}_{1 \leq j \leq n}$ in $[-1, 1]$ that minimizes $\max_{-1 \leq x \leq 1} \sum_{j=1}^{n} |l_{n-1}(x; x_j)|$ seems a difficult, open problem (Rivlin, 1974), although the set that minimizes $\max_{-1 \leq x \leq 1} \sum_{j=1}^{n} l_{n-1}^2(x; x_j)$ is known[5]. Erdős (1961) demonstrates[6] that there exists a positive real constant c such that $\max_{-1 \leq x \leq 1} \sum_{j=1}^{n} |l_{n-1}(x; x_j)| > \frac{2}{\pi} \log n - c$, for any set $\{x_j\}_{1 \leq j \leq n}$. Consequently, given the set $\{x_j\}_{1 \leq j \leq n}$, when $n \to \infty$, there exists a continuous function $f(x)$ on

---

[5] Rivlin (1974, p. 52) mentions that this optimal set $\{x_j\}_{1 \leq j \leq n}$ consists of zeros of $\int_x^1 P_{n-1}(t) \, dt$, where $P_n(x)$ is the n-th Legendre polynomial, for which $\max_{-1 \leq x \leq 1} \sum_{j=1}^{n} l_{n-1}^2(x; x_j) = 1$.

[6] Rivlin (1974, p. 18) proves that, with the Euler constant $\gamma = 0.5772$,

$$\frac{2}{\pi} \log n + \frac{2}{\pi} \left(\log \frac{8}{\pi} + \gamma \right) < \max_{-1 \leq x \leq 1} \sum_{j=1}^{n} |l_{n-1}(x; x_j)| \leq \frac{2}{\pi} \log n + 1$$

where each point x_j in the set $\{x_j\}_{1 \leq j \leq n}$ is a zero of the Chebyshev polynomial $T_n(x) = \cos(n \arccos x)$.

$[-1, 1]$ for which the Lagrange polynomial (B.20) does *not* converge uniformly to $f(x)$.

Hermite interpolation requires that the interpolating polynomial $h_{2n-1}(x)$ of degree $2n - 1$ satisfies, besides the function values $y_j = h_{2n-1}(x_j)$ at x_j, also the first derivative $y'_j = h'_{2n-1}(x_j)$ for each $1 \le j \le n$, so that

$$h_{2n-1}(x) = \sum_{j=1}^{n} \left\{ y_j \left(1 - \frac{F''_n(x_j)}{F'_n(x_j)} (x - x_j) \right) + y'_j (x - x_j) \right\} l_{n-1}^2(x; x_j)$$

Whereas Lagrange interpolation failed, Rivlin (1974, p. 27) demonstrates that Hermite interpolation at the zeros of the Chebyshev polynomial leads to a sequence of polynomials that converges to the function $f(x)$. Consequently, Hermite interpolation also proves Weierstrass's famous approximation theorem:

Theorem 84 (Weierstrass's approximation theorem) *For any continuous real-valued function $f(x)$, defined on the real interval $[a, b]$, and for every $\epsilon > 0$, there exists a polynomial $p(x)$ such that $|f(x) - p(x)| < \epsilon$, for all $x \in [a, b]$.*

Also Bernstein polynomials $b_{k,n}(x) = \binom{n}{k} x^k (1 - x)^{n-k}$ of degree n and integer $k \in [0, n]$ provide a constructive proof of Weierstrass's approximation theorem. Indeed, it can be shown that the polynomial $\sum_{k=0}^{n} f\left(\frac{k}{n}\right) b_{k,n}(x)$ converges to $f(x)$ uniformly for any $x \in [0, 1]$.

Since $p_{n-1}(x_j) = f(x_j)$ for $1 \le j \le n$, the remainder $r_n(x) = f(x) - p_{n-1}(x)$ is zero at each interpolation point x_j and we can write with the definition (B.17) of $F_n(x)$ that

$$r_n(x) = F_n(x) g(x)$$

where $g(x)$ is a function related to $f(x)$. Consider the auxiliary function

$$w_y(t) = f(t) - p_{n-1}(t) - F_n(t) g(y)$$

which is zero at $t = x_j$ for $1 \le j \le n$ and also, by definition of the remainder $r_n(y)$, at $t = y$. If $y \in [a, b]$ and $y \ne x_j$, then $w_y(t)$ has at least $n + 1$ different zeros in $[a, b]$, $\frac{dw_y(t)}{dt}$ has at least n zeros lying in between those of $w_y(t)$ since the interpolation points and y are different, and so on. Thus, $\frac{d^n w_y(t)}{dt^n}$ has at least one zero ξ lying inside the interval (x_n, x_1) and satisfying

$$\left. \frac{d^n w_y(t)}{dt^n} \right|_{t=\xi} = 0 \quad \Leftrightarrow \quad f^{(n)}(\xi) - n! g(y) = 0$$

In summary, assuming that f has continuous derivatives up to order n in $[a, b]$, we arrive at the Lagrange interpolating polynomial with remainder

$$f(x) = \sum_{k=1}^{n} \frac{F_n(x)}{(x - x_k) F'_n(x_k)} f(x_k) + \frac{f^{(n)}(\xi)}{n!} F_n(x) \tag{B.21}$$

where $x_n < \xi < x_1$.

The error $\tilde{r}_n(x) = f(x) - h_{2n-1}(x)$ of Hermite interpolation is zero at each interpolation point x_j and each zero has multiplicity two because both $f(x_j) = h_{2n-1}(x_j)$ and $f'(x_j) = h'_{2n-1}(x_j)$ so that $\tilde{r}_n(x) = F_n^2(x)g(x)$. By a similar argument that led to (B.21), the Hermite interpolating polynomial with remainder is

$$f(x) = h_{2n-1}(x) + \frac{f^{(2n)}(\xi)}{(2n)!}F_n^2(x) \tag{B.22}$$

305. *Lagrange interpolation and the Vandermonde matrix.* There is a noteworthy relation with the Vandermonde matrix (**art.** 224), when we write the set of equations, $y_j = p_{n-1}(x_j)$ for $1 \le j \le n$, using $p_{n-1}(x) = \sum_{k=0}^{n-1} a_k x^k$, as

$$\begin{cases} a_0 + a_1 x_1 + a_2 x_1^2 + \ldots + a_{n-1} x_1^{n-1} = y_1 \\ a_0 + a_1 x_2 + a_2 x_2^2 + \ldots + a_{n-1} x_2^{n-1} = y_2 \\ \vdots \\ a_0 + a_1 x_n + a_2 x_n^2 + \ldots + a_{n-1} x_n^{n-1} = y_n \end{cases}$$

which is

$$\begin{bmatrix} 1 & x_1 & x_1^2 & \cdots & x_1^{n-1} \\ 1 & x_2 & x_2^2 & \cdots & x_2^{n-1} \\ \vdots & \vdots & \vdots & \vdots & \vdots \\ 1 & x_n & x_n^2 & \cdots & x_n^{n-1} \end{bmatrix} \begin{bmatrix} a_0 \\ a_1 \\ \vdots \\ a_{n-1} \end{bmatrix} = \begin{bmatrix} y_1 \\ y_2 \\ \vdots \\ y_n \end{bmatrix}$$

In matrix form, the interpolation problem becomes $V_n(x) a = y$, where the coefficient vector $a = (a_0, a_1, \ldots, a_{n-1})$ is transformed to the ordinate vector $y = (y_1, y_2, \ldots, y_n)$ by the Vandermonde matrix $V_n(x)$ of the abscissa vector $x = (x_1, x_2, \ldots, x_n)$. Using Cramer's rule (**art.** 220), the coefficient a_k reads

$$a_k = \frac{1}{\det V_n^T(x)} \begin{vmatrix} 1 & \cdots & x_1^{k-2} & y_1 & x_1^k & \cdots & x_1^{n-1} \\ 1 & \cdots & x_2^{k-2} & y_2 & x_2^k & \cdots & x_2^{n-1} \\ \vdots & \vdots & \vdots & \vdots & \vdots & \vdots & \vdots \\ 1 & \cdots & x_n^{k-2} & y_n & x_n^k & \cdots & x_n^{n-1} \end{vmatrix}$$

After expanding the determinant in cofactors of the k-th column (**art.** 212) and recalling that $\text{cofactor}_{mk} V_n^T(x) = \text{cofactor}_{km} V_n(x)$, we find

$$a_k = \frac{1}{\det V_n(x)} \sum_{m=1}^{n} y_m \text{cofactor}_{km} V_n(x) \tag{B.23}$$

The coefficient vector a can be found after multiplying out the product of $l_{n-1}(x; x_m)$

in (B.19). Let[7]

$$\prod_{j=1; j\neq m}^{n} (x - x_j) = \sum_{k=0}^{n-1} b_k (m) \, x^k \tag{B.24}$$

and introduction into the Lagrange polynomial (B.20) results in

$$p_{n-1} (x) = \sum_{m=1}^{n} \frac{y_m}{\prod_{j=1; j\neq m}^{n} (x_m - x_j)} \sum_{k=0}^{n-1} b_k (m) \, x^k$$

Equating corresponding powers of x in $p_{n-1} (x) = \sum_{k=0}^{n-1} a_k x^k$ and the above form yields

$$a_k = \sum_{m=1}^{n} \frac{y_m}{\prod_{j=1; j\neq m}^{n} (x_m - x_j)} b_k (m) \tag{B.25}$$

Combining (B.23) and (B.25) leads to an explicit expression for the cofactor of the Vandermonde matrix,

$$\frac{\text{cofactor}_{km} V_n (x)}{\det V_n (x)} = \frac{b_k (m)}{\prod_{j=1; j\neq m}^{n} (x_m - x_j)} \tag{B.26}$$

306. *Newton interpolation.* Before Lagrange, Newton has proposed to construct the interpolating polynomial passing through n different points $\{(x_j, y_j)\}_{1 \leq j \leq n}$ based on Newton polynomials of degree j,

$$n_j (x) = \prod_{k=1}^{j} (x - x_k) \tag{B.27}$$

where $n_0 (x) = 1$, $n_1 (x) = x - x_1$, and so on. With the definition (B.17) of $F_n (x)$, we recognize that $n_j (x) = F_j (x)$ and that $F_j (x_m) = 0$ if $j \geq m$. Similarly as the Lagrange polynomial (B.20), the Newton interpolating polynomial is

$$p_{n-1} (x) = \sum_{k=0}^{n-1} a_k F_k (x) \tag{B.28}$$

---

[7] After relabeling the set $z_j = x_j$ for $1 \leq j \leq m - 1$ and $z_j = x_{j+1}$ for $m \leq j \leq n - 1$, we write

$$\prod_{j=1; j\neq m}^{n} (x - x_j) = \prod_{j=1}^{n-1} (x - z_j) = \sum_{k=0}^{n-1} b_k (m) \, x^k$$

and by using Vieta's Theorem 83, we have

$$b_k (m) = (-1)^{n-1-k} \sum_{j_1=1}^{n-1} \sum_{j_2=j_1+1}^{n-1} \cdots \sum_{j_{n-1-k}=j_{n-k-2}+1}^{n-1} \prod_{i=1}^{n-1-k} z_{j_i} \qquad (j_0 = 0) \tag{$B.29$}$$

Thus, $b_k (m)$ is equal to the elementary symmetric polynomial of degree k in $n - 1$ variables $z_1, z_2, \ldots, z_n$ (**art. 297**).

where the coefficients a_k can be found from the relation that $p_{n-1}(x_j) = y_j$, so that, for $1 \le j \le n$,

$$y_j = \sum_{k=0}^{n-1} a_k F_k(x_j) = \sum_{k=0}^{j-1} a_k F_k(x_j) \tag{B.29}$$

Explicitly, executing the relation $p_{n-1}(x_j) = y_j$ for all n coordinates leads to a set of linear equations that determine $a_0, a_1, \ldots, a_{n-1}$ as

$$\begin{bmatrix} 1 & 0 & 0 & \cdots & 0 \\ 1 & (x_2 - x_1) & 0 & \cdots & 0 \\ \vdots & \vdots & \vdots & \ddots & \vdots \\ 1 & (x_n - x_1) & (x_n - x_1)(x_n - x_2) & \cdots & \prod_{k=1}^{n-1}(x_n - x_k) \end{bmatrix} \begin{bmatrix} a_0 \\ a_1 \\ \vdots \\ a_{n-1} \end{bmatrix} = \begin{bmatrix} y_1 \\ y_2 \\ \vdots \\ y_n \end{bmatrix}$$

Hence, we find that $a_0 = y_1$, $a_1 = \frac{y_2 - y_1}{x_2 - x_1}$, $a_2 = \frac{\frac{y_3 - y_2}{x_3 - x_2} - \frac{y_2 - y_1}{x_2 - x_1}}{x_3 - x_1}$, which is also written as

$$a_2 = \frac{y_1}{(x_1 - x_2)(x_1 - x_3)} + \frac{y_2}{(x_2 - x_1)(x_2 - x_3)} + \frac{y_3}{(x_3 - x_1)(x_3 - x_2)}$$

By iterating (B.29) further, we observe that

$$a_j = \sum_{k=1}^{j+1} \frac{y_k}{\prod_{m=1;m \ne k}^{j+1}(x_k - x_m)} = \sum_{k=1}^{j+1} \frac{y_k}{F'_{j+1}(x_k)} \tag{B.30}$$

where (B.18) has been used. Indeed, substitution of (B.30) into (B.29) justifies the correctness of (B.30),

$$y_j = \sum_{k=0}^{n-1} \sum_{m=1}^{k+1} \frac{y_m F_k(x_j)}{F'_{k+1}(x_m)} = \sum_{k=1}^{n} \sum_{m=1}^{k} y_m \frac{F_{k-1}(x_j)}{F'_k(x_m)} = \sum_{m=1}^{n} y_m \sum_{k=m}^{n} \frac{F_{k-1}(x_j)}{F'_k(x_m)}$$

Since $F_{k-1}(x_j) = \prod_{q=1}^{k-1}(x_j - x_q)$, it holds that $F_{k-1}(x_j) = 0$ if $k \ge j + 1$. Also, $F_{k-1}(x_j) = \prod_{q=1}^{m}(x_j - x_q) \prod_{q=m+1}^{k-1}(x_j - x_q)$ illustrates that, if $j < m$, then $F_{k-1}(x_j) = 0$. Hence, $k = j = m$ and

$$\frac{F_{k-1}(x_j)}{F'_k(x_j)} = \lim_{x \to x_j} \frac{F_k(x)}{(x - x_j) F'_k(x_j)} = l(x_j; x_j) = 1$$

The expression (B.30) for a_j is called the *divided difference* of $(y_1, y_2, \ldots, y_{j+1})$ and denoted as $a_j = [y_1, y_2, \ldots, y_{j+1}]$. The divided difference possesses interesting recursive properties, such as

$$[y_1, y_2, \ldots, y_{j+1}] = \frac{[y_2, \ldots, y_{j+1}] - [y_1, y_2, \ldots, y_j]}{x_{j+1} - x_1}$$

similar to forward and backward differences, for which we refer to Lanczos (1988).

In conclusion, the Newton interpolating polynomial that passes through the n points $\{(x_j, y_j)\}_{1 \leq j \leq n}$ is

$$p_{n-1}(x) = \sum_{k=0}^{n-1} \left\{ \sum_{m=1}^{k+1} \frac{y_m}{F'_{k+1}(x_m)} \right\} F_k(x) \tag{B.31}$$

Explicitly,

$$p_{n-1}(x) = y_1 + \left\{ \frac{y_2 - y_1}{x_2 - x_1} \right\}(x - x_1) + \left\{ \frac{\frac{y_3 - y_2}{x_3 - x_2} - \frac{y_2 - y_1}{x_2 - x_1}}{x_3 - x_1} \right\}(x - x_1)(x - x_2) + \dots$$

which shows that each term in k-sum in (B.31) is a polynomial of degree k, whereas each term in j-sum in the Lagrange interpolating polynomial (B.20) is of degree $n-1$. Since the polynomial (B.31) is unique, the Lagrange and Newton interpolating polynomials are equal, but just different in representation.

307. *Equidistant interpolation.* When the set of abscissae $\{x_k\}_{1 \leq k \leq n}$ is chosen in an equidistant way as $x_k = k\Delta x$ for $1 \leq k \leq n$, then the Lagrange interpolating polynomial (B.20) reduces to

$$p_{n-1}(x) = F_n(x) \sum_{j=1}^{n} \frac{p_{n-1}(j\Delta x)}{(x - j\Delta x) \prod_{m=1; m \neq j}^{n} (j - m)\Delta x}$$

Using $\prod_{m=1; m \neq j}^{n}(j - m) = (-1)^{n-j}(j-1)!(n-j)!$, we obtain

$$p_{n-1}(x) = \frac{\prod_{k=1}^{n}(x - k\Delta x)}{(\Delta x)^{n-1}(n-1)!} \sum_{j=1}^{n} \binom{n-1}{j-1} \frac{(-1)^{n-j} p_{n-1}(j\Delta x)}{x - j\Delta x}$$

It is often more convenient to interpolate the polynomial $p_n(x)$ from $x_1 = 0$ with steps of $\Delta x = y$ up to $x_n = (n-1)y$, in which case we arrive at the classical equidistant interpolating polynomial:

$$p_n(xy) = \frac{\prod_{k=0}^{n}(x - k)}{n!} \sum_{j=0}^{n} \binom{n}{j} \frac{(-1)^{n-j} p_n(jy)}{x - j} \tag{B.32}$$

In particular, the special case where $x = -1$ leads to

$$p_n(-y) = \sum_{j=0}^{n} \binom{n+1}{j+1} (-1)^j p_n(jy) \tag{B.33}$$

which expresses the negative argument values of a polynomial in terms of positive argument values.

Finally, we present

$$p_n(xy) = \sum_{q=0}^{n} \Delta^q p_n(0) \binom{y}{q} \tag{B.34}$$

where

$$\Delta^q p_n(0) = \sum_{k=0}^{q} (-1)^{q-k} \binom{q}{k} p_n(kx)$$

is the q-th difference obeying $\Delta^q f_m = \Delta^{q-1} f_{m+1} - \Delta^{q-1} f_m$ for all $q \in \mathbb{N}_0$, thus $\Delta f_m = f_{m+1} - f_m$. By iteration, we have that $\Delta^q f_m = \sum_{k=0}^{q} \binom{q}{k}(-1)^{q-k} f_{m-q+k}$, which we apply to the set $\{f_0, f_1, \ldots, f_n\} = \{p_n(0), p_n(x), \ldots, p_n(nx)\}$. Substituting the polynomial form $p_n(x) = \sum_{k=0}^{n} a_k x^k$ into the q-th difference yields

$$\frac{\Delta^q p_n(0)}{q!} = \sum_{m=q}^{n} \mathcal{S}_m^{(q)} a_m x^m$$

where $\mathcal{S}_m^{(q)}$ are the Stirling numbers of the second kind (Abramowitz and Stegun, 1968, Section 24.1.4). The relation (B.34) is commonly known for $x = 1$ as Newton's equidistant difference expansion for polynomials. If both sides of (B.34) converge in the limit for $n \to \infty$, the left-hand side converges to the Taylor series around $y = 0$ of a complex function f and the right-hand side then equals the difference expansion for f,

$$f(y) = \sum_{k=0}^{\infty} f_k y^k = \sum_{k=0}^{\infty} \binom{y}{k} \Delta^k f_0 \tag{B.35}$$

The series (B.35) first appeared in Newton's famous book Philosophiae Naturalis Principia Mathematica (Newton, 1687). If the Taylor coefficients $\{f_k(z_0)\}_{k \geq 0}$ are known around z_0, then Newton's series (B.35) generalizes to

$$f(z) = \sum_{k=0}^{\infty} f_k(z_0)(z - z_0)^k = \sum_{k=0}^{\infty} \binom{z - z_0}{k} \Delta^k f_0(z_0)$$

Carlson's theorem[8] indicates that Newton's series (B.35) is unique.

Using the formula

$$\sum_{j=k}^{n} (-1)^{k+j} \binom{j}{k} \binom{y}{j} = \frac{(-1)^{k+n} \Gamma(1+y)}{(y-k)\,k!(n-k)!\,\Gamma(-n+y)} \tag{B.36}$$

$$= \delta_{n,k} \qquad \text{if } y = n$$

where $\Gamma(x)$ is the Gamma function, we may verify that Newton's difference expansion (B.34) for polynomials is equivalent to the equidistant interpolating formula (B.32).

---

[8] A special case of the Phragmén-Lindelöf theorem (Titchmarsh, 1964, p. 176).

308. *Inequalities for derivatives.* The Lagrange interpolation polynomial (B.20), applied to the derivative $p'_n(x)$ of a polynomial $p_n(x)$, is written with $y_j = p'_{n-1}(x_j)$ for $1 \leq j \leq n$ and the definition (B.19) of $l_{n-1}(x; x_k)$ as

$$\frac{p'_n(x)}{F_n(x)} = \sum_{j=1}^{n} \frac{\frac{p'_n(x_j)}{F'_n(x_j)}}{x - x_j}$$

while the logarithmic derivative of $F_n(x)$ is $\frac{F'_n(x)}{F_n(x)} = \sum_{j=1}^{n} \frac{1}{x - x_j}$. The derivative

$$\frac{d}{dx}\left(\frac{p'_n(x)}{F_n(x)}\right) = \frac{p''_n(x)F_n(x) - p'_n(x)F'_n(x)}{F_n^2(x)} = -\sum_{j=1}^{n} \frac{\frac{p'_n(x_j)}{F'_n(x_j)}}{(x - x_j)^2}$$

illustrates that, at a zero $x = \zeta$ of $F'_n(x)$,

$$\left|\frac{p''_n(\zeta)}{F_n(\zeta)}\right| = \left|-\sum_{j=1}^{n} \frac{\frac{p'_n(x_j)}{F'_n(x_j)}}{(\zeta - x_j)^2}\right| \leq \sum_{j=1}^{n} \frac{\left|\frac{p'_n(x_j)}{F'_n(x_j)}\right|}{|\zeta - x_j|^2}$$

If $|p'_n(x_j)| \leq |F'_n(x_j)|$ and all $x_1, x_2, \ldots, x_n$ are real and distinct, then Rolle's theorem[9] states that also ζ is real so that

$$\left|\frac{p''_n(\zeta)}{F_n(\zeta)}\right| \leq \sum_{j=1}^{n} \frac{\left|\frac{p'_n(x_j)}{F'_n(x_j)}\right|}{(\zeta - x_j)^2} \leq \sum_{j=1}^{n} \frac{1}{(\zeta - x_j)^2} = \left|\frac{F''_n(\zeta)}{F_n(\zeta)}\right|$$

from which we conclude that $|p''_n(\zeta)| \leq |F''_n(\zeta)|$. The last equality only holds provided all $\{x_j\}_{1 \leq j \leq n}$ are real and distinct. Since the argument holds for any zero of $F'_n(x)$, we find that, if $|p'_n(x_j)| \leq |F'_n(x_j)|$ for each zero x_j of $F_n(x)$, then $|p''_n(\zeta)| \leq |F''_n(\zeta)|$ at each zero ζ of $F'_n(x)$, provided all $x_1, x_2, \ldots, x_n$ are real and distinct.

The analysis can be extended by observing that $F'_n(x) = q \prod_{j=1}^{n-1}(x - \zeta_j)$, where $q \in \mathbb{R}_0$, all zeros ζ_j are real and distinct and, in addition, different from those of $F_n(x)$, by Rolle's theorem. Generalizing this observation, we find that $F_n^{(m)}(x) = q_m \prod_{j=1}^{n-m}\left(x - \zeta_j^{(m)}\right)$, where all zeros $\zeta_j^{(m)}$ are real and distinct and different from those in the sets $\left\{\zeta_j^{(k)}\right\}_{1 \leq j \leq n-k}$ where $0 \leq k < m$ and $\zeta_j^{(0)} = x_j$. The corresponding Lagrange interpolation polynomial (B.20) becomes

$$\frac{p_n^{(m+1)}(x)}{F_n^{(m)}(x)} = \sum_{j=1}^{n-m} \frac{\frac{p_n^{(m+1)}\left(\zeta_j^{(m)}\right)}{F_n^{(m+1)}\left(\zeta_j^{(m)}\right)}}{x - \zeta_j^{(m)}}$$

---

[9] If a real, continuous function $f(x)$ on the interval $[a, b]$, that is differentiable inside the interval (a, b), vanishes at $x = a$ and $x = b$, $f(a) = f(b) = 0$, then there exists at least one point $\xi \in (a, b)$ for which $f'(\xi) = 0$.

while the logarithmic derivative of $F_n^{(m)}(x)$ is

$$\frac{F_n^{(m+1)}(x)}{F_n^{(m)}(x)} = \sum_{j=1}^{n-m} \frac{1}{x - \zeta_j^{(m)}}$$

We proceed by induction. Assume that $\left| p_n^{(m+1)}\left(\zeta_j^{(m)}\right) \right| \le \left| F_n^{(m+1)}\left(\zeta_j^{(m)}\right) \right|$ for each real zero $\zeta_j^{(m)}$ of $F_n^{(m)}(x)$. Differentiation as above yields

$$\left| \frac{p_n^{(m+2)}(x) F_n^{(m)}(x) - p_n^{(m+1)}(x) F_n^{(m+1)}(x)}{\left(F_n^{(m)}(x)\right)^2} \right| = \sum_{j=1}^{n-m} \frac{\left| \begin{matrix} p_n^{(m+1)}\left(\zeta_j^{(m)}\right) \\ F_n^{(m+1)}\left(\zeta_j^{(m)}\right) \end{matrix} \right|}{\left(x - \zeta_j^{(m)}\right)^2}$$

$$\le \sum_{j=1}^{n-m} \frac{1}{\left(x - \zeta_j^{(m)}\right)^2} = \left| \frac{d}{dx} \frac{F_n^{(m+1)}(x)}{F_n^{(m)}(x)} \right|$$

which holds for any real x different from $\zeta_j^{(m)}$. By choosing $x = \zeta_j^{(m+1)}$ for which $F_n^{(m+1)}(x) = 0$, we find that $\left| p_n^{(m+2)}\left(\zeta_j^{(m+1)}\right) \right| \le \left| F_n^{(m+2)}\left(\zeta_j^{(m+1)}\right) \right|$, which establishes the induction for all m, because the case for $m = 0$ has been demonstrated above. In conclusion, we have proved:

Lemma 15 *Let all $x_1, x_2, \ldots, x_n$ be real and distinct. Define $\zeta_j^{(0)} = x_j$ for $1 \le j \le n$. Further, consider a polynomial $p_n(z)$ and the Lagrange product $F_n(z)$, defined in (B.17). If $|p_n'(x_j)| \le |F_n'(x_j)|$ for each $1 \le j \le n$, then it holds for any integer $m \ge 0$ that*

$$\left| p_n^{(m+1)}\left(\zeta_j^{(m)}\right) \right| \le \left| F_n^{(m+1)}\left(\zeta_j^{(m)}\right) \right|$$

where $\zeta_j^{(m)}$ is a real zero of $F_n^{(m)}(x) = q_m \prod_{j=1}^{n-m} \left(x - \zeta_j^{(m)}\right)$.

11.4 The Euclidean algorithm

309. Consider two polynomials[10] $s_0(z) = \sum_{k=0}^{n} a_k z^k$ and $s_1(z) = \sum_{k=0}^{m} b_k z^k$, both with complex coefficients and where the degree n of $s_0(z)$ is larger than or equal to m. Then, there always exists a polynomial $q_1(z)$, called the quotient, such that

$$s_0(z) = q_1(z) s_1(z) + s_2(z)$$

and the degree of the remainder polynomial $s_2(z)$ is smaller than m. Indeed, we can

---

[10] The indices in $s_0(z)$ and $s_1(z)$ here deviate from the general definition (B.1) and do not reflect the degree, but the sequence order of polynomials in the Euclidean iteration scheme.

always remove the highest degree term $a_n z^n$ in $s_0(z)$ by subtracting $\frac{a_n}{b_m} z^{n-m} s_1(z)$. This first step in the long division yields

$$s_0(z) - \frac{a_n}{b_m} z^{n-m} s_1(z) = \sum_{j=0}^{n-1} \left(a_j - \frac{a_n}{b_m} b_{j-n+m} \right) z^j = r(z)$$

where the convention $b_{-j} = 0$ for $j > 0$ and where the degree of $r(z) = \sum_{j=0}^{n-1} r_j z^j$ is at most $n-1$. If the degree of the remainder $r(z)$ is larger than m, we repeat the process and subtract $\frac{r_{n-1}}{b_m} z^{n-1-m} s_1(z)$ from $r(z)$, resulting in a remainder with degree at most $n-2$. As long as the degree of the remainder polynomial exceeds m, we repeat the process of subsequent lowering the highest degree. Eventually, we arrive at a remainder $s_2(z)$ with degree smaller than m. This operation is the well-known long division.

Next, we can rewrite the equation as

$$\frac{s_0(z)}{s_1(z)} = q_1(z) + \frac{s_2(z)}{s_1(z)} = q_1(z) + \frac{1}{\frac{s_1(z)}{s_2(z)}}$$

and apply the same recipe to $\frac{s_1(z)}{s_2(z)} = q_2(z) + \frac{s_3(z)}{s_2(z)}$, where the degree of $s_3(z)$ is smaller than that of $s_2(z)$. Thus,

$$\frac{s_0(z)}{s_1(z)} = q_1(z) + \frac{1}{q_2(z) + \frac{1}{\frac{s_2(z)}{s_3(z)}}}$$

We can repeat the recipe to $\frac{s_2(z)}{s_3(z)} = q_3(z) + \frac{s_4(z)}{s_3(z)}$, where again the degree of $s_4(z)$ is smaller than that of $s_3(z)$. Hence, we can always reduce the degree of the remainder and eventually it will be equal to zero. The result is a finite continued fraction for $\frac{s_0(z)}{s_1(z)}$ in terms of the subsequent quotients $q_1(z), q_2(z), \ldots, q_m(z)$,

$$\frac{s_0(z)}{s_1(z)} = q_1(z) + \cfrac{1}{q_2(z) + \cfrac{1}{q_3(z) + \cfrac{1}{\ddots + \cfrac{1}{q_m(z)}}}}$$

Alternatively, we obtain a system of polynomial equations

$$
\begin{array}{ll}
s_0(z) = q_1(z) s_1(z) + s_2(z) & (0 < \deg s_2 < \deg s_1) \\
s_1(z) = q_2(z) s_2(z) + s_3(z) & (0 < \deg s_3 < \deg s_2) \\
s_2(z) = q_3(z) s_3(z) + s_4(z) & (0 < \deg s_4 < \deg s_3) \\
\cdots & \cdots \\
s_{m-2}(z) = q_{m-1}(z) s_{m-1}(z) + s_m(z) & (0 < \deg s_m < \deg s_{m-1}) \\
s_{m-1}(z) = q_m(z) s_m(z) &
\end{array}
$$

which is known as Euclid's algorithm. The last equation shows that $s_m(z)$ divides $s_{m-1}(z)$. The penultimate polynomial equation, written as $s_{m-2}(z) = q_{m-1}(z) q_m(z) s_m(z) + s_m(z)$, indicates that $s_m(z)$ also divides $s_{m-2}(z)$. Continuing upwards, we see that the polynomial $s_m(z)$ divides all polynomials $s_k(z)$ with

$0 \leq k \leq m$. Hence, $s_m(z)$ cannot be zero for all z, otherwise all $s_k(z)$ would be zero.

Thus, $s_m(z)$ is the greatest common divisor polynomial of both $s_0(z)$ and $s_1(z)$. Indeed, any divisor polynomial $d(z)$ of $s_0(z)$ and $s_1(z)$ obeys $d|s_0$ and $d|s_1$, then the first Euclidean equation indicates that $d|s_2$, and subsequently, $d|s_k$. Since the degree of the sequence of polynomials $s_k(z)$ strictly decreases, $s_m(z)$ is the largest possible common divisor polynomial. Consequently, the functions $f_k(z) = \frac{s_k(z)}{s_m(z)}$ are again polynomials.

310. *Minimal polynomial.* The minimal polynomial associated to a polynomial $p_n(z) = a_n \prod_{k=1}^{l}(z - z_k)^{m_k}$, where m_k denotes the multiplicity of zero z_k, is defined as

$$m_{p_n}(z) = a_n \prod_{k=1}^{l}(z - z_k) \tag{B.37}$$

The minimal polynomial divides $p_n(z)$ and is the lowest degree polynomial possessing the same zeros of $p_n(z)$, all with multiplicity 1. If $p_n(z)$ has only simple zeros, i.e., $m_k = 1$ for all $1 \leq k \leq n$, then $m_{p_n}(z) = p_n(z)$.

The minimal polynomial plays an important role in matrix polynomials (**art.** 229).

311. *Division by a first degree polynomial.* The division of $p_n(z) = \sum_{k=0}^{n} a_k z^k$ by the polynomial $z - \xi$ of degree 1 can be computed explicitly. In the notation of **art.** 309, the long division of $s_0(z) = p_n(z)$ by $s_1(z) = z - \xi$ gives the remainder $s_2(z)$ and the quotient

$$q_1(z) = \sum_{k=0}^{n-1}\left\{\frac{1}{\xi^{k+1}} \sum_{j=k+1}^{n} a_j \xi^j\right\} z^k \tag{B.38}$$

It is instructive to relate the long division with Taylor series. We assume that $\xi \neq 0$. With the convention that $a_k = 0$ if $k > n$ and $a_n \neq 0$, execution of the Cauchy product of two Taylor series around $z = 0$ yields, for $|z| < |\xi|$,

$$\frac{p_n(z)}{z - \xi} = -\sum_{k=0}^{\infty} a_k z^k \sum_{k=0}^{\infty} \frac{z^k}{\xi^{k+1}} = -\sum_{k=0}^{\infty}\left\{\frac{1}{\xi^{k+1}} \sum_{j=0}^{k} a_j \xi^j\right\} z^k$$

We split the series into two parts and take into account that $a_k = 0$ if $k > n$,

$$\frac{p_n(z)}{z - \xi} = -\sum_{k=0}^{n-1}\left\{\frac{1}{\xi^{k+1}} \sum_{j=0}^{k} a_j \xi^j\right\} z^k - \sum_{k=n}^{\infty}\left\{\frac{1}{\xi^{k+1}} \sum_{j=0}^{n} a_j \xi^j\right\} z^k$$

Observing that $p_n(\xi) = \sum_{j=0}^{n} a_j \xi^j$, the last sum equals

$$\sum_{k=n}^{\infty} \left\{ \frac{1}{\xi^{k+1}} \sum_{j=0}^{n} a_j \xi^j \right\} z^k = p_n(\xi) \sum_{k=n}^{\infty} \frac{z^k}{\xi^{k+1}} = p_n(\xi) \left(\sum_{k=0}^{\infty} \frac{z^k}{\xi^{k+1}} - \sum_{k=0}^{n-1} \frac{z^k}{\xi^{k+1}} \right)$$

$$= -\frac{p_n(\xi)}{z - \xi} - p_n(\xi) \sum_{k=0}^{n-1} \frac{z^k}{\xi^{k+1}}$$

such that

$$\frac{p_n(z)}{z - \xi} = \sum_{k=0}^{n-1} \left\{ \frac{1}{\xi^{k+1}} \left(p_n(\xi) - \sum_{j=0}^{k} a_j \xi^j \right) \right\} z^k + \frac{p_n(\xi)}{z - \xi} \qquad (\text{B.39})$$

where the k-sum is the quotient $q_1(z) = \frac{p_n(z) - p_n(\xi)}{z - \xi}$ in (B.38) obtained by the long division. It follows from (B.39) that $q_1(\xi) = p'_n(\xi)$. The latter is, indeed, deduced from (B.38) as

$$q_1(\xi) = \sum_{k=0}^{n-1} \sum_{j=k}^{n-1} a_{j+1} \xi^j = \sum_{j=0}^{n-1} a_{j+1} \xi^j \sum_{k=0}^{j} 1 = \sum_{j=0}^{n-1} (j+1) a_{j+1} \xi^j = p'_n(\xi)$$

We rewrite the quotient $q_1(z)$ in (B.38) with (B.39) as

$$\frac{p_n(z) - p_n(\xi)}{z - \xi} = \sum_{k=1}^{n} \left\{ \frac{1}{\xi^k} \sum_{j=k}^{n} a_j \xi^j \right\} z^{k-1} = \sum_{k=1}^{n} \left\{ \sum_{j=0}^{n-k} a_{j+k} \xi^j \right\} z^{k-1}$$

After letting $l = n - k$, we obtain

$$\frac{p_n(z) - p_n(\xi)}{z - \xi} = \sum_{l=0}^{n-1} \left\{ \sum_{j=0}^{l} a_{j+n-l} \xi^j \right\} z^{n-1-l}$$

$$= a_n z^{n-1} + \{a_{n-1} + a_n \xi\} z^{n-2} + \cdots + \left\{ \sum_{j=0}^{n-1} a_{j+1} \xi^j \right\}$$

Let us denote the set of polynomials

$$f_l(x) = \sum_{j=0}^{l} a_{j+n-l} x^j \qquad \text{for } 0 \le l \le n \qquad (\text{B.40})$$

then $f_0(x) = a_n$, $f_1(x) = a_{n-1} + a_n x, \ldots,$ and $f_n(x) = \sum_{j=0}^{n} a_j x^j = p_n(x)$. With definition (B.40), the quotient $q_1(z)$ becomes

$$\frac{p_n(z) - p_n(\xi)}{z - \xi} = \sum_{l=0}^{n-1} f_l(\xi) z^{n-1-l} \qquad (\text{B.41})$$

which is also valid if ξ is a zero of $p_n(z)$ and $p_n(\xi) = 0$. Writing

$$f_{l+1}(x) = \sum_{j=0}^{l+1} a_{j+n-l-1} x^j = \sum_{j=-1}^{l} a_{j+n-l} x^{j+1} = x \sum_{j=0}^{l} a_{j+n-l} x^j + a_{n-(l+1)}$$

shows that the polynomials $\{f_l(x)\}_{0 \le l < n}$ obey the recursion

$$f_{l+1}(x) = x f_l(x) + a_{n-(l+1)} \qquad \text{for } 0 \le l \le n-1 \tag{B.42}$$

312. *Application of art. 311.* Perron (1907) considers a monic polynomial $p_n(z) = \sum_{k=0}^{n} a_k z^k$, i.e. with $a_n = 1$ and $a_0 \ne 0$, and defines, for the polynomials in (B.40),

$$\lambda = \sum_{l=1}^{n-1} |f_l(\xi)|$$

After introducing the recursion (B.42)

$$\lambda = \frac{1}{|\xi|} \sum_{l=1}^{n-1} \left| f_{l+1}(\xi) - a_{n-(l+1)} \right| \le \frac{1}{|\xi|} \sum_{l=2}^{n} (|f_l(\xi)| + |a_{n-l}|)$$

$$= \frac{1}{|\xi|} \left(|f_n(\xi)| + \lambda - |f_1(\xi)| + \sum_{l=2}^{n} |a_{n-l}| \right)$$

and using $|f_1(\xi)| = |a_{n-1} + a_n \xi|$, we find the inequality

$$\lambda(|\xi| - 1) \le |f_n(\xi)| + \sum_{k=0}^{n-2} |a_k| - |a_{n-1} + a_n \xi|$$

The sharpest bound is achieved if ξ is a zero of $p_n(z)$, for which $0 = p_n(\xi) = f_n(\xi)$. If $|\xi| = 1$, then

$$\sum_{k=0}^{n-2} |a_k| \ge |a_{n-1} + a_n| \tag{B.43}$$

If $|\xi| > 1$, then $\lambda \le \frac{\sum_{k=0}^{n-2} |a_k| - |a_{n-1} + a_n \xi|}{|\xi| - 1}$. Perron (1907) now constrains the right-hand side to be smaller than 1, which leads to the inequality

$$\sum_{k=0}^{n-2} |a_k| + 1 - |\xi| < |a_{n-1} + a_n \xi| \tag{B.44}$$

Perron (1907) remarks that this constraint is not compatible with $|\xi| = 1$, because it violates (B.43).

Let β be another zero of $p_n(z)$, different from the zero ξ, then the quotient polynomial in (B.41) becomes $0 = \sum_{l=0}^{n-1} f_l(\xi) \beta^{n-1-l}$, which is equivalent, with $f_0(x) = a_n = 1$, to $\beta^{n-1} = -\sum_{l=1}^{n-1} f_l(\xi) \beta^{n-1-l}$. If we assume that $|\beta| > 1$, then

$$|\beta|^{n-1} \le \sum_{l=1}^{n-1} |f_l(\xi)| |\beta|^{n-1-l} < |\beta|^{n-2} \sum_{l=1}^{n-1} |f_l(\xi)| = |\beta|^{n-2} \lambda$$

which implies that $|\beta| < \lambda$. However, the constraint (B.44) implies that $\lambda < 1$, and thus, that $|\beta| < 1$, in contrast to the assumption. In summary, we have proved

Theorem 85 (Perron) *If a zero ξ of the monic polynomial $p_n(z) = \sum_{k=0}^{n} a_k z^k$, i.e. with $a_n = 1$ and $a_0 \neq 0$, obeys $|\xi| \geq 1$ and inequality (B.44), then it holds that $|\xi| > 1$, while all $n-1$ other zeros are smaller, in absolute value, than 1.*

Since $|a_{n-1}| - |\xi| < |a_{n-1} + \xi|$ and requiring that $\sum_{k=0}^{n-2} |a_k| + 1 - |\xi| < |a_{n-1}| - |\xi|$, then leads to a more stringent, but easier Perron constraint than (B.44),

$$\sum_{k=0}^{n} |a_k| < 2 |a_{n-1}| \tag{B.45}$$

Theorem 85 with constraint (B.44) replaced by (B.45) still holds.

On the other hand, it also holds that $|\xi| - |a_{n-1}| < |a_{n-1} + \xi|$ and then a second more stringent constraint than (B.44) is

$$\sum_{k=0}^{n-2} |a_k| + 1 - |\xi| < |\xi| - |a_{n-1}| < |a_{n-1} + a_n \xi|$$

from which a second lower bound for the zero $1 < |\xi|$ follows,

$$\frac{1}{2} \left(\sum_{k=0}^{n-1} |a_k| + 1 \right) < |\xi|$$

Theorem 85 then states that, for a monic polynomial $p_n(z)$, whose zero ξ obeys $|\xi| > \min\left(1, \frac{1}{2}\left(\sum_{k=0}^{n-1} |a_k| + 1\right)\right)$, any other zero β of $p_n(z)$ satisfies $|\beta| < 1$. If $\alpha = \frac{1}{2}\left(\sum_{k=0}^{n-1} |a_k| + 1\right) > 1$, then $|\xi| > 1$ and ξ is the only zero outside the unit circle. If all coefficients a_k of the polynomial $p_n(z)$ are real, then the zero ξ must[11] be real. Thus, if the zero $\xi > 0$ is positive, then ξ is the only real zero lying between α and $+\infty$, so that $p_n(\alpha) < 0$, because $a_n = 1$. If ξ is negative, then ξ is the only real zero lying between $-\alpha$ and $-\infty$, so that $(-1)^n p_n(-\alpha) < 0$. In conclusion, Perron's Theorem 85 can be rephrased as:

Theorem 86 (Perron) *Let $p_n(z) = \sum_{k=0}^{n} a_k z^k$ be a monic polynomial with real coefficients a_k with $a_n = 1$ and $a_0 \neq 0$ and define $\alpha = \frac{1}{2}\left(\sum_{k=0}^{n-1} |a_k| + 1\right)$. If either $p_n(\alpha) < 0$ or $(-1)^n p_n(-\alpha) < 0$ holds, then $p_n(z)$ has a real zero ξ with $|\xi| > 1$, while all $n-1$ other zeros are smaller, in absolute value, than 1.*

---

[11] If ξ were complex, then also its complex conjugate ξ^* with $|\xi^*| = |\xi| > 1$ would be a zero, but there is only one zero outside the unit disk.

313. *Division by an m-degree polynomial.* By using the Taylor series in case $m = 2$,

$$\frac{1}{(z - \xi)(z - \zeta)} = \frac{1}{\xi \zeta} \sum_{k=0}^{\infty} \frac{z^k}{\xi^k} \sum_{k=0}^{\infty} \frac{z^k}{\zeta^k} = \frac{1}{\xi \zeta} \sum_{k=0}^{\infty} \left\{ \frac{1}{\zeta^k} \sum_{j=0}^{k} \left(\frac{\zeta}{\xi} \right)^j \right\} z^k$$

$$= \frac{1}{\xi - \zeta} \sum_{k=0}^{\infty} \left(\frac{1}{\zeta^{k+1}} - \frac{1}{\xi^{k+1}} \right) z^k$$

a similar series expansion manipulation as in **art. 311** leads to

$$\frac{p_n(z)}{(z - \xi)(z - \zeta)} = \frac{1}{\zeta - \xi} \sum_{k=0}^{n-2} \left\{ \frac{1}{\zeta^{k+1}} \sum_{j=k+1}^{n} a_j \zeta^j - \frac{1}{\xi^{k+1}} \sum_{j=k+1}^{n} a_j \xi^j \right\} z^k$$

$$+ \frac{p_n(\zeta)}{\zeta - \xi} \frac{1}{z - \zeta} + \frac{p_n(\xi)}{\xi - \zeta} \frac{1}{z - \xi}$$

When $\xi = \zeta$, differentiation yields

$$\frac{p_n(z)}{(z - \xi)^2} = \sum_{k=0}^{n-2} \left\{ \frac{1}{\zeta^{k+2}} \sum_{j=k+1}^{n} (j - k - 1) a_j \zeta^j \right\} z^k + \frac{p'_n(\xi)}{z - \xi} + \frac{p_n(\xi)}{(z - \xi)^2}$$

The general result is elegantly deduced from the k-th derivative of the Cauchy integral,

$$\frac{1}{k!} \frac{d^k f(z)}{dz^k} = \frac{1}{2\pi i} \int_{C(z)} \frac{f(w)}{(w - z)^{k+1}} dw \qquad \text{(B.46)}$$

where the contour $C(z)$ encloses the point $w = z$. Let $F_m(z) = \prod_{j=1}^{m} (z - \xi_j)$ in (B.17), where $x_k = \xi_k$, and assuming that all zeros ξ_j of $F_m(z)$ are different, then the Cauchy integral with $k = 1$ in (B.46) becomes

$$\frac{1}{F_m(z)} = \frac{1}{2\pi i} \int_{C(z)} \frac{1}{\displaystyle\prod_{j=1}^{m} (w - \xi_j)(w - z)} dw$$

Since $\lim_{z \to \infty} \frac{1}{|F_m(z)|} = 0$, we can deform the contour $C(z)$ to enclose the entire complex plane except for an arbitrary small region around the point $w = z$. The function $\frac{1}{F_m(z)}$ is analytic everywhere, except for the simple poles at $w = \xi_j$ and does not possess zeros for finite z. Cauchy's residue theorem (Titchmarsh, 1964) then leads to the partial fraction expansion

$$\frac{1}{F_m(z)} = -\sum_{l=1}^{m} \lim_{w \to \xi_l} \frac{w - \xi_l}{\displaystyle\prod_{j=1}^{m} (w - \xi_j)(w - z)}$$

and

$$\frac{1}{F_m(z)} = \sum_{l=1}^{m} \frac{1}{\prod_{j=1; j \neq l}^{m} (\xi_l - \xi_j)} \frac{1}{z - \xi_l}$$

If not all zeros are simple, a tedious computation using (B.46) leads to the general partial fraction expansion

$$\frac{1}{\prod_{j=1}^{l} (z - \xi_j)^{m_j}} = \sum_{q=1}^{l} \sum_{r=1}^{m_q} \frac{\chi_q(r-1; \xi_q)}{(z - \xi_q)^r} \tag{B.47}$$

$$= \sum_{q=1}^{l} \left\{ \frac{\chi_q(0; \xi_q)}{(z - \xi_q)} + \frac{\chi_q(1; \xi_q)}{(z - \xi_q)^2} + \cdots + \frac{\chi_q(m_q - 1; \xi_q)}{(z - \xi_q)^{m_q}} \right\}$$

where the residues are

$$\chi_q(r; w) = \frac{1}{(m_q - 1 - r)!} \frac{d^{m_q-1-r}}{dw^{m_q-1-r}} \left(\frac{1}{\prod_{j=1; j \neq q}^{l} (w - \xi_j)^{m_j}} \right) \tag{B.48}$$

With $\frac{1}{z - \xi_l} = \sum_{k=0}^{\infty} \frac{z^k}{\xi_l^{k+1}}$ for $|z| < |\xi_l|$ and proceeding with distinct zeros, we find the Taylor expansion, for $|z| < \min_{1 \leq l \leq m} |\xi_l|$,

$$\frac{1}{F_m(z)} = \frac{1}{\prod_{j=1}^{m} (z - \xi_j)} = -\sum_{k=0}^{\infty} \left\{ \sum_{l=1}^{m} \frac{1}{\xi_l^{k+1}} \prod_{j=1; j \neq l}^{m} \frac{1}{\xi_l - \xi_j} \right\} z^k \tag{B.49}$$

Similarly as in **art. 311** by computing the Cauchy product of $p_n(z)$ and $\frac{1}{F_m(z)}$, we arrive at

$$\frac{p_n(z)}{\prod_{j=1}^{m} (z - \xi_j)} = \sum_{k=0}^{n-m} \left\{ \sum_{l=1}^{m} \frac{\sum_{q=k+1}^{n} a_q \xi_l^q}{\xi_l^{k+1} \prod_{j=1; j \neq l}^{m} (\xi_l - \xi_j)} \right\} z^k + \sum_{l=1}^{m} \frac{p_0(\xi_l)}{\prod_{j=1; j \neq l}^{m} (\xi_l - \xi_j)} \frac{1}{z - \xi_l} \tag{B.50}$$

We recognize that $\prod_{j=1; j \neq l}^{m} (\xi_l - \xi_j) = F_m'(\xi_l)$ and (B.50) reduces, when $m = n$, to

$$p_n(z) = \sum_{l=1}^{n} p_n(\xi_l) \frac{F_n(z)}{(z - \xi_l) F_n'(\xi_l)}$$

which is the Lagrange interpolation polynomial (B.20) corresponding to the set of n points $\{(\xi_l, p_n(\xi_l))\}_{1 \leq l \leq n}$. The first sum in (B.50), which reduces to the $k = 0$ term when $n = m$, vanishes because

$$\sum_{l=1}^{n} \frac{p_n(\xi_l) - p_n(0)}{\xi_l F_n'(\xi_l)} = \sum_{l=1}^{n} \frac{p_n(\xi_l)}{\xi_l F_n'(\xi_l)} - p_n(0) \sum_{l=1}^{n} \frac{1}{\xi_l F_n'(\xi_l)} = 0$$

which follows from the Taylor expansion (B.49) and the Lagrange polynomial, both evaluated at $z = 0$.

We have implicitly assumed that not all ξ_l are zeros of $p_n(z)$. If all $\{\xi_l\}_{1 \le l \le m}$ are simple zeros of $p_n(z)$, then (B.50), written in terms of the polynomial in (B.17), reduces to the quotient polynomial

$$q_m(z) = \frac{p_n(z)}{F_m(z)} = \sum_{k=0}^{n-m} \left\{ \sum_{l=1}^{m} \frac{\sum_{q=0}^{n-k-1} a_{q+k+1}\xi_l^q}{F_m'(\xi_l)} \right\} z^k \tag{B.51}$$

Compared to the quotient polynomial (B.38) for $m = 1$, the coefficients in the general version (B.51) of the quotient polynomial only require m similar polynomial evaluations $\sum_{q=0}^{n-k-1} a_{q+k+1}\xi_l^q$ as in (B.38) and m additional $F_m'(\xi_l)$ computations.

11.5 Descartes' rule of signs

314. A famous theorem due to René Descartes is:

Theorem 87 (Descartes' rule of signs) *Let C denote the number of changes of sign in the sequence of real coefficients $a_0, a_1, \ldots, a_n$ of a polynomial $p_n(z) = \sum_{k=0}^{n} a_k x^k$ and let Z denote the number of positive real zeros of $p_n(z)$, then*

$$C - Z = 2k \ge 0$$

where k is a non-negative integer.

Before proving Theorem 87, we make the following observation. Since the polynomial $p_n(-x)$ has coefficients $(-1)^k a_k$, Descartes' rule of signs indicates that the number of negative real zeros of $p_n(x)$ is not larger than the number of changes in signs of $a_0, -a_1, \ldots, (-1)^n a_n$ in $p_n(-x)$.

The product form in (B.1) for a real polynomial with v real and $2m$ complex zeros (m conjugate pairs), such that $n = v + 2m$, can be written as

$$p_n(z) = a_n \prod_{j=1}^{v} (z - x_j) \prod_{j=1}^{m} (z - \operatorname{Re} z_j)^2 + (\operatorname{Im} z_j)^2$$

from which

$$a_0 = p_n(0) = a_n \prod_{j=1}^{v} (-x_j) \prod_{j=1}^{m} (\operatorname{Re} z_j)^2 + (\operatorname{Im} z_j)^2$$

shows that the sign of a_0 does not depend on the complex zeros. For example, a_0 has the sign of a_n if all real zeros x_j are negative. The sequence $a_0, a_1, \ldots, a_n$ in which $\operatorname{sign}(a_0) = \operatorname{sign}(a_n)$, equivalent to $a_0 a_n > 0$, has an even number of changes in sign. This is verified when the sequence is plotted as a piece-wise linear function through the points (k, a_k) for $0 \le k \le n$, similarly as for the random walk in **art. 184**. The number of sign changes equals the number of k-axis crossings. Zero coefficients do not contribute to a change in sign. For example, the sequence $\{1, 2, 0, 0, -1, 0, 1\}$ has two sign changes.

Generalizing this observation, if the polynomial $p_n(z)$ with real coefficients has an even number of real, positive zeros such that $\text{sign}(a_0) = \text{sign}(a_n)$, the number C of sign changes in $a_0, a_1, \ldots, a_n$ is even, whereas, if $p_n(z)$ has an odd number of real, positive zeros such that $\text{sign}(a_0) = -\text{sign}(a_n)$, the number C of sign changes is odd. Hence, C and Z are both even or odd, which demonstrates that $C - Z = 2k$ is even. To show that k is non-negative, a deeper argument is needed.

Proof[12] **(by Laguerre):** Let $z = e^x$, then the number of real zeros of the function $p_n(e^x) = \sum_{k=0}^{n} a_k e^{kx}$ is the same as the number of positive zeros of $p_n(x)$, because $x = \log z$ is monotonous increasing for $z > 0$. Laguerre actually proves a more general result by considering the entire function

$$F(x) = \sum_{k=0}^{n} a_k e^{\lambda_k x}$$

where the real numbers obey $\lambda_0 < \lambda_1 < \cdots < \lambda_n$. Clearly, if we choose $\lambda_k = k$, we obtain $p_n(e^x)$. Let C denote the number of changes in sign in the sequence $a_1, a_2, \ldots, a_n$ and let Z denote the number of real zeros of the entire function $F(x)$. For $x \to \infty$, the term $a_n e^{\lambda_n x}$ dominates, while for $x \to -\infty$, the term $a_0 e^{\lambda_0 x}$ is dominant; by the argument above, therefore, $C - Z$ is even. The proof that $C - Z \geq 0$ is by induction. If there are no changes of sign ($C = 0$), then there are no zeros ($Z = 0$) and $C \geq Z$. Assume that the Theorem holds for $C-1$ changes of sign (hypothesis). Suppose that $F(x)$ has $C > 0$ changes of sign and let $\beta + 1$ be an index of change, i.e., $a_\beta a_{\beta+1} < 0$ for $1 \leq \beta < n$. Consider now the related function

$$G(x) = \sum_{k=0}^{n} a_k (\lambda_k - \lambda) e^{\lambda_k x}$$

then, for $\lambda_\beta < \lambda < \lambda_{\beta+1}$, the number of changes of sign in the sequence

$$-a_0 (\lambda - \lambda_0), -a_1 (\lambda - \lambda_1), \ldots, -a_\beta (\lambda - \lambda_\beta), a_{\beta+1} (\lambda_{\beta+1} - \lambda), \ldots, a_n (\lambda_n - \lambda)$$

is precisely $C_G = C - 1$ because now $-a_\beta (\lambda - \lambda_\beta) a_{\beta+1} (\lambda_{\beta+1} - \lambda) > 0$, where all other consecutive products remain unchanged. Further,

$$G(x) = e^{\lambda x} \frac{d}{dx} \left(e^{-\lambda x} F(x) \right)$$

and, since $e^{-\lambda x} > 0$ for all real x, both $e^{-\lambda x} F(x)$ and $F(x)$ have the same real zeros. Rolle's Theorem (Hardy, 2006) states that the derivative $f'(x)$ has not less than $Z - 1$ zeros in the same interval where $f(x)$ has Z zeros. Hence, $G(x)$ has at

---

[12] In 1828, Gauss proved Descartes' rule, which was published in 1637 in his *Géométrie*. Laguerre studied and extended Descartes' rule in several papers, combined by Hermite *et al.* (1972) in the part on Algebra.

least $Z_G \geq Z - 1$ zeros. On the other hand, $G(x)$ has at most $Z - 1$ zeros[13]. Thus, $C_G - Z_G = C - 1 - (Z - 1)$ and, by the induction argument, $C - 1 \geq Z - 1$, we arrive at $C_G \geq Z_G$. Introducing $C_G = C - 1$ and $Z_G = Z - 1$, we finally obtain that $C \geq Z$, which completes the induction. $\qquad\square$

Since the set of exponents $\{\lambda_k\}_{1 \leq k \leq n}$ can be real numbers, Laguerre's proof thus extends Descartes' rule of signs to a finite sum of non-integer powers of x. For example, $x^3 - x^2 + x^{1/3} + x^{1/7} - 1 - x^{-2} = 0$ has $C = 3$ sign changes, and thus at most $Z \leq 3$ positive (real) zeros.

Descartes' rule of signs is only exact if $C < 2$ because $k \geq 0$; thus, in case there is no $(C = 0)$ or only one $(C = 1)$ sign variation, which corresponds to no or exactly one positive real zero. The reverse of the $C = 0$ case holds: if all zeros of a real polynomial have negative real part, then all coefficients are positive and there is no change in sign. However, the reverse implication, $\{Z = 1\} \implies \{C = 1\}$ does not hold in general as the example $x^3 - x^2 + x - 1 = (x - 1)(x - i)(x + i)$ shows.

Example 1 The polynomial $p_5(x) = 2x^5 - x^4 + x^3 + 11x^2 - x + 2$ has four changes in sign, while $p_5(-x) = -2x^5 - x^4 - x^3 + 11x^2 + x + 2$ only has one change in sign. Hence, while there are in total precisely five zeros, there is at most one negative real zero and at most four real positive. Since complex zeros appear in pairs, there can be either four, two or zero real positive zeros, but precisely one negative zero.

Example 2 Milovanović *et al.* (1994) mention the remarkable inequality, valid for all real x and even integers $n = 2m > 0$,

$$q_n(x) = x^n - nx + n - 1 \geq 0$$

with equality only if $x = 1$. Since $q_n(-x)$ has only positive coefficients, $q_n(-x) > 0$. For $x > 0$, there are $C = 2$ changes in sign and Descartes' rule of signs states that there are at most two real zeros. Since $q_n(1) = q_n'(1) = 0$, the polynomial $q_n(x)$ has a double zero at $x = 1$, which is thus the only real zero and this implies $q_n(x) \geq 0$.

315. *Number of sign changes in the sequence of the differences.* Let C be the number of sign changes in the sequence $a_0, a_1, \ldots, a_n$ and assume that these sign changes occur between the elements

$$(a_{k_1}, a_{m_1}), (a_{k_2}, a_{m_2}), \ldots, (a_{k_C}, a_{m_C})$$

where $k_j \leq m_j - 1$ and the equality sign only occurs if there are no zero elements between a_{k_j} and a_{m_j}. We denote $a_{m_0} = a_0$ and $a_{k_{C+1}} = a_n$, which has the same sign as a_{m_C} and $a_{k_{C+1}+1} = 0$. Assume, without loss of generality, that $a_{m_0} > 0$. Then, we have that $\text{sign}(a_{k_j}) = (-1)^{j-1}$, $\text{sign}(a_{m_j}) = (-1)^j$ for $1 \leq j \leq C$ and

$$(-1)^{j-1} a_{m_j - 1} \geq 0$$

---

[13] If $f(z)$ is an analytic function in the interior of a single closed contour C defined by $|f(z)| = M$, where M is a constant, then the number of zeros of $f(z)$ in this region exceeds the number of zeros of the derivative $f'(z)$ in that same region by unity (Whittaker and Watson, 1996, p. 121).

Consider now the sequence of the differences $a_0, a_1 - a_0, a_2 - a_1, \ldots, a_n - a_{n-1}$. We denote the difference by $\Delta a_j = a_j - a_{j-1}$ for $1 \leq j \leq n$ and $\Delta a_0 = a_0$. The sign of the $1 \leq j \leq C$ elements,

$$(-1)^i \left(a_{m_j} - a_{m_j - 1} \right) = (-1)^i \Delta a_{m_j} > 0$$

is known. Since Δa_{m_j-1} and Δa_{m_j} have opposite sign for $1 \leq j \leq C$ by construction, the changes in sign of all differences between them is odd; an odd number of k-axis crossings. The last subsequence between Δa_{m_C} and $\Delta a_{n+1} = a_{n+1} - a_{k_{C+1}} = -\text{sign}(a_{m_C}) = -\text{sign}(\Delta a_{m_C})$ also has an odd number of sign changes. Summing the sign changes in all $C + 1$ subintervals equals C plus an odd number of sign changes. Thus, we have proved:

Lemma 16 *If C is the number of sign changes in the sequence $a_0, a_1, \ldots, a_n$, then the number of sign changes in the sequence of the differences $\Delta a_0, \Delta a_1, \ldots, \Delta a_n$, where $\Delta a_j = a_j - a_{j-1}$ for $1 \leq j \leq n$ and $\Delta a_0 = a_0$, equals C plus an odd positive number.*

316. *Application of Lemma 16.* Consider the polynomial $q_{n+1}(x) = (x - \xi) p_n(x)$,

$$q_{n+1}(x) = a_n x^{n+1} + \sum_{k=1}^{n} (a_{k-1} - \xi a_k) x^k - \xi a_0 = \sum_{j=0}^{n+1} (a_{j-1} - \xi a_j) x^j$$

with the convention that $a_{-1} = a_{n+1} = 0$. If $\xi > 0$, the number of sign changes in the coefficients $b_j = a_{j-1} - \xi a_j$ of $q_{n+1}(x)$ equals the number of sign changes in the difference $\Delta \left(\xi^j a_j \right) = \xi^j a_j - \xi^{j-1} a_{j-1} = -\xi^{j-1} b_j$. Lemma 16 shows that the number of sign changes in the difference sequence equals that in the polynomial $p_n(z)$ plus an odd positive number. Descartes' rule of signs in Theorem 87 states that $C_{p_n(z)} = Z_{p_n(z)} + 2k$ and, hence,

$$C_{q_{n+1}(z)} = Z_{p_n(z)} + 2k + 2m + 1 = Z_{q_{n+1}(z)} + 2(k + m)$$

The argument and Lemma 16 provide a second proof of Descartes' rule of signs, Theorem 87, because we have just shown the inductive step: if the rule holds for $p_n(z)$, it also holds for $q_{n+1}(z)$. Descartes' rule of signs definitely holds for $n = 0$ and this completes the second proof.

If $\xi > 0$ and the number of changes in sign in $p_n(x)$ is zero, which implies by Descartes' rule of signs in Theorem 87 that $p_n(x)$ has no positive real zeros, then ξ is the largest real zero of $q_{n+1}(x)$. If $\xi < 0$ and the coefficients of $p_n(x)$ are alternating (equivalent to the fact that $p_n(x)$ does not have negative real zeros (**art. 314**)), then ξ is the smallest real zero of $q_{n+1}(x)$.

317. *Laguerre's extension of Descartes' rule of signs.* Laguerre (see Hermite et al. (1972)) has elegantly and ingeniously extended Descartes' rule of signs. As in **art. 314**, Laguerre considers, as a generalization of the polynomial $p_n(z) =$

$\sum_{k=0}^{n} a_k z^k$, the entire function

$$F(z) = \sum_{k=0}^{n} a_k z^{\beta_k} = a_n z^{\beta_n} + a_{n-1} z^{\beta_{n-1}} + \ldots + a_0 z^{\beta_0} \qquad (B.52)$$

where $\beta_n > \beta_{n-1} > \ldots > \beta_0$ are real numbers.

Theorem 88 (Laguerre) *The number of real zeros Z of the entire function $F(z)$, defined in (B.52), that are larger than a positive number ξ, is at most equal the number C of changes in signs of the partial sum sequence*

$$a_n \xi^{\beta_n}, a_n \xi^{\beta_n} + a_{n-1} \xi^{\beta_{n-1}}, a_n \xi^{\beta_n} + a_{n-1} \xi^{\beta_{n-1}} + a_{n-2} \xi^{\beta_{n-2}}, \ldots, F(\xi)$$

and $C - Z = 2k \geq 0$.

Proof: We start from the polynomial identity (B.39),

$$\frac{p_n(z)}{z - \xi} = \sum_{k=0}^{n-1} \left(\sum_{j=k+1}^{n} a_j \xi^{j-k-1} \right) z^k + \frac{p_n(\xi)}{z - \xi}$$

Using the expansion of $(z - \xi)^{-1}$ for $z > \xi$ results in

$$\frac{p_n(z)}{z - \xi} = \sum_{k=0}^{n-1} \left(\sum_{j=k+1}^{n} a_j \xi^{j-k-1} \right) z^k + \sum_{k=0}^{\infty} \frac{\xi^k p_n(\xi)}{z^{k+1}}$$

Since $\xi > 0$, all terms $\xi^k p_n(\xi)$ for $k > 0$ have the same sign as $p_n(\xi)$, which implies that the number C of sign changes is equal to the number of sign changes of the coefficients in the first k-sum. Each of these coefficients has the same sign as the partial sum $\sum_{j=k+1}^{n} a_j \xi^j$ of $p_n(\xi)$, because $\xi > 0$. This proves the theorem in case $F(z)$ is a polynomial.

We can always reduce $F(z)$ to a polynomial form. If $\beta_0 < 0$ and all exponents β_k are integers, $z^{-\beta_0} F(x)$ is a polynomial and the above argument applies. If $\beta_k \in \mathbb{Q}^+$, then $F(z^w)$ is a polynomial provided w is the least common multiple of the denominators of the set $\{\beta_k\}_{1 \leq k \leq n}$. Since each real number can be approximated arbitrarily close by a rational number, so can $F(z)$ with real exponents approximated arbitrarily close by a polynomial, which proves Theorem 88. □

Theorem 88 is modified when we want the number of *positive* zeros smaller than ξ. In that case, we may verify by following the same steps as in the proof above that the Theorem 88 also holds for the number of *positive* zeros smaller than ξ, provided the order of terms (B.52) is written according to increasing exponents, i.e., $\beta_n < \beta_{n-1} < \ldots < \beta_0$. As a corollary, the number of zeros in $[0, 1]$ of $F(z)$ is at most equal to the number of sign changes in the sequence

$$a_n, a_n + a_{n-1}, \ldots, F(1)$$

Another application is $F(z) = p_n(z + h)$, which we expand by Taylor's theorem

$$p_n(z + h) = \sum_{j=0}^{n} \frac{p_n^{(j)}(h)}{j!} z^j = p_n(h) + z p_n'(h) + z^2 \frac{p_n''(h)}{2} + \ldots + z^n \frac{p_n^{(n)}(h)}{n!}$$

into a polynomial, written with exponents of z in increasing order. The number of real zeros of $F(z)$ between $[0, \xi]$, and thus the real zeros of $p_n(z)$ between $[h, h+\xi]$, is at most equal to the number of sign changes in the sequence

$$\left\{ p_n(h), p_n(h) + \xi p_n'(h), p_n(h) + \xi p_n'(h) + \xi^2 \frac{p_n''(h)}{2}, \ldots, p_n(\xi + h) \right\}$$

and their difference is an even integer (possibly zero).

The whole idea can subsequently be applied to $\frac{p_n(z)}{(z-\xi)^m}$ using **art. 313**. Since the number of sign changes in $\frac{p_n(z)}{(z-\xi)^m}$ is at most equal to that in $\frac{p_n(z)}{(z-\xi)^{m-1}}$ (**art. 316**), but not smaller than the number of real zeros of $p_n(z)$ larger than ξ, we may expect to deduce, by choosing an appropriate m, an exact way to determine the number of such real zeros. In fact, Laguerre succeeded (Hermite *et al.*, 1972, p. 24-25) to propose an exact method, that involves the discriminant (**art. 298**), which is hard to compute. In summary, his method turns out to be less attractive than that of Sturm, discussed in **art. 326**.

318. We present another nice approach due to Laguerre (Hermite *et al.*, 1972, p. 26-41). Consider the polynomial

$$f_n(z) = \sum_{j=1}^{m} A_j p_n(\xi_j z) = \sum_{k=0}^{n} \left(a_k \sum_{j=1}^{m} A_j \xi_j^k \right) z^k$$

where $0 < \xi_m < \xi_{m-1} < \ldots < \xi_1$ and $p_n(z) = \sum_{k=0}^{n} a_k z^k$. Descartes' rule of signs in Theorem 87 states that the number Z of positive zeros is at most equal to the number C of variations in sign in the sequence

$$\left\{ a_0 \sum_{j=1}^{m} A_j, a_1 \sum_{j=1}^{m} A_j \xi_j, \ldots, a_n \sum_{j=1}^{m} A_j \xi_j^n \right\}$$

That number C is also equal to the number C_1 of sign changes in the sequence $(l < n)$

$$S_1 = \left\{ a_0 \sum_{j=1}^{m} A_j, a_1 \sum_{j=1}^{m} A_j \xi_j, \ldots, a_l \sum_{j=1}^{m} A_j \xi_j^l \right\}$$

plus the number C_2 of changes of sign in the remaining sequence

$$S_2 = \left\{ a_l \sum_{j=1}^{m} A_j \xi_j^l, a_{l+1} \sum_{j=1}^{m} A_j \xi_j^{l+1}, \ldots, a_n \sum_{j=1}^{m} A_j \xi_j^n \right\}$$

$$= \left\{ a_l \phi(0), a_{l+1} \phi(1), \ldots, a_n \phi(n - l) \right\}$$

where

$$\phi(x) = \sum_{j=1}^{m} A_j \xi_j^l \xi_j^x = \sum_{j=1}^{m} A_j \xi_j^l \left(e^x\right)^{\log \xi_j}$$

If we suppose that all $a_k \geq 0$, then the number C_2 of variations in sign in S_2 is at most equal to the number Z_ϕ of positive zeros of $\phi(x)$, because even if $\phi(k)\phi(k-1) > 0$, there can be an even number of zeros in the interval $(k-1,k)$. The number Z_ϕ is also equal to the number of real zeros of $\phi(\log z) = 0$, which is greater than 1. Theorem 88 in **art. 317** shows that the number Z_ϕ of real zeros of $\phi(\log z)$ larger than 1 is at most equal to the number C_ϕ of sign changes in the sequence

$$\left\{A_1 \xi_1^l, A_1 \xi_1^l + A_2 \xi_2^l, \dots, \phi(0)\right\}$$

Hence, $Z \leq C \leq C_1 + C_\phi$. Since the above holds for all $0 \leq l < n$, the simplest choice is $l = 0$. Thus, we have proved

Theorem 89 (Laguerre) *The number of real roots Z of the equation*

$$\sum_{j=1}^{m} A_j p_n\left(\xi_j z\right) = 0$$

where $0 < \xi_m < \xi_{m-1} < \dots < \xi_1$ and $p_n(z) = \sum_{k=0}^{n} a_k z^k$, is at most equal to the number of changes in sign of the sequence $\left\{A_1, A_1 + A_2, \dots, \sum_{j=1}^{m} A_j\right\}$.

Theorem 89 holds for $f(z) = \lim_{n \to \infty} p_n(z)$ provided the polynomial sum converges. Let us consider $f(z) = e^z$. The equation $\sum_{j=0}^{m} A_j \exp\left(\xi_j z\right) = 0$ possesses the same roots as $\sum_{j=0}^{m} A_j \exp\left(\left(\xi_j + k\right) z\right) = 0$, where k is a finite real number such that the restriction $0 < \xi_m$ can be removed. Let $\xi_j = a + j\Delta t$ and $\xi_m = b > a$, such that $m = \frac{b-a}{\Delta t}$, then we obtain the Riemann sum,

$$\lim_{\Delta t \to 0} \sum_{j=1}^{\frac{b-a}{\Delta t}} A_j e^{(a+j\Delta t)z} \Delta t = \int_a^b e^{xz} \psi(z)\, dz$$

where $\psi(z)$ is an arbitrary function, because the coefficients $A_0, A_1, \dots, A_m$ are arbitrary. The number of sign changes in $\left\{A_1, A_1 + A_2, \dots, \sum_{j=1}^{m} A_j\right\}$ is, in that limit, at least equal to the number of zeros of $\int_a^x \psi(z)\, dz = 0$ in the interval (a,b). For example, let

$$\psi(z) = \sum_{k=0}^{n} a_k \frac{z^{\beta_k + w - 1}}{\Gamma\left(\beta_k + w\right)}$$

where all $\beta_k > 0$, $w \geq 0$ and $\Gamma(x)$ is the Gamma function. For $a = 0$ and $b = \infty$, the equation $\int_0^\infty e^{xz} \psi(z)\, dz = 0$ becomes

$$\sum_{k=0}^{n} \frac{a_k}{x^{\beta_k + w}} = 0$$

whose number of positive zeros is, after the transformation $x \to x^{-1}$, precisely equal to those of $\sum_{k=0}^{n} a_k x^{\beta_k + w} = 0$ and, thus, of $\sum_{k=0}^{n} a_k x^{\beta_k} = 0$. On the other hand, the number of positive zeros of the equation $\int_0^x \psi(z)\,dz = 0$, computed as

$$\sum_{k=0}^{n} a_k \frac{x^{\beta_k + w}}{\Gamma(\beta_k + w + 1)} = 0$$

is at least equal to those of $\sum_{k=0}^{n} a_k x^{\beta_k} = 0$. Now, for $\beta_k = k$ the equations reduce to polynomials and we observe that, after a transform $x \to -x$, the number of negative zeros of the polynomial $p_n(x) = \sum_{k=0}^{n} a_k x^k$ is at most equal to those of the polynomial $\sum_{k=0}^{n} a_k \frac{x^k}{\Gamma(k+w+1)}$. Consequently, we arrive at

Theorem 90 (Laguerre) *If all zeros of the polynomial $p_n(x) = \sum_{k=0}^{n} a_k x^k$ are real, then the zeros of the related polynomial $q_n(x;w) = \sum_{k=0}^{n} a_k \frac{x^k}{\Gamma(k+w+1)}$ are also all real, for any real number $w \geq 0$.*

Many extensions, so-called *zero mapping transformations*, have been deduced of Laguerre's Theorem 90. Consider the set of real numbers $\{\gamma_k\}_{k \geq 0}$, which is a zero mapping transformation, satisfying certain properties. If all zeros of the polynomial $p_n(x) = \sum_{k=0}^{n} a_k x^k$ are real, then the zeros of the related, transformed polynomial $t_n(x;\gamma) = \sum_{k=0}^{n} \gamma_k a_k x^k$ are also all real. A large list of particular sequences $\{\gamma_k\}_{k \geq 0}$ is presented in Milovanović *et al.* (1994).

319. Theorem 90 in **art. 318** can be extended,

Theorem 91 (Laguerre) *Let $p_n(z) = \sum_{k=0}^{n} a_k z^k$ be a polynomial with real zeros and let $f(z)$ be an entire function (of genus 0 or 1), which is real for real z and all the zeros are real and negative. Then, the polynomial $g_n(z) = \sum_{k=0}^{n} a_k f(k) z^k$ has all real zeros, and as many positive, zero and negative zeros as $p_n(z)$.*

Proof: See Hermite *et al.* (1972, p. 200) or Titchmarsh (1964, pp. 268-269). $\square$

It can be shown (Titchmarsh, 1964, pp. 269-270) that, if $n \to \infty$ and $p(z) = \lim_{n \to \infty} p_n(z)$ is an entire function, then $g(z) = \lim_{n \to \infty} g_n(z)$ is entire, all of whose zeros are real and negative. Hence, applied to $p(z) = e^z$, Laguerre's theorem 91 (extended to $n \to \infty$) shows that the Taylor series

$$g(z) = \sum_{k=0}^{\infty} \frac{f(k)}{k!} z^k$$

is an entire function $g(z)$ with negative, real zeros.

320. *Application of Descartes' rule of signs.* The polynomial

$$r_n(z) = |a_n| z^n - \sum_{k=0}^{n-1} |a_k| z^k$$

where $|a_n| > 0$ and $\sum_{k=0}^{n-1} |a_k| > 0$, has precisely one positive real zero. Descartes' rule of signs in Theorem 87 tells us that there is at most one positive real zero, because there is one change of sign. Since the k-sum in

$$r_n(z) = |a_n| z^n \left(1 - \sum_{k=0}^{n-1} \frac{|a_k|}{|a_n|} z^{k-n} \right)$$

is monotone decreasing from ∞ to 0 when z increases from 0 to ∞ along the real axis, there is precisely one point $z = \xi$ at which the k-sum equals one and $r_n(\xi) = 0$. Moreover, $r_n(z) < 0$ if $z < \xi$ and $r_n(z) > 0$ if $z > \xi$. If z_0 is a zero of the polynomial $p_n(z) = \sum_{k=0}^{n-1} a_k z^k$, then

$$|a_n| \, |z_0^n| = \left| -\sum_{k=0}^{n-1} a_k z_0^k \right| \leq \sum_{k=0}^{n-1} |a_k| \, |z_0|^k = |a_n| \, |z_0|^n - r_n(|z_0|)$$

which shows that $r_n(|z_0|) \leq 0$, implying that $|z_0| \leq \xi$. Hence, we have proved

Theorem 92 *If z_0 is a zero of $p_n(z) = \sum_{k=0}^{n} a_k z^k$ and ξ is the only positive zero of $r_n(z) = |a_n| z^n - \sum_{k=0}^{n-1} |a_k| z^k$, then $|z_0| \leq \xi$.*

In other words, the absolute values of all zeros of $p_n(z)$ are smaller than or equal to the only positive zero of $r_n(z)$. Theorem 92 is related to, but different from Perron's Theorem 85 in **art. 312**.

321. *Cauchy's rule.* We derive an upperbound $\zeta \geq 0$ for any positive zero of the real polynomial $p_n(z) = \sum_{k=0}^{n} a_k z^k$, without resorting to Descartes' rule.

Theorem 93 (Cauchy's rule) *No zero of the real polynomial $p_n(z)$ is larger in absolute value than*

$$\zeta = \max_{0 \leq k \leq n-1 \ and \ \frac{a_k}{a_n} < 0} \left(\left(\frac{|a_k|}{c_k |a_n|} \right)^{1/(n-k)} \right) \tag{B.53}$$

where

$$\sum_{k=0 \ and \ \frac{a_k}{a_n} < 0}^{n-1} c_k \leq 1 \tag{B.54}$$

.

Proof: The upperbound $\zeta \geq 0$ satisfies

$$0 \leq \left| \frac{p_n(\zeta)}{a_n} \right| = \left| \zeta^n + \sum_{k=0}^{n-1} \frac{a_k}{a_n} \zeta^k \right| \leq \zeta^n + \sum_{k=0}^{n-1} \left| \frac{a_k}{a_n} \right| \zeta^k$$

Since the coefficients of $p_n(z)$ are real, we rewrite the latter bound as

$$\zeta^n \left(1 - \sum_{k=0 \ and \ \frac{a_k}{a_n} < 0}^{n-1} \frac{a_k}{a_n} \zeta^{k-n} \right) + \sum_{k=0 \ and \ \frac{a_k}{a_n} \geq 0}^{n-1} \frac{a_k}{a_n} \zeta^k \geq 0$$

Let us denote $c_k = \left|\frac{a_k}{a_n}\right| \zeta^{k-n} \geq 0$ for all k indices for which $\frac{a_k}{a_n} < 0$. Then,
$\zeta = \left(\frac{|a_k|}{c_k |a_n|}\right)^{1/(n-k)}$ and the above inequality reduces to (B.54). □

Examples Let $\mu = \sum_{k=0 \text{ and } \frac{a_k}{a_n}<0}^{n-1} 1$ be the number of negative coefficients of $\frac{p_n(z)}{a_n}$, then the choice $c_k = \frac{1}{\mu}$ satisfies the condition (B.54), leading to

$$\zeta = \max_{0 \leq k < n \text{ and } \frac{a_k}{a_n}<0} \left(\mu \left|\frac{a_k}{a_n}\right|\right)^{\frac{1}{n-k}} \tag{B.55}$$

A weaker bound $\zeta = \max_{0 \leq k \leq n-1} \left(\left(n\frac{|a_k|}{|a_n|}\right)^{1/(n-k)}\right)$, derived from the inequality in (B.53), follows from the choice $c_k = \frac{1}{n} \leq \frac{1}{\mu}$. Another choice, that satisfies (B.54) for all k, is $c_k = \frac{\binom{n-1}{k}}{2^{n-1}}$.

322. *Rescaling.* The equation $p_n(z) = 0$ where $a_n \neq 0$ can be transformed by the substitution $z = bx$ into $x^n + \sum_{k=0}^{n-1} \frac{a_k}{a_n} b^{k-n} x^k = 0$. Let us confine to odd n. Odd polynomials with real coefficients have at least one real zero. We now choose b such that $\frac{a_0}{a_n} b^{-n} = -1$ or $b = \left(-\frac{a_0}{a_n}\right)^{1/n}$. This choice reduces the original equation $p_n(z) = 0$ into

$$q_n(x) = x^n - \sum_{k=1}^{n-1} \frac{a_k}{a_0} \left(-\frac{a_0}{a_n}\right)^{k/n} x^k - 1 = 0$$

Since $q_n(0) = -1 < 0$ and $\lim_{x \to \infty} q_n(x) > 0$, there must lie at least one real root in the interval $(0, \infty)$. If $q_n(1) > 0$, the root must lie between 0 and 1; if $q_n(1) < 0$, then the root lies in the interval $(1, \infty)$. By the transform $x = y^{-1}$, the interval $(1, \infty)$ can be changed to $(0, 1)$. Alternatively, **art. 291** shows that $\prod_{k=1}^n z_k = 1$ which indicates that not all zeros can lie in $(0, 1)$ nor in $(1, \infty)$. Hence, we have reduced the problem to find a real zero of $p_n(z)$ with odd degree n, into a new problem of finding the real root of $q_n(x)$ in $(0, 1)$. We refer to Lanczos (1988) for a scheme of successively lowering the order of the polynomial $q_n(x)$ by shifted Chebyshev polynomials.

323. *Isolation of real zeros via continued fractions.* Let $m_1 \in \mathbb{N}$ and $m_k \in \mathbb{N}_0$ for all $k > 1$. Akritas (1989, p. 367-371) has proved:

Theorem 94 (Vincent-Uspensky-Akritas) *There exists a continued fraction transform with a non-negative m_1 and further positive integer partial quotients $\{m_k\}_{2 \leq k \leq l}$,*

$$z = \frac{A_l w + A_{l-1}}{B_l w + B_{l-1}} = m_1 + \cfrac{1}{m_2 + \cdots + \cfrac{1}{m_l + w}} \tag{B.56}$$

that transforms the polynomial $p_n(z)$ with rational coefficients a_k and simple zeros into the function $p_n\left(\frac{A_l w + A_{l-1}}{B_l w + B_{l-1}}\right) = (B_l w + B_{l-1})^{-n} \tilde{p}_n(w)$ such that the polynomial

$\tilde{p}_n(w)$ *has either zero or one sign variation. The integer l is the smallest integer such that $F_{l-1}\frac{d}{2} > 1$ and $F_{l-1}F_l d > 1 + \varepsilon_n^{-1}$, where d is the minimum distance between any two zeros, F_m is the m-th Fibonacci number that obeys $F_m = F_{m-1} + F_{m-2}$ for $m > 1$ and with $F_0 = F_1 = 1$ and where $\varepsilon_n = \left(1 + \frac{1}{n}\right)^{\frac{1}{n-1}} - 1$.*

While the converse of Descartes' Theorem 87 in case $C = 0$, implying that there is no positive real zero, is generally true, the converse of the case $C = 1$ is not generally true as demonstrated in **art. 314**. The part of Theorem 94 that details the determination of the integer l guarantees that, if there is one zero with positive real part and all others have negative real part and lying in an ε_n-disk around -1, the corresponding polynomial has exactly one change in sign. The Fibonacci numbers F_m enter the scene because they are the denominators of the m-th convergent of the continued fraction of the golden mean (see e.g. Govers *et al.* (2008, p. 316)),

$$\frac{1 + \sqrt{5}}{2} = 1 + \cfrac{1}{1 + \cdots + \cfrac{1}{1 + \cfrac{1}{\ddots}}}$$

in the limit case where all $m_k = 1$ for $k \geq 1$. The continued fraction transform (B.56) roughly maps one zero to the interval $(0, \infty)$ and all others in clusters around -1 with negative real part. The continued fraction (B.56) is equivalent to a series of successive substitutions of the form $z = m_j + \frac{1}{w}$ for $1 \leq j \leq l$. The best way to choose the set of integers $\{m_k\}_{1 \leq k \leq l}$ is still an open issue. Akritas (1989) motivates to choose m_j in each substitution round equal to Cauchy's estimate (B.55). Finally, Akritas (1989) claims that his method for isolating a zero is superior in computational effort to Sturm's classical bisection method based on Theorem 96.

11.6 The number of real zeros in an interval

324. *The Cauchy index.* Consider the rational function $r(z) = \frac{p_m(z)}{p_n(z)}$ that has at most n poles: the zeros of the polynomial $p_n(z)$ that are not zeros of the numerator polynomial $p_m(z)$. We further assume $n > m$, else we can always reduce the rational function as the sum of a polynomial and a rational function, where the numerator polynomial has a smaller degree than the denominator polynomial as explained in **art. 309**.

If ξ_k is a zero with multiplicity m_k of $p_n(z)$ but not of $p_m(z)$, then $r(z) = p_m(z) \prod_{k=1}^{l} (z - \xi_k)^{-m_k}$ and the partial fraction expansion (B.47) shows that the behavior of $r(z)$ around a pole ξ_q of order m_q is dominated by $b_q(z - \zeta_q)^{-m_q}$, where $b_q = p_m(\xi_q)\chi_q(m_q - 1; \xi_q) = p_m(\xi_q)\prod_{j=1;j \neq q}^{l}(\xi_q - \xi_j)^{-m_j} \neq 0$.

The Cauchy index of a rational function $r(z)$ at a *real* pole y is defined to be $+1$ if

$$\lim_{x \to y^-} r(x) = -\infty \quad \text{and} \quad \lim_{x \to y^+} r(x) = \infty$$

and the Cauchy index is -1 if

$$\lim_{x \to y^-} r(x) = \infty \quad \text{and} \quad \lim_{x \to y^+} r(x) = -\infty$$

while the Cauchy index is zero if both limits are the same. Hence, the Cauchy index at a real zero ξ_q of $p_n(z)$ equals 0 if m_q is even and $\text{sign}(b_q)$ if m_q is odd. The Cauchy index of a rational function r for the interval $[a, b]$, denoted by $I_a^b r(x)$, is defined as the sum of the Cauchy indices at all real poles y between a and b, such that $a < y < b$ and a and b are not poles of r.

The logarithmic derivative of $p_n(z) = \prod_{k=1}^{l} (z - \zeta_k)^{-m_k}$ is

$$\frac{d \log p_n(z)}{dz} = \frac{p_n'(z)}{p_n(z)} = \sum_{k=1}^{l} \frac{m_k}{z - \zeta_k} = \sum_{k=1}^{s} \frac{m_k}{z - \zeta_k} + r_1(z)$$

where only the first s zeros are real in the interval $[a, b]$. The Cauchy index for the interval $[a, b]$ is $I_a^b \frac{p_n'(x)}{p_n(x)} = s$, which is equal to the number of distinct real zeros of $p_n(z)$ in the interval $[a, b]$. Since $p_n(z)$ has a finite number of zeros, $I_{-\infty}^{+\infty} \frac{p_n'(x)}{p_n(x)}$ equals all distinct real zeros of $p_n(z)$.

Sturm's classical Theorem 95 in **art.** 325 is a method to compute the Cauchy index for the logarithmic derivative, which determines the number of real zeros of a polynomial in a possibly infinite interval $[a, b]$.

325. *A Sturm sequence.* A sequence of real polynomials $f_1(x), f_2(x), \ldots, f_m(x)$ is a Sturm sequence on the interval (a, b) if it obeys for each $a < x < b$ two properties: (i) $f_m(x) \neq 0$ and (ii) $f_{k-1}(x) f_{k+1}(x) < 0$ for any k where $f_k(x) = 0$.

Let $V(x)$ denote the number of changes in sign of the sequence $f_1(x), f_2(x), \ldots, f_m(x)$ at a fixed $x \in (a, b)$. The value of $V(x)$ can only change when x varies from a to b, if one of the functions $f_k(x)$ passes through zero. However, for a Sturm sequence, property (ii) shows that, when $f_k(x) = 0$ for any $2 \leq k \leq m-1$, the value of $V(x)$ versus x does not change. Only if $f_1(x)$ passes through a zero $\xi \in (a, b)$, then $V(x)$ changes by ± 1 according to the Cauchy index of $\frac{f_2(x)}{f_1(x)}$ at $x = \xi$. Hence, we have shown:

Theorem 95 (Sturm) *If $f_1(x), f_2(x), \ldots, f_m(x)$ is a Sturm sequence on the interval (a, b), then $I_a^b \frac{f_2(x)}{f_1(x)} = V(a) - V(b)$.*

326. An interesting property of a Sturm sequence is its connection to the Euclidean algorithm (**art.** 309), which we modify (all remainders have negative sign) into

$$\begin{aligned}
s_0(z) &= q_1(z) s_1(z) - s_2(z) & (0 < \deg s_2 < \deg s_1) \\
s_1(z) &= q_2(z) s_2(z) - s_3(z) & (0 < \deg s_3 < \deg s_2) \\
s_2(z) &= q_3(z) s_3(z) - s_4(z) & (0 < \deg s_4 < \deg s_3) \\
&\cdots & \cdots \\
s_{m-2}(z) &= q_{m-1}(z) s_{m-1}(z) - s_m(z) & (0 < \deg s_m < \deg s_{m-1}) \\
s_{m-1}(z) &= q_m(z) s_m(z)
\end{aligned}$$

The sequence $\{s_k(x)\}_{0 \leq k \leq m}$ is a Sturm sequence if the largest common divisor polynomial $s_m(x)$ does not change sign in the interval (a,b). By the modified Euclidean construction, we observe that property (ii) in **art.** 325 is always fulfilled. Indeed, in the modified Euclidean algorithm for any $0 < k < m$ and $x \in (a,b)$ relation $s_{k-1}(x) = q_k(x)s_k(x) - s_{k+1}(x)$ shows that, if $s_k(x) = 0$, both $s_{k-1}(x)$ and $s_{k+1}(x)$ have opposite sign and do not contribute to changes in $V(x)$.

The Euclidean algorithm, applied to the logarithmic derivative $r(z) = \frac{p'(z)}{p(z)}$ where $s_0(z) = p(x)$ and $s_1(z) = p'(z)$, provides information about the multiplicity of zeros of the polynomial $p(z)$. If ξ is a zero with multiplicity m of $p(z)$, then it is a zero with multiplicity $m-1$ of $p'(z)$. Hence, both $p(z)$ and $p'(z)$ have the factor $(z-\xi)^{m-1}$ in common, and since, by construction, $s_m(z)$ is the largest common divisor polynomial, $s_m(z)$ also must possess the factor $(z-\xi)^{m-1}$.

In summary, applying the (modified) Euclidean algorithm to the logarithmic derivative $r(x) = \frac{p'(x)}{p(x)}$ of a polynomial $p(x)$, **art.** 324 with Theorem 95 leads to:

Theorem 96 (Sturm) *Let $p(z)$ be a polynomial with real coefficients and let $\{p_k\}$ be the sequence of polynomials generated by the (modified) Euclidean algorithm starting with $s_0(z) = p(z)$ and $s_1(z) = p'(z)$. The polynomial $p(z)$ has exactly $V(a) - V(b)$ distinct real zeros in (a,b), where $V(x)$ denotes the number of changes of sign in the sequence $\{s_k(x)\}$. A complex number ξ is a zero of multiplicity m of $p(z)$ if and only if ξ is a zero of multiplicity $m-1$ of $s_m(z)$. Thus, all zeros of $p(z)$ in (a,b) are simple if and only if $s_m(z)$ has no zeros in (a,b).*

Example Let $p(z) = z^4 - 2z^2 + z + 1$. Descartes' rule of signs in Theorem 87 states that there are either 2 or 0 real positive zeros. The (modified) Euclidean algorithm yields, with $s_0(z) = p(z)$ and $s_1(z) = p'_0(z)$,

$$s_0(z) = z^4 - 2z^2 + z + 1 = \left(\frac{z}{4}\right)s_1(z) - \left(z^2 - \frac{3}{4}z - 1\right)$$

$$s_1(z) = 4z^3 - 4z + 1 = (4z+3)s_2(z) - \left(-\frac{9}{4}z - 4\right)$$

$$s_2(z) = z^2 - \frac{3}{4}z - 1 = \left(-\frac{4z}{9} + \frac{91}{81}\right)s_3(z) - \frac{283}{81}$$

$$s_3(z) = -\frac{9}{4}z - 4 = \left(\frac{729}{1132}z + \frac{324}{283}\right)s_4(z)$$

$$s_4(4) = -\frac{283}{81}$$

The corresponding continued fraction of the modified Euclidean algorithm is

$$\frac{s_0(z)}{s_1(z)} = q_1(z) - \cfrac{1}{q_2(z) - \cfrac{1}{q_3(z) - \cfrac{1}{\ddots \, - \cfrac{1}{q_m(z)}}}}$$

and, here,

$$\frac{s_0(z)}{s_1(z)} = \frac{z}{4} - \cfrac{1}{4z + 3 - \cfrac{1}{-\frac{4z}{9} + \frac{91}{81} - \cfrac{1}{\frac{729}{1132}z + \frac{324}{283}}}}$$

The sequence of signs in $s_0(z), s_1(z), \ldots, s_4(z)$ at $z = 0$ is $+, +, -, -, -$ such that $V(0) = 1$. For $z \to \infty$, the signs of the leading coefficients are $+, +, +, -, -$ and $V(\infty) = 1$, while $V(-\infty) = 3$. There is no positive real zero, but two negative zeros. The zeros are simple because $s_4(z)$ is a constant. The zeros of $p(z)$ are $z_1 = -1.49$, $z_2 = -0.52$, $z_{3,4} = 1.01 \pm 0.51i$.

11.7 Real zeros and the sequence of coefficients

We discuss a beautiful result of Newton on the sequence of the real coefficients $a_0, a_1, \ldots, a_n$ of a polynomial with real zeros. Instead of starting with the usual definition $p_n(z) = \sum_{k=0}^{n} a_k z^k$ in (B.1) of a polynomial, Newton considers the polynomial in two variables

$$t_n(x, y) = \sum_{k=0}^{n} a_k x^k y^{n-k} = a_n x^n + a_{n-1} x^{n-1} y + \ldots + a_0 y^n = y^n p_n\left(\frac{x}{y}\right)$$

whose zeros $z = \frac{x}{y}$ are all real.

327. If the zeros of the polynomial $t_n(x, y)$ are real, then also the polynomials $\frac{\partial t_n}{\partial x}$ and $\frac{\partial t_n}{\partial y}$ possess real zeros by Rolle's theorem (Hardy, 2006), provided $n > 1$. Applying Rolle's theorem recursively leads to the conclusion that any polynomial $\frac{\partial^{m+l} t_n}{\partial x^m \partial y^l}$ with $m + l < n$ has real zeros. In particular, all polynomials $\frac{\partial^{m+l} t_n}{\partial x^m \partial y^l}$ of degree $n - (m + l) = 2$,

$$\frac{\partial^{n-2} t_n}{\partial x^{m-1} \partial y^{n-1-m}} = \sum_{k=0}^{n} a_k \frac{k!}{(k-m+1)!} \frac{(n-k)!}{(m-k+1)!} x^{k-m+1} y^{m-k+1}$$

$$= \sum_{k=m-1}^{m+1} a_k \frac{k!}{(k-m+1)!} \frac{(n-k)!}{(m-k+1)!} x^{k-m+1} y^{m-k+1}$$

$$= a_{m-1} \frac{(m-1)!\,(n-m+1)!}{2} y^2 + a_m m!\,(n-m)! xy$$

$$\quad + a_{m+1} \frac{(m+1)!\,(n-m-1)!}{2} x^2$$

$$= \frac{n!}{2}\left(\frac{a_{m-1}}{\binom{n}{m-1}} y^2 + \frac{2a_m}{\binom{n}{m}} xy + \frac{a_{m+1}}{\binom{n}{m+1}} x^2\right)$$

possess real zeros, which is equivalent to the fact that all these polynomials for $1 \leq m \leq n - 1$ have a non-negative discriminant (**art. 298**):

$$\frac{a_m^2}{\binom{n}{m}^2} \geq \frac{a_{m-1}}{\binom{n}{m-1}} \frac{a_{m+1}}{\binom{n}{m+1}}$$

Hence, we have shown

Theorem 97 (Newton) *If $p_n(z) = \sum_{k=0}^{n} a_k z^k$ is a polynomial with real coefficients and real zeros, then the coefficients satisfy the inequality*

$$a_m^2 \geq a_{m-1} a_{m+1} \frac{m+1}{m} \frac{n-m+1}{n-m} \tag{B.57}$$

For example, the inequality for $m = n - 1$ in (B.57) yields $a_{n-1}^2 \geq a_{n-2} a_n \frac{2n}{n-1}$. Using a different argument, **art. 300** concludes that, if $a_{n-1} = 0$, then a_n and a_{n-2} must have opposite sign.

328. *Unimodal sequences.* A real sequence $v_0, v_1, \ldots, v_n$ is unimodal (Comtet, 1974) if there exist two integers l and m such that

$$\begin{cases} v_k \leq v_{k+1} & \text{for } 0 \leq k \leq l - 2 \\ v_k \geq v_{k+1} & \text{for } k \geq m + 1 \end{cases}$$

and with an intermediate region where $v_{l-1} < v_l = v_{l+1} = \ldots = v_m > v_{m+1}$. If $l < m$, there is a plateau, else ($l = m$), there is a peak separating the non-decreasing and non-increasing subsequence. A real sequence $v_k \geq 0$ with $0 \leq k \leq n$ is logarithmically convex on $[a, b]$ if $v_k^2 \leq v_{k-1} v_{k+1}$ for $a + 1 \leq k \leq b - 1$, while a real sequence $w_0, w_1, \ldots, w_n$ is convex on $[a, b]$ if $w_k \leq \frac{1}{2}(w_{k-1} + w_k)$ for $a + 1 \leq k \leq b - 1$. The transform $w_k = \log v_k$ explains the logarithmic convexity inequality, that is also rewritten as

$$\frac{v_k}{v_{k-1}} \leq \frac{v_{k+1}}{v_k}$$

demonstrating that $y_k = \frac{v_k}{v_{k-1}}$ is increasing in k on $[a + 1, b]$. Similarly, the logarithmic concavity inequality $v_k^2 \geq v_{k-1} v_{k+1}$ implies that $y_k = \frac{v_k}{v_{k-1}}$ is decreasing on $[a + 1, b]$. If the sequence is logarithmically concave and $y_b \geq 1$, then v_k is increasing in k, while if $y_{a+1} \leq 1$, then v_k is decreasing. If $y_{a+1} > 1$ and $y_b < 1$, then v_k is unimodal. Finally, if the sequence is logarithmically strictly concave obeying the inequality $v_k^2 > v_{k-1} v_{k+1}$ so that y_k is strictly decreasing, then there is at most one value of k where $y_k = \frac{v_k}{v_{k-1}} = 1$, which results in a plateau of two points. If there is no such value of k, then the unimodal sequence has a peak.

This preparation is needed to conclude from Newton's Theorem 97 that if all coefficients $a_k \geq 0$ and all zeros are real (and non-positive by Decartes' Theorem 87), then the sequence $a_k \geq 0$ is unimodal with either a plateau of two points or a peak because

$$a_m^2 \geq a_{m+1} a_{m-1} \frac{m+1}{m} \frac{n-m+1}{n-m} > a_{m-1} a_{m+1}$$

Comtet (1974) illustrates that generating functions of many positive combinatorial numbers, such as binomial and Stirling numbers of the second kind, are polynomials with real zeros and the sequences of such combinatorial numbers are unimodal with either a plateau of two points or a peak.

329. *Interlacing polynomials.* A polynomial $g_{n-1}(x) = \prod_{k=1}^{n-1}(x - \gamma_k)$ interlaces a polynomial $p_n(x) = \prod_{k=1}^{n}(x - z_k)$ if their real zeros interlace

$$z_n \le \gamma_{n-1} \le z_{n-1} \le \gamma_{n-2} \le \cdots \le \gamma_1 \le z_1$$

A set of (monic) polynomials $_1p_n(x), _2p_n(x), \ldots, _mp_n(x)$ have a common interlacing if there is a single polynomial $g_{n-1}(x)$ that interlaces each of them. If $_jp_n(x) = \prod_{k=1}^{n}(x - z_{j;k})$, then the polynomials $_1p_n(x), _2p_n(x), \ldots, _mp_n(x)$ possess a common interlacing if there exist numbers $\gamma_n \le \gamma_{n-1} \le \cdots \le \gamma_1 \le \gamma_0$ so that $z_{j;k} \in [\gamma_{k-1}, \gamma_k]$ for all $k \in [1, n]$ and all $j \in [1, m]$. The numbers $\gamma_{n-1} \le \cdots \le \gamma_1$ can represent the zeros of a polynomial $g_{n-1}(x)$, while γ_n (γ_0) is smaller (larger) than any of the zeros of any polynomial $_jp_n(x)$. Marcus *et al.* (2015, Lemma 4.2) prove

Lemma 17 *If the monic polynomials $_1p_n(x), _2p_n(x), \ldots, _mp_n(x)$ have a common interlacing, then there exists a polynomial $_ip_n(x)$ with $i \in [1, m]$ for which the largest zero $z_{i;1}$ is at most the largest zero ϕ_1 of the sum polynomial $f_n(x) = \sum_{j=1}^{m} {}_jp_n(x)$.*

Proof: The monic polynomial $g_{n-1}(x) = \prod_{k=1}^{n-1}(x - \gamma_k)$ interlaces all monic polynomials $_1p_n(x), _2p_n(x), \ldots, _mp_n(x)$, implying that $\gamma_1 \le z_{j;1} \le \gamma_0$ and for $x \ge z_{j;1}$, $_jp_n(x) > 0$ because the leading coefficient is 1 for monic polynomials. Since each polynomial $_jp_n(x)$ has exactly one zero $\gamma_1 \le z_{j;1}$, it holds that $_jp_n(\gamma_1) \le 0$ for all $1 \le j \le m$. Hence, $f_n(\gamma_1) = \sum_{j=1}^{m} {}_jp_n(\gamma_1) \le 0$ and $f_n(x)$ becomes eventually positive for $x > \gamma_1$. In other words, the sum polynomial $f_n(x)$ has a zero $\phi_1 \ge \gamma_1$. Furthermore, there must be some $i \in [1, m]$ for which polynomial $_ip_n(\phi_1) \ge 0$, else $f_n(\phi_1)$ were negative, contradicting that ϕ_1 is a zero of $f_n(x)$. Hence, there exists a polynomial $_ip_n(x)$ with largest zero $\gamma_1 \le z_{i;1} \le \phi_1$. □

A similar argument can be deduced for the second largest zero and, further, for the k-th largest zero. Consequently, Lemma 17 implies that the sum polynomial $f_n(x) = \sum_{j=1}^{m} {}_jp_n(x)$ also interlaces the polynomial $g_{n-1}(x)$ and thus possesses all real zeros.

In general, a sum polynomial of real-rooted polynomials does not possess necessarily all real zeros, which underlines the strong property of interlacing. However, even if all zeros of the sum polynomial are real, but interlacing is violated, Marcus *et al.* (2015) consider the sum $f_3(x)$ of the polynomials $(x + 5)(x - 9)(x - 10)$ and $(x + 6)(x - 1)(x - 8)$, whose zeros are approximately $-5.3, 6.4$ and 7.4, indicating that both the largest zeros 10 and 8 are larger than 7.4 of the sum polynomial, in contrast to Lemma 17.

11.8 Locations of zeros in the complex plane

330. *Center of gravity.* Consider the real numbers $\nu_1 > 0, \nu_2 > 0, \ldots, \nu_n > 0$ that obey $\sum_{j=1}^{n} \nu_j = 1$, and let $\{z_k\}_{1 \le k \le n}$ denote the n complex zeros of a polynomial

$p_n(z)$, then the center of gravity is defined as

$$z = \sum_{j=1}^{n} \nu_j z_j \qquad (\text{B.58})$$

and the number ν_j can be interpreted as a mass placed at the position z_j. If we consider all possible sets $\{\nu_j\}_{1 \le j \le n}$ of masses at the fixed points $\{z_k\}_{1 \le k \le n}$ in the complex plane, then the corresponding centers of gravity cover the interior of a convex polygon, the smallest one containing the points $z_1, z_2, \ldots, z_n$. The only exception occurs if all zeros lie on a straight line. In that case, all the centers of gravity lie in the smallest line segment that contains all the points $z_1, z_2, \ldots, z_n$.

Any straight line through the center of gravity[14] separates the set $\{z_k\}_{1 \le k \le n}$ into parts, one on each side of the line, except if all the points $z_1, z_2, \ldots, z_n$ lie on a line. Indeed, since $\sum_{j=1}^{n} \nu_j = 1$, we can write (B.58) with $w_j = \nu_j(z_j - z)$ as $\sum_{j=1}^{n} w_j = 0$. If all the points $w_1, w_2, \ldots, w_n$ are on the same side of a straight line passing through the origin, then $\sum_{j=1}^{n} w_j \ne 0$ and $\sum_{j=1}^{n} \frac{1}{w_j} \ne 0$. Indeed, we can always rotate the coordinate axis such that the imaginary axis coincides with the straight line through the origin. If all points are on one side, then they lie in either the positive or negative half plane and $\sum_{j=1}^{n} \operatorname{Re}(w_j) = \operatorname{Re}\left(\sum_{j=1}^{n} w_j\right)$ and $\sum_{j=1}^{n} \operatorname{Re}(w_j^{-1})$ is non-zero. The argument shows that not all the points $\nu_j(z_j - z)$ lie on the same side of a line. Translate the origin from the center of gravity z to any other point in the plane and verify that the property still holds.

Theorem 98 (Gauss) *No zero of the derivative $p_n'(z)$ of a polynomial $p_n(z)$ lies outside the smallest convex polygon that contains all the zeros of $p_n(z)$.*

Proof: Let $z_1, z_2, \ldots, z_n$ denote the zeros of $p_n(z)$ and let w be a zero of $p_n'(z)$, different from $z_1, z_2, \ldots, z_n$, then

$$\frac{p_n'(w)}{p_n(w)} = \sum_{j=1}^{n} \frac{1}{w - z_j} = 0$$

Since also the complex conjugate $\sum_{j=1}^{n} \frac{1}{(w-z_j)^*} = 0$, we have that $\sum_{j=1}^{n} \frac{w-z_j}{|w-z_j|^2} = 0$. This is equivalent to $w \sum_{j=1}^{n} \frac{1}{|w-z_j|^2} = \sum_{j=1}^{n} \frac{1}{|w-z_j|^2} z_j$. With $V = \sum_{j=1}^{n} \frac{1}{|w-z_j|^2}$, we arrive at

$$w = \sum_{j=1}^{n} \frac{1}{V|w - z_j|^2} z_j$$

which expresses a center of gravity if $\nu_j = \frac{1}{V|w-z_j|^2}$ in (B.58) and, by construction, $\sum_{j=1}^{n} \nu_j = 1$. As shown above, any center of gravity lies inside the smallest convex polygon formed by the points $z_1, z_2, \ldots, z_n$. □

---

[14] We may also interpret the vector $z_j - z$ as a force directed from z to z_j with magnitude $\nu_j|z_j - z|$. Then z represents an equilibrium position of a material point subject to repellant forces exerted by the points $z_1, z_2, \ldots, z_n$. If z were outside the smallest convex polygon that contains the z_j's, the resultant of the several forces acting on z could not vanish: no equilibrium is possible.

Any smallest convex polygon containing all zeros can be enclosed by a circular disk C, because all zeros are finite. If c is a point lying on the boundary of the circle C, then the Möbius transform in **art.** 301 $s(z) = \frac{1}{z-c}$ maps the disk into a half-plane containing the point at infinity, since $s(c) = \infty$. Further considerations of Gauss's Theorem 98 and the Möbius transform are discussed in Henrici (1974).

331. *Apolar polynomials.* There exists a quite remarkable result that relates the zeros of two polynomials, that satisfy the *apolar* condition (B.59). Two polynomials $p_n(z) = \sum_{k=0}^{n} a_k z^k$ and $q_n(z) = \sum_{k=0}^{n} b_k z^k$ are called *apolar* if they satisfy

$$\sum_{k=0}^{n} (-1)^k \frac{a_k b_{n-k}}{\binom{n}{k}} = 0 \tag{B.59}$$

Let $\beta_k = \frac{(-1)^k b_k}{\binom{n}{k}}$, then the Cauchy product of the polynomials $p_n(z)$ and $\tilde{q}_n(z) = \sum_{k=0}^{n} \beta_k z^k$ is

$$p_n(z)\, \tilde{q}_n(z) = \sum_{k=0}^{2n} \left(\sum_{j=0}^{k} a_j \beta_{k-j} \right) z^k$$

which shows that the apolar condition (B.59) implies that the n-th coefficient or n-th derivative at $z = 0$ of the product $p_n(z)\, \tilde{q}_n(z)$ is zero.

Theorem 99 (Grace) *Let $p_n(z) = \sum_{k=0}^{n} a_k z^k$ and $q_n(z) = \sum_{k=0}^{n} b_k z^k$ be apolar, thus satisfying (B.59). If all zeros of $p_n(z)$ lie in a circular region R, then $q_n(z)$ has at least one zero in R.*

Proof: See, e.g., Szegő (1922), Henrici (1974, pp. 469-472). □

Example Consider $q_n(z) = z^n + b_{n-k} z^{n-k}$, whose coefficient b_{n-k} is chosen to satisfy the apolar condition (B.59), such that $a_0 + (-1)^k \binom{n}{k}^{-1} a_k b_{n-k} = 0$. Thus, for $b_{n-k} = (-1)^{k+1} \binom{n}{k} \frac{a_0}{a_k}$, the zeros of $q_n(z)$ are $[0]^{n-k}$ and $(b_{n-k})^{1/k} e^{2\pi l/k}$ for $0 \le l < k$. All zeros of $q_n(z)$ lie at the origin or on the circle R around the origin with radius $\left| \binom{n}{k} \frac{a_0}{a_k} \right|^{1/k}$. Grace's Theorem 99 states that, there is at least one zero of $p_n(z)$ that lies inside that circle R.

The example shows that, by choosing an appropriate polynomial $q_n(z)$ whose zeros are known and that can be made apolar to $p_n(z)$, valuable information about the locations of some zeros of $p_n(z)$ can be derived. Related to Grace's Theorem 99 is:

Theorem 100 (Szegő's Composition Theorem) *Suppose that all the zeros of $p_n(z) = \sum_{k=0}^{n} a_k \binom{n}{k} z^k$ lie in a circular region R. If η is a zero of $q_n(z) = \sum_{k=0}^{n} b_k \binom{n}{k} z^k$, then each zero ξ of $w_n(z) = \sum_{k=0}^{n} a_k b_k \binom{n}{k} z^k$ can be written as $\xi = -s\eta$, where s is a point belonging to R.*

Proof: See Szegő (1922). □

332. *A variation on Cauchy's rule.* Let us assume that there is no zero of the polynomial $p_n(z) = \sum_{k=0}^n a_k z^k$ in a disk around z_0 with radius ρ. After transforming $z \to z - z_0$, we obtain the polynomial expansion $p_n(z) = \sum_{k=0}^n b_k(z_0)(z - z_0)^k$ around z_0, where $b_0(z_0) = p_n(z_0) \neq 0$, by the assumption. Further, we bound $p_n(z)$ for $|z - z_0| < \rho$ as

$$|p_n(z)| = \left| b_0(z_0) + \sum_{k=1}^n b_k(z_0)(z - z_0)^k \right| > |b_0(z_0)| - \sum_{k=1}^n |b_k(z_0)|(z - z_0)^k$$

$$> |b_0(z_0)| - \sum_{k=1}^n |b_k(z_0)| \rho^k = |b_0(z_0)| \left\{ 1 - \sum_{k=1}^n \frac{|b_k(z_0)|}{|b_0(z_0)|} \rho^k \right\}$$

Cauchy's rule in **art. 321** shows that we may deduce a sharper bound if all coefficients $b_k(z_0)$ are real. There is exactly one positive solution for ρ of $|b_0(z_0)| = \sum_{k=1}^n |b_k(z_0)| \rho^k$ because the right-hand side is monotonously increasing from zero at $\rho = 0$ on. Since finding such solution is generally not easy, we proceed as in **art. 321**. Let $\beta_k = \frac{|b_k(z_0)|}{|b_0(z_0)|} \rho^k > 0$, for each k where $|b_k(z_0)| > 0$, then

$$|p_n(z)| > |b_0(z_0)| \left\{ 1 - \sum_{k=1;|b_k(z_0)|>0}^n \beta_k \right\} \geq |b_0(z_0)| \left\{ 1 - \sum_{k=1}^n \beta_k \right\}$$

It suffices to require that $\sum_{k=1}^n \beta_k \leq 1$ to obtain $|p_n(z)| > 0$. Hence, given a set of positive numbers β_k satisfying $\sum_{k=1}^n \beta_k \leq 1$, then there are no zeros in a disk around z_0 with radius

$$\rho = \min_{1 \leq k \leq n; |b_k(z_0)|>0} \beta_k^{1/k} \left| \frac{b_k(z_0)}{b_0(z_0)} \right|^{1/k}$$

Example 1 If $\beta_k = 2^{-k}$ for which $\sum_{k=1}^n \beta_k = \sum_{k=1}^n 2^{-k} < \sum_{k=1}^\infty 2^{-k} = 1$, then a zero free disk around z_0 has radius

$$\rho = \frac{1}{2} \min_{1 \leq k \leq n; |b_k(z_0)|>0} \left| \frac{b_k(z_0)}{b_0(z_0)} \right|^{1/k}$$

Example 2 If $\beta_k = \binom{n}{k} y^k (1-y)^{n-k}$, then $\sum_{k=1}^n \beta_k = \sum_{k=1}^n \binom{n}{k} y^k (1-y)^{n-k} = 1$ and

$$\rho = \frac{y}{1-y} \min_{1 \leq k \leq n; |b_k(z_0)|>0} (1-y)^{\frac{n}{k}} \left| \binom{n}{k} \frac{b_k(z_0)}{b_0(z_0)} \right|^{1/k}$$

If $0 < y < 1$, then $(1-y)^{\frac{n}{k}} < (1-y)^n$ such that

$$\rho \geq y(1-y)^{n-1} \min_{1 \leq k \leq n; |b_k(z_0)|>0} \left| \binom{n}{k} \frac{b_k(z_0)}{b_0(z_0)} \right|^{1/k}$$

Finally, the maximum of $y(1-y)^{n-1}$ occurs at $y = \frac{1}{n}$ and is $\frac{1}{n}\left(1 - \frac{1}{n}\right)^{n-1} > \frac{1}{ne}$.

Thus, a zero free disk around z_0 has radius

$$\rho = \frac{1}{ne} \min_{1 \leq k \leq n; |b_k(z_0)| > 0} \left| \binom{n}{k} \frac{b_k(z_0)}{b_0(z_0)} \right|^{1/k}$$

Example 2 has another interesting property: Vieta's formula (B.13) applied to $p_n(z) = \sum_{k=0}^{n} b_k(z_0)(z - z_0)^k$ shows that

$$\frac{b_m(z_0)}{b_0(z_0)} = (-1)^m \sum_{j_1=1}^{n} \sum_{j_2=j_1+1}^{n} \cdots \sum_{j_m=j_{m-1}+1}^{n} \prod_{i=1}^{m} \frac{1}{z_{j_i} - z_0}$$

where the multiple sum contains $\binom{n}{m}$ terms as shown in **art. 296**. Now, let $d = \min_{1 \leq k \leq n} |z_k - z_0|$ denote the distance of z_0 to the nearest zero of $p_n(z)$, then $|z_k - z_0|^{-1} \leq d^{-1}$ for all $1 \leq k \leq n$. Introduced in the above Vieta formula yields, for $1 \leq m \leq n$,

$$\left| \frac{b_m(z_0)}{b_0(z_0)} \right| = \sum_{j_1=1}^{n} \sum_{j_2=j_1+1}^{n} \cdots \sum_{j_m=j_{m-1}+1}^{n} \prod_{i=1}^{m} \left| \frac{1}{z_{j_i} - z_0} \right| \leq \binom{n}{m} d^{-m}$$

from which

$$d \leq \min_{1 \leq k \leq n; |b_k(z_0)| > 0} \left| \binom{n}{k} \frac{b_k(z_0)}{b_0(z_0)} \right|^{1/k} = ne\rho$$

Thus, we have shown that there is at least one zero in the disk around z_0 with radius $ne\rho$, while Example 2 demonstrates that there are no zeros in the disk with the same center z_0 but radius ρ. Finally, we use the bound $\left| \frac{b_m(z_0)}{b_0(z_0)} \right| \leq \binom{n}{m} d^{-m}$ into $|b_0(z_0)| = \sum_{k=1}^{n} |b_k(z_0)| \rho^k$ and find

$$1 = \sum_{k=1}^{n} \frac{|b_k(z_0)|}{|b_0(z_0)|} \rho^k \leq \sum_{k=1}^{n} \binom{n}{k} \left(\frac{\rho}{d} \right)^k = \left(1 + \frac{\rho}{d} \right)^n - 1$$

such that $d \leq \frac{\rho}{2^{1/n}-1}$. Given the solution ρ of $|b_0(z_0)| = \sum_{k=1}^{n} |b_k(z_0)| \rho^k$, the disk around z_0 with radius $\frac{\rho}{2^{1/n}-1}$ contains at least one zero of $p_n(z)$.

There exist theorems, for which we refer to Henrici (1974, pp. 457-462), that give conditions for the radius of a disk to enclose at least m zeros.

333. *If $a_0 > a_1 > \cdots > a_n > 0$, then the polynomial $p_n(z) = \sum_{k=0}^{n} a_k z^k$ does not have a zero in the unit disk $|z| \leq 1$ nor on the positive real axis.*

Proof: If $z = r$ is real and positive, $p_n(r) > 0$. For the other cases where $z = re^{i\theta}$ and $\theta \neq 0$, consider

$$|(1-z)p_n(z)| = \left| a_0 - \left(\sum_{k=1}^{n} (a_{k-1} - a_k) z^k + a_n z^{n+1} \right) \right|$$

$$\geq a_0 - \left| \sum_{k=1}^{n} (a_{k-1} - a_k) z^k + a_n z^{n+1} \right|$$

Further, with $r \leq 1$,

$$\left| \sum_{k=1}^{n} (a_{k-1} - a_k) z^k + a_n z^{n+1} \right| = \left| \sum_{k=1}^{n} (a_{k-1} - a_k) r^k e^{ik\theta} + a_n r^{n+1} e^{i(n+1)\theta} \right|$$

$$< \sum_{k=1}^{n} (a_{k-1} - a_k) + a_n = a_0$$

where the inequality stems from the fact that not all arguments $e^{ik\theta}$ are equal, because $\theta \neq 0$. Hence, $|(1-z)p_n(z)| > 0$ for $|z| \leq 1$. $\qquad\square$

Art. 333 also holds for a polynomial with alternating coefficients, $t_n(z) = \sum_{k=0}^{n}(-1)^k a_k z^k$, where $a_0 > a_1 > \cdots > a_n > 0$, because a zero x of $t_n(z)$ is also a zero of $p_n(-z)$ for which $|-x| > 1$. If $a_n > a_{n-1} > \cdots > a_0 > 0$, then all the zeros of the polynomial $p_n(z) = \sum_{k=0}^{n} a_k z^k$ lie within the unit disk $|z| < 1$. This case is a consequence of **art.** 333 and (B.2) in **art.** 291.

334. *Two extensions of art. 333 due to Aziz and Zargar (2012).*

Theorem 101 (Aziz-Zargar) *If $xa_n \geq a_{n-1} \geq \cdots \geq a_1 \geq ya_0 \geq 0$ where $x \geq 1$ and $0 < y \leq 1$, then all the zeros of the polynomial $p_n(z) = \sum_{k=0}^{n} a_k z^k$ lie in the closed disk $|z + x - 1| \leq x + \frac{2a_0}{a_n}(1-y)$.*

Proof: We rewrite $q(z) = (1-z)p_n(z)$ as

$$q(z) = -a_n z^{n+1} + (a_n - a_{n-1}) z^n + \sum_{k=2}^{n-1} (a_k - a_{k-1}) z^k + (a_1 - a_0) z + a_0$$

$$= -a_n z^{n+1} + a_n z^n - x a_n z^n + (x a_n - a_{n-1}) z^n + \sum_{k=2}^{n-1} (a_k - a_{k-1}) z^k$$

$$+ (a_1 - y a_0) z + (y - 1) a_0 z + a_0$$

and further $q(z) = -A + B$, where $A = a_n z^n (z + x - 1)$ and

$$B = (x a_n - a_{n-1}) z^n + \sum_{k=2}^{n-1} (a_k - a_{k-1}) z^k + (a_1 - y a_0) z + (y - 1) a_0 z + a_0$$

Hence,

$$|q(z)| \geq |A| - |B|$$

$$\geq |a_n| |z|^n \left\{ |z + x - 1| - \frac{1}{|a_n|} \left((x a_n - a_{n-1}) + \sum_{k=2}^{n-1} (a_k - a_{k-1}) |z|^{k-n} \right. \right.$$

$$\left. \left. + (a_1 - y a_0) \frac{1}{|z|^{n-1}} + |(y-1)| \frac{a_0}{|z|^{n-1}} + \frac{a_0}{|z|^n} \right) \right\}$$

where in the last step, the inequality of the coefficients in the theorem has been used. For $|z| > 1$, we have that

$$|q(z)| > |a_n| |z^n| \left\{ |z + x - 1| - \frac{1}{a_n} \left(xa_n - a_{n-1} + \sum_{k=2}^{n-1} a_k - \sum_{k=2}^{n-1} a_{k-1} \right) \right.$$

$$\left. + a_1 - ya_0 + (1-y) a_0 + a_0) \right\}$$

$$= |a_n| |z^n| \left\{ |z + x - 1| - \frac{1}{a_n} (xa_n - ya_0 + (1-y) a_0 + a_0) \right\}$$

If $|z + x - 1| > \frac{xa_n + 2(1-y)a_0}{a_n}$ and $|z| > 1$, then $|q(z)| > 0$. Therefore, all the zeros of $q(z)$ with modulus larger than 1 lie in the closed disk $|z + x - 1| \leq x + 2(1-y) \frac{a_0}{a_n}$. Now, the zeros of $q(z)$ with modulus smaller than or equal to 1, also satisfy $|z + x - 1| \leq x + 2(1-y) \frac{a_0}{a_n}$, because $|z + x - 1| \leq |z| + x - 1 \leq x$. Since all the zeros of $p_n(z)$ are also zeros of $q(z)$, the theorem is proved. □

When $x = y = 1$, Theorem 101 reduces to **art. 333**. **Art. 333** cannot be applied to the polynomial $p_n(z) = az^n + (a-1) \sum_{k=1}^{n-1} x^k + a$ with $a > 1$, whereas Theorem 101 with $x = 1$ and $y = \frac{a-1}{a}$ shows that all the zeros of $p_n(z)$ lie in the disk $|z| \leq 1 + \frac{2}{a}$.

Theorem 102 (Aziz-Zargar) *If $a_n \leq a_{n-1} \leq \cdots \leq a_{l+1} \leq a_l \geq a_{l-1} \geq \cdots \geq a_1 \geq ya_0$ where $0 \leq l \leq n-1$ and $0 < y \leq 1$, then all the zeros of the polynomial $p_n(z) = \sum_{k=0}^{n} a_k z^k$ lie in the closed disk*

$$\left| z + \frac{a_{n-1}}{a_n} - 1 \right| \leq \frac{2a_l - a_{n-1} + (2-y) |a_0| - ya_0}{|a_n|}$$

Proof: Similar as the one above and omitted (see Aziz and Zargar (2012)). □

335. *If the polynomial $p_n(z) = \sum_{k=0}^{n} a_k z^k$ has real, positive coefficients, then all its zeros lie in the annulus $\min_{1 \leq k \leq n} \left(\frac{a_{k-1}}{a_k} \right) \leq |z| \leq \max_{1 \leq k \leq n} \left(\frac{a_{k-1}}{a_k} \right)$.*

Proof: Consider $p_n \left(\frac{z}{x} \right) = \sum_{k=0}^{n} a_k x^{-k} z^k$ and we can always choose x such that $a_k x^{-k} < a_{k-1} x^{1-k}$ for each $1 \leq k \leq n$. Indeed, it suffices that $x^{-1} < \frac{a_{k-1}}{a_k}$ for each k or that $x^{-1} = \min_{1 \leq k \leq n} \left(\frac{a_{k-1}}{a_k} \right)$. For those x, **art. 333** shows that $\left| p_n \left(\frac{z}{x} \right) \right| > 0$ for $|z| \leq 1$, which implies that $p_n(z)$ has no zeros within the disk with radius x^{-1}, thus $|z_k| > x^{-1}$. Applying the same method to $z^n p_n \left(\frac{y}{z} \right) = \sum_{k=0}^{n} a_{n-k} y^{n-k} z^k$ and choose y such that $a_{n-k} y^{n-k} < a_{n-k+1} y^{n-k+1}$ for each $1 \leq k \leq n$, or $y = \max_{1 \leq k \leq n} \left(\frac{a_{k-1}}{a_k} \right)$. For those y, **art. 333** indicates that $\left| p_n \left(\frac{y}{z} \right) \right| > 0$ for $|z| \leq 1$, which implies that all zeros of $z^n p_n \left(\frac{1}{z} \right)$ lie outside the disk with radius y. In view of **art. 291**, the zeros z_k of $p_n(z)$ lie within that disk with radius y, thus $|z_k| < y$. Combining both bounds completes the proof. □

336. *Upper bound for the number of real zeros of a polynomial $p_n(z)$.* The square of the distance between the complex numbers $z = e^{i\varphi}$ and $w = \rho e^{i\theta}$ equals

$$|z - w|^2 = 1 + \rho^2 - 2\rho \cos(\theta - \varphi)$$

from which

$$\frac{|z - w|^2}{|w|} = \rho + \frac{1}{\rho} - 2\cos(\theta - \varphi)$$

and the right-hand side is minimal when $\rho = 1$, so that

$$\frac{|z - w|^2}{|w|} \geq 2 - 2\cos(\theta - \varphi) = \left| z - e^{i\theta} \right|^2$$

After applying this inequality to the n zeros $z_k = r_k e^{i\theta_k}$ for $1 \leq k \leq n$ of a polynomial $p_n(z)$, we obtain for z on the unit circle, i.e. $|z| = 1$,

$$\frac{\left| \prod_{k=1}^n (z - z_k) \right|^2}{\left| \prod_{k=1}^n z_k \right|} \geq \left| \prod_{k=1}^n \left(z - e^{i\theta_k} \right) \right|^2$$

With the definition (B.1) of $p_n(z)$ and defining the polynomial

$$q_n(z) = \prod_{k=1}^n \left(z - e^{i\theta_k} \right)$$

whose zeros are all on the unit circle and each zero $e^{i\theta_k}$ of $q_n(z)$ possesses precisely the same phase θ_k as the zero $z_k = r_k e^{i\theta_k}$ of $p_n(z)$, we find

$$|q_n(z)|^2 \leq \frac{|p_n(z)|^2}{|a_n|^2 \left| \prod_{k=1}^n z_k \right|} = \frac{|p_n(z)|^2}{|a_n a_0|} \qquad \text{for any complex } z \text{ with } |z| = 1$$

Since $|p_n(z)|^2 = \left| p_n\left(e^{i\varphi}\right) \right|^2$ for any real φ and $\left| p_n\left(e^{i\varphi}\right) \right| = \left| \sum_{k=0}^n a_k e^{ik\varphi} \right| \leq \sum_{k=0}^n |a_k|$, we arrive at Shur's inequality, according to Erdős and Turán (1950),

$$|q_n(z)| \leq \frac{\sum_{k=0}^n |a_k|}{\sqrt{|a_n a_0|}} \qquad \text{for any complex } z \text{ with } |z| = 1$$

The zeros of $q_n(z)$ lie on a known interval $[0, 2\pi]$ and, if $p_n(z)$ has m positive real zeros, then $q_n(z)$ has a zero at $z = 1$ with multiplicity m. By using extremal properties of orthogonal polynomials, Erdős and Turán (1950) derived a lower bound for $|q_n(z)|$ and established

Theorem 103 (Schmidt-Schur-Erdős-Turán) *The number r of real zeros of the polynomial $p_n(z) = \sum_{k=0}^n a_k z^k$ is upper bounded by*

$$r^2 \leq 4n \log \frac{\sum_{k=0}^n |a_k|}{\sqrt{|a_n a_0|}} \tag{B.60}$$

11.9 Iterative algorithms for the zeros

337. *Method of Newton-Raphson.* Assume that z_0 is a reasonably good approximation of a zero ζ of $f(z)$, so that $\zeta - z_0 = h$ is sufficiently small. Then, Taylor's theorem $f(z) = \sum_{k=0}^{\infty} f_k(z_0)(z - z_0)^k$ with $f_k(z_0) = \frac{1}{k!}\frac{d^k f(u)}{du^k}\Big|_{u=z_0}$ shows that

$$f(z_0 + h) = f(z_0) + f_1(z_0)h + O(h^2)$$

Since $f(z_0 + h) = 0$, a good approximation of h up to $O(h^2)$ can be computed as $h = -\frac{f(z_0)}{f_1(z_0)}$ and the approximation of the zero ζ is $z_1 = z_0 + h$. Newton observed that repeating the argument increasingly leads to a better approximation for the zero ζ. If the first derivative can be computed in a range around z_0, then the Newton-Raphson iteration scheme for the zero is

$$z_k = z_{k-1} - \frac{f(z_{k-1})}{f_1(z_{k-1})} \tag{B.61}$$

and the sequence $z_0, z_1, z_2, \ldots, z_m$ converges to the correct zero ζ of $f(z)$. Indeed, Taylor's theorem indicates that $f(z_k) = f\left(z_{k-1} - \frac{f(z_{k-1})}{f_1(z_{k-1})}\right)$ is

$$f(z_k) = f_2(z_{k-1})\left(\frac{f(z_{k-1})}{f_1(z_{k-1})}\right)^2 + O\left(\left(\frac{f(z_{k-1})}{f_1(z_{k-1})}\right)^3\right)$$

which implies, provided that $h_k = -\frac{f(z_{k-1})}{f_1(z_{k-1})}$ is small enough to ignore terms of order 3 and higher, that

$$f(z_k) \simeq \frac{f_2(z_{k-1})}{f_1^2(z_{k-1})}(f(z_{k-1}))^2$$

If the derivatives $f_1(z_{k-1})$ are not too small, nor $f_2(z_{k-1})$ is too large, then the sequence $\{f(z_k)\}_{k \geq 0}$ converges *quadratically*: if $f(z_{k-1}) = 10^{-a}$ is small, then $f(z_k) \simeq 10^{-2a}$ and each iteration doubles the number of correct digits, which is amazing!

338. *Weierstrass's iterative method.* Weierstrass argues similarly. Ideally, all $w_j + \Delta w_j = z_j$ for all zeros $1 \leq j \leq n$ such that the product form (B.1) of the polynomial equals

$$p_n(z) = a_n \prod_{j=1}^{n}(z - w_j + \Delta w_j)$$

Taylor's Theorem in **art.** 200 of the n-dimensional function $p_n(z; z_1, \ldots, z_n)$ in the vector $(z_1, z_2, \ldots, z_n)$ around the vector $(w_1, w_2, \ldots, w_n)$ yields

$$p_n(z) = p_n(z; w_1, \ldots, w_n) + \sum_{j=1}^{n} \frac{\partial p_n(z)}{\partial z_j}\Big|_{z_j=w_j} \Delta w_j + r$$

where the remainder r contains higher order terms in Δw_j such as $(\Delta w)^T H \Delta w$, where H is the Hessian. Ignoring the remainder as in Newton-Raphson's rule (**art.** 337) and computing the derivative yields

$$p_n(z) \simeq a_n \prod_{j=1}^{n} (z - w_j) - a_n \sum_{j=1}^{n} \prod_{k=1; k \neq j}^{n} (z - w_k) \Delta w_j$$

All increments Δw_j for $1 \leq j \leq n$ are solved from this relation by subsequently letting $z = w_m$ for $1 \leq m \leq n$, resulting in

$$p_n(w_m) \simeq -a_n \sum_{j=1}^{n} \Delta w_j \prod_{k=1; k \neq j}^{n} (w_m - w_k) = -a_n \Delta w_m \prod_{k=1; k \neq m}^{n} (w_m - w_k)$$

from which Weierstrass' increments for $1 \leq m \leq n$ are obtained:

$$\Delta w_m = \frac{-p_n(w_m)}{a_n \prod_{k=1; k \neq m}^{n} (w_m - w_k)} \tag{B.62}$$

Iterations of $w_m^{(k+1)} = w_m^{(k)} + \Delta w_m^{(k)}$ for $1 \leq m \leq n$ in $k = 0, 1, \ldots$ converge also quadratically in k to all the $1 \leq m \leq n$ zeros z_m under much milder conditions than the Newton-Raphson rule. McNamee (2007) demonstrates that Weierstrass's scheme nearly always converges, irrespective of the initial guesses $w_m^{(0)}$ for $1 \leq m \leq n$.

There is an interesting alternative derivation of the Weierstrass increments (B.62). The application of the Newton-Raphson rule (B.61) to the coefficients (B.11) of Vieta's formula expressed in terms of the zeros yields a set of linear equations in Δw_m that leads to (B.62). The simplest linear equation, $\frac{a_{n-1}}{a_n} = -\sum_{k=1}^{n} z_k$ for $k = n - 1$ in (B.11), is linear in $z_k = w_k + \Delta w_k$ and shows that, at each iteration $\sum_{m=1}^{n} w_m^{(k+1)} = -\frac{a_{n-1}}{a_n}$, meaning that the sum of approximations equals the exact sum. Just as there are many improvements of the Newton-Raphson rule to enhance the convergence towards the root, so are there many variants that improve Weierstrass's rule. Moreover, there are conditions for the initial values $w_m^{(0)}$ to guarantee convergence, which are discussed in McNamee (2007).

11.10 Zeros of complex functions

339. *The argument principle.*

Theorem 104 *If $f(z)$ is analytic on and inside the contour C, then the number of zeros of $f(z)$ inside C is*

$$N = \frac{1}{2\pi i} \int_C \frac{f'(z)}{f(z)} dz = \frac{1}{2\pi} \Delta_C \arg f(z)$$

where Δ_C denotes the variation of the argument of f round the contour C.

Proof: See Titchmarsh (1964, p. 116). □

Since polynomials are analytic in the entire complex plane, Theorem 104 is valid for any contour C and can be used to compute the number of zeros inside a certain contour as shown in Section 11.6.

340. *Theorem of Rouché.* The famous and simple theorem of Rouché is very powerful.

Theorem 105 (Rouché) *If $f(z)$ and $g(z)$ are analytic inside and on a closed contour C, and $|g(z)| < |f(z)|$ on C, then $f(z)$ and $f(z) + g(z)$ have the same number of zeros inside C.*

Proof: See Titchmarsh (1964, p. 116). □

Corollary 5 *If at all points of a contour C around the origin, it holds that $|a_k z^k| > \left| \sum_{j=0; j \neq k}^{n} a_j z^j \right|$, then the contour C encloses k zeros of $p_n(z) = \sum_{k=0}^{n} a_k z^k$.*

Proof: A proof not directly based on Rouché's Theorem is given in Whittaker and Watson (1996, p. 120). The result directly follows from Rouché's Theorem 105 with $f(z) = a_k z^k$, which has a k-multiple zero at the origin and $g(z) = \sum_{j=0; j \neq k}^{n} a_j z^j$. □

We give another application of Rouché's Theorem to a polynomial $p_n(z)$ with real coefficients $a_0 > a_1 > \cdots > a_n > 0$. If R is such that $a_0 > \sum_{k=1}^{n} a_k R^k$, then $p_n(z)$ has no zeros in the disk around the origin with radius R. If $R > 1$, an improvement of **art. 333** is obtained.

341. *Theorem of Jensen and bounds of Mahler.*

Theorem 106 (Jensen) *Let $f(z)$ be analytic for $|z| < R$. Suppose that $f(0) \neq 0$, and let $r_1 \leq r_2 \leq \ldots \leq r_n \leq \ldots$ be the moduli of the zeros of $f(z)$ in the circle $|z| < R$. Then, if $r_n \leq r \leq r_{n+1}$,*

$$n \log r + \log |f(0)| - \sum_{j=1}^{n} \log r_j = \frac{1}{2\pi} \int_0^{2\pi} \log \left| f\left(re^{i\theta}\right) \right| d\theta$$

Proof: See Titchmarsh (1964, p. 125). □

Consider the polynomial $p_n(z) = \sum_{k=0}^{n} a_k z^k$ with zeros, ordered as $|z_1| > |z_2| > \ldots > |z_m| > 1 \geq |z_{m+1}| > \ldots > |z_n|$. Assuming that $a_0 \neq 0$, then Jensen's Theorem 106 states for $r = 1$ that $\frac{1}{2\pi} \int_0^{2\pi} \log \left| p_n\left(e^{i\theta}\right) \right| d\theta = - \log \left(\frac{1}{|a_0|} \prod_{j=m+1}^{n} |z_j| \right)$.

Using $a_0 = (-1)^n a_n \prod_{k=1}^{n} z_k$ in **art. 291** yields

$$\frac{1}{2\pi} \int_0^{2\pi} \log \left| p_n\left(e^{i\theta}\right) \right| d\theta = \log \left(|a_n| \prod_{k=1}^{m} |z_k| \right) \tag{B.63}$$

With $\left|p_n\left(e^{i\theta}\right)\right| = \left|\sum_{k=0}^n a_k e^{ik\theta}\right| \leq \sum_{k=0}^n |a_k|$, we obtain the inequality of Mahler (1960),

$$|a_n| \prod_{k=1}^m |z_k| \leq \sum_{k=0}^n |a_k| \tag{B.64}$$

Mahler (1960) also derives a lower bound for $|a_n| \prod_{k=1}^m |z_k|$. Since $|z_1| > |z_2| > \ldots > |z_m| > 1 \geq |z_{m+1}| > \ldots > |z_n|$, it holds, for $1 \leq j_k \leq n$ and $0 \leq k \leq n$, that $\prod_{i=1}^k |z_{j_i}| \leq \prod_{l=1}^m |z_l|$. Vieta's formula (B.11) shows that, for each $0 \leq k \leq n$,

$$\left|\frac{a_{n-k}}{a_n}\right| = \left|\sum_{j_1=1}^n \sum_{j_2=j_1+1}^n \cdots \sum_{j_k=j_{k-1}+1}^n \prod_{i=1}^k z_{j_i}\right| \leq \sum_{j_1=1}^n \sum_{j_2=j_1+1}^n \cdots \sum_{j_k=j_{k-1}+1}^n \left|\prod_{i=1}^k z_{j_i}\right|$$

$$\leq \prod_{l=1}^m |z_l| \sum_{j_1=1}^n \sum_{j_2=j_1+1}^n \cdots \sum_{j_k=j_{k-1}+1}^n 1 = \binom{n}{k} \prod_{l=1}^m |z_l|$$

Multiplying by ρ^{n-k} and summing over all k results in

$$\sum_{k=0}^n |a_{n-k}| \rho^{n-k} = \sum_{k=0}^n |a_k| \rho^k \leq |a_n| \prod_{l=1}^m |z_l| \sum_{k=0}^n \binom{n}{k} \rho^{n-k} = (1+\rho)^n |a_n| \prod_{l=1}^m |z_l|$$

which gives Mahler's lower bound when $\rho = 1$,

$$2^{-n} \sum_{k=0}^n |a_k| \leq |a_n| \prod_{l=1}^m |z_l| \tag{B.65}$$

342. *Lagrange's series for the inverse of a function.* Let the function $w = f(z)$ be analytic around the point z_0 and $f'(z_0) \neq 0$. Then, there exists a region around $w_0 = f(z_0)$, in which each point has a unique inverse $z = f^{-1}(w)$ belonging to the analytic region around z_0. The Lagrange series for the inverse of a function (Markushevich, 1985, II, pp. 88),

$$f^{-1}(w) = z_0 + \sum_{n=1}^\infty \frac{1}{n!} \frac{d^{n-1}}{dz^{n-1}} \left(\frac{z-z_0}{f(z)-f(z_0)}\right)^n \Bigg|_{z=z_0} (w - f(z_0))^n \tag{B.66}$$

is a special case (for $G(z) = z$) of the more general result

$$G(f^{-1}(w)) = G(z_0) + \sum_{n=1}^\infty \frac{1}{n!} \frac{d^{n-1}}{dz^{n-1}} \left[G'(z)\left(\frac{z-z_0}{f(z)-f(z_0)}\right)^n\right]\Bigg|_{z=z_0} (w - f(z_0))^n \tag{B.67}$$

Provided that $G(z)$ is analytic inside the contour C around z_0, that encloses a region where $f(z) - w_0$ has only a single zero, then the last series (B.67) follows from expanding the integral definition of an inverse function

$$G(f^{-1}(w)) = \frac{1}{2\pi i} \int_C G(z) \frac{f'(z)}{f(z)-w} \, dz$$

in a Taylor series around $w_0 = f(z_0)$,

$$G(f^{-1}(w)) = \sum_{n=0}^{\infty} \left[\frac{1}{2\pi i} \int_C G(z) \frac{f'(z)}{(f(z) - f(z_0))^{n+1}} \, dz \right] (w - f(z_0))^n$$

Applying integration by parts for $n > 0$ gives

$$\frac{1}{2\pi i} \int_C G(z) \frac{f'(z)}{(f(z) - f(z_0))^{n+1}} \, dz = \frac{1}{2\pi i n} \int_C \frac{G'(z)}{(f(z) - f(z_0))^n} \, dz$$

After rewriting,

$$\frac{1}{2\pi i} \int_C \frac{G'(z)}{(f(z) - f(z_0))^n} \, dz = \frac{1}{2\pi i} \int_C \left(\frac{(z - z_0)}{(f(z) - f(z_0))} \right)^n \frac{G'(z)}{(z - z_0)^n} \, dz$$

and invoking Cauchy's integral formula (B.46) for the k-th derivative, we obtain (B.67).

A zero ζ of a function $w = f(z)$, whose inverse is $z = f^{-1}(w)$, satisfies $\zeta = f^{-1}(0)$. If the Taylor series $f(z) = \sum_{k=0}^{\infty} f_k(z_0)(z - z_0)^k$ is known, the Lagrange series (B.66) can be computed formally using characteristic coefficients (Van Mieghem, 2007) to any desired order. Explicitly, up to order five in $\frac{f_0(z_0)}{f_1(z_0)}$, we have

$$\zeta(z_0) \approx z_0 - \frac{f_0(z_0)}{f_1(z_0)} - \frac{f_2(z_0)}{f_1(z_0)} \left(\frac{f_0(z_0)}{f_1(z_0)} \right)^2 + \left[-2 \left(\frac{f_2(z_0)}{f_1(z_0)} \right)^2 + \frac{f_3(z_0)}{f_1(z_0)} \right] \left(\frac{f_0(z_0)}{f_1(z_0)} \right)^3$$

$$+ \left[-5 \left(\frac{f_2(z_0)}{f_1(z_0)} \right)^3 + 5 \frac{f_3(z_0)}{f_1(z_0)} \frac{f_2(z_0)}{f_1(z_0)} - \frac{f_4(z_0)}{f_1(z_0)} \right] \left(\frac{f_0(z_0)}{f_1(z_0)} \right)^4$$

$$+ \left[-14 \left(\frac{f_2(z_0)}{f_1(z_0)} \right)^4 + 21 \frac{f_3(z_0)}{f_1(z_0)} \left(\frac{f_2(z_0)}{f_1(z_0)} \right)^2 - 3 \left(\frac{f_3(z_0)}{f_1(z_0)} \right)^2 \right.$$

$$\left. -6 \frac{f_4(z_0)}{f_1(z_0)} \frac{f_2(z_0)}{f_1(z_0)} + \frac{f_5(z_0)}{f_1(z_0)} \right] \left(\frac{f_0(z_0)}{f_1(z_0)} \right)^5 \tag{B.68}$$

from which we observe that the two first terms are Newton-Raphson's correction (B.61) in **art. 338**. The formal Lagrange expansion, where only a few terms in (B.68) are presented, only converges provided z_0 is chosen sufficiently close to the zero $\zeta(z_0)$, which underlines the importance of the choice for z_0. Another observation is that, if all Taylor coefficients $f_k(z_0)$ are real as well as z_0, the Lagrange series only possesses real terms such that the zero $\zeta(z_0)$ is real. Thus, for any polynomial $f(z) = p_n(z)$ with given real coefficients, a converging Lagrange series for some real z_0 identifies a real zero $\zeta(z_0)$.

11.11 Bounds on values of a polynomial

Milovanović *et al.* (1994) have collected a large amount of bounds on values of polynomials. Here, we only mention the first contributions to the field by Pavnuty

Chebyshev and refer to Borwein and Erdélyi (1995) and Rivlin (1974) for the deeper theory. Properties of Chebyshev polynomials are studied in Section 12.7.

343. Chebyshev proved the following theorem:

Theorem 107 (Chebyshev) *Let $p_n(x) = \sum_{k=0}^{n} a_k x^k$ be a monic polynomial with real coefficients and $a_n = 1$, then, for $n \geq 1$,*

$$\max_{-1 \leq x \leq 1} |p_n(x)| \geq \frac{1}{2^{n-1}}$$

The equality sign is obtained for $p_n(z) = \frac{T_n(z)}{2^{n-1}}$, where $T_n(z) = \cos(n \arccos z)$ are the Chebyshev polynomials of the first kind.

Proof: See Aigner and Ziegler (2003, Chapter 18) and Rivlin (1974, p. 56). □

An immediate consequence of Chebyshev's Theorem 107 is:

Corollary 6 *If a real and monic polynomial $p_n(z)$ obeys $|p_n(x)| \leq c$ for all $x \in [a, b]$, then $b - a \leq 4 \left(\frac{c}{2}\right)^{1/n}$.*

Proof: We map the x-interval $[a, b]$ onto the y-interval $[-1, 1]$ by the linear transform $y = \frac{2}{b-a}(x - a) - 1$. The polynomial $q_n(y) = p_n\left(\frac{b-a}{2}(y + 1) + a\right)$ has leading coefficient $\left(\frac{b-a}{2}\right)^n$ and satisfies $\max_{-1 \leq y \leq 1} |q_n(y)| = \max_{a \leq x \leq b} |p_n(x)|$. Chebyshev's Theorem 107 states that $\max_{-1 \leq y \leq 1} |q_n(y)| \geq \left(\frac{b-a}{2}\right)^n \frac{1}{2^{n-1}}$. Hence,

$$2 \left(\frac{b - a}{4}\right)^n \leq \max_{a \leq x \leq b} |p_n(x)| \leq c$$

such that $b - a \leq 4 \left(\frac{c}{2}\right)^{1/n}$.

□

11.12 Bounds for the spacing between zeros

344. *Minimum distance between zeros.* Mahler (1964) proved a beautiful theorem that bounds the minimum spacing or distance between any pair of simple zeros of a polynomial. Only if all zeros are simple or distinct, the discriminant $\Delta(p_n)$ is non-zero as shown in **art.** 298. Moreover, Mahler's lower bound (B.69) is the best possible.

Theorem 108 (Mahler) *For any polynomial $p_n(x)$ with degree $n \geq 2$ and distinct zeros, ordered as $|z_1| > |z_2| > \ldots > |z_m| > 1 \geq |z_{m+1}| > \ldots > |z_n|$, the minimum distance between any pair of zeros is bounded from below by*

$$\min_{1 \leq k < j \leq n} |z_k - z_j| > \frac{\sqrt{3 |\Delta(p_n)|}}{n^{\frac{n}{2}+1} \left(|a_n| \prod_{j=1}^{m} |z_j|\right)^{n-1}} \geq \frac{\sqrt{3 |\Delta(p_n)|}}{n^{\frac{n}{2}+1} \left(\sum_{k=0}^{n} |a_k|\right)^{n-1}} \qquad \text{(B.69)}$$

where $\Delta\left(p_{n}\right)$ is the discriminant, defined in **art. 298**.

Proof: The relation (B.15) between the discriminant and the Vandermonde determinant suggests us to start considering the Vandermonde matrix $V_{n}\left(z\right)$ in (A.75) of the zeros, ordered as in Theorem 108. Subtract row s from row r and use the algebraic formula $x^{k}-y^{k}=\left(x-y\right)\sum_{j=0}^{k-1}x^{k-1-j}y^{j}$ such that

$$
\det V_{n}\left(z\right)=\left(z_{r}-z_{s}\right)
\begin{bmatrix}
1 & z_{1} & z_{1}^{2} & z_{1}^{3} & \cdots & z_{1}^{n-1} \\
1 & z_{2} & z_{2}^{2} & z_{2}^{3} & \cdots & z_{2}^{n-1} \\
\vdots & \vdots & \vdots & \vdots & \vdots & \vdots \\
0 & 1 & z_{r}+z_{s} & z_{r}^{2}+z_{r}z_{s}+z_{s}^{2} & \cdots & \sum_{j=0}^{n-2}z_{r}^{n-2-j}z_{s}^{j} \\
\vdots & \vdots & \vdots & \vdots & \vdots & \vdots \\
1 & z_{n} & z_{n}^{2} & z_{n}^{3} & \cdots & z_{n}^{n-1}
\end{bmatrix}
$$

We now proceed similarly as in **art.** 225 by dividing the first m rows, corresponding to the components with absolute value larger than 1, by z_{j}^{n-1} for $1\leq j\leq m$. The only difference lies in row r, that consists of the elements

$$
\begin{array}{cccccl}
0 & 1 & z_{r}+z_{s} & z_{r}^{2}+z_{r}z_{s}+z_{s}^{2} & \cdots & \sum_{j=0}^{n-2}z_{r}^{n-2-j}z_{s}^{j} & \text{if } r>m \\
0 & z_{r}^{-(n-2)} & \frac{z_{r}+z_{s}}{z_{r}^{n-1}} & \frac{z_{r}^{2}+z_{r}z_{s}+z_{s}^{2}}{z_{r}^{n-1}} & \cdots & \frac{\sum_{j=0}^{n-2}z_{r}^{n-2-j}z_{s}^{j}}{z_{r}^{n-1}} & \text{if } r\leq m
\end{array}
$$

Since $r<s$, the ordering tells us that $\left|z_{r}\right|>\left|z_{s}\right|$. If $r>m$, then $1\geq\left|z_{r}\right|>\left|z_{s}\right|$, and the k-th element in row r is bounded by $\left|\sum_{j=0}^{k-2}z_{r}^{k-2-j}z_{s}^{j}\right|\leq k-1$, while if $r\leq m$, then $\left|z_{r}\right|>1$ and the k-th element in row r is bounded by $\left|\sum_{j=0}^{k-2}\frac{z_{r}^{k-2-j}z_{s}^{j}}{z_{r}^{n-1}}\right|\leq k-1$. Hadamard's inequality (A.78) shows that

$$
\left|\det V_{n}\left(z\right)\right|\leq\left|z_{r}-z_{s}\right|n^{\frac{n-1}{2}}\prod_{j=1}^{m}\left|z_{j}\right|^{n-1}\sqrt{\sum_{k=1}^{n}\left(k-1\right)^{2}}
$$

Using $\sum_{k=1}^{n-1}k^{2}=\frac{n(n-1)(2n-1)}{6}<\frac{n^{3}}{3}$ (Abramowitz and Stegun, 1968, Section 23.1.4), we have

$$
\left|\det V_{n}\left(z\right)\right|\leq\frac{\left|z_{r}-z_{s}\right|}{\sqrt{3}}n^{\frac{n+2}{2}}\prod_{j=1}^{m}\left|z_{j}\right|^{n-1} \tag{B.70}
$$

This inequality (B.70) is nearly the best possible, because equality is attained if $z_{j}=e^{2\pi i\frac{j}{n}}$ as shown in **art.** 225 and **art.** 242. Choosing $z_{r}=1$ and $z_{s}=e^{\frac{2\pi i}{n}}$, $\left|z_{r}-z_{s}\right|=2\sin\frac{\pi}{n}$, while we know from (A.79) that $\left|\det V_{n}\left(z\right)\right|=n^{n/2}$ such that $\frac{\left|\det V_{n}(z)\right|}{\left|z_{r}-z_{s}\right|}=\frac{n^{n/2}}{2\sin\frac{\pi}{n}}=\frac{\frac{\pi}{n}}{\sin\frac{\pi}{n}}\left(n^{\frac{n+2}{2}}\frac{}{2\pi}\right)$, which tends to $\frac{\left|\det V_{n}(z)\right|}{\left|z_{r}-z_{s}\right|}\to\frac{n^{\frac{n+2}{2}}}{2\pi}$ for large n. This illustrates that (B.70) cannot be improved, except perhaps for a slightly smaller prefactor than $\frac{1}{\sqrt{3}}$. Since the inequality (B.70) holds for any r and s, Theorem 108 now follows from the definition of the discriminant (B.15). The last inequality in (B.69) follows from (B.64). $\qquad\square$

Usually, the discriminant $\Delta(p_n)$ is not easy to determine. However, for a polynomial with integer coefficients and thus also rational coefficients because $\frac{p_n(z)}{\alpha}$ and $p_n(z)$ have the same zeros for any complex number $\alpha \neq 0$, **art. 298** shows that $\Delta(p_n)$ is a function of the coefficients a_k and, hence, an integer. In addition $\Delta(p_n) \neq 0$, such that $|\Delta(p_n)| \geq 1$. Thus, the minimum spacing between the simple zeros of a polynomial with rational coefficients $a_k \in \mathbb{Q}$ is lower bounded by

$$\min_{1 \leq k < j \leq n} |z_k - z_j| > \frac{\sqrt{3}}{n^{\frac{n}{2}+1} \left(\sum_{k=0}^{n} |a_k|\right)^{n-1}} \tag{B.71}$$

345. *Lupas' upper bound for the minimum spacing.* An upper bound for the spacing (Milovanović *et al.*, 1994, p. 106) due to Lupas is

$$\min_{1 \leq k < j \leq n} |z_k - z_j| \leq 2\sqrt{\frac{3 \operatorname{Var}[z]}{n^2 - 1}} \tag{B.72}$$

where the variance of the zeros of a real polynomial is defined as

$$\operatorname{Var}[z] = \frac{1}{n} \sum_{k=1}^{n} z_k^2 - \left(\frac{1}{n} \sum_{k=1}^{n} z_k\right)^2$$

Using the Newton identities in **art. 294** yields $\operatorname{Var}[z] = \frac{1}{n^2} \left\{ (n-1) \frac{a_{n-1}^2}{a_n^2} - \frac{2n a_{n-2}}{a_n} \right\}$. Equality in the upper bound (B.72) is attained for the polynomial

$$p_n(z) = \prod_{k=1}^{n} \left(z - E[z] - (n - 2k + 1) \sqrt{\frac{3 \operatorname{Var}[z]}{n^2 - 1}} \right)$$

where the mean of the zeros $E[z] = \frac{1}{n} \sum_{k=1}^{n} z_k = -\frac{a_{n-1}}{n a_n}$.

11.13 Bounds on the zeros of a polynomial

346. *Bounds on the largest zero.* Let w be a zero of the polynomial $p_n(z)$, then $-a_n w^n = \sum_{k=0}^{n-1} a_k w^k$ implies $|a_n| |w|^n \leq \sum_{k=0}^{n-1} |a_k| |w|^k$. If $|w| \geq 1$, then we can further bound as $|a_n| |w|^n \leq |w|^{n-1} \sum_{k=0}^{n-1} |a_k|$. Thus, unless all zeros lie within the unit disk and $|w| < 1$, the zero $\zeta = \max_{1 \leq k \leq n} |z_k|$ with largest modulus of the polynomial $p_n(z)$ obeys $1 \leq \zeta \leq \frac{1}{|a_n|} \sum_{k=0}^{n-1} |a_k|$. The Newton equation $Z_1 = \sum_{k=1}^{n} z_k = -\frac{a_{n-1}}{a_n}$ in (B.8) leads to $\left| \frac{a_{n-1}}{a_n} \right| \leq \sum_{k=1}^{n} |z_k| \leq n\zeta$. Hence, the zero ζ with largest modulus is bounded by

$$\max\left(1, \frac{1}{n} \frac{|a_{n-1}|}{|a_n|} \right) \leq \zeta \leq \frac{|a_{n-1}|}{|a_n|} + \frac{1}{|a_n|} \sum_{k=0}^{n-2} |a_k|$$

The difference between upper and lower bound illustrates that the bounds are rather loose.

McNamee (2007) provides a long list of bounds on the modulus of the largest zero $\zeta = \max_{1 \leq k \leq n} |z_k|$ of $p_n(z)$, where the coefficients $a_k \in \mathbb{C}$ and $a_n = 1$. By testing over a large number of polynomials, he mentions that the relatively simple formula, due to Deutsch (1970),

$$\zeta \leq |a_{n-1}| + \max_{1 \leq k \leq n-1 \text{ and } a_k \neq 0} \left| \frac{a_{k-1}}{a_k} \right|$$

which is an extension of **art. 291** to complex coefficients a_k derived from the companion matrix (**art. 242**) using matrix norms, ended up as second best. The best one is due to Kalantari,

$$\zeta \leq \max_{4 \leq k \leq n+3} \left| a_{n-1}^2 a_{n-k+3} - a_{n-1} a_{n-k+2} - a_{n-2} a_{n-k+3} + a_{n-k+1} \right|^{\frac{1}{k-1}}$$

347. *Euler's bounds.* Let $\zeta = \max_{1 \leq k \leq n} |z_k|$ denote the zero with largest modulus of the polynomial $p_n(z) = \sum_{k=0}^{\infty} a_k z^k$ and define $\sigma_j = \sum_{k=1}^{n} |z_k|^j \geq |Z_j|$, where the Newton equations (B.4) determine Z_j in terms of the coefficients $\{a_k\}$. Evidently, $\zeta^j \leq \sigma_j$. On the other hand, since $\zeta \geq |z_k|$ for any $1 \leq k \leq n$,

$$\sigma_{j+1} = \sum_{k=1}^{n} |z_k|^j |z_k| \leq \zeta \sum_{k=1}^{n} |z_k|^j = \zeta \sigma_j$$

Combining both inequalities yields the bounds, for any $j > 0$,

$$\frac{\sigma_{j+1}}{\sigma_j} \leq \zeta \leq (\sigma_j)^{\frac{1}{j}}$$

In the limit for $j \to \infty$ and for functions f with only real positive zeros where $Z_m = \sum_{k=1}^{n} z_k^m = \sigma_m$, both bounds tend to each other, because the radius of convergence R of the Taylor series $\frac{f'(z)}{f(z)} = \frac{f'(0)}{f(0)} - \sum_{m=1}^{\infty} \left(\sum_{k=1}^{\infty} \frac{1}{z_k^{m+1}} \right) z^m$, which is nothing else than ζ, can be calculated from (Buck, 1978, pp. 240) as $\frac{1}{R} = \lim_{m \to \infty} \sup |Z_m|^{\frac{1}{m}}$ and $\frac{1}{R} = \lim_{m \to \infty} \left| \frac{Z_{m+1}}{Z_m} \right|$, when the latter exists. Watson (1995, pp. 500) mentions that Euler, already in 1781, has devised a similar method to calculate the three smallest zeros of the Bessel function $J_0(2\sqrt{z})$.

348. *Inequalities for $\sigma_j = \sum_{k=1}^{n} |z_k|^j$.* For any integer m and j, we may write $\sigma_j = \sum_{k=1}^{n} \left| z_k^{j-m} \right| |z_k^m|$. Applying the Hölder inequality (A.10) gives

$$\sigma_j \leq \left(\sum_{k=1}^{n} |z_k|^{p(j-m)} \right)^{\frac{1}{p}} \left(\sum_{k=1}^{n} |z_k|^{\frac{pm}{p-1}} \right)^{1 - \frac{1}{p}}$$

where we require that $p(j-m) = l$ and $\frac{pm}{p-1} = h$ and both l and m are integers. All solutions satisfy $m(l-h) = (l-j)h$ with $l > 0$ and $p = \frac{l}{j-m}$ and we obtain the recursion inequality

$$\sigma_j \leq (\sigma_l)^{\frac{1}{p}} (\sigma_h)^{1 - \frac{1}{p}} \qquad \text{with } l = p(j-m) \text{ and } h = \frac{pm}{p-1}$$

For example, the solution $h = l = j$ and $p = \frac{j}{j-m}$ returns, for any m and j, an equality, namely the definition of σ_j. The case $p = 2$ is

$$\sigma_j^2 \le \left(\sum_{k=1}^n |z_k|^{2(j-m)} \right) \left(\sum_{k=1}^n |z_k|^{2m} \right) = \sigma_{2(j-m)} \sigma_{2m} \tag{B.73}$$

which is particularly useful in the case where j is even and all zeros are real.

349. The next theorem sharpens the bounds in **art. 347**:

Theorem 109 *If $z_1, \ldots, z_n$ are the real zeros of a polynomial $p_n(z) = \sum_{k=0}^n a_k z^k$ for which $Z_1 = \sum_{k=1}^n z_k = 0$, then any zero $\zeta \in \{z_1, \ldots, z_n\}$ is bounded for positive integers $1 \le m$ by*

$$-\left(\frac{Z_{2m}}{1 + \frac{1}{(n-1)^{2m-1}}} \right)^{\frac{1}{2m}} \le \zeta \le \left(\frac{Z_{2m}}{1 + \frac{1}{(n-1)^{2m-1}}} \right)^{\frac{1}{2m}} \tag{B.74}$$

where $Z_j = \sum_{k=1}^n z_k^j$ for $1 \le j \le n$ is uniquely expressed via the Newton recursion (B.7) in terms of the coefficients a_k.

As shown in **art. 300**, the condition $Z_1 = 0$, which is equivalent to the requirement that $a_{n-1} = 0$ by (B.8), is not confining.

Proof: Let z_1 denote an arbitrary zero of $p_n(z)$, because we can always relabel the zeros. Applying the Hölder inequality (A.10) to $x_j = 1$ and $y_j = z_j \in \mathbb{R}$ for $2 \le j \le n$, yields for even $q = 2m > 1$,

$$\frac{1}{(n-1)^{2m-1}} \left(\sum_{j=2}^n |z_j| \right)^{2m} \le \sum_{j=2}^n |z_j|^{2m} \tag{B.75}$$

Since $\left| \sum_{j=2}^n z_j \right| \le \sum_{j=2}^n |z_j|$ and $\sum_{j=2}^n z_j = -\frac{a_{n-1}}{a_n} - z_1$, the inequality (B.75) becomes for real zeros only,

$$\frac{1}{(n-1)^{2m-1}} \left(\frac{a_{n-1}}{a_n} + z_1 \right)^{2m} \le Z_{2m} - z_1^{2m} \tag{B.76}$$

Using the assumption that $a_{n-1} = 0$, we finally arrive, for any integer $1 \le m$, at our bounds (B.74) for any zero of $p_n(z)$, and thus also for the largest real zero. $\square$

Theorem 109 actually generalizes a famous theorem due to Laguerre in which $m = 1$ and where the condition that $Z_1 = 0$ is not needed:

Theorem 110 (Laguerre) *If all the zeros $z_1, \ldots, z_n$ of a polynomial $p_n(x) = \sum_{k=0}^n a_k x^k$ with $a_n = 1$ are real, then they all lie in the interval $[y_-, y_+]$ where*

$$y_\pm = -\frac{a_{n-1}}{n} \pm \frac{n-1}{n} \sqrt{a_{n-1}^2 - \frac{2n}{n-1} a_{n-2}}$$

Proof: Laguerre's theorem follows immediately from (B.76) and the Newton identities in **art. 294** for $m = 1$. See also Aigner and Ziegler (2003, p. 101). $\square$

Since the quartic equation ($m = 2$ in (B.76)) is still solvable exactly, that case can be expressed in closed form without the condition $Z_1 = 0$, as in the proof of Laguerre's Theorem 110. However, all other $m \geq 2$ cases are greatly simplified by the condition $Z_1 = 0$, that relieves us from solving a polynomial equation of order $2m$.

Theorem 111 *If $z_1, \ldots, z_n$ are the real zeros of a polynomial $p_n(z) = \sum_{k=0}^{n} a_k z^k$, then any zero $\zeta \in \{z_1, \ldots, z_n\}$ is upper bounded, for positive integers m, by*

$$\zeta \leq \left(\frac{Z_{2m}}{n} + \sqrt{(n-1)} \sqrt{\frac{Z_{4m}}{n} - \left(\frac{Z_{2m}}{n} \right)^2} \right)^{\frac{1}{m}} \tag{B.77}$$

and lower bounded for odd integer values of m, by

$$\zeta \geq \left(\frac{Z_{2m}}{n} - \sqrt{(n-1)} \sqrt{\frac{Z_{4m}}{n} - \left(\frac{Z_{2m}}{n} \right)^2} \right)^{\frac{1}{m}} \tag{B.78}$$

Proof: In a similar vein, application of (B.73) for $j = 2m$ and $m = l$, gives

$$\left(Z_{2m} - z_1^{2m} \right)^2 \leq \left(Z_{2(2m-l)} - z_1^{2(2m-l)} \right) \left(Z_{2l} - z_1^{2l} \right)$$

If $l = \frac{m}{2}$ or $l = \frac{3m}{2}$, then $z_1^{3m} Z_m - 2z_1^{2m} Z_{2m} + z_1^m Z_{3m} + Z_{2m}^2 - Z_{3m} Z_m \leq 0$, whose exact solution via Cardano's formula is less attractive. For $l = 0$, the quadratic inequality

$$z_1^{4m} - \frac{2Z_{2m}}{n} z_1^{2m} + \frac{Z_{2m}^2 - (n-1) Z_{4m}}{n} \leq 0$$

is obeyed for any z_1^m lying in between

$$z_{\pm} = \frac{Z_{2m}}{n} \pm \sqrt{(n-1)} \sqrt{\frac{Z_{4m}}{n} - \left(\frac{Z_{2m}}{n} \right)^2}$$

The Cauchy–Schwarz inequality (A.12) shows that $\frac{Z_{4m}}{n} - \left(\frac{Z_{2m}}{n} \right)^2 \geq 0$, implying that the roots are real. Thus, we find the upper bound (B.77) and the lower bound (B.78), that always exists for odd m. $\square$

12

Orthogonal polynomials

The classical theory of orthogonal polynomials is reviewed from an algebraic point of view. The book by Szegő (1978) is regarded as the standard text, although it approaches the theory of orthogonal polynomials via complex function theory and differential equations. The classical theory of orthogonal polynomials is remarkably beautiful, and powerful at the same time. Moreover, as shown in Section 6.13, we found interesting relations with graph theory.

An overview and properties of the classical orthogonal polynomials such as Legendre, Chebyshev, Jacobi, Laguerre and Hermite polynomials is found in Abramowitz and Stegun (1968, Chapter 22) and Rainville (1971). A general classification scheme of orthogonal polynomials is presented by Koekoek *et al.* (2010) and by Koornwinder *et al.* in Olver *et al.* (2010, Chapter 18). The theory of expanding an arbitrary function in terms of solutions of a second-order differential equation, initiated by Sturm and Liouville, and treated by Titchmarsh (1962) and by Titchmarsh (1958) for partial differential equations, can be regarded as the generalization of orthogonal polynomial expansions.

12.1 Definitions

350. The usual scalar or inner product between two vectors x and y, that is denoted as $x^T y$, is generalized to real functions f and g as the Stieltjes-Lebesgue integral[1] over the interval $[a, b]$

$$(f, g) = \int_a^b f(u) g(u) \, dW(u) \tag{B.79}$$

---

[1] As mentioned in the introduction of Van Mieghem (2014, Chapter 2), the Stieltjes integral unifies both continuous and differentiable distribution functions as well as discrete ones, in which case, the integral reduces to a sum. If W is differentiable, then (B.79) simplifies to

$$(f, g) = \int_a^b f(u) g(u) w(u) \, du$$

where the non-negative function $w(u) = W'(u)$ is often called a weight function. A broader discussion is given by Szegő (1978, Section 1.4).

where the distribution function W is a non-decreasing, non-constant function in $[a, b]$. As in linear algebra, the functions f and g are called orthogonal if $(f, g) = 0$ and, likewise, the norm of f is $\|f\| = \sqrt{(f, f)}$. Moreover, the generalization (B.79) is obviously linear, $(\alpha f + \beta h, g) = \alpha (f, g) + \beta (h, g)$ and commutative $(f, g) = (g, f)$. The definition thus assumes the knowledge of both the interval $[a, b]$ as well as the distribution function W. All functions f, for which the integral $\int_a^b |f(u)|^2 \, dW(u)$ in (B.79) exists, constitute the space $L^2_{[a,b]}$.

351. An orthogonal set of real functions $f_0(x), f_1(x), \ldots, f_m(x)$ is defined, for any $k \neq l \in \{0, 1, \ldots, m\}$, by

$$(f_k, f_l) = \int_a^b f_k(u) f_l(u) \, dW(u) = 0 \tag{B.80}$$

When $(f_k, f_k) = 1$ for all $k \in \{0, 1, \ldots, m\}$, the set is normalized and called an orthonormal set of functions. Just as in linear algebra, these functions $\{f_k\}_{0 \leq k \leq m}$ are linearly independent. Since polynomials are special types of functions, an orthogonal set of polynomials $\{\pi_k\}_{0 \leq k \leq m}$ is also defined by (B.80), and we denote, an orthogonal polynomial of degree n, by π_n or $\pi_n(x)$. In addition, the general polynomial expression (B.1) is

$$\pi_n(x) = \sum_{k=0}^n c_{k;n} x^k \tag{B.81}$$

The special scalar product $m_k = \left(x^k, 1\right)$, or in integral form

$$m_k = \int_a^b u^k \, dW(u) \tag{B.82}$$

is called the moment of order k, and is further studied in **art. 354.**

If $\pi_n(x)$ is an orthogonal polynomial, then $\widetilde{\pi}_n(x) = \frac{\pi_n(x)}{\sqrt{(\pi_n, \pi_n)}}$ is an orthonormal polynomial. Although the highest coefficients $c_{n;n}$ can be any real number, we may always choose $c_{n;n} > 0$ since any polynomial can be multiplied by a number without affecting the zeros. The fact that $c_{n;n} > 0$ is sometimes implicitly assumed.

352. *The Gram-Schmidt orthogonalization process.* Analogous to linear algebra, where a set of n linearly independent vectors that span the n-dimensional space are transformed into an orthonormal set of vectors, the Gram-Schmidt orthogonalization process can also be used to construct a set of orthonormal polynomials, defined by the scalar product (B.79). First, the constant polynomial of degree zero $\pi_0(x) = \pi_0$ is chosen to obey

$$1 = (\pi_0, \pi_0) = \pi_0^2 \int_a^b dW(u) = \pi_0^2 m_0$$

where the moment of order zero in (B.82) equals

$$m_0 = W(b) - W(a) > 0$$

because the distribution function $W(x)$ is non-decreasing in x. Thus, $\tilde{\pi}_0(x) = \frac{1}{\sqrt{m_0}}$.

The degree one polynomial, $\pi_1(x) = c_{1;1}x + c_{0;1}$ must be orthogonal to $\pi_0(x)$ and orthonormal, $(\pi_1, \pi_1) = 1$. These two requirements result in

$$(\pi_1, \pi_0) = c_{1;1}(x, \pi_0) + c_{0;1}(1, \pi_0) = 0$$

such that $c_{0;1} = -\frac{c_{1;1}(x,\pi_0)}{(1,\pi_0)}$, while normalization requires that $\tilde{\pi}_1(x) = \frac{c_{1;1}x + c_{0;1}}{\sqrt{(\pi_1,\pi_1)}}$. Combining both leads to

$$\tilde{\pi}_1(x) = \frac{c_{1;1}}{\sqrt{(\pi_1, \pi_1)}}\left(x - \frac{(x, \pi_0)}{(1, \pi_0)}\right)$$

Both π_1 and π_0 are real polynomials.

We can continue the process and compute the degree two polynomial that is orthogonal to both π_1 and π_0, and that is also orthonormal. Suppose now that we have constructed a sequence of orthonormal polynomials $\pi_0, \pi_1, \ldots, \pi_{n-1}$, which are all real, obey the orthogonality condition (B.80) and are normalized, $(\pi_k, \pi_k) = 1$. Next, we construct the polynomial $\pi_n(x)$ that is orthogonal to all lower degree orthonormal polynomials by considering

$$\pi_n(x) = c_{n;n}x^n - \sum_{k=0}^{n-1} a_k \pi_k(x)$$

Orthogonality requires for $j < n$ that

$$0 = (\pi_n, \pi_j) = c_{n;n}(x^n, \pi_j) - \sum_{k=0}^{n-1} a_k(\pi_k, \pi_j) = c_{n;n}(x^n, \pi_j) - a_j(\pi_j, \pi_j)$$

such that

$$a_j = c_{n;n}\frac{(x^n, \pi_j)}{(\pi_j, \pi_j)}$$

After normalization $(\pi_n, \pi_n) = 1$, we obtain the real, orthonormal polynomial of degree n:

$$\tilde{\pi}_n(x) = \frac{c_{n;n}}{\sqrt{(\pi_n, \pi_n)}}\left(x^n - \sum_{k=0}^{n-1}\frac{(x^n, \pi_k)}{(\pi_k, \pi_k)}\pi_k(x)\right)$$

By induction on n, it follows that there exists an orthonormal set of polynomials belonging to the scalar product (B.79).

12.2 Properties

353. *Key orthogonality property.* Let $p_n(x)$ be an arbitrary polynomial of degree n, then $p_n(x)$ can be written as a linear combination of the linearly independent, orthogonal polynomials $\{\pi_k\}_{0 \le k \le m}$, provided $m \ge n$,

$$p_n(x) = \sum_{k=0}^{n} b_{k;n} \pi_k(x) \tag{B.83}$$

After multiplying both sides by $\pi_l(x)$, taking the scalar product, and using the orthogonality definition in (B.80), we find that, for all $0 \le l \le n$,

$$b_{l;n} = \frac{(p_n, \pi_l)}{(\pi_l, \pi_l)}$$

Hence, any polynomial of degree $n \le m$ can be expressed in a unique way as a linear combination of the set of orthogonal polynomials $\{\pi_k\}_{0 \le k \le m}$. Because the polynomial $p_n(x)$ is of degree n, the fundamental theorem of algebra states that the coefficients $b_{l;n} = 0$ when $l > n$. In summary, a key property of orthogonality is

$$(p_n, \pi_l) = \begin{cases} b_{l;n} \|\pi_l\|^2 & \text{if } 0 \le l \le n \\ 0 & \text{if } l > n \end{cases} \tag{B.84}$$

Example If $p_n(x) = \pi_n(x) = \sum_{k=0}^{n} c_{k;n} x^k$, then $(p_n, \pi_l) = \sum_{k=0}^{n} c_{k;n} (x^k, \pi_l) = \sum_{k=l}^{n} c_{k;n} (x^k, \pi_l)$, because (B.84) indicates that $(x^k, \pi_l) = 0$ if $l > k$. By orthogonality (B.80), it holds that $(p_n, \pi_l) = (\pi_n, \pi_l) = \delta_{l,n} = b_{l;n} \|\pi_l\|^2$. Thus, $\sum_{k=l}^{n} c_{k;n} (x^k, \pi_l) = 0$ for $l < n$ and, for $n = l$, we find that $b_{n;n} = 1$, such that

$$c_{n;n} = \frac{(\pi_n, \pi_n)}{(x^n, \pi_n)} \tag{B.85}$$

354. A first interesting consequence of **art. 353** arises for the special class of polynomials $p_n(x) = x^n$. In that case, if $l > n$, then $(x^n, \pi_l) = 0$, but $(x^n, \pi_n) \ne 0$. Introducing the polynomial form (B.81) and using $(x^n, x^k) = (x^{n+k}, 1) = m_{n+k}$ in (B.82) yields, for $n \le l$,

$$(x^n, \pi_l) = \sum_{k=0}^{l} c_{k;l} m_{n+k}$$

which is written in matrix form, taking into account that $(x^n, \pi_l) = (x^n, \pi_n) \delta_{l,n}$ for $0 \le n \le l$, as

$$\begin{bmatrix} m_0 & m_1 & \cdots & m_l \\ m_1 & m_2 & \cdots & m_{l+1} \\ \vdots & \vdots & \vdots & \vdots \\ m_l & m_{l+1} & \cdots & m_{2l} \end{bmatrix} \begin{bmatrix} c_{0;l} \\ c_{1;l} \\ \vdots \\ c_{l;l} \end{bmatrix} = \begin{bmatrix} 0 \\ 0 \\ \vdots \\ (x^l, \pi_l) \end{bmatrix}$$

The symmetric moment matrix

$$M_l = \begin{bmatrix} m_0 & m_1 & \cdots & m_l \\ m_1 & m_2 & \cdots & m_{l+1} \\ \vdots & \vdots & \vdots & \vdots \\ m_l & m_{l+1} & \cdots & m_{2l} \end{bmatrix}$$

is an $(l+1) \times (l+1)$ Hankel matrix[2]. The Gram-Schmidt orthogonalization process (**art. 352**) shows that there always exists an orthogonal set of polynomials, such that a set of not-all-zero coefficients $\{c_{k;l}\}_{0 \le k \le l}$ exists. This implies that the determinant of the moment matrix M_l is non-zero. By Cramer's rule (A.68), the solution is

$$c_{k;l} = \frac{\left(x^l, \pi_l\right) \text{cofactor}_{l+1,k} M_l}{\det M_l}$$

In particular, the definition (A.36) of a cofactor in **art. 212** shows that

$$c_{l;l} = \frac{\det M_{l-1}}{\det M_l} \left(x^l, \pi_l\right) \tag{B.86}$$

which is always different from zero.

By applying (B.86) for $n = l$, any polynomial $p_n(x) = \sum_{k=0}^{n} a_k x^k$ can be written as

$$(p_n, \pi_n) = \sum_{k=0}^{n} a_k \left(x^k, \pi_n\right) = a_n \left(x^n, \pi_n\right) = a_n \frac{c_{n;n} \det M_n}{\det M_{n-1}}$$

and the particular choice $p_n(x) = \pi_n(x)$ shows that

$$\frac{\det M_n}{\det M_{n-1}} = \frac{(\pi_n, \pi_n)}{c_{n;n}^2} > 0 \tag{B.87}$$

355. Another consequence of the orthogonality property in **art. 353** is that for any monic polynomial $p_n(x) = \sum_{k=0}^{n} a_k x^k$ with $a_n = 1$ and any set of monic orthogonal polynomials $\pi_n(x)$ with $c_{n;n} = 1$, it holds that

$$\int_a^b p_n^2(x) \, dW(x) \ge \int_a^b \pi_n^2(x) \, dW(x) = \|\pi_n\|^2 \tag{B.88}$$

with equality only if $p_n = \pi_n$. Among all monic polynomials of degree n, an orthogonal polynomial has the smallest integral of its square with respect to its weight function $W(x)$ and its orthogonality interval $[a, b]$.

Proof of (B.88): We consider

$$\int_a^b (p_n(x) - \pi_n(x))^2 \, dW(x) = \int_a^b p_n^2(x) \, dW(x) + \int_a^b \pi_n^2(x) \, dW(x)$$
$$- 2 \int_a^b p_n(x) \pi_n(x) \, dW(x)$$

---

[2] We refer for properties of the Hankel matrix to Gantmacher (1959a, pp. 338-348).

and the last integral equals with **art. 351** and **art. 353**

$$\int_a^b p_n(x)\,\pi_n(x)\,dW(x) = (p_n, \pi_n) = b_{n;n}\,\|\pi_n\|^2$$

Equating the coefficient of x^n in $p_n(x) = \sum_{k=0}^n b_{k;n}\pi_k(x)$ and (B.81) shows, with the definition (B.81), that $a_n = b_{n;n}c_{n;n}$ and, thus, $b_{n;n} = 1$. With $\int_a^b \pi_n^2(x)\,dW(x) = (\pi_n, \pi_n) = \|\pi_n\|^2$, we arrive at

$$\int_a^b (p_n(x) - \pi_n(x))^2\,dW(x) = \int_a^b p_n^2(x)\,dW(x) - \int_a^b \pi_n^2(x)\,dW(x)$$

and since the left-hand side is non-negative, we have established (B.88). □

The proof is generalized from a polynomial $p_n(x)$ to a real function $f(x) \in L^2_{[a,b]}$, defined in **art. 350**. Similarly as above, we minimize $\int_a^b (q_n(x) - f(x))^2\,dW(x)$, where $q_n(x)$ is a polynomial of degree n with real coefficients, which can be expanded by **art. 353** as $q_n(x) = \sum_{k=0}^n a_k \tilde{\pi}_k(x)$. Hence,

$$\int_a^b (q_n(x) - f(x))^2\,dW(x) = \int_a^b f^2(x)\,dW(x) - 2\sum_{k=0}^n a_k \int_a^b \tilde{\pi}_k(x)\,f(x)\,dW(x)$$

$$+ \sum_{k=0}^n a_k \sum_{j=0}^n a_j \int_a^b \tilde{\pi}_j(x)\,\tilde{\pi}_k(x)\,dW(x)$$

By orthonormality (**art. 351**), $\int_a^b \tilde{\pi}_j(x)\,\tilde{\pi}_k(x)\,dW(x) = \delta_{jk}$, and defining

$$h_k = (f, \tilde{\pi}_k) = \int_a^b f(x)\,\tilde{\pi}_k(x)\,dW(x) \tag{B.89}$$

we find

$$\int_a^b (q_n(x) - f(x))^2\,dW(x) = \int_a^b f^2(x)\,dW(x) - 2\sum_{k=0}^n a_k h_k + \sum_{k=0}^n a_k^2$$

$$= \int_a^b f^2(x)\,dW(x) + \sum_{k=0}^n (a_k - h_k)^2 - \sum_{k=0}^n h_k^2 \geq 0 \tag{B.90}$$

illustrating that the right-hand side is minimal when $a_k = h_k$, thus when $q_n(x)$ is replaced by $s_n(x) = \sum_{k=0}^n h_k \tilde{\pi}_k(x)$, which leads to *Bessel's inequality*,

$$\sum_{k=0}^n h_k^2 \leq \int_a^b f^2(x)\,dW(x) \tag{B.91}$$

Since the right-hand side of (B.91) is independent of n and finite, $\sum_{k=0}^\infty h_k^2$ converges, implying that $\lim_{n\to\infty} h_n = 0$. Moreover, if $f \in L^2_{[a,b]}$, then equality holds in Bessel's inequality (B.91) when $n \to \infty$ by Weierstrass's approximation theorem (**art. 304**): for $\varepsilon > 0$, there exists a degree n such that the minimizer polynomial $s_n(x)$ obeys $|f(x) - s_n(x)| \leq \varepsilon$ for all $x \in [a,b]$, so that

$\int_a^b \left(s_n\left(x\right) - f\left(x\right)\right)^2 dW\left(x\right) \le \varepsilon^2$. For ε arbitrarily small, thus n arbitrarily large, relation (B.90) reduces to *Parseval's equality*

$$\sum_{k=0}^{\infty} h_k^2 = \int_a^b f^2\left(x\right) dW\left(x\right)$$

12.3 The three-term recursion

356. *The three-term recursion.* Another, even more important application of **art.** 353 follows from the polynomial $p_n\left(x\right) = x\pi_{n-1}\left(x\right)$, that has an orthogonal polynomial expansion

$$x\pi_{n-1}\left(x\right) = \sum_{l=0}^{n} b_{l;n}\pi_l\left(x\right)$$

where the coefficients are

$$b_{l;n} = \frac{\left(x\pi_{n-1}, \pi_l\right)}{\left(\pi_l, \pi_l\right)} = \frac{\left(\pi_{n-1}, x\pi_l\right)}{\left(\pi_l, \pi_l\right)}$$

Since $x\pi_l$ is a polynomial of degree $l+1$, we know from **art.** 353 that $b_{l;n} = 0$ when $n - 1 > l + 1$, thus when $l < n - 2$. Hence, we find that any set of orthogonal polynomials possesses a three-term recursion for $2 \le n \le l$

$$x\pi_{n-1}\left(x\right) = b_{n;n}\pi_n\left(x\right) + b_{n-1;n}\pi_{n-1}\left(x\right) + b_{n-2;n}\pi_{n-2}\left(x\right) \tag{B.92}$$

When $n = 1$, then $\pi_0\left(x\right)$ is a constant and $\pi_{-1}\left(x\right) = 0$. Observe that any other polynomial $p_n\left(x\right) = x^j\pi_{n-j}\left(x\right)$ with $j > 1$ will result in a recursion with $2j + 1$ terms because $\frac{\left(x^j\pi_{n-j}, \pi_l\right)}{\left(\pi_l, \pi_l\right)} = \frac{\left(x^j\pi_l, \pi_{n-j}\right)}{\left(\pi_l, \pi_l\right)} = 0$ if $n - j > l + j$, thus $l < n - 2j$.

The coefficients $b_{l;n}$ for $n - 2 \le l \le n$ in the three-term recursion (B.92) can be related to the coefficients $c_{k;l}$ in **art.** 354 of the moment expansion. Taking the scalar product in (B.92) with x^{n-2} yields

$$\left(x\pi_{n-1}, x^{n-2}\right) = b_{n;n}\left(\pi_n, x^{n-2}\right) + b_{n-1;n}\left(\pi_{n-1}, x^{n-2}\right) + b_{n-2;n}\left(\pi_{n-2}, x^{n-2}\right)$$

Since $\left(x\pi_{n-1}, x^{n-2}\right) = \left(\pi_{n-1}, x^{n-1}\right)$ while $\left(\pi_j, x^{n-2}\right) = 0$ for $j > n - 2$, we find, beside $b_{n-2;n} = \frac{\left(\pi_{n-1}, x\pi_{n-2}\right)}{\left(\pi_{n-2}, \pi_{n-2}\right)}$, that

$$b_{n-2;n} = \frac{\left(\pi_{n-1}, x^{n-1}\right)}{\left(\pi_{n-2}, x^{n-2}\right)} = \frac{c_{n-1;n-1} \det M_{n-1} \det M_{n-3}}{c_{n-2;n-2}\left(\det M_{n-2}\right)^2} \tag{B.93}$$

where the last formula follows from (B.86). It demonstrates that $b_{n-2;n} \ne 0$. Moreover, (B.87) shows that for monic polynomials, i.e., if $c_{n;n} = 1$, that $b_{n-2;n} > 0$ and, thus, $\left(\pi_{n-1}\left(x\right), x\pi_{n-2}\right) > 0$. Substituting (B.81) in

$$b_{n-2;n} = \frac{\left(x\pi_{n-2}, \pi_{n-1}\right)}{\left(\pi_{n-2}, \pi_{n-2}\right)} = \frac{\sum_{k=1}^{n-1} c_{k-1;n-2}\left(x^k, \pi_{n-1}\right)}{\left(\pi_{n-2}, \pi_{n-2}\right)} = \frac{c_{n-2;n-2}\left(x^{n-1}, \pi_{n-1}\right)}{\left(\pi_{n-2}, \pi_{n-2}\right)}$$

leads with (B.85) to

$$b_{n-2;n} = \frac{c_{n-2;n-2}\left(\pi_{n-1}, \pi_{n-1}\right)}{c_{n-1;n-1}\left(\pi_{n-2}, \pi_{n-2}\right)} \tag{B.94}$$

Further,

$$b_{n-1;n} = \frac{\left(x\pi_{n-1}, \pi_{n-1}\right)}{\left(\pi_{n-1}, \pi_{n-1}\right)} = \frac{1}{\left(\pi_{n-1}, \pi_{n-1}\right)} \int_a^b u\pi_{n-1}^2\left(u\right) dW\left(u\right) \tag{B.95}$$

shows that $b_{n-1;n}$ is positive if $b > a \geq 0$ and negative, if $b < a \leq 0$. It can only be zero provided symmetry holds, $a = -b$ and $w\left(u\right) = \frac{dW}{du} = w\left(-u\right)$.

The coefficient $b_{n;n}$ can be rewritten as

$$b_{n;n} = \frac{\left(x\pi_{n-1}\left(x\right), \pi_n\right)}{\left(\pi_n, \pi_n\right)} = \frac{\sum_{k=1}^n c_{k-1;n-1}\left(x^k, \pi_n\right)}{\left(\pi_n, \pi_n\right)} = \frac{c_{n-1;n-1}\left(x^n, \pi_n\right)}{\left(\pi_n, \pi_n\right)}$$

Using (B.85) leads to

$$b_{n;n} = \frac{c_{n-1;n-1}}{c_{n;n}} \tag{B.96}$$

The expressions (B.93) and (B.96) simplify for monic polynomials where $c_{n;n} = 1$.

357. Often, the three-term recursion (B.92) is rewritten in normalized form with $\pi_n\left(x\right) = \tilde{\pi}_n\left(x\right)\sqrt{\left(\pi_n, \pi_n\right)}$ as

$$\tilde{\pi}_n\left(x\right) = \left(x - b_{n-1;n}\right)\tilde{\pi}_{n-1}\left(x\right)\frac{\sqrt{\left(\pi_{n-1}, \pi_{n-1}\right)}}{b_{n;n}\sqrt{\left(\pi_n, \pi_n\right)}} - \tilde{\pi}_{n-2}\left(x\right)\frac{b_{n-2;n}\sqrt{\left(\pi_{n-2}, \pi_{n-2}\right)}}{b_{n;n}\sqrt{\left(\pi_n, \pi_n\right)}}$$

Substituting the expressions (B.96), (B.95) and (B.93) for the b-coefficients yields

$$\tilde{\pi}_n\left(x\right) = \left(x - b_{n-1;n}\right)\tilde{\pi}_{n-1}\left(x\right)\frac{\tilde{c}_{n;n}}{\tilde{c}_{n-1;n-1}} - \tilde{\pi}_{n-2}\left(x\right)\frac{\tilde{c}_{n;n}\tilde{c}_{n-2;n-2}}{\tilde{c}_{n-1;n-1}^2}$$

where $\tilde{c}_{n;n} = \frac{c_{n;n}}{\sqrt{\left(\pi_n, \pi_n\right)}}$. Thus, the normalized three-term recursion is

$$\tilde{\pi}_n\left(x\right) = \left(A_n x + B_n\right)\tilde{\pi}_{n-1}\left(x\right) - C_n\tilde{\pi}_{n-2}\left(x\right) \tag{B.97}$$

where $A_n = \frac{\tilde{c}_{n;n}}{\tilde{c}_{n-1;n-1}}$, $B_n = -b_{n-1;n}\frac{\tilde{c}_{n;n}}{\tilde{c}_{n-1;n-1}}$ and $C_n = \frac{\tilde{c}_{n;n}\tilde{c}_{n-2;n-2}}{\tilde{c}_{n-1;n-1}^2} = \frac{A_n}{A_{n-1}}$. The major advantage of the normalized expression is the relation $C_n = \frac{A_n}{A_{n-1}}$, as illustrated in **art.** 358.

The converse is proven by Favard: if a set of polynomials satisfies a three-term recursion as (B.92), then the set of polynomials is orthogonal. Favard's theorem is proven in Chihara (1978, p. 22) for monic polynomials, where $A_n = 1$ and $C_n > 0$.

358. *Christoffel-Darboux formula.* Multiplying both sides of the normalized three-term recursion (B.97) by $\tilde{\pi}_{n-1}\left(y\right)$,

$$\tilde{\pi}_{n-1}\left(y\right)\tilde{\pi}_n\left(x\right) = \left(A_n x + B_n\right)\tilde{\pi}_{n-1}\left(x\right)\tilde{\pi}_{n-1}\left(y\right) - C_n\tilde{\pi}_{n-2}\left(x\right)\tilde{\pi}_{n-1}\left(y\right)$$

Similarly, letting $x \to y$ in (B.97) and multiplying both sides by $\tilde{\pi}_{n-1}\left(x\right)$ yields

$$\tilde{\pi}_{n-1}\left(x\right)\tilde{\pi}_n\left(y\right) = \left(A_n y + B_n\right)\tilde{\pi}_{n-1}\left(x\right)\tilde{\pi}_{n-1}\left(y\right) - C_n\tilde{\pi}_{n-2}\left(y\right)\tilde{\pi}_{n-1}\left(x\right)$$

Subtracting the second equation from the first results in

$$\tilde{\pi}_{n-1}(y)\,\tilde{\pi}_n(x) - \tilde{\pi}_{n-1}(x)\,\tilde{\pi}_n(y) = A_n(x-y)\,\tilde{\pi}_{n-1}(x)\,\tilde{\pi}_{n-1}(y)$$
$$- C_n\{\tilde{\pi}_{n-2}(x)\,\tilde{\pi}_{n-1}(y) - \tilde{\pi}_{n-2}(y)\,\tilde{\pi}_{n-1}(x)\}$$

At this stage, we employ the relation $C_n = \frac{A_n}{A_{n-1}}$, that only holds for the normalized three-term recursion (B.97) and not for (B.92). Defining

$$g_n = \frac{\tilde{\pi}_{n-1}(y)\,\tilde{\pi}_n(x) - \tilde{\pi}_{n-1}(x)\,\tilde{\pi}_n(y)}{A_n}$$

leads to

$$(x-y)\,\tilde{\pi}_{n-1}(x)\,\tilde{\pi}_{n-1}(y) = g_n - g_{n-1}$$

Summing both sides over n,

$$(x-y)\sum_{n=1}^{m+1}\tilde{\pi}_{n-1}(x)\,\tilde{\pi}_{n-1}(y) = \sum_{n=1}^{m+1} g_n - \sum_{n=1}^{m+1} g_{n-1} = g_{m+1} - g_0 = g_{m+1}$$

because $\tilde{\pi}_{-1} = 0$. Hence, we arrive at the famous Christoffel-Darboux formula,

$$\sum_{n=0}^{m}\tilde{\pi}_n(x)\,\tilde{\pi}_n(y) = \frac{1}{A_{m+1}}\frac{\tilde{\pi}_m(y)\,\tilde{\pi}_{m+1}(x) - \tilde{\pi}_m(x)\,\tilde{\pi}_{m+1}(y)}{x-y} \tag{B.98}$$

which can also be written as

$$\sum_{n=0}^{m}\tilde{\pi}_n(x)\,\tilde{\pi}_n(y) = \frac{\tilde{\pi}_m(y)}{A_{m+1}}\frac{\tilde{\pi}_{m+1}(x) - \tilde{\pi}_{m+1}(y)}{x-y} - \frac{\tilde{\pi}_{m+1}(y)}{A_{m+1}}\frac{\tilde{\pi}_m(x) - \tilde{\pi}_m(y)}{x-y}$$

The special case, where $x = y$, follows, after invoking the definition of the derivative, as

$$\sum_{n=0}^{m}\tilde{\pi}_n^2(x) = \frac{\tilde{\pi}_m(x)\,\tilde{\pi}'_{m+1}(x) - \tilde{\pi}'_m(x)\,\tilde{\pi}_{m+1}(x)}{A_{m+1}} = \frac{\tilde{\pi}_m^2(x)}{A_{m+1}}\frac{d}{dx}\left(\frac{\tilde{\pi}_{m+1}(x)}{\tilde{\pi}_m(x)}\right) \tag{B.99}$$

359. *Associated orthogonal polynomials.* Similar to the derivation of the Christoffel-Darboux formula in **art. 358**, we consider the difference at two arguments of the three-term recursion (B.92) for $n > 1$,

$$x\pi_{n-1}(x) - y\pi_{n-1}(y) = b_{n;n}[\pi_n(x) - \pi_n(y)] + b_{n-1;n}[\pi_{n-1}(x) - \pi_{n-1}(y)]$$
$$+ b_{n-2;n}[\pi_{n-2}(x) - \pi_{n-2}(y)]$$

We rewrite the left-hand side as

$$x\pi_{n-1}(x) - y\pi_{n-1}(y) = x[\pi_{n-1}(x) - \pi_{n-1}(y)] + (x-y)\pi_{n-1}(y)$$

and obtain

$$(x-y)\pi_{n-1}(y) = b_{n;n}[\pi_n(x) - \pi_n(y)] + (b_{n-1;n} - x)[\pi_{n-1}(x) - \pi_{n-1}(y)]$$
$$+ b_{n-2;n}[\pi_{n-2}(x) - \pi_{n-2}(y)]$$

After multiplying both sides by $\frac{dW(y)}{x-y}$ and integrating over $[a, b]$, we have

$$\int_a^b \pi_{n-1}(y)\, dW(y) = b_{n;n} \int_a^b \frac{\pi_n(x) - \pi_n(y)}{x-y} dW(y)$$

$$+ (b_{n-1;n} - x) \int_a^b \frac{\pi_{n-1}(x) - \pi_{n-1}(y)}{x-y} dW(y)$$

$$+ b_{n-2;n} \int_a^b \frac{\pi_{n-2}(x) - \pi_{n-2}(y)}{x-y} dW(y)$$

Since $\int_a^b \pi_{n-1}(y)\, dW(y) = (\pi_{n-1}, 1) = 0$ for $n > 1$ by orthogonality (**art. 353**), we arrive, with the definition

$$\sigma_n(x) = \int_a^b \frac{\pi_n(x) - \pi_n(y)}{x-y} dW(y) \tag{B.100}$$

at the same three-term recursion as (B.92) for $n > 1$,

$$x\sigma_{n-1}(x) = b_{n;n}\sigma_n(x) + b_{n-1;n}\sigma_{n-1}(x) + b_{n-2;n}\sigma_{n-2}(x)$$

If $n = 0$, in which case $\pi_0(x)$ is a constant, then (B.100) shows that $\sigma_0(x) = 0$. For $n = 1$ where $\pi_1(x) = c_{1;1}x + c_{0;1}$, the integral (B.100) with (B.82) gives $\sigma_1(x) = c_{1;1}m_0$. By introducing (B.81) in (B.100), we have

$$\sigma_n(x) = \sum_{k=0}^n c_{k;n} \int_a^b \frac{x^k - y^k}{x-y} dW(y) = \sum_{k=0}^n c_{k;n} \sum_{j=0}^{k-1} x^j \int_a^b y^{k-1-j} dW(y)$$

Using (B.82) yields $\sigma_n(x) = \sum_{k=0}^n c_{k;n} \sum_{j=0}^{k-1} x^j m_{k-1-j}$. After reversal of the summation, we find that

$$\sigma_n(x) = \sum_{j=0}^{n-1} \left(\sum_{k=j+1}^n c_{k;n} m_{k-1-j} \right) x^j$$

is a polynomial of order $n - 1$. Since the polynomials $\sigma_n(x)$ satisfy a three-term recursion, Favard's theorem (**art. 356**) states that these polynomials are also orthogonal. The polynomials $\sigma_n(x)$, defined by the integral (B.100), are called *orthogonal polynomials of the second kind* or *associated orthogonal polynomials*. The analysis shows that, by choosing other initial conditions, another set of orthogonal polynomials can be obtained from the three-term recursion (B.92).

360. The integral (B.100) of $\sigma_n(x)$ cannot be split into two integrals when $x \in [a, b]$, due to the pole at x. However, when $x \in \mathbb{C} \setminus [a, b]$, then we can write,

$$\sigma_n(x) = \pi_n(x) \int_a^b \frac{dW(y)}{x-y} - \int_a^b \frac{\pi_n(y)}{x-y} dW(y)$$

For $|x| > \max(|a|, |b|)$, we expand $\frac{1}{x-y} = \sum_{k=0}^\infty \frac{y^k}{x^{k+1}}$ and interchange the integration and summation, which is valid when assuming absolute convergence. The first integral,

$$F(x) = \int_a^b \frac{dW(y)}{x - y} \tag{B.101}$$

becomes

$$F(x) = \sum_{k=0}^{\infty} \frac{1}{x^{k+1}} \int_a^b y^k dW(y) = \sum_{k=0}^{\infty} \frac{m_k}{x^{k+1}}$$

while the second integral,

$$\rho_n(x) = \int_a^b \frac{\pi_n(y)}{x - y} dW(y) \tag{B.102}$$

reads

$$\rho_n(x) = \sum_{k=0}^{\infty} \frac{1}{x^{k+1}} \int_a^b y^k \pi_n(y) dW(y) = \sum_{k=n}^{\infty} \frac{(y^k, \pi_n)}{x^{k+1}}$$

because $(y^k, \pi_n) = 0$ for $k < n$, by orthogonality (**art.** 353). Hence, for large x, we rewrite $\sigma_n(x) = \pi_n(x) F(x) - \rho_n(x)$ as

$$\frac{\sigma_n(x)}{\pi_n(x)} = F(x) - \frac{\rho_n(x)}{\pi_n(x)} = F(x) + O\left(x^{-2n-1}\right) \tag{B.103}$$

whose consequences are further explored in **art.** 367. Convergence considerations when $n \to \infty$ are discussed in Gautschi (2004).

361. *Computing the weight function $w(x)$.* The two functions $F(z)$ and $\rho_n(z)$ are analytic in $\mathbb{C} \backslash [a, b]$ and both integral representations resemble the Cauchy integral, $f(z) = \frac{1}{2\pi i} \int_{C(z)} \frac{f(w)}{w-z} dw$, where $C(z)$ is a contour that encloses the point $z \in \mathbb{C}$. A general theorem (Markushevich, 1985, p. 312) states that the integral of the Cauchy type,

$$f(z) = \frac{1}{2\pi i} \int_L \frac{\varphi(w)}{w - z} dw \tag{B.104}$$

satisfies

$$\lim_{z \to z_0; z \in I(z_0)} f(z) - \lim_{w \to z_0; w \in E(z_0)} f(w) = \varphi(z_0) \tag{B.105}$$

where L is a not necessarily closed path in the complex plane and on which $|\varphi(z) - \varphi(z_0)| \le c(z - z_0)^\beta$ for any point $z, z_0 \in L$, and where $c, \beta > 0$ are constants. The interior $I(z_0)$ is a region enclosed[3] by a closed contour $C_1(z_0)$ around $z_0 \in L$ in which $\varphi(w)$ is analytic. The exterior $E(z_0)$ is the region that is not enclosed by a contour $C_2(z_0)$. The contours $C_1(z_0)$ and $C_2(z_0)$ are here formed by a circle around z_0 with a radius such that it intersects the path L in two points z_1 and z_2 and such that $\varphi(w)$ is analytic in the enclosed region. The first contour $C_1(z_0)$ follows the path L in the positive direction from z_1 to z_2 and returns to

---

[3] An observer traveling in the direction of the contour around z_0 in counter-clockwise sense finds the interior on his left and the exterior on his right.

z_1 along the circle in positive direction, whereas contour $C_2(z_0)$ similarly follows the path L in the positive direction from z_1 to z_2, but it returns to z_1 along the circle around z_0 in negative direction. Hence, any point z lying inside the contour $C_1(z_0)$ is enclosed in positive direction, whereas any point w lying inside the contour $C_2(z_0)$ is enclosed in negative direction. Finally, consider the circle $C(z_0)$ around z_0 that passes through z_1 and z_2. By Cauchy's theorem and the fact that $\varphi(w)$ is analytic inside the circle $C(z_0)$, we have that $\varphi(z_0) = \frac{1}{2\pi i} \int_{C(z_0)} \frac{\varphi(w)}{w-z} dw$. By deforming the circle into the contour $C(z_0) = C_1(z_0) - C_2(z_0)$, we arrive at (B.105).

We apply this theorem to the integral (B.102). The path L is the segment $[a, b]$ on the real axis and $z_0 = x_0 \in [a, b]$. The contour $C(z_0)$ around $z_0 = x_0$ is the path from $x_0 - r > a$ to $x_0 + r < b$ along the real axis and the circle segment lying above the real axis (with positive imaginary part). The contour $C'(z_0)$ follows the same segment from $x_0 - r > a$ to $x_0 + r < b$, but returns along the semicircle below the real axis. Hence,

$$\lim_{z \to z_0; z \in I(z_0)} \rho_n(z) = \lim_{y \to 0} \rho_n(x_0 + iy)$$

$$\lim_{w \to z_0; w \in E(z_0)} \rho_n(w) = \lim_{y \to 0} \rho_n(x_0 - iy)$$

Since the complex conjugate $\rho_n^*(z) = \rho_n(z^*)$, by the reflection principle (Titchmarsh, 1964, p. 155) because $\rho_n(z)$ is real on the real axis, we have that

$$\rho_n(x_0 - iy) = \operatorname{Re} \rho_n(x_0 - iy) + i \operatorname{Im} \rho_n(x_0 - iy)$$
$$= \operatorname{Re} \rho_n(x_0 + iy) - i \operatorname{Im} \rho_n(x_0 + iy)$$

Finally, (B.105) shows that

$$\frac{1}{\pi} \lim_{y \to 0} \operatorname{Im} \rho_n(x_0 + iy) = -\pi_n(x_0) w(x_0)$$

Similarly, from the integral (B.101), the density function is found at $x_0 \in [a, b]$ by

$$w(x_0) = -\frac{1}{\pi} \lim_{y \to 0} \operatorname{Im} F(x_0 + iy)$$

362. *Cauchy transform.* When the path of integration L in (B.104) coincides with the real axis, **art. 361** has demonstrated that the Cauchy transform $\Gamma_f(z)$ of a function f,

$$\Gamma_f(z) = \int_a^b \frac{f(x)}{z - x} dx$$

possesses, for any $x \in [a, b]$, the inverse

$$f(x) = -\frac{1}{\pi} \lim_{y \to 0} \operatorname{Im} \Gamma_f(x + iy)$$

A short formal, but different demonstration of the inverse Cauchy transform is based on the Dirac function. Indeed, for any $x \in [a, b]$, the characteristic property of the Dirac function (see **art.** 172) shows that

$$f(x) = \int_a^b f(u) \delta(x - u) \, du$$

Substituting the representation (7.3) of the Dirac function yields

$$f(x) = -\frac{1}{\pi} \lim_{y \to 0} \operatorname{Im} \int_a^b \frac{f(u)}{x - u + iy} \, du = -\frac{1}{\pi} \lim_{\eta \to 0} \operatorname{Im} \Gamma_f(x + iy)$$

12.4 Zeros of orthogonal polynomials

We illustrate that a lot of information about the zeros of orthogonal polynomials can be deduced. Both the orthogonal polynomial $\pi_n(x)$ and its normalized version $\tilde{\pi}_n(x) = \frac{\pi_n(x)}{\sqrt{(\pi_n, \pi_n)}}$ possess the same zeros.

363. *Zeros of orthogonal polynomials.*

Theorem 112 *All zeros of the orthogonal polynomial $\pi_l(u)$ are real, simple and lie inside the interval $[a, b]$.*

Proof: Art. 354 has shown that $(x^n, \pi_l) = 0$ if $n < l$. The particular case $n = 0$ and $l \geq 1$, written with the scalar product (B.79) as

$$\int_a^b \pi_l(u) \, dW(u) = 0$$

indicates that there must exist at least one point within the interval (a, b) at which $\pi_l(u)$ changes sign, because $W(u)$ is a distribution function with positive density. The change in sign implies that such a point is a zero with odd multiplicity. Let $z_1, z_2, \ldots, z_k$ be all such points and consider the polynomial $q_k(x) = \prod_{j=1}^k (x - z_j)$. **Art.** 353 shows that $(\pi_l, q_l) \neq 0$ but that $(\pi_l, q_k) = 0$ if $l > k$,

$$\int_a^b \pi_l(u) \prod_{j=1}^k (u - z_j) \, dW(u) = 0$$

By construction, $\pi_l(u) \prod_{j=1}^k (u - z_j)$ does not change sign for any $u \in [a, b]$ and, hence, the integral cannot vanish. Orthogonality shows that the non-vanishing of (π_l, q_k) is only possible provided $k = l$. The fundamental theorem of algebra (**art.** 291) together with the odd multiplicity of each zero z_j then implies that all zeros are simple. $\qquad \square$

Szegő (1978, p. 45) presents other proofs. For example, the simplicity of the zeros can be deduced by applying the Sturm sequence (**art.** 325) to the tree-term recursion (B.97) assuming $c_{n;n} > 0$. Theorem 112 shows that any orthogonal polynomial only possesses real and simple zeros. An arbitrary polynomial with real coefficients possesses zeros that, with high probability, do not all lie on a line segment in the complex plane, which illustrates the peculiar nature of orthogonal polynomials.

Let $a \leq z_{n;n} < z_{n-1;n} < \cdots < z_{1;n} \leq b$ denote the zeros of the orthogonal polynomial $\pi_n(x)$. Combining Theorem 112 and (B.1) yields

$$\pi_n(x) = c_{n;n} \prod_{j=1}^{n} (x - z_{j;n}) \tag{B.106}$$

Finally, if $b < 0$ and $c_{n;n} > 0$, then all coefficients $\{c_{k;n}\}_{0 \leq k \leq n}$ of $\pi_n(x) = \sum_{k=0}^{n} c_{k;n} x^k$ in (B.81) are non-negative. As a consequence of Newton's Theorem 97 in **art.** 328 and Theorem 112, the sequence $\{c_{k;n} \geq 0\}_{0 \leq k \leq n}$ of the coefficients in (B.81) is unimodal with either a plateau of two points or a peak.

364. *Interlacing property of zeros of orthogonal polynomials.* The main observations are derived from the Christoffel-Darboux formula (B.99), which implies, assuming $A_{n+1} > 0$,

$$\tilde{\pi}_n(x) \tilde{\pi}'_{n+1}(x) - \tilde{\pi}'_n(x) \tilde{\pi}_{n+1}(x) \geq \frac{1}{m_0} > 0 \tag{B.107}$$

because $\tilde{\pi}_0(x) = \frac{1}{\sqrt{m_0}}$. The simplicity of the zeros (Theorem 112) implies that the derivative $\tilde{\pi}'_m(x)$ cannot have the same zero as $\tilde{\pi}_m(x)$. Hence, the above inequality indicates that $\tilde{\pi}_m(x)$ and $\tilde{\pi}_{m+1}(x)$ cannot have a same zero.

Theorem 113 (Interlacing) *Let $a < z_{n;n} < z_{n-1;n} < \cdots < z_{1;n} < b$ be the zeros of the orthogonal polynomial $\pi_n(x)$. The zeros of $\pi_n(x)$ and $\pi_{n+1}(x)$ are interlaced,*

$$a < z_{n+1;n+1} < z_{n;n} < z_{n;n+1} < z_{n-1;n} < \cdots < z_{1;n} < z_{1;n+1} < b$$

In other words (**art.** 329), between each pair of consecutive zeros of $\pi_n(x)$, there lies a zero of $\pi_{n+1}(x)$, thus, $z_{k;n} < z_{k;n+1} < z_{k-1;n}$ for all $1 \leq k \leq n$, while the smallest and largest zero obey $a < z_{n+1;n+1} < z_{n;n}$ and $z_{1;n} < z_{1;n+1} < b$.

Proof: Theorem 112 shows that the zeros are simple and real such that

$$\pi'_n(z_{k;n}) \pi'_n(z_{k-1;n}) < 0$$

On the other hand, the inequality (B.107) implies that

$$-\pi'_n(z_{k;n}) \pi_{n+1}(z_{k;n}) > 0 \quad \text{and} \quad -\pi'_n(z_{k-1;n}) \pi_{n+1}(z_{k-1;n}) > 0$$

Multiplying both and taking $\pi'_n(z_{k;n}) \pi'_n(z_{k-1;n}) < 0$ into account yields

$$\pi_{n+1}(z_{k;n}) \pi_{n+1}(z_{k-1;n}) < 0$$

which means that there is at least one zero $z_{j;n+1}$ between $z_{k;n} < z_{j;n+1} < z_{k-1;n}$. Since the inequalities hold for all $1 \leq k \leq n$, the argument accounts for at least $n-1$ zeros of $\pi_{n+1}(x)$. With the convention that $c_{n;n} > 0$ for all $n \geq 0$, we know that $\pi_n(x)$ is increasing at least from the largest zero on, $\pi'_n(z_{1;n}) > 0$. The inequality

(B.107) indicates that $\pi_{n+1}(z_{1;n}) < 0$. By the convention $c_{n;n} > 0$ for all $n \geq 0$, we have that $\pi_{n+1}(b) > 0$ such that there must be a zero, in fact the largest $z_{1;n+1}$ of $\pi_{n+1}(x)$ in the interval $[z_{1;n}, b]$. A similar argument applies for the smallest zero $z_{n+1;n+1}$, thereby proving the theorem. $\qquad\square$

If $z_{n;n} = a$, the interlacing Theorem 113 implies that the set $\{\widetilde{\pi}_k(x)\}_{0 \leq k \leq n}$ is finite and that $\widetilde{\pi}_n(x)$ is the highest order polynomial of that finite orthogonal set with a zero equal to $z_{n;n} = a$. All other smallest zeros are larger, i.e., $z_{k;k} > a$ for $1 \leq k \leq n-1$.

Another noteworthy consequence of the interlacing Theorem 113 is the partial fraction decomposition

$$\frac{\pi_n(x)}{\pi_{n+1}(x)} = \sum_{k=1}^{n+1} \frac{\beta_{k;n+1}}{x - z_{k;n+1}}$$

where the coefficients, in general, obey

$$\beta_{k;n+1} = \lim_{x \to z_{k;n+1}} \frac{\pi_n(x)(x - z_{k;n+1})}{\pi_{n+1}(x)} = \frac{\pi_n(z_{k;n+1})}{\pi'_{n+1}(z_{k;n+1})}$$

Inequality (B.107) shows that all $\beta_{k;n+1} > 0$. We include here a sharpening of the interlacing property whose proof relies on the Gaussian quadrature Theorem 115 derived in Section 12.5.

Theorem 114 *Between two zeros of $\pi_n(x)$, there is at least one zero of $\pi_l(x)$ with $l > n$.*

Proof: Assume the contrary, namely $\pi_l(x)$ has no zero between $z_{k;n}$ and $z_{k-1;n}$ for some $k \in [1, n]$. Then, the polynomial $p_m(x) = \pi_n(x) q_{n-2}(x)$ of degree $m = 2n - 2$, where

$$q_{n-2}(x) = \frac{\pi_n(x)}{(x - z_{k;n})(x - z_{k-1;n})}$$

is everywhere non-zero in $[a, b]$, except in the interval $(z_{k;n}, z_{k-1;n})$, where $p_m(x)$ is negative. The Gaussian quadrature Theorem 115 shows, for $m = 2n - 2 < 2l$, that

$$\int_a^b p_m(x)\, dW(x) = \sum_{j=1}^l p_m(z_{j;l}) \lambda_{j;l} > 0$$

because (a) the Christoffel numbers $\lambda_{j;l}$ are positive (**art. 366**), and (b) $p_m(z_{j;l})$ cannot vanish at every zero $z_{j;l}$ of $\pi_l(x)$ and $p_m(z_{j;l}) \geq 0$ since, by hypothesis, $z_{j;l} \notin [z_{k;n}, z_{k-1;n}]$. But this contradicts the basic orthogonality property,

$$\int_a^b p_m(x)\, dW(x) = (\pi_n, q_{n-2}) = 0$$

established in **art. 353**. $\qquad\square$

Szegő (1978, p. 112) mentions the following distance result between consecutive zeros. If the density or weight function $\frac{dW(x)}{dx} = w(x) \geq w_{\min} > 0$ and the zeros are written as $z_{k;n} = \frac{a+b}{2} + \frac{b-a}{2} \cos \theta_{k;n}$, where $0 < \theta_{k;n} < \pi$, for $1 \leq k \leq n$, then it holds that

$$\theta_{k+1;n} - \theta_{k;n} < \alpha \frac{\log n}{n}$$

where the constant α depends on $w_{\min}$, a and b. If stronger constraints are imposed on the weight function w, the $\log n$ factor in the numerator can be removed. More precise results on the location of zeros only seem possible in specific cases and/or when the differential equation of the set of orthogonal polynomials is known.

12.5 Gaussian quadrature

Lanczos (1988, pp. 396-414) nicely explains Gauss's genial idea to compute the integral $\int_{-1}^{1} f(u)\, du$ with "double order accuracy" compared to other numerical integration methods. The underlying principle of Gauss's renowned quadrature method is orthogonality and properties of orthogonal polynomials. Before giving an example, we first focus on the theory.

365. We consider the Lagrange polynomial q_{n-1} of degree $n-1$ (**art.** 303) that coincides at n points, defined by their finite coordinates (x_j, y_j) for $1 \leq j \leq n$, with the arbitrary polynomial $p_m(x)$ of degree $m > n$,

$$q_{n-1}(x) = \sum_{j=1}^{n} y_j l_{n-1}(x; x_k) = \sum_{j=1}^{n} y_j \frac{F_n(x)}{(x - x_j) F_n'(x_j)}$$

where $F_n(x) = \prod_{j=1}^{n}(x - x_j)$ and $y_j = p_m(x_j)$. We further assume that the abscissae coincide with the distinct zeros of the orthogonal polynomial $\pi_n(x)$, thus $x_j = z_{j;n}$. Then, from (B.106), it follows that $\frac{F_n(x)}{F_n'(z_{j;n})} = \frac{\pi_n(x)}{\pi_n'(z_{j;n})}$ for all $1 \leq j \leq n$ and we obtain

$$q_{n-1}(x) = \sum_{j=1}^{n} p_m(z_{j;n}) \frac{\pi_n(x)}{(x - z_{j;n}) \pi_n'(z_{j;n})}$$

The difference polynomial $r_m(x) = p_m(x) - q_{n-1}(x)$ has degree m and $r_m(x)$ vanishes at the n points $x_j = z_{j;n}$, taken as the zeros of $\pi_n(x)$. Thus,

$$r_m(x) = t_{m-n}(x) \pi_n(x)$$

where $t_{m-n}(x)$ is some polynomial of degree $m - n$. Taking the scalar product $(r_m, 1)$ or multiplying both sides by $dW(x)$ and integrating over $[a, b]$ shows that

$$(r_m, 1) = (t_{m-n}, \pi_n)$$

which, by **art. 353**, vanishes provided $n > m - n$, or $2n > m$. In the case that m is at most $2n - 1$ and $(r_m, 1) = (p_m - q_{n-1}, 1) = 0$, we find that

$$\int_a^b p_m(x)\, dW(x) = \int_a^b q_{n-1}(x)\, dW(x)$$

$$= \sum_{j=1}^n p_m(z_{j;n}) \int_a^b \frac{\pi_n(x)\, dW(x)}{(x - z_{j;n})\, \pi_n'(z_{j;n})}$$

In summary, we have demonstrated Gauss's famous quadrature formula,

Theorem 115 (Gauss's quadrature formula) *Let* $a < z_{n;n} < z_{n-1;n} < \cdots < z_{1;n} < b$ *be the zeros of the orthogonal polynomial* $\pi_n(x)$ *on the interval* $[a, b]$ *with respect to the distribution function* $W(x)$. *For any polynomial* $p_m(x)$ *of degree* m *at most* $2n - 1$,

$$\int_a^b p_m(x)\, dW(x) = \sum_{j=1}^n p_m(z_{j;n}) \lambda_{j;n} \tag{B.108}$$

where the Christoffel numbers are

$$\lambda_{j;n} = \int_a^b \frac{\pi_n(x)\, dW(x)}{(x - z_{j;n})\, \pi_n'(z_{j;n})} \tag{B.109}$$

The extension from a polynomial $p_m(x)$ to a real function $f(x) \in L_{[a,b]}^2$ may suggest us to consider the remainder $R_n(x) = f(x) - q_{n-1}(x)$, which has n zeros in $[a, b]$ since $f(z_{j;n}) = q_{n-1}(z_{j;n})$ for $1 \le j \le n$ and which, as in **art. 304**, can be written as $R_n(x) = \pi_n(x) g(x)$. However, since Gauss's quadrature formula (B.108) with n evaluations is exact for polynomials with degree at most $2n - 1$, a sharper error estimate is achieved by considering Hermite interpolation (**art. 304**). After integration of the Hermite interpolating polynomial (B.21) at the interpolation points $x_j = z_{j;n}$ for $1 \le j \le n$ and replacing $F_n(x) = \frac{\pi_n(x)}{c_{n;n}}$ by (B.106), we obtain

$$\int_a^b f(x)\, dW(x) = \int_a^b h_{2n-1}(x)\, dW(x) + \frac{f^{(2n)}(\xi)}{(2n)! c_{n;n}^2} \int_a^b \pi_n^2(x)\, dW(x)$$

Using (B.108), we arrive at Gauss's quadrature formula with remainder

$$\int_a^b f(x)\, dW(x) = \sum_{j=1}^n f(z_{j;n}) \lambda_{j;n} + \frac{f^{(2n)}(\xi)}{(2n)! c_{n;n}^2} \|\pi_n\|^2 \tag{B.110}$$

where $a < \xi < b$.

366. *Christoffel numbers.* The Christoffel numbers in (B.109) possess interesting properties. First, let $p_m(x) = \left\{ \frac{\pi_n(x)}{(x - z_{j;n})\pi_n'(z_{j;n})} \right\}^2$ such that $p_m(z_{k;n}) = \delta_{kj}$, then Gauss's quadrature formula (B.108) reduces to

$$\lambda_{j;n} = \int_a^b \left\{ \frac{\pi_n(x)}{(x - z_{j;n})\, \pi_n'(z_{j;n})} \right\}^2 dW(x) \tag{B.111}$$

demonstrating that all Christoffel numbers $\lambda_{j;n}$ are positive.

The integral (B.102), corresponding to the associated orthogonal polynomials $\sigma_n(x)$ and valid for $x \in \mathbb{C} \backslash [a, b]$, actually is finite at the zeros of $\pi_n(x)$. Comparison with (B.109) shows that

$$\rho_n(z_{j;n}) = -\lambda_{j;n} \pi_n'(z_{j;n})$$

Next, the Christoffel-Darboux formula (B.98), with $y = z_{j;n}$, is

$$\sum_{k=0}^{n-1} \widetilde{\pi}_k(x) \widetilde{\pi}_k(z_{j;n}) = \frac{1}{A_{n+1}} \frac{-\widetilde{\pi}_n(x) \widetilde{\pi}_{n+1}(z_{j;n})}{x - z_{j;n}}$$

Taking the scalar product $(.,1)$ of both sides yields

$$\sum_{k=0}^{n-1} (\widetilde{\pi}_k, 1) \widetilde{\pi}_k(z_{j;n}) = -\frac{\widetilde{\pi}_{n+1}(z_{j;n})}{A_{n+1}} \int_a^b \frac{\widetilde{\pi}_n(x)}{x - z_{j;n}} dW(x)$$

Art. 353 shows that $(\widetilde{\pi}_k, 1) = 0$ except when $k = 0$. In that case, $\widetilde{\pi}_0(x) = \frac{1}{\sqrt{m_0}}$ and $(\widetilde{\pi}_0, 1) = \int_a^b \widetilde{\pi}_0(x) dW(x) = \sqrt{m_0}$ such that $\sum_{k=0}^{n-1} (\widetilde{\pi}_k, 1) \widetilde{\pi}_k(z_{j;n}) = 1$. The definition (B.109) of the Christoffel numbers shows that

$$\lambda_{j;n} = \int_a^b \frac{\pi_n(x) dW(x)}{(x - z_{j;n}) \pi_n'(z_{j;n})} = \int_a^b \frac{\widetilde{\pi}_n(x) dW(x)}{(x - z_{j;n}) \widetilde{\pi}_n'(z_{j;n})}$$

such that

$$1 = -\frac{\widetilde{\pi}_{n+1}(z_{j;n})}{A_{n+1}} \int_a^b \frac{\widetilde{\pi}_n(x)}{x - z_{j;n}} dW(x) = -\frac{\widetilde{\pi}_{n+1}(z_{j;n}) \widetilde{\pi}_n'(z_{j;n})}{A_{n+1}} \lambda_{j;n}$$

Thus, the Christoffel numbers obey

$$\lambda_{j;n} = -\frac{A_{n+1}}{\widetilde{\pi}_{n+1}(z_{j;n}) \widetilde{\pi}_n'(z_{j;n})} = \frac{A_n}{\widetilde{\pi}_{n-1}(z_{j;n}) \widetilde{\pi}_n'(z_{j;n})} \tag{B.112}$$

where the latter follows from (B.97).

Finally, the Christoffel-Darboux formula (B.99) evaluated at $x = z_{j;n}$ combined with (B.112) gives

$$\lambda_{j;n} = \frac{1}{\sum_{k=0}^{n-1} \widetilde{\pi}_k^2(z_{j;n})} \tag{B.113}$$

which again illustrates that all Christoffel numbers $\lambda_{j;n}$ are positive.

367. *Partial fraction decomposition of* $\frac{\sigma_n(x)}{\pi_n(x)}$. The associated orthogonal polynomials (**art. 359**) are of degree $n - 1$, such that the fraction $\frac{\sigma_n(x)}{\pi_n(x)}$ can be expanded as

$$\frac{\sigma_n(x)}{\pi_n(x)} = \sum_{k=1}^n \frac{a_{k;n}}{x - z_{k;n}}$$

where we need to determine the coefficients $a_{k;n}$. Substituting in (B.103)

$$\frac{\sigma_n (x)}{\pi_n (x)} - F (x) = O \left(x^{-2n-1} \right)$$

the partial fraction expansion and the integral (B.101) of $F (x)$ yields

$$\sum_{k=1}^{n} \frac{a_{k;n}}{x - z_{k;n}} - \int_a^b \frac{dW (y)}{x - y} = O \left(x^{-2n-1} \right)$$

After expanding the left-hand side in a power series in x^{-1} and after equating the corresponding power of x^{-m}, we obtain, for $0 \le m \le 2n - 1$,

$$\sum_{k=1}^{n} a_{k;n} \left(z_{k;n} \right)^m - \int_a^b y^m dW (y) = 0$$

Gauss's quadrature formula (B.108) applied to $p_m (x) = y^m$ for $0 \le m \le 2n - 1$ gives

$$\int_a^b y^m dW (y) = \sum_{k=1}^{n} \lambda_{k;n} z_{k;n}^m$$

whence $a_{k;n} = \lambda_{k;n}$. In summary, the partial fraction decomposition becomes

$$\frac{\sigma_n (x)}{\pi_n (x)} = \sum_{k=1}^{n} \frac{\lambda_{k;n}}{x - z_{k;n}}$$

from which the Christoffel numbers follow as

$$\lambda_{j;n} = \lim_{x \to z_{j;n}} \frac{\sigma_n (x) (x - z_{j;n})}{\pi_n (x)} = \frac{\sigma_n (z_{j;n})}{\pi_n' (z_{j;n})} \qquad \text{(B.114)}$$

368. *Parameterized weight functions.* Suppose that the distribution function W is differentiable at any point of $[a, b]$, and that W depends on a parameter t. In addition, we assume that the density or weight function $w (x, t) = \frac{dW(x,t)}{dx}$ is positive and that $w (x, t)$ is also continuous and differentiable in t. The explicit dependence on the parameter t in Gauss's quadrature formula (B.108) is written as

$$\int_a^b p_m (x) w (x, t) \, dx = \sum_{j=1}^{n} p_m (z_{j;n} (t)) \lambda_{j;n} (t)$$

Differentiation with respect to t yields

$$\int_a^b p_m (x) \frac{\partial w (x, t)}{\partial t} dx = \sum_{j=1}^{n} p_m' (z_{j;n} (t)) z_{j;n}' (t) \lambda_{j;n} (t) + \sum_{j=1}^{n} p_m (z_{j;n} (t)) \lambda_{j;n}' (t)$$

For the particular choice of $p_m (x) = \frac{\tilde{\pi}_n^2 (x,t)}{x - z_{k;n}(t)}$, we have that $p_m (z_{j;n} (t)) = 0$ and that

$$p_m' (z_{j;n} (t)) = \left. \frac{dp_m (x)}{dx} \right|_{x = z_{j;n}(t)} = \left(\tilde{\pi}_n' (z_{j;n} (t) , t) \right)^2 \delta_{jk}$$

such that

$$\int_a^b \frac{\widetilde{\pi}_n^2\,(x,t)}{x - z_{k;n}\,(t)}\,\frac{\partial w\,(x,t)}{\partial t}\,dx = \left(\widetilde{\pi}_n'\,(z_{k;n}\,(t)\,,t)\right)^2 z_{k;n}'\,(t)\,\lambda_{k;n}\,(t)$$

On the other hand,

$$\left(\widetilde{\pi}_n, \frac{\widetilde{\pi}_n}{x - z_{k;n}\,(t)}\right) = \int_a^b \frac{\widetilde{\pi}_n^2\,(x,t)}{x - z_{k;n}\,(t)}\,w\,(x,t)\,dx = 0$$

by orthogonality (**art. 353**). Subtraction from the previous integral yields

$$\left(\widetilde{\pi}_n'\,(z_{k;n}\,(t)\,,t)\right)^2 z_{k;n}'\,(t)\,\lambda_{k;n}\,(t) = \int_a^b \frac{\widetilde{\pi}_n^2\,(x,t)}{x - z_{k;n}\,(t)}\left\{\frac{\partial w\,(x,t)}{\partial t} - \xi w\,(x,t)\right\}\,dx$$

$$= \int_a^b \frac{\widetilde{\pi}_n^2\,(x,t)}{x - z_{k;n}\,(t)}\left\{\frac{\frac{\partial w(x,t)}{\partial t}}{w\,(x,t)} - \xi\right\}\,w\,(x,t)\,dx$$

If the constant ξ is chosen equal to $\xi = \left.\frac{1}{w(x,t)}\frac{\partial w(x,t)}{\partial t}\right|_{x=z_{k;n}(t)}$, then the function

$$\frac{\left\{\frac{1}{w(x,t)}\frac{\partial w(x,t)}{\partial t} - \left.\frac{1}{w(x,t)}\frac{\partial w(x,t)}{\partial t}\right|_{x=z_{k;n}(t)}\right\}}{x - z_{k;n}\,(t)} \geq 0$$

provided that $\frac{1}{w(x,t)}\frac{\partial w(x,t)}{\partial t}$ is increasing in x. In that case, the integral at the right-hand side is positive (because it cannot vanish at any point $x \in [a,b]$) and, hence, $z_{k;n}'\,(t) > 0$: the zero $z_{k;n}\,(t)$ of $\widetilde{\pi}_n\,(x,t)$ is increasing in the parameter t.

An interesting application is the choice $w\,(x,t) = (1-t)\,w_1\,(x) + t w_2\,(x)$, where w_1 and w_2 are two weight functions on $[a,b]$, both positive and continuous for $x \in (a,b)$. In addition,

$$\frac{1}{w\,(x,t)}\frac{\partial w\,(x,t)}{\partial t} = \frac{w_2\,(x) - w_1\,(x)}{(1-t)\,w_1\,(x) + t w_2\,(x)} = \frac{1}{t}\left(1 - \frac{1}{t\frac{w_2(x)}{w_1(x)} + 1 - t}\right)$$

is increasing if $\frac{w_2(x)}{w_1(x)}$ is increasing for $0 < t < 1$. Then, we have shown above that the zero $z_{k;n}\,(t)$ of $\widetilde{\pi}_n\,(x,t)$ is increasing in t. Let $\{z_{1;k;n}\}_{1\leq k\leq n}$ and $\{z_{2;k;n}\}_{1\leq k\leq n}$ denote the set of zeros of the orthogonal polynomials corresponding to w_1 and w_2, respectively. Thus, $z_{2;k;n} = z_{k;n}\,(1)$ is larger than $z_{1;k;n} = z_{k,n}\,(0)$ for all $1 \leq k \leq n$, because $w\,(x,0) = w_1\,(x)$ and $w\,(x,1) = w_2\,(x)$. In summary, if the ratio $\frac{w_2(x)}{w_1(x)}$ of two weight functions is increasing on $x \in [a,b]$, then the respective zeros obey $z_{2;k;n} > z_{1;k;n}$ for all $1 \leq k \leq n$.

369. *Numerical integration.* Let us consider the integral $\int_a^b f\,(x)\,dW\,(x)$, which we evaluate by the Gaussian quadrature formula (B.110) with remainder. The Christoffel numbers $\{\lambda_{j;n}\}_{1\leq j\leq n}$ and the zeros $\{z_{j;n}\}_{1\leq j\leq n}$ of the orthogonal polynomial $\pi_n\,(x)$ are independent of the function $f\,(x)$. Theorem 115 states that (B.110) is exact for any polynomial $f\,(x)$ of degree at most $2n - 1$. In fact, it can be shown (see Gautschi (2004)) that the Gaussian quadrature formula is the only interpolating quadrature rule with n function evaluations with the largest possible precision of $2n - 1$.

Since $dW(u) = du$ for Legendre polynomials $P_n(x)$, where $a = -1$ and $b = 1$, the most straightforward numerical computation of the integral $\int_a^b f(u)\,du$ uses Legendre's orthogonal polynomials. After substitution $u = \frac{b+a}{2} + \frac{b-a}{2}x$,

$$\int_a^b f(u)\,du = \frac{b-a}{2} \int_{-1}^1 f\left(\frac{b+a}{2} + \frac{b-a}{2}x\right) dx$$

Gauss's quadrature formula (B.110) gives us

$$\int_a^b f(u)\,du \simeq \frac{b-a}{2} \sum_{j=1}^n f\left(\frac{b+a}{2} + \frac{b-a}{2}\omega_{j;n}\right) \int_{-1}^1 \frac{P_n(x)\,dx}{(x - \omega_{j;n})\,P_n'(\omega_{j;n})}$$

where $\omega_{j;n}$ is the j-th zero of $P_n(z)$. For Chebyshev polynomials $T_n(x)$ studied in Section 12.7, the Gaussian quadrature formula (B.110) simplifies to

$$\int_{-1}^1 \frac{f(x)}{\sqrt{1-x^2}}\,dx = \int_0^\pi f(\cos\theta)\,d\theta = \frac{\pi}{n} \sum_{j=1}^n f\left(\cos\frac{(2j-1)\pi}{2n}\right) + \frac{\pi}{2^{2n-1}} \frac{f^{(2n)}(\xi)}{(2n)!}$$

because the zeros $z_{j;n}$ are given in (B.126), $c_{n;n}$ is specified in (B.124), $(T_n, T_n) = \frac{\pi}{2}$ in **art.** 381 and the Christoffel numbers (B.109) are all equal to $\lambda_{j;n} = \frac{\pi}{n}$. We refer to Lanczos (1988, p. 400-404) for a numerical example that illustrates the power of the Gaussian quadrature formula.

12.6 The Jacobi matrix

370. *The Jacobi matrix.* The three-term recursion (B.92) is written in matrix form by defining the vector $\tau(x) = \begin{bmatrix} \pi_0(x) & \pi_1(x) & \cdots & \pi_{n-1}(x) \end{bmatrix}^T$ as

$$x\begin{bmatrix} \pi_0(x) \\ \pi_1(x) \\ \vdots \\ \pi_{n-2}(x) \\ \pi_{n-1}(x) \end{bmatrix} = \begin{bmatrix} b_{0;1} & b_{1;1} & & & \\ b_{0;2} & b_{1;2} & b_{2;2} & & \\ & \ddots & \ddots & \ddots & \\ & & b_{n-3;n-1} & b_{n-2;n-1} & b_{n-1;n-1} \\ & & & b_{n-2;n} & b_{n-1;n} \end{bmatrix} \begin{bmatrix} \pi_0(x) \\ \pi_1(x) \\ \vdots \\ \pi_{n-2}(x) \\ \pi_{n-1}(x) \end{bmatrix} + \begin{bmatrix} 0 \\ 0 \\ \vdots \\ 0 \\ b_n\pi_n(x) \end{bmatrix}$$

Thus,

$$x\tau(x) = \Upsilon\tau(x) + b_n\pi_n(x)\,e_n \tag{B.115}$$

where the basic vector $e_n = \begin{bmatrix} 0 & 0 & \cdots & 0 & 1 \end{bmatrix}^T$ and the $n \times n$ matrix

$$\Upsilon = \begin{bmatrix} b_{0;1} & b_{1;1} & & & & \\ b_{0;2} & b_{1;2} & b_{2;2} & & & \\ & \ddots & \ddots & & \ddots & \\ & & b_{n-3;n-1} & b_{n-2;n-1} & b_{n-1;n-1} \\ & & & b_{n-2;n} & b_{n-1;n} \end{bmatrix}$$

We observe that, when $x = z_k$ is a zero of $\pi_n(x)$, then (B.115) reduces to the eigenvalue equation

$$\Upsilon \tau(z_k) = z_k \tau(z_k)$$

such that the zero z_k is an eigenvalue of Υ belonging to the eigenvector $\tau(z_k)$. This eigenvector is never equal to the zero vector because the first component $\pi_0(x) = c_{0;0} \neq 0$.

There must be a similarity transform to make the matrix Υ symmetric, since all zeros of $\pi_n(x)$ are real (Theorem 112). A similarity transform (**art. 239**) preserves the eigenvalues. The simplest similarity transform is $H = \mathrm{diag}(h_1, h_2, \ldots, h_n)$ such that

$$\widetilde{\Upsilon} = H \Upsilon H^{-1} = \begin{bmatrix} b_{0;1} & \frac{h_1}{h_2} b_{1;1} & & & & \\ \frac{h_2}{h_1} b_{0;2} & b_{1;2} & \frac{h_2}{h_3} b_{2;2} & & & \\ & \ddots & \ddots & & \ddots & \\ & & \frac{h_{n-1}}{h_{n-2}} b_{n-3;n-1} & b_{n-2;n-1} & \frac{h_{n-1}}{h_n} b_{n-1;n-1} \\ & & & \frac{h_n}{h_{n-1}} b_{n-2;n} & b_{n-1;n} \end{bmatrix}$$

In order to produce a symmetric tri-band matrix $\widetilde{\Upsilon} = \widetilde{\Upsilon}^T$, we need to require that $\left(\widetilde{\Upsilon}\right)_{i,i-1} = \left(\widetilde{\Upsilon}\right)_{i-1,i}$ for all $1 \leq i \leq n$, implying for $i \geq 2$ that $\frac{h_i}{h_{i-1}} b_{i-2;i} = \frac{h_{i-1}}{h_i} b_{i-1;i-1}$, whence $\left(\frac{h_i}{h_{i-1}}\right)^2 = \frac{b_{i-1;i-1}}{b_{i-2;i}}$. **Art. 356** shows that $b_{i-1;i-1}$ and $b_{i-2;i}$ have the same sign. Thus, $h_i = \sqrt{\frac{b_{i-1;i-1}}{b_{i-2;i}}} h_{i-1}$ for $1 \leq i \leq n$ and we can choose $h_1 = 1$ such that

$$h_i = \sqrt{\prod_{k=1}^{i-1} \frac{b_{k;k}}{b_{k-1;k+1}}}$$

The eigenvector belonging to the zero z_k equals $\widetilde{\tau}(z_k) = H\tau(z_k)$. After the similarity transform H, the resulting symmetric matrix $\widetilde{\Upsilon}$ is

$$\begin{bmatrix} b_{0;1} & \sqrt{b_{0;2}b_{1;1}} & & & & \\ \sqrt{b_{0;2}b_{1;1}} & b_{1;2} & \sqrt{b_{1;3}b_{2;2}} & & & \\ & \ddots & \ddots & & \ddots & \\ & & \sqrt{b_{n-3;n-1}b_{n-2;n-2}} & b_{n-2;n-1} & \sqrt{b_{n-2;n}b_{n-1;n-1}} \\ & & & \sqrt{b_{n-2;n}b_{n-1;n-1}} & b_{n-1;n} \end{bmatrix}$$

371. Similarly as in **art.** 370, the three-term recursion (B.97) of the normalized polynomials $\{\widetilde{\pi}_j(x)\}_{0 \le j \le n-1}$ is written in matrix form as

$$
x\begin{bmatrix} \widetilde{\pi}_0(x) \\ \widetilde{\pi}_1(x) \\ \vdots \\ \widetilde{\pi}_{n-2}(x) \\ \widetilde{\pi}_{n-1}(x) \end{bmatrix} = \begin{bmatrix} -\frac{B_1}{A_1} & \frac{1}{A_1} & & & \\ \frac{1}{A_1} & -\frac{B_2}{A_2} & \frac{1}{A_2} & & \\ & \ddots & \ddots & \ddots & \\ & & \frac{1}{A_{n-2}} & -\frac{B_{n-1}}{A_{n-1}} & \frac{1}{A_{n-1}} \\ & & & \frac{1}{A_{n-1}} & -\frac{B_n}{A_n} \end{bmatrix} \begin{bmatrix} \widetilde{\pi}_0(x) \\ \widetilde{\pi}_1(x) \\ \vdots \\ \widetilde{\pi}_{n-2}(x) \\ \widetilde{\pi}_{n-1}(x) \end{bmatrix} + \begin{bmatrix} 0 \\ 0 \\ \vdots \\ 0 \\ \frac{1}{A_n}\widetilde{\pi}_n(x) \end{bmatrix}
$$

Thus, in the normalized case where $C_n = \frac{A_n}{A_{n-1}}$, the matrix $\widetilde{\Upsilon}$ is symmetric,

$$
x\widetilde{\tau}(x) = \widetilde{\Upsilon}\widetilde{\tau}(x) + \frac{1}{A_n}\widetilde{\pi}_n(x)\,e_n
$$

where the vector $\widetilde{\tau}(x) = \operatorname{diag}\left(\|\pi_j\|^{-1}\right)\tau(x)$.

If there exist two different similarity transforms H_1 and H_2 that transform a matrix A into two different symmetric matrices, $B_1 = H_1 A H_1^{-1}$ and $B_2 = H_2 A H_2^{-1}$, then $H_1^T H_1 = H_2^T H_2$. Indeed, $A = H_1^{-1} B_1 H_1 = H_2^{-1} B_2 H_2$ from which $B_1 = H_1 H_2^{-1} B_2 H_2 H_1^{-1}$. Since $B_1 = B_1^T$ and $B_2 = B_2^T$, we have that $B_1 = \left(H_2 H_1^{-1}\right)^T B_2 \left(H_1 H_2^{-1}\right)^T$. Hence, $H_1 H_2^{-1} = \left(H_2 H_1^{-1}\right)^T$ and $H_2 H_1^{-1} = \left(H_1 H_2^{-1}\right)^T$, which lead to $H_1^T H_1 = H_2^T H_2$. If H_1 and H_2 are, in addition, also symmetric as in the case of a diagonal matrix, then $H_1^2 = H_2^2$ or $H_1 = \pm H_2$. Since $\widetilde{\pi}_j(x) = \pi_j(x)\|\pi_j\|^{-1}$, both similarity transforms H_1 and H_2 must be the same. This implies that $H = \operatorname{diag}(\sqrt{b_j}) = \operatorname{diag}\left(\|\pi_j\|^{-1}\right)$, thus $b_j = \|\pi_j\|^{-2} = \frac{1}{(\pi_j, \pi_j)}$. In addition, in agreement with **art.** 356, we have

$$
b_{j-1;j} = -\frac{B_j}{A_j}
$$

$$
\sqrt{b_{j-1;j+1}b_{j;j}} = \frac{1}{A_j}
$$

Hence, transforming Υ by a similarity transform H to a symmetric matrix $\widetilde{\Upsilon}$ corresponds to normalizing the orthogonal polynomials.

372. Gerschgorin's Theorem 65 tells us that there lies a zero z_k of $\pi_n(x)$ in a disk centered around $b_{j-1;j} = -\frac{B_j}{A_j}$ with radius $\frac{1}{A_{j-1}} + \frac{1}{A_j}$. Overall, the symmetric matrix $\widetilde{\Upsilon}$ leads to the sharpest bounds on the eigenvalues/zeros of $\pi_n(x)$ because the above similarity transform H minimizes the off-diagonal elements. However, not always. In particular, ignoring the attempt to symmetrize Υ, we may choose h_1 and h_n in such a way that $\left(H\Upsilon H^{-1}\right)_{12}$ and $\left(H\Upsilon H^{-1}\right)_{n;n-1}$ are arbitrarily small but not zero. But, by making h_1 and h_n very small, we increase the radius around $b_{0;1}$ and $b_{n-1;n}$. Gerschgorin's Theorem 65 indicates that there is a zero z_k close to $b_{0;1}$ and another zero close to $b_{n-1;n}$.

373. *Continued fraction associated to orthogonal polynomials.* By systematic row multiplication and subtraction from the next one, we can eliminate the lower diagonal elements $\widetilde{\Upsilon}_{j-1;j}$ in the determinant $\det\left(\widetilde{\Upsilon} - xI\right)$, which eventually results in a continued fraction expansion of $W_n = \det\left(\widetilde{\Upsilon} - xI\right) = \left|\widetilde{\Upsilon} - xI\right|$,

$$
W_n = \begin{vmatrix}
-\frac{B_1}{A_1} - x & \frac{1}{A_1} & & & \\
\frac{1}{A_1} & -\frac{B_2}{A_2} - x & \frac{1}{A_2} & & \\
& \ddots & \ddots & & \ddots \\
& & \frac{1}{A_{n-2}} & -\frac{B_{n-1}}{A_{n-1}} - x & \frac{1}{A_{n-1}} \\
& & & \frac{1}{A_{n-1}} & -\frac{B_n}{A_n} - x
\end{vmatrix}
$$

We write the determinant W_n in block form,

$$
W_n = \begin{vmatrix}
-\frac{A_1 x + B_1}{A_1} & \frac{1}{A_1} e_1^T \\
\frac{1}{A_1} e_1 & W_{1;n}
\end{vmatrix}
$$

where the basis vector is $e_1 = (1, 0, \ldots)$ and where the matrix $W_{1;n}$ is obtained by deleting the first row and the first column in $\widetilde{\Upsilon} - xI$,

$$
W_{1;n} = \begin{bmatrix}
-\frac{A_2 x + B_2}{A_2} & \frac{1}{A_2} & & & \\
\frac{1}{A_2} & -\frac{A_3 x + B_3}{A_3} & \frac{1}{A_3} & & \\
& \ddots & \ddots & & \ddots \\
& & \frac{1}{A_{n-2}} & -\frac{A_{n-1} x + B_{n-1}}{A_{n-1}} & \frac{1}{A_{n-1}} \\
& & & \frac{1}{A_{n-1}} & -\frac{A_n x + B_n}{A_n}
\end{bmatrix}
$$

Invoking the Schur complement (A.57) yields

$$
W_n = -\frac{A_1 x + B_1}{A_1} \left| W_{1;n} + \frac{1}{A_1} \frac{e_1 e_1^T}{A_1 x + B_1} \right|
$$

and $e_1 e_1^T = \widehat{O}$ equals the zero matrix with same dimensions as W_{n-1}, except for the element $\widehat{O}_{11} = 1$. Thus,

$$
W_n = -\frac{A_1 x + B_1}{A_1} \begin{vmatrix}
-\frac{A_2 x + B_2}{A_2} + \frac{1}{A_1} \frac{1}{A_1 x + B_1} & \frac{1}{A_2} e_1^T \\
\frac{1}{A_2} e_1 & W_{2;n}
\end{vmatrix}
$$

where we denote by $W_{j;n}$ the matrix obtained by deleting the first j rows and the first j columns in $\widetilde{\Upsilon} - xI$. Again invoking (A.57) yields, with $C_n = \frac{A_n}{A_{n-1}}$,

$$
W_n = \frac{A_1 x + B_1}{A_1 A_2} \left(A_2 x + B_2 - \frac{C_2}{A_1 x + B_1} \right) \begin{vmatrix}
\frac{1}{A_2} \frac{1}{A_2 x + B_2 - \frac{C_2}{A_1 x + B_1}} - \frac{A_3 x + B_3}{A_3} & \frac{1}{A_3} e_1^T \\
\frac{1}{A_3} e_1 & W_{3;n}
\end{vmatrix}
$$

The next iteration

$$W_n = -\frac{A_1 x + B_1}{A_1 A_2 A_3}\left((A_2 x + B_2) - \frac{C_2}{A_1 x + B_1}\right)\left((A_3 x + B_3) - \frac{C_3}{A_2 x + B_2 - \frac{C_2}{A_1 x + B_1}}\right)$$

$$\times \begin{vmatrix} -\frac{A_4 x + B_4}{A_4} + \frac{1}{A_3} & \frac{1}{A_3 x + B_3 - \frac{C_3}{A_2 x + B_2 - \frac{C_2}{A_1 x + B_1}}} & \frac{1}{A_4} e_1^T \\ \frac{1}{A_4} e_1 & & W_{4;n} \end{vmatrix}$$

reveals the structure $W_n = \frac{(-1)^n}{\prod_{k=1}^n A_k}\prod_{k=1}^n \theta_k(x)$, where the continued fraction $\theta_k(x)$ equals

$$\theta_k(x) = A_k x + B_k - \cfrac{C_k}{A_{k-1} x + B_{k-1} - \cfrac{C_{k-1}}{A_{k-2} x + B_{k-2} - \cfrac{C_{k-2}}{\ddots - \cfrac{C_2}{A_2 x + B_2 - \frac{C_2}{A_1 x + B_1}}}}} \tag{B.116}$$

The continued fraction thus satisfies the recursion[4]

$$\theta_k(x) = A_k x + B_k - \frac{C_k}{\theta_{k-1}(x)} \tag{B.117}$$

Art. 370 shows that the characteristic polynomial $W_n = \det\left(\widetilde{\Upsilon} - xI\right)$ has the same zeros as $\pi_n(x)$, such that $W_n = \frac{(-1)^n}{c_{n;n}}\pi_n(x)$, and

$$\pi_n(x) = \frac{c_{n;n}}{\prod_{k=1}^n A_k}\prod_{k=1}^n \theta_k(x)$$

from which,

$$\theta_n(x) = \frac{A_n c_{n-1;n-1}}{c_{n;n}}\frac{\pi_n(x)}{\pi_{n-1}(x)} = \frac{\widetilde{\pi}_n(x)}{\widetilde{\pi}_{n-1}(x)}$$

Introducing $\theta_n(x) = \frac{\widetilde{\pi}_n(x)}{\widetilde{\pi}_{n-1}(x)}$ into the recursion (B.117) again leads to the normalized three-term recursion (B.97).

More results on continued fractions are presented in Gautschi (2004) and in Chihara (1978).

---

[4] In most textbooks, a finite continued fraction is written in a differently labeled form as

$$\theta_n = a_0 - \cfrac{b_1}{a_1 - \cfrac{b_2}{a_2 - \cfrac{b_3}{\ddots - \cfrac{b_n}{a_n}}}}$$

from which the recursive structure is less naturally observed. If the determinant T_n is expanded by the last row and last column, up to the first one, a same labeling would have been found. The main purpose in classical treatment to use the highest index in the deepest fraction is to study the convergence of $\lim_{n\to\infty}\theta_n$.

374. If $\widetilde{\Upsilon}$ is positive semidefinite, then $\widetilde{\Upsilon}$ can be considered as a Gram matrix (art. 280), i.e. $\widetilde{\Upsilon} = A^T A$ where $\widetilde{\Upsilon}_{i,i} \geq 0$. **Art.** 370 demonstrates that $\widetilde{\Upsilon}$ is positive semidefinite if all zeros of the orthogonal polynomials are non-negative. Theorem 112 guarantees semidefiniteness when the orthogonality interval $[a, b]$ lies on the non-negative real axis, i.e., if $b > a \geq 0$.

Since $\widetilde{\Upsilon}$ is a three-band matrix, A is a two-band matrix with diagonal elements $A_{jj} = a_j$ for $1 \leq j \leq n$ and upper diagonal elements $A_{j,j+1} = b_j$ for $1 \leq j \leq n-1$. Indeed,

$$\widetilde{\Upsilon}_{ij} = \sum_{k=1}^{n} \left(A^T\right)_{ik} A_{kj} = \sum_{k=1}^{n} A_{ki} A_{kj}$$
$$= A_{ii} A_{ij} + A_{i-1,i} A_{i-1,j} = a_i A_{ij} + b_{i-1} A_{i-1,j}$$

and

$$\widetilde{\Upsilon}_{ij} = \begin{cases} a_{i-1} b_{i-1} & \text{if } j = i-1 \\ a_i^2 + b_{i-1}^2 & \text{if } j = i \\ a_i b_i & \text{if } j = i+1 \end{cases}$$

Hence, if $i = 1$, comparison shows that $\widetilde{\Upsilon}_{11} = a_1^2$ and $\widetilde{\Upsilon}_{12} = a_1 b_1$ such that $a_1 = \sqrt{-\frac{B_1}{A_1}}$ and $b_1 = \frac{1}{\sqrt{-B_1 A_1}}$. For the i-th row, we find the equations

$$\begin{cases} \widetilde{\Upsilon}_{i,i-1} = a_{i-1} b_{i-1} = \frac{1}{A_{i-1}} \\ \widetilde{\Upsilon}_{i,i} = a_i^2 + b_{i-1}^2 = -\frac{B_i}{A_i} \\ \widetilde{\Upsilon}_{i,i+1} = a_i b_i = \frac{1}{A_i} \end{cases}$$

whose solution, by iteration from $i = 1$, is a continued fraction

$$a_j^2 = -\frac{B_j}{A_j} + \cfrac{1}{A_{j-1}B_{j-1} - \cfrac{A_{j-1}^2}{A_{j-2}B_{j-2} - \cfrac{A_{j-2}^2}{A_{j-3}B_{j-3} - \cfrac{A_{j-3}^2}{\ddots \quad \cfrac{\ddots}{A_2 B_2 - \frac{A_2^2}{A_1 B_1}}}}}}$$

$$b_j^2 = \cfrac{1}{-A_j B_j + \cfrac{A_j^2}{A_{j-1}B_{j-1} - \cfrac{A_{j-1}^2}{A_{j-2}B_{j-2} - \cfrac{A_{j-2}^2}{A_{j-3}B_{j-3} - \cfrac{A_{j-3}^2}{\ddots \quad \cfrac{\ddots}{A_2 B_2 - \frac{A_2^2}{A_1 B_1}}}}}}}$$

satisfying the recursion $a_j^2 = -\frac{B_j}{A_j} - \frac{1}{A_{j-1}^2 a_{j-1}^2}$. Either the positive square root $\left(\sqrt{a_j^2}, \sqrt{b_j^2}\right)$ or the negative square root $\left(-\sqrt{a_j^2}, -\sqrt{b_j^2}\right)$ are solutions. By comparison with the continued fraction (B.116) where $C_n = \frac{A_n}{A_{n-1}}$, we verify that

$$a_j^2 = -\frac{\theta_j(0)}{A_j} \quad \text{and} \quad b_j^2 = -\frac{1}{A_j \theta_j(0)}$$

In summary, the matrix A, which satisfies $\widetilde{\Upsilon} = A^T A$, is

$$A = \begin{bmatrix} a_1 & b_1 & & & & \\ 0 & a_2 & b_2 & & & \\ & \ddots & \ddots & & \ddots & \\ & & 0 & a_{n-1} & b_{n-1} \\ & & & 0 & a_n \end{bmatrix}$$

The eigenvalues of A are its diagonal elements a_i. The eigenvector x_i of A belonging to $\lambda = a_i$ can be written explicitly: just write out $Ax_i = a_i x_i$, starting with the last component $(x_i)_n = 1_{\{i=n\}}$, and iterate upwards. Thus, $A = X \text{diag}(\lambda(A)) X^{-1}$ can be explicitly written, where X is the matrix with its eigenvectors as columns, and

$$\widetilde{\Upsilon} = A^T A = \left(X^{-1}\right)^T \text{diag}\left(\lambda(A)\right) X^T X \text{diag}\left(\lambda(A)\right) X^{-1}$$

After eigenvalue decomposition, the symmetric matrix

$$\widetilde{\Upsilon} = U \text{diag}\left(\lambda_k\left(\widetilde{\Upsilon}\right)\right) U^{-1}$$

where $U^T = U^{-1}$ is an orthogonal matrix (**art. 247**). The latter is a property of symmetric matrices and does not hold in general. Hence, X is not necessarily orthogonal, although the eigenvectors $x_1, x_2, \ldots, x_n$ of A are linearly independent. Since $\widetilde{\Upsilon}$ is positive semidefinite, $\lambda_k\left(\widetilde{\Upsilon}\right) \geq 0$ and, thus $\sqrt{\lambda_k\left(\widetilde{\Upsilon}\right)}$ is real such that

$$\widetilde{\Upsilon} = U \text{diag}\left(\lambda_k\left(\widetilde{\Upsilon}\right)\right) U^{-1} = U \text{diag}\left(\sqrt{\lambda_k(\widetilde{\Upsilon})}\right) Y^T Y \text{diag}\left(\sqrt{\lambda_k(\widetilde{\Upsilon})}\right) U^T$$

$$= U \text{diag}\left(\sqrt{\lambda_k(\widetilde{\Upsilon})}\right) Y^T \left(U \text{diag}\left(\sqrt{\lambda_k(\widetilde{\Upsilon})}\right) Y^T\right)^T$$

where Y is an orthogonal matrix. Hence, we can construct the matrix $A = Y \text{diag}\left(\sqrt{\lambda_k(\widetilde{\Upsilon})}\right) U^T$, which is a singular value decomposition[5]. Obviously, the simplest choice is $Y = I$, in which case, $A = \text{diag}\left(\sqrt{\lambda_k(\widetilde{\Upsilon})}\right) U^T$. However, multiplication by a diagonal matrix only multiplies row j in U^T by $\sqrt{\lambda_j(\widetilde{\Upsilon})}$ and the resulting structure should be the two-band structure of A. Since the two-band structure of A is not orthogonal (the column vectors in A are not orthogonal), $Y = I$

---

[5] Although the singular values are unique, the singular vectors are not, and, hence $A = U \Sigma V^T$ is not unique.

is not a correct choice. Also, $Y \neq U$, because A is not symmetric. Applying QR-decomposition (see, e.g., Golub and Van Loan (1996)) to $A = X\text{diag}(\lambda(A))X^{-1}$ with $X = QR$ and $X^{-1} = R_1 Q_1^T$ yields

$$A = QR\text{diag}(\lambda(A))R_1 Q_1^T$$

Since $A = Y\text{diag}\left(\sqrt{\lambda_k(\widetilde{\Upsilon})}\right)U^T$, it remains to show that $R\text{diag}(\lambda(A))R_1$ is a diagonal matrix. Unfortunately, the major difficulty is to find an orthogonalization process for the eigenvectors X such that $A = X\text{diag}(\lambda(A))X^{-1}$ has a singular value decomposition $A = Y\text{diag}\left(\sqrt{\lambda_k(\widetilde{\Upsilon})}\right)U^T$. There does not seem to exist a general method to achieve this result. If it existed, we would have, at least for the class of orthogonal polynomials with zeros on the positive real axis, a general method to compute the exact zeros!

12.7 Chebyshev polynomials

Instead of the Legendre, Hermite or Jacobi polynomials, we have chosen the Chebyshev polynomials to exemplify an orthogonal set of polynomials, because Chebyshev polynomials appear in the spectrum of the small-world graph (Section 6.2.2), the cycle (Section 6.3) and the path (Section 6.4).

375. *Definition.* The Chebyshev polynomial of degree n is defined by

$$T_n(x) = \cos(n\theta) \quad \text{with } x = \cos\theta \tag{B.118}$$

For real θ, $\cos\theta$ ranges between -1 and 1 and (B.118) defines the Chebyshev polynomial $T_n(x)$ for x in the interval $[-1,1]$. The compact definition for $x \in [-1,1]$, corresponding to $0 \leq \theta \leq \pi$, is

$$T_n(x) = \cos(n\arccos x)$$

For complex $\theta = iy$, then $\cos\theta = \cosh y$ which is larger than 1 for real $y \neq 0$ and the corresponding compact definition for $x > 1$ is

$$T_n(x) = \cosh(n\,\text{arccosh}x)$$

A direct consequence of the compact definition is

$$T_n(T_m(x)) = \cos(n\arccos\cos(m\arccos x)) = \cos(nm\arccos x)$$

demonstrating the semi-group property or commutativity under composition of the Chebyshev polynomials,

$$T_n(T_m(x)) = T_{mn}(x) \tag{B.119}$$

Rivlin (1974, p. 161) shows that no other polynomial than T_k itself can commute with T_n if $n \geq 2$. Among all polynomials $p_n(x)$, only the powers of x^n and $T_n(x)$ obey $p_n(p_m(x)) = p_{mn}(x)$.

376. *Polynomial form for* $T_n(x)$. The Taylor expansion of the Chebyshev polynomial $T_n(x)$ around x_0 is

$$T_n(x) = \sum_{k=0}^{n} t_{k;n}(x_0)(x - x_0)^k \tag{B.120}$$

Writing $x - x_0 = (x - x_1) + (x_1 - x_0)$, substituting the binomial series in (B.120), reversing the summations and equating corresponding powers in $x - x_0$ lead to

$$t_{k;n}(x_1) = \sum_{k=j}^{n} t_{k;n}(x_0)\binom{k}{j}(x_1 - x_0)^{k-j}$$

with the obvious inverse after replacing x_1 and x_2. Choosing $x_0 = 0$ thus expresses the Taylor coefficients $t_{k;n}(x_1)$ around x_1 in terms of the Taylor coefficients $t_{k;n}(0)$ around $x_0 = 0$. The Taylor coefficients $t_{k;n}(0)$ of $T_n(x)$ around $x_0 = 0$ in (B.120) are elegantly derived from Euler's formula $e^{i\theta} = \cos\theta + i\sin\theta$. Indeed, from

$$e^{in\theta} = \cos n\theta + i\sin n\theta = (\cos\theta + i\sin\theta)^n$$

the binomial expansion $(\cos\theta + i\sin\theta)^n = \sum_{k=0}^{n}\binom{n}{k}i^k\cos^{n-k}\theta\sin^k\theta$ is split in even and odd powers of i^k using the general formula

$$\sum_{k=1}^{n} f(k) = \sum_{k=1}^{\left[\frac{n}{2}\right]} f(2k) + \sum_{k=1}^{\left[\frac{n+1}{2}\right]} f(2k-1) \tag{B.121}$$

as $e^{in\theta} = \cos^n\theta + \sum_{k=1}^{\left[\frac{n}{2}\right]}\binom{n}{2k}\frac{\cos^{n-2k}\theta\sin^{2k}\theta}{i^{-2k}} + \sum_{k=1}^{\left[\frac{n+1}{2}\right]}\binom{n}{2k-1}\frac{\cos^{n-2k+1}\theta\sin^{2k-1}\theta}{i^{1-2k}}$. Equating the real and imaginary part of both sides yields

$$\begin{cases} \cos n\theta = \cos^n\theta + \sum_{k=1}^{\left[\frac{n}{2}\right]}\binom{n}{2k}(-1)^k\cos^{n-2k}\theta\sin^{2k}\theta \\ \frac{\sin n\theta}{\sin\theta} = \frac{1}{\sin^2\theta}\sum_{k=1}^{\left[\frac{n+1}{2}\right]}\binom{n}{2k-1}(-1)^{k-1}\cos^{n-2k+1}\theta\sin^{2k}\theta \end{cases}$$

Only even powers of $\sin\theta$ occur. With $\sin^{2k}\theta = \left(1 - \cos^2\theta\right)^k = \sum_{j=0}^{k}\binom{k}{j}(-1)^j\cos^{2j}\theta$, we obtain $\cos n\theta = \sum_{k=0}^{\left[\frac{n}{2}\right]}\sum_{j=0}^{k}\binom{k}{j}\binom{n}{2k}(-1)^{k+j}\cos^{n-2(k-j)}\theta$ and, after reversing the sums, $\cos n\theta = \sum_{j=0}^{\left[\frac{n}{2}\right]}\sum_{k=j}^{\left[\frac{n}{2}\right]}\binom{k}{j}\binom{n}{2k}(-1)^{k+j}\cos^{n-2(k-j)}\theta$. Let $q = k - j$, then $0 \le q \le \left[\frac{n}{2}\right]$. Moreover, $j = k - q \ge 0$ and $j \le k$, so that $k \ge q$, while $k \le \left[\frac{n}{2}\right]$. Hence, we arrive at

$$\cos n\theta = \sum_{q=0}^{\left[\frac{n}{2}\right]}(-1)^q\left(\sum_{k=q}^{\left[\frac{n}{2}\right]}\binom{k}{q}\binom{n}{2k}\right)\cos^{n-2q}\theta$$

The $\frac{\sin n\theta}{\sin\theta}$ counterpart can be treated similarly. However, we will later in **art. 378** see a more convenient way. With the definition (B.118), we find the Taylor expansion around $x_0 = 0$,

$$T_n(x) = \sum_{q=0}^{\left[\frac{n}{2}\right]} (-1)^q \left(\sum_{k=q}^{\left[\frac{n}{2}\right]} \binom{k}{q} \binom{n}{2k} \right) x^{n-2q} \tag{B.122}$$

which demonstrates that $T_n(x) = \sum_{q=0}^{n} t_{q;n}(0) x^q$ is a polynomial of degree n in x, valid for any complex number x, with coefficients

$$t_{n-2q;n}(0) = (-1)^q \sum_{k=q}^{\left[\frac{n}{2}\right]} \binom{k}{q} \binom{n}{2k} \tag{B.123}$$

Since (B.122) only contains powers of x^{n-2q}, we observe that

$$T_n(-x) = (-1)^n T_n(x)$$

The Chebyshev polynomials $T_n(x)$ for first few degrees n are $T_0(x) = 1$ and

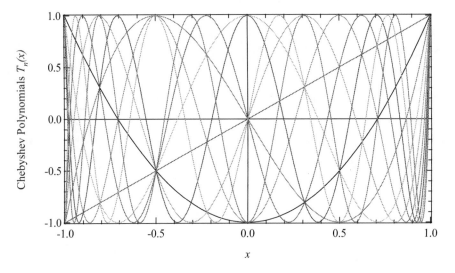

Fig. 12.1. The Chebyshev polynomials $T_n(x)$ for $n = 1, 2, \ldots 10$ in the interval $[-1, 1]$.

$$
\begin{array}{ll}
T_1(x) = x & T_4(x) = 8x^4 - 8x^2 + 1 \\
T_2(x) = 2x^2 - 1 & T_5(x) = 16x^5 - 20x^3 + 5x \\
T_3(x) = 4x^3 - 3x & T_6(x) = 32x^6 - 48x^4 + 18x^2 - 1
\end{array}
$$

The coefficient $t_{n;n}(0)$ of x^n in (B.123) equals, for $n > 0$,

$$t_{n;n}(0) = \sum_{k=0}^{\left[\frac{n}{2}\right]} \binom{k}{0} \binom{n}{2k} = \sum_{k=0}^{\left[\frac{n}{2}\right]} \binom{n}{2k}$$

The binomial sum $(1+x)^n = \sum_{k=0}^{n} \binom{n}{k} x^k$ indicates that

$$(1+x)^n + (1-x)^n = \sum_{k=0}^{n} \binom{n}{k} \left(1+(-1)^k\right) x^k = 2 \sum_{k=0}^{\left[\frac{n}{2}\right]} \binom{n}{2k} x^{2k}$$

from which the coefficient $c_{n;n} = t_{n;n}(0)$ of the highest power in x in (B.81) equals

$$t_{n;n}(0) = 2^{n-1} \tag{B.124}$$

377. *Closed form for $T_n(x)$.* A closed form for $T_n(x)$ follows from $\cos\theta = \frac{e^{i\theta}+e^{-i\theta}}{2}$ and Euler's formula as

$$\cos n\theta = \frac{1}{2} \left\{ (\cos\theta + i\sin\theta)^n + (\cos\theta - i\sin\theta)^n \right\}$$
$$= \frac{1}{2} \left\{ \left(\cos\theta + i\sqrt{1-\cos^2\theta}\right)^n + \left(\cos\theta - i\sqrt{1-\cos^2\theta}\right)^n \right\}$$

For real θ, it holds that $x = \cos\theta$ is in absolute value smaller than or equal to 1, i.e. $|x| \le 1$ and the definition (B.118) of $T_n(x)$ shows that

$$T_n(x) = \frac{1}{2} \left\{ \left(x + i\sqrt{1-x^2}\right)^n + \left(x - i\sqrt{1-x^2}\right)^n \right\}$$

which can be written, for $|x| < 1$, as

$$T_n(x) = \frac{1}{2} \left\{ \left(x + \sqrt{x^2-1}\right)^n + \left(x - \sqrt{x^2-1}\right)^n \right\} \tag{B.125}$$

The closed form (B.125) straightforwardly extends to $|x| > 1$ as well and illustrates that $T_n(1) = 1$, $T_n(-1) = (-1)^n$ and $T_n(0) = \frac{e^{i\frac{\pi}{2}n} + e^{-i\frac{\pi}{2}n}}{2} = \cos\frac{\pi n}{2} = (-1)^n$.

378. *Zeros and extrema and Chebyshev polynomial $U_n(x)$.* The definition (B.118) with $0 \le \theta \le \pi$ shows that $\cos n\theta = 0$ for $\theta = \frac{(2k-1)\pi}{2n}$ for $k = 1, 2, \ldots, n$, so that the zeros $z_{1;n} > z_{2;n} > \ldots > z_{n;n}$ of $T_n(x)$ are

$$z_{k;n} = \cos\frac{(2k-1)\pi}{2n} \quad \text{for } 1 \le k \le n \tag{B.126}$$

which indicates that all zeros are real, different and lying in the interval $(-1, 1)$. With $t_{n;n}(0) = 2^{n-1}$ in (B.124) and $T_n(x) = \sum_{q=0}^{n} t_{q;n}(0) x^q$, the product form in (B.1) becomes

$$T_n(x) = 2^{n-1} \prod_{m=1}^{n} \left(x - \cos\left(\frac{\pi(2m-1)}{2n}\right) \right) \tag{B.127}$$

The zeros of $\cos(n \arccos z) - 1 = 0$ are $z_m = \cos\left(\frac{2m\pi}{n}\right)$, for $m = 1, \ldots, n$, so that the definition $T_n(x) = \cos(n \arccos x)$ and (B.1) leads to an alternative product form

$$T_n(x) - 1 = 2^{n-1} \prod_{m=1}^{n} \left(x - \cos\left(\frac{2m\pi}{n}\right) \right) \tag{B.128}$$

Similarly, the extrema of $\cos n\theta$ occur at $n\theta = k\pi$, for $k = 0, 1, \ldots, n$ so that the extrema of $T_n(x)$, obeying $T_n(\eta_{k;n}) = (-1)^k$, are

$$\eta_{k;n} = \cos \frac{k\pi}{n} \text{ for } 0 \leq k \leq n$$

which also lie inside the interval $(-1, 1)$, except for $\eta_{0;n} = 1$ and $\eta_{n;n} = -1$. Those extrema of $T_n(x)$ inside the interval $(-1, 1)$ are the zeros of $T'_n(x)$, which is a polynomial of degree $n - 1$. By differentiating (B.118) with respect to $x = \cos\theta$,

$$T'_n(x) = \frac{d}{d\theta}(\cos n\theta) \frac{d\theta}{dx} = -n\sin n\theta \left(\frac{1}{-\sin\theta}\right) = n\frac{\sin n\theta}{\sin\theta} \qquad \text{(B.129)}$$

and the polynomial of degree $n - 1$, with zeros $\eta_{k;n} = \cos\frac{k\pi}{n}$ for $1 \leq k \leq n - 1$,

$$U_{n-1}(x) = \frac{1}{n}T'_n(x) = \frac{\sin n\theta}{\sin\theta} \qquad \text{(B.130)}$$

is called the *Chebyshev polynomial of the second kind*.

379. *Differential equations of $T_n(x)$.* Differentiation of (B.129) yields

$$T''_n(x) = n\frac{d}{d\theta}\left(\frac{\sin n\theta}{\sin\theta}\right)\frac{d\theta}{dx} = -\frac{n}{\sin^2\theta}\left(n\cos n\theta - \cos\theta\frac{\sin n\theta}{\sin\theta}\right)$$

from which we deduce with (B.129) that $T_n(x)$ satisfies the second-order linear differential equation

$$\left(1 - x^2\right)T''_n(x) - xT'_n(x) + n^2 T_n(x) = 0 \qquad \text{(B.131)}$$

Remarkably, we can integrate this second-order linear differential equation once. After multiplying both sides in (B.131) by $T'_n(x)$,

$$\left(1 - x^2\right)T''_n(x)T'_n(x) = x\left(T'_n(x)\right)^2 - n^2 T_n(x)T'_n(x)$$

and substituting $T''_n(x)T'_n(x) = \frac{1}{2}\frac{d}{dx}\left(T'_n(x)\right)^2$ and $T_n(x)T'_n(x) = \frac{1}{2}\frac{d}{dx}\left(T_n(x)\right)^2$, we obtain

$$\left(1 - x^2\right)\frac{d}{dx}\left(T'_n(x)\right)^2 = 2x\left(T'_n(x)\right)^2 - n^2\frac{d}{dx}\left(T_n(x)\right)^2$$

Let $y = T_n(x)$, then $\left(1 - x^2\right)\frac{d}{dx}(y')^2 - 2x(y')^2 = -n^2\frac{d}{dx}(y^2)$ and observing that $\frac{d}{dx}\left\{\left(1 - x^2\right)(y')^2\right\} = \left(1 - x^2\right)\frac{d}{dx}\left((y')^2\right) - 2x\left((y')^2\right)$ leads to

$$\frac{d}{dx}\left\{\left(1 - x^2\right)(y')^2\right\} = -n^2\frac{d}{dx}(y^2)$$

After integrating both sides

$$\left(1 - x^2\right)(y')^2 = -n^2 y^2 + c$$

and using $T_n(1) = 1$, we find that the constant of integration is $c = n^2$. In summary, the Chebyshev polynomials $T_n(x)$ also obey the first order non-linear differential equation

$$\left(1 - x^2\right)\left(T'_n(x)\right)^2 = n^2\left(1 - T_n^2(x)\right) \qquad \text{(B.132)}$$

Introducing the definition (B.130) of the Chebyshev polynomial of the second kind $U_n(x)$, the first order non-linear differential equation (B.132) becomes

$$T_n^2(x) - (x^2 - 1) U_{n-1}^2(x) = 1$$

where we recognize the famous Pell diophantine equation $z^2 - my^2 = 1$ in the unknown integers z and y, given the integer m. Hence, the integer $z = T_n(x)$ and $y = U_{n-1}(x)$, given that $m = x^2 - 1$ is an integer, solves the Pell equation in number theory. On the other hand, standard solution techniques for the Pell equation can generate integer solutions for the pair $(T_n(x), U_{n-1}(x))$.

380. Coefficients of $T_n(x)$. We deduce a recursion for the Taylor coefficients $t_{q;n}(0)$ around $x_0 = 0$ in (B.120) from the second-order linear differential equation (B.131) in **art. 379**. By substitution of Taylor expansion (B.120) into the differential equation (B.131) and simplifying $t_{q;n}(0)$ by t_q, we obtain

$$\left(1 - x^2\right) \sum_{q=0}^{n} q(q-1) t_q x^{q-2} - x \sum_{q=0}^{n} q t_q x^{q-1} + n^2 \sum_{q=0}^{n} t_q x^q = 0$$

or

$$\sum_{q=0}^{n-2} (q+2)(q+1) t_{q+2} x^q + \sum_{q=0}^{n} \left(n^2 - q^2\right) t_q x^q = 0$$

Equating corresponding powers of x yields,

$$(q+2)(q+1) t_{q+2} + \left(n^2 - q^2\right) t_q = 0 \quad \text{for } 0 \le q \le n-2$$

and $\left(n^2 - (n-1)^2\right) t_{n-1} = 0$. Hence, for $0 \le q \le n-2$, we find the recursion

$$t_{q+2} = -\frac{(n-q)(n+q)}{(q+2)(q+1)} t_q$$

and $t_{n-1} = 0$. The recursion illustrates that all odd coefficients $t_{2j-1} = 0$. After iterating the recursion p times, we have

$$t_q = (-1)^p \frac{(q-2p)!\,(n+q-2)\ldots(n+q-2p)(n-q+2)\ldots(n-q+2p)}{q!} t_{q-2p}$$

With $t_n = 2^{n-1}$ in (B.124), we find with $q = n$ that

$$t_{n-2p} = (-1)^p \frac{n!}{(n-2p)!\,(2n-2)(2n-4)\ldots(2n-2p)\,2.4\ldots 2p} 2^{n-1}$$

$$= n(-1)^p \frac{(n-p-1)!}{(n-2p)!p!} 2^{n-2p-1} = (-1)^p \frac{n}{n-p} \frac{(n-p)!}{(n-2p)!p!} 2^{n-2p-1}$$

In conclusion, for $p = 0, 1, \ldots, \left[\frac{n}{2}\right]$, the differential equation (B.131) leads to the more concise[6] Chebyshev coefficient of x^{n-2p}

$$t_{n-2p;n}(0) = (-1)^p \frac{n}{n-p} \binom{n-p}{p} 2^{n-2p-1} \tag{B.133}$$

and the Taylor expansion (B.120) around $x_0 = 0$ is

$$T_n(x) = \cos(n \arccos x) = \frac{1}{2} \sum_{k=0}^{\left[\frac{n}{2}\right]} (-1)^k \frac{n}{n-k} \binom{n-k}{k} (2x)^{n-2k} \tag{B.134}$$

The Chebyshev polynomial of the second kind, defined by (B.130), is

$$U_n(x) = \frac{\sin((n+1)\arccos x)}{\sin \arccos x} = \sum_{k=0}^{\left[\frac{n}{2}\right]} (-1)^k \frac{(n-k)!}{k!(n-2k)!} (2x)^{n-2k} \tag{B.135}$$

and has zeros at $z_m = \cos \frac{m\pi}{n+1}$ for $m = 1, \ldots, n$ such that

$$U_n(x) = 2^n \prod_{m=1}^{n} \left(x - \cos \frac{m\pi}{n+1} \right) \tag{B.136}$$

381. *Orthogonality and three-term recursion.* The well-known orthogonality property (B.80) in the theory of Fourier series is

$$\int_0^\pi \cos m\theta \cos k\theta d\theta = 0 \quad \text{for } k \neq m$$

and

$$\int_0^\pi \cos^2 k\theta d\theta = \begin{cases} \frac{\pi}{2} & \text{for } k \neq 0 \\ \pi & \text{for } k = 0 \end{cases}$$

If we substitute $\theta = \cos x$ in these integrals and invoke the definition (B.118), then we find the orthogonal relations for the Chebyshev polynomials

$$\int_{-1}^{1} T_m(x) T_k(x) \frac{dx}{\sqrt{1-x^2}} = 0 \quad \text{for } k \neq m$$

and

$$\int_0^\pi T_k^2(x) \frac{dx}{\sqrt{1-x^2}} = \begin{cases} \frac{\pi}{2} & \text{for } k \neq 0 \\ \pi & \text{for } k = 0 \end{cases}$$

which shows that the set $\{T_n(x)\}_{n \geq 0}$ is a sequence of orthogonal polynomials on the interval $[-1, 1]$ with respect to the weight function $(1-x^2)^{-1/2}$.

---

[6] Comparison with (B.123) incidentally establishes the identity (Riordan, 1968)

$$\sum_{k=p}^{\left[\frac{n}{2}\right]} \binom{k}{p} \binom{n}{2k} = \frac{n}{n-p} \binom{n-p}{p} 2^{n-1-2p}$$

From the trigonometric identity, $\cos n\theta + \cos (n-2)\theta = 2\cos (n-1)\theta \cos \theta$, the definition (B.118) directly leads to the three-term recursion (**art.** 356)

$$T_n (x) = 2xT_{n-1} (x) - T_{n-2} (x)$$

with $T_0 (x) = 1$ and $T_1 (x) = x$. Favard's theorem (**art.** 357) demonstrates again that $T_n (x)$ is an orthogonal polynomial.

382. *Generating functions.* From the geometric series $\sum_{k=0}^{\infty} z^k e^{ik\theta} = \frac{1}{1-ze^{i\theta}}$ for $|z| < 1$, we obtain for real z, after equating the real and imaginary part of both sides,

$$\begin{cases} \sum_{k=0}^{\infty} z^k \cos k\theta = \frac{1-z\cos\theta}{1-2z\cos\theta+z^2} \\ \sum_{k=0}^{\infty} z^k \sin k\theta = \frac{z\sin\theta}{1-2z\cos\theta+z^2} \end{cases}$$

With $x = \cos\theta$, the definition (B.118) provides us with the generating function

$$\sum_{k=0}^{\infty} T_k (x) z^k = \frac{1 - zx}{1 - 2zx + z^2} \tag{B.137}$$

while the definition $U_{k-1} (x) = \frac{\sin k\theta}{\sin\theta}$ in (B.130) of Chebyshev polynomial of the second kind indicates

$$\sum_{k=0}^{\infty} U_k (x) z^k = \frac{1}{1 - 2zx + z^2} \tag{B.138}$$

The real and imaginary part of both sides in $\sum_{k=0}^{\infty} \frac{z^k e^{ik\theta}}{k!} = e^{ze^{i\theta}}$, that converges for all $z = re^{i\omega}$, yields

$$\begin{cases} \sum_{k=0}^{\infty} \frac{z^k \cos k\theta}{k!} = e^{r\cos(\theta+\omega)} \cos (r\sin (\theta + \omega)) \\ \sum_{k=0}^{\infty} \frac{z^k \sin k\theta}{k!} = e^{r\cos(\theta+\omega)} \sin (r\sin (\theta + \omega)) \end{cases}$$

With $x = \cos\theta$ and (B.118), we obtain the exponential generating functions in the real r,

$$\sum_{k=0}^{\infty} \frac{r^k T_k (x)}{k!} = e^{rx} \cos \left(r\sqrt{1 - x^2}\right)$$

and

$$\sum_{k=0}^{\infty} \frac{r^{k+1} U_k (x)}{(k + 1)!} = e^{rx} \frac{\sin \left(r\sqrt{1 - x^2}\right)}{\sqrt{1 - x^2}}$$

From the generating function (B.137), the Cauchy integral representation for the k-th derivative (B.46) gives

$$T_k (x) = \frac{1}{2\pi i} \int_{C(0)} \frac{1 - zx}{1 - 2zx + z^2} \frac{dz}{z^{k+1}}$$

where the contour $C(0)$ encloses the origin. Since the integrand vanishes at $|z| \rightarrow$

∞, we deform the contour to enclose the entire z-plane, except for the origin. Cauchy's residue theorem (Titchmarsh, 1964) yields

$$T_k\left(x\right) = -\lim_{z \to z_1} \frac{\left(1-zx\right)\left(z-z_1\right)}{1-2zx+z^2}\frac{1}{z^{k+1}} - \lim_{z \to z_2} \frac{\left(1-zx\right)\left(z-z_2\right)}{1-2zx+z^2}\frac{1}{z^{k+1}}$$

where $1-2zx+z^2 = \left(z-z_1\right)\left(z-z_2\right)$ with $z_1 = x+\sqrt{x^2-1}$ and $z_2 = x-\sqrt{x^2-1}$, from which we find again the closed form (B.125).

Bibliography

Abramowitz, M. and Stegun, I. A. (1968). *Handbook of Mathematical Functions*. (Dover Publications, Inc., New York).

Aigner, M. and Ziegler, G. M. (2003). *Proofs from THE BOOK*, third edn. (Springer-Verlag, Berlin).

Akritas, A. G. (1989). *Elements of Computer Algebra with Applications*. (John Wiley & Sons, New York).

Albert, R. and Barabási, A.-L. (2002). Statistical mechanics of complex networks. *Reviews of Modern Physics 74*, 47–97.

Alon, N. (1986). Eigenvalues and expanders. *Combinatorica 6*, 2, 83–96.

Alon, N. and Milman, V. D. (1985). λ_1, isoperimetric inequalities for graphs and supercontractors. *Journal of Combinatorial Theory, Series B 38*, 73–88.

Anderson, G. W., Guionnet, A., and Zeitouni, O. (2010). *An Introduction to Random Matrices*. (Cambridge University Press, Cambridge, U.K.).

Anderson, W. N. and Morley, T. D. (1985). Eigenvalues of the Laplacian of a graph. *Linear and Multilinear Algebra 18*, 141–145.

Aouchiche, M. and Hansen, P. (2014). Distance spectra of graphs: A survey. *Linear Algebra and its Applications 458*, 301–386.

Aziz, A. and Zargar, B. A. (2012). Bounds for the zeros of a polynomial with restricted coefficients. *Applied Mathematics 3*, 30–33.

Bapat, R. B. (2004). Resistance matrix of a weighted graph. *Match, Communications in Mathematical and in Computer Chemistry 50*, 73–82.

Bapat, R. B. (2013). On the adjacency matrix of a threshold graph. *Linear Algebra and its Applications 439*, 3008–3015.

Barabási, A. L. (2002). *Linked, The New Science of Networks*. (Perseus, Cambridge, MA).

Barabási, A. L. (2016). *Network Science*. (Cambridge University Press, Cambridge, U.K.).

Barabási, A. L. and Albert, R. (1999). Emergence of scaling in random networks. *Science 286*, 509–512.

Barrat, A., Bartelemy, M., and Vespignani, A. (2008). *Dynamical Processes on Complex Networks*. (Cambridge University Press, Cambridge, U.K.).

Batson, J., Spielman, D. A., Srivastava, N., and Teng, S.-H. (2013). Spectral sparsification of graphs: Theory and algorithms. *Communications of the ACM 56*, 8 (August), 87–94.

Behzad, M. and Chartrand, G. (1967). No graph is perfect. *The American Mathematical Monthly 74*, 8, 962–963.

Biggs, N. (1996). *Algebraic Graph Theory*, 2nd edn. (Cambridge University Press, Cambridge, U.K.).

Bollobás, B. (1998). *Modern Graph Theory*. Graduate Texts in Mathematics, (Springer-Verlag, New York).

Bollobás, B. (2001). *Random Graphs*, 2nd edn. (Cambridge University Press, Cambridge).

Bollobás, B. (2004). *Extremal Graph Theory*. (Courier Dover Publishers, New York).

Borghs, G., Bhattacharyya, K., Deneffe, K., Van Mieghem, P., and Mertens, R. (1989). Band-gap narrowing in highly doped n- and p-type GaAs studied by photoluminescence spectroscopy. *Journal of Applied Physics 66*, 9, 4381–4386.

Borwein, P. and Erdélyi, T. (1995). *Polynomials and Polynomial Inequalities*. (Springer-Verlag, New York).

Brankov, V., Hansen, P., and Stevanović, D. (2006). Automated conjectures on upper bounds for the largest Laplacian eigenvalue of graphs. *Linear Algebra and its Applications 414*, 407–424.

Brouwer, A. E. and Haemers, W. H. (2008). A lower bound for the Laplacian eigenvalues of a graph: Proof of a conjecture by Guo. *Linear Algebra and its Applications 429*, 2131–2135.

Brouwer, A. E. and Haemers, W. H. (2012). *Spectra of Graphs*. Universitext, (Springer, New York).

Buck, R. C. (1978). *Advanced Calculus*. (McGraw-Hill Book Company, New York).

Budel, G. and Van Mieghem, P. (2021). Detecting the number of clusters in a network. *Journal of Complex Networks 8*, 6 (December), cnaa047.

Buldyrev, S. V., Parshani, R., Paul, G., Stanley, H. E., and Havlin, S. (2010). Catastrophic cascade of failures in interdependent networks. *Nature Letters 464*, 1025–1028.

Chandra, A. K., Raghavan, P., Ruzzo, W. L., and Smolensky, R. (1997). The electrical resistance of a graph captures its commute and cover times. *Computational Complexity 6*, 312–340.

Chihara, T. S. (1978). *An Introduction to Orthogonal Polynomials*. (Gordon and Breath, New York).

Chu, T., Gao, Y., Peng, R., Sachdeva, S., Sawlani, S., and Wang, J. (2020). Graph sparsification, spectral sketches, and faster resistance computation via short cycle decomposition. *SIAM Journal on Computing 0*, 0, FOCS18-85–FOCS18-157.

Chung, F. R. K. (1989). Diameters and eigenvalues. *Journal of the American Mathematical Society 2*, 2 (April), 187–196.

Chung, F. R. K., Faber, V., and Manteuffel, T. A. (1994). An upper bound on the diameter of a graph from eigenvalues associated with its Laplacian. *SIAM Journal of Discrete Mathematics 7*, 3 (August), 443–457.

Cioabă, S. M. (2006). Sums of powers of the degrees of a graph. *Discrete Mathematics 306*, 1959–1964.

Cioabă, S. M. (2007). The spectral radius and the maximum degree of irregular graphs. *Electronic Journal of Combinatorics 14*, R38.

Cioabă, S. M., van Dam, E. R., Koolen, J. H., and Lee, J.-H. (2010). A lower bound for the spectral radius of graphs with fixed diameter. *European Journal of Combinatorics 31*, 6, 1560–1566.

Cohen-Tannoudji, C., Diu, B., and Laloë, F. (1977). *Mécanique Quantique*. Vol. I and II. (Hermann, Paris).

Comtet, L. (1974). *Advanced Combinatorics*, revised and enlarged edn. (D. Riedel Publishing Company, Dordrecht, Holland).

Coppersmith, D., Feige, U., and Shearer, J. (1996). Random walks on regular and irregular graphs. *SIAM Journal of Discrete Mathematics 9*, 2 (May), 301–308.

Cormen, T. H., Leiserson, C. E., and Rivest, R. L. (1991). *An Introduction to Algorithms*. (MIT Press, Boston).

Courant, R. and Hilbert, D. (1953). *Methods of Mathematical Physics*, first English edition, translated and revised from the German original of Springer in 1937 edn. Wiley Classic Library, vol. I. (Interscience, New York).

Cvetković, D., Rowlinson, P., and Simić, S. (1997). *Eigenspaces of graphs*. (Cambridge University Press, Cambridge, U.K.).

Cvetković, D., Rowlinson, P., and Simić, S. (2004). *Spectral Generalizations of Line Graphs*. (Cambridge University Press, Cambridge, U.K.).

Cvetković, D., Rowlinson, P., and Simić, S. (2007). Signless Laplacian of finite graphs. *Linear Algebra and its Applications 432*, 155–171.

Cvetković, D., Rowlinson, P., and Simić, S. (2009). *An Introduction to the Theory of Graph Spectra*. (Cambridge University Press, Cambridge, U.K.).

Cvetković, D. M., Doob, M., and Sachs, H. (1995). *Spectra of Graphs, Theory and Applications*, third edn. (Johann Ambrosius Barth Verlag, Heidelberg).

Das, K. C. (2004). The Laplacian spectrum of a graph. *Computers and Mathematics with Applications 48*, 715–724.

Das, K. C. and Kumar, P. (2004). Some new bounds on the spectral radius of graphs. *Discrete Mathematics 281*, 149–161.

de Abreu, N. M. M. (2007). Old and new results on algebraic connectivity of graphs. *Linear Algebra and its Applications 423*, 53–73.

de Bruijn, N. G. (1960). Opgave 12. *Wiskundige Opgaven met de oplossingen, edited by Het wiskundig genootschap, Amsterdam and published by Noordhoff, Groningen Deel 21*, 1, 12–14.

Dehmer, M. and Emmert-Streib, F. (2009). *Analysis of Complex Networks*. (Wiley-VCH Verlag GmbH& Co., Weinheim).

Deutsch, E. (1970). Matricial norms and the zeros of polynomials. *Linear Algebra and its Applications 3*, 483–489.

Devriendt, K. (2022a). Effective resistance is more than distance: Laplacians, Simplices and the Schur complement. *Linear Algebra and its Applications 639*, 24–49.

Devriendt, K. (2022b). *Graph geometry from effective resistances*. (PhD thesis, Mansfield College, University of Oxford).

Devriendt, K. and Lambiotte, R. (2022). Discrete curvature on graphs from the effective resistance. *Journal of Physics: Complexity 3*, 2 (June), 025008.

Devriendt, K. and Van Mieghem, P. (2019a). The Simplex Geometry of Graphs. *Journal of Complex Networks 7*, 4 (August), 469–490.

Devriendt, K. and Van Mieghem, P. (2019b). Tighter spectral bounds for the cut-set based on Laplacian eigenvectors. *Linear Algebra and its Applications 572*, 68–91.

Diestel, R. (2010). *Graph Theory*, fourth edn. (Springer-Verlag, Heidelberg).

Dorogovtsev, S. N., Goltsev, A. V., Mendes, J. F. F., and Samukhin, A. N. (2003). Spectra of complex networks. *Physical Review E 68*, 046109.

Dorogovtsev, S. N. and Mendes, J. F. F. (2003). *Evolution of Networks, From Biological Nets to the Internet and WWW*. (Oxford University Press, Oxford).

Doyle, P. G. and Snell, J. L. (1984). *Random Walks and Electric Networks*. Carus mathematical monographs; updated in arXiv:math/0001057, Jan. 2000, (Mathematical Association of America, Washington).

Edelman, A. and Raj Rao, N. (2005). Random matrix theory. *Acta Numerica*, 1–65.

Ellens, W., Spieksma, F. A., Van Mieghem, P., Jamakovic, A., and Kooij, R. E. (2011). Effective graph resistance. *Linear Algebra and its Applications 435*, 2491–2506.

Erdős, P. (1961). Problems and results on the theory of interpolation. ii. *Acta Mathematica Hungarica 12*, 1-2 (March), 235–244.

Erdős, P. and Turán, P. (1950). On the distribution of roots of polynomials. *Annals of Mathematics 51*, 1 (January), 105–119.

Estrada, E. (2012). Path Laplacian matrices: Introduction and applications to the analysis of consensus in networks. *Linear Algebra and its Applications 436*, 3373–3391.

Feller, W. (1970). *An Introduction to Probability Theory and Its Applications*, 3rd edn. Vol. 1. (John Wiley & Sons, New York).

Feynman, R. P., Leighton, R. B., and Sands, M. (1963). *The Feynman Lectures on Physics*. Vol. 2. (Addison-Wesley, Massachusetts).

Fiedler, M. (1972). Bounds for eigenvalues of doubly stochastic matrices. *Linear Algebra and its Applications 5*, 299–310.

Fiedler, M. (1973). Algebraic connectivity of graphs. *Czechoslovak Mathematical Journal 23*, 2, 298–305.

Fiedler, M. (1975). A property of eigenvectors of nonnegative symmetric matrices and its application to graph theory. *Czechoslovak Mathematical Journal 25*, 4, 619–633.

Fiedler, M. (2009). *Matrices and Graphs in Geometry*. (Cambridge University Press, Cambridge, U.K.).

Fiedler, M. and Pták, V. (1962). On matrices with non-positive off-diagonal elements and positive principal minors. *Czechoslovak Mathematical Journal 12*, 3, 382–400.

Fiol, M. A. and Garriga, E. (2009). Number of walks and degree powers in a graph. *Discrete Mathematics 309*, 2613–2614.

Foster, R. M. (1949). The average impedance of an electrical network. *Contributions to Applied Mechanics (Reissner Anniversary Volume), Edwards Brothers, Ann Arbor, Michigan*, 333–340.

Füredi, Z. and Komlós, J. (1981). The eigenvalues of random symmetric matrices. *Combinatorica 1*, 3, 233–241.

Gama, F., Isufi, E., Leus, G., and Ribeiro, A. (2020). Graphs, convolutions and neural networks. *IEEE Signal Processing Magazine*, 128–138.

Gantmacher, F. R. (1959a). *The Theory of Matrices*. Vol. I. (Chelsea Publishing Company, New York).

Gantmacher, F. R. (1959b). *The Theory of Matrices*. Vol. II. (Chelsea Publishing Company, New York).

Gauss, C. F. (1801). *Disquisitiones Arithmeticae*. (translated in English by A. A. Clark (1966), Yale University Press, New Haven and London).

Gautschi, W. (2004). *Orthogonal Polynomials. Computation and Approximation*. (Oxford University Press, Oxford, U.K.).

Ghosh, A., Boyd, S., and Saberi, A. (2008). Minimizing effective resistance of a graph. *SIAM Review 50*, 1 (February), 37–66.

Gilbert, W. J. and Nicholson, W. K. (2004). *Modern Algebra with Applications*, 2nd edn. (John Wiley & Sons, Hoboken, New Jersey).

Godsil, C. D. and McKay, B. D. (1982). Constructing cospectral graphs. *Aequationes Mathematicae 25*, 257–268.

Godsil, C. D. and Royle, G. (2001). *Algebraic Graph Theory*. (Springer, New York).

Golub, G. H. and Van Loan, C. F. (1996). *Matrix Computations*, 3rd edn. (The John Hopkins University Press, Baltimore).

Goulden, I. P. and Jackson, D. M. (1983). *Combinatorial Enumeration*. (John Wiley & Sons, New York).

Govers, T., Barrow-Green, J., and Leader, I. (2008). *The Princeton Companion to Mathematics*. (Princeton University Press, Princeton and Oxford).

Grone, R. and Merris, R. (1994). The Laplacian spectrum of a graph π^*. *SIAM Journal on Discrete Mathematics 7*, 2 (May), 221–229.

Gvishiani, A. D. and Gurvich, V. A. (1987). Metric and ultrametric spaces of resistances. *Uspekhi Mat. Nauk 2*, 6, 187–188.

Haemers, W. H. (1995). Interlacing eigenvalues and graphs. *Linear Algebra and its Applications 227-228*, 593–616.

Haemers, W. H. and Spence, E. (2004). Enumeration of cospectral graphs. *European Journal of Combinatorics 25*, 199–211.

Hammer, P. L. and Kelmans, A. K. (1996). Laplacian spectra and spanning trees of threshold graphs. *Discrete Applied Mathematics 65*, 255–273.

Harary, F. (1962). The determinant of the adjacency matrix of a graph. *SIAM Review 4*, 3 (July), 202–210.

Harary, F., King, C., Mowshowitz, A., and Read, R. C. (1971). Cospectral graphs and digraphs. *Bulletin of the London Mathematical Society 3*, 321–328.

Harary, F. and Nash-Williams, C. S. J. A. (1965). On Eulerian and Hamiltonian graphs and line graphs. *Canadian Mathematical Bulletin 8*, 6, 701–709.

Harary, F. and Palmer, E. M. (1973). *Graphical Enumeration*. (Academic Press, New York and London).

Hardy, G. H. (2006). *A Course of Pure Mathematics*, 10th edn. (Cambridge University Press).

Hardy, G. H., Littlewood, J. E., and Polya, G. (1999). *Inequalities*, 2nd edn. (Cambridge University Press, Cambridge, U.K.).

Hardy, G. H. and Wright, E. M. (2008). *An Introduction to the Theory of Numbers*, 6th edn. (Oxford University Press, London).

Heilbronner, E. (1953). Das Kompositions-Prinzip: Eine anschauliche Methode zur elektronentheoretischen Behandlung nicht oder niedrig symmetrischer Molekuln im Rahmen der MO-Theorie. *Helvetica Chimica Acta 36*, 170–188.

Henrici, P. (1974). *Applied and Computational Complex Analysis*. Vol. 1. (John Wiley & Sons, New York).

Hermite, C., Poincaré, H., and Rouché, E. (1972). *Œuvres de Laguerre*, 2nd edn. Vol. I. (Chelsea Publishing Company, Bronx, New York).

Hofmeister, M. (1988). Spectral radius and degree sequence. *Mathematische Nachrichten 139*, 37–44.

Horn, A. (1962). Eigenvalues of sums of Hermitian matrices. *Pacific Journal of Mathematics 12*, 1, 225–241.

Horn, R. A. and Johnson, C. R. (1991). *Topics in Matrix Analysis*. (Cambridge University Press, Cambridge, U.K.).

Horn, R. A. and Johnson, C. R. (2013). *Topics in Matrix Analysis*, 2nd edn. (Cambridge University Press, Cambridge, U.K.).

Kelmans, A. K. and Chelnokov, V. M. (1974). A certain polynomial of a graph and graphs with an extremal number of trees. *Journal of Combinatorial Theory (B) 16*, 197–214.

Kirchhoff, G. (1847). Ueber die Auflösung der Gleichungen, auf welche man bei der Untersuchung der linearen Verteheilung galvanischer Ströme gefürt wird. *Annalen der Physik und Chemie 72*, 12, 497–508.

Klein, D. J. (2002). Resistance-distance sum rules. *Croatica Chemica Acta 75*, 2, 633–649.

Klein, D. J. and Randić, M. (1993). Resistance distance. *Journal of Mathematical Chemistry 12*, 81–95.

Knutson, A. and Tao, T. (2001). Honeycombs and sums of Hermitian matrices. *Notices of the AMS 48*, 2 (February), 175–186.

Koekoek, R., Lesky, P. A., and Swarttouw, R. F. (2010). *Hypergeometric Orthogonal Polynomials and Their q-Analogues*. Springer Monographs in Mathematics, (Springer-Verlag, New York).

Kollár, A. J. and Sarnak, P. (2021). Gap sets for the spectra of cubic graphs. *arXiv:2005.05379v4*.

Kolokolnikov, T. (2015). Maximizing algebraic connectivity for certain families of graphs. *Linear Algebra and its Applications 471*, 122–140.

Krivelevich, M., Sudakov, B., Vu, V. H., and Wormald, N. C. (2001). Random regular graphs of high degree. *Random Structures and Algorithms 18*, 4 (July), 346–363.

Lanczos, C. (1988). *Applied Analysis*. (Dover Publications, Inc., New York).

Lehot, P. G. H. (1974). An optimal algorithm to detect a line graph and output its root graph. *Journal of the Association for Computing Machinery 21*, 4 (October), 569–575.

Li, C., Wang, H., and Van Mieghem, P. (2012). Bounds for the spectral radius of a graph when nodes are removed. *Linear Algebra and its Applications 437*, 319–323.

Li, L., Alderson, D., Doyle, J. C., and Willinger, W. (2006). Towards a theory of scale-free graphs: Definition, properties, and implications. *Internet Mathematics 2*, 4, 431–523.

Liu, D., Trajanovski, S., and Van Mieghem, P. (2015). ILIGRA: An efficient inverse line graph algorithm. *Journal of Mathematical Modelling and Algorithms in Operations Research 14*, 1, 13–33.

Liu, D., Wang, H., and Van Mieghem, P. (2010). Spectral perturbation and reconstructability of complex networks. *Physical Review E 81*, 1 (January), 016101.

Lovász, L. (1993). Random walks on graphs: A survey. *Combinatorics 2*, 1–46.

Lovász, L. (2003). *Combinatorial Problems and Exercises*, 2nd edn. (Elsevier, Amsterdam).

Lovász, L. and Pelikán, J. (1973). On the eigenvalues of trees. *Periodica Mathematica Hungarica 3*, 1-2, 175–182.

Lubotzky, A., Philips, R., and Sarnak, P. (1988). Ramanujan graphs. *Combinatorica 8*, 261–277.

Lyons, R. and Peres, Y. (2016). *Probability on Trees and Networks*. Cambridge Series in Statistical and Probabilistic Mathematics, (Cambridge University Press, Cambridge, U.K.).

Mahler, K. (1960). An application of Jensen's formula to polynomials. *Mathematika 7*, 98–100.

Mahler, K. (1964). An inequality for the discriminant of a polynomial. *Michigan Mathematical Journal 11*, 3, 257–262.

Marcus, A. W., Spielman, D. A., and Srivastava, N. (2015). Interlacing families i: Bipartite ramanujan graphs of all degrees. *Annals of Mathematics 182*, 307–325.

Marcus, M. and Ming, H. (1964). *A Survey of Matrix Theory and Matrix Inequalities*. (Allyn and Bacon, Inc., Boston).

Markushevich, A. I. (1985). *Theory of Functions of a Complex Variable*. Vol. I – III. (Chelsea Publishing Company, New York).

Marshall, A. W., Olkin, I., and Arnold, B. C. (2011). *Inequalities: Theory of Majorization and Its Applications*, 2nd edn. (Springer, New York).

Martin-Hernandez, J., Li, Z., and Van Mieghem, P. (2014). Weighted betweenness and algebraic connectivity. *Journal of Complex Networks 2*, 3, 272–287.

Marčenko, V. A. and Pastur, L. A. (1967). Distribution of eigenvalues for some sets of random matrices. *Math. USSR - Sbornik 1*, 4, 457– 483.

McKay, B. D. (1981). The expected eigenvalue distribution of a large regular graph. *Linear Algebra and its Applications 40*, 203–216.

McKay, B. D. and Piperno, A. (2014). Practical graph isomorphism, II. *Journal of Symbolic Computation 60*, 94–112.

McKay, B. D. and Wormald, N. C. (1990). Asymptotic enumeration by degree sequence of graphs of high degree. *European Journal of Combinatorics 11*, 565–580.

McNamee, J. M. (2007). *Numerical Methods for Roots of Polynomials, Part I*. (Elsevier, Amsterdam).

Mehta, M. L. (1991). *Random Matrices*, 2nd edn. (Academic Press, Boston).

Merris, R. (1994). Laplacian matrices of graphs: A survey. *Linear Algebra and its Applications 197-198*, 143–176.

Merris, R. (1999). Note: An upper bound for the diameter of a graph. *SIAM Journal of Discrete Mathematics 12*, 8, 412.

Meyer, C. D. (2000). *Matrix Analysis and Applied Linear Algebra*. (Society for Industrial and Applied Mathematics (SIAM), Philadelphia).

Micchelli, C. A. and Willoughby, R. A. (1979). On functions which preserve the class of Stieltjes matrices. *Linear Algebra and its Applications 23*, 141–156.

Milovanović, G. V., Mitrinović, D. S., and Rassias, T. M. (1994). *Topics in Polynomials: Extremal Problems, Inequalities and Zeros*. (World Scientific Publishing, Singapore).

Minc, H. (1970). On the maximal eigenvector of a positive matrix. *SIAM Journal on Numerical Analysis* 7, 3 (September), 424–427.

Mirsky, L. (1982). *An Introduction to Linear Algebra.* (Dover Publications, Inc., New York).

Mohar, B. (1989). Isoperimetric numbers of graphs. *Journal of Combinatorial Theory, Series B 47*, 274–291.

Mohar, B. (1991). Eigenvalues, diameter and mean distance in graphs. *Graphs and Combinatorics 7*, 53–64.

Morse, P. M. and Feshbach, H. (1978). *Methods of Theoretical Physics.* (McGraw-Hill Book Company, New York).

Motzkin, T. and Straus, E. G. (1965). Maxima for graphs and a new proof of a theorem of Turán. *Canadian Journal of Mathematics 17*, 533–540.

Mowshowitz, A. (1972). The characteristic polynomial of a graph. *Journal of Combinaotiral Theory (B) 12*, 177–193.

Muir, T. (1906-1911-1920-1923-1930). *The Theory of Determinants in the Historical Order of Development.* Vol. I-V. (MacMillan and co., London).

Myers, C. R. (2003). Software systems as complex networks: Structure, function, and evolvability of software collaboration graphs. *Physical Review E 68*, 4, 046116.

Nash-Williams, C. S. J. A. (1959). Random walk and electric currents in networks. *Mathematical Proceedings of the Cambridge Philosophical Society 55*, 2 (April), 181–194.

Newman, M. E. J. (2003a). Mixing patterns in networks. *Physical Review E 67*, 026126.

Newman, M. E. J. (2003b). The structure and function of complex networks. *SIAM Review 45*, 2, 167–256.

Newman, M. E. J. (2006). Modularity and community structure in networks. *Proceedings of the National Academy of Sciences of the United States of America (PNAS) 103*, 23 (June), 8577–8582.

Newman, M. E. J. (2010). *Networks: An Introduction.* (Oxford University Press, Oxford, U.K.).

Newman, M. E. J. (2018). *Networks*, second edn. (Oxford University Press, Oxford, U.K.).

Newman, M. E. J. and Girvan, M. (2004). Finding and evaluating community structure in networks. *Physical Review E 69*, 026113.

Newman, M. E. J., Strogatz, S. H., and Watts, D. J. (2001). Random graphs with arbitrary degree distributions and their applications. *Physical Review E 64*, 026118.

Newton, I. S. (1687). *Philosophiae Naturalis Principia Mathematica*, first edn. (Londini).

Nikiforov, V. (2002). Some inequalities for the largest eigenvalue of a graph. *Combinatorics, Probability and Computing 11*, 2 (March), 179–189.

Nikiforov, V. (2021). On a theorem of Nosal. *arXiv:2104.1217.*

Nikiforov, V., Tait, M., and Timmons, C. (2018). Degenerate Turán problems for hereditary properties. *The Electronic Journal of Combinatorics 25*, 4, 4.39.

Nikoloski, Z., Deo, N., and Kucera, L. (2005). Degree-correlation of a scale-free random graph process. *Discrete Mathematics and Theoretical Computer Science (DMTCS) proc. AE, EuroComb 2005*, 239–244.

Nilli, A. (1991). On the second eigenvalue of a graph. *Discrete Mathematics 91*, 207–210.

Noldus, R. and Van Mieghem, P. (2015). Assortativity in complex networks. *Journal of Complex Networks 3*, 4, 507–542.

Olver, F. W. J., Lozier, D. W., Boisvert, R. F., and Clark, C. W. (2010). *NIST Handbook of Mathematical Functions.* (Cambridge University Press, New York).

Ortega, A. (2022). *Introduction to Graph Signal Processing.* (Cambridge University Press, Cambridge, U.K.).

Ostrowski, A. M. (1960). On the eigenvector belonging to the maximal root of a non-negative matrix. *Proceedings of the Edinburgh Mathematical Society (Series 2) 12*, 2 (December), 107–112.

Pastor-Satorras, R., Castellano, C., Van Mieghem, P., and Vespignani, A. (2015). Epidemic processes in complex networks. *Review of Modern Physics 87*, 3 (September), 925–979.

Perron, O. (1907). Neue Kriterien für die Irreduzililität algebraischer Gleichungen. *Journal für die reine und angewandte Mathematik 132*, 288–307.

Petrović, M. and Gutman, I. (2002). The path is the tree with smallest greatest Laplacian eigenvalue. *Kragujevac Journal of Mathematics 24*, 67–70.

Pinkus, A. (2010). *Totally Positive Matrices.* Cambridge Tracts in Mathematics, (Cambridge University Press, Cambridge, U.K.).

Pothen, A., Simon, H. D., and Liou, K.-P. (1990). Partitioning sparse matrices with eigenvectors of graphs. *SIAM Journal of Matrix Analysis and Applications 11*, 3 (July), 430–452.

Powell, P. D. (2011). Calculating determinants of block matrices. *arXiv:1112.4379.*

Powers, D. L. (1988). Graph partitioning by eigenvectors. *Linear Algebra and its Applications 101*, 121–133.

Prasse, B., Devriendt, K., and Van Mieghem, P. (2021). Clustering for epidemics on networks: a geometric approach. *Chaos: An Interdisciplinary Journal of Nonlinear Science 31*, 6, 063115.

Press, W. H., Teukolsky, S. A., Vetterling, W. T., and Flannery, B. P. (1992). *Numerical Recipes in C*, 2nd edn. (Cambridge University Press, New York).

Rademacher, H. (1973). *Topics in Analytic Number Theory*. (Springer-Verlag, Berlin).

Rainville, E. D. (1971). *Special Functions*, reprinted edn. (Chelsea Publishing Company, New York).

Raj Rao, N. and Newman, M. E. J. (2013). Spectra of random graphs with arbitrary expected degrees. *Physical Review E 87*, 012803.

Restrepo, J. G., Ott, E., and Hunt, B. R. (2005). Onset of synchronization in large networks of coupled oscillators. *Physical Review E 71*, 036151, 1–12.

Riordan, J. (1968). *Combinatorial Identities*. (John Wiley & Sons, New York).

Rivlin, T. J. (1974). *The Chebyshev Polynomials*. (John Wiley & Sons, New York).

Roussopoulos, N. D. (1973). A max m, n algorithm for determining the graph h from its line graph g. *Information Processing Letters 2*, 108–112.

Rowlinson, P. (1988). On the maximal index of graphs with a prescribed number of edges. *Linear Algebra and its Applications 110*, 43–53.

Rudelson, M. and Vershynin, R. (2007). Sampling from large matrices: An approach through geometric functional analysis. *Journal of the ACM 54*, 4 (July), Article 21.

Ruiz, L., Gama, F., and Ribeiro, A. (2021). Graph neural networks: Architectures, stability and transferability. *Proceedings of the IEEE 109*, 5 (May), 660–682.

Rulkov, N. F., Sushchik, M. M., Tsimring, L. S., and Abarbanel, H. D. (1995). Generalized synchronization of chaos in directionally coupled chaotic systems. *Physical Review E 51*, 2, 980–994.

Sahneh, F. D., Scoglio, C., and Van Mieghem, P. (2015). Coupling threshold for structural transition in interconnected networks. *Physical Review E 92*, 040801.

Schnakenberg, J. (1976). Network theory of microscopic and macroscopic behavior of master equation systems. *Review of Modern Physics 48*, 4 (October), 571–585.

Schoone, A. A., Bodlaender, H. L., and van Leeuwen, J. (1987). Diameter increase caused by edge deletion. *Journal of Graph Theory 11*, 3, 409–427.

Sheil-Small, T. (2002). *Complex Polynomials*. Cambridge Studies in Advanced Mathematics 73, (Cambridge University Press, Cambridge, U.K.).

Shi, L. (2009). The spectral radius of irregular graphs. *Linear Algebra and its Applications 431*, 189–196.

Shilov, G. E. (1977). *Linear Algebra*. (Dover Publications, Inc., New York).

Simić, S. K., Belardo, F., Li Marzi, E. M., and Tosić, D. V. (2010). Connected graphs of fixed order and size with maximal index: Some spectral bounds. *Linear Algebra and its Applications 432*, 2361–2372.

Sinkhorn, R. (1964). A relationship between arbitrary positive matrices and doubly stochastic matrices. *The Annals of Mathematical Statistics 35*, 2 (June), 876–879.

Spielman, D. A. and Srivastava, N. (2011). Graph sparsification by effective resistances. *SIAM Journal on Computing 40*, 6, 1913–1926.

Spielman, D. A. and Teng, S.-H. (2007). Spectral partitioning works: Planar graphs and finite element meshes. *Linear Algebra and its Applications 421*, 284–305.

Stam, C. J. and Reijneveld, J. C. (2007). Graph theoretical analysis of complex networks in the brain. *Nonlinear Biomedical Physics 1*, 3 (July), 1–19.

Stanić, Z. (2015). *Inequalities for Graph Eigenvalues*. London Mathematical Society Lecture Note Series: 423, (Cambridge University Press, Cambridge, U.K.).

Stevanović, D. (2004). The largest eigenvalue of nonregular graphs. *Journal of Combinatorial Theory, Series B 91*, 1, 143–146.

Stevanović, D. (2015). *Spectral Radius of Graphs*. (Academic Press, London).

Stewart, I. (2004). *Galois Theory*, 3rd edn. (Chapman & Hall/CRC, Boca Raton).

Stoer, M. and Wagner, F. (1997). A simple min-cut algorithm. *Journal of the ACM 44*, 4 (July), 585–591.

Strogatz, S. H. (2001). Exploring complex networks. *Nature 410*, 8 (March), 268–276.

Subak-Sharpe, G. E. (1990). On the structure of well-conditioned positive resistance networks. *IEEE Proceedings of the 33rd Midwest Symposium on Circuits and Systems*, 293–296.

Sun, L., Wang, W., Zhou, J., and Bu, C. (2015). Some results on resistance distances and resistance matrices. *Linear and Multilinear Algebra 63*, 3, 523–533.

Szegő, G. (1922). Bemerkungen zu einem Satz von J.H. Grace über die Wurzeln algebraischer Gleichungen. *Mathematische Zeitschrift 13*, 28–55.

Szegő, G. (1978). *Orthogonal Polynomials*, 4th edn. (American Mathematical Society, Providence, Rhode Island).

Tao, T. and Vu, V. (2011). Random matrices: Universality of local eigenvalue statistics. *Acta Mathematica 206*, 127–204.

Tao, T. and Vu, V. H. (2010). Random matrices: Universality of ESDs and the circular law. *The Annals of Probability 38*, 5, 2023–2065.

Tewarie, P., Prasse, B., Meier, J., Byrne, A., Domenico, M., Stam, C. J., M., B., Hillebrand, A., Daffertshofer, A., Coombes, S., and Van Mieghem, P. (2021). Interlayer connectivity reconstruction for multilayer brain networks using phase oscillator models. *New Journal of Physics 23*, 063065.

Titchmarsh, E. C. (1958). *Eigenfunction Expansions Associated with Second-order Differential Equations, Part II*. (Oxford University Press, Amen House, London).

Titchmarsh, E. C. (1962). *Eigenfunction Expansions Associated with Second-order Differential Equations, Part I*, 2nd edn. (Oxford University Press, Amen House, London).

Titchmarsh, E. C. (1964). *The Theory of Functions*. (Oxford University Press, Amen House, London).

Titchmarsh, E. C. and Heath-Brown, D. R. (1986). *The Theory of the Zeta-function*, 2nd edn. (Oxford Science Publications, Oxford).

van Dam, E. R. (2007). Graphs with given diameter maximizing the spectral radius. *Linear Algebra and its Applications 426*, 545–457.

van Dam, E. R. and Haemers, W. H. (1995). Eigenvalues and the diameter of graphs. *Linear and Multilinear Algebra 39*, 1-2, 33–44.

van Dam, E. R. and Haemers, W. H. (2003). Which graphs are determined by their spectrum? *Linear Algebra and its Applications 373*, 241–272.

van Dam, E. R. and Kooij, R. E. (2007). The minimal spectral radius of graphs with a given diameter. *Linear Algebra and its Applications 423*, 2-3, 408–419.

van Lint, J. H. and Wilson, R. M. (1996). *A Course in Combinatorics*. (Cambridge University Press, Cambridge, U.K.).

Van Mieghem, P. (1996). The asymptotic behaviour of queueing systems: Large deviations theory and dominant pole approximation. *Queueing Systems 23*, 27–55.

Van Mieghem, P. (2006). *Performance Analysis of Communications Networks and Systems*. (Cambridge University Press, Cambridge, U.K.).

Van Mieghem, P. (2007). A new type of lower bound for the largest eigenvalue of a symmetric matrix. *Linear Algebra and its Applications 427*, 1 (November), 119–129.

Van Mieghem, P. (2010). *Data Communications Networking*, 2nd edn. (Piet Van Mieghem, ISBN 978-94-91075-01-8, Delft).

Van Mieghem, P. (2013). Decay towards the overall-healthy state in SIS epidemics on networks. Delft University of Technology, Report20131016 (www.nas.ewi.tudelft.nl/people/Piet/TUDelftReports) *and arXiv:1310.3980*.

Van Mieghem, P. (2014). *Performance Analysis of Complex Networks and Systems*. (Cambridge University Press, Cambridge, U.K.).

Van Mieghem, P. (2015a). Graph eigenvectors, fundamental weights and centrality metrics for nodes in networks. Delft University of Technology, Report20150808 (www.nas.ewi.tudelft.nl/people/Piet/TUDelftReports); *arXiv:1401.4580*.

Van Mieghem, P. (2015b). A Lagrange series approach to the spectrum of the kite graph. *Electronic Journal of Linear Algebra 30*, 934–943.

Van Mieghem, P. (2016). Interconnectivity structure of a general interdependent network. *Physical Review E 93*, 4 (April), 0423405.

Van Mieghem, P. (2021). Some Laplacian eigenvalues can be computed by matrix perturbation. Delft University of Technology, report20211109 (www.nas.ewi.tudelft.nl/people/Piet/TUDelftReports).

Van Mieghem, P., Devriendt, K., and Cetinay, H. (2017). Pseudo-inverse of the Laplacian and best spreader node in a network. *Physical Review E 96*, 3 (September), 032311.

Van Mieghem, P., Ge, X., Schumm, P., Trajanovski, S., and Wang, H. (2010). Spectral graph analysis of modularity and assortativity. *Physical Review E 82*, 5 (November), 056113.

Van Mieghem, P. and Kuipers, F. A. (2004). Concepts of exact quality of service algorithms. *IEEE/ACM Transaction on Networking 12*, 5 (October), 851–864.

Van Mieghem, P., Omic, J., and Kooij, R. E. (2009). Virus spread in networks. *IEEE/ACM Transactions on Networking 17*, 1 (February), 1–14.

Van Mieghem, P., Stevanović, D., Kuipers, F. A., Li, C., van de Bovenkamp, R., Liu, D., and Wang, H. (2011). Decreasing the spectral radius of a graph by link removals. *Physical Review E 84*, 1 (July), 016101.

Van Mieghem, P. and van de Bovenkamp, R. (2013). Non-Markovian infection spread dramatically alters the SIS epidemic threshold in networks. *Physical Review Letters 110*, 10 (March), 108701.

Van Mieghem, P. and Wang, H. (2009). Spectra of a new class of graphs with extremal properties. Delft University of Technology, Report20091010 (www.nas.ewi.tudelft.nl/people/Piet/TUDelftReports).

Van Mieghem, P., Wang, H., Ge, X., Tang, S., and Kuipers, F. A. (2010). Influence of assortativity and degree-preserving rewiring on the spectra of networks. *The European Physical Journal B 76*, 4, 643–652.

van Rooij, A. C. M. and Wilf, H. S. (1965). The interchange graph of a finite graph. *Acta Mathematica Hungarica 16*, 3-4, 263–269.

Vu, V. H. (2007). Spectral norm of random matrices. *Combinatorica 27*, 6, 721–736.

Walker, S. G. and Van Mieghem, P. (2008). On lower bounds for the largest eigenvalue of a symmetric matrix. *Linear Algebra and its Applications 429*, 2-3 (July), 519–526.

Wang, H., Douw, L., Hernandez, J. M., Reijneveld, J. C., Stam, C. J., and Van Mieghem, P. (2010). Effect of tumor resection on the characteristics of functional brain networks. *Physical Review E 82*, 2 (August), 021924.

Wang, H., Kooij, R. E., and Van Mieghem, P. (2010). Graphs with given diameter maximizing the algebraic connectivity. *Linear Algebra and its Applications Vol. 433*, 11, 1889–1908.

Wang, H., Martin Hernandez, J., and Van Mieghem, P. (2008). Betweenness centrality in weighted networks. *Physical Review E 77*, 4 (April), 046105.

Wang, X., Dubbeldam, J. L. A., and Van Mieghem, P. (2017). Kemeny's constant and the effective graph resistance. *Linear Algebra and its Applications 535*, 232–244.

Watson, G. N. (1995). *A Treatise on the Theory of Bessel Functions*, Cambridge Mathematical Library edn. (Cambridge University Press, Cambridge, U.K.).

Watts, D. J. (1999). *Small Worlds, The Dynamics of Networks between Order and Randomness.* (Princeton University Press, Princeton, New Jersey).

Watts, D. J. and Strogatz, S. H. (1998). Collective dynamics of "small-worlds" networks. *Nature 393*, 440–442.

Weil, A. (1984). *Number Theory, An Approach through History - From Hammurapi to Legendre.* (Birkhäuser, Boston).

Weyl, H. (1912). Das asymptotische Verteilungsgesetz der Eigenwerte linearer partieller Differentialgleichungen (mit einer Anwendung auf die Theorie der Holhraumstrahlung). *Mathematische Annalen 71*, 4, 441–479.

Whitney, H. (1932). Congruent graphs and the connectivity of graphs. *American Journal of Mathematics 54*, 1, 150–168.

Whittaker, E. T. and Watson, G. N. (1996). *A Course of Modern Analysis*, Cambridge Mathematical Library edn. (Cambridge University Press, Cambridge, U.K.).

Wigner, E. P. (1955). Characteristic vectors of bordered matrices with infinite dimensions. *Annals of Mathematics 62*, 3 (November), 548–564.

Wigner, E. P. (1957). Characteristic vectors of bordered matrices with infinite dimensions ii. *Annals of Mathematics 65*, 2 (March), 203–207.

Wigner, E. P. (1958). On the distribution of the roots of certain symmetric matrices. *Annals of Mathematics 67*, 2 (March), 325–327.

Wilf, H. S. (1986). Spectral bounds for the clique and independence numbers of graphs. *Journal of Combinatorial Theory, Series B 40*, 113–117.

Wilkinson, J. H. (1965). *The Algebraic Eigenvalue Problem.* (Oxford University Press, New York).

Xing, R. and Zhou, B. (2013). On least eigenvalues and least eigenvectors of real symmetric matrices and graph. *Linear Algebra and its Applications 438*, 2378–2384.

Xiong, L. (2001). The Hamiltonian index of a graph. *Graphs and Combinatorics 17*, 775–784.

Zhou, B. and Trinasjtić. (2008). A note on Kirchhoff index. *Chemical Physics Letters 455*, 120–123.

Zhou, J., Wang, Z., and Bu, C. (2016). On the resistance matrix of a graph. *The Electronic Journal of Combinatorics 23*, 1 (March), P1.41.

Zhu, X., Ghahramani, Z., and Lafferty, J. (2003). Semi-supervised learning using Gaussian fields and harmonic functions. *Proceedings of the Twentieth International Conference on Machine Learning (ICML-2003, Washington D.C.)* 912-919 (August).

Index

Printed in the United States
by Baker & Taylor Publisher Services